NUCLEAR AND PARTICLE PHYSICS SOURCE BOOK

THE McGRAW-HILL SCIENCE REFERENCE SERIES

Acoustics Source Book
Communications Source Book
Computer Science Source Book
Fluid Mechanics Source Book
Meteorology Source Book
Optics Source Book
Physical Chemistry Source Book
Solid-State Physics Source Book
Spectroscopy Source Book

For more information about other McGraw-Hill materials, call 1-800-2-MCGRAW in the United States. In other countries, call your nearest McGraw-Hill office.

NUCLEAR AND PARTICLE PHYSICS SOURCE BOOK

Sybil P. Parker, Editor in Chief

McGRAW-HILL BOOK COMPANY

New York   St. Louis   San Francisco
Auckland   Bogotá   Caracas   Colorado Springs   Hamburg
Lisbon   London   Madrid   Mexico   Milan   Montreal
New Delhi   Oklahoma City   Panama   Paris   San Juan
São Paulo   Singapore   Sydney   Tokyo   Toronto

Cover: Central part of the collider detector at Fermilab, used to observe collisions in the tevatron proton-antiproton collider having a total energy of 1.8 TeV (1.8×10^{12} eV). The detector stands over 20 ft (6 m) tall, weighs 4500 tons (4000 metric tons), and was built by a collaboration of more than 170 physicists.

This material has appeared previously in the McGRAW-HILL ENCYCLOPEDIA OF SCIENCE AND TECHNOLOGY, 6th Edition, copyright © 1987 by McGraw-Hill, Inc. All rights reserved.

NUCLEAR AND PARTICLE PHYSICS SOURCE BOOK, copyright © 1988 by McGraw-Hill, Inc. All rights reserved. Printed in the United States of America. Except as permitted under the United States Copyright Act of 1976, no part of this publication may be reproduced or distributed in any form or by any means, or stored in a data base or retrieval system, without prior written permission of the publisher.

1 2 3 4 5 6 7 8 9 0   DOC/DOC   8 9 5 4 3 2 1 0 9 8

ISBN 0-07-045509-0

Library of Congress Cataloging in Publication Data:

Nuclear and particle physics source book.

   (The McGraw-Hill science reference series)
   Bibliography: p.
   Includes index.
   1. Nuclear physics.   2. Particles (Nuclear physics).
I. Parker, Sybil P.   II. Series.
QC776.N77   1988   539.7   88-13483
ISBN 0-07-045509-0

For more information about other McGraw-Hill materials, call 1-800-2-MCGRAW in the United States. In other countries, call your nearest McGraw-Hill office.

TABLE OF CONTENTS

Introduction	1
Nuclear Properties and Structure	5
Types of Nuclei	39
Nuclear Probes and Phenomena	51
Nuclear Collisions and Reactions	103
Energetic Particles and Radiation	165
Elementary Particles	195
Hadrons	241
Exotic Atoms	283
Fundamental Interactions and Fields	291
Symmetry Laws and Conserved Quantities	347
Kinematics and Scattering Theory	377
Particle Accelerators	397
Particle Sources and Detectors	455
Contributors	511
Index	517

NUCLEAR AND PARTICLE PHYSICS SOURCE BOOK

INTRODUCTION

NUCLEAR AND particle physics are disciplines that study the ultimate structure of matter. The two fields are closely related: They test and extend the same fundamental symmetries and laws of nature, and they share much of the same instrumentation and experimental techniques. Elementary particles are the fundamental constituents of matter, and particle physics probes their properties and interactions. Nuclear physics studies the structure of atomic nuclei and their interactions with each other, with their constituent particles, and with the spectrum of elementary particles that is provided by particle accelerators. As the only system in which all the known natural forces can be studied simultaneously, the atomic nucleus provides a natural laboratory that yields results which complement those of particle physics.

Nuclear physics. The origins of nuclear physics are bound up with those of atomic physics, relativity, and quantum theory in the early twentieth century. Major early advances include the discovery of radioactivity (1896), the discovery of polonium and radium (1898), the discovery of the atomic nucleus by interpretation of alpha-particle scattering results (1911), the identification of isotopes and isobars (1911), the establishment of the displacement laws governing the changes in atomic number accompanying radioactive decay (1913), the production of nuclear transmutations by alpha-particle bombardment (1919) and by artificially accelerated particles (1932), the formulation of the theory of beta decay (1933), the production of radioactive nuclides by accelerated particles (1934), and the discovery of nuclear fission (1938).

The nuclear domain occupies a central position between the atomic range of forces and sizes and those of elementary-particle physics, characteristically within the nucleons themselves. Containing a reasonably large yet manageable number of strongly interacting components, the nucleus also occupies a central position in the universal many-body problems of physics, falling between the few-body problems characteristic of elementary-particle interactions and the extreme many-body situations of plasma physics and condensed matter in which statistical approaches dominate. The nucleus provides the scientist with a rich range of phenomena to investigate—with the hope of understanding these phenomena at a microscopic level.

Activity in the field centers on three broad and interdependent subareas. The first is referred to as classical nuclear physics, wherein the structural and dynamic aspects of nuclear behavior are probed in numerous laboratories and in many nuclear systems, with the use of a broad range of experimental and theoretical techniques. The second is higher-energy nuclear physics (referred to as medium-energy physics in the United States), which emphasizes the nuclear interior and nuclear interactions with mesonic probes. The third is heavy-ion physics, internationally the most rapidly growing subfield, wherein accelerated beams of nuclei spanning the periodic table are used to study previously inaccessible nuclear phenomena.

Nuclear physics is unique in the extent to which it merges the most fundamental and the most applied topics. Its instrumentation has found broad applicability throughout science, technology, and medicine; nuclear engineering and nuclear medicine are two very important areas of applied specialization.

Particle physics. While elementary-particle physics can be said to have originated with the identification or discovery of the electron (1897), proton (1919), neutron (1932), positron (1932), and muon (1936), the discovery and classification of large numbers of particles were carried out in the years 1947–1964, first through studies of cosmic-ray events and then with increasingly powerful particle accelerators. Although the resulting "zoo" of over 100 "particles" seemed to defy attempts at constructing a theory of their behavior, different sets of hypotheses were successful in organizing various aspects of the data, and it was possible to classify both the particles and their interactions with the aid of symmetry principles.

The interactions could be classified into four types—strong nuclear, electromagnetic, weak nuclear, and gravitational—that differed radically in terms of their strengths, ranges, and other characteristics. The particles could be classified into two principal groups, the hadrons and leptons, according to whether or not they responded to the strong interactions. The list of leptons remained limited to the electron, muon, and (later discovered) tau, and their corresponding neutrinos; it was among the hadrons that the proliferation of particles took place.

The application of symmetry principles to the study of hadrons led to the hypothesis that they were composed of constituents of fractional electric charge called quarks, and were thus not truly elementary particles. The over 100 hadrons could thus be understood in terms of the configuration of their constituent quarks, much as the atoms had been understood in terms of the configuration of their electrons, protons, and neutrons. However, although the quark model received support from high-energy electron-proton scattering data, several puzzling issues, including the failure of all attempts to find free quarks, raised serious doubts as to whether it could form the basis of a physical theory.

This picture of particle physics has changed radically as the result of both theoretical advances and experimental results. Central to the theoretical work has been the concept of gauge invariance, whereby interactions are derived from symmetry principles and are mediated by force-carrying particles with a spin of 1 called gauge bosons. The simplest example of a gauge theory is that of the electromagnetic interactions, mediated by the photon. In 1967, S. Weinberg and A. Salam constructed a gauge theory that unified the electromagnetic and weak forces in the so-called electroweak interaction, in which the photon is joined by three massive gauge bosons, the W^+, W^-, and Z^0. The application of the gauge-invariance principle to the strong interactions of quarks led to the creation of quantum chromodynamics, in which the quarks exchange massless gauge bosons called gluons; the puzzles surrounding the quark model were largely resolved in this theory.

Experimental results have provided dramatic support for these theories. The results include the discovery of the so-called neutral-current weak interactions (1973) predicted by the Weinberg-Salam theory; the discovery of the J/ψ particle (1974), a heavy hadron of unusually long lifetime that was recognized to be a bound system of a new massive quark with its antiquark; the similar discovery of upsilon particles (1977); the observation of the W and Z bosons (1983); and the observation, in high-energy collisions, of emerging jets of particles believed to emanate from quarks and gluons.

The picture that emerges, known as the standard model, is one of great simplicity and generality. All matter appears to be composed of quarks and leptons, which are pointlike, structureless particles with a spin of ½. If gravitation (which is a negligible perturbation at the energy scales usually considered) is set aside, the interactions among these particles are the strong and electroweak, both described by gauge theories and mediated by spin-1 gauge bosons. The quarks and leptons are arranged in families, and there is a hint of extended families containing both quarks and leptons.

While the standard model provides a wide-ranging synthesis, it cannot be the final answer. It does not predict the number of families of quarks and leptons, their masses or mass ratios, or numerous other parameters in the theory. Perhaps most important, the mechanism that causes the W and Z bosons to acquire large masses while leaving the photon massless is poorly understood. Current research is directed toward extending the standard model to remedy these defects and to unify the strong and electroweak interactions, and perhaps even the gravitational interaction, in a single theory.

1
NUCLEAR PROPERTIES AND STRUCTURE

Atomic nucleus	6
Nuclear structure	6
Atomic number	15
Mass number	16
Isotope	16
Isotone	20
Isobar	20
Atomic weight	21
Mass defect	22
Nuclear binding energy	22
Garvey-Kelson mass relations	24
Magic numbers	25
Analog states	26
Nuclear isomerism	29
Giant nuclear resonances	31
Nuclear moments	35
Magneton	37

ATOMIC NUCLEUS
Henry E. Duckworth

The central region of an atom. Atoms are composed of negatively charged electrons, positively charged protons, and electrically neutral neutrons. The protons and neutrons (collectively known as nucleons) are located in a small central region known as the nucleus. The electrons move in orbits which are large in comparison with the dimensions of the nucleus itself. Protons and neutrons possess approximately equal masses, each roughly 1840 times that of an electron. The number of nucleons in a nucleus is given by the mass number A and the number of protons by the atomic number Z. Nuclear radii r are given approximately by $r = 1.4 \times 10^{-13} A^{1/3}$ cm. See Nuclear Structure.

NUCLEAR STRUCTURE
F. Iachello

The atomic nucleus is at the center of the atom and contains 99.975% of the total mass of the atom. Its average density is about 3×10^{11} kg/cm^3; its diameter is about 10^{-12} cm, and thus much smaller than the diameter of the atom, which is about 10^{-8} cm.

The nucleus is composed of protons and neutrons. The number of protons is usually denoted by Z, while that of neutrons is denoted by N. The number of protons is equal to the number of electrons in the atom. Since the proton is positively charged, the electron is negatively charged, and the neutron is neutral, the atom as a whole is neutral under normal conditions. The total number of protons and neutrons in a nucleus is called the mass number and denoted by $A = N + Z$. Nuclei having the same proton number but different neutron number are called isotopes. Nuclei having the same neutron number but different proton number are called isotones. Finally, nuclei with the same mass number are called isobars. See Electron; Isotope; Proton.

Bulk properties. The bulk properties of nuclei include their sizes, density distributions, and masses.

Nuclear densities and sizes. The average radius of the proton distribution of the nucleus can be easily measured by using the scattering of fast electrons and the energies of x-rays emitted by μ-mesic atoms. When a high-velocity electron approaches a nucleus, it is deflected from its path because of the electromagnetic interaction with the protons in the nucleus. A measurement of this deflection permits the determination of the charge distribution of the nucleus. When a negative μ-meson (of mass $m_\mu \simeq 207\, m_e$, where m_e is the mass of the electron) is brought to rest in matter, it is attracted by the positive charge of the nucleus and captured into an orbit around it, thus forming a μ-mesic atom. The μ-meson can then jump from one orbit to another, emitting x-rays. The lowest orbits of the μ-meson have diameters comparable to the size of the nucleus, and thus the x-ray energies are greatly affected by the actual charge distribution of the nucleus. A measurement of these energies provides information on the proton distribution in nuclei.

From both methods, it has been possible to determine that the charge distribution in nuclei can be quite accurately described by Eq. (1), where $\rho(r)$ is the proton density at a distance r from

$$\rho(r) = \rho(0)[1 + e^{(r-R)/a}]^{-1} \qquad (1)$$

the center, and $\rho(0)$ is the density at the center. This bell-shaped distribution (called a Woods-Saxon distribution) is shown in **Fig. 1** for nuclei of cobalt and bismuth. The quantity R is the point at which the nuclear density has fallen to one-half of its central value, and is referred to as the nuclear radius. It is usually measured in fermis (1 F = 10^{-13} cm = 10^{-15} m = 1 fm). The dependence of R on the mass number is approximately given by Eq. (2). The quantity a is called the surface thickness and is given by Eq. (3). The average radius R is proportional to $A^{1/3}$, which

$$R = (1.07 \pm 0.02)\, A^{1/3}\ \text{F} \qquad (2) \qquad\qquad a = (0.55 \pm 0.07)\ \text{F} \qquad (3)$$

implies that the volume is proportional to A and thus that the mean density is independent of the size of the nucleus. The phenomenon is usually referred to as saturation.

It is much more difficult to measure the neutron distribution. Since the neutrons have no

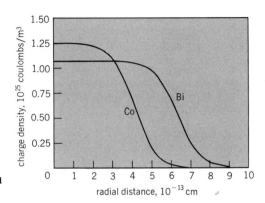

Fig. 1. Charge density in nuclei of cobalt and bismuth.

electric charge, a measurement of their distribution must be done with strongly interacting particles, such as alpha particles. However, unlike electrons, these particles are not pointlike, but have a finite extent of their own. Moreover, the strong interaction is not so well known as the electromagnetic interaction, and this introduces further uncertainties in the measurement of the neutron distribution. Although most theories predict a neutron radius in heavy nuclei larger than the proton radius, measurements are consistent with a neutron radius approximately equal to the proton radius ($R_n - R_p \cong 0.1$–0.2 F, where R_n is the neutron radius and R_p the proton radius). A method of measurement based on the scattering of positively and negatively charged π-mesons may provide more definitive information on the difference $R_n - R_p$. SEE SCATTERING EXPERIMENTS (NUCLEI).

Nuclear masses. Nuclear masses are usually measured in unified mass units (symbol u). One mass unit equals one-twelfth of the mass of the carbon atom, which has $A = 12$. The observed masses of the proton and neutron are 1.007277 and 1.008665 u, respectively, while that of the hydrogen atom is 1.007825 u. By convention, whenever one speaks of a nuclear mass, one includes the mass of the Z electrons, thus quoting the mass of the corresponding neutral atom and not of the nucleus alone.

Atomic masses are, to a good approximation, described by a semiempirical mass formula. The following terms contribute to this formula. First, there is the contribution of the Z protons, N neutrons, and Z electrons, altogether given by $1.007825 Z + 1.008665 N$. There is then a term which is roughly proportional to the volume of the nucleus. Since from Eq. (2), R is proportional to $A^{1/3}$, this gives a contribution proportional to A, $-c_v A$. Next, there is a term proportional to the area of the surface of the nucleus, $c_a A^{2/3}$. Another term comes from the Coulomb repulsion between protons. This term can be written in the form $c_c Z^2/A^{1/3}$. The Coulomb repulsion makes it more favorable to have in the nucleus more neutrons than protons. However, this repulsion is counterbalanced by the strong nuclear interaction, that which holds nuclei together. This interaction favors the situation in which the number of protons and neutrons is equal, and contributes $c_s(N - Z)^2/A$. Finally there is a term which takes into account the fact that nucleons in nuclei tend to pair together, gaining energy when doing so. This last term is of the form $\pm c_p A^{-3/4}$, where the plus sign is used for odd-odd nuclei and the minus sign is used for even-even nuclei. The term is assumed to be zero in even-odd nuclei. Inserting the appropriate numerical constants, the mass of an atom, $M(A,Z,N)$ is given by Eq. (4), called the Bethe-Weizsacker mass formula.

$$M(A,Z,N) = 1.007825\,Z + 1.008665\,N - 0.015\,A \\ + 0.014\,A^{2/3} + 0.021\,\frac{(N-Z)^2}{A} \\ + 0.000627\,\frac{Z^2}{A^{1/3}} \pm 0.036\,A^{-3/4} \quad (4)$$

A convenient way to present miscellaneous data concerning nuclei is offered by the chart of nuclei, in which each nucleus is represented by a unit square in a plot of Z versus N. The general layout of this chart is shown in **Fig. 2**. Stable nuclei are found along a stability valley. This valley departs more and more from the line $Z = N$ the heavier the nucleus. Nuclei away from the stability line decay either by converting protons into neutrons (or vice versa), creating in the process an electron and a neutrino (beta decay), or by boiling off particles, such as neutrons (n decay), protons (p decay), and He nuclei (alpha decay), or finally by dividing into two or more pieces (fission). SEE NUCLEAR FISSION; RADIOACTIVITY.

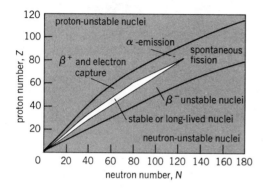

Fig. 2. General arrangement of the chart of nuclei, with lines of stability against various break-up modes.

Nuclei far from stability. The number of nuclei which have been investigated experimentally has increased considerably since the early 1970s. This has been possible because of the development of heavy-ion accelerators. When a heavy ion collides with a target, it produces new nuclear species. These subsequently decay by emitting gamma rays. Measurements of the energies of the emitted gamma rays provide information on nuclei far from stability. For each isotopic chain (constant Z) there are on the average about 20 nuclei known. A more detailed portion of the chart of nuclei is shown in **Fig. 3**. Nuclei with an even number of protons and neutrons, whose

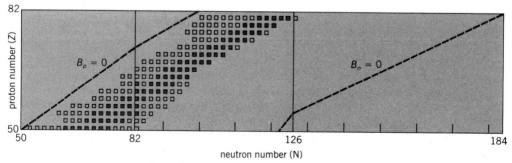

Fig. 3. Detailed portion of the chart of the nuclei.

spectroscopic properties are known with some confidence, are shown by open squares. Stable or long-lived nuclei (with half-life greater than 10^{20} s) are shown by dark squares. The lines at which nuclei become unstable under proton or neutron emission are denoted by $B_p = 0$ and $B_n = 0$. These are sometimes called proton and neutron drip lines. Since heavy-ion collisions produce mostly neutron-deficient isotopes, more information is available on nuclei to the left of the valley of stability.

In addition, attempts have been made to produce heavier elements in the laboratory, and elements up to $Z = 109$ and $A = 262$ have been identified. Preliminary attempts to produce even heavier elements with special stability properties, called superheavy elements ($Z = 114$ and $A = 298$), have not yet succeeded. SEE ELEMENT 109; SUPERTRANSURANICS; TRANSURANIUM ELEMENTS.

Detailed properties. The detailed properties of a nucleus include the energies, angular momenta, and parities of its quantum states, and its magnetic and electric moments.

Energy-level diagram. The nucleus is a quantum-mechanical system. Its properties are best described by a diagram, called the energy-level diagram, in which its quantum states are listed, together with the expectation values of all measurable quantities. Three of these are particularly important: the energy, the angular momentum, and the parity. Energies are usually measured from the lowest state (called the ground state) and are given in millions of electronvolts (1 MeV = 1.60219×10^{-13} joules). Angular momenta are labeled by a quantity J which is half-integer for odd-even nuclei and integer for even-even and odd-odd nuclei. In terms of this quantity the square of the angular momentum is given by $P_J^2 = J(J + 1)\hbar^2$, where $\hbar = 1.05459 \times 10^{-34}$ J·s. With no exception, the ground states of even-even nuclei have been found to have $J = 0$. Parity, denoted by π, is a purely quantum-mechanical concept, which describes the transformation of the wave function of the system under reflection $\vec{x} \to -\vec{x}$. It is either plus or minus. SEE PARITY.

Magnetic and electric moments. In addition to energies, angular momenta, and parities, other properties are often measured. Among these, especially important are the intensities of the electromagnetic transitions between two energy levels. Nuclei, like atoms, in a state of higher energy can decay to states of lower energy by emitting photons. The corresponding transitions can be either electric (E) or magnetic (M), and have multipolarity $l = 0, 1, 2, \ldots$. Electromagnetic transitions satisfy certain selection rules which are related to the angular momenta and parities of the energy levels between which the transition is taking place. For electric transitions of multipolarity l, the parity of the initial, π_i, and final, π_f, states are related by $\pi_i \pi_f = (-)^l$. For magnetic transitions, they are related by $\pi_i \pi_f = (-)^{l+1}$. The multipolarity of the transition l is limited by $|J_i - J_f| \leq l \leq |J_i + J_f|$. Thus an electric-dipole, $l = 1$, transition ($E1$) between a state with $J_i^{\pi_i} = 2^+$ and $J_f^{\pi_f} = 0^+$ is not allowed. Conversely, the same transition is allowed between $J_i^{\pi_i} = 1^-$ and $J_f^{\pi_f} = 0^+$. Other measured properties of nuclear states are their static magnetic and electric moments. Magnetic dipole moments are measured in nuclear magnetons (1 nuclear magneton = 1/1836.15 Bohr magneton = 5.0508×10^{-27} J/T). In the scale of nuclear magnetons, the proton has a magnetic moment of 2.79285 nuclear magnetons, and the neutron of -1.91304 nuclear magnetons. Electric quadrupole moments are usually measured in electron-barns (1 eb = $10^{-24} e$ cm^2, where e is the magnitude of the charge of the electron). SEE MAGNETON; MULTIPOLE RADIATION; NUCLEAR MOMENTS; SELECTION RULES.

Nuclear models. A number of models have been developed to account for the properties of nuclei.

Shell model. The basic model for the description of nuclear properties is the shell model. The strong nuclear force binds together protons and neutrons, called by the single name nucleons. Each individual nucleon moves in the average potential generated by all the others. This potential has the same shape as the nuclear matter distribution of Fig. 1, but upside down. The states of a single nucleon in the average potential cluster together into layers or shells, much like the single-particle states in atoms. Attempts to predict the location of the shells were not very successful until 1949, when M. G. Mayer and J. H. Jensen introduced a new term in the average potential field. The term, spin-orbit interaction, describes an interaction of the intrinsic spin of the nucleon with its orbital angular momentum, and it can be written in the form of Eq. (5), where $f(r)$ is a

$$V_{so} = f(r) (\vec{s} \cdot \vec{l}) \qquad (5)$$

function of the radial distance r, and $\vec{s}$ and $\vec{l}$ are the intrinsic and orbital angular momenta of the nucleon. The introduction of this term gives rise to the single-particle structure shown in **Fig. 4**. In this figure the closed shells appear at nucleon numbers 2, 8, 20, 28, 50, 82, 126, in agreement with experiment. The numbers at which closed shells occur are called magic numbers. SEE MAGIC NUMBERS; SPIN.

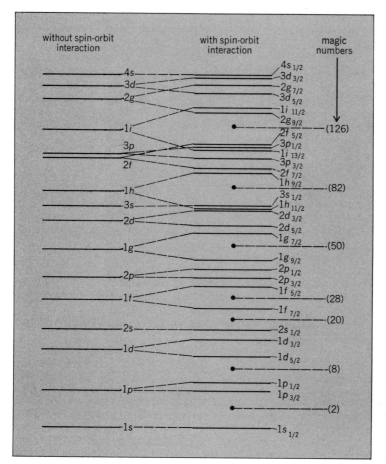

Fig. 4. Single-particle levels according to the shell model. Symbols at right indicate their spectroscopic notation. The letter indicates orbital angular momentum (s, p, d, f, . . ., for $l = 0, 1, 2, 3,$ . . .), the integer indicates the order of levels with the same l, and the subscript indicates the j (total angular momentum) value. Numbers in parentheses indicate magic numbers.

Nuclei with few valence particles. Closed-shell nuclei behave, in many respects, as inert. Most properties of nuclei can be described by considering as active only nucleons outside the closed shells. These are called valence nucleons. In addition to the average potential in which they move, it is assumed that there is a residual interaction between valence nucleons. When the number of valence nucleons is small (up to about four), the effects of the residual interaction can be calculated easily. **Figure 5** shows a comparison between the calculated energy-level diagram, called a spectrum, of an even-even nucleus with two valence nucleons, and the experimental scheme. The agreement between theory and experiment is usually found to be good.

Nuclei with many valence particles. When the number of valence nucleons is large (greater than 5–6), some striking regularities develop in the level scheme. These regularities, first recognized by J. Rainwater, A. Bohr, and B. R. Mottelson, indicate the presence of collective features in nuclei. The collective features arise from two special properties of the residual interaction between valence nucleons: first, its pairing property which tends to pair off nucleons, one with angular momentum up and one with angular momentum down, to give zero resultant; and second, a quadrupole property which favors a distortion of the nuclear surface in the shape of an ellipsoid.

A large variety of collective spectra of quadrupole type have been observed in nuclei. These spectra can be discussed in terms of the geometry of the nuclear surface. The nuclear surface can

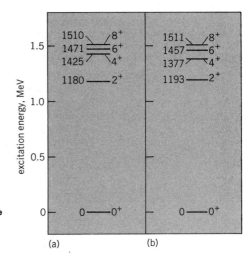

Fig. 5. Comparison of (a) experimental level scheme of $^{210}_{84}\text{Po}_{126}$ ($A = 210$, $Z = 84$, $N = 126$) and (b) that calculated using the shell model. The numbers on the left of each level are the excitation energies in kiloelectronvolts, and the symbols on the right are the angular momenta and parities.

be characterized by giving its radius R. For nuclei with quadrupole distortions, R can be written as in Eq. (6), where $Y_{2\mu}(\theta,\phi)$ is the spherical harmonic of order two and the quantities α_μ ($\mu =$

$$R = R_0 \left(1 + \sum_{\mu=-2}^{+2} \alpha_\mu Y_{2\mu}(\theta,\phi)\right) \quad (6)$$

0, ±1, ±2) are called deformation parameters. When $\alpha_\mu = 0$, the nucleus is said to be spherical; when $\alpha_\mu \neq 0$, the nucleus is said to be deformed. For spherical nuclei, the radius R is given by Eq. (2). Instead of α_μ, it has become customary to introduce another set of variables, related to the α_μ's by Eqs. (7). These are called Bohr variables. Collective spectra can then be classified

$$\alpha_0 = \beta \cos \gamma \qquad \alpha_2 = \alpha_{-2} = \frac{1}{\sqrt{2}} \beta \sin \gamma \qquad \alpha_1 = \alpha_{-1} = 0 \quad (7)$$

according to the values of β and γ which describe the shape of a nucleus. The observed spectra fall into three major categories:

1. Rotational spectra of deformed nuclei with axial symmetry. These correspond to shapes characterized by a value of $\beta = \beta_{\text{equ}} \neq 0$, called equilibrium deformation, and $\gamma = 0^0$. The spectra consist of a series of energy levels, connected by large electric quadrupole ($E2$) transitions, called bands. Within each band, the excitation energies are approximately given by Eq. (8), where κ is

$$E = \kappa J(J + 1) \quad (8)$$

a constant typical of a given nucleus and J is the angular momentum of the level. In even-even nuclei, the band built on the ground state is composed of levels with $J = 0, 2, 4, \ldots$; levels with odd angular momenta, $J = 1, 3, \ldots$, are missing. Other bands, with higher excitation energy, may or may not begin with angular momentum $J = 0$. If they do, the angular momentum sequence is the same as in the ground-state band. Otherwise, both even and odd angular momenta appear. An example is shown in **Fig. 6**.

2. Rotational spectra of deformed nuclei with γ instability. These correspond to shapes characterized by a value $\beta = \beta_{\text{equ}} \neq 0$, but no fixed value of γ. Here again, the spectrum consists of a series of bands, but the band structure is very different from the previous case, and energies within a band are given by Eq. (9), where A is a constant typical of a given nucleus and τ is a

$$E = A\tau(\tau + 3) \quad (9)$$

quantum number which labels the energy levels ($\tau = 0, 1, 2, \ldots$). For each value of τ, there can

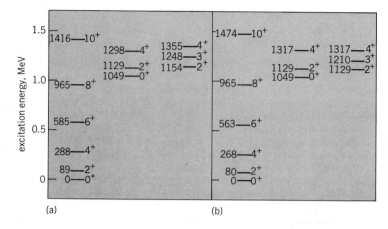

Fig. 6. Comparison of (a) experimental level scheme of $^{156}_{64}Gd_{92}$ and (b) that calculated using the collective model. The numbers on the left of each level are the excitation energies in kiloelectronvolts, and the symbols on the right are the angular momenta and parities.

be one or more levels with different J. The values of J belonging to each τ are given, for the lowest levels, by $\tau = 0$, $J = 0$; $\tau = 1$, $J = 2$; $\tau = 2$, $J = 4, 2$; and $\tau = 3$, $J = 6, 4, 3, 0$.

3. Vibrational spectra of spherical nuclei. These correspond to shapes characterized by a value of $\beta = \beta_{equ} = 0$ and no fixed value of γ. The spectrum consists of a series of vibrational multiplets, whose energy is approximately given by Eq. (10), where ϵ is the vibrational energy,

$$E = \epsilon n \qquad (10)$$

typical of a given nucleus, and n is a vibrational quantum number ($n = 0, 1, 2, \ldots$). Here again, for each value of n, there can be one or more levels with different J. The values of J belonging to each n are given, for the lowest levels, by $n = 0$, $J = 0$; $n = 1$, $J = 2$; $n = 2$, $J = 4, 2, 0$; $n = 3$, $J = 6, 4, 3, 2, 0$.

An alternative description of collective spectra in nuclei has been developed by A. Arima, F. Iachello, and I. Talmi. In this description collective features arise from the fact that protons and neutrons inside nuclei bind together into pairs. These pairs, called interacting bosons, have angular momenta $J = 0$ and $J = 2$. The three major categories of observed spectra discussed above appear here as special cases of a more general description, corresponding to dynamic symmetries of the hamiltonian operator describing the nuclear system. A classification of nuclear shapes is then obtained in terms of symmetry groups similar to the classification of shapes of crystals in terms of point groups. The three symmetry groups of the collective quadrupole motion in nuclei are called SU(3), SO(6), and SU(5), respectively. In Fig. 6, the level scheme on the right-hand side has been calculated by using the SU(3) symmetry. SEE SYMMETRY LAWS.

Even-even nuclei with several valence particles show spectra which are either of these three types or intermediate between them. A similar, but more complex, classification scheme of collective spectra has also been developed for odd-even and odd-odd nuclei.

Statistical model. The number of energy levels of a nucleus below an excitation energy of 2 MeV is usually small. As shown above, these levels can be described by using the shell model and allowing only for excitations of the valence nucleons from some single-particle levels to others nearby in energy. At higher excitation energies, the number of observed states increases considerably. These states arise from the excitation of the valence nucleons to higher single-particle levels and from excitations of nucleons from the closed shells to the valence shell. Because of their large number, it is no longer possible to describe properties of individual states. Thus, statistical methods are employed to describe average properties of states of this sort. An important statistical property is the average density of states ρ as a function of the excitation energy E. This is given by Eq. (11), where a is a coefficient which varies from nucleus to nucleus. For a nucleus with

$$\rho(E) = b \frac{1}{E^2} \exp\{2(aE)^{1/2}\} \qquad (11)$$

mass number $A = 150$, $a \cong 16$ MeV^{-1}. The constant b depends on the spin J of the level, but not on the parity π, since it is assumed that at a given excitation energy E there are an equal number of states with parity $+$ as with parity $-$. The principal source of information on the quantities b and a comes from the study of the resonances observed in the interaction of slow neutrons with nuclei. SEE NEUTRON SPECTROMETRY.

Simple modes. It is an interesting property of nuclei that simple collective modes of excitation occur even at higher excitation energy. Unlike the low-lying collective modes, these are no longer necessarily associated with pairing and quadrupole correlations. They arise from some coherent excitation of nucleons to shells other than the valence shell. These modes are called giant resonances, and are observed in inelastic scattering of strongly interacting particles off nuclei (alpha scattering, for example), in inelastic scattering of electrons off nuclei, and in photoabsorption. The resonances appear as bumps in the cross section, indicating that the special properties of these collective states are spread over many states in a certain energy region. The spreading occurs because, in the same energy region, there exist many states of the type described above (random excitations). The collective state then mixes with these underlying states, sharing its properties with them. In addition to its energy, angular momentum, and parity, a giant resonance is characterized by its spreading width, Γ, that is, the interval in energy over which its properties are spread (**Fig. 7**). The first giant mode to be observed in nuclei was the $J^\pi = 1^-$ mode, called giant dipole resonance. The energy of this mode follows approximately Eq. (12).

$$E = 79A^{-1/3} \text{ MeV} \qquad (12)$$

Other modes observed are the giant quadrupole resonance, $J^\pi = 2^+$, and the giant monopole resonance, $J^\pi = 0^+$. SEE GIANT NUCLEAR RESONANCES.

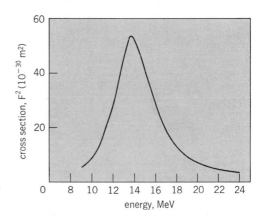

Fig. 7. Total photoabsorption cross section for $^{197}_{79}$Au$_{11}$. The resonance has energy $E = 13.9$ MeV, width $\Gamma = 4.2$ MeV, and spin and parity $J\pi = 1^-$.

Nuclear forces. The nucleons which form the nucleus are held together by the strong nuclear forces.

Free interaction. The forces which act between two nucleons when they are free (not inside the nucleus) have been studied by performing scattering experiments of one nucleon on the other, and by analyzing the properties of the deuteron, the only bound state of two nucleons. Because the proton and the neutron share many properties, it is convenient to consider them as two different states of the same particle, the nucleon. In order to distinguish protons from neutrons, the nucleon is given a sort of intrinsic angular momentum, called isospin. Since there are only two states, the isospin of the nucleon is taken to be $t = \frac{1}{2}$, and the proton and the neutron correspond to the two components $t_z = \pm\frac{1}{2}$ of t. Nuclear forces depend on the total isospin T of the nucleons. This isospin can be $T = 0$ or $T = 1$. A study of low-energy scattering of nucleons on nucleons has revealed that the proton-proton, neutron-neutron, and proton-neutron $T = 1$ forces are practically identical. This property is called charge independence. The proton-neutron

$T = 0$ force is rather different. The free nucleon-nucleon force strongly depends also on the total intrinsic angular momentum S of the two particles. Since each nucleon has spin $s = \frac{1}{2}$, the total spin can be $S = 0$ or $S = 1$, called singlet and triplet, respectively. SEE DEUTERON; ISOBARIC SPIN.

To a good approximation, the free nucleon-nucleon force can be written as a sum of a spin-independent interaction, $V_c(r)$, sometimes called a central interaction, an interaction which depends on the total intrinsic, S, and orbital, L, angular momentum, $V_{LS}(r)\,(\vec{S} \cdot \vec{L})$, called a spin-orbit interaction, and a more complex interaction which involves both spins simultaneously, $V_T(r)[3(\vec{s}_1 \cdot \hat{r}) \cdot (\vec{s}_2 \cdot \hat{r}) - (\vec{s}_1 \cdot \vec{s}_2)]$, called a tensor interaction. Here r is the relative distance between the two nucleons. The form of $V_c(r)$ is shown in **Fig. 8**. It consists of a short-range repulsion ($r \leqslant 0.6$ F) and a long-range attraction ($r \geqslant 0.6$ F). It is rather similar to the interaction between two neutral atoms but, on the contrary, very different from the Coulomb interaction between two pointlike charges $V_{Coul}(r) \propto 1/r$. The nucleon-nucleon interaction decreases as $e^{-\mu r}/r$ for large distances, $r \to \infty$. The inverse length μ is of the order of 0.7 F^{-1}.

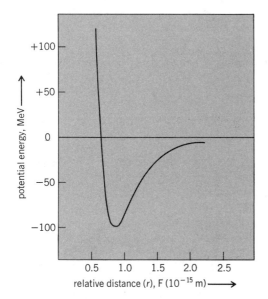

Fig. 8. Shape of the nucleon-nucleon potential for two nucleons in a singlet $S = 0$ state and isotopic spin $T = 1$ as a function of the relative distance r.

Forces inside nuclei. Nuclear structure calculations require the knowledge of the residual interaction between nucleons. This interaction is different from the free nucleon-nucleon interaction because of the modifications introduced by the presence of the other nucleons in the nucleus. Most calculations are done by assuming a strong, attractive interaction of zero range, usually written as $-V_0\delta(\vec{r} - \vec{r}')$. The function $\delta(\vec{r} - \vec{r}')$, called the Dirac delta function, states that the interaction is different from zero only when the relative distance between the two nucleons is zero. In order to take into account the dependence on intrinsic angular momenta, $\vec{s}_1$, and $\vec{s}_2$, and isospins, $\vec{t}_1$, $\vec{t}_2$, the full interaction is written as in Eq. (13), where W, B, H, and M are constants

$$V(\vec{r} - \vec{r}') = -V_0\delta(\vec{r} - \vec{r}')[W + B(\vec{s}_1 \cdot \vec{s}_2) + H(\vec{t}_1 \cdot \vec{t}_2) + M(\vec{s}_1 \cdot \vec{s}_2)(\vec{t}_1 \cdot \vec{t}_2)] \tag{13}$$

which describe the spin-isospin dependence of the interaction and V_0 is an overall strength. The interaction inside nuclei is called the effective interaction, and its relationship to the free interaction is not fully understood.

One property of the effective interaction is its charge independence. This property arises from the charge independence of the free nucleon-nucleon interaction. If charge independence were exactly true, the spectra of the two nuclei $(Z, N + 1)$ and $(Z + 1, N)$, called mirror nuclei,

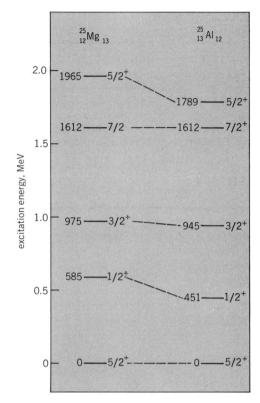

Fig. 9. Comparison of the experimental level schemes of the mirror nuclei $^{25}_{12}Mg_{13}$ and $^{25}_{13}Al_{12}$. The Coulomb energy difference has been removed by placing the two ground states at the same energy. The numbers on the left of each level are the excitation energies in kiloelectronvolts, and the symbols on the right are the spins and parities.

would be exactly identical. This cannot be actually correct since, in addition to the nuclear interaction, there is the electrostatic repulsion between nucleons. However, if this Coulomb piece is removed, the two spectra should be identical. In the example of **Fig. 9**, the two spectra are, to a large extent, identical. The observed differences are due to high-order effects of the electrostatic interaction between the protons which are not removed by subtracting the Coulomb energy. SEE ANALOG STATES; NUCLEAR REACTION; NUCLEAR SPECTRA; STRONG NUCLEAR INTERACTIONS.

Bibliography. A. Bohr and B. R. Mottelson, *Nuclear Structure*, vols. 1 and 2, 1969–1975; A. deShalit and H. Feshbach, *Theoretical Nuclear Physics*, vol. 1: *Nuclear Structure*, 1974; H. A. Enge, *Introduction to Nuclear Physics*, 1966; K. S. Krane, *Introductory Nuclear Physics*, 1987.

ATOMIC NUMBER
HENRY E. DUCKWORTH

The number of elementary positive charges (protons) contained within the nucleus of an atom. It is denoted by the letter Z. For an electrically neutral atom, the number of planetary electrons is also given by the atomic number. Atoms with the same Z (isotopes) belong to the same element. The lightest element, hydrogen, has $Z = 1$. The heaviest naturally occurring element, plutonium, has $Z = 94$. All elements up to and including $Z = 107$, as well as $Z = 109$, either occur in nature or have been created artificially. The atomic number of an atom is altered during radioactive decay: For α-emission, $Z \to Z - 2$; for β^- emission, $Z \to Z + 1$; for β^+ emission or electron capture, $Z \to Z - 1$. When specifically written, the atomic number is usually placed before and below the elemental symbol, for example, $_1H$, $_{92}U$. SEE MASS NUMBER; RADIOACTIVITY; TRANSURANIUM ELEMENTS.

MASS NUMBER
HENRY E. DUCKWORTH

The mass number A of an atom is the total number of its nuclear constituents, or nucleons, as the protons and neutrons are collectively called. The mass number is placed before and above the elemental symbol, as in ^{238}U. Because of the approximate equality of the proton and neutron masses, and the relative insignificance of that of the electron, the mass number gives a useful rough figure for the atomic mass; for example, H^1 = 1.00814 atomic mass units (amu), ^{238}U = 238.124 amu, and so on. The mass number is reduced by four during alpha-emission, but it is not altered during beta-decay or electron capture. SEE ATOMIC NUMBER.

ISOTOPE
DANIEL J. HOREN

One member of a (chemical-element) family of atomic species which has two or more nuclides with the same number of protons (Z) but a different number of neutrons (N). Because the atomic mass is determined by the sum of the number of protons and neutrons contained in the nucleus, isotopes differ in mass. Since they contain the same number of protons (and hence electrons), isotopes have the same chemical properties. However, the nuclear and atomic properties of isotopes can be different. The electronic energy levels of an atom depend upon the nuclear mass. Thus, corresponding atomic levels of isotopes are slightly shifted relative to each other. A nucleus can have a magnetic moment which can interact with the magnetic field generated by the electrons and lead to a splitting of the electronic levels. The number of resulting states of nearly the same energy depends upon the spin of the nucleus and the characteristics of the specific electronic level. SEE HYPERFINE STRUCTURE; ISOTOPE SHIFT.

Of the 108 elements reported thus far, 81 have at least one stable isotope whereas the others exist only in the form of radioactive nuclides. Some radioactive nuclides (for example, ^{115}In, ^{232}Th, ^{235}U, ^{238}U) have survived from the time of formation of the elements. Several thousand radioactive nuclides produced through natural or artificial means have been identified. SEE RADIOISOTOPE.

Of the 83 elements which occur naturally in significant quantities on Earth, 20 are found as a single isotope (mononuclidic), and the others as admixtures containing from 2 to 10 isotopes. Isotopic composition is mainly determined by mass spectroscopy.

Nuclides with identical mass number (that is, $A = N + Z$) but differing in the number of protons in the nucleus are called isobars. Nuclides having different mass number but the same number of neutrons are called isotones. SEE ISOBAR; ISOTONE.

Nuclear stability. The stability of a nuclide is governed by its total energy E as given by the Einstein relation $E = Mc^2$, where M is the nuclidic mass and c is the velocity of light. If E is less than the combined energies of possible decay products, the nuclide will be stable. A major factor in determining stability is the relative strength of the nuclear force which acts to attract nucleons and the coulombic force (repulsive) which arises from the electric charge on the protons. Nuclides with an even number of protons or neutrons are prevalent in the **table**. Of the 287 nuclides tabulated, 168 are even-even (that is, an even number of both neutrons and protons), 110 are odd, and only 9 are odd-odd. This demonstrates the increased attraction of the nuclear force between pairs of nucleons of the same type (the pairing effect). Nuclides for which the number of either protons or neutrons (or both) comprises so-called magic numbers (for example, 8, 20, 50, 82, etc.) have increased stability. SEE NUCLEAR STRUCTURE.

Isotopic abundance. The term isotopic abundance refers to the isotopic composition of an element found in its natural terrestrial state. The isotopic composition for most elements does not vary much from sample to sample. This is true even for samples of extraterrestrial origin such as meteorites and lunar materials brought back to Earth by United States crewed and Soviet uncrewed missions. However, there are a few exceptional cases for which variations of up to several percent have been observed. There are several phenomena that can account for such variations, the most likely being some type of nuclear process which changes the abundance of one isotope relative to the others. For some of the lighter elements, the processes of distillation or

Natural isotopic compositions of the elements

Atomic no.	Element symbol	Mass no.	Isotopic abundance, %	Atomic no.	Element symbol	Mass no.	Isotopic abundance, %
1	H*	1	99.985	27	Co	59	100
		2	0.015	28	Ni	58	68.27
2	He*	3	0.000138			60	26.10
		4	99.999862			61	1.13
3	Li*	6	7.5			62	3.59
		7	92.5			64	0.91
4	Be	9	100	29	Cu	63	69.17
5	B*	10	19.9			65	30.83
		11	80.1	30	Zn	64	48.6
6	C*	12	98.90			66	27.9
		13	1.10			67	4.1
7	N*	14	99.634			68	18.8
		15	0.366			70	0.6
8	O*	16	99.762	31	Ga	69	60.1
		17	0.038			71	39.9
		18	0.200	32	Ge	70	20.5
9	F	19	100			72	27.4
10	Ne*†	20	90.51			73	7.8
		21	0.27			74	36.5
		22	9.22			76	7.8
11	Na	23	100	33	As	75	100
12	Mg	24	78.99	34	Se	74	0.9
		25	10.00			76	9.0
		26	11.01			77	7.6
13	Al	27	100			78	23.5
14	Si	28	92.23			80	49.6
		29	4.67			82	9.4
		30	3.10	35	Br	79	50.69
15	P	31	100			81	49.31
16	S*	32	95.02	36	Kr†	78	0.35
		33	0.75			80	2.25
		34	4.21			82	11.6
		36	0.02			83	11.5
17	Cl	35	75.77			84	57.0
		37	24.23			86	17.3
18	Ar†	36	0.337	37	Rb	85	72.165
		38	0.063			87	27.835
		40	99.600	38	Sr†	84	0.56
19	K	39	93.2581			86	9.86
		40	0.0117			87	7.00
		41	6.7302			88	82.58
20	Ca	40	96.941	39	Y	89	100
		42	0.647	40	Zr	90	51.45
		43	0.135			91	11.27
		44	2.086			92	17.17
		46	0.004			94	17.33
		48	0.187			96	2.78
21	Sc	45	100	41	Nb	93	100
22	Ti	46	8.0	42	Mo	92	14.84
		47	7.3			94	9.25
		48	73.8			95	15.92
		49	5.5			96	16.68
		50	5.4			97	9.55
23	V	50	0.250			98	24.13
		51	99.750			100	9.63
24	Cr	50	4.35	44	Ru	96	5.52
		52	83.79			98	1.88
		53	9.50			99	12.7
		54	2.36			100	12.6
25	Mn	55	100			101	17.0
26	Fe	54	5.8			102	31.6
		56	91.72			104	18.7
		57	2.2	45	Rh	103	100
		58	0.28				

Natural isotopic compositions of the elements (cont.)

Atomic no.	Element symbol	Mass no.	Isotopic abundance, %	Atomic no.	Element symbol	Mass no.	Isotopic abundance, %
46	Pd	102	1.02	60	Nd	142	27.13
		104	11.14			143	12.18
		105	22.33			144	23.80
		106	27.33			145	8.30
		108	24.46			146	17.19
		110	11.72			148	5.76
47	Ag	107	51.839			150	5.64
		109	48.161	62	Sm	144	3.1
48	Cd	106	1.25			147	15.0
		108	0.89			148	11.3
		110	12.49			149	13.8
		111	12.80			150	7.4
		112	24.13			152	26.7
		113	12.22			154	22.7
		114	28.73	63	Eu	151	47.8
		116	7.49			153	52.2
49	In	113	4.3	64	Gd	152	0.20
		115	95.7			154	2.18
50	Sn	112	1.0			155	14.80
		114	0.7			156	20.47
		115	0.4			157	15.65
		116	14.7			158	24.84
		117	7.7			160	21.86
		118	24.3	65	Tb	159	100
		119	8.6	66	Dy	156	0.06
		120	32.4			158	0.10
		122	4.6			160	2.34
		124	5.6			161	18.9
51	Sb	121	57.3			162	25.5
		123	42.7			163	24.9
52	Te	120	0.096			164	28.2
		122	2.60	67	Ho	165	100
		123	0.908	68	Er	162	0.14
		124	4.816			164	1.61
		125	7.14			166	33.6
		126	18.95			167	22.95
		128	31.69			168	26.8
		130	33.80			170	14.9
53	I	127	100	69	Ta	169	100
54	Xe†	124	0.10	70	Yb	168	0.13
		126	0.09			170	3.05
		128	1.91			171	14.3
		129	26.4			172	21.9
		130	4.1			173	16.12
		131	21.2			174	31.8
		132	26.9			176	12.7
		134	10.4	71	Lu	175	97.40
		136	8.9			176	2.60
55	Cs	133	100	72	Hf	174	0.16
56	Ba	130	0.106			176	5.2
		132	0.101			177	18.6
		134	2.417			178	27.1
		135	6.592			179	13.74
		136	7.854			180	35.2
		137	11.23	73	Ta	180	0.012
		138	71.70			181	99.988
57	La	138	0.09	74	W	180	0.13
		139	99.91			182	26.3
58	Ce	136	0.19			183	14.3
		138	0.25			184	30.67
		140	88.48			186	28.6
		142	11.08	75	Re	185	37.40
59	Pr	141	100			187	62.60

Natural isotopic compositions of the elements (cont.)

Atomic no.	Element symbol	Mass no.	Isotopic abundance, %	Atomic no.	Element symbol	Mass no.	Isotopic abundance, %
76	Os†	184	0.02	80	Hg	196	0.15
		186	1.58			198	10.1
		187	1.6			199	17.0
		188	13.3			200	23.1
		189	16.1			201	13.2
		190	26.4			202	29.65
		192	41.0			204	6.8
77	Ir	191	37.3	81	Tl	203	29.524
		193	62.7			205	70.467
78	Pt	190	0.01	82	Pb†	204	1.4
		192	0.79			206	24.1
		194	32.9			207	22.1
		195	33.8			208	52.4
		196	25.3	83	Bi	209	100
		198	7.2	90	Th	232	100
79	Au	197	100	92	U*	234	0.0055
						235	0.7200
						236	99.2745

*Isotopic composition may vary with sample depending upon geological or biological origin.
†Isotopic composition may vary with sample because some of the isotopes may be formed as a result of radioactive decay or nuclear reactions.

chemical exchange between different chemical compounds could be responsible for isotopic differences. *See* NUCLEAR REACTION; RADIOACTIVITY.

The lead isotopes ^{206}Pb, ^{207}Pb, and ^{208}Pb are stable and are end products of naturally decaying ^{238}U, ^{235}U, and ^{232}Th, respectively, whereas ^{204}Pb is not produced by any long-lived decay chain. Thus, the isotopic composition of lead samples will depend upon their prior contact with thorium and uranium.

The potassium isotope ^{40}K has a half-life of 1.28×10^9 years and decays by beta-ray emission to ^{40}Ar and by electron capture to ^{40}Ca, which are both stable. This can cause the argon in potassium-bearing minerals to differ in isotopic abundance from that found in air. It is possible to determine the age of rocks by measuring the ratio of their ^{40}K/^{40}Ar content. This ratio technique can also be used for rock samples which bear other long-lived, naturally occurring isotopes such as ^{87}Rb (rubidium), thorium, and uranium.

An interesting example of anomalous isotopic compositions has been observed in the Oklo uranium deposit in Gabon (western Africa). Based upon extensive research, it has been concluded that this is the site of a natural chain reaction that took place about 1.8×10^9 years ago. Much of the uranium in this formation has been depleted of the fissionable isotope ^{235}U. The isotopic composition of some of the other elements found at or near this deposit has also been altered as a result of fission, neutron absorption, and radioactive decay.

Use of separated isotopes. The areas in which separated (or enriched) isotopes are utilized have become fairly extensive, and a partial list includes nuclear research, nuclear power generation, nuclear weapons, nuclear medicine, and agricultural research.

Various methods are employed to prepare separated isotopes. Mass spectroscopy is used in the United States and the Soviet Union to prepare inventories of separated stable isotopes. Distillative, exchange, and electrolysis processes have been used to produce heavy water (enriched in deuterium, ^{2}H), which is used as a neutron moderator in some reactors. The uranium enrichment of ^{235}U, which is used as a fuel in nuclear reactors, has mainly been accomplished by using the process of gaseous diffusion in uranium hexafluoride gas in very large plants. This method has the disadvantage of requiring large power consumption. Techniques which can overcome this problem and which are finding increasing favor include centrifugal separation and laser isotope separation.

For many applications there is a need for separated radioactive isotopes. These are usually obtained through chemical separations of the desired element following production by means of a suitable nuclear reaction. Separated radioactive isotopes are used for a number of diagnostic studies in nuclear medicine, including the technique of positron tomography.

Studies of metabolism, drug utilization, and other reactions in living organisms can be done with stable isotopes such as ^{13}C, ^{15}N, ^{18}O, and ^{2}H. Molecular compounds are "spiked" with these isotopes, and the metabolized products are analyzed by using a mass spectrometer to measure the altered isotopic ratios. The use of separated isotopes as tracers for determining the content of a particular element in a sample by dilution techniques had broad applicability.

Atomic mass. Atomic masses are given in terms of an internationally accepted standard which at present defines an atomic mass unit (amu) as exactly equal to one-twelfth the mass of a neutral atom of ^{12}C in its electronic and nuclear ground states. On this scale the mass of a ^{12}C atom is equal to 12.0 amu. Atomic masses can be determined quite accurately by means of mass spectroscopy, and also by the use of nuclear reaction data with the aid of the Einstein mass-energy relation.

Atomic weight is the average mass per atom per amu of the natural isotopic composition of an element. Practically all atomic weights are now calculated by taking the sum of the products of the fractional isotopic abundances times their respective masses in amu. The atomic weight of a neutral atom of a nuclide is nearly equal to an integer value A, because the mass of both the neutron and proton is almost identically 1 amu. SEE ATOMIC WEIGHT.

Bibliography. E. Anders and M. Ebihara, Solar-system abundances of the elements, *Geochim. Cosmochim. Acta*, 46:2363–2380, 1982; E. Browne and R. B. Firestone, *Table of Radioactive Isotopes*, 1986; H. E. Duckworth et al., *Mass Spectroscopy*, 1986; N. E. Holden, R. L. Martin, and I. L. Barnes, Isotopic compositions of the elements 1981, *Pure Appl. Chem.*, 55:1119–1136, 1983; F. W. Walker, D. G. Mitler, and F. Feiner, *Knolls Atomic Power Laboratory Chart of the Nuclides*, General Electric Nuclear Energy Division, San Jose, California, December 1983.

ISOTONE
Henry E. Duckworth

One of two or more atoms which display a constant difference $A - Z$ between their mass number A and their atomic number Z. Thus, despite differences in the total number of nuclear constituents, the numbers of neutrons in the nuclei of isotones are the same.

The numbers of naturally occurring isotones provide useful evidence concerning the stability of particular neutron configurations. For example, the relatively large number (six and seven, respectively) of naturally occurring 50- and 82-neutron isotones suggests that these nuclear configurations are especially stable. On the other hand, from the fact that most atoms with odd numbers of neutrons are anisotonic, one may conclude that odd-neutron configurations are relatively unstable. SEE NUCLEAR STRUCTURE.

ISOBAR
Henry E. Duckworth

One of two or more atoms which have a common mass number A but which differ in atomic number Z. Thus, although isobars possess approximately equal masses, they differ in chemical properties; they are atoms of different elements. Isobars whose atomic numbers differ by unity cannot both be stable; one will inevitably decay into the other by β^- emission ($Z \rightarrow Z + 1$), β^+ emission ($Z \rightarrow Z - 1$), or electron capture ($Z \rightarrow Z - 1$). There are many examples of stable isobaric pairs, such as ^{50}Ti ($Z = 24$) and ^{50}Cr ($Z = 26$), and four examples of stable isobaric triplets. At most values of A the number of known radioactive isobars exceeds the number of stable ones. SEE ELECTRON CAPTURE; RADIOACTIVITY.

ATOMIC WEIGHT
CORNELIUS P. BROWNE

A measure of the average mass of the atoms of a chemical element. Most elements consist of a mixture of isotopes, which are atoms having the same chemical properties but different masses. Although the mixture of isotopes may depend on the origin of the sample, especially when some of the material is the product of radioactive decay, these variations are small in most natural materials. The assumption that all atoms of a given isotope have the same mass is a fundamental part of atomic theory. It is confirmed by the existence of chemical combining weights, that is, the fact that the ratio of the weights of elements which combine to form compounds is very nearly the ratio of small whole numbers. These ratios are indeed the ratios of the numbers of atoms in the molecules. This verifies Avogadro's hypothesis that a mass of any material having a weight equal to its molecular weight contains the same number of molecules. This number is called the Avogadro number. SEE ISOTOPE.

Relative weights. The relative weight of elements may be determined by measuring the chemical combining weights. For example, if it is found that 1 g of carbon combines with 1.332 g of oxygen to form CO, then the ratio of the atomic weight of oxygen to that of carbon is 1.332 (approximately 16/12). A second method of determining atomic weights is based on the fact that equal volumes of ideal gases at the same temperature and pressure contain equal numbers of molecules (Avogadro's law). By measuring the ratio of the densities of two real gases at a series of decreasing pressures and extrapolating to zero density, the ratio of molecular weights may be determined.

Relative atomic weights may be used to form a table of chemical atomic weights if it is agreed to assign a value for the weight of one element. For example, chemists used to choose a value of exactly 16 for the average atomic weight of naturally occurring oxygen. Carbon then had a weight of about 12, hydrogen a weight of about 1, and so forth. To obtain the mass of a single atom from the relative weights, it is necessary to determine the Avogadro number. There are many methods of doing this, one of the most accurate being the use of x-ray crystallography to measure the distance between atoms in a crystal lattice. This spacing, together with the overall dimensions of the crystal, gives the number of atoms in the crystal, whereas the weight of the crystal gives the fraction of an atomic weight. The ratio of the two is the Avogadro number.

Unified scale. To the physicist, the mass of an atom of a given isotope is the significant quantity. Since it is now possible to measure the masses of individual atoms, and in order to remove the uncertainties of isotopic composition, a Unified Scale of Atomic Masses has been adopted by international agreement. The chemical and physical scales are unified by defining the mass of the abundant isotope of carbon to be exactly 12 **u**, where **u** is the symbol for the unified atomic mass unit. On this scale all atoms have mass values that are nearly integers. The difference, called the mass deficit, represents the difference in mass between the sum of the masses of the constituent protons, neutrons, and electrons and the mass of the atom. This difference represents the binding energy of the nucleus. SEE NUCLEAR BINDING ENERGY.

Accurate mass values. Results from two methods of accurate measurements of individual atomic masses are combined to give the best values. These methods are mass spectroscopy and nuclear reaction energy measurements.

Mass spectroscopy. In mass spectroscopy a beam of ionized atoms is sent through a combination of electric and magnetic fields. For an ion of mass m and charge q moving perpendicular to an electric field of strength E, with velocity v, the radius r of the path is given by Eq. (1). In a magnetic field of strength B the radius r' is given by Eq. (2).

$$r = \frac{m}{q}\frac{v^2}{E} \qquad (1) \qquad\qquad r' = \frac{m}{q}\frac{v}{B} \qquad (2)$$

By proper choice of the pattern of E and B fields, ions of a given m/q ratio may be focused along a given line on a detector. In practice, rather than relying on absolute measurements of the fields and radii, the small differences in position of ions of nearly equal m/q are measured. Thus mass differences between ions of the unknown mass and known mass are determined. For example, the masses of $^{12}C^1H_4{}^+$ and $^{16}O^+$ are both nearly 16 u. Given the mass of carbon (12 u) and the mass of hydrogen, a very precise value for the mass of ^{16}O may then be found.

Nuclear reaction energies. Differences in atomic masses are measured very accurately by using the conversion of mass to energy in a nuclear reaction. From special relativity theory $E = (\Delta m)c^2$, where E is the energy released or absorbed in the reaction, c is a fundamental constant (the velocity of light), and Δm is the difference between the total mass before and after the reaction. The energy may appear in the form of kinetic energy of the recoiling product masses or as electromagnetic energy in the form of gamma rays or x-rays, or a combination of these forms. For example, if a hydrogen atom ^1H and a neutron n combine to form a deuterium atom ^2H, some 2.2 MeV of energy is released as a gamma ray. This energy may be very accurately measured by diffraction techniques or with solid-state detectors. Again, if ^{12}C atoms are bombarded by high-speed ^2H ions, the reaction $^{12}C + {}^2H \rightarrow {}^{13}C + {}^1H$ occurs. The kinetic energies of the ^2H ions before the reaction and of the ^1H ions produced may be precisely measured by deflections in magnetic fields. From these and the direction of the ^1H ions, the energy of the ^{13}C recoil may be deduced and thus the net change in kinetic energy. From this, Δm is calculated. SEE NUCLEAR REACTION; NUCLEAR SPECTRA.

Mass tables. When the mass differences from enough different reactions and mass spectroscopy ion pairs are available, a set of equations having the masses as unknowns may be written. There are actually many more known differences than masses, so the equations may be solved for all the unknown masses. All the nuclear reaction and mass spectroscopic data are combined to give the Table of Atomic Masses. There are thousands of input values, and great computing power is required. The table is periodically revised as new measurements are made and new unstable nuclear species are discovered. At present the masses of most stable atoms are known to 1 part in 10^7, and the masses of a great many artificially produced unstable atoms are known to 1 part in 10^6 or better.

MASS DEFECT
W. W. Watson

The difference between the mass of an atom and the sum of the masses of its individual components in the free (unbound) state. The mass of an atom is always less than the total mass of its constituent particles; this means, according to Albert Einstein's well-known formula, that an energy of $E = mc^2$ has been released in the process of combination, where m is the difference between the total mass of the constituent particles and the mass of the atom, and c is the velocity of light.

The mass defect, when expressed in energy units, is called the binding energy, a term which is perhaps more commonly used. SEE NUCLEAR BINDING ENERGY.

NUCLEAR BINDING ENERGY
Henry E. Duckworth and D. H. Wilkinson
D. H. Wilkinson wrote the last paragraph.

The amount by which the mass of an atom is less than the sum of the masses of its constituent protons, neutrons, and electrons expressed in units of energy. This energy difference accounts for the stability of the atom. In principle, the binding energy is the amount of energy which was released when the several atomic constituents came together to form the atom. Most of the binding energy is associated with the nuclear constituents (protons and neutrons), or nucleons, and it is customary to regard this quantity as a measure of the stability of the nucleus alone. SEE NUCLEAR STRUCTURE.

A widely used term, the binding energy (BE) per nucleon, is defined by the equation below,

$$\text{BE/nucleon} = \frac{[ZH + (A - Z)n - {}_zM^A]c^2}{A}$$

where $_ZM^A$ represents the mass of an atom of mass number A and atomic number Z, H and n are the masses of the hydrogen atom and neutron, respectively, and c is the velocity of light. The binding energies of the orbital electrons, here practically neglected, are not only small, but increase with Z in a gradual manner; thus the BE/nucleon gives an accurate picture of the variations and tends in nuclear stability. The **illustration** shows the BE/nucleon (in megaelectronvolts) plotted against mass number for $A > 40$.

The BE/nucleon curve at certain values of A suddenly changes slope in such a direction as to indicate that the nuclear stability has abruptly deteriorated. These turning points coincide with particularly stable configurations, or nuclear shells, to which additional nucleons are rather loosely bound. Thus there is a sudden turning of the curve over $A = 52$ (28 neutrons); the maximum occurs in the nickel region (28 protons, $\sim A = 60$); the stability rapidly deteriorates beyond $A = 90$ (50 neutrons); there is a slightly greater than normal stability in the tin region (50 protons, $\sim A = 118$); the stability deteriorates beyond $A = 140$ (82 neutrons) and beyond $A = 208$ (82 protons plus 126 neutrons).

The BE/nucleon is remarkably uniform, lying for most atoms in the range 5–9 MeV. This near constancy is evidence that nucleons interact only with near neighbors; that is, nuclear forces are saturated.

The binding energy, when expressed in mass units, is known as the mass defect, a term sometimes incorrectly applied to quantity $M - A$, where M is the mass of the atom. S*ee* M*ass*
D*efect*.

The term binding energy is sometimes also used to describe the energy which must be supplied to a nucleus in order to remove a specified particle to infinity, for example, a neutron, proton, or alpha particle. A more appropriate term for this energy is the separation energy. This quantity varies greatly from nucleus to nucleus and from particle to particle. For example, the binding energies for a neutron, a proton, and a deuteron in O^{16} are 15.67, 12.13, and 20.74 MeV, respectively, while the corresponding energies in O^{17} are 4.14, 13.78, and 14.04 MeV, respectively. The usual order of neutron or proton separation energy is 7–9 MeV for most of the periodic table.

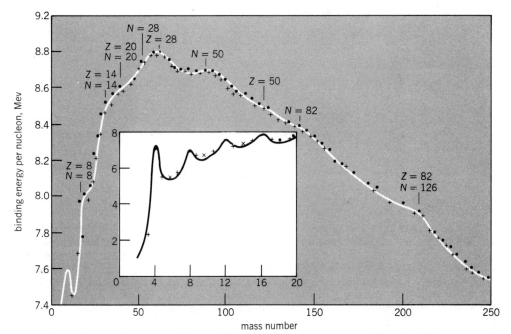

Graph of binding energy per nucleon (in megaelectronvolts) plotted against mass number. N = number of neutrons. (After A. H. Wapstra, Isotopic measure, part 1, where A is less than 34, Physica, 21:367–384, 1955)

GARVEY-KELSON MASS RELATIONS
GERALD T. GARVEY

A set of equations relating the masses of atomic nuclei with slightly different numbers of neutrons and protons. These relationships are derived from general physical arguments and are observed to be well obeyed. They can be employed to predict the masses of exotic nuclei which occur in stellar processes but which are presently impossible to create in the laboratory.

The fact that the masses of nuclei which differ slightly in their number of neutrons and protons are very closely related was quantitatively characterized and explained by G. T. Garvey and I. Kelson. The mass of a nucleus, $M(N,Z)$, is the total energy content of the lowest energy state of the system with N neutrons and Z protons. This is the state in which the nucleus usually occurs. In addition to the straightforward mass contribution coming from the mass of individual constituents, there is a not precisely understood contribution from the interaction energy between the constituent neutrons and protons.

The simplest form that the Garvey-Kelson mass relations take is given by the equation below, where N_0 and Z_0 can take any value provided that the absolute value of $N_0 - Z_0$ is equal

$$M(N_0, Z_0 - 1) - M(N_0 - 1, Z_0) + M(N_0 - 1, Z + 1)$$
$$- M(N_0 + 1, Z_0 - 1) + M(N_0 + 1, Z) - M(N_0, Z_0 + 1) = 0$$

to or greater than 2. The equation is written so that all neutron-neutron, neutron-proton, and proton-proton interactions cancel. A simple depiction of the form of the equation is shown in the **illustration**. The equation is observed to be obeyed to within a standard deviation of 0.18×10^6 eV when it is applied to the several hundred cases in which all six nuclear masses are known. A characteristic size of the mass difference of neighboring nuclei is typically 5×10^6 eV.

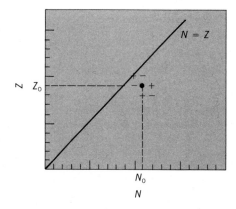

Plot of the coefficients of the masses $M(N,Z)$ in the equation which appears in the text.

If all masses in a specific case of the above equation are known except one, the equation can be used to predict that unknown mass. Thus, starting with relatively few known masses, the equation can be applied over and over again to fill in a complete mass table. Thus, the particle stability of such unusual species as ^{8}He, ^{11}Li, and ^{14}B has been predicted and subsequently observed in laboratory experiments. Further, the stability of species as neutron-rich as ^{70}Ca are also predicted. While it is unlikely that such species can be found in the laboratory, their potential existence in very neutron-rich stellar environments has significant consequences for nucleosynthesis of the elements. *SEE NUCLEAR STRUCTURE.*

Bibliography. G. T. Garvey et al., A set of nuclear mass relations and a resultant mass table, *Rev. Mod. Phys.*, 41:510–580, 1969.

MAGIC NUMBERS
D. Allan Bromley

Numbers of neutrons or protons in nuclei which correspond to particularly stable structures and closed shells. In nature the magic numbers for both neutrons and protons are 2, 8, 14, 20, 50, and 82. The next neutron magic numbers are 126 and 184; the next proton magic numbers are expected to be 114 and perhaps 164. Nuclei having magic numbers of both neutrons and protons have been found to have spherical equilibrium shapes and special stability; they have anomalously low capture probabilities for additional neutrons and protons respectively. *See* Nuclear structure.

Magic number effects were first observed as systematic trends in neutron-capture cross sections (see **illus**.), in nuclear electric quadrupole moments, in isotopic and isotonic abundances, and in nuclear magnetic dipole moments. The illustration shows neutron-capture cross sections for a wide range of nuclei; these data provide clear evidence for the magic character of neutron numbers 50, 82, and 126. These data led in turn to the establishment of the nuclear shell model in analogy to the electronic shell structure of atoms; in both cases the magic numbers of nucleons (neutrons and protons) and of electrons are those required to close the shells, that is, to complete particularly stable orbital structures. In the nuclear case, however, the magic numbers are quite different from those in the atomic case because of the strong nuclear spin-orbit effect; the energy of a nuclear configuration depends sensitively upon the relative orientations of the orbital and the spin angular momenta. *See* Neutron spectrometry; Nuclear moments.

The abundant isotopes of helium, oxygen, silicon, calcium, tin, and lead owe their special nuclear stability to magic numbers of protons and (in all cases except tin) to magic numbers of neutrons as well. *See* Isotope.

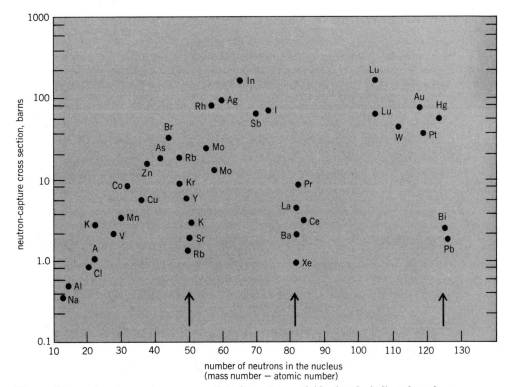

Measured thermal neutron-capture cross sections for various nuclei having the indicated number of neutrons already present. Arrows indicate magic neutron numbers 50, 82, and 126.

The hope of finding stable or quasistable superheavy nuclear species well beyond any occurring naturally, and beyond the transuranic species (between elements number 92 and 109) thus far studied with heavy-ion accelerators, depends critically upon the assumed stability of the doubly magic species having 114 protons and 184 neutrons. SEE SUPERTRANSURANICS.

ANALOG STATES
NELSON STEIN

Certain states belonging to neighboring nuclear isobars and possessing identical structure except for the transformation of one or more neutrons into the same number of protons. Hundreds of examples of analog states (also known as isobaric analog states) have been observed throughout the periodic table, and their existence is considered to be a fundamental property of nuclear structure. SEE ISOBAR; NUCLEAR STRUCTURE.

Starting with a nucleus of N number of neutrons and Z number of protons in a particular state of energy, angular momentum, parity, and so forth, an analog state may be constructed in the neighboring isobar with $N-1$ neutrons and $Z+1$ protons by replacing a neutron with a proton in the same orbit and then coupling the proton to the remaining nucleons in exactly the same fashion as the original neutron. Under the assumption of charge independence of nuclear forces, the only difference between a state and its analog should arise from the increase in the Coulomb interactions attributed to the increased number of protons in the nucleus. Since the total energy of a state has a Coulomb contribution, the most outstanding difference between analog states is their Coulomb energy. When this difference is taken into account, then the total energies of corresponding analog states become nearly equal. Another, smaller effect which must also be considered is the neutron-proton mass difference of 0.782 MeV. However, even then, all properties of analog states do not become precisely identical because there remain subtle differences in the behavior of neutrons and protons due to electromagnetic effects.

Isobaric spin. The simplest nucleus is hydrogen, which consists of a single proton; its analog is a free neutron. In the framework of the nuclear shell model, states of composite nuclei are built up from the individual neutrons and protons that are considered to be moving in orbits which can be calculated by assuming a suitable nuclear potential energy well. Such a framework proves ideal for understanding analog states, when one introduces the concept of isobaric spin, also termed isotopic spin or isospin. The neutron and proton are endowed with a quantum number for isospin of $t = \frac{1}{2}$ (just as they have $s = \frac{1}{2}$ spin angular momentum). The z-projection of isospin is then defined as $t_z = +\frac{1}{2}$ for neutrons and $t_z = -\frac{1}{2}$ for protons. The isospin of a nuclear state with N neutrons and Z protons is constructed by combining the individual neutron and proton isospins, so that the total z-projection is $T_z = (N-Z)/2$, and by analogy with angular momentum, the isospin T must be greater than or equal to T_z up to a maximum of $(N + Z)/2$. States which are analogs have the same total isospin T and differ only in their value of T_z. Interchange of a single neutron and proton either raises or lowers T_z by one unit. SEE ISOBARIC SPIN.

Nuclei with N=Z. In nuclei with $N=Z$, the value of T_z is 0, and the ground and lowest excited states are characterized by $T = 0$. These states have no analogs, since replacement of a neutron by a proton, or the reverse, leads to $T_z = -1$ or $+1$, respectively, for which a value of $T=0$ cannot occur. Physically, this means that when a neutron is transformed into a proton in a state with $T = 0$, the structure cannot be preserved merely by correcting for the difference in Coulomb energy. The neutron-proton exchange must also change one or more additional measurable properties of the nuclear state.

Figure 1 shows a portion of the energy level diagrams for three mass-12 nuclei: ^{12}B, ^{12}C, and ^{12}N. All of the states up to 15.11-MeV excitation in ^{12}C have $T = 0$. The first $T = 1$ state at 15.11 MeV has spin 1 and positive parity. There should also exist two analog states with $T = 1$ and $T_z = +1$ and -1 respectively, which along with the 15.11-MeV state in ^{12}C form an isobaric analog triplet. These states are just the ground states of ^{12}B and ^{12}N, which are also known from experiments to have spin 1 and positive parity. In Fig. 1 the energy scale has been adjusted by the Coulomb energy differences of ^{12}B, ^{12}C, and ^{12}N and for the neutron-proton mass difference. It

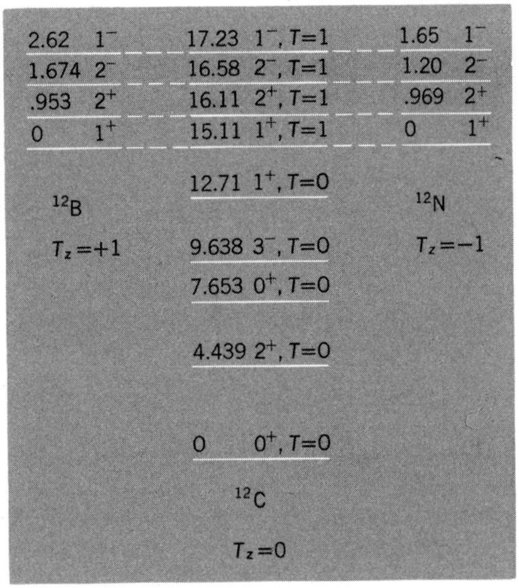

2.62 1⁻	17.23 1⁻, T=1	1.65 1⁻
1.674 2⁻	16.58 2⁻, T=1	1.20 2⁻
.953 2⁺	16.11 2⁺, T=1	.969 2⁺
0 1⁺	15.11 1⁺, T=1	0 1⁺
	12.71 1⁺, T=0	
¹²B		¹²N
$T_z = +1$	9.638 3⁻, T=0	$T_z = -1$
	7.653 0⁺, T=0	
	4.439 2⁺, T=0	
	0 0⁺, T=0	
	¹²C	
	$T_z = 0$	

Fig. 1. Partial level diagram of mass-12 isobars. Broken lines indicate analog states that are members of isospin triplets. Numbers at left of levels give energy in MeV, and symbols at right give spin and parity.

is assumed that all states of $T = 1$ in ^{12}C have corresponding analog states in ^{12}B and ^{12}N, forming a series of isobaric analog triplets. In a similar fashion at higher excitation energy, $T = 2$ states should occur in ^{12}C. Since $T = 2$ is compatible with $T_z = -2, -1, 0, +1$, and $+2$, such states should be members of isobaric analog quintets ranging over five isobars with mass 12. SEE PARITY; SPIN.

Nuclei with neutron excess (N>Z). Above ^{3}He, all stable nuclei have $N \geq Z$ because of the Coulomb repulsion of the protons. In cases such as ^{13}C and ^{27}Al where $N = Z + 1$, the isospin and its z-component have the value of ½, and all states are members of isobaric analog doublets with the second member residing in the mirror nuclei ^{13}N and ^{27}Si respectively, both with $T_z = -½$.

Heavy nuclei. Until about 1961, it was thought that analog states were a property only of light-mass nuclei with mass number A less than about 40. This belief was based on the supposition that the increased Coulomb energy that results from the increased number of protons in heavy nuclei would finally destroy the isobaric symmetry so that the analog states at high excitation energy would not remain as distinct measurable entities. However, in a number of experiments not specifically designed to search for analog states, dramatic evidence was found for their existence throughout the periodic table up to the heaviest known nuclei. This discovery was one of the most important in nuclear physics during the decade of the 1960s. It led to a large number of experiments and a great increase in knowledge of nuclear structure.

Mass 208 system. The mass 208 system provides an example of analog states in isobars with large neutron excess. **Figure 2** shows a partial level diagram of ^{208}Pb and its neighbor ^{208}Bi. The z-component of isospin for ^{208}Pb is $T_z = 22$, and all states up to about 23 MeV of excitation also have $T = 22$. The value of T_z for ^{208}Bi is 21, which is also the value of T for all states up to 15.21 MeV. Just at that energy, however, an excitation is observed in the ^{208}Bi spectrum which possesses many of the properties of the ground state of ^{208}Pb. Moreover, if the Coulomb energy difference of 18.86 MeV is subtracted from ^{208}Bi, then the total energy of each of the two states is the same except for the neutron-proton mass difference. This leads to the identification of ths excitation at 15.21 MeV in ^{208}Bi as the $T_z = 21$ analog of the $T = T_z = 22$ ground state of ^{208}Pb. Similarly, higher excitations of ^{208}Bi can be identified as analog states of excited states of ^{208}Pb, with nearly equal energies after correction for Coulomb energy and neutron-proton mass differences.

In neutron excess nuclei, the states with $T = T_z$ are often referred to as the parent states and the states with $T > T_z$ as the analog states. The microscopic structure of this parent-analog

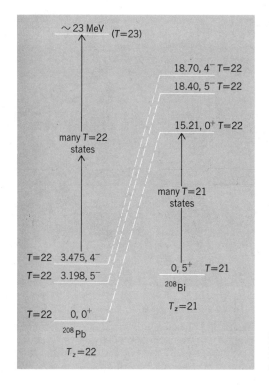

Fig. 2. Partial level diagram of ^{208}Pb and ^{208}Bi. Broken lines indicate parent-analog state pairs. Numbers at left of levels give energy in MeV, and symbols at right give spin and parity.

state pair in terms of the nuclear shell model is shown in **Fig. 3**. The ground state of ^{208}Pb is characterized by neutrons and protons filling all available orbits up to the closed shells at $Z = 82$ and $N = 126$. For the analog state in ^{208}Bi, one of the neutrons of ^{208}Pb is replaced by a proton in the same orbit. However, not all of the neutrons of ^{208}Pb are allowed to undergo this change. The Pauli exclusion principle limits the number of protons that may occupy a particular orbit, so that only those neutrons occupying orbits unfilled by protons are available for the interchange. Moreover, of those neutrons that are available, all have equal probability, so that the structure of the analog state consists of a superposition of many microscopic configurations (shown in Fig. 3b) that look almost like ^{208}Pb. Each microscopic configuration differs from ^{208}Pb by a proton occupying an unfilled orbit and a corresponding neutron hole in a neutron orbit. The numbers under the square root signs in Fig. 3b are the amplitudes which multiply each configuration. The amplitudes are determined by the number of neutrons that fill each orbit. Thus the structure of an analog state in a neutron excess nucleus is characterized by each neutron, which occupies an orbit unfilled by protons, appearing part of the time as a proton in that orbit. The greater the neutron excess, the more neutrons available and the less time needed for each to spend as a proton.

Width of states. An important distinction between analog states in light- and heavy-mass nuclei is their width. In nuclei with low mass, analog states are usually very narrow, not more than tens or hundreds of electron volts wide, and their widths are often comparable to those of their corresponding parent states. However, with increasing nuclear mass, analog state widths also increase, attaining values of about 200,000 eV in the lead region, while their corresponding parent states remain very narrow. Associated with the broadening of analog states in medium- and heavy-mass nuclei is the fact that they mix with, and their properties become fragmented over, many narrow states with one unit less of isospin which reside in the region of the analog state. In this way the analog state ceases to be an individual nuclear state, but rather it develops fine structures and is characterized by a distribution of its properties over the underlying states which span the region of the spectrum corresponding to the energy and width of the analog state.

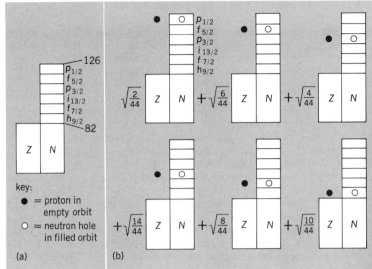

Fig. 3. Shell model structure of a parent-analog state pair. Letters indicate orbital angular momentum of neutron orbits ($s, p, d, f, g, \ldots$, for $l = 0, 1, 2, 3, 4, \ldots$); the subscript gives the j value. (a) The parent state, the ground state of ^{208}Pb. (b) Its analog state in ^{208}Bi.

Applications. The discovery of analog states in heavy nuclei has provided a powerful tool for studying nuclear structure that was previously unforeseen. First, it has enabled the study of Coulomb energies to be extended from low-mass nuclei to the heaviest nuclei; and second, because of their high excitation energy, analog states normally are able to decay by particle emission, unlike their corresponding parent states which usually decay only by γ-ray emission to the ground state. By measuring the properties of the particles that are emitted from the analog state, unique information is derived about the parent state which often cannot be obtained in any other fashion. *See* RADIOACTIVITY.

Bibliography. J. D. Anderson et al., *Nuclear Isospin*, 1969; A. Bohr and B. R. Mottelson, *Nuclear Structure*, vol. 1, 1969, vol. 2, 1976; B. L. Cohen, *Concepts of Nuclear Physics*, 1971; K. S. Krane, *Introductory Nuclear Physics*, 1987; D. H. Wilkinson (ed.), *Isospin in Nuclear Physics*, 1969.

NUCLEAR ISOMERISM
RUSSELL BETTS

The existence of excited states of atomic nuclei with unusually long lifetimes. A nucleus may exist in an excited quantum state with well-defined excitation energy (E_x), spin (J), and parity (π). Such states are unstable and decay, usually by the emission of electromagnetic radiation (gamma rays), to lower excited states or to the ground state of the nucleus. The rate at which this decay takes place is characterized by a half-life ($\tau_{1/2}$), the time in which one-half of a large number of nuclei, each in the same excited state, will have decayed. If the lifetime of a specific excited state is unusually long, compared with the lifetimes of other excited states in the same nucleus, the state is said to be isomeric. The definition of the boundary between isomeric and normal decays is arbitrary, and the term is therefore used loosely. *See* PARITY; SPIN.

Spin isomerism. The predominant decay mode of excited nuclear states is by gamma-ray emission. The rate at which this process occurs is determined largely by the spins, parities, and excitation energies of the decaying state and of those to which it is decaying. In particular, the rate is extremely sensitive to the difference in the spins of initial and final states and to the difference in excitation energies. Both extremely large spin differences and extremely small energy differences can result in a slowing of the gamma-ray emission by many orders of magnitude, resulting in some excited states having unusually long lifetimes and therefore being termed isomeric.

Occurrence. Isomeric states have been observed to occur in almost all known nuclei. However, they occur predominantly in regions of nuclei with neutron numbers N and proton numbers Z close to the so-called magic numbers at which shell closures occur. This observation is taken as important evidence for the correctness of the nuclear shell model which predicts that high-spin, and therefore isomeric, states should occur at quite low excitation energies in such nuclei. SEE MAGIC NUMBERS.

Examples. Three examples of isomeric states are shown in the **illustration**. In the case of ^{90}Zr (illus. a), the 2.319-MeV, $J^\pi = 5^-$ state has a half-life of 809 milliseconds, compared to the much shorter lifetimes of 93 femtoseconds, and 61 nanoseconds for the 2.186-MeV and 1.761-MeV states, respectively. The spin difference of 5 between the 2.319-MeV state and the ground and 1.761 $J^\pi = 0^+$ states, together with the spin difference of 3 and small energy difference between the 2.319-MeV state and the 2.186-MeV, $J^\pi = 2^+$ state, produce this long lifetime.

For ^{42}Sc (illus. b), the gamma decay of the $J^\pi = 7^+$ state at 0.617 MeV is so retarded by the large spin changes involved that the intrinsically much slower β^+ decay process can take place, resulting in a half-life of 62 s.

Yet another process takes place in the case of the high-spin isomer in ^{212}Po (illus. c), which decays by alpha-particle emission with a half-life of 45 s rather than gamma decays.

The common feature of all these examples is the slowing of the gamma-ray emission process due to the high spin of the isomeric state.

Other mechanisms. Not all isomers are the spin isomers described above. Two other types of isomers have been identified. The first of these arises from the fact that some excited nuclear states represent a drastic change in shape of the nucleus from the shape of the ground state. In many cases this extremely deformed shape displays unusual stability, and states with this shape are therefore isomeric. A particularly important class of these shape isomers is observed in the decay of heavy nuclei by fission, and the study of such fission isomers has been the subject of intensive effort. The possibility that nuclei may undergo sudden changes of shape at high rotational velocities has spurred searches for isomers with extremely high spin which may also be termed shape isomers. SEE NUCLEAR FISSION.

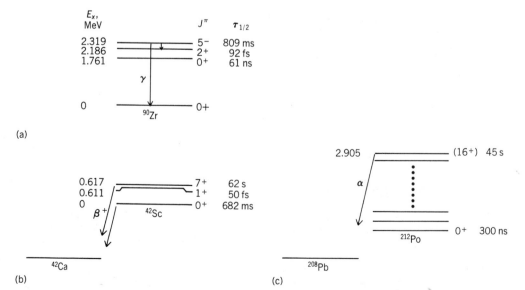

Examples of nuclear isomerism in (a) ^{90}Zr, (b) ^{42}Sc, and (c) ^{212}Po. (After C. M. Lederer and V. S. Shirley, *Table of Isotopes*, 7th ed., John Wiley and Sons, 1978)

A more esoteric form of isomer has also been observed, the so-called pairing isomer, which results from differences in the microscopic motions of the constituent nucleons in the nucleus. A state of this type has a quite different character from the ground state of the nucleus, and is therefore also termed isomeric. SEE NUCLEAR STRUCTURE.

Bibliography. B. L. Cohen, *Concepts of Nuclear Physics*, 1971; K. S. Krane, *Introductory Nuclear Physics*, 1987; C. M. Lederer and V. S. Shirley, *Table of Isotopes*, 7th ed., 1978; K. Siegbahn (ed.), *Alpha, Beta and Gamma-Ray Spectroscopy*, vols. 1 and 2, 1965.

GIANT NUCLEAR RESONANCES
STANLEY S. HANNA

Systematic excitations of the atomic nucleus which occur with great strength in a concentrated energy region. When high-energy electromagnetic radiation (gamma radiation) impinges on a nucleus, it can be strongly absorbed into a number of high-lying resonances: one in which the nucleus is excited in a dipole mode of oscillation that is electric in nature (E1), a second dipole model that is magnetic in nature (M1), and a third excitation that can be identified as an electric quadrupole oscillation (E2). When excited into these resonances, the nucleus can deexcite by emitting gamma radiation or, if energetically allowed, by emitting particles, especially protons and neutrons. If proton and neutron emissions are allowed, it is possible to study these resonances by the inverse process in which the proton or the neutron is captured in the giant resonance and gamma radiation is emitted. These giant resonances can also be excited by inelastic scattering of electrons and other particles such as protons, deuterons, ^{3}He, and alpha particles. For inelastic excitation with particles, greater momentum can be imparted to the nucleus, and higher modes such as octupole oscillations can be excited. The particles can also excite the electric monopole vibration (E0), which corresponds to a "breathing" mode. SEE GAMMA RAYS; MULTIPOLE RADIATION; NUCLEAR MOMENTS.

Giant E1 resonances. The giant E1 resonance has long been the object of intensive study. The three important properties which characterize it are its systematic occurrence in all nuclei, its great strength, and its localized nature. The reactions that have been used to study the giant E1 resonances are the following:

1. The photoneutron process (γ,n). In this case, a gamma ray is absorbed by the nucleus, and the subsequent neutron emission from the nucleus is measured as a function of the gamma-ray energy. A typical E1 giant resonance obtained with this method is shown in **Fig. 1**.

2. The inverse proton capture reaction (p,γ). In this type of experiment, the gamma-ray yield is measured as a function of the incident proton energy. A typical giant resonance is shown in **Fig. 2**.

3. Inelastic electron scattering (e,e'). In this case, the energy distribution of scattered electrons is measured at a fixed energy of the incident electrons. In general, a detailed analysis is required to separate the giant E1 resonance from other giant excitations.

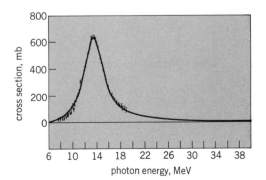

Fig. 1. Giant E1 resonance in ^{208}Pb observed with the photonuclear reaction, 1 millibarn = 10^{-31} m^2. (*After B. L. Berman and S. C. Fultz, Measurements of the giant dipole resonance with monoenergetic photons, Rev. Mod. Phys., 47:713–761, 1975*)

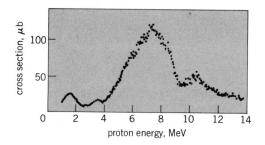

Fig. 2. Giant E1 resonance at 7.2 MeV in ^{12}C observed with the "inverse" proton capture reaction, ^{11}B$(p,\gamma)^{12}$C. 1 microbarn = 10^{-34} m^2. (*After R. G. Allas et al., Radiative capture of protons by B^{11} and the giant dipole resonance in C^{12}, Nucl. Phys., 58:122–144, 1964*)

The dominant features of the E1 resonance may be summarized as follows (see **table**):

1. In nuclei of medium and heavy mass, the E1 resonance occurs at an excitation energy E_x of $77/A^{1/3}$ MeV, where A is the mass number of the nucleus. However, in the light nuclei below ^{40}Ca, the energy of the resonance falls off to values of $50/A^{1/3}$ for the very lightest nuclei.

2. The giant E1 resonance "exhausts" the theoretically allowed limit for absorption of gamma radiation by an E1 mode.

3. Perhaps the most impressive feature of the E1 resonance is its localized nature, despite the fact that it occurs at a high excitation energy where the nucleus can decay in many ways. From the lightest to the heaviest nuclei, the width Γ is given by $\Gamma/E_x \simeq 1/5$, with some important exceptions. The best-established broadening of the resonance is that caused by the deformation of the nucleus.

The giant E1 resonance can be described in terms of characteristic single-particle excitations in the nucleus which absorb most of the E1 strength. Alternatively, it can be attributed to collective oscillations of the nucleus. In the latter picture, the basic mode is one in which all the protons in the nucleus vibrate against all the neutrons. Nuclear theory has been successful in showing the equivalence of these two pictures and in accounting for the prominent features of the E1 resonance.

Summary of giant resonance properties*

Property	E1	M1†	E2
E_x	$77/A^{1/3}$	$40/A^{1/3}$	$63/A^{1/3}$
Strength	Theoretical limit	≃Theoretical limit	≤ Theoretical limit
Γ/E_x	0.2	≤ 0.2	≃ 0.2

*E_x is the excitation energy in MeV and Γ is the width of the resonance. A is the number of protons and neutrons in the nucleus.
†Properties established for light nuclei.

Giant M1 resonances. Information on the giant M1 resonances has become rather extensive and exists for nuclei from mass 8 to 208. The methods that have been used to study the M1 resonance can exists for nuclei from mass 8 to 208. However, the resonance and its properties have been well established only for the light nuclei. The methods that have been used to study the M1 resonance can be summarized as follows:

1. The inverse proton capture reaction (p,γ). Some resonances have also been studied by reactions of the type $(X,Y\gamma)$ where X and Y stand for nuclear particles.

2. Gamma-ray fluorescence (γ,γ'). In these experiments the nucleus is excited by means of inelastic gamma-ray scattering.

3. Inelastic electron scattering at 180°. The use of 180° scattering is necessary to sort out magnetic from electric multipoles. A typical case of an M1 resonance observed with electron scattering is shown in **Fig. 3**.

4. The photoneutron process (γ,n). This process has been used to give valuable information just above the neutron threshold in heavy nuclei. Information comes also from (n,γ) results.

The properties of the M1 resonances are summarized in the table. The giant M1 resonance can be described in terms of single-particle excitations in the nucleus which produce a change in the direction of the particle spin. In the collective picture, the oscillation can be thought of as one in which particles with one spin direction vibrate against those with the opposite spin direction.

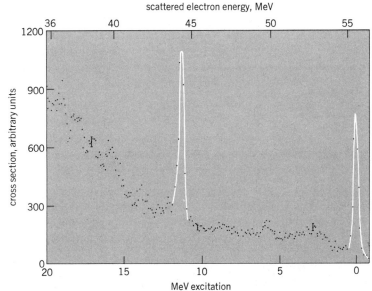

Fig. 3. Giant M1 resonance at 11.3 MeV in ^{20}Ne observed with inelastic electron scattering at 180°. Energy of incident electrons is 56.0 MeV. (*After L. W. Fagg, Electroexcitation of nuclear magnetic dipole transitions, Rev. Mod. Phys., 47:683–711, 1975*)

Giant E2 resonances. Interest in the study of the giant E2 resonances stemmed from the observation (in 1971) of these resonances below the giant E1 resonance in electron scattering and their identification in inelastic proton scattering. E2 strength had been seen earlier in proton capture experiments in the lighter nuclei. However, the later observations established the E2 resonance as a compact, systematic excitation occurring in all nuclei. The methods which have been used to study E2 strength can be classified as follows:

1. Inelastic electron scattering (e,e'). Since the momentum imparted to the nucleus can be easily varied, electron scattering provides a sensitive method for studying the excitations.

2. Inelastic scattering by nuclear particles (X,X'). the E2 resonances have been observed in the inelastic scattering of a variety of particles, such as protons, deuterons, ^{3}He, alpha particles, heavy ions, and pions. Systematic properties have been developed throughout the whole nuclear table of isotopes, as given in the table. A typical E2 resonance is shown in **Fig. 4**.

3. The inverse capture reaction (X,γ). The (p,γ) work has become much more definitive by the use of polarized protons. Important information has been obtained from the (γ,p) process. A great deal of evidence has also been accumulated from the (α,γ) reaction.

As for the dipole resonances, the E2 resonances can be successfully described by means of single-particle excitations or with a collective picture in which the nucleus undergoes shape oscillations of a quadrupole nature.

Giant E0 resonances. The importance of these resonances lies in the fact that the energy at which they occur determines the incompressibility of a nucleus, a property that is basic

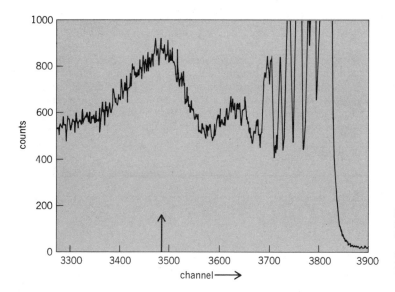

Fig. 4. Giant E2 resonance at channel 3490 (equal to an excitation energy of 13.3 MeV) observed in the inelastic scattering of 152-MeV alpha particles from ^{120}Sn.

to understanding the force between neutrons and protons. These resonances became well established through a series of measurements on the inelastic excitation of a nucleus by alpha particles. It was necessary to observe the scattered alpha particles at very small angles in order to separate the E0 resonance from the other resonances discussed above. An illustration of this separation is shown in **Fig. 5**, where there is a marked increase in the E0 (0^+) peak relative to the E2 (2^+) peak in going from a scattering angle $\theta = 4°$ to $\theta = 0°$. Many of the properties of the E0 resonance

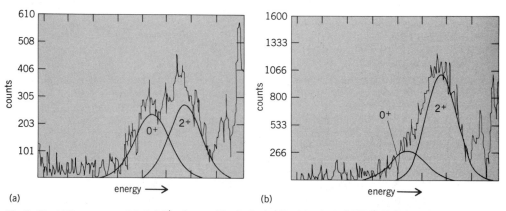

Fig. 5. Giant E0 resonance, labeled 0^+, observed in the inelastic scattering of 127-MeV alpha particles from ^{208}Pb. (a) Scattering angle $\theta = 0°$. (b) $\theta = 4°$. (*After F. E. Bertrand, ed., Giant Multipole Resonances, vol. 1, pp. 113–137, Harwood Academic Publishers, 1980*)

have been established, but active research has continued to extend and quantify the results. The resonances are being studied by inelastic excitation by nuclear particles and electrons, and their decay properties are being investigated. *See* NUCLEAR REACTION; NUCLEAR SPECTRA; NUCLEAR STRUCTURE.

Bibliography. B. L. Berman and S. C. Fultz, Measurements of the giant dipole resonance with monoenergetic photons, *Rev. Mod. Phys.*, 47:713–761, 1975; F. E. Bertrand, Excitation of giant multipole resonances through inelastic scattering, *Annu. Rev. Nucl. Sci.*, 26:457–509, 1976; G. E. Bertsch and R. A. Broglia, Giant resonances in hot nuclei, *Phys. Today*, 39(8):44–52, August 1986; L. W. Fagg, Electroexcitation of nuclear magnetic dipole transitions, *Rev. Mod. Phys.*, 47:683–711, 1975; S. S. Hanna, Giant multipole resonances, in *Photonuclear Reactions*, International School on Electro- and Photonuclear Reactions, vol. 1, pp. 275–339, 1976; K. A. Snover, Giant resonances in excited nuclei, *Ann. Rev. Nucl. Part. Sci.*, 36:545–603, 1986; D. H . Youngblood, The giant monopole resonance: An experimental review, in *Topical Conference on Giant Multipole Resonances*, Oak Ridge, Tennessee, 1979.

NUCLEAR MOMENTS
Noémie Koller

Intrinsic properties of atomic nuclei; electric moments result from deviations of the nuclear charge distribution from spherical symmetry; magnetic moments are a consequence of the intrinsic spin and the rotational motion of nucleons within the nucleus. The classical definitions of the magnetic and electric multipole moments are written in general in terms of multipole expansions. SEE NUCLEAR STRUCTURE; SPIN.

Parity conservation allows only even-rank electric moments and odd-rank magnetic moments to be nonzero. The most important terms are the magnetic dipole, given by Eq. (1), and the electric monopole, quadrupole, and hexadecapole, given by Eq. (2), for $l = 0, 2, 4$. Here m is

$$\vec{\mu} = \int \vec{M}(\vec{r})\, dv \qquad (1) \qquad\qquad Q = \frac{1}{e} \int r^l Y_{lm}(\theta, \phi) \rho(\vec{r})\, dv \qquad (2)$$

the projection of the orbital angular momentum l on a z axis appropriately chosen in space, $\vec{M}\vec{r}$ is the magnetization density of the nucleus and depends on the space coordinates $\vec{r}$, e is the electronic charge, $\rho\vec{r}$ is the charge density in the nucleus, and Y_{lm} are normalized spherical harmonics that depend on the angular coordinates θ and ϕ. SEE PARITY.

Quantum mechanically only the z components of the effective operators have nonvanishing values. Magnetic dipole and electric quadrupole moments are usually expressed in terms of the nuclear spin I through Eqs. (3) and (4), where g, the nuclear gyromagnetic factor, is a measure of

$$\mu/\mu_N = gI \qquad (3) \qquad\qquad Q(m_I) = \frac{[3m_I^2 - I(I+1)]Q}{I(2I-1)} \qquad (4)$$

the coupling of nuclear spins and orbital angular momenta, $\mu_N = eh/4\pi M_p c = 5.0508 \times 10^{-24}$ erg/gauss = 5.0508×10^{-27} joule/tesla, is the nuclear magneton, M_p is the proton mass, h is Planck's constant, e is the electron charge, and c is the speed of light. $Q(m_I)$ is the effective quadrupole moment in the state m_I, and Q is the quadrupole moment for the state $m_I = I$. All angular momenta are expressed in units of $h/2\pi$. The magnitude of g varies between 0 and 1.8, and Q is of the order of 10^{-25} cm^2. SEE MAGNETON.

In special cases nuclear moments can be measured by direct methods involving the interaction of the nucleus with an external magnetic field or with an electric field gradient produced by the scattering of high-energy charged particles. In general, however, nuclear moments manifest themselves through the hyperfine interaction between the nuclear moments and the fields or field gradients produced by either the atomic electrons' currents and spins, or the molecular or crystalline electronic and lattice structures. SEE HYPERFINE STRUCTURE.

Effects of nuclear moments. In a free atom the magnetic hyperfine interaction between the nuclear spin $\vec{I}$ and the effective magnetic field $\vec{H}_e$ associated with electronic angular momentum $\vec{J}$ results in an energy $W = -\vec{\mu} \cdot \vec{H}_e = ha\vec{I} \cdot \vec{J}$, which appears as a splitting of the energy levels of the atom. The magnetic field at the nucleus due to atomic electrons can be as large as 10–100 T for a neutral atom. The constant a is of the order of 1000 MHz.

The electric monopole moment is a measure of the nuclear charge and does not give rise to hyperfine interactions. The quadrupole moment Q reflects the deviation of the nuclear charge distribution from a spherical charge distribution. It is responsible for a quadrupole hyperfine interaction energy W_Q, which is proportional to the quadrupole moment Q and to the spatial derivative of the electric field at the nucleus due to the electronic charges, and is given by Eq. (5), where q

$$W_Q = \frac{e^2 Qq}{2I(2I-1)J(2J-1)} \cdot [3(\vec{I} \cdot \vec{J})^2 + 3/2\,(\vec{I} \cdot \vec{J}) - I(I+1)] \tag{5}$$

is the average of expression (6). Here, r_i is the radius vector from the nucleus to the ith electron,

$$\sum_i (3\cos^2\theta - 1)/r_i^{-3} \tag{6}$$

and θ_i is the angle between r_i and the z axis. SEE MÖSSBAUER EFFECT.

In free molecules the hyperfine couplings are similar to those encountered in free atoms, but as the charge distributions and the spin coupling of valence electrons vary widely, depending on the nature of the molecular bonding, a greater diversity of magnetic dipole and quadrupole interactions is met.

In crystals the hyperfine interaction patterns become extremely complex, because the crystalline electric field is usually strong enough to compete with the spin orbit coupling of the electrons in the ion. Nevertheless, the energy-level structure can often be resolved by selective experiments at low temperatures on dilute concentrations of the ion of interest.

Measurement. The hyperfine interactions affect the energy levels of either the nuclei or the atoms, molecules, or ions, and therefore can be observed either in nuclear parameters or in the atomic, molecular, or ionic structure. The many different techniques that have been developed to measure nuclear moments can be grossly grouped in three categories: the conventional techniques based mostly on spectroscopy of energy levels, the methods based on the detection of nuclear radiation from aligned excited nuclei, and techniques involving the interactions of fast ions with matter or of fast ions with laser beams.

Hyperfine structure of spectral lines. The hyperfine interaction causes a splitting of the electronic energy levels which is proportional to the magnitude of the nuclear moments, to the angular momenta I and J of the nuclei and their electronic environment, and to the magnetic field or electric field gradient at the nucleus. The magnitude of the splitting is determined by the nuclear moments, and the multiplicity of levels is given by the relevant angular momenta I or J involved in the interaction. The energy levels are identified either by optical or microwave spectroscopy.

Optical spectroscopy (in the visible and ultraviolet) has the advantage of allowing the study of atomic excited states and of atoms in different states of ionization. Furthermore, optical spectra provide a direct measure of the monopole moments, which are manifested as shifts in the energy levels of atoms of different nuclear isotopes exhibiting different nuclear radii and charge distributions. Optical spectroscopy has a special advantage over other methods in that the intensity of the lines often yields the sign of the interaction constant a. SEE ISOTOPE SHIFT.

Microwave spectroscopy is a high-resolution technique involving the attenuation of a signal in a waveguide containing the absorber in the form of a low-pressure gas. The states are identified by the observation of electric dipole transitions of the order of 20,000 MHz. The levels are split by quadrupole interactions of the order of 100 MHz. Very precise quadrupole couplings are obtained, as well as vibrational and rotational constants of molecules, and nuclear spins.

Atomic and molecular beams and nuclear resonance. Atomic and molecular beams passing through inhomogeneous magnetic fields are deflected by an amount depending on the nuclear moment. However, because of the small size of the nuclear moment, the observable effect is very small. The addition of a radio-frequency magnetic field at the frequency corresponding to the energy difference between hyperfine electronic states has vastly extended the scope of the technique. For nuclei in solids, liquids, or gases, the internal magnetic fields and gradients of the electric fields may be quenched if the pairing of electrons and the interaction between the nuclear magnetic moment and the external field dominate. The molecular beam apparatus is designed to detect the change in orientation of the nuclei, while the nuclear magnetic resonance system is designed to detect absorbed power (resonance absorption) or a signal induced at resonance in a pick-up coil around the sample (nuclear induction). The required frequencies for fields of about 0.5 tesla are of the order of 1–5 MHz. The principal calibration of the field is accomplished in relation to the resonant frequency for the proton whose g-factor is accurately known. Sensitivities of 1 part in 10^8 are possible under optimum experimental conditions. The constant a for ^{133}Cs has been measured to 1 part in 10^{10}, and this isotope is used as a time standard.

The existence of quadrupole interactions produces a broadening of the resonance line above the natural width and a definite structure determined by the value of the nuclear spin.

In some crystals the electric field gradient at the nucleus is large enough to split the energy levels without the need for an external field, and pure quadrupole resonance spectra are observed. This technique allows very accurate comparison of quadrupole moments of isotopes.

Atomic and molecular beams with radioactive nuclei. The conventional atomic and molecular beam investigations can be applied to radioactive nuclei if the beam current measurement is replaced by the much more sensitive detectors of radiations emitted in a radioactive decay. Moments of nuclei with half-lives down to the order of minutes have been determined. SEE RADIOACTIVITY.

Perturbed angular correlations. The angular distribution and the polarization of radiation emitted by nuclei depend on the angle between the nuclear spin axis and the direction of emissions. In radioactive sources in which the nuclei have been somewhat oriented by a nuclear reaction or by static or dynamic polarization techniques at low temperatures, the ensuing nonisotropic angular correlation of the decay radiation can be perturbed by the application of external magnetic fields, or by the hyperfine interaction between the nuclear moment and the electronic or crystalline fields acting at the nuclear site. Magnetic dipole and electric quadrupole moments of ground and excited nuclear states with half-lives as short as 10^{-9} have been measured. SEE DYNAMIC NUCLEAR POLARIZATION.

Techniques involving interactions of fast ions. Techniques involving the interaction of intense light beams from tuned lasers with fast ion beams have extended the realm of resonance spectroscopy to the study of exotic species, such as nuclei far from stability, fission isomers, and ground-state nuclei with half-lives shorter than minutes. The hyperfine interactions in beams of fast ions traversing magnetic materials result from the coupling between the nuclear moments and unpaired polarized s-electrons, and are strong enough (H_e is of the order of 2×10^3 teslas) to extend the moment measurements to excited states with lifetimes as short as 10^{-12} s. Progress in atomic and nuclear technology has contributed to the production of hyperfine interactions of increasing strength, thus allowing for the observation of nuclear moments of nuclei and nuclear states of increasing rarity.

Bibliography. P. Averbuch (ed.), *Magnetic Resonance and Radiofrequency Spectroscopy*, 1969; H. Kopferman, *Nuclear Moments*, 1958; *Proceedings of the International Conference on Nuclear Moments and Nuclear Structure*, vol. 43, Physics Society, Japan, 1973; N. F. Ramsey, *Molecular Beams*, 1985.

MAGNETON
McAllister H. Hull, Jr.

A unit of magnetic moment used for atomic, molecular, or nuclear magnets.

The Bohr magneton μ_B has the value of the classical magnetic moment of the electron, which can theoretically be calculated as shown in Eq. (1), where e and m are the electronic charge

$$\mu_B = \mu_0 = \frac{e\hbar}{2mc}$$
$$= (9.274078 \pm 0.00004) \times 10^{-24} \text{ joule/tesla} \quad (1)$$

and mass, $\hbar$ is Planck's constant divided by 2π, and c is the velocity of light. A consistent relativistic treatment of the magnetic moment of the free electron shows that corrections to the classical calculation are necessary, so that the electron moment is about 0.1% larger than μ_0.

The magnetic moment of an atom or molecule results from contributions from both the orbital angular momentum of the atomic electrons and the electronic moments themselves (attributed to the electron spin). When certain groups of atoms are compared, their moments show simple ratios. Observation of this fact led to the definition of the Weiss magneton (before the Bohr magneton) on a purely experimental basis as the unit for these moments. Its value is given by Eq. (2).

$$\mu_W = 1.853 \times 10^{-24} \text{ joule/tesla} \qquad (2)$$

The nuclear magneton is obtained from the Bohr magneton by replacing m by the proton mass; it is thus 1836.15152 times smaller than μ_B; and is thus given by Eq. (3). However, mea-

$$\mu_N = (5.050787 \pm 0.000002) \times 10^{-27} \text{ joule/tesla} \qquad (3)$$

sured nuclear magnetic moments are not closely approximated by this value (only the order of magnitude is given). Thus the proton and neutron moments are given in terms of μ_N by Eqs. (4).

$$\mu_p = (2.79284739 \pm 0.00000006)\, \mu_N \qquad \mu_n = (-1.9130428 \pm 0.0000004)\, \mu_N \qquad (4)$$

The ratio is $\mu_p/\mu_n = -1.4599$. Since the neutron is uncharged, it has no classical magnetic moment at all. *See* NEUTRON; NUCLEAR MOMENTS; PROTON.

According to current understanding, nucleons (and hadrons in general) are not elementary particles (while the electron is). Remarkable confirmation of this view is provided by treating the neutron and proton as composed of three quarks each and calculating the magnetic moments of these composite systems in the lowest order of quantum chromodynamics. The result of the calculation is to give a theoretical value for μ_p/μ_n of -1.5, which differs by only about 3% from the measured value. For higher mass hadrons (Λ, Σ, Ξ) the agreement is only at the 15% level, with all the signs of the moments correctly given. Contributions of quark orbital angular momenta improve agreement with measured values for the hyperons, but do not yield as good agreement as is achieved with the μ_p/μ_n ratio. *See* ELEMENTARY PARTICLE; QUANTUM CHROMODYNAMICS; QUARKS.

TYPES OF NUCLEI

Deuteron	40
Triton	40
Transuranium elements	40
Supertransuranics	46
Hypernuclei	50

DEUTERON
Henry E. Duckworth

The nucleus of the atom of heavy hydrogen, ^{2}H (deuterium). The deuteron d is composed of a proton and a neutron. As the simplest multinucleon nucleus, the deuteron has been the subject of extensive study. Its binding energy is 2.225 MeV; that is, this is the amount of energy which must be added to a deuteron for it to dissociate into a proton and a neutron. The accurate determination of this dissociation energy provides the means of calculating the mass of the neutron, the mass of the deuteron (2.0141018 amu) and proton being known from other experiments.

The intrinsic angular momenta, or spins, of the proton and neutron combine to produce a deuteron spin of unity; hence, the deuteron obeys the type of quantum statistics known as Bose-Einstein statistics. The deuteron possesses a magnetic moment (0.8574382 nuclear magneton) and an electric quadrupole moment (2.738×10^{-27} cm^2).

Deuterons are much used as projectiles in nuclear bombardment experiments, especially to produce (d,p), (d,n), and (d,α) reactions. In the first two reactions, because of the low binding energy of the deuteron, the neutron n or proton p is stripped from it and captured by the target nucleus. Meanwhile, the other half of the deuteron (that is, the proton or neutron) carries away the excess energy. The ^{1}H/^{2}H abundance ratio in nature is 6700. SEE NUCLEAR REACTION.

TRITON
Henry E. Duckworth

The nucleus of $_1$H^3 (tritium); it is the only known radioactive nuclide belonging to hydrogen. The triton is produced in nuclear reactors by neutron absorption in deuterium ($_1$H^2 + $_0$n^1 → $_1$H^3 + γ), and decays by β^- emission to $_2$He^3 with a half-life of 12.4 years. The spin of the triton is ½, its magnetic moment is 2.9788 nuclear magnetons, and its mass is 3.0160493 atomic mass units. Much of the interest in producing $_1$H^3 arises from the fact that the fusion reaction $_1$H^3 + $_1$H^1 → $_2$He^4 releases about 20 MeV of energy. Tritons are also used as projectiles in nuclear bombardment experiments. SEE NUCLEAR REACTION.

TRANSURANIUM ELEMENTS
Glenn T. Seaborg

Those synthetic elements with atomic numbers larger than that of uranium (atomic number 92). They are the members of the actinide series, from neptunium (atomic number 93) through lawrencium (atomic number 103), and the transactinide elements (with higher atomic numbers than 103). Of these elements, plutonium, an explosive ingredient for nuclear weapons and a fuel for nuclear power because it is fissionable, has been prepared on the largest (ton) scale, while some of the others have been produced in kilograms (neptunium, americium, curium) and in much smaller quantities (berkelium, californium, and einsteinium).

The concept of atomic weight in the sense applied to naturally occurring elements is not applicable to the transuranium elements, since the isotopic composition of any given sample depends on its source. In most cases the use of the mass number of the longest-lived isotope in combination with an evaluation of its availability has been adequate. Good choices at present are neptunium, 237; plutonium, 242; americium, 243; curium, 248; berkelium, 249; californium, 249; einsteinium, 254; fermium, 257; mendelevium, 258; nobelium, 259; lawrencium, 260; rutherfordium, 261; hahnium, 262; and element 106,263.

The actinide elements are chemically similar and have a strong chemical resemblance to the lanthanide, or rare-earth, elements (atomic numbers 57–71). The transactinide elements, with atomic numbers 104 to 118, should be placed in an expanded periodic table under the row of elements beginning with hafnium, number 72, and ending with radon, number 86. This arrangement allows prediction of the chemical properties of these elements and suggests that they will have an element-by-element chemical analogy with the elements which appear immediately above them in the periodic table.

The transuranium elements up to and including fermium (atomic number 100) are produced in largest quantity through the successive capture of neutrons in nuclear reactors. The yield decreases with increasing atomic number, and the heaviest to be produced in weighable quantity is einsteinium (number 99). Many additional isotopes are produced by bombardment of heavy target isotopes with charged atomic projectiles in accelerators; beyond fermium all elements are produced by bombardment with heavy ions. Brief descriptions of transuranium elements follow. They are listed according to increasing atomic number.

Neptunium. Neptunium (Np, atomic number 93, named after the planet Neptune) was the first transuranium element discovered. In 1940 E. M. McMillan and P. H. Abelson at the University of California, Berkeley, identified the isotope ^{239}Np (half-life 2.35 days), which was produced by the bombardment of uranium with neutrons according to reaction (1). The element as

$$^{238}U(n,\gamma)^{239}U \rightarrow ^{239}Np \qquad (1)$$

^{237}Np was first isolated as a pure compound, the oxide, in 1944 by L. G. Magnusson and T. J. La Chapelle. Neptunium in trace amounts is found in nature, the element being produced in nuclear reactions in uranium ores caused by the neutrons present. Kilogram and larger quantities of ^{237}Np (half-life 2.14×10^6 years), used for chemical and physical investigations, are being produced as a by-product of the production of plutonium in nuclear reactors. Isotopes from mass number 227 to 244 have been synthesized by various nuclear reactions.

Neptunium displays five oxidation states in aqueous solution: Np^{3+} (pale purple), Np^{4+} (yellow-green), NpO_2^+ (green-blue), NpO_2^{2+} (pink), and NpO_5^{3-} (green). The ion NpO_2^+, unlike corresponding ions of uranium, plutonium, and americium, can exist in aqueous solution at moderately high concentrations. The element forms tri- and tetrahalides such as NpF_3, NpF_4, $NpCl_3$, $NpCl_4$, $NpBr_3$, NpI_3, as well as NpF_6 and oxides of various compositions such as those found in the uranium-oxygen system, including Np_3O_8 and NpO_2.

Neptunium metal has a silvery appearance, is chemically reactive, and melts at 637°C (1179°F); the solid metal has at least three crystalline forms between room temperature and its melting point.

Plutonium. Plutonium (Pu, atomic number 94, named after the plant Pluto) in the form of ^{238}Pu was discovered in late 1940 and early 1941 by G. T. Seaborg, McMillan, J. W. Kennedy, and A. C. Wahl at the University of California, Berkeley. The element was produced in the bombardment of uranium with deuterons according to reaction (2). The important isotope ^{239}Pu was dis-

$$^{238}U(d,2n)^{238}Np \xrightarrow[2.1 \text{ days}]{\beta^-} {}^{238}Pu \qquad (2)$$

covered by Kennedy, Seaborg, E. Segrè, and Wahl in 1941. Plutonium-239 (half-life 24,400 years), because of its property of being fissionable with neutrons, is used as the explosive ingredient in nuclear weapons and is a key material in the development of nuclear energy for industrial purposes, 1 lb (0.45 kg) being equivalent to about 10,000,000 kWh of heat energy; it is produced in ton quantities in nuclear reactors. The alpha radioactivity and physiological behavior of this isotope make it one of the most dangerous poisons known, but means for handling it safely have been devised. Plutonium as ^{239}Pu was first isolated as a pure compound, the fluoride, in 1942 by B. B. Cunningham and L. B. Werner. Minute amounts of plutonium formed in much the same way as naturally occurring neptunium are present in nature. Much smaller quantities of the longer-lived isotope ^{244}Pu (half-life 83,000,000 years) have been found in nature; in this case it may represent the small fraction remaining from a primordial source or it may be caused by cosmic rays. Isotopes of mass number 232–246 are known. The longer-lived isotopes ^{242}Pu (half-life 390,000 years) and ^{244}Pu, produced in nuclear reactors, are more suitable than ^{239}Pu for chemical and physical investigation because of their longer half-lives and lower specific activities.

Plutonium has five oxidation states in aqueous solution: Pu^{3+} (blue to violet), Pu^{4+} (yellow-brown), PuO_2^+ (pink), PuO_2^{2+} (pink-orange), and PuO_5^{3-} (blue-green). The ions Pu^{4+} and PuO_2^+ undergo extensive disproportionation to the ions of higher and lower oxidation states. Four oxidation states (III, IV, V, and VI) can exist simultaneously at appreciable concentrations in equilibrium with each other, an unusual situation that leads to complicated solution phenomena.

Plutonium forms binary compounds with oxygen (PuO, PuO_2, and intermediate oxides of variable composition); with the halogens (PuF_3, PuF_4, PuF_6, $PuCl_3$, $PuBr_3$, PuI_3); with carbon, nitrogen, and silicon (including PuC, PuN, $PuSi_2$); in addition, oxyhalides are well known ($PuOCl$, $PuOBr$, $PuOI$).

The metal is silvery in appearance, is chemically reactive, melts at 640°C (1184°F), and has six crystalline modifications between room temperature and its melting point.

Americium. Americium (Am, atomic number 95, named after the Americas) was the fourth transuranium element discovered. The element as ^{241}Am (half-life 433 years) was produced by the intense neutron bombardment of plutonium and was identified by Seaborg, R. A. James, L. O. Morgan, and A. Ghiorso in late 1944 and early 1945 at the wartime Metallurgical Laboratory at the University of Chicago. By using the isotope ^{241}Am, the element was first isolated as a pure compound, the hydroxide, in 1945 by B. B. Cunningham. Isotopes of mass numbers 237–247 have been prepared. Kilogram quantities of ^{241}Am are being produced in nuclear reactors. The less radioactive isotope ^{243}Am (half-life 7400 years), also produced in nuclear reactors, is more suitable for use in chemical and physical investigation.

Americium exists in four oxidation states in aqueous solution: Am^{3+} (light salmon), AmO_2^{+} (light tan), AmO_2^{2+} (light tan), and a fluoride complex of the IV state (pink). The trivalent state is highly stable and difficult to oxidize. AmO_2^{+}, like plutonium, is unstable with respect to disproportionation into Am^{3+} and AmO_2^{2+}. The ion Am^{4+} may be stabilized in solution only in the presence of very high concentrations of fluoride ion, and tetravalent solid compounds are well known. Divalent americium has been prepared in solid compounds; this is consistent with the presence of seven 5f electrons in americium (enhanced stability of half-filled 5f electron shell) and is similar to the analogous lanthanide, europium, which can be reduced to the divalent state.

Americium dioxide, AmO_2, is the important oxide; Am_2O_3 and, as with previous actinide elements, oxides of variable composition between $AmO_{1.5}$ and AmO_2 are known. The halides AmF_2 (in CaF_2), AmF_3, AmF_4, $AmCl_2$ (in $SrCl_2$), $AmCl_3$, $AmBr_3$, AmI_2, and AmI_3 have also been prepared.

Metallic americium is silvery-white in appearance, is chemically reactive, and has a melting point of 1176°C (2149°F). It has two crystalline modifications between room temperature and its melting point.

Curium. The third transuranium element to be discovered, curium (Cm, atomic number 96, named after Pierre and Marie Curie), as the isotope ^{242}Cm, was identified by Seaborg, James, and Ghiorso in 1944 at the wartime Metallurgical Laboratory of the University of Chicago. This was produced by the helium-ion bombardment of ^{239}Pu in the University of California 60-in. (152 cm) cyclotron. Curium was first isolated, using the isotope ^{242}Cm, in the form of a pure compound, the hydroxide, in 1947 by L. B. Werner and I. Perlman. Isotopes of mass number 238–251 are known. Chemical investigations with curium have been performed using ^{242}Cm (half-life 163 days) and ^{244}Cm (half-life 18 years), but the higher-mass isotopes ^{247}Cm and ^{248}Cm with much longer half-lives (1.6 × 10^7 and 3.5 × 10^5 years, respectively) are more satisfactory for this purpose; these are all produced by neutron irradiation in nuclear reactors.

Curium exists solely as Cm^{3+} (colorless to yellow) in the uncomplexed state in aqueous solution. This behavior is related to its position as the element in the actinide series in which the 5f electron shell is half filled; that is, it has the especially stable electronic configuration $5f^7$, analogous to its lanthanide homolog, gadolinium. A curium IV fluoride complex ion exists in aqueous solution. Solid compounds include Cm_2O_3, CmO_2 (and oxides of intermediate composition), CmF_3, CmF_4, $CmCl_3$, $CmBr_3$, and CmI_3.

The metal is silvery and shiny in appearance, is chemically reactive, melts at 1340°C (2444°F), and resembles americium metal in its two crystal modifications.

Berkelium. Berkelium (Bk, atomic number 97, named after Berkeley, California) was produced and identified by S. G. Thompson, Ghiorso, and Seaborg in late 1949 at the University of California, Berkeley, and was the fifth transuranium element discovered. The isotope ^{243}Bk (half-life 4.6 h) was synthesized by helium-ion bombardment of ^{241}Am. The first isolation of berkelium in weighable amount, as ^{249}Bk (half-life 314 days), produced by neutron irradiation, was accomplished in 1958 by Thompson and Cunningham; this isotope, produced in nuclear reactors, is used in the chemical and physical investigation of berkelium. Isotopes of mass number 242–251 are known.

Berkelium exhibits two ionic oxidation states in aqueous solution, Bk^{3+} (yellow-green) and somewhat unstable Bk^{4+} (yellow) as might be expected by analogy with its rare-earth homolog, terbium. Solid compounds include Bk_2O_3, BkO_2 (and oxides of intermediate composition), BkF_3, BkF_4, $BkCl_3$, $BkBr_3$, and BkI_3.

Berkelium metal is chemically reactive, exists in two crystal structure modifications, and melts at 986°C (1807°F).

Californium. The sixth transuranium element to be discovered, californium (Cf, atomic number 98, named after the state and University of California), in the form of the isotope ^{245}Cf (half-life 44 min), was first prepared by the helium-ion bombardment of microgram quantities of ^{242}Cm at the University of California, Berkeley. The element was discovered by Thompson, K. Street, Jr., Ghiorso, and Seaborg at the University of California, Berkeley, early in 1950. Cunningham and Thompson, at Berkeley, isolated californium in weighable quantities for the first time in 1958 using a mixture of the isotopes ^{249}Cf, ^{250}Cf, ^{251}Cf, and ^{252}Cf, produced by neutron irradiation. Isotopes of mass number 239–256 are known. The best isotope for the investigation of the chemical and physical properties of californium is ^{249}Cf (half-life 350 years), produced in pure form as the beta-particle decay product of ^{249}Bk.

Californium exists mainly as Cf^{3+} in aqueous solution (emerald green), but it is the first of the actinide elements in the second half of the series to exhibit the II state, which becomes progressively more stable on proceeding through the heavier members of the series. It also exhibits the IV oxidation state in CfF_4 and CfO_2, which can be prepared under somewhat intensive oxidizing conditions. Solid compounds also include Cf_2O_3 (and higher intermediate oxides), CfF_3, $CfCl_3$, $CfBr_2$, $CfBr_3$, CfI_2, and CfI_3.

Californium metal is chemically reactive, is quite volatile, and can be distilled at temperatures of the order of 1100 to 1200°C (2010–2190°F). It appears to exist in three different crystalline modifications between room temperature and its melting point, 900°C (1652°F).

Einsteinium. The seventh transuranium element to be discovered, einsteinium (Es, atomic number 99, named after Albert Einstein), was found by Ghiorso and coworkers in the debris from the "Mike" thermonuclear explosion staged by the Los Alamos Scientific Laboratory in November 1952. Very heavy uranium isotopes were formed by the action of the intense neutron flux on the uranium in the device, and these decayed into isotopes of elements 99, 100, and other transuranium elements of lower atomic number. Chemical investigation of the debris in late 1952 by workers at the University of California Radiation Laboratory, Argonne National Laboratory, and Los Alamos Scientific Laboratory revealed the presence of element 99 as the isotope ^{253}Es. Einsteinium was isolated in a macroscopic (weighable) quantity for the first time in 1961 by Cunningham, J. C. Wallman, L. Phillips, and R. C. Gatti at Berkeley; they used the isotope ^{253}Es, produced in nuclear reactors, working with only a few hundredths of a microgram. The macroscopic property that they determined in this case was the magnetic susceptibility. Isotopes of mass number 243–256 have been synthesized. Einsteinium is the heaviest transuranium element to be isolated in weighable form. Most of the investigations have used the short-lived ^{253}Es (half-life 20.5 days) because of its greater availability, but the use of ^{254}Es (half-life 276 days) will increase as it becomes more available as the result of production in nuclear reactors.

Einsteinium exists in normal aqueous solution essentially as Es^{3+} (green), although Es^{2+} can be produced under strong reducing conditions. Solid compounds such as Es_2O_3, $EsCl_3$, $EsOCl$, $EsBr_2$, $EsBr_3$, EsI_2, and EsI_3 have been made.

Einsteinium metal is chemically reactive, is quite volatile, and melts at 860°C (1580°F); one crystal structure is known.

Fermium. Fermium (Fm, atomic number 100, named after Enrico Fermi), the eighth transuranium element discovered, was isolated as the isotope ^{255}Fm (half-life 20 h) from the heavy elements formed in the "Mike" thermonuclear explosion. The element was discovered in early 1953 by Ghiorso and coworkers during the same investigation which resulted in the discovery of element 99. Fermium isotopes of mass number 242–259 have been prepared.

No isotope of fermium has yet been isolated in weighable amounts, and thus all the investigations of this element have been done with tracer quantities. The longest-lived isotope is ^{257}Fm (half-life about 100 days) whose production in high-neutron-flux reactors is extremely limited because of the very long sequence of neutron-capture reactions that is required.

Despite its very limited availability, fermium, in the form of the 3.24-h ^{254}Fm isotope, has been identified in the "metallic" zero-valent state in an atomic-beam magnetic resonance experiment. This established the electron structure of elemental fermium in the ground state as $5f^{12}7s^2$ (beyond the radon structure).

Fermium exists in normal aqueous solution almost exclusively as Fm^{3+}, but strong reducing conditions can produce Fm^{2+}, which has greater stability than Es^{2+} and less stability than Md^{2+}.

Mendelevium. Mendelevium (Md, atomic number 101, named after Dmitri Mendeleev), the ninth transuranium element discovered, was identified by Ghiorso, B. G. Harvey, G. R. Choppin, Thompson, and Seaborg at the University of California, Berkeley, in 1955. The element as ^{256}Md (half-life 1.5 h) was produced by the bombardment of extremely small amounts (approximately 10^9 atoms) of ^{253}Es with helium ions in the 60-in. (152-cm) cyclotron. The first identification of mendelevium was notable in that only one or two atoms per experiment were produced. (This served as the prototype for the discovery of all heavier transuranium elements, which have been first synthesized and identified on a one-atom-at-a-time basis.) Isotopes of mass numbers 247–259 are known. Although the isotope ^{258}Md (half-life 56 days) is sufficiently long-lived, it cannot be produced in nuclear reactors, and hence it will be very difficult and perhaps impossible to isolate it in weighable amount.

The chemical properties have been investigated on the tracer scale, and the element is found to behave in aqueous solution as a typical tripositive actinide ion; it can be reduced to the II state with moderately strong reducing agents.

Nobelium. The discovery of nobelium (No, atomic number 102, named after Alfred Nobel), the tenth transuranium element to be discovered, has a complicated history. For the first time scientists from countries other than the United States embarked on serious efforts to compete with the United States in this field. The reported discovery of element 102 in 1957 by an international group of scientists working at the Nobel Institute for Physics in Stockholm, who suggested the name nobelium, has never been confirmed and must be considered to be erroneous. Working at the Kurchatov Institute of Atomic Energy in Moscow, G. N. Flerov and coworkers in 1958 reported a radioactivity which they thought might be attributed to element 102, but a wide range of half-lives was suggested and no chemistry was performed. As the result of more definitive work performed in 1958, Ghiorso, T. Sikkeland, J. R. Walton, and Seaborg reported an isotope of the element, produced by bombarding a mixture of curium isotopes with ^{12}C ions in the then-new Heavy Ion Linear Accelerator (HILAC) at Berkeley. They described a novel "double recoil" technique which permitted identification by chemical means, one atom at a time, of any daughter isotope of element 102 that might have been formed. The isotope ^{250}Fm was identified conclusively by this means, indicating that its parent should be the isotope of element 102 with mass number 254 produced by the reaction of ^{12}C ions with ^{246}Cm. However, another isotope of element 102, with half-life 3 s, also observed indirectly in 1958, and whose alpha particles were shown to have an energy of 8.3 MeV by Ghiorso and coworkers in 1959, was shown later by Flerov and coworkers (working at the Dubna Laboratory near Moscow) to be of an isotope of element 102 with mass number 252 rather than 254; in other words, two isotopes of element 102 were discovered by the Berkeley group in 1958, but the correct mass number assignments were not made until later. On the basis that they identified the atomic number correctly, the Berkeley scientists probably have the best claim to the discovery of element 102; they suggest the retention of nobelium as the name for this element.

All known isotopes (mass numbers 250–259) of nobelium are short-lived and are produced by the bombardment of lighter elements with charged particles (heavy ions); the longest lived is ^{259}No with a half-life of 58 min. All of the chemical investigations have been, and presumably must continue to be, done on the tracer scale. These have demonstrated the existence of No^{3+} and No^{2+} in aqueous solution, with the latter much more stable than the former. The stability of No^{2+} is consistent with the expected presence of the completed shell of fourteen $5f$ electrons in this ion.

Lawrencium. Lawrencium (Lr, atomic number 103, named after Ernest O. Lawrence) was discovered in 1961 by Ghiorso, Sikkeland, A. E. Larsh, and R. M. Latimer using the HILAC at the University of California, Berkeley. A few micrograms of a mixture of ^{249}Cf, ^{250}Cf, ^{251}Cf, and ^{252}Cf (produced in a nuclear reactor) were bombarded with ^{10}B and ^{11}B ions to produce single atoms of an isotope of element 103 with a half-life measured as 8 s and decaying by the emission of alpha particles of 8.6 MeV energy. Ghiorso and coworkers suggested at that time that this radioactivity might be assigned the mass number 257. G. N. Flerov and coworkers have disputed this discovery on the basis that their later work suggests a greatly different half-life for the isotope with the mass number 257. Subsequent work by Ghiorso and coworkers proves that the correct assignment of mass number to the isotope discovered in 1961 is 258, and this later work gives 4 s as a better value for the half-life.

All known isotopes of lawrencium (mass numbers 253–260) are short-lived and are produced by bombardment of lighter elements with charged particles (heavy ions); chemical investigations have been, and presumably must be, performed on the tracer scale. Work with ^{260}Lr (half-life 3 min) has demonstrated that the normal oxidation state in aqueous solution is the III state, corresponding to the ion Lr^{3+}, as would be expected for the last member of the actinide series.

Element 104. Rutherfordium (Rf, atomic number 104, after Lord Rutherford), the first transactinide element to be discovered, was probably first identified in a definitive manner by Ghiorso, M. Nurmia, J. Harris, K. Eskola, and P. Eskola in 1969 at Berkeley. Flerov and coworkers have suggested the name kurchatovium (named after Igor Kurchatov with symbol Ku) on the basis of an earlier claim to the discovery of this element; they bombarded, in 1964, ^{242}Pu with ^{22}Ne ions in their cyclotron at the Joint Institute for Nuclear Research in Dubna and reported the production of an isotope, suggested to be ^{260}Ku, which was held to decay by spontaneous fission with a half-life of 0.3 s. After finding it impossible to confirm this observation, Ghiorso and coworkers reported definitive proof of the production of alpha-particle-emitting ^{257}Rf and ^{259}Rf (half-lives 4.5 and 3 s, respectively), demonstrated by the identification of the previously known ^{253}No and ^{255}No as decay products, by means of the bombardment of ^{249}Cf with ^{12}C and ^{13}C ions in the Berkeley HILAC.

All known isotopes of rutherfordium (mass numbers 253–262) are short-lived and are produced by bombardment of lighter elements with charged heavy-ion particles. The isotope ^{261}Rf (half-life 65 s) has made it possible, by means of rapid chemical experiments, to demonstrate that the normal oxidation state of rutherfordium in aqueous solution is the IV state corresponding to the ion Rf^{4+}. This is consistent with expectations for this first "transactinide" element which should be a homolog of hafnium, an element that is exclusively tetrapositive in aqueous solution.

Hahnium. Hahnium (Ha, atomic number 105, named after Otto Hahn), the second transactinide element to be discovered, was probably first identified in a definitive manner in 1970 by Ghiorso, Nurmia, K. Eskola, Harris, and P. Eskola at Berkeley. They reported the production of alpha-particle-emitting ^{260}Ha (half-life 1.6 s), demonstrated through the identification of the previously known ^{256}Lr as the decay product, by bombardment of ^{249}Cf with ^{15}N ions in the Berkeley HILAC. Again the Berkeley claim to discovery is disputed by Flerov and coworkers, who earlier in 1970 reported the discovery of an isotope held to be element 105, decaying by the less definitive process of spontaneous fission, produced by the bombardment of ^{243}Am with ^{22}Ne ions in the Dubna cyclotron; in later work Flerov and coworkers may have also observed the alpha-particle-emitting isotope of element 105 reported by Ghiorso and workers. Flerov has suggested nielsbohrium (named after Niels Bohr, symbol Ns) as the name for element 105.

The known isotopes of element 105 (mass numbers 257–262) are short-lived and are produced by bombardment of lighter elements with charged heavy-ion particles. The isotope ^{262}Ha (half-life 40 s) makes it possible, with rapid chemical techniques, to study the chemical properties of hahnium. It should exhibit the V oxidation state like its homolog tantalum.

Element 106. The discovery of element 106 took place in 1974 simultaneously as the result of experiments by Ghiorso and coworkers at Berkeley and Flerov, Y. T. Oganessian, and coworkers at Dubna. The Ghiorso group used the SuperHILAC (the rebuilt HILAC) to bombard a target of californium (the isotope ^{249}Cf) with ^{18}O ions. This resulted in the production and positive identification of the alpha-particle-emitting isotope ^{263}10b, which decays with a half-life of 0.9 ± 0.2 s by the emission of alpha particles of principal energy 9.06 MeV. The definitive identification consisted of the establishment of the genetic link between the element 106 alpha-particle-emitting isotope (^{263}106) and previously identified daughter (^{259}104) and granddaughter (^{255}102) nuclides, that is, the demonstration of the decay sequence: $^{263}106 \xrightarrow{\alpha} {}^{259}Rf \xrightarrow{\alpha} {}^{255}No \xrightarrow{\alpha}$. A total of seventy-three ^{263}106 alpha particles and approximately the expected corresponding number of ^{259}Rf daughter and ^{255}No granddaughter alpha particles were recorded.

The Dubna group chose lead (atomic number 82) as their target because, they believe, its closed shells of protons and neutrons and consequent small relative mass leads to minimum excitation energy for the compound nucleus and therefore an enhancement in the cross section for the production of the desired product nuclide. They bombarded ^{207}Pb and ^{208}Pb with ^{54}Cr ions (atomic number 24) in their cyclotron to find a product that decays by the spontaneous fission mechanism (a total of 51 events), with the very short half-life of 7 ms, which they assign to the isotope ^{259}106. Later work at Dubna and the G.S.I. laboratory in Darmstadt, Germany, has shown that this assignment is not correct.

On the basis of its projected position in the periodic table, element 106 is expected to have chemical properties similar to those of tungsten (atomic number 74).

Elements 107, 108, 109. G. Münzenberg and coworkers, working at the G.S.I. laboratory, Darmstadt, have synthesized and identified isotopes of elements 107, 108, and 109 which have relatively long half-lives in the millisecond range, indicating an unexpected relative stability in this region of high atomic numbers. They used targets in the region of closed nucleon shells, ^{208}Pb and ^{209}Bi, which when bombarded with heavy ions led to compound nuclei of minimum excitation ("cold" compound systems) in order to enhance the yields, and identified the alpha-particle-emitting products by establishment of the genetic links with their known alpha-particle-emitting descendants, as was done by Ghiorso and coworkers in their discovery of element 106. They observed six atoms of 5-ms (the time interval for its decay) ^{262}107, produced by the ^{209}Bi (^{54}Cr,n) reaction, in 1981; three atoms of 2-ms ^{265}108, produced by the ^{208}Pb (^{58}Fe,n) reaction, in 1984; and one atom of 5-ms ^{266}109, produced by the ^{209}Bi (^{58}Fe,n) reaction, in 1982. Element 107 is expected to have chemical properties similar to those of rhenium, 108 similar to those of osmium, and 109 similar to those of iridium.

Superheavy elements. Although the transactinide elements immediately beyond element 109 are predicted to have very short half-lives, theoretical considerations suggest increased nuclear stability, compared with preceding and succeeding elements, for a range of elements with larger atomic numbers because of increased stability predicted to result from closed nucleon shells, especially a closed neutron shell at 184 neutrons. The element with atomic number 114 (which should have chemical properties similar to those of lead) seems to show special promise of such relative stability, that is, relatively long half-life. It should be possible to synthesize isotopes of such "superheavy" elements through bombardments of heavy-element targets with intense beams of very heavy ions accelerated in specially constructed heavy-ion accelerators. SEE SUPERTRANSURANICS.

It should be possible to predict the chemical properties of all the transactinide elements with the help of the periodic table. Rutherfordium should chemically be like hafnium, hahnium like tantalum, element 106 like tungsten, 107 like rhenium, and so on across the periodic table to element 118, which should be a noble gas like radon. Beyond element 118, the elements 119, 120, and 121 should fit into the periodic table under the elements francium, radium, and actinium (atomic numbers 87, 88, and 89). At about this point there should start another series, but a special kind of inner transition series, perhaps similar in some respects to the actinide series. However, this series, which may be termed the superactinide series, will be different in that it will contain 32 elements, corresponding to the filling of 18-member and 14-member inner electron shells. After the filling of these shells at element 153, the still higher elements should again be placed in the main body of the periodic table leading to the next noble gas at element 168. The larger atomic numbers are far beyond the predicted region of nuclear stability and hence presumably are not accessible to experimentation. SEE NUCLEAR FISSION; NUCLEAR REACTION. RARE-EARTH ELEMENTS.

Bibliography. G. R. Choppin and J. Rydberg, *Nuclear Chemistry: Theory and Application*, 1980; A. J. Freeman et al. (eds.), *Handbook on the Physics and Chemistry of the Actinides*, vols. 1–4, 1984–1986; J. J. Katz et al. (eds.), *The Chemistry of the Actinide Elements*, 2 vols., 1986; Max Planck Society for the Advancement of Science, *Transurane-Transuranium Elements*, 8th ed., 1975; G. T. Seaborg, The new elements, *Amer. Sci.*, 68:3, 1980; G. T. Seaborg, (ed.), *Transuranium Elements: Products of Modern Alchemy*, 1978.

SUPERTRANSURANICS
J. RAYFORD NIX

A group of relatively stable elements, with atomic numbers around 114 and mass numbers around 298, that are predicted to exist beyond the present periodic table of known elements. Also called superheavy elements, they have not yet been discovered experimentally, probably because of difficulties in synthesizing them rather than their instability once they are produced. SEE ATOMIC NUMBER; MASS NUMBER.

Nuclear stability. There are approximately 300 naturally occurring nuclei, representing isotopes of elements containing from 1 to 94 protons. Some 2500 additional nuclei have been made artificially since around 1925; the heaviest nucleus produced thus far has 109 protons and a mass number of 266. This so-called peninsula of known nuclei terminates because of the disruptive electrostatic forces between positively charged protons, which increase faster than the cohesive nuclear forces along the peninsula in the direction toward heavier nuclei. The large electrostatic forces cause heavy nuclei to decay rapidly by the emission of alpha particles and by spontaneous fission. *See* A*lpha particles;* A*tomic nucleus;* N*uclear fission.*

The possibility of an island of supertransuranics past the end of the peninsula is associated with the extra stability that arises from the closing of a proton or neutron shell. As protons or neutrons are added to a nucleus, they go into definite single-particle orbits. When a given shell of protons or neutrons is completely filled, that nucleus has relatively lower energy, or extra binding. This increases the stability of the nucleus against both alpha emission and spontaneous fission. *See* N*eutron;* P*roton.*

Because of the extra stability and other properties of nuclei with completely filled proton or neutron shells, such nuclei are said to be magic. From experiments it has been determined that the proton magic numbers are 2, 8, 14, 28, 50, and 82, and that the neutron magic numbers are 2, 8, 14, 28, 50, 82, and 126. It is possible to reproduce these magic numbers theoretically by solving the Schrödinger equation of quantum mechanics with an appropriate single-particle potential to describe the motion of the protons and neutrons. When such calculations are extrapolated to heavier nuclei, the next predicted proton magic number is 114 and neutron magic number, 184. The difference between the next predicted proton magic number of 114 and the corresponding neutron magic number of 126 arises because of the electrostatic forces, which become increasingly important for heavy nuclei. *See* M*agic numbers.*

The predicted island of supertransuranics, separated from the peninsula of known nuclei by a sea of instability, is shown in the **illustration**. The mountains, representing nuclei that are especially stable against alpha emission and spontaneous fission, occur near the proton and neu-

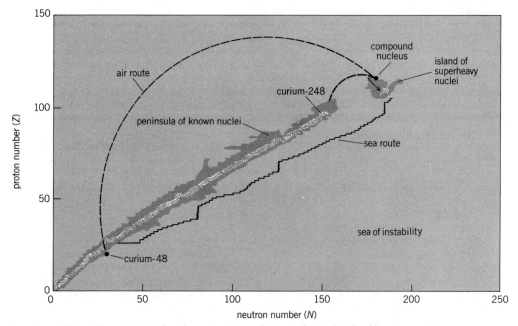

Location of the predicted island of supertransuranics relative to the peninsula of known nuclei. (*After P. Möller and J. R. Nix, Calculation of fission barriers, Proceedings of the 3d IAEA Symposium on the Physics and Chemistry of Fission, Rochester, New York, 1973, vol. 1, pp. 103–143, International Atomic Energy Agency, 1974*)

tron magic numbers. These magic numbers also lead to the irregularities that are seen along the coastlines. The ridge along the center of the peninsula, representing nuclei that are stable against the emission or capture of negatively charged electrons, corresponds to nuclei with an increasing ratio of neutrons to protons in the direction of heavier nuclei.

Widespread interest in supertransuranics began in 1965 when it was estimated that the fission barrier of such a nucleus should be even higher than that of a conventional nucleus like uranium. Shortly thereafter, an improved method for calculating the potential energy of a nucleus as a function of its shape was developed. This method is a two-part approach, with the smooth trends of the potential energy taken from the liquid-drop model of the nucleus and the local fluctuations from the shell model of the nucleus. It was used to make the first systematic survey of the expected stability of supertransuranics. More refined calculations have subsequently been made by several other groups.

Since supertransuranics can decay by spontaneous fission, the emission of alpha particles, or the emission or capture of electrons, all of these possible decay modes must be considered in determining their half-lives. The partial half-life for spontaneous fission is longest for the doubly magic nucleus with 114 protons and 184 neutrons, and decreases away from this nucleus in any direction. The partial half-life for alpha emission is longer for nuclei with a smaller number of protons. Only certain nuclei with an appropriate ratio of neutrons to protons are stable against electron emission or capture. *See* ELECTRON; HALF-LIFE.

In many early calculations the supertransuranic nucleus with 110 protons and 184 neutrons had the longest total half-life, corresponding to decay by alpha emission, with values ranging from 10^5 to 10^9 years. In another calculation this nucleus is unstable against electron emission, and the longest-lived supertransuranic nucleus has 110 protons and 178 neutrons, with a calculated half-life of 200 days. The **table** lists this and other predicted half-lives of supertransuranic nuclei, corresponding to decay by alpha emission.

Predicted half-lives of several supertransuranic nuclei

Atomic number (Z)	Neutron number (N)	Mass number (A)	Half-life ($T_{1/2}$)
110	178	288	200 days
112	184	296	20 days
114	184	298	3 hours
116	190	306	0.3 second

Production methods. There are two possible methods for producing supertransuranics, the multiple capture of neutrons and a reaction between two fairly heavy nuclei. The first method, which is like trying to sail over to the island through the sea of instability, would produce nuclei in the southeastern or neutron-rich shore of the island (see illus.). In this process, a nucleus successively moves east by capturing one or more neutrons, and northwest by converting a neutron into a proton through the emission of an electron. Many naturally occurring nuclei were made in this way from the enormous quantities of neutrons that were released in supernovas. Other nuclei have been produced artificially by the multiple capture of neutrons in nuclear explosions and in nuclear reactors. However, it is unlikely that supertransuranics can be produced by this process because the intermediate nuclei formed in the sea of instability between the peninsula and the island terminate the process by undergoing fission.

The second method, which is like trying to fly over to the island, would produce nuclei in the northwestern or proton-rich side of the island by means of a collision between two fairly heavy nuclei (see illus.). There are two promising ways to do this. The first approach is to fire a projectile such as calcium-48 at a target such as curium-248 with sufficient energy to produce a nearly spherical compound nucleus. In most cases the system will undergo fission, either before or after a compound nucleus is formed; but there is a small probability that the compound nucleus, which is produced with some excitation energy, can deexcite by the emission of neutrons and gamma rays to produce an unexcited supertransuranic nucleus. This nucleus can in turn decay by successively emitting alpha particles and capturing electrons to produce some final supertransuranic nuclei that live for days. In the second approach, a somewhat heavier projectile is used, part of which may combine with the target nucleus to form a supertransuranic nucleus while the remaining portion flies on by.

Searches in nature. Since early calculations suggested that some supertransuranics may live as long as the age of the solar system and they could possibly have been produced in supernovas by the multiple capture of neutrons, many scientists have searched for them in nature since 1968. Their quest has led them to some unusual spots—from California platinum and gold mines to meteorites and Moon rocks, from a fourteenth-century Russian orthodox church to the ocean floor off the Fiji Islands, and from monazite minerals on Madagascar to hot-water springs on the Cheleken Peninsula in the Caspian Sea.

The searches have utilized minerals and ores of known elements whose chemical properties are similar to those of the supertransuranics being sought. The chemical properties of the supertransuranics can be predicted by performing self-consistent calculations for the electrons surrounding the nuclei by use of relativistic quantum mechanics. This permits the extension of the periodic table to elements heavier than those currently known. In this way it is found that the supertransuranics with atomic numbers 110, 111, 112, 113, 114, and 115 should be similar to Pt, Au, Hg, Tl, Pb, and Bi, respectively, although there are some differences arising from relativistic effects.

The first search for supertransuranics in nature was made in 1968 by S. G. Thompson and his coworkers at Berkeley, who looked for element 110, eka-platinum. To detect the possible presence of supertransuranics in the samples considered, they used the predicted large number of neutrons per fission of a supertransuranic, as well as activation analysis, mass spectrometry, x-ray fluorescence, and direct spontaneous-fission counting. Later searches were carried out by G. N. Flerov's group at Dubna, G. Herrmann's group at Mainz, and many others.

No clear evidence for supertransuranics has been found, although a tantalizing possibility for their prior existence was presented in 1975 by E. Anders and his group at Chicago. They analyzed the isotopic distribution of the element Xe in the Allende meteorite by radiochemical neutron activation, finding an anomalous distribution of neutron-rich Xe isotopes. This could have resulted from the spontaneous fission of an extinct volatile supertransuranic nucleus with atomic number 115, 114, or 113. The failure to find supertransuranics in nature is probably due to difficulties in producing them through the multiple capture of neutrons and—if they were initially produced—to their subsequent decay with half-lives that are much shorter than the age of the solar system.

Searches at accelerators. Attempts to synthesize supertransuranics artificially were first made at Berkeley in 1968 by Thompson's group and by A. Ghiorso's group, who used reactions between such nuclei as argon-40 and curium-248. Later attempts by M. Lefort's group at Orsay involved reactions between krypton-84 and thorium-232, whereas Flerov, Yu. Ts. Oganessian, and their coworkers at Dubna bombarded uranium-238 targets with projectiles as heavy as xenon-136. Secondary reactions induced by energetic heavy recoils resulting from the collision of high-energy protons with suitable targets have also been utilized at Geneva by A. Marinov and his coworkers. More recent attempts have included collisions between two uranium nuclei at Darmstadt by Herrmann's group, as well as reactions between calcium-48 and curium-248 at Berkeley by an international collaboration. In these experiments, various methods that could detect nuclei with half-lives as short as 10^{-5} s and as long as 100 years have been used.

None of these attempts has provided any evidence for the production of nuclei in the island of supertransuranics. This probably means that the system of two fusing nuclei is undergoing fission because of the strong electrostatic and dissipative forces that are present, rather than reaching a spherical shape and deexciting. However, it could also mean that the half-lives of any supertransuranics that are produced are much shorter than predicted.

The heaviest nucleus produced thus far contained 109 protons and 157 neutrons, corresponding to a mass number of 266. Formed at Darmstadt by P. Armbruster's group through a reaction involving iron-58 and bismuth-209 , with the release of a neutron, it was detected by observing a sequence of correlated decays by alpha emission, electron capture, and spontaneous fission. To increase the number of neutrons in the system, G. T. Seaborg and his coworkers were planning to bombard einsteinium-254 with calcium-48 at Berkeley in a future attempt to produce supertransuranics. SEE TRANSURANIUM ELEMENTS.

Bibliography. P. Armbruster, On the production of heavy elements by cold fusion: The elements 106 to 109, *Ann. Rev. Nucl. Part. Sci.*, 35:135–194, 1985; G. Herrmann, Superheavy-element research, *Nature*, 280:543–549, 1979; M. A. K. Lodhi (ed.), *Proceedings of the International Symposium on Superheavy Elements*, Lubbock, Texas, pp. 1–572, 1978; P. Möller, G. A. Leander, and J. R. Nix, On the instability of the transeinsteinium elements, *Z. Phys. A*, 323:41–45, 1986.

HYPERNUCLEI
BOGDAN POVH

Nuclei that consist of protons, neutrons, and one or more strange particles such as lambda particles. The lambda particle is the lightest strange baryon (hyperon); its lifetime is 2.6×10^{-10} s. Because strangeness is conserved in strong interactions, the lifetime of the lambda particle remains essentially unchanged in the nucleus also. Lambda hypernuclei live long enough to permit detailed study of their properties. SEE BARYON; ELEMENTARY PARTICLE; HYPERON; STRONG NUCLEAR INTERACTIONS.

There is no bound lambda-nucleon system, demonstrating that the lambda-nucleon force is weaker than the force between nucleons which can bind two nucleons into deuterium. The lightest bound Λ hypernucleus is $^3_\Lambda$H—lambda hypertriton—which is composed of a proton, a neutron, and a lambda particle. Lambda hypernuclei up to $^{16}_\Lambda$O have been identified experimentally.

Two cases of the double lambda hypernuclei have been found so far; one is $^6_{\Lambda\Lambda}$He, which is composed of two lambda particles coupled to the ^{4}He nucleus. The bound double-lambda system is still being sought. In some theoretical models of the elementary particles a strong binding for the two lambda particles is predicted.

The lambda particle in the nucleus experiences an attractive potential, the strength of which is about two-thirds that for the nucleon. The spin-orbit force, which is strong in the case of the nucleon as it causes splitting comparable to the energy differences between the nucleon shells, is negligibly small in the case of the lambda particle. Theoretically, the difference between the lambda-nucleus and nucleon-nucleus interaction is explained as reflecting the differences in the internal quark structure of the two baryons.

Sigma, xi, and omega particles, all with a lifetime of about 10^{-10} s as free particles, convert, through the strong interaction, into lambda particles in the nucleus. Nevertheless, sigma hypernuclei have been experimentally observed, the lifetime being of the order of 10^{-21} s. This lifetime is long enough to permit determination of the most important parameters of the sigma-nucleus interaction. There is a good chance that xi and omega hypernuclei could also be investigated experimentally. SEE NUCLEAR STRUCTURE.

NUCLEAR PROBES AND PHENOMENA

Radioactivity	52
Radioisotope	78
Half-life	78
Transmutation	79
Nuclear spectra	79
Multipole radiation	81
Sum rules	82
Angular correlations	83
Mössbauer effect	84
Neutron spectrometry	87
Time-of-flight spectrometers	92
Nuclear orientation	94
Dynamic nuclear polarization	95
Isotope shift	99
Hyperfine structure	100

RADIOACTIVITY
Joseph H. Hamilton

A phenomenon resulting from an instability of the atomic nucleus in certain atoms whereby the nucleus experiences a spontaneous but measurably delayed nuclear transition or transformation with the resulting emission of radiation. The discovery of radioactivity by H. Becquerel in 1896 was an indirect consequence of the discovery of x-rays a few months earlier by W. Roentgen, and marked the birth of nuclear physics. Studies of the radioactive decays of new isotopes far from the stable ones in nature continue as a major frontier in nuclear research. *See Isotope.*

In 1934 I. Curie and F. Joliot demonstrated that radioactive nuclei can be made in the laboratory. All chemical elements may be rendered radioactive by adding or by subtracting (except for hydrogen and helium) neutrons from the nucleus of the stable ones. The availability of this wide variety of radioactive isotopes has stimulated their use in science and technology in an enormous number of applications. *See Alpha particles; Beta particles; Gamma rays.*

A particular radioactive transition may be delayed by less than a microsecond or by more than a billion years, but the existence of a measurable delay or lifetime distinguishes a radioactive nuclear transition from a so-called prompt nuclear transition, such as is involved in the emission of most gamma rays. The delay is expressed quantitatively by the radioactive decay constant, or by the mean life, or by the half-period for each type of radioactive atom.

The first five types of radioactivity given in **Table 1** are the most commonly found with

Table 1. Types of radioactivity

Type and symbol	Particles emitted	Change in atomic number, ΔZ	Change in atomic mass number, ΔA	Example[a]
Alpha, α	Helium nucleus	-2	-4	$^{226}_{86}\text{Ra} \rightarrow {}^{222}_{84}\text{Rn} + \alpha$
Beta negatron, β^-	Negative electron and antineutrino	$+1$	0	$^{24}_{11}\text{Na} \rightarrow {}^{24}_{12}\text{Mg} + e^- + \bar{\nu}$
Beta positron, β^+	Positive electron and neutrino	-1	0	$^{22}_{11}\text{Na} \rightarrow {}^{22}_{10}\text{Ne} + e^+ + \nu$
Electron capture, EC	Neutrino	-1	0	$^{7}_{4}\text{Be} + e^- \rightarrow {}^{7}_{3}\text{Li} + \nu$
Isomeric transition, IT	Gamma rays or conversion electrons or both (and positive-negative electron pair)[b]	0	0	$^{137m}_{56}\text{Ba} \rightarrow {}^{137}_{56}\text{Ba} + \gamma$ or c.e.
Proton, p	Proton	-1	-1	$^{151}_{71}\text{La} \rightarrow {}^{150}_{70}\text{Yb} + p$ $^{53m}_{27}\text{Co} \rightarrow {}^{52}_{26}\text{Fe} + p$
Spontaneous fission, SF	Heavy fragments and neutrons	Various	Various	$^{238}_{92}\text{U} \rightarrow {}^{133}_{50}\text{Sn} + {}^{103}_{42}\text{Mo} + 2n$
Isomeric spontaneous fission, ISF	Heavy fragments and neutrons	Various	Various	$^{244f}_{95}\text{Am} \rightarrow {}^{134}_{53}\text{I} + {}^{107}_{42}\text{Mo} + 3n$
Beta-delayed spontaneous fission, (EC + β^+)SF	Positive electron, neutrino, heavy fragments, and neutrons	Various	Various	$^{246}_{99}\text{Es} \rightarrow \beta^+ + \nu + {}^{246f}_{98}\text{Cf} \rightarrow$ $^{138}_{54}\text{Xe} + {}^{107}_{44}\text{Ru} + n$
β^-SF	Negative electron, antineutrino, heavy fragments, and neutrons	Various	Various	$^{236}_{91}\text{Pa} \rightarrow \beta^- + \bar{\nu} + {}^{236f}_{92}\text{U} \rightarrow$ $^{139}_{53}\text{I} + {}^{94}_{39}\text{Y} + 3n$

NUCLEAR PROBES AND PHENOMENA

Table 1. Types of radioactivity (cont.)

Type and symbol	Particles emitted	Change in atomic number, ΔZ	Change in atomic mass number, ΔA	Example[a]
Beta-delayed neutron, $\beta^- n$	Negative electron, and antineutrino, neutron	+1	−1	$^{11}_{3}\text{Li} \rightarrow \beta^- + \bar{\nu} + {}^{11}_{4}\text{Be}^* \rightarrow {}^{10}_{4}\text{Be} + n$
Beta-delayed two-neutron (three-neutron), $\beta^- 2n$ (3n)	Negative electron, antineutrino, and two (three) neutrons	+1	−2 (−3)	$^{11}_{3}\text{Li} \rightarrow \beta^- + \bar{\nu} + {}^{11}_{4}\text{Be}^* \rightarrow {}^{9(8)}_{4}\text{Be} + 2n \ (3n)$
Beta-delayed proton, $\beta^+ p$ or $(\beta^+ + EC)p$	Positive electron, neutrino, and proton	−2	−1	$^{114}_{55}\text{Cs} \rightarrow \beta^+ + \nu + {}^{114}_{54}\text{Xe}^* \rightarrow {}^{113}_{53}\text{I} + p$
Beta-delayed two-proton, $\beta^+ 2p$	Positive electron, neutrino, and two protons	−3	−2	$^{22}_{13}\text{Al} \rightarrow \beta^+ + \nu + {}^{22}_{12}\text{Mg}^* \rightarrow {}^{20}_{10}\text{Ne} + 2p$
Beta-delayed triton, $\beta^- {}^3_1\text{H}$	Negative electron, antineutrino and triton	0	−3	$^{11}_{3}\text{Li} \rightarrow \beta^- + \bar{\nu} + {}^{11}_{4}\text{B}^* \rightarrow {}^{8}_{3}\text{Li} + {}^{3}_{1}\text{H}$
Beta-delayed alpha, $\beta^+ \alpha$	Positive electron, neutrino, and alpha	−3	−4	$^{114}_{55}\text{Cs} \rightarrow \beta^+ + \nu + {}^{114}_{54}\text{Xe}^* \rightarrow {}^{110}_{52}\text{Te} + \alpha$
$\beta^- \alpha$	Negative electron, antineutrino, and alpha	−1	−4	$^{214}_{83}\text{Bi} \rightarrow \beta^- + \bar{\nu} + {}^{114}_{84}\text{Po}^* \rightarrow {}^{210}_{82}\text{Pb} + \alpha$
Beta-delayed alpha-neutron, $\beta^- \alpha, n$	Negative electron, antineutrino, alpha, and neutron	−1	−5	$^{11}_{3}\text{Li} \rightarrow \beta^- + \bar{\nu} + {}^{11}_{4}\text{B}^* \rightarrow {}^{6}_{2}\text{He} + \alpha + n$
Double beta decay [c], $\beta^- \beta^-$	Two negative electrons and two antineutrinos	+2	0	[d] $^{130}_{52}\text{Te} \rightarrow {}^{130}_{54}\text{Xe} + 2\beta^- + 2\bar{\nu}$
$\beta^+ \beta^+$	Two positive electrons and two neutrinos	−2	0	[c] $^{130}_{56}\text{Ba} \rightarrow {}^{130}_{54}\text{Xe} + 2\beta^+ + 2\nu$
Double electron capture [c], EC EC	Two neutrinos	−2	0	[c] $^{130}_{56}\text{Ba} + 2e^- \rightarrow {}^{130}_{54}\text{Xe} + 2\nu$
Two-proton [c], 2p	Two protons	−2	−2	[c] $^{114}_{55}\text{Cs} \rightarrow {}^{112}_{53}\text{I} + 2p$
Neutron [c], n	Neutron	0	−1	
Two-neutron [c], 2n	Two neutrons	0	−2	
Heavy clusters [e], $^{14}_{6}\text{C}$	$^{14}_{6}\text{C}$ nucleus	−6	−14	$^{223}_{88}\text{Ra} \rightarrow {}^{209}_{82}\text{Pb} + {}^{14}_{6}\text{C}$
$^{20}_{8}\text{O}$	$^{20}_{8}\text{O}$ nucleus	−8	−20	[c] $^{227}_{89}\text{Ac} \rightarrow {}^{207}_{81}\text{Tl} + {}^{20}_{8}\text{O}$
$^{24}_{10}\text{Ne}$	$^{24}_{10}\text{Ne}$ nucleus	−10	−24	[c] $^{232}_{92}\text{U} \rightarrow {}^{108}_{82}\text{Pb} + {}^{24}_{10}\text{Ne}$

[a] Excited states with relatively long measured half-lives are called isomeric and are identified by placing the symbol m for metastable after the mass number, as in ^{137m}Ba. Excited states with essentially prompt decay are identified by asterisks, as in ^{11}Be*.
[b] Occurs as an additional decay mode when the decay energy exceeds 1.022 MeV.
[c] Theoretically predicted but not established experimentally.
[d] Some indirect evidence for this particular decay has been reported, but one cannot say double beta decay is established.
[e] There are other possible clusters in addition to those shown.

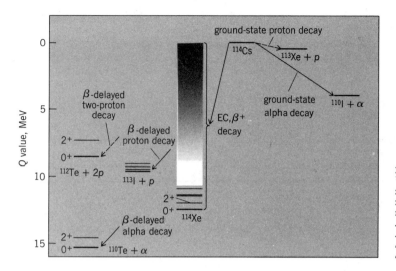

Fig. 1. Decay modes of ^{114}Cs based on Q values from the droplet-model formula for nuclear masses. (After J. H. Hamilton et al., eds., *Future Directions in Studies of Nuclei Far from Stability*, 1980)

each characterized by the particular type of nuclear radiation which is emitted by the transforming parent nucleus. In addition, there are several other decay modes that are observed more rarely in specific regions of the periodic table, as shown in the lower half of Table 1. Several of these rarer processes are in fact two-step processes, as shown in **Figs. 1** and **2**. In addition, there are several other processes predicted theoretically that remain to be verified. There is some indirect evidence for double beta decay of ^{130}Te.

In the first type in Table 1, in alpha radioactivity the parent nucleus spontaneously emits an alpha ray; then the atomic number, or nuclear charge Z, of the decay product is 2 units less than that of the parent, and the nuclear mass A of the product is 4 atomic mass units less than that of the parent, because the emitted alpha particle carries away this amount of nuclear charge and mass. This decrease of 2 units of atomic number or nuclear charge between parent and

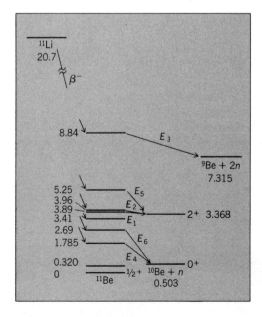

Fig. 2. Observed decay modes of ^{11}Li. Energies in MeV. (After J. H. Hamilton et al., eds., *Future Directions in Studies of Nuclei Far from Stability*, 1980)

product means that the decay product will be a different chemical element, displaced by 2 units to the left in a periodic table of the elements. For example, radium has atomic number 88 and is found in column II of the periodic table. Its decay product after the emission of an alpha ray is a different chemical element, radon, whose atomic number is 86 and whose position is in column 0 of the periodic table.

TRANSITION RATES AND DECAY LAWS

This section covers radioactive decay constant, dual decay, exponential decay law, mean life, and half-period.

Radioactive decay constant. The rate of radioactive transformation, or the activity, of a source equals the number A of identical radioactive atoms present in the source, multiplied by their characteristic radioactive decay constant λ. Thus Eq. (1) holds, where the decay constant λ

$$\text{Activity} = A\lambda \text{ disintegrations per second} \tag{1}$$

has dimensions of s^{-1}. The numerical value of λ expresses the statistical probability of decay of each radioactive atom in a group of identical atoms, per unit time. For example, if $\lambda = 0.01\ s^{-1}$ for a particular radioactive species, then each atom has a chance of 0.01 (1%) of decaying in 1 s, and a chance of 0.99 (99%) of not decaying in any given 1-s interval. The constant λ is one of the most important characteristics of each radioactive nuclide: λ is essentially independent of all physical and chemical conditions such as temperature, pressure, concentration, chemical combination, or age of the radioactive atoms. There are a few cases where measurable effects are observed for different chemical combinations. One of the largest observed is a 3.2% change in λ for the 24-s isomer in ^{90}Nb. The half-period is inversely proportional to λ.

The identification of some radioactive samples can be made simply by measuring λ, which then serves as an equivalent of qualitative chemical analysis. For the most common radioactive nuclides, the range of λ extends from $3 \times 10^6\ s^{-1}$ (for thorium C') to $1.6 \times 10^{-18}\ s^{-1}$ (for thorium).

Dual decay. Many radioactive nuclides have two or more independent and alternative modes of decay. For example, ^{238}U can decay either by alpha-ray emission or by spontaneous fission. A single atom of ^{64}Cu can decay in any of three competing independent ways: negatron beta-ray emission, positron beta-ray emission, or electron capture. When two or more independent modes of decay are possible, the nuclide is said to exhibit dual decay.

The competing modes of decay of any nuclide have independent partial decay constants given by the probabilities $\lambda_1, \lambda_2, \lambda_3, \ldots$ per second, and the total probability of decay is represented by the total decay constant λ, defined by Eq. (2). If there are A identical atoms present,

$$\lambda = \lambda_1 + \lambda_2 + \lambda_3 + \cdots \tag{2}$$

the partial activities, as measured by the different modes of decay, are $A\lambda_1, A\lambda_2, A\lambda_3, \ldots$, and the total activity $A\lambda$ is given by Eq. (3). The partial activities, $A\lambda_1, \ldots$, such as positron beta-

$$A\lambda = A\lambda_1 + A\lambda_2 + A\lambda_3 + \cdots \tag{3}$$

rays from ^{64}Cu, are proportional to the total activity, $A\lambda$, at all times.

The branching ratio is the fraction of the decaying atoms which follow a particular mode of decay, and equals $A\lambda_1/A\lambda$ or λ_1/λ. For example, in the case of ^{64}Cu the measured branching ratios are $\lambda_1/\lambda = 0.40$ for negatron beta decay, $\lambda_2/\lambda = 0.20$ for positron beta-decay, and $\lambda_3/\lambda = 0.40$ for electron capture. The sum of all the branching ratios for a particular nuclide is unity.

Exponential decay law. The total activity, $A\lambda$, equals the rate of decrease $-dA/dt$ in the number of radioactive atoms A present. Because λ is independent of the age t of an atom, integration of the differential equation of radioactive decay, $-dA/dt = A\lambda$, gives Eq. (4), where

$$\ln \frac{A}{A_0} = -\lambda(t - t_0) \tag{4}$$

ln represents the natural logarithm to the base e, and A atoms remain at time t if there were A_0 atoms initially present at time t_0. If $t_0 = 0$, then Eq. (4) can be rewritten as the exponential law of radioactive decay in its most common form, Eq. (5). The initial activity at $t = 0$ was $A_0\lambda$, and

$$A = A_0 e^{-\lambda t} \tag{5}$$

the activity at t, when only A atoms remain untransformed, is $A\lambda$. Because λ is a constant, the fractional activity $A\lambda/A_0\lambda$ at time t and the fractional amount of radioactive atoms A/A_0 are given by Eq. (6). In cases of dual decay, the partial activities $A\lambda_1$, $A\lambda_2$, ... also decrease with time as

$$\frac{A\lambda}{A_0\lambda} = \frac{A}{A_0} = e^{-\lambda t} \tag{6}$$

$e^{-\lambda t}$, not as $e^{-\lambda_1 t}$, ..., because $A\lambda_1/A_0\lambda_1 = A/A_0 = e^{-\lambda t}$ where λ is the total decay constant. This is because the decrease of each partial activity with time is due to the depletion of the total stock of atoms A, and this depletion is accomplished by the combined action of all the competing modes of decay.

Mean life. The actual life of any particular atom can have any value between zero and infinity. The average or mean life of a large number of identical radioactive atoms is, however, a definite and important quantity.

If there are A_0 atoms present initially at $t = 0$, then the number remaining undecayed at a later time t is $A = A_0 e^{-\lambda t}$, by Eq. (5). Each of these A atoms has a life longer than t. In an additional infinitesimally short time interval dt, between time t and $t + dt$, the absolute number of atoms which will decay on the average is $A\lambda dt$, and these atoms had a life-span t. The total L of the life-spans of all the A_0 atoms is the sum or integral of $tA\lambda\, dt$ from $t = 0$ to $t = \infty$, which is given by Eq. (7). Then the average lifetime L/A_0, which is called the mean life τ, is given by

$$L = \int_0^\infty tA\lambda\, dt = \int_0^\infty tA_0\lambda e^{-\lambda t}\, dt = \frac{A_0}{\lambda} \tag{7}$$

Eq. (8), where λ is the total radioactive decay constant of Eq. (2). Substitution of $t = \tau = 1/\lambda$ into

$$\tau = 1/\lambda \tag{8}$$

Eq. (6) shows that the mean life is the time required for the number of atoms, or their activity, to fall to $e^{-1} = 0.368$ of any initial value.

Half-period (half-life). The time interval over which the chance of survival of a particular radioactive atom is exactly one-half is called the half-period T (also called half-life, written $T_{1/2}$). From Eq. (4), Eq. (9) is obtained. Then the half-period T is related to the total radioactive decay

$$-\ln(A/A_0) = \ln(A_0/A) = \ln 2 = 0.693 = \lambda T \tag{9}$$

constant λ, and to the mean life τ, by Eq. (10). For mnemonic reasons, the half-period T is much

$$T = 0.693/\lambda = 0.693\tau \tag{10}$$

more frequently employed than the total decay constant λ or the mean life τ. For example, it is more common to speak of ^{232}Th as having a half-period of 1.4×10^{10} years than to speak of its mean life of 2.0×10^{10} years or its total decay constant of 1.6×10^{-18} s^{-1}, although all three are equivalent statements of the average longevity of ^{232}Th atoms.

The relationships between T, τ, and λ are summarized graphically in **Fig. 3**. Any initial activity $A_0\lambda$ is reduced to $\frac{1}{2}$ in 1 half-period T, to $1/e$ in 1 mean life τ, to $\frac{1}{4}$ in 2 half-periods $2T$, and so on. The slope of the activity curve, or rate of decrease of activity, is $d(A\lambda)/dt = -\lambda dA/dt = -\lambda(A\lambda)$. Thus the initial slope is $-(A_0\lambda) = -(A_0\lambda)\tau$. The area under the activity curve, if integrated to $t = \infty$, is simply A_0, the total initial number of radioactive atoms. Also, the initial activity $A_0\lambda$, if it could continue at a constant value for one mean life τ, would exactly destroy all the atoms because $(A_0\lambda)\tau = A_0$.

RADIOACTIVE SERIES DECAY

In a number of cases a radioactive nuclide A decays into a nuclide B which is also radioactive; the nuclide B decays into C which is also radioactive, and so on. For example, $^{232}_{90}$Th decays into a series of 10 successive radioactive nuclides. Substantially all the primary products of nuclear fission are negatron beta-ray emitters which decay through a chain or series of two to six successive beta-ray emitters before a stable nuclide is reached as an end product. *See Nuclear fission.*

Let the initial part of such a series be represented by reaction (11), where radioactive atoms

$$A \xrightarrow{\lambda_A} B \xrightarrow{\lambda_B} C \xrightarrow{\lambda_C} D \xrightarrow{\lambda_D} \cdots \tag{11}$$

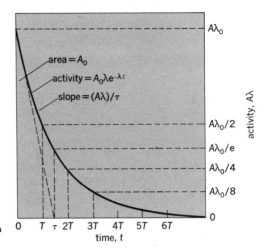

Fig. 3. Graphical representation of relationships in decay of a single radioactive nuclide.

of types $A, B, C, D, \ldots$ have radioactive decay constants given by $\lambda_A, \lambda_B, \lambda_C, \lambda_D, \ldots$. Then if there are initially present, at time $t = 0$, A_0 atoms of type A, the numbers $A, B, C, \ldots$ of atoms of types $A, B, C, \ldots$, which will be present at a later time t, are given by Eqs. (12)–(14), and

$$A = A_0 e^{-\lambda_A t} \qquad (12) \qquad B = A_0 \frac{\lambda_A}{\lambda_B - \lambda_A}(e^{-\lambda_A t} - e^{-\lambda_B t}) \qquad (13)$$

$$C = A_0 \left(\frac{\lambda_A}{\lambda_C - \lambda_A} \frac{\lambda_B}{\lambda_B - \lambda_A} e^{-\lambda_A t} + \frac{\lambda_A}{\lambda_A - \lambda_B} \frac{\lambda_B}{\lambda_C - \lambda_B} e^{-\lambda_B t} + \frac{\lambda_A}{\lambda_A - \lambda_C} \frac{\lambda_B}{\lambda_B - \lambda_C} e^{-\lambda_C t} \right) \qquad (14)$$

the activities of $A, B, C, \ldots$ are $A\lambda_A, B\lambda_B, C\lambda_C, \ldots$. General equations describing the amounts and activities of any number of radioactive decay products are more complicated and are given in standard texts.

Figure 4 illustrates the growth and decay of the activity of a short series of radioactive decay products in accord with Eqs. (12)–(14).

Radioactive equilibrium. In Fig. 4 the ratio $B\lambda_B/A\lambda_A$ of the activities of the parent A and the daughter product B change with time. The activity $B\lambda_B$ is zero initially and also after a very long time, when all the atoms have decayed. Thus $B\lambda_B$ passes through a maximum value, and it can be shown that this occurs at a time t_m given by Eq. (15). The situation in which the

$$t_m = \frac{\ln(\lambda_B/\lambda_A)}{(\lambda_B - \lambda_A)} \qquad (15)$$

activities $A\lambda_A$ and $B\lambda_B$ are exactly equal to each other is called ideal equilibrium, and exists only at the moment t_m.

If the parent A is longer-lived than the daughter B, as occurs in many cases, then at a time which is long compared with the mean life τ_B of B, the activity ratio approaches a constant value given by Eq. (16), where T_A and T_B are the half-periods of A and of B. When the activity ratio

$$\frac{B\lambda_B}{A\lambda_A} = \frac{\lambda_B}{\lambda_B - \lambda_A} = \frac{T_A}{T_A - T_B} \qquad (16)$$

$B\lambda_B/A\lambda_A$ is constant, a particular type of radioactive equilibrium exists. This is spoken of as secular equilibrium if the activity ratio is experimentally indistinguishable from unity, as occurs when T_A is very much greater than T_B.

Equilibrium concepts are applied also between a long-lived parent and any of its decay products in a long series. For example, in a sufficiently old uranium ore, radium ($T = 1620$ years) is in secular equilibrium with its ultimate parent uranium ($T = 4.5 \times 10^9$ years) although there

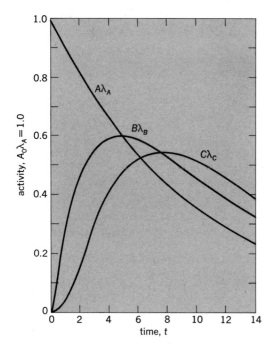

Fig. 4. Growth and decay of the activity $B\lambda_B$ of the daughter product, and $C\lambda_C$ of the granddaughter product, in an initially pure source of a radioactive parent whose activity at $t = 0$ is $A_0\lambda_A$.

are four intermediate radioactive substances intervening in the series between uranium and radium. Here, secular equilibrium expresses the fact that the activities of radium and uranium continue to be equal to each other even though the activity of the parent uranium is decreasing with time.

When T_B is comparable with T_A, Eq. (16) shows that the equilibrium ratio will clearly exceed unity; this situation is spoken of as transient equilibrium. For example, in fission-product decay series (17) the half-period of ^{140}Ba is 307 h and that of ^{140}La is 40 h. In an initially pure

$$^{140}\text{Ba} \rightarrow {}^{140}\text{La} \rightarrow {}^{140}\text{Ce} \qquad (17)$$

source of ^{140}Ba the activity of ^{140}La starts at zero, rises to a maximum at $t_m = 135$ h [Eq. (15)], then decreases, and after a few hundred hours is in transient equilibrium with its parent, when the ^{140}La activity [by Eq. (16)] is $307/(307-40) = 1.15$ times the activity of its parent ^{140}Ba.

Radioactivity in the Earth. A number of isotopes of elements found in the Earth are radioactive. All known or theoretically predicted isotopes of elements above bismuth are radioactive. Because the Earth is composed of atoms which were believed to have been created more than 3×10^9 years ago, the naturally occurring parent radioactive isotopes are those which have such long half-periods that detectable residual activity is still observable today. As a general rule, one can detect the presence of a radioactive substance for about 10 half-lives. Therefore activities with $T \lesssim 0.3 \times 10^9$ years should not be found in the Earth. For example, present-day uranium is an isotopic mixture containing 99.3% ^{238}U, whose half-period is 4.5×10^9 years, and only 0.7% of the shorter-lived uranium isotope ^{235}U, whose half-period is 0.7×10^9 years, whereas these isotopes presumably were produced in roughly equal amounts in the Earth a few billion years ago. Geophysical evidence indicates that originally some ^{236}U was present also, but none is found in nature now as expected with its half-period of 0.02×10^9 years. The elements technetium ($Z = 43$) and promethium ($Z = 61$) are not found in the Earth's crust because all their isotopes are radioactive with much shorter half-periods (their longest-lived are $T = 2.6 \times 10^6$ years for ^{97}Tc and $T = 17.7$ years for ^{145}Pm).

Uranium-238 decays through a long series of 14 radioactive decay products before ending as a stable isotope of lead, ^{206}Pb. Some of these members of the ^{238}U decay chain have very short half-periods, so their existence in nature is entirely dependent on the presence of their long-lived parent, and thus is a genealogical accident. For example, radium occurs in nature only in the minerals of its parent, uranium. The decay series of ^{235}U supports 14, and the decay series of ^{232}Th supports 10, short-lived radioactive substances found in nature.

A few of the common elements contain long-lived, naturally radioactive isotopes. For example, all terrestrial potassium contains 0.012% of the radioactive isotope ^{40}K, which has a half-period of 1.3×10^9 years, and emits negatron or positron beta-rays and gamma rays in a dual decay to stable ^{40}Ca and ^{40}A. This isotope is the principal source of radioactivity in a normal human being; each human contains about 0.1 microcurie (3.7×10^3 becquerel) of the radioactive potassium isotope ^{40}K.

Table 2 summarizes the radioactive properties of all the well-established cases of radioactivities found in the Earth's surface. Geological age measurements are based on the accumulation of decay products of these long-lived isotopes, especially in the cases of ^{40}K, ^{87}Rb, ^{232}Th, ^{235}U, and ^{238}U.

Table 2. Parent radioactive nuclides found in nature

Nuclide		Percent abundance in nature	Half-period, years	Radioactive transitions observed	Disintegration energy, MeV
Atomic number, Z	Mass number, Z				
19 K	40	0.0117	1.3×10^9	β^-, EC	β^- 1.3 − EC 1.5
37 Rb	87	27.83	4.8×10^{10}	β^-	0.3
48 Cd	113	12.2	9×10^{15}	β^-	0.3
49 In	115	95.77	5.1×10^{14}	β^-	0.5
52 Te	130	34.49	2×10^{21}	Growth of $^{130}_{54}$Xe†	1.6
57 La	138	0.089	1.1×10^{11}	β^-, EC	β^- 1.0 − EC 1.75
60 Nd	144	23.8	2.1×10^{15}	α	1.9
62 Sm	147	15.07	1.1×10^{11}	α	2.3
62 Sm	148	11.3	8×10^{15}	α	1.99
64 Gd	152	0.20	1.1×10^{14}	α	2.2
71 Lu	176	2.6	3.6×10^{10}	β^-, γ	0.6
72 Hf	174	0.16	2×10^{15}	α	2.5
75 Re	187	62.6	4×10^{10}	β^-	0.003
78 Pt	190	0.013	6×10^{11}	α	3.24
90 Th	232	100.	1.4×10^{10}	α	4.08
92 U	235	0.715	7.0×10^8	α	4.68
92 U	238	99.28	4.5×10^9	α	4.27

†Indirect evidence for $\beta^-\beta^-$ decay.

Laboratory produced radioactive nuclei. With particle accelerators and nuclear reactors, over 1650 radioactive isotopes not found in detectable quantities in the Earth's crust have been produced in the laboratory since 1935, including those of at least 17 new chemical elements up to element 109. Earlier titles of induced or artificial radioactivities for these isotopes are misnomers. Many of these now have been identified in meteorites and in stars, and others are produced in the atmosphere by cosmic rays. There are over 5000 isotopes theoretically predicted to exist. As one approaches the place where a proton or neutron is no longer bound in a nucleus of an element (the limits of the existence of that element), the half-periods become extremely short.

For example, carbon-14 is a negatron beta-ray emitter, with a half-period of about 5600 years, which can be produced in the laboratory as the product of a variety of different nuclear transmutation experiments. Nuclear bombardment of ^{11}B nuclei by alpha rays (helium nuclei) can produce excited compound nuclei of ^{15}N which promptly emit a proton (hydrogen nucleus), leaving ^{14}C as the end product of the transmutation. The same end-product ^{14}C can be produced by bombarding ^{14}N with neutrons, resulting in nuclear reaction (18). This reaction is easily carried

$$^{14}\text{N} + \text{neutron} \rightarrow ^{15}\text{N*} \rightarrow ^{14}\text{C} + \text{proton} \tag{18}$$

out by using neutrons from nuclear accelerators or a nuclear reactor. This particular transmutation reaction is one which occurs in nature also, because the nitrogen in the Earth's atmosphere is continually bombarded by neutrons which are produced by cosmic rays, thus producing radioactive ^{14}C. Mixing of ^{14}C with stable carbon provides the basis for radiocarbon dating of systems that absorb carbon for times up to about 50,000 years ago (10 half-lives). SEE NUCLEAR REACTION; PARTICLE ACCELERATOR.

Radioactive hydrogen, ^{3}H, is also formed in the atmosphere from the ^{14}N + neutron → ^{12}C + ^{3}H reaction. Also, ^{3}H is produced in the Sun, and the Earth's water as well as satellites shows an additional concentration of ^{3}H from the Sun. Over two dozen radioactive products, ranging in half-life from a few days to millions of years, have been identified in meteorites that have fallen to Earth. The carbon and hydrogen burning cycles that produce energy for stars produce radioactive ^{13}N, ^{15}O, ^{3}H. At higher temperatures the radioactivities ^{7}Be and even ^{8}Be ($T \approx 10^{-16}$ s) help burn hydrogen and helium. In addition to the production of radioactive as well as stable isotopes prior to the formation of the solar system, nucleosynthesis continues to go on in stars with the production of many short-lived radioactive atoms by different processes. SEE CARBON-NITROGEN-OXYGEN CYCLES; PROTON-PROTON CHAIN.

The yield of any radioactivity produced in the laboratory is the initial rate of the activity under the particular conditions of nuclear bombardment. When a target material A is bombarded to produce a radioactive product B whose radioactive decay constant is λ_B, the number of atoms B which are present after a bombardment of duration t, and their activity $B\lambda_B$, are given by Eq. (19), where the yield Y has dimensions equivalent to curies of activity produced per second of

$$B\lambda_B = \frac{Y}{\lambda_B}(1 - e^{-\lambda_B t}) \qquad (19)$$

bombardment. The yield Y depends on the number of atoms A present in the target, the intensity of the beam of bombarding particles, and the cross section, or probability of the reaction per bombarding particle under the conditions of bombardment.

Radioactive transformation series. As noted in Eqs. (12)–(14), many radioactive substances have decay products which are also radioactive. Thus many long chains or series of radioactive transformations are known.

The three naturally occurring transformation series are headed by ^{232}Th, ^{235}U, and ^{238}U. Their genealogical relationships are summarized in **Fig. 5** and **Table 3**.

Each of the naturally occurring radioactive isotopes in these transformation series has two synonymous names. For example, the commercially important radioisotope whose classical name is mesothorium-1 is now known to be an isotope of radium with mass number of 228 and is designated as radium-228 (^{228}Ra). Table 3 summarizes the names, symbols, and some radioactive properties of these three transformation series. However, these chains are not complete, and their uniqueness or importance as chains is an accident of the very long half-lives of ^{232}Th, ^{235}U, and ^{238}U. For example, element 105 of mass 260 has a succession of seven alpha decays and one electron capture and positron decay to ^{232}Th. The special importance of the chains in Table 3 is related to the fact that they were essentially the only early sources of radioactive materials, and more recently to their role in nuclear power.

Transformation series are now known for every element in the periodic table except hydrogen. Chains of neutron-rich isotopes have been produced and studied among the products of nuclear fission. Heavy-ion-induced reactions and high-flux reactors have been used to extend knowledge of the elements beyond uranium. The elements from number 93 (neptunium) out to 109 (as yet un-named), which have so far not been found on Earth, were made in the laboratory. Both proton- and heavy-ion-induced reactions have extended knowledge of chains and neutron-deficient isotopes of the stable elements.

ALPHA-PARTICLE DECAY

Alpha-particle decay is that type of radioactivity in which the parent nucleus expels an alpha particle (a helium nucleus). The alpha particle is emitted with a speed of the order of 1 to 2 × 10^7 m/s (10^4 mi/s), that is, about 1/20 of the velocity of light.

In the simplest case of alpha decay, every alpha particle would be emitted with exactly the

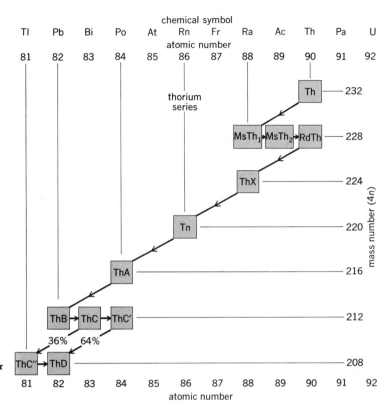

Fig. 5. Main line of decay of uranium series, or $4n + 2$ series, of heavy radioactive nuclides, headed in nature by uranium-238. Each member has a mass number given by $4n + 2$, where n is an integer.

same velocity and hence the same kinetic energy. However, in most cases there are two or more discrete energy groups called lines, as shown in **Fig. 6** in the spectrum of alpha particles from ^{184}Tl and ^{184}Hg decays. For example, in the alpha decay of a large group of ^{238}U atoms, 77% of the alpha decays will be by emission of alpha particles whose kinetic energy is 4.20 MeV, while 23% will be by emission of 4.15-MeV alpha particles. When the 4.20-MeV alpha particle is emitted, the decay product nucleus is formed in its ground (lowest energy) level. When a 4.15-MeV alpha particle is emitted, the decay product is produced in an excited level, 0.05 MeV above the ground level. This nucleus promptly transforms to its ground level by the emission of a 0.05-MeV gamma ray or alternatively by the emission of the same amount of energy in the form of a conversion electron and the associated spectrum of characteristic x-rays. Thus in all alpha-particle spectra, the alpha particles are emitted in one or more discrete and homogeneous energy groups, and alpha-particle spectra are accompanied by gamma-ray and conversion electron spectra whenever there are two or more alpha-particle groups in the spectrum.

Geiger-Nuttall rule. Among all the known alpha-particle emitters, most alpha-particle energy spectra lie in the domain of 4–6 MeV, although a few extend as low as 2 MeV ($^{147}_{62}$Sm) and as high as 10 MeV (ThC'). There is a systematic relationship between the kinetic energy of the emitted alpha particles and the half-period of the alpha emitter. The highest-energy alpha particles are emitted by short-lived nuclides, and the lowest-energy alpha particles are emitted by the very-long-lived alpha-particle emitters. H. Geiger and J. M. Nuttall showed that there is a linear relationship between log λ and the energy of the alpha particle.

The Geiger-Nuttall rule is inexplicable by classical physics, but emerges clearly from quantum, or wave, mechanics. In 1928 the hypothesis of transmission through nuclear potential barriers, as introduced by G. Gamow and independently by R. W. Gurney and E. U. Condon, was shown to give a satisfactory account of the alpha-decay data, and it has been altered subsequently

Table 3. Names, symbols, and radioactive properties of members of the three naturally occurring radioactive transformation series[†]

Conventional name	Conventional symbol	Atomic number	Mass number	Isotopic symbol	Half-period	Type of decay
Uranium (4n + 2) series						
Uranium I	UI	92	238	^{238}U	4.5×10^9 y	α
Uranium X_1	UX_1	90	234	^{234}Th	24 d	β^-
Uranium X_2	UX_2	91	234	^{234m}Pa	1.2 m	It,β^-
Uranium Z	UZ	91	234	^{234}Pa	6.7 h	β^-
Uranium II	UII	92	234	^{234}U	2.5×10^5 y	α
Ionium	Io	90	230	^{230}Th	8×10^4 y	α
Radium	Ra	88	226	^{226}Ra	1600 y	α
Radon	Rn	86	222	^{222}Rn	3.8 d	α
Radium A	RaA	84	218	^{218}Po	3.0 m	α
Radium B	RaB	82	214	^{214}Pb	27 m	β^-
Radium C	RaC	83	214	^{214}Bi	20 m	β^-,α
Radium C'	RaC'	84	214	^{214}Po	1.6×10^{-4} s	α
Radium C''	RaC''	81	210	^{210}Tl	1.3 m	β^-
Radium D	RaD	82	210	^{210}Pb	22 y	β^-
Radium E	RaE	83	210	^{210}Bi	5.0 d	β^-
Radium F	RaF	84	210	^{210}Po	138 d	α
Polonium	Po	84	210	^{210}Po	138 d	α
Radium G	RaG	82	206	^{206}Pb	Stable	Stable
Thorium (4n) series						
Thorium	Th	90	232	^{232}Th	1.4×10^{10} y	α
Mesothorium$_1$	MsTh$_1$	88	228	^{228}Ra	5.8 y	β^-
Mesothorium$_2$	MsTh$_2$	89	228	^{228}Ac	6.1 h	β^-
Radiothorium	RdTh	90	228	^{228}Th	1.9 y	α
Thorium X	ThX	88	224	^{224}Ra	3.7 d	α
Thoron	Tn	86	220	^{220}Rn	56 s	α
Thorium A	ThA	84	216	^{216}Po	0.15 s	α
Thorium B	ThB	82	212	^{212}Pb	10.6 h	β^-
Thorium C	ThC	83	212	^{212}Bi	1.0 h	β^-,α
Thorium C'	ThC'	84	212	^{212}Po	3×10^{-7} s	α
Thorium C''	ThC''	81	208	^{208}Tl	3.1 m	β^-
Thorium D	ThD	82	208	^{208}Pb	Stable	Stable
Actinium (4n + 3) series						
Actinouranium	AcU	92	235	^{235}U	7.0×10^8 y	α
Uranium Y	UY	90	231	^{231}Th	26 h	β^-
Protactinium	Pa	91	231	^{231}Pa	3.3×10^4 y	α
Actinium	Ac	89	227	^{227}Ac	22 y	β^-,α
Radioactinium	RdAc	90	227	^{227}Th	19 d	α
Actinium K	AcK	87	223	^{223}Fr	22 m	β^-,α
Actinium X	AcX	88	223	^{223}Ra	11 d	α
Astatine	At	85	219	^{219}At	0.9 m	α,β^-
Actinon	An	86	219	^{219}Rn	4.0 s	α
Actinium A	AcA	84	215	^{215}Po	1.8×10^{-3} s	α
Actinium B	AcB	82	211	^{211}Pb	36 m	β^-
Actinium C	AcC	83	211	^{211}Bi	2.2 m	α,β^-
Actinium C'	AcC'	84	211	^{211}Po	0.5 s	α
Actinium C''	AcC''	81	207	^{207}Tl	4.8 m	β
Actinium D	AcD	82	207	^{207}Pb	Stable	Stable

[†]Radon-223 has been shown to have a very weak radioactive ^{14}C decay branch to ^{209}Pb. Several of the isotopes in these chains, such as $^{234, 235, 238}U$, ^{226}Ra, and others are predicted to have such very weak, heavy cluster decay branches.

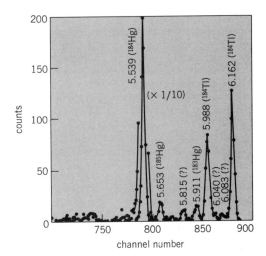

Fig. 6. Alpha groups in the decay of ^{184}Tl ($T = 11$ s) and ^{184}Hg and weak groups from ^{183}Hg and ^{185}Hg, very far off stability (17 neutrons less than the lightest stable thallium isotope). Energies in MeV. (*After K. S. Toth et al., Observation of α-decay in thallium nuclei, including the new isotopes ^{184}Tl and ^{185}Tl, Phys. Lett., 63B:150–153, 1976*)

only in details. The form of the barrier-penetration equations is such that correlation plots of log λ against $1/\sqrt{E}$ give nearly straight lines.

Nuclear potential barrier. At distances r which are large compared with the nuclear radius, the potential energy of an alpha particle, whose charge is $2e$, in the field of a residual nucleus, whose charge is $(Z - 2)e$, is $2(Z - 2)e^2/r$. At very close distances this electrostatic repulsion is opposed and overcome by short-range, specifically nuclear, attractive forces. The net potential energy U as a function of the separation r between the alpha particle and its residual nucleus is the nuclear potential barrier.

One of several operating definitions of the nuclear radius R is the distance $r = R$ at which the attractive nuclear forces just balance the repulsive electrostatic forces. At this distance, called the top of the nuclear barrier, the potential energy is about 25–30 MeV for typical cases of heavy, alpha-emitting nuclei, as indicated in **Fig. 7**. SEE NUCLEAR STRUCTURE.

Inside the nucleus the alpha particle is represented as a de Broglie matter wave. According to wave mechanics, this wave has a very small but finite probability of being transmitted through the nuclear potential energy barrier and thus of emerging as an alpha particle emitted from the nucleus. The transmission of a particle through such an energy barrier is completely forbidden in classical electrodynamics but is possible according to wave mechanics. This transmission of a

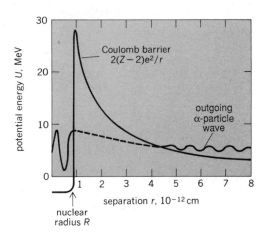

Fig. 7. Schematic of nuclear potential barrier, illustrating emission of an alpha particle as a wave which can be transmitted through the barrier.

matter wave through an energy barrier is analogous to the familiar case of the transmission of ordinary visible light through an opaque metal such as gold: if the gold is thin enough, some light does get through, as in the case of the thin gold leaf which is sometimes used for lettering signs on store windows.

The wave-mechanical probability of the transmission of an alpha particle through the nuclear potential barrier is very strongly dependent upon the energy of the emitted alpha particle. Analytically the probability of transmission T depends exponentially upon a barrier transmission exponent γ according to Eq. (20). To a good approximation, Eq. (21) holds, where $h = 6.626 \times$

$$T = e^{-\gamma} \qquad (20)$$

$$\gamma = \left(\frac{4\pi^2}{h}\right)\frac{(Z-2)2e^2}{V} - \left(\frac{8\pi}{h}\right)[2(Z-2)2e^2MR]^{1/2} \qquad (21)$$

10^{-34} joule-second is Planck's constant, and M is the so-called reduced mass of the alpha particle. For the alpha decay of ^{226}Ra, the numerical value of γ is about 71: hence $T = e^{-71} = 10^{-31}$. The first term on the right side of Eq. (21) is about 154 and is therefore the dominant term. When this term is taken alone, $e^{-(4\pi^2/h)(Z-2)2e^2/V}$ is called the Gamow factor for barrier penetration.

Inspection of Eq. (21) shows that the barrier transmission decreases with increasing nuclear charge $(Z-2)e$, increases with increasing velocity V of emission of the alpha particle, and increases with increasing radius R of the nucleus. When the experimentally known values of alpha-decay energy are substituted into Eq. (21), with R about 10^{-12} cm and Z about 90, the transmission coefficient $T = e^{-\gamma}$ is found to extend over a domain of about 10^{-20} to 10^{-40}. This range of about 10^{20} is just what is needed to relate the alpha-disintegration energy to the broad domain of known alpha-decay half-periods. Equation (21) thus explains the Geiger-Nuttall rule very successfully. **Figure 8** presents a modern form of the Geiger-Nuttall relationship. The individual points show the measured half-periods and alpha-disintegration energies (alpha-particle energy plus recoil energy) for a number of high-Z emitters of alpha particles. The smooth curves are drawn using the wave-mechanical theory of transmission through nuclear barriers, with a nuclear radius of $R = 1.48 \times 10^{-15} A^{1/3}$ m, where A is the mass number of the alpha-particle decay product. The agreement between experiment and theory is good.

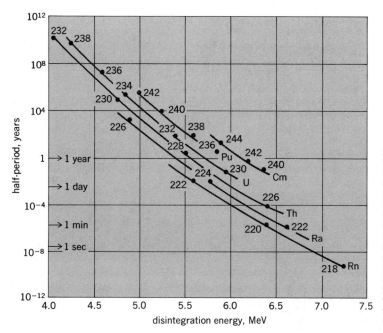

Fig. 8. Systematics of the broad range of half-periods for alpha-particle decay and their strong dependence on alpha-decay energy and weaker dependence on nuclear charge. Numbers beside experimental points are mass numbers of parent alpha-particle emitters. Lines connect parent isotopes and are drawn using wave-mechanical theory of alpha-particle transmission through nuclear potential barriers.

NUCLEAR PROBES AND PHENOMENA

Since 1970, knowledge of alpha-emitting isotopes has been greatly enlarged through the identification of many isotopes far off stability in the region just above tin and in the broad region from neodymium all the way to uranium. For example, fusion reactions between 290-MeV ^{58}Ni ions and ^{58}Ni and ^{63}Cu targets have been used to produce and study very-neutron-deficient radioactive isotopes, including 12 alpha emitters between tin and cesium. These results provide important data on the atomic masses of nuclei far from the stable ones in nature. These data test understanding of nuclear mass formulas and their validity in new regions of the periodic table.

BETA-PARTICLE DECAY

Beta-particle decay is a type of radioactivity in which the parent nucleus emits a beta particle. There are two types of beta decay established: in negatron beta decay (β^-) the emitted beta particle is a negatively charged electron (negatron); in positron beta decay (β^+) the emitted beta particle is a positively charged electron (positron). In beta decay the atomic number shifts by one unit of charge, while the mass number remains unchanged (Table 1). In contrast to alpha decay, when beta decay takes place between two nuclei which have a definite energy difference, the beta particles from a large number of atoms will have a continuous distribution of energy.

The continuous number-versus-energy distribution of emitted beta particles is illustrated in **Fig. 9**. For each beta-particle emitter, there is a definite maximum or upper limit to the energy spectrum of beta particles. This maximum energy, E_{max}, corresponds to the change in nuclear

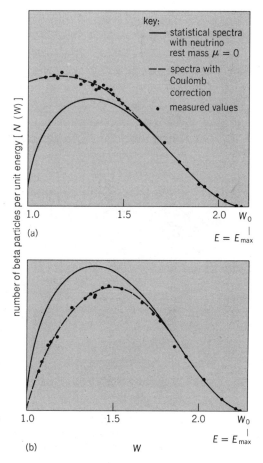

Fig. 9. Spectra of beta particles. (a) β^- decay of ^{64}Cu. (b) β^+ decay of ^{64}Cu. (*After L. M. Langer, R. D. Moffat, and H. C. Price, The beta-spectrum of Cu^{64}, Phys. Rev., 76:1725–1726, 1949*)

energy in the beta decay. Thus $E_{max} = 0.57$ MeV for β^- decay of ^{64}Cu, and $E_{max} = 0.66$ MeV for β^+ decay of ^{64}Cu. For positron decay to occur, the total decay energy must exceed 1.022 MeV (twice the rest energy of the electron). The total decay energy for β^+ decay is then $E_{max}(\beta^+)$ plus 1.022 MeV. As in the case of alpha decay, most beta-particle spectra are not this simple, but include additional continuous spectra which have less maximum energy and which leave the product nucleus in an excited level from which gamma rays are then emitted.

For nuclei very far from stability, the energies of these excited states populated in beta decay are so large that the excited states may decay by proton, two-proton, neutron, two-neutron, three-neutron or alpha emission, or spontaneous fission. In some cases, the energies are so great that the number of excited states to which beta decay can occur is so large that only the gross strength of the beta decays to many states can be studied.

Neutrinos. The continuous spectrum of beta-particle energies shown in Fig. 9 implies the simultaneous emission of a second particle besides the beta particle, in order to conserve energy and angular momentum for each decaying nucleus. This particle is the neutrino. The sum of the kinetic energy of the neutrino and the beta particle equals E_{max} for the particular transition involved except in the rare cases where internal bremsstrahlung or shake-off electrons are emitted along with the beta particle and neutrino. The neutrino has zero charge and presumably zero rest mass, travels at the same speed as light (3×10^8 m/s or 1.86×10^5 mi/s), and is emitted as a companion particle with each beta particle.

Earlier careful measurements of the beta spectra of ^{3}H established an upper limit for the neutrino rest mass as less than 0.0005 times the rest energy of the electron (**Fig. 10**). In 1980, however, some indirect evidence and a new ^{3}H beta spectra measurement yielded evidence for a rest mass much smaller than this limit, but finite. If the neutrino does have a nonzero rest mass, this will have many consequences, such as the size of the total mass of the universe but will not radically change the general features of the beta decay as presented here.

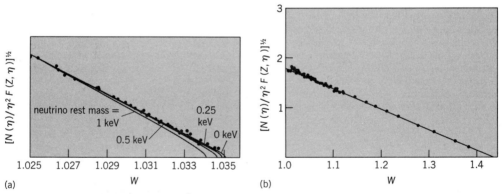

Fig. 10. Kurie plots. (a) Allowed decay of ^{3}H (after L. M. Langer and R. J. D. Moffat, The beta-spectrum of tritium and the mass of the neutrino, Phys. Rev., 88:689–694, 1952). (b) Once-forbidden decay of ^{147}Pm (after J. H. Hamilton, L. M. Langer, and W. G. Smith, The shape of the ^{143}Pr spectrum, Phys. Rev., 112:2010–2019, 1958).

Two forms of neutrinos are distinguished in beta decay. In positron beta decay, a proton p in the nucleus transforms into a neutron n in the nucleus, thus reducing the nuclear charge by 1 unit. At the time of this transition, two particles, the positron β^+ and the neutrino ν, are created and emitted. The emitted β^+ and ν together carry away the energy E_{max} of the transition and provide for conservation of energy, momentum, angular momentum, charge, and statistics. Thus positron beta decay is represented by reaction (22). Negatron beta decay is a closely related

$$p \rightarrow n + \beta^+ + \nu \tag{22}$$

process, except that a neutron n changes to a proton p in the nucleus, and a negatron beta-particle β^-

and its characteristic companion particle, the antineutrino $\bar{\nu}$, are emitted. Thus reaction (23)

$$n \to p + \beta^- + \bar{\nu} \qquad (23)$$

is written. The antineutrino is the antiparticle of the neutrino as the β^+ is the antiparticle of the β^-. The ν and $\bar{\nu}$ have the same properties of zero charge and zero rest mass, and differ only with respect to the direction of alignment of their intrinsic spin along their direction of motion. In most beta-decay contexts, the term neutrino includes both its forms, neutrino and antineutrino. Because of the fact that the neutron rest mass is greater than the proton rest mass, free neutrons can undergo beta decay (23), but protons must use part of the nuclear energy available inside a nucleus to make up the rest mass difference.

The interaction of neutrinos with matter is exceedingly feeble. A neutrino can pass all the way through the Sun with little chance of collision. The thickness of lead required to attenuate neutrinos by the factor ½ is about 10^{18} m (10^{15} mi), or 100 light-years of lead! SEE ANTIMATTER; NEUTRINO; POSITRON.

Average beta energy. Charged particles, such as beta particles or alpha particles, are easily absorbed in matter, and their kinetic energy is thereby converted into heat. In beta decay the average energy E_{av} of the beta particles is far less than the maximum energy E_{max} of the particular beta-particle spectrum. The detailed shape of beta-particle spectra and hence the exact value of the ratio E_{av}/E_{max} varies somewhat with Z, E_{max}, the degree of forbiddenness of the transition, and the sign of charge of the emitted beta particle. A rough rule of thumb which covers many practical cases is $E_{av} = (0.40 \pm 0.05) E_{max}$, with slightly higher values for positron beta-particle spectra than for negatron beta-particle spectra. The remaining disintegration energy is emitted as kinetic energy of neutrinos and is not recoverable in finite absorbers.

There are other processes that carry off part of the energy of beta decay, including internal bremsstrahlung (gamma rays) and shake-off electrons (atomic electrons). The total probabilities for these additional two processes are the order of 1% or much less per beta decay, and the probability of their emission decreases rapidly with increasing energy so they are mainly low-energy (less than about 50 keV) radiations. In internal bremsstrahlung, through an interaction of the beta particle and the emitting nucleus, part of the decay energy is emitted as a gamma ray. In the shake-off process, part of the beta-decay energy is given to one of the atomic electrons. The gamma rays are not absorbed in matter as easily as the beta particles. In addition, if one tries to absorb the beta particles in matter, the beta particles can interact with the atoms and give off external bremsstrahlung (gamma rays). The number of these gamma rays again is a strongly decreasing function of energy, but their emission extends up to the maximum energies of the beta particles.

Fermi theory. By postulating the simultaneous emission of a beta particle and a neutrino, as in reaction (22), E. Fermi developed in 1934 a quantum-mechanical theory which satisfactorily gives the shape of beta-particle spectra (Fig. 9), and the relative half-periods of beta-particle emitters for allowed beta decays. The energy distribution of beta particles in allowed transitions is then given by Eq. (24).

$$N(W) \, dW = \frac{|P|^2}{\tau_0} F(Z,W) \, (W^2 - 1)^{1/2} (W_0 - W)^2 W \, dW \qquad (24)$$

$N(W) \, dW$ = number of beta particles in energy range W to $W + dW$
$W = 1 + E/(m_0 c^2)$ = total energy of beta particle in units of rest energy $m_0 c^2 = 0.51$ MeV for an electron (m_0 = electron mass, c = velocity of light)
$W_0 = 1 + E_{max}/(m_0 c^2)$ = maximum energy of the beta-particle spectrum
$|P|^2$ = squared matrix element for the transition, and is of the order of unity for allowed transitions
τ_0 = time constant $\cong 7000$ s
$F(Z,W)$ = complex, dimensionless function involving the nuclear radius, nuclear charge, beta-particle energy, and whether the decay is β^- or β^+

Physically this distribution function involves the product of the energy W and momentum $(W^2 - 1)^{1/2}$ of the beta particle times the energy $(W_0 - W)$ and the momentum $(W_0 - W)/c$ of the neutrino. The product of these factors gives a "statistical" distribution for the number of beta particles as a function of energy as shown in Fig. 9. The observed spectra show an excess of low-energy β^- and a deficiency of low-energy β^+ particles. This arises because of the Coulomb attraction and repulsion of the nucleus for β^- and β^+. The statistical spectrum is corrected by the Fermi function, $F(Z,W)$, and the new distribution agrees with experiments as shown in Fig. 9.

Equation (24) essentially matches the energy spectra of allowed beta-particle transitions and therefore furnishes one type of experimental verification of the properties of neutrinos. Its counterpart in terms of the beta-particle momentum spectrum is often used for the analysis of spectra, and is given by Eq. (25). The momentum distribution is much more nearly symmetric than its corresponding energy spectrum.

$$N(\eta)\, d\eta = \frac{|P|^2}{\tau_0} F(Z,\eta)(W_0 - W)^2 \eta^2\, d\eta \qquad (25)$$

$N(\eta)\, d\eta$ = number of beta particles in the momentum interval from η to $\eta + d\eta$
$\eta = (W^2 - 1)^{1/2}$ = momentum of the beta particle in units of $m_0 c$
$F(Z,\eta) = F(Z,W)$ of Eq. (24)

Konopinski-Uhlenbeck theory. After the work of Fermi which explained allowed decay, E. J. Konopinski and G. E. Uhlenbeck in 1941 developed the theory of forbidden beta decay. Allowed decays occur between nuclear states which differ in spin by 0 or 1 unit and which have the same parity. Konopinski and Uhlenbeck developed a theory to describe beta decays where energy is available for decay but the allowed selection rules on spin or parity or both are violated. These beta transitions occur at a slower rate and are called forbidden transitions. In 1949 the theory of forbidden beta decay was confirmed by L. M. Langer and H. C. Price. The orders of forbiddenness, which retard the rate of decay, are: once-forbidden decay when the change in nuclear spin ΔI is again 0 or 1 as in allowed decay, but a parity charge $\Delta \pi$ occurs; once-forbidden unique decay when $\Delta \pi$ changes and $\Delta J = 2$; n-times forbidden decay when $\Delta J = n$, $\Delta \pi = (-)^n$, where $\Delta \pi = -$ indicates a parity change; and n-times forbidden unique decay when $\Delta J = n + 1$, $\Delta \pi = (-)^n$. These are illustrated in **Table 4**. In forbidden decays the first-order allowed matrix elements of the Fermi theory in Eq. (24) vanish because of the selection rules on angular momentum and spin. Then the much smaller higher-order matrix elements that can be neglected compared to the large allowed matrix elements come into play. SEE PARITY; SELECTION RULES.

Table 4. Selection rules for beta decay and log fT values

Type	ΔJ	$\Delta \pi$	Log fT	Examples
Allowed (favored)	0 or 1	No	3	n, ^{3}H
Allowed (normal)	0 or 1	No	4 to 7	^{35}S, ^{30}P
Allowed (l-forbidden)	1	No	6 to 9	^{32}P, ^{65}Ni
Once-forbidden	0 or 1	Yes	6 to 8	^{111}Ag, ^{143}Pr
Once-forbidden (unique)	2	Yes	8 to 9	^{42}K, ^{91}Y
Twice-forbidden	2	No	11 to 14	^{36}Cl, ^{59}Fe
Twice-forbidden (unique)	3	No	12 to 14	^{22}Na, ^{60}Co
Third-forbidden	3	Yes	17 to 19	^{87}Rb, ^{138}La
Third-forbidden (unique)	4	Yes	(~18)	^{40}K
Fourth-forbidden	4	No	~24	^{115}In
Fourth-forbidden (unique)	5	No		

Comparative half-lives. The half-period T of beta decay can be derived from Eq. (24) because the radioactive decay constant $\lambda = 0.693/T$ is simply the total probability of decay, or $N(W)\, dW$ integrated over all possible values of the beta-particle energy from $W = 1$ to $W = W_0$.

For allowed decays, the matrix elements are not functions of the beta energy and can be factored out of Eq. (24), so Eq. (26) is valid, where f is given by Eq. (27), and the constants include

$$\lambda = 0.693/T = \text{constants} \times f \qquad (26)$$

$$f = \int_1^W F(Z,W)\,(W^2 - 1)^{1/2}(W_0 - W)^2 W\,dW \qquad (27)$$

$|P|^2$ of Eq. (24). Equation (26) can be rearranged as Eq. (28). For different beta decays, T varies

$$fT = \frac{0.693}{\text{constants}} = \text{comparative half-life} \qquad (28)$$

over a range greater than 10^{18} and inversely depends on the beta-decay energy in analogy to the Geiger-Nuttall rule for alpha decay. However, Eq. (28) says that the comparative half-life should be a constant. Indeed it is found experimentally that different classes of beta decay do have very similar fT values. It is generally easier to give the $\log_{10} fT$ for comparison. The groups are illustrated in Table 4 and include, in addition to the forbidden decays, three classes of allowed decays: the favored or superallowed decays of nuclei whose structures are very similar so that the matrix element in the denominator of Eq. (28) is large and $\log fT$ is small; normal allowed; and allowed l-forbidden where the total angular momentum selection rule holds, but the individual particle that is undergoing beta decay has a change of 2 units of orbital angular momentum. The matrix elements for each degree of forbiddenness get progressively smaller and so $\log fT$ values increase sharply with each degree of forbiddenness. The ranges of these fT values for each degree of forbiddenness are in general so well established that measurements of fT values can be used to establish changes in spins and parities between nuclear states in beta decay.

Kurie plots. For allowed transitions, the transition matrix element $|P|^2$ is independent of the momentum η. Then Eq. (25) can be put in the form of Eq. (29). Therefore a straight line results

$$\left(\frac{N(\eta)}{\eta^2 F(Z,\eta)}\right)^{1/2} = \text{const}\,(W_0 - W) \qquad (29)$$

when the quantity $\sqrt{N/\eta^2 F}$ is plotted against beta-particle energy, either as W or as E, on a linear scale. Such graphs are called Kurie plots, Fermi plots, or Fermi-Kurie plots. These are especially useful for revealing deviations from the allowed theory and for obtaining the upper energy limit E_{max} as the extrapolated intercept of $\sqrt{N/\eta^2 F}$ on the energy axis. Practically all results on the shapes of beta-particle spectra are published as Kurie plots, rather than as actual momentum or energy spectra.

Figure 10 shows representative Kurie plots for ^{3}H and ^{147}Pm. When spectral data give a straight line, such as these, then $N(\eta)$ is in agreement with the Fermi momentum distribution, Eq. (25); and the intercept of this straight line, on the energy axis, gives the disintegration energy E_{max}. For β^+ decay the total decay energy is E_{max} (β^+) plus 1.022 MeV. In Fig. 10a, theoretical curves are given for various values of the neutrino rest mass, and the data points, which are experimental values, lie on the curve corresponding to zero mass.

In addition to allowed decays, all but one known once-forbidden decays have Kurie plots that are essentially linear in energy (Fig. 10b). The once-forbidden unique decays have a pronounced characteristic energy dependence for their matrix elements, and thus the conventional Kurie plot has a characteristic shape that differs from a straight line (**Fig. 11**). When the data are corrected by the unique shape factor, given by Konopinski and Uhlenbeck, a linear Kurie plot is again obtained. This unique shape was the key to the discovery of forbidden beta decay by Langer and Price, and Fig. 11 is, in fact, from the data with which they made this discovery. The higher-order forbidden spectra each show different strong energy dependences in their Kurie plots, each characteristic of their degree of forbiddenness.

Double beta decay. When the ground state of a nucleus differing by 2 units of charge from nucleus A has lower energy than A, then it is theoretically possible for A to emit two beta particles, either $\beta^+\beta^+$ or $\beta^-\beta^-$ as the case may be, and two neutrinos or antineutrinos, and go from Z to $Z \pm 2$. Here two protons decay into two neutrons, or vice versa. This is a second-order process and so should go much slower than beta decay. There are a number of cases where such decays should occur, but their half-lives are of the order of 10^{20} years or greater. These are obviously very difficult to detect and have not been seen directly. There is indirect evidence for

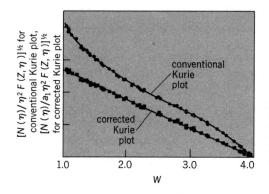

Fig. 11. Once-forbidden spectrum of ^{91}Y: conventional Kurie plot and Kurie plot corrected by the unique shape factor, $a_1 = W^2 - 1 + (W_0 - W)^2$, given by Konopinski and Uhlenbeck, which linearizes the data. (After L. M. Langer and H. C. Price, Shape of the beta spectrum of the forbidden transition of yttrium 91, Phys. Rev., 75:1109, 76:641, 1949)

double beta decay in one case, that of $^{130}_{52}$Te, from the observed buildup of $^{130}_{54}$Xe in samples. However, it cannot be said with certainty that this process has been shown to occur in nature.

Electron-capture transitions. Whenever it is energetically allowed by the mass difference between neighboring isobars, a nucleus Z may capture one of its own atomic electrons and transform to the isobar of atomic number $Z - 1$ (Table 1). Usually the electron-capture (EC) transition involves an electron from the K shell of atomic electrons, because these innermost electrons have the greatest probability density of being in or near the nucleus. SEE ELECTRON CAPTURE.

In EC transitions, a proton p bound in the parent nucleus absorbs an electron e^- and changes to a bound neutron n. The disintegration energy is carried away by an emitted neutrino ν as in Eq. (30). The residual nucleus may be left either in its ground level or in an excited level

$$p + e^- \to n + \nu \tag{30}$$

from which gamma-ray emission follows.

EC transitions compete with all cases of positron beta-particle decay. EC has an energetic advantage over β^+ decay equivalent to the mass of two electrons, or 1.02 MeV, because in Eq. (30) one electron mass e^- enters the reaction and is available, whereas in Eq. (22) one electron mass β^+ must be produced as a product of the positron beta-particle decay. For example, in the radioactive decay of $^{64}_{29}$Cu, twice as many transitions go by EC to $^{64}_{28}$Ni as go by positron beta decay to the same decay product. In the heavy, high-Z elements, EC is greatly favored over the competing β^+ decay, and examples of measurable β^+ decay are practically unknown for Z greater than 80, although there are a large number of examples of electron capture. As the energy for decay increases beyond 1.02 MeV, the probability of β^+ decay increases relative to EC and dominates at several MeV of energy.

Several examples are known of completely pure EC radioactivity in which there is insufficient nuclear energy to allow any positron beta-particle decay (total decay energy is less than 1.022 MeV). For example, $^{55}_{26}$Fe emits no positron beta particles, but transforms with a half-period of 2.6 years entirely by EC to the ground level of $^{55}_{25}$Mn. This radioactivity is detectable through the K-series x-particles which are emitted from ^{55}Mn when the atomic electron vacancy, produced by nuclear capture of a K electron, refills from the L shell of atomic electrons. Also, double electron capture, analogous to double beta decay, is theoretically predicted to exist. Here two atomic electrons are captured and two neutrinos emitted.

GAMMA-RAY DECAY

Gamma-ray decay involves a transition between two excited levels of a nucleus, or between an excited level and the ground level. A nucleus in its ground level cannot emit any gamma radiation. Therefore gamma-ray decay occurs only as a sequel of one of the processes in Table 1 or of some other process whereby the product nucleus is left in an excited state. Such additional processes include gamma rays observed following the fusion of two nuclei, as occurs in bombarding ^{58}Ni with ^{16}O to form an excited compound nucleus of ^{74}Kr. This compound nucleus first promptly gives off a few particles like two neutrons to leave ^{72}Kr* or two protons to leave ^{72}Se*, both of which will be in excited states which will emit gamma rays. Or one may excite states in a nucleus by the Coulomb force between two nuclei when they pass close to each other but do not touch (their separation is greater than the sum of the radii of the two nuclei). There are also other nuclear reactions such as induced nuclear fission that leave nuclei in excited states to undergo gamma decay. SEE COULOMB EXCITATION.

A gamma ray is high-frequency electromagnetic radiation (a photon) in the same family with radiowaves, visible light, and x-rays. The energy of a gamma ray is given by $h\nu$, where h is Planck's constant and ν is the frequency of oscillation of the wave in hertz. The gamma-ray or photon energy $h\nu$ lies between 0.05 and 3 MeV for the majority of known nuclear transitions. Higher-energy gamma rays are seen in neutron capture and some reactions.

Gamma rays carry away energy, linear momentum, and angular momentum, and account for changes of angular momentum, parity, and energy between excited levels in a given nucleus. This leads to a set of gamma-ray selection rules for nuclear decay and a classification of gamma-ray transitions as "electric" or as "magnetic" multipole radiation of multipole order 2^l, where $l = 1$ is called dipole radiation, $l = 2$ is quadrupole radiation, and $l = 3$ is octupole, l being the vector change in nuclear angular momentum. The most common type of gamma-ray transition in nuclei is the electric quadrupole (E2). There are cases where several hundred gamma rays with different energies are emitted in the decays of atoms of only one isotope. SEE MULTIPOLE RADIATION.

Mean life for transitions. A reasonably successful approximate theory of the mean life for gamma-ray decay was developed by V. F. Weisskopf in 1951, using the single-particle shell model of nuclei. **Figure 12** summarizes the numerical consequences of this theory. An E2 transition of about 1 MeV is expected to take place with a mean life, τ_{el}, or mean delay in the upper level, of about 10^{-11} s. Thus most gamma-ray transitions are prompt transitions, in which the mean life of the excited level is too short to be measured easily. Figure 12 is for electric multipole transitions. The mean life τ_{mag} for magnetic multipoles is of the order of 30 (for $A = 20$) to 150 (for $A = 200$) times longer than τ_{el}.

At low energies or high Z, or both, the internal conversion process becomes a very important additional mode of decay that markedly shortens the mean lives of the nuclear levels. In addition, in many cases the structure of the nucleus comes into play and alters the observed mean lives considerably compared to those in Fig. 12. Electric dipole (E1) transitions are generally retarded (longer mean lives) by factors of 10^6 over the Weisskopf estimates of Fig. 12. On the other hand, A. Bohr and B. Mottelson developed a model of collective nuclear motions where E2 transitions are enhanced by factors of 100 or more (shorter τ) over the Weisskopf single-particle estimates, and these predictions are confirmed by experiments. The magnetic dipole (M1) transitions are also often hindered by factors of 100 or more. Measurements of the mean lives for gamma-ray decay provide important tests of nuclear models.

Internal conversion. An alternative type of deexcitation which always competes with gamma-ray emission is known as internal conversion. Instead of the emission of a gamma ray, the nuclear excitation energy can be transferred directly to a bound electron of the same atom. Then the nuclear energy difference is converted to energy of an atomic electron, which is ejected from the atom with a kinetic energy E_i given by Eq. (31). Here B_i is the original atomic binding

$$E_i = W - B_i \qquad (31)$$

energy of the particular electron, which is ejected, and W is the nuclear transition energy which would otherwise have been emitted as a gamma-ray photon having energy $h\nu = W$.

The spectrum of internal conversion electrons is then a series of discrete energies, or "lines," each corresponding to an individual value of B_i, for the K, L (L_1, L_2, L_3) M, . . . electrons in each shell and subshell of the atom. Thus conversion electron spectra are much more complex than gamma spectra. From the spacing of the E_i values in this conversion electron spectrum, it is possible to assign definitely the atomic number Z of the atom in which the nuclear transition W took place. In this way it is known that the conversion electron and the competing gamma-ray emission are sequels and not antecedents of alpha decay, beta decay, and electron-capture transitions. Partial electron spectra, showing K, L, and M shell conversion, and gamma-ray spectra are shown in **Fig. 13** for the decay of ^{186}Tl. By comparing the K and L electron intensities of the 402 + 405 and 522 keV transitions with the gamma-ray spectrum, it can be seen that the strong 522-keV electron transition has no gamma ray associated with it. The strong 511-keV gamma ray

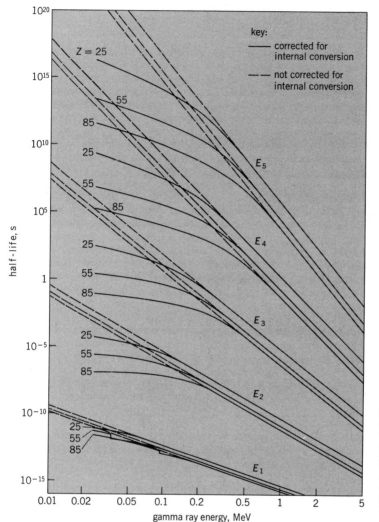

Fig. 12. Theoretical values of half-lives for decay of nuclear levels by emission of gamma rays and conversion electrons, for electric multipoles. (*After A. H. Wapstra, G. J. Nijgh, and R. van Lieshout, Nuclear Spectroscopy Tables, North-Holland, 1959*)

is from the annihilation of positrons and is not a nuclear transition, and so has no conversion electrons of this energy. One can improve the energy resolution by factors of 100–1000 over that in Fig. 13 with magnetic spectrometers so that, for example, one can separate the lines with different energies from even the five M subshells.

The total internal conversion coefficient α_T is the ratio of the number of transitions proceeding by internal conversion, N_{e_T}, to the number going by gamma-ray emission, N_γ, for any particular nuclear transformation from an excited level to a lower-lying level, as in Eq. (32). The

$$\alpha_T = \frac{N_{e_T}}{N_\gamma} \tag{32}$$

total internal conversion coefficient is a sum of the conversion coefficients for each shell [K, $L(L_1 + L_2 + L_3)$, $M(M_1 + \cdots)$, and so forth], and is given by Eq. (33), where $\alpha_K = N_{e_K}/N_\gamma$, $\alpha_L =$

$$\alpha_T = \alpha_K + \alpha_L + \alpha_M + \cdots \tag{33}$$

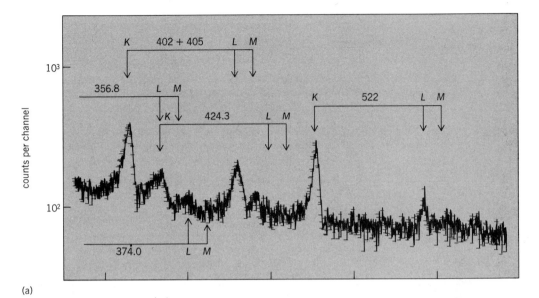

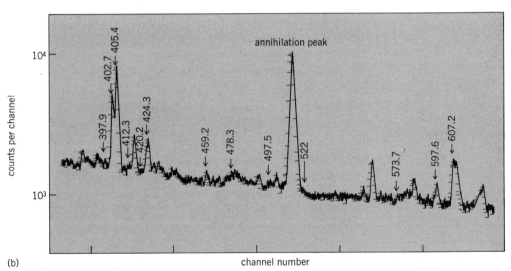

Fig. 13. Spectra from the decay of 30-s half-life ^{186}Tl far off stability (15 neutrons less than the lightest stable thallium isotope). (a) Internal conversion electrons. (b) Gamma rays. Nuclear transition energies are given in keV. (After J. H. Hamilton et al., Shape coexistence in ^{186}Hg and the decay of ^{186}Tl, Phys. Rev., C16:2010–2018, 1977)

$N_{e_L}/N_\gamma = \alpha_{L_1} + \alpha_{L_2} + \alpha_{L_3}$ ($\alpha_{L_1} = N_{e_{L1}}/N_\gamma, \ldots$), and N_{e_K}, N_{e_L} ($N_{e_{l1}}, \ldots$), and so forth are the numbers of electrons ejected from the K, L (L_1, L_2, L_3), . . . shells, respectively. In general, this probability of internal conversion relative to gamma-ray emission increases with increasing atomic number Z, with increasing multipole order 2^l, and with decreasing nuclear deexcitation energy W. In middle-weight elements, for $W = 1$ MeV, α is of the order of 10^{-2} to 10^{-4}; while for $W = 0.2$ MeV, α is of the order of 0.1 for electric $l = 2$ transitions, and 10 or larger for electric $l = 5$ transitions. When internal conversion electron decay occurs, this process is always followed by the emission of characteristic x-rays of the element and Auger electrons from outer shells when the inner shell vacancy is filled. This emission can include K x-rays, L x-rays and x-rays from higher shells, and K Auger electrons, L Auger electrons, and so forth.

Radiationless transitions. There are cases where gamma-ray emission is strictly forbidden and conversion electron emission allowed. This occurs when both nuclear states have zero spin and the same parity. The conversion electrons are called electric monopole radiations, E0. These transitions occur because of the penetration of the atomic electrons into the nuclear volume where they interact directly with the nucleus. An example of the E0 decay is shown in Fig. 13. E0 radiation can occur in principle whenever two states have the same spin and same parity, but in practice, E0 decays are found to be very, very small in these cases. There are some exceptions in well-deformed nuclei and in nuclei which have states with quite different shapes. In these cases, E0 decays can totally dominate the electron emission for transitions that have no charge in spin and that involve decays between states with large differences in their nuclear shapes (Fig. 13).

The E0 decays which arise because of the penetration of the atomic electrons into the nuclear volume are thus sensitive measures of changes in shape between two nuclear states, and have played important roles in establishing vibrations of the nuclear shape and the coexistence of states with quite different deformation in the same nucleus. There also are other circumstances where the penetration of the atomic electron into the nuclear volume gives rise to additional contributions to the conversion-electron decay. Again these penetration effects probe details of the structure of the nucleus.

Internal pair formation. When the energy between two states in the same nucleus exceeds 1.022 MeV, twice the rest mass energy of an electron, it is possible for the nucleus to give up its excess energy to an electron-positron pair—a pair creation process. This is a third alternate mode to gamma decay and conversion electron decay. This process becomes more important as the gamma-ray energy increases. It is relatively unimportant below 2–3 MeV of decay energy. SEE ELECTRON-POSITRON PAIR PRODUCTION.

Isomeric transitions. Measurably delayed radioactive transitions from an excited level of a nucleus are known as isomeric transitions. The measurably long-lived excited level is called an isomeric or metastable level or an isomer of the ground level. What constitutes an isomer is not well defined. The terminology grew up when it was difficult to measure mean lives shorter than 10^{-7} s. States with longer mean lives were isomers. Now mean lives down to 10^{-13} s can be measured for many transitions in different nuclei, but these are not generally called isomers. The break point is simply not defined.

Figure 12 shows that if the excitation energy is small (say, 0.5 MeV or less) and the angular momentum difference l is large (say, $l = 3$ or more) then the mean life of an excited level for gamma-ray or conversion-electron emission can be of the order of 1 s up to several years.

Most of the long-lived isomers occur in nuclei which have odd mass number A. Then either the number of protons Z in the nucleus is odd, or the number of neutrons N in the nucleus is odd. The frequency distribution of odd-A isomeric pairs, excited level and ground level, displays so-called islands of isomerism in which the odd-proton or odd-neutron number is less than 50, or less than 82. The distribution is one of several lines of evidence for closed shells of identical nucleons at N or $Z = 50$ or 82 in nuclei, and it plays an important role in the so-called shell model of nuclei. SEE MAGIC NUMBERS; NUCLEAR ISOMERISM.

SPONTANEOUS FISSION

This involves the spontaneous breakup of a nucleus into two heavy fragments (two intermediate atomic number elements, for example, with $Z = 42$ and 50) and neutrons, as shown in Table 1. Spontaneous fission can occur when the sum of the masses of the two heavy fragments and the neutrons is less than the mass of the parent undergoing decay. After the discovery of fission in 1939, it was subsequently discovered that isotopes like ^{238}U had very weak decay branches for spontaneous fission, with branching ratios on the order of 10^{-6}. New isotopes subsequently identified like ^{252}Cf have large (3.1%) spontaneous fission branching. In these cases, the nucleus can go to a lower energy state by spontaneously splitting apart into two heavy fragments of rather similar mass plus a few neutrons. This process liberates a large amount of energy compared to any other decay mode. Thus, ^{252}Cf has become important in many applications in medicine and industry as a compact energy source or as a source of nuclear radiation, since the fragments themselves are left in excited states and so emit gamma rays.

An important isomeric decay mode was discovered in the early 1960s in the very heavy elements, spontaneous fission isomers. Here the nucleus in an excited state, rather than emit a gamma ray or conversion electron, spontaneously breaks apart into two heavy fragments plus neutrons exactly as in spontaneous fission. To identify these isomers, the symbol f is often placed after their atomic mass, for example, $^{244f}_{95}$Am. Their half-lives are generally short, 10^{-3} to 10^{-9} s. It is now understood that these fission isomers are states with much larger deformation than the ground states of these isotopes. The Coulomb barrier against fission is in fact a double-hump barrier with the fission isomers in the valley at large deformation (**Fig. 14**). The study of these fission isomers has provided important tests of understanding of the behavior and structure of nuclei with very large deformation.

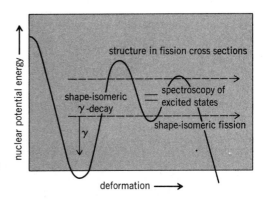

Fig. 14. Observable consequences of a double-humped nuclear potential barrier against fission. The potential well at the larger deformation gives rise to isomeric spontaneous fission.

HEAVY-CLUSTER DECAYS

In 1980 through the extension of their two-centered shell model, W. Greiner and coworkers predicted that a certain class of heavy nuclei should exhibit a new form of radioactivity intermediate between alpha decay and spontaneous fission. This mode is a new manifestation of the nuclear shell structure. It should be seen when the emission of a heavy cluster results in the decay of the parent to nucleus at or very near the strong double closed shells, $Z = 82$, and $N = 126$, at $^{208}_{82}$Pb$_{126}$. Then there is much energy available because of the very strong nuclear binding around this double magic closed shell nucleus.

In 1984 $^{223}_{88}$Ra was reported to decay back to $^{209}_{82}$Pb$_{127}$ through the emission of $^{14}_{6}$C$_{8}$ with a decay energy of 32 MeV. The branching ratio for emission of ^{12}C to α particles is $(8.5 \pm 2.5) \times 10^{-10}$. Greiner and coworkers have noted several other modes of heavy cluster decays such as $^{227}_{89}$Ac → $^{207}_{81}$Tl$_{126}$ + $^{20}_{8}$O$_{12}$, $^{232}_{92}$U → $^{208}_{82}$Pb$_{126}$ + $^{24}_{10}$Ne$_{14}$, and $^{237}_{93}$Np → $^{207}_{81}$Tl$_{126}$ + $^{30}_{12}$Mg$_{18}$. While quite rare, these modes provide very interesting tests of the understanding of the structure of heavy nuclei, including shell effects, deformation, and clustering of many protons and neutrons in a nucleus.

PROTON RADIOACTIVITY

Proton radioactivity is a mode of radioactive decay that is generally expected to arise in proton-rich nuclei far from the stable isotopes, in which the parent nucleus changes its chemical identity by emission of a proton in a single-step process. Its physical interpretation parallels almost exactly the quantum-mechanical treatment of alpha-ray decay. It is also theoretically predicted that one can have the simultaneous emission of two protons—two-proton radioactivity. Although proton radioactivity has been of considerable theoretical interest since 1951 and is expected to be a general phenomenon, so far only a few examples of this decay mode have been observed, because of the narrow range of half-lives and decay energies where this mode can compete with other modes. **Figure 15** presents the decay scheme of the first nuclide found, in 1970, to decay by proton radioactivity. It is $^{53m}_{27}$Co, where the m (metastate) denotes a (relatively) long-lived iso-

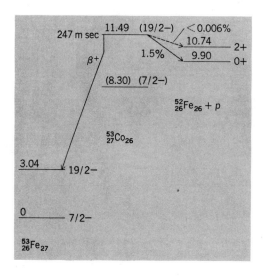

Fig. 15. Decay scheme of $^{53}_{27}$Co*. Numbers to left of levels represent energies in MeV, relative to ground state of $^{53}_{26}$Fe$_{27}$. Symbols to right of levels are spin and parity. (After J. Cerny et al., Further results on the proton radioactivity of ^{53}Co*, Nucl. Phys., A188:666–672, 1972)

meric state. Because of its very high angular momentum of 19/2 and odd parity, gamma decay is highly forbidden. This mode of decay is essentially the same as that of β-delayed proton emission discussed below, except that the energy of the excited nuclear level is low, and angular momentum selection rules highly forbid gamma-ray decay so the state lives a relatively long time in comparison to those states populated in beta decay. It was produced in the laboratory by the compound nucleus reactions $^{16}_{8}$O + $^{40}_{20}$Ca → $^{53m}_{27}$Co + p + 2n and $^{54}_{26}$F + p → $^{53m}_{27}$Co + 2n. This 247-ms isomer exhibits two different decay modes: though it predominantly decays by positron (β$^+$) emission to a similar 19/2$^-$ level in $^{53}_{26}$Fe, a 1.5% branch in its decay occurs via direct emission of a 1.59-MeV proton to the $^{52}_{26}$Fe ground state. The calculated half-life that $^{53m}_{27}$Co would possess if proton radioactivity were the only decay mode (its partial half-life for this decay branch) is the surprisingly long time of 17 s.

The first two cases of ground-state proton radioactivities were reported in 1982. These were ^{151}Lu → $^{150}_{70}$Yb + p($T_{1/2}$ = 85 ms) and $^{147}_{69}$Tm → $^{146}_{68}$Er + p ($T_{1/2}$ = 0.56 s) produced in the reactions ^{58}Ni + ^{96}Ru and ^{58}Ni + ^{92}Mo. Others have now been found as heavy-ion experiments have reached isotopes still further off stability, for example, $^{113}_{55}$Cs. Nevertheless, the windows for the observation of direct proton decays are small and, therefore, such decays are very difficult to identify. Searches for two-proton radioactivities from ground states of nuclei are being carrried out.

NEUTRON RADIOACTIVITY

In very neutron-rich nuclei near the boundary line of nucleus stability, one may find nuclei with ground states which are unstable to the emission of one or two neutrons. Here there is no Coulomb barrier, but one can have a centrifugal barrier that may give rise to one- or even two-neutron radioactivity. These processes for ground states would be very near the limits where nuclei become totally unstable to the addition of a neutron, the neutron drip line, and very difficult to even make much less measure. However, there may be neutron-rich nuclei with high-spin isomeric states where the high spin analogous to the one in ^{53m}Co gives rise to a large centrifugal barrier. Such isomeric states may undergo one- or two-neutron radioactivity.

DELAYED PARTICLE EMISSIONS

As shown in Table 1, seven types of beta-delayed particle emissions have been observed. Beta-delayed deuteron (^{2}H) and triton (^{3}H) emissions, which are not shown there, might also be expected. Over 100 beta-delayed particle radioactivities are now known. Theoretically, the number of isotopes which can undergo beta-delayed particle emission could exceed 1000. Thus, this mode, which was observed in only a few cases prior to 1965, is among the important ones in nuclei very far from the stable ones in nature. Studies of these decays can provide insights into the nucleus which can be gained in no other way.

Beta-delayed alpha radioactivity. The β^- decay of ^{214}Bi to ^{214}Po leaves the nucleus in such a high-energy excited state that it can emit an alpha particle and go to ^{210}Pb as an alternative to gamma-ray decay to lower levels in ^{214}Po. This is a two-step process with beta decay the first step. After beta decay the nucleus is in such a highly excited state that it can emit either an alpha particle or gamma ray.

The β^- delayed alpha emission has been found relatively rarely, but many cases of beta (β^+, EC) delayed alpha emission have been discovered. In proton-rich nuclei far from stability, the conditions are more favorable for beta (β^+, EC) delayed alpha emission because of the excess of nuclear charge, and number of such beta-delayed alpha emitters are not known.

Beta-delayed neutron radioactivity. In 1939, shortly after the discovery of nuclear fission, it was proposed that the delayed neutrons observed following fission were in fact beta-delayed neutrons. That is, after the nucleus fissioned, the beta decay of the neutron-rich fission fragments populated high-energy excited states that could promptly undergo dual decay, emitting either a gamma ray or neutron. Beta-delayed neutron emission is illustrated in Fig. 2. The processes of beta-delayed two- and three-neutron emission were discovered in 1979 and 1980 in the decay of ^{11}Li. The former is shown in Fig. 2, and β^-2n decays were subsequently observed in other nuclei.

Beta-delayed proton radioactivities. In addition to proton radioactivity, one can have beta-delayed proton and beta-delayed two-proton radioactivities which again ultimately result in emission of protons from the nucleus. These latter processes also occur in quite proton-rich nuclei with very high decay energies; however, they are complex two-step decay modes whose fundamental first step is beta decay.

Over 40 nuclei ranging from $^{9}_{6}$C to $^{183}_{80}$Hg have been identified to decay by the two-step mode of beta-delayed proton radioactivity. **Figure 16** presents the observed proton energy spectrum arising in the decay of $^{33}_{18}$Ar, with a half-life of 173 ms; it was produced by the $^{32}_{16}$S + $^{3}_{2}$He → $^{33}_{18}$Ar + 2n reaction. This isotope decays by superallowed and allowed β^+ decay to a number of

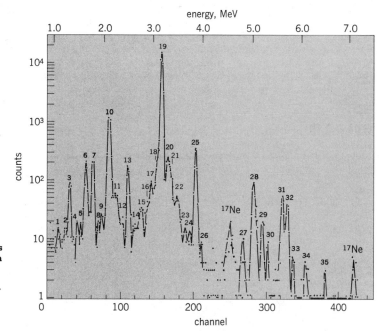

Fig. 16. Spectrum of β+-delayed protons from the decay of $^{33}_{18}$Ar as observed in a counter telescope; the proton laboratory energy is indicated at the top. Proton groups are numbered 1 through 35. (After J. C. Hardy et al., Isospin purity and delayed-proton decay: ^{17}Ne and ^{33}Ar, Phys. Rev., C3:700–718, 1971)

levels in its daughter nucleus $^{33}_{17}$Cl, which immediately (in less than 10^{-17} s) breaks up into $^{32}_{16}$S and a proton. More than 30 proton groups arising from the decay of $^{33}_{18}$Ar are observed, ranging in energy from 1 to approximately 6 MeV and varying in intensity over four orders of magnitude. Although it is normally very difficult to study many β-decay branches in the decay of a particular nuclide—because of the continuous nature of the energy spectrum of the emitted beta particles—it is possible to do so when investigating beta-delayed proton emitters. The observed proton group energies and intensities can be correlated with the levels fed in the preceding beta decay and their transition rates, thereby permitting sensitive tests via beta decay of nuclear wave functions arising from different models of the nucleus. β$^+$-delayed two-proton decay has also been discovered.

Beta-delayed spontaneous fission. There are also observed beta-decay processes where the excited nucleus following beta decay has a probability of undergoing spontaneous fission rather than gamma-ray decay. This is the same process as in spontaneous or isomeric spontaneous fission. The excitation energy of the nuclear level provides the extra energy to make fission possible. The nucleus splits into two nearly equal fragments plus some neutrons. This process is like isomeric spontaneous fission except that the lifetime of the nuclear level is so short that the level would not normally be called an isomer.

Bibliography. R. Bock (ed.), *Heavy Ion Reactions*, vol. 3, 1982; J. M. Eisenberg and W. Greiner, *Nuclear Theory*, 3 vols., 1976; J. H. Hamilton et al. (eds.), *Future Directions in Studies of Nuclei Far from Stability*, 1980; J. H. Hamilton, P. G. Hansen, and E. F. Zganjar, Advances in studies of nuclei far from stability, *Rep. Prog. Phys.*, 48:631–708, 1985; *International Conference on Nuclei Far from Stability*, CERN 81-09, 1981; K. S. Krane, *Introductory Nuclear Physics*, 1987; C. M. Lederer and V. S. Shirley (eds.), *Table of Isotopes*, 7th ed., 1978; P. B. Price et al., Discovery of radioactive decay of ^{222}Ra and ^{224}Ra by ^{14}C emission, *Phys. Rev. Lett.*, 54:297–299, 1985; F. Rösel et al., *Atomic and Nuclear Data Tables*, 21:91, 1978.

RADIOISOTOPE
HENRY E. DUCKWORTH

A radioactive isotope (as distinguished from a stable isotope) of an element. Atomic nuclei are of two types, unstable and stable. Those in the former category are said to be radioactive and eventually are transformed, by radioactive decay, into the latter. One of the three types of radioactive ray (alpha, beta, and gamma rays) is emitted during each stage of the decay.

The term radioisotope is also loosely used to refer to any radioactive atomic species. Whereas approximately a dozen radioisotopes are found in nature in appreciable amounts, hundreds of different radioisotopes have been artificially produced by bombarding stable nuclei with various atomic projectiles. *See Isotope; Radioactivity.*

HALF-LIFE
ROBLEY D. EVANS

The average time interval required for one-half of any quantity of identical radioactive atoms to undergo radioactive decay.

The activity of a source of any single radioactive substance decreases to one-half in 1 half-period, because the activity is always proportional to the number of radioactive atoms present. For example, the half-period of ^{60}Co (cobalt-60) is $t_{1/2} = 5.3$ years. Then a ^{60}Co source whose initial activity was 100 curies will decrease to 50 curies in 5.3 years. The activity of any radioactive source decreases exponentially with time t, in proportion to exp $-0.693\ t/t_{1/2}$. After 1 half-period (when $t = t_{1/2}$) the activity will be reduced by the factor $e^{-0.693} = 1/2$. In 1 additional half-period this activity will be further reduced by the factor 1/2. Thus, the fraction of the initial activity which remains is 1/2 after 1 half-period, 1/4 after 2 half-periods, 1/8 after 3 half-periods, 1/16 after 4 half-periods, and so on.

The half-period is sometimes also called the half-value time or, with less justification, but frequently, the half-life. The half-period is 0.693 times the mean life or average life of a group of identical radioactive atoms. The probability is exactly 1/2 that the actual life-span of one individual radioactive atom will exceed its half-period. *See* Radioactivity.

TRANSMUTATION
Frank H. Rockett

The nuclear change of one element into another, either naturally, in radioactive elements, or artificially, by bombarding an element with electrons, deuterons, or alpha particles in particle accelerators or with neutrons in atomic piles.

Natural transmutation was first explained by Marie Curie about 1900 as the result of the decay of radioactive elements into others of lower atomic weight. Ernest Rutherford produced the first artificial transmutation (nitrogen into oxygen and hydrogen) in 1919. Artificial transmutation is the method of origin of the heavier, artificial transuranium elements, and also of hundreds of radioactive isotopes of most of the chemical elements in the periodic table. Practically all of these elements also have been artificially transmuted into neighboring elements under experimental conditions. *See* Nuclear reaction.

NUCLEAR SPECTRA
D. J. Horen

The distribution in energy or momentum of radiations emitted by radioactive nuclei or in a nuclear reaction; also, the graphical display of data from instruments used to measure such radiations (for example, magnetic spectrometers and scintillation detectors).

Radiations can occur when the total energy of a nuclear system (or state) is higher than that of a different configuration (or state) to which a transition can take place. The specific type of radiation which is emitted in such a process is determined by the characteristics of the initial and final nuclear systems as well as the properties of the interaction mechanism. The radiations not only remove energy from the initial system but can also lead to changes in mass, charge, angular momentum (spin), and parity (or symmetry characteristic) of the system.

Nuclear spectra are widely used in both basic and applied research. For the former, it is usually the information which can be inferred from studies of such spectra rather than the radiations themselves which are of primary interest. From experimental determinations of the type of radiation, its energy, and the changes that it causes in angular momentum and party, one is able to deduce information pertaining to the static and dynamic properties of the initial and final nuclear configurations. This, in turn, enables one to gain insight into the structure of nuclei as well as the forces between nucleons. In applied research, it is the radiations themselves or the effects which they produce that are of most importance. *See* Nuclear structure.

The distribution in energy (momentum) of radiations emitted during transitions between nuclear configurations can be discrete or continuous. Transitions which give rise to discrete spectra are those in which a single type of radiation is emitted. The emitted particle then has a specific energy determined by the two nuclear configurations and type of radiation involved. On the other hand, if two or more radiations are emitted during a transition, the energy will be shared among them. In this case, the energy of each particle emitted can take on a continuum of values between well-defined limits. This then gives rise to a continuous spectrum.

Mathematically a spectrum is described as the number of particles with a given energy (that is, relative intensity) as a function of energy. Graphically, the relative intensity is usually plotted along the ordinate and the energy along the abscissa. A spectrum can be composed of a series of discrete lines (or peaks), a smooth curve, or a combination of these, depending upon the specifics of the transitions involved. In practice, the spectra of particles emitted during nuclear transitions are distorted by the detection instruments used to measure them. For discrete spectra, the instruments are usually calibrated so that the central position of the peak along the abscissa defines the energy of the detected radiation, and the area under the peak is proportional to its relative intensity. The full width at half maximum of the peak is a measure of the effective resolution. A continuous spectrum has an end point corresponding to the energy change of the transition, and it often has a definite shape that is related to the characteristics of the radiation.

In early years, the study of nuclear spectra was mainly confined to naturally occurring radioisotopes (unstable nuclei). The development of nuclear particle accelerators greatly broadened the scope of such research so that nuclear spectra characteristics of over 1600 nuclei have been measured. Some types and characteristics of nuclear spectra are described below, with those associated with the decay of radioisotopes being discussed first. SEE ISOTOPE; PARTICLE ACCELERATOR; RADIOISOTOPE.

Beta-ray spectrum. Beta rays are electrons emitted by a nucleus of atomic number Z which, in its ground state, is unstable with respect to one of its neighboring isobars of charge $Z + 1$ or $Z - 1$. Beta decay can also take place from an excited state (isomer) if radiative decay to the ground state is greatly hindered. The charged-particle emission consists of a negative electron to the $Z + 1$ nucleus or a positive electron (positron) to the $Z - 1$ nucleus, and the transition can take place to either the ground or to an excited state of the daughter nucleus.

A transition involving the emission of a beta ray is accompanied by the simultaneous emission of a neutrino, a particle that has neither mass nor charge and is exceedingly difficult to detect. The total energy of such a transition is a fixed quantity, and since this energy is shared between the beta ray and the neutrino, the beta ray can emerge with any energy from zero up to the maximum transition energy. Thus the spectrum of beta rays emitted from an ensemble of nuclei which undero such a transition is continuous. As a result of the statistical manner in which the energy is shared between the beta ray and neutrino, the spectrum of beta-ray intensity plotted versus energy has a bell shape with a broad maximum located at somewhat less than half the maximum energy.

Beta-ray spectra are usually analyzed by a so-called Fermi-Kurie plot. Such a plot converts the experimental intensity-versus-energy spectrum (corrected for source thickness and instrumental effects) into a straight-line distribution which intersects the abscissa at the maximum energy (end-point energy). For negative beta-ray emitters, the end-point energy is equal to the transition energy, except for a nuclear recoil correction. In most instances, this correction is negligible due to the small mass of the beta ray relative to the nucleus. For a positron emitter, the transition energy is equal to the end-point energy plus 1.02 MeV. If the Fermi-Kurie plot of the experimental data, under the assumption of a purely statistical intensity-versus-energy distribution, is not a straight line, the beta-ray transition is said to have a nonstatistical shape and to be of a forbidden type. The degree of forbiddenness is determined by the correction factors needed to straighten out the Fermi-Kurie plot. Whether a beta-ray transition has the allowed or forbidden shape depends upon the characteristics of the initial and final nuclear states that are involved. SEE BETA PARTICLES.

Alpha-particle spectra. The emission of an alpha particle (equivalent to the nucleus of a mass-4 helium atom) can occur when the state of a nucleus with charge Z, mass A, is unstable with respect to a state of a nucleus with charge $Z - 2$, mass $A - 4$. In order for the alpha particle to emerge, it must overcome the Coulomb-charge-potential barrier of the nucleus and also a centrifugal barrier which depends upon the angular momentum removed. Alpha-particle spectra of radioisotopes are discrete, and the peak energies are less than the corresponding transition energies by an amount equal to the nuclear recoil energy. SEE ALPHA PARTICLES.

Spontaneous fission. Some radioisotopes decay by spontaneous fission; that is, they break up into two fragments (occasionally three). The primary fission fragments and a number of their daughters are themselves radioactive and give rise to nuclear spectra. SEE NUCLEAR FISSION.

Gamma-ray spectra. Gamma rays are emitted when a transition takes place from an excited state to a lower state in the same nucleus. Of the various types of nuclear radiation, gamma rays produce the least amount of nuclear recoil, although in certain kinds of experiments the recoil energy shift is observable. Thus, the gamma-ray energy is almost exactly equal to the energy difference between the states. A process which sometimes competes with gamma-ray emission is internal electron conversion. This process creates holes in the atomic shell structure which, when filled, are accompanied by the emission of x-rays. Hence, it is not unusual to find x-rays in the low-energy regions of gamma-ray spectra.

When gamma rays interact with matter, they produce secondary radiations by means of the photoelectric and Compton effects as well as by pair production (the emission of a positron-electron pair). Instruments (such as ionization chambers, scintillators, and magnetic spectrometers) that measure gamma rays are based upon the detection of these secondary radiations. Of the various types of gamma-ray detectors, it is the scintillator (especially with LiGe crystals) which has been responsible for the gathering of enormous quantities of gamma-ray data. From a determination of gamma-ray energies and intensities, it has been possible to construct nuclear level schemes for many nuclei. Since there is little probability that the level structure of any two nuclei will be identical, precise measurements of gamma-ray transitions can serve as a means to uniquely identify a particular isotope. This feature is used extensively in applied research. SEE COMPTON EFFECT; ELECTRON-POSITRON PAIR PRODUCTION; GAMMA RAYS.

Nuclear spectra from reactions. There is a myriad of nuclear reactions by which one can produce nuclear spectra of various types (for example, gamma rays, neutrons, protons, and many nuclei). Nuclear reactions can be categorized roughly into three types: (1) elastic and inelastic scattering; (2) transfer reactions in which one or more nucleons (that is, protons or neutrons) are transferred between the projectile and target nucleus; and (3) reactions in which the projectile and target nucleus coalesce to form a compound system which then decays in some manner. In past years, the particles that could be used as projectiles were limited to the lighter elements. However, the development of heavy-ion accelerators has been undertaken to make possible the use of even the heaviest elements as projectiles.

The spectra of particles produced in a nuclear reaction depend upon the target nucleus, the reaction energy release, and the kinematics involved (including geometric and recoil effects). Usually, more than one type of radiation is emitted, and, in practice, the researcher chooses which particles to measure on the basis of the nuclear structure or reaction properties under study. Whether the spectrum of particles of any given type is discrete or continuous depends upon the specifics of the reaction used.

As an example, consider the case of inelastic scattering. Inelastic scattering takes place when the projectile gives up some of its energy to a target nucleus and is scattered. If the energy transferred to the target is sufficient only to excite discrete levels, the spectrum of the scattered particles will consist of a series of peaks corresponding to each level excited. However, if the energy transferred were sufficient to excite the nucleus up to the point where the density of states is continuous, then the corresponding portion of the spectrum would also be continuous. In the same reaction, nuclear radiations of one type or another (such as gamma rays and neutrons) will also be emitted following the decay of the states which have been excited. In most cases (but not always), the decay of discrete states excited in such reactions takes place by the emission of gamma rays and the spectra involved are also discrete. The decay of excited states in the continuum region usually takes place by neutron or proton emission, and these spectra are usually continuous. SEE NUCLEAR REACTION; RADIOACTIVITY.

Bibliography. G. F. Bertsch (ed.), *Nuclear Spectroscopy: Proceedings*, 1980; J. Cerny (ed.), *Nuclear Spectroscopy and Reactions*, pts. A, B, C, and D, 1974; H. Ejiri and T. Fukuda, *Nuclear Spectroscopy and Nuclear Interactions: New Development of Inbeam Spectroscopy and Exotic Nuclear Spectroscopy*, 1984; J. H. Hamilton and J. C. Manthuruthil (eds.), *Radioactivity in Nuclear Spectroscopy*, vols. 1 and 2, 1972; K. Siegbahn (ed.), *Alpha-, Beta-, and Gamma-Ray Spectroscopy*, vols. 1 and 2, 1965.

MULTIPOLE RADIATION
DAVID E. ALBURGER

Gamma rays, internal conversion electrons, or positron-electron pairs of defined characteristics emitted from an atom when the nucleus makes a transition between two energy states. The multipole order is the number of units of angular momentum removed by the radiation. This number is not necessarily equal to the difference between the spins of the nucleus in its initial and final states because the nuclear spin direction may change. Thus a quadrupole radiation will result when a state of spin 2 makes a transition to a state of spin 0, but a transition from a state of spin 2 to one of spin 1 may also result in quadrupole radiation if there is an appropriate change in nuclear spin direction. If I_i and I_f are the spins of the nucleus in its initial and final states, then the multipole order ΔI must be as in the equation shown below in order to conserve angular momentum.

$$|I_i - I_f| \leq \Delta I \leq |I_i + I_f|$$

From multipole radiation measurements, the static and dynamic properties of nuclear energy states may be determined, and this information may be used in theories of nuclear structure.

In addition to the energy and angular momentum values, a third characteristic of a nuclear state is the parity. The parity of a given state is odd ($-$) or even ($+$), depending on whether the quantum-mechanical wave function describing the nucleus in that state changes sign upon transposing the function to a reflected coordinate system. There are two classes of multipole radiation, the electric and the magnetic, and the designation of a given radiation depends upon both the angular momentum change and whether the parities of the initial **table** illustrates the spin and parity changes for the various multipoles. SEE PARITY; SPIN.

Spin and parity changes for various multipoles

ΔI	0	1	2	3	4	5
Multipole	Monopole	Dipole	Quadrupole	Octupole	2^4-pole	2^5-pole
Electric	E0 no	E1 yes	E2 no	E3 yes	E4 no	E5 yes
Magnetic		M1 no	M2 yes	M3 no	M4 yes	M5 no

The characteristics of multipole radiations may be determined experimentally by measurements of one or more of the following: the angular correlation of the gamma ray with another coincident or cascade gamma ray; the relative intensities of internal conversion electrons and gamma rays; the relative intensities of internal conversion electrons from various electron shells of the atom; the characteristics of internal positron-electron pair formation; and the half-life of the transition. Nuclear transitions generally take place with emission of the lowest order of multipole radiation, because the higher the order the longer the half-life. SEE RADIOACTIVITY; SELECTION RULES.

SUM RULES
GEORGE F. BERTSCH

Formulas in quantum mechanics for transitions between energy levels, in which the sum of the transition strengths is expressed in a simple form. Sum rules are used to describe the properties of many physical systems, including solids, atoms, atomic nuclei, and nuclear constituents such as protons and neutrons. The sum rules are derived from quite general principles, and are useful in situations where the behavior of individual energy levels is too complex to describe by a precise quantum-mechanical theory.

In atomic physics an important sum rule relates to the probability for an atom to absorb and emit photons, the quanta of light. In one formulation the transition probability between any two levels is expressed in terms of the wavelength of the photon, the fundamental constants of nature, and an overall dimensionless scale factor f. According to the sum rule, the f factors for all the possible transitions from a given energy level add up to a number equal to the number of electrons in the atom.

In nuclear physics there is an analogous sum rule for describing the absorption of gamma-ray photons by nuclei. The formula is written with the sum over transitions replaced by an integral over energy. The sum rule is given by the equation below, where $\sigma(E)$ is the cross section for a

$$\int_0^\infty \sigma(E) \, dE = \frac{2\pi^2 e^2 \hbar}{Mc} \frac{NZ}{A}$$

photon of energy E to be absorbed by a nucleus of mass, charge, and neutron numbers A, Z, and N; M is the mass of a proton; e is the electron charge; $\hbar$ is Planck's constant divided by 2π; and c is the speed of light.

Sum rules are used in nuclear physics to describe other kinds of transitions occurring in nuclear reactions, such as the scattering of particles from nuclei, or reactions which add or remove nucleons. In beta radioactivity, limitations on decay rates are provided by sum rules. In elementary particle physics, which treats the properties of mesons and baryons, the decays of these particles and their reactions with low-energy pions are connected by sum rules to the electromagnetic properties of the particles. SEE BARYON; ELEMENTARY PARTICLE; MESON; NUCLEAR REACTION; RADIOACTIVITY.

In general, sum rules are derived by using Heisenberg's quantum-mechanical algebra to construct operator equalities, which are then applied to particles or the energy levels of a system.

ANGULAR CORRELATIONS
A. JOHN FERGUSON

A technique of nuclear experimentation for measuring spins of nuclear states, the angular momentum mixtures of incoming or outgoing particles, and the multipole mixing of emitted gamma rays. The technique is to measure the dependence of the intensity or the cross section of a nuclear reaction on the directions of two or more radiations. The experiments to which the technique is applicable are reactions excited by a high-energy incident beam, in which the incident beam and an emitted reaction product define two directions, or radioactive decay, in which two emitted radiations, electrons or gamma rays from a decaying nucleus, are measured in coincidence. Angular correlations between an incident and an emitted beam are also called angular distributions, and have been used since the inception of nuclear physics to study the interaction potential between nucleons or nuclei and to investigate direct interaction models. A closely related class of experiments are the polarization–direction correlations, which involve the correlation between an incoming and an outgoing particle or gamma ray coupled with polarization of the incident particle or detection of the polarization of the outgoing radiation, or sometimes both. Here the polarization is equivalent to defining an additional direction in space. SEE MULTIPOLE RADIATION; NUCLEAR REACTION; SCATTERING EXPERIMENTS (NUCLEI).

Compound nucleus interpretation. The term angular correlation is normally applied to experiments that can be interpreted in terms of the compound nucleus model. It is assumed that the intermediate state has sharp spin and parity or is a mixture of a small number of overlapping states. This requirement is ordinarily satisfied for experiments with radioactive sources. For bombardment experiments the work must be carried out at a strong resonance. The correlation function is expressed by the expansion below, where $W(\theta)$ is the intensity of the reaction at an

$$W(\theta) = \sum_{k=0}^{N} a_k P_k (\cos \theta)$$

angle θ between the two radiations, $P_k (\cos \theta)$ is the Legendre polynomial of degree k for the argument $\cos \theta$, and the a_k are expansion coefficients that can be measured experimentally and also evaluated theoretically if the spins and angular momenta of the reaction are known or can be assumed. N, the highest degree of expansion, is limited by the spin of the intermediate state and the incoming and outgoing orbital momenta and gamma-ray multipolarities. If the intermediate state has definite parity, only even values of k are present; otherwise both odd and even values occur. In a typical reaction the expansion has only the terms $k = 0, 2, 4$. SEE PARITY.

$W(\theta)$ is measured at a series of angular positions in the range $0° \leq \theta \leq 180°$. Comparison with theory is made either by extracting the a_k coefficients of the above expansion and then seeking spins and mixing parameters that reproduce these, or by seeking spins and mixing parameters that reproduce the measured $W(\theta)$. A unique assignment is sometimes obtained from either analysis, but frequently two or more theoretical assignments give acceptable fits, leaving the problem unresolved.

Bombardment experiments. Bombardment experiments in medium- and heavy-mass nuclei are usually performed at energies well above the Coulomb barrier where the compound state is not resonant, so that the requirements for angular correlation analysis are not satisfied. The final nucleus is normally left in an excited, bound state of sharp spin and parity by the outgoing particle. If the outgoing particle is undetected or is detected in a counter axially symmetric about the incident beam axis, then the orientation of the excited final nucleus is described by only a few parameters. Double or triple angular correlation measurements can then be performed between the incident beam and one or two cascade gamma rays emitted by this state whose orientation parameters can be found from the correlation data.

Ambiguities in spin and mixing parameter assignments may be resolved by additional information obtained from the measurement of the correlation of adjacent transitions; the measurement of the polarization of the transition of interest; and triple angular correlations as described above. Triple correlations, which intrinsically provide much more information than double correlations, are found to be only marginally more effective than experiments in which correlation of adjacent transitions or the polarizations of transitions of interest are measured. Several cases of ambiguity cannot be resolved by any type of correlation experiment.

Radioactive decay. For radioactive sources, angular correlations are measured by using coincidences between electrons or gamma rays. Gamma–gamma correlations are analyzed by using the compound nucleus interpretation discussed above. If a gamma-ray transition is strongly or totally converted, a gamma–conversion electron correlation may be measured and will provide additional independent information. Beta-gamma polarization–direction correlations, in which the gamma-ray circular polarization is measured, provide detailed information about the beta-decay matrix elements. *See* RADIOACTIVITY.

Perturbations. Nuclear orientations can be perturbed by their own atomic electrons or by the fields of nearby atoms. For angular correlation measurements used to determine nuclear parameters, perturbations must be absent. This requirement will be satisfied for nuclear lifetimes less than 10^{-11} s since atomic processes are in general slower than this. Alternatively, perturbations have been extensively used to measure nuclear gyromagnetic ratios g and static quadrupole couplings of excited nuclear states. *See* NUCLEAR MOMENTS.

Bibliography. A. J. Ferguson, *Angular Correlation Methods in Gamma-Ray Spectroscopy*, 1965; R. D. Gill, *Gamma-Ray Angular Correlations*, 1975; K. Siegbahn (ed.), *Alpha-, Beta- and Gamma-Ray Spectroscopy*, vols. 1 and 2, 1965; P. J. Twin, Inherent ambiguities in gamma-ray angular correlation experiments, *Nucl. Instrum. Meth.*, 106:481–492, 1973.

MÖSSBAUER EFFECT
ROLFE H. HERBER

Recoil-free gamma-ray resonance absorption. The Mössbauer effect, also called nuclear gamma resonance fluorescence, has become the basis for a type of spectroscopy which has found wide application in nuclear physics, structural and inorganic chemistry, biological sciences, the study of the solid state, and many related areas of science.

Theory of effect. The fundamental physics of this effect involves the transition (decay) of a nucleus from an excited state of energy E_e to a ground state of energy E_g with the emission of a gamma ray of energy E_γ. If the emitting nucleus is free to recoil, so as to conserve momentum, the emitted gamma ray energy is $E_\gamma = (E_e - E_g) - E_r$, where E_r is the recoil energy of the nucleus. The magnitude of E_r is given classically by the relationship $E_r = E_\gamma^2/2mc^2$, where m is the mass of the recoiling atom. Since E_r is a positive number, the E_γ will always be less than the difference $E_e - E_g$, and if the gamma ray is now absorbed by another nucleus, its energy is insufficient to promote the transition from the nuclear ground state E_g to the excited state E_e.

In 1957 R. L. Mössbauer discovered that if the emitting nucleus is held by strong bonding forces in the lattice of a solid, the whole lattice takes up the recoil energy, and the mass in the recoil energy equation given above becomes the mass of the whole lattice. Since this mass typically corresponds to that of 10^{10} to 10^{20} atoms, the recoil energy is reduced by a factor of 10^{-10} to 10^{-20}, with the important result that $E_r \sim 0$ so that $E_\gamma = E_e - E_g$; that is, the emitted gamma-ray energy is exactly equal to the difference between the nuclear ground-state energy and the excited-state energy. Consequently, absorption of this gamma ray by a nucleus which is also firmly bound to a solid lattice can result in the "pumping" of the absorber nucleus from the ground state to the excited state. The newly excited nucleus remains, on the average, in its upper energy state for a time given by its mean lifetime τ (a quantity dependent on energy, spin, and parity of the nuclear states involved in the deexcitation process) and then falls back to the ground state by reemission of the gamma ray. An important feature of this reemission process is the fact that it is essentially isotropic; that is, it occurs with equal probability in all directions. SEE GAMMA RAYS.

Energy modulation. Before this phenomenon of resonance fluorescence can be turned into a spectroscopic technique, it is necessary to provide an appropriate energy modulation of the gamma ray emitted in the initial decay process. An estimate of the energy needed to accomplish this can be calculated from a knowledge of the inherent width or sharpness of the excited-state nuclear level. This is given by the Heisenberg uncertainty principle as $\Gamma = h/2\pi\tau$ (h is Planck's constant and τ is the mean lifetime of the excited state). In the case of ^{57}Fe, a nucleus for which resonance fluorescence is especially easy to observe experimentally, $\Gamma = 4.6 \times 10^{-12}$ keV. In order to modulate the emitted gamma-ray energy, which in this case corresponds to 14.4 keV, one can take advantage of the Doppler phenomenon which states that if a radiation source has a velocity relative to an observer of v, its energy will be shifted by an amount equal to $E = (v/c)E_\gamma$. Setting the required Doppler energy equal to the width of the nuclear level and E_γ equal to the nuclear transition energy leads to the equation below. Relative velocities of this order of magni-

$$v = c\frac{\Gamma}{E_\gamma} = 3 \times 10^{10} \text{ cm/s} \times \frac{4.6 \times 10^{-12}}{14.4}$$
$$= 0.0096 \text{ cm/s} = 0.0038 \text{ in./s}$$

tude can be used to modulate the gamma ray emitted in a typical Mössbauer transition, that is, to "sweep through" the energy width of the nuclear transition. SEE RADIOACTIVITY.

Experimental realization. The experimental realization of gamma-ray resonance fluorescence can be achieved with the arrangement illustrated schematically in **Fig. 1**. In a typical Mössbauer experiment the radioactive source is mounted on a velocity transducer which imparts a smoothly varying motion (relative to the absorber, which is held stationary), up to a maximum of several centimeters per second, to the source of the gamma rays. These gamma rays are incident on the material to be examined (the absorber). Some of the gamma rays (those for which E_γ is exactly equal to $E_e - E_g$) are absorbed and reemitted in all directions, while the remainder of the gamma rays traverse the absorber and are registered in an appropriate detector which causes one or more pulses to be stored in a multichannel analyzer. The electronics are so arranged that the location (address) in the multichannel analyzer, where the transmitted pulses are stored, is synchronized with the magnitude of the relative motion of source and absorber.

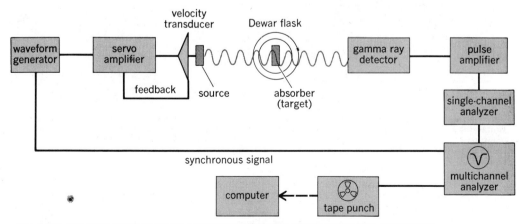

Fig. 1. Experimental arrangement for performing Mössbauer effect spectroscopy. This typical Mössbauer experiment is with ^{57}Fe or ^{119}Sn. (*After R. H. Herber, Mössbauer spectroscopy, Sci. Amer., 225(4):86–95, October 1971*)

A typical display of a Mössbauer spectrum, which is the result of many repetitive scans through the velocity range of the transducer, is shown in **Fig. 2**. Such a Mössbauer spectrum is characterized by a position δ of the resonance maximum (corresponding to a maximum in the isotropic scattering, and thus a minimum in the intensity of the transmitted radiation), a line width Γ, and a resonance effect magnitude ε corresponding to the total area A under the resonance curve. SEE GAMMA-RAY DETECTORS.

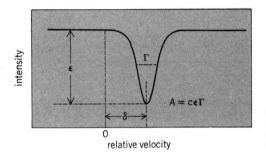

Fig. 2. Mössbauer spectrum of an absorber which gives an upsplit resonance line. The spectrum is characterized by a position δ, a line with Γ, and an A related to the effect megnitude ε.

In the case of the Mössbauer active nuclides ^{57}Fe and ^{119}Sn, among others, two additional features which are of great interest to chemists and physicists may be experimentally elucidated. One of these is the quadrupole coupling which is observed if the Mössbauer nuclide is located in an environment where the electric charge distribution does not have cubic (that is, tetrahedral or octahedral) symmetry. Such a spectrum is shown in **Fig. 3**, in which the magnitude of the quadrupole interaction Δ is equal to $e^2qQ/2$, where e is the electron charge, q is the gradient of the electrostatic field at the nucleus, and Q is the nuclear quadrupole moment. Finally, a Mössbauer spectrum can also give information on the magnitude of the magnetic field H_0 acting on the nucleus through the magnetic hyperfine interaction. This is illustrated in **Fig. 4**, where only a single resonance line would be observed in the absence of a magnetic interaction. SEE HYPERFINE STRUCTURE; NUCLEAR MOMENTS.

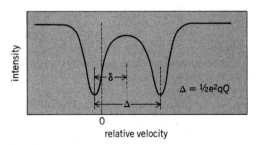

Fig. 3. Mössbauer spectrum of an absorber (containing for example ^{57}Fe or ^{119}Sn) which shows quadrupole splitting Δ.

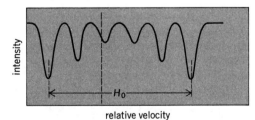

Fig. 4. Mössbauer spectrum of metallic iron showing the splitting of the resonance line by the internal magnetic field (H_0 = 33 T = 330 kG at room temperature).

Moreover, all of these parameters—δ, Δ, Γ, A, and H_0—are temperature-dependent quantities, and their study over a range of temperatures and conditions can shed a great deal of light on the nature of the environment in which the Mössbauer nuclide is located in the sample under investigation. More than one hundred Mössbauer transitions, involving 43 different elements, have been experimentally observed and reported.

Bibliography. G. M. Bancroft, *Mössbauer Spectroscopy*, 1973; T. E. Cranshaw, *Mössbauer Spectroscopy and Its Applications*, 1986; T. C. Gibb, *Principles of Mössbauer Spectroscopy*, 1976; U. Gonser (ed.), *Mössbauer Spectroscopy*, 1975; R. H. Herber (ed.), *Chemical Mössbauer Spectroscopy*, 1984; G. J. Long (ed.), *Mössbauer Spectroscopy Applied to Inorganic Chemistry*, 1984; G. K. Shenoy and F. E. Wagner (eds.), *Mössbauer Isomer Shifts*, 1978.

NEUTRON SPECTROMETRY
JOHN A. HARVEY

A generic term applied to experiments in which neutrons are used as the probe for measuring excited states of nuclides and for determining the properties of these states. The term neutron spectroscopy is also used. The strength of the interaction between a neutron and a target nuclide can vary rapidly as a function of the energy of the incident neutron, and it is different for every nuclide. At particular neutron energies the interaction strength for a specific nuclide can be very strong; these narrow energy regions of strong interactions are called resonances. The strength of the interaction, expressing the probability that an interaction of a given kind will take place, can be considered as the effective cross-sectional area presented by a nucleus to an incident neutron. This cross-sectional area is expressed in barns (1 barn = 10^{-28} m^2) and is represented by the symbol σ. The neutron total cross section of the nuclide ^{231}Pa from 0.01 to 10 eV is shown in **Fig. 1**. Even though the neutron has zero charge, neutron energies are measured in electronvolts (1 eV = 1.60×10^{-19} joule). Neutron spectroscopy covers the vast energy range from 10^{-3} eV to 10^3 MeV.

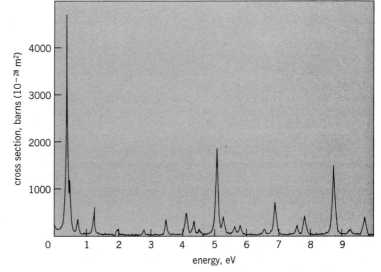

Fig. 1. Neutron total cross section of ^{231}Pa + neutron. The variation in sizes of the resonances and the nonuniform spacing of resonances are apparent.

Unbound and bound states of nuclides. Each resonance corresponds to an unbound excited state of the compound nucleus (**Fig. 2**) at an excitation energy that is the sum of the energy of the neutron and the binding energy of the neutron (4–11 MeV) which has been added to the target nuclide. The compound nucleus has a mass number which is one more than that of the target nuclide. Near the ground state of the compound nuclide the spacing of energy levels may be 10^4 to 10^6 eV. However, for a heavy nuclide, such as the compound nucleus ^{232}Pa, the excited states at an excitation energy just above the binding energy of approximately 5.5 MeV

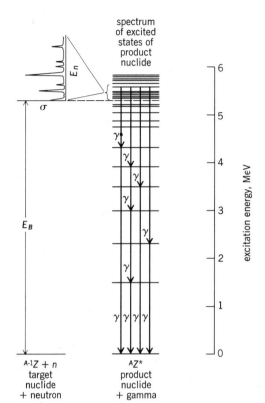

Fig. 2. Energy-level diagram for the product nucleus $^AZ^*$. The asterisk emphasizes that the product nucleus is in an excited state.

are less than 1 eV apart. To observe the individual states, the neutron energy resolution must be smaller than the level spacing. This can be achieved with low-energy neutrons, because they provide the requisite resolution, and there is no Coulomb repulsion to prevent them from entering the target nucleus. Neutron spectroscopy is presently the only technique which can provide this detailed information. For light nuclei (atomic weights ≤ 40) the spacing between the excited states can be many keV, and resonances can be resolved up to neutron energies of several MeV (**Fig. 3**). Lower-energy bound excited states of the compound nucleus below the neutron binding energy can also be studied by gamma-ray spectroscopy, observing the energy of the gamma rays emitted after the capture of neutrons at resonances or at thermal energy (Fig. 2).

Nuclear energy levels of the target nuclide can also be determined by measuring the energy spectrum of neutrons which are inelastically scattered by the target under bombardment from monoenergetic MeV incident neutrons (**Fig. 4**). The energy of an excited state is equal to the difference in energies between the incident and scattered neutrons, $E_n - E_{n'}$. If the incident neutron in Fig. 4 has energy E_{n1}, it has enough energy to excite any of the six lowest levels and emit a neutron of lesser energy than E_{n1}. A neutron of energy E_{n2} could excite only the two lowest levels and emit a neutron of lesser energy than E_{n2}. Information on these same low-energy states can be obtained by measuring the energies and intensities of the gamma rays from the deexcitation of these states excited by inelastically scattered neutrons.

Neutron reactions and resonance parameters. The abbreviated notation for neutron reactions, (n,n) and so on, lists the bombarding particle before the comma, and the emitted particle or particles after the comma. The standard symbols are: n (neutron), p (proton), d (deuteron), α (alpha particle), γ (gamma ray), f (fission), and T (total). A more complete description of the reaction lists also the target and product nuclides, for example, $^AZ(n,\gamma)$ ^{A+1}Z. The reactions most useful for neutron spectroscopy are: the total interaction; elastic scattering (n,n); radiative capture

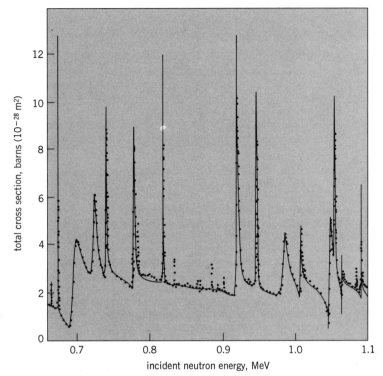

Fig. 3. Experimental neutron total cross section of sulfur compared to a theoretical fit. The fit does not include contributions from small resonances and minor isotopes of sulfur. The asymmetry of some resonances arises from the interference of resonance and potential scattering.

(n,γ); fission (n,f); inelastic scattering (n,n'); charged particle emission (n,p), (n,α), and (n,d); and three-body breakup or sequential decay $(n,2n)$ and (n,np). SEE NUCLEAR REACTION.

The resonances observed in these various reactions can be fitted by a theoretical formula to give parameters of the resonances (E_0, Γ, Γ_f, Γ_γ, Γ_n, and so forth) which correspond to detailed properties of the excited states in the compound nucleus. For example, E_0 is the resonance energy; the fission width, Γ_f, is obtained from the fission cross section; the radiation width, Γ_γ, from the capture cross section, and so forth. The neutron width, Γ_n, can be obtained from the scattering cross section or the total cross section. The total width, Γ, can be obtained if the energy resolution is less than or equal to Γ. In addition, two other properties, the angular momentum of the neutron forming the resonance and the spin, J, of the state can often be determined. For narrow resonances where Γ (in eV) $\leq 0.05 \sqrt{E_0}$ (in eV), it is necessary to consider the Doppler broadening of resonances due to the thermal motion of the target nuclides.

Neutron cross sections. The measurement of a cross section for a particular reaction consists of measuring the number of such reactions produced by a known number of neutrons incident on a known number of target nuclides. When the probability of all neutron interactions with the target nucleus is small, the number of reactions of a particular process i, per unit area and unit time using a beam of neutrons equals $(nv)(Nx)\sigma_i$. The quantity nv is the number of incident neutrons per unit area normal to their direction per unit time, N is the number of target nuclei per unit volume, x is the thickness of the target in the direction of the incident neutrons, and σ_i is the cross section per target nucleus for a particular reaction expressed in units of area. If the probability of all interactions is not small, the incident beam will be attenuated exponentially, as $\exp(-Nx\sigma_T)$, in passing through the sample, where Nx is the number of target nuclei per unit area normal to the beam, and σ_T is the neutron total cross section.

The most common type of neutron cross-section measurement, which can usually be made with the highest neutron energy resolution and usually with the most accuracy, is that of the total cross section. This measurement consists of measuring the transmission of a well-collimated beam

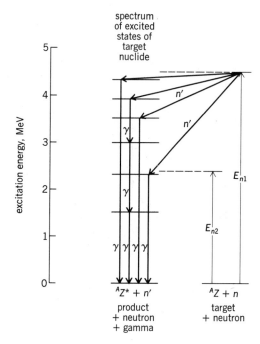

Fig. 4. Energy diagram of the target nuclide showing excitation of levels by the inelastic scattering of MeV neutrons.

of neutrons through a sample of known thickness; the transmission through the sample is simply the ratio of the intensity of the beam passing through the sample to that incident on the sample. The intensity of the incident beam is reduced in passing through the sample because the incident neutrons are absorbed or scattered by the target particles. The total cross section, σ_T, is determined from the equation $\sigma_T = -[\log_e \text{(transmission)}]/Nx$.

In order to measure partial cross sections, more elaborate equipment is needed, in general, than for total cross-section measurements, and the measurements are considerably more difficult. For example, to measure the differential elastic scattering cross section, it is necessary to measure the number of elastically scattered neutrons as a function of angle of the scattered neutron relative to that of the incident neutron. MeV-energy neutrons, in addition to being elastically scattered, can also lose energy when scattered from a target nucleus [inelastic scattering, $(n,n'\gamma)$]. Several techniques have been developed for determining inelastic cross sections, both by measuring the energy spectrum of the inelastically scattered neutrons and by measuring the energies and intensities of the gamma rays emitted from the excited nuclei. These cross sections can also be measured as a function of the angle relative to the direction of the incident neutron.

Techniques for neutron spectroscopy. Neutron spectroscopy can be carried out by two different techniques (or a combination): (1) by the use of a time-pulsed neutron source which emits neutrons of many energies simultaneously, combined with the time-of-flight technique to measure the velocities of the neutrons; this time-of-flight technique can be used for neutron measurements from 10^{-3} eV to about 200MeV; (2) by the use of a beam of nearly monoenergetic neutrons whose energy can be varied in small steps approximately equal to the energy spread of the neutron beam; however, useful "monoenergetic" neutron sources are not available from about 10 eV to about 10 keV.

Time-of-flight neutron spectrometers. Time-of-flight neutron spectrometers are the most widely used spectrometers for most neutron cross-section measurements. The time-of-flight technique requires an intense pulse of neutrons which contains neutrons of many energies and a flight path to measure the velocities of the neutrons. Various detectors are placed at the end of the flight path depending on the type of cross-section measurement. Burst widths from 10^{-5} to 10^{-9} s and flight paths from 3 to 3000 ft (1 to 1000 m) have been used. The resolution of a time-of-flight spectrometer is often quoted in microseconds or nanoseconds per meter. The energy resolution ($\Delta E/E$) is equal to $2\Delta t/t$, where Δt is the time width of the neutron burst plus the time spread in the detector, and t is the time of flight of the neutron [t (in microseconds) = 72.3 × path length (in meters)/$\sqrt{E}$ (in eV)]. The time between pulses, the flight path length, and filters in the beam must be selected so that low-energy neutrons from previous pulses do not interfere with the high-energy neutrons from following pulses. By the use of multichannel storage, usually a computer, the complete neutron spectrum (or cross section) can be obtained in one measurement with good energy resolution over a broad energy range.

The most valuable neutron source for neutron time-of-flight spectroscopy is an electron or other charged-particle accelerator capable of producing intense pulses of neutrons of short duration (on the order of 10^{-9} s). Excellent neutron cross-section measurements can be made with these spectrometers from 0.01 eV to 200 MeV. For example, a beam of 140-MeV electrons incident on a tantalum target produces neutrons with an energy distribution that has a peak at about 1 MeV and extends up to about 80 MeV. The peak of the neutron distribution for protons or deuterons incident on a heavy target occurs at a higher neutron energy than for electrons; for deuterons a broad peak occurs at about half the energy of the deuterons. Lower-energy neutrons from these accelerators are obtained by placing a moderator (about 0.8 in. or 2 cm thick) around or near the target. The duration of these moderated neutron pulses in nanoseconds is approximately $2/\sqrt{E}$ (in MeV). The flux distribution of these moderated neutrons approximately follows the relation $E^{-0.8}$ down to thermal neutron energies. With a moderated neutron source the pulse repetition must be sufficiently low, depending on the flight path length, beam filter, and energy range, to prevent overlap of neutrons from successive pulses. Typical high-resolution results obtained using a moderated source are shown in Figs. 1 and 3. *See* Particle accelerator.

Before the development of short-pulse accelerators, a mechanical chopper rotating at high speed in a well-collimated beam from a moderated fission reactor was used to produce bursts of electronvolt-energy neutrons. The neutron pulses were sufficiently short (about 10^{-6} s) and of sufficient intensity for measurements to be made up to a neutron energy of about 10^4 eV using neutron flight paths up to about 300 ft (100 m) in length. In order to produce pulses of only 10^{-6} s duration, the neutron beam had to be collimated to narrow slits (0.02 in. or 0.05 cm) to match the narrow slits through the chopper. Only when the rotation of the chopper was such that the slits in the rotor lined up with the slits in the collimator was a neutron beam with a broad energy spread passed. A fast-chopper time-of-flight spectrometer is particularly useful for transmission measurements on samples which are available in small amounts, since the sample only needs to be large enough to cover the beams passing through the narrow slits in the collimator.

For time-of-flight measurements in the energy region from about 10 keV to 1 MeV, a pulsed electrostatic accelerator using the ^{7}Li (p,n) reaction is capable of producing neutron pulses of short duration (10^{-9} s) with sufficient intensity for measurements with flight paths of a few meters. By selecting the proton-bombarding energy and a suitable target thickness, neutrons produced in the reaction at a given angle can have well-defined upper and lower energy limits. With no low-energy neutrons and short flight paths, rather high repetition rates of 10^6 Hz can be used and an energy resolution of about 1% can be realized.

The most intense pulsed neutron source used for neutron time-of-flight spectroscopy is that achieved from an underground nuclear explosion. The burst duration is about 80 nanoseconds, and the neutron distribution extends down to about 20 eV. Fission cross-section measurements have been made on very small samples of many radioactive heavy nuclides using such a source and an approximately 980-ft (300-m) flight path length. The availability of this source is obviously rather restricted, but it is unique for measurements on highly radioactive samples.

Neutron time-of-flight measurements have also been made using a pulsed fission reactor where the duration of burst is about 40 μs. Finally, subcritical boosters have been used to multiply the intensity of the neutron pulses from electron accelerators by factors of 10–200, which results in pulse durations of 0.08–4 μs.

Monoenergetic neutron spectrometers. The best technique for obtaining an intense beam of low-energy neutrons (≤10 eV) with an energy spread of only about 1% is to use a crystal monochromator placed in a well-collimated beam of neutrons from a high-flux moderated fission reactor. If a single crystal (such as beryllium, copper, or lead) is properly oriented in a collimated neutron beam, neutrons of a discrete energy, E, will be elastically scattered from a particular set of planes of atoms in the crystal through an angle 2θ given by Bragg's law $n\lambda = 2d\sin\theta$. In this equation, the integer n is the order of the reflection, λ is the neutron wavelength, d is the spacing between the planes of atoms of the particular set in the crystal, and θ is the angle of incidence between the direction of the neutron beam and the set of planes of atoms being considered. The neutron wavelength λ in centimeters equal $0.286 \times 10^{-8}/\sqrt{E}$, where E is in eV. The energy of the diffracted beam can be continuously varied by changing the angle of the crystal. Measurements of many rare-earth nuclides and heavy nuclides have been made with crystal spectrometers up to 10 eV neutron energy. Capture gamma-ray spectra have also been studied as a function of neutron energy, specifically from different neutron resonances.

"Monoenergetic" neutrons in the energy range from a few keV to 20 MeV can be obtained by bombarding various thin targets with protons or deuterons from a variable-energy accelerator such as an electrostatic accelerator. The most useful (p,n) reactions to cover the energy range from a few keV to a few MeV are those on lithium and tritium targets. The (d,n) reaction on deuterium is useful from about 1 to 10 MeV, and the (d,n) reaction on tritium from 10 to 20 MeV. In the energy range up to 1 MeV an energy resolution of about 1 keV is possible, but this resolution is usually not adequate for neutron spectroscopy for neutrons with energies less than 10^4 eV. The measurement of a complete cross-section spectrum up to 1 MeV may require 1000 sequential measurements at slightly different neutron energies. Monoenergetic neutron sources are also useful for measurements such as activation, which cannot be done with the time-of-flight technique.

Applications. Neutron spectroscopy has yielded a mass of valuable information on nuclear systematics for almost all nuclides. The distribution of the spacings between nuclear levels and the average of these spacings have provided valuable tests for various nuclear theories. The properties of these levels, that is, the probabilities that they decay by neutron or gamma-ray emission, or by fission, and the averages and distribution of these probabilities have stimulated much theoretical effort.

In addition, knowledge of neutron cross sections is fundamental for the optimum design of thermal fission power reactors and fast neutron breeder reactors, as well as fusion power reactors now in the conceptual stage. Cross sections are needed for nuclear fuel materials such as ^{235}U or ^{239}Pu, for fertile materials such as ^{238}U, for structural materials such as iron and chromium, for coolants such as sodium, for moderators such as beryllium, for shielding materials such as concrete. The optimum choice of materials for the energy region under consideration is critical to the success of the project and is of great economic significance. SEE николаев NUCLEAR STRUCTURE.

Bibliography. D. I. Garber and R. R. Kinsey, *Neutron Cross Sections*, 3d ed., vol. 2: *Curves*, 1976; J. A. Harvey (ed.), *Experimental Neutron Resonance Spectroscopy*, 1970; D. J. Hughes, *Neutron Cross Sections*, 1957; J. E. Lynn, *The Theory of Neutron Resonance Reactions*, 1968; J. B. Marion and J. L. Fowler (eds.), *Fast Neutron Physics*, vol. 1, 1959, and vol. 2, 1963; S. F. Mughabghsab, S. F. Divadeenam, and N. E. Holden, *Neutron Cross Sections*, 4th ed., vol. 1, *Resonance Parameters and Thermal Cross Sections, Part A: Z = 1–60*, 1981, Part B: Z = 61–100, 1984.

TIME-OF-FLIGHT SPECTROMETERS
Frank W. K. Firk

A general class of instruments in which the speed of a particle is determined directly by measuring the time that it takes to travel a measured distance. By knowing the particle's mass, its energy can be calculated. If the particles are uncharged (for example, neutrons), difficulties arise because standard methods of measurement (such as deflection in electric and magnetic fields) are not possible. The time-of-flight method is a powerful alternative, suitable for both uncharged and charged particles, that involves the measurement of the time t that a particle takes to travel a distance l. If the rest mass of the particle is m_0, its kinetic energy E_T can be calculated from its measured speed, $v = l/t$, using the equation below, where c is the speed of light.

$$E_T = m_0 c^2 \{[1 - (v/c)^2]^{-1/2} - 1\}$$
$$\approx m_0 v^2 / 2 \quad \text{if } v \ll c$$

Some idea of the time scales involved in measuring the energies of nuclear particles can be gained by noting that a slow neutron of kinetic energy $E_T = 1$ eV takes 72.3 microseconds to travel 1 m. Its flight time along a 10-m path (typical of those found in practice) is therefore 723 microseconds, whereas a 4-MeV neutron takes only 361.5 nanoseconds.

The time intervals are best measured by counting the number of oscillations of a stable oscillator that occur between the instants that the particle begins and ends its journey (see **illus.**). Oscillators operating at 100 MHz are in common use. If the particles from a pulsed source have different energies, those with the highest energies arrive at the detector first. Digital information from the "gated" oscillator consists of a series of pulses whose number $N(t)$ is proportional to the time-of-flight t. These pulses can be counted and stored in an on/line computer that provides many thousands of sequential "time channels," t_0, $t_0 + \Delta t$, $t_0 + 2\Delta t$, $t_0 + 3\Delta t$, . . . , where t_0 is the time at which the particles are produced and Δt is the period of the oscillator. To store an event in channel $N(t)$, the contents of memory address $N(t)$ are updated by "adding 1."

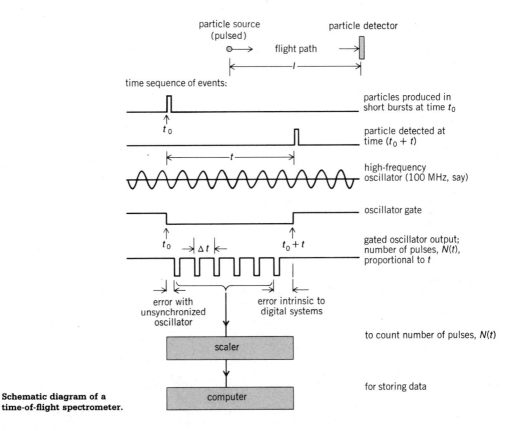

Schematic diagram of a time-of-flight spectrometer.

Time-of-flight spectrometers have been used for energy measurements of uncharged and charged elementary particles, electrons, atoms, and molecules. Their popularity is due to the broad energy range that can be covered, their high resolution ($\Delta E_T/E_T \approx 2\Delta t/t$, where δE_T and Δt are the uncertainties in the energy and time measurements, respectively), their adaptability for studying different kinds of particles, and their relative simplicity. *See* NEUTRON SPECTROMETRY.

NUCLEAR ORIENTATION
Harvey Marshak

The directional ordering of an assembly of nuclear spins I with respect to some axis in space. Under normal conditions nuclei are not oriented; that is, all directions in space are equally probable. For a system of nuclear spins with rotational symmetry about an axis, the degree of orientation is completely characterized by the relative populations a_m of the $2I + 1$ magnetic sublevels m ($m = I, I - 1, \ldots, -I$). There are just $2I$ independent values of a_m, since they are normalized to unity, namely,

$$\sum_m a_m = 1$$

Rather than specify these populations directly, it turns out to be more useful to form the moments

$$\sum_m m^\nu a_m$$

since these occur in the theoretical calculations. There are $2I$ independent linear combinations of these moments which are called orientation parameters, $f_k(I)$, and are defined by the equation below. Here $f_0(I) = 1$ and all $f_k(I)$ with $k \geq 2I + 1$ are zero.

$$f_k(I) = \binom{2k}{k}^{-1} I^{-k} \sum_m \sum_{\nu=0}^{k} (-1)^\nu \cdot \frac{(I-m)!(I+m)!}{(I-m-\nu)!(I+m-k+\nu)!} \binom{k}{\nu}^2 a_m$$

Nuclear polarization and alignment. Nuclear polarization is said to be present when one or more $f_k(I)$ with k-odd is not zero, regardless of the even $f_k(I)$ values. In this case the nuclear spin system is polarized. If all the $f_k(I)$ for k-odd are zero and at least one $f_k(I)$ for k-even is not zero, nuclear alignment is said to be present; that is, the nuclear spin system is aligned. Simply stated, if the z axis is the axis of quantization of the nuclear spin system, polarization represents a net moment along the z axis, whereas alignment does not. Unfortunately, the term nuclear polarization is usually associated with $f_1(I)$, and nuclear alignment with $f_2(I)$, although their meanings are in fact much more general. There are other definitions of nuclear orientation parameters; they are mathematically equivalent to the one above. If the nuclear spin system does not have cylindrical symmetry, a more general definition of nuclear orientation is needed leading to the statistical tensors. SEE SPIN.

Production. Nuclear orientation can be achieved in various ways. The most obvious way is to modify the energies of the $2I + 1$ magnetic sublevels so as to remove their degeneracy and thereby change the populations of these sublevels. The spin degeneracy can be removed by a magnetic field interacting with the nuclear magnetic dipole moment, or by an inhomogeneous electric field interacting with the nuclear electric quadrupole moment. Significant differences in the populations of the sublevels can be established by cooling the nuclear sample to low temperatures T such that T is in the region around $\Delta E/k$, where ΔE is the energy separation of adjacent magnetic sublevels of energy E_m, and k is the Boltzmann constant. If the nuclear spin system is in thermal equilibrium, the populations a_m are given by the normalized Boltzmann factor

$$\exp(-E_m/kT) \Big/ \sum_m \exp(-E_m/kT)$$

This means of producing nuclear orientation is called the static method. In contrast, there is the dynamic method, which is related to optical pumping in gases. There are other ways to produce oriented nuclei; for example, in a nuclear reaction such as the capture of polarized neutrons (produced by magnetic scattering) by unoriented nuclei, the resulting compound nuclei could be polarized. In addition to polarized neutron beams, polarized beams of protons, deuterons, tritons, helium-3, lithium-6, and other nuclei have been produced.

Applications. Oriented nuclei have proved to be very useful in various fields of physics. They have been used to measure nuclear properties, for example, magnetic dipole and electric quadrupole moments, spins, parities, and mixing rations of nuclear states. Oriented nuclei have been used to examine some of the fundamental properties of nuclear forces, for example, nonconservation of parity in the weak interaction. Measurement of hyperfine fields, electric-field gradients, and other properties relating to the environment of the nucleus have been made by using oriented nuclei. Nuclear orientation thermometry is one of the few sources of a primary temperature scale at low temperatures. Oriented nuclear targets used in conjunction with beams of polarized and unpolarized particles have proved very useful in examining certain aspects of the nuclear force. SEE HYPERFINE STRUCTURE; NUCLEAR MOMENTS; NUCLEAR STRUCTURE; PARITY.

DYNAMIC NUCLEAR POLARIZATION
JAMES M. DANIELS AND B. SHIVAKUMAR
B. Shivakumar wrote the section Laser-Induced Polarization.

The creation of assemblies of nuclei whose spin axes are not oriented at random, and which are in a steady state that is not a state of thermal equilibrium. Under commonly occurring conditions, the spin axes of nuclei (with nonzero spin) are oriented at random; where this is not so, the nuclei are said to be polarized. Assemblies of polarized nuclei are not in a state of thermal equilibrium except under rather extreme conditions (for example, temperatures below 10 millikelvins or 0.02°F above absolute zero, and magnetic fields greater than several teslas), and therefore schemes have been devised to produce polarized assemblies, in a steady state which is not a state of thermal equilibrium, under less extreme conditions of temperature and so forth. Such schemes constitute dynamic nuclear polarization.

Among the many applications of polarized nuclei are the following. Nuclear forces are spin-dependent, and although the spin-dependent part can be found by using unpolarized assemblies, the experiments are simpler and their interpretation is clearer if polarized nuclei are used. Assemblies of polarized nuclei have a lower geometrical symmetry than assemblies of randomly oriented nuclei, and so these have been used to investigate the fundamental symmetries of nature. Polarized nuclei have been used to enhance the signal in free precession magnetometers and similar instruments, and the use of an assembly of polarized nuclei as a gyroscope has also been suggested. SEE NUCLEAR ORIENTATION; PARITY; SPIN.

The simplest and most common parameter used to define the state of polarization is called the "polarization" or vector polarization **P**. It is a vector whose component along any given axis (for example, the z axis) is the average value of the projection on that axis of a unit vector parallel to the nuclear spin axis. Often, nuclei lie with their spins pointing along the z axis in equal numbers in both directions, but none lies in the x-y plane. Such a system is not random, but **P** = 0. The parameter used to quantify this arrangement is the average value of $(3/2 \cos^2 \Theta - 1/2)$ and is one component of the so-called tensor polarization. In general, the average value of any zonal harmonic is a measure of a kind of polarization. It is a property of nuclear spin operators that harmonics of order greater than $2J$ vanish identically, where J is the nuclear spin. In particular, spin 0 nuclei cannot be polarized and spin $1/2$ nuclei can have a vector polarization only.

There are, in general, two common features of all dynamic nuclear polarization schemes. There is some external system or interaction, for example, an oscillating magnetic field, which disturbs the system; this is called the pumping mechanism. There are also processes which tend to restore the system to thermal equilibrium; these are called relaxation processes. However, the functions of these two processes are not easily distinguished, for, in many schemes, the relaxation mechanisms actually play a part in setting up the nuclear polarization, this being one of the stages back to thermal equilibrium.

Microwave pumping. Consider an electron spin and a nucleus of spin $1/2$ (for example, a proton) in close association. Typically, the electron spin is part of a paramagnetic ion, and the proton is a hydrogen nucleus in a nearby water molecule. In a magnetic field of about 1 T, the energy levels of this combination are as shown in **Fig. 1**. M_e and m_p are, respectively, the electron and proton spin magnetic quantum numbers which specify whether the electron and proton spins are oriented in the same direction as the magnetic field or in the opposite direction. An oscillating magnetic field is applied to induce transitions between pairs of levels. The two transitions A and B, in which only an electron spin is reversed, constitute the "allowed" transition. The two weaker transitions, C and D, in which a nuclear spin is reversed, occurring one at a higher and the other

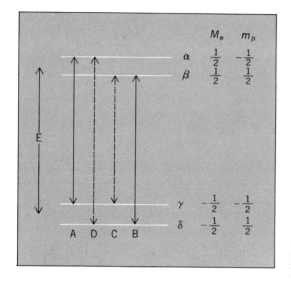

Fig. 1. Energy levels of an electron and a proton in a magnetic field.

at a lower field (or frequency) than the allowed transition, are called forbidden transitions. *See* Selection rules.

In the most common and most successful mechanism of this type for polarizing protons, one of the forbidden transitions (C, for instance) is excited strongly enough to saturate it. Then the populations of levels β and γ are equal, and are kept equal by the oscillating magnetic field. The strongest relaxation processes, which try to bring the system into thermal equilibrium, are associated with the strongest transitions A and B. As a result of these processes, fewer systems occupy level α than level γ, and similarly, fewer systems occupy level β than level δ. Thus more proton spins are oriented in one direction ($m_p = +\frac{1}{2}$) than in the opposite direction ($m_p = -\frac{1}{2}$), and a nuclear polarization results. By saturating the other forbidden transition, an equal and opposite polarization results.

In another method the allowed transition is saturated, and so the populations of levels α and γ are equal, and similarly the populations of levels β and δ are equal. The strong relaxation processes, which involve only the inversion of an electron spin, are ineffective since they are swamped by the radio-frequency field. The "cross relaxation" processes, which connect level α with level δ (called the flip-flop relaxation), and which connect level β with level γ (called the flip-flip relaxation), are of different strengths. This is because in the flip-flop transition, spin angular momentum is conserved, but not so in the flip-flip transition. The relative populations of the levels are thus determined by the relative strengths of the two processes. If, as in the case of metals, the flip-flip transition can be neglected, the flip-flop process results in fewer systems occupying level α than level δ, and a nuclear polarization results. If the electron spin and the nuclear spin are coupled by magnetic dipole interaction, and direct nuclear spin lattice relaxation can be neglected, the flip-flip process is twice as strong as the flip-flop process, and a smaller nuclear polarization of the opposite sign results.

Although this explanation describes very effectively what happens, it is oversimplified. Nuclei do not interact in turn with individual electron spins, and the latter are not independent. Especially when protons are polarized in organic materials, the effect of the oscillating magnetic field is to cool the assembly of interacting electron spins, which then cool the nuclei by thermal contact. These processes are known by the acronym DONKEY effects.

In order to produce large polarizations, the external magnetic field should be of the order of 2 T, the frequency of the oscillating field should be about 70 GHz, and the temperature should be 1 K (2°F above absolute zero) or less. Values of polarization of 0.8 or more (corresponding to 90% of the spins oriented in one direction and only 10% in the opposite direction) are regularly obtainable for protons polarized in $La_2Mg_3(NO_3)_{12} \cdot 24H_2O$ in which 1% of the lanthanum, La, is replaced by neodymium, Nd; in an empirically determined mixture consisting mainly of butanol, C_4H_9OH, and about 1% of porphyrexide, a free radical; or in ethylene glycol or propanediol with about 1% of an organic Cr^V complex. Deuteron polarizations of 0.4 have been obtained by using C_4D_9OD and porphyrexide at temperatures of 0.3 K (0.5°F above absolute zero).

Butanol and diol systems are favored to make targets of polarized protons for experiments in high-energy physics. The beam passing through the target causes radiation damage which adversely affects thepolarization. To a first approximation, the polarization P falls off as an exponential function of the dose ϕ, given by the equation below, where P_0 is the initial polarization

$$P = P_0 \exp(-\phi/\phi_0)$$

and ϕ_0 is approximately 2.5×10^{15} minimum ionizing particles per square inch (4×10^{14} minimum ionizing particles per square centimeter); but this levels off to a constant value of approximately $0.7 P_0$. This effect is caused by the production of free radicals, and free electrons which are trapped to form color centers, both of which are paramagnetic. Most of the radiation damage can be annealed out by raising the temperature of the target material to about $-208°F$ (140 K) for butanol, or $-136°F$ (180 K) for diols, and keeping it there for about 15 min.

Polarization by rotation. This method was most spectacularly demonstrated by rotating a crystal of yttrium ethylsulfate. $Y(C_2H_5SO_4)_3 \cdot 9H_2O$, in which a few percent of yttrium, Y, was replaced by ytterbium, Yb, in a magnetic field of 1 T at a temperature of 1.4 K (2.5°F above absolute zero), at a rate of 60 revolutions per second. The proton polarization was found to be 0.19 (corresponding to 59.5% of the spins oriented in one direction and 40.5% in the other direction); higher values can be obtained by modifying the experimental conditions. The splitting of the two energy levels of the protons in the external magnetic field is independent of the orientation of the crystal, but the splitting of the two energy levels of the Yb^{3+} ion depends strongly on the orientation of the magnetic field relative to the c axis of the crystal, being quite large when these are parallel and almost zero when they are perpendicular. The Yb^{3+} ions are strongly coupled to the lattice when this angle is 45°, but almost isolated when it is 0 or 90°. Thus, as the crystal is rotated, when it reaches the 45° position, the Yb^{3+} ions quickly come to equilibrium at the temperature of the lattice. Then, as the crystal is rotated to the 90° position, the energy levels of the Yb^{3+} ions move together; this is equivalent to adiabatic demagnetization, and the spin temperature of the YB^{3+} ions drops. Near the 90° position, the Yb^{3+} ions can exchange energy with the protons, because the two splittings are the same, and a flip-flop transition involving a Yb^{3+} ion and a proton conserves energy. The protons and the Yb^{3+} ions thus come to thermal equilibrium at the temperature of the Y^{3+} ions after demagnetization. Continuous rotation repeats the cycle, and the ideal steady state is one in which the proton polarization is the same as the polarization of the Yb^{3+} spins at the 45° position.

Optical pumping. This method, pioneered by Alfred Kastler and Jean Brossel, makes use of the fact that circularly polarized light carries angular momentum, and when it is absorbed, that angular momentum is given to the absorber. Typically circularly polarized resonance radiation (such as sodium-D radiation) is incident on, for example, sodium vapor at room temperature (at a pressure of about 10^{-6} torr or 10^{-8} psi or 10^{-4} pascal) in a small magnetic "guide" field directed along the direction of the light beam. It excites only the $\sigma+$ transitions in which the magnetic quantum number of the absorbing atom, which specifies its angular momentum along the direction of the magnetic field, increases by 1. When the excited state decays, on the average there is no change in the magnetic quantum number. Thus an equilibrium is set up in which the ground-state atom spins are polarized. Polarizations of almost 1.0 can easily be obtained for the alkali metals.

The absorbing atoms are usually mixed with a gas such as argon, called a buffer gas, at a pressure of $1-100$ torr ($0.01-1$ psi or 10^2-10^4 Pa) to delay their diffusion to the walls of the container where they may be depolarized. If some other gas is present instead of, or in addition to, the buffer gas, the angular momentum of the absorber is transferred in collisions to the atoms of this gas and can end up as a nuclear polarization.

This method has been very successfully applied to ^3He, which acts as a buffer gas. The absorbing atoms are metastable 2^3S_1 ^3He atoms produced by striking a weak electrodeless discharge in the ^3He gas. Polarizations of up to 0.4 have been achieved, and under suitable conditions, when the polarizing mechanism is turned off, the polarization decays with a time constant of the order of several days.

Laser-induced polarization. Nuclear orientation through optical pumping can also be achieved by using lasers. This method has been refined with the availability of high-power tunable continuous-wave dye lasers. The notion of a preferred direction in space is passed on to an ensemble of randomly oriented nuclei through atomic transitions induced by a polarized laser beam.

The energy levels in an atom exhibit fine and hyperfine structure. The hyperfine structure arises through the interaction of the magnetic and quadrupole moment of the nucleus with the atomic electrons. The hyperfine energy levels are classified in terms of the total angular momentum of the electron-nuclear system $\vec{F}$. $\vec{F} = \vec{I} + \vec{J}$, where $\vec{I}$ and $\vec{J}$ are the electron and nuclear angular momenta respectively. F can take the smaller of $2I + 1$ or $2J + 1$ values. The projection of $\vec{F}$ on an arbitrary (quantization) axis, labeled M_F, can take on values in the range $-F$ to F in integral steps. In the absence of electromagnetic fields, the various M_F levels associated with a given F have the same energy and, under normal conditions in a large ensemble of atoms, are equally likely to be populated. Optical transitions can be induced by using a polarized laser to alter the random distribution of population among the M_F states; this constitutes nuclear orientation. The transitions are governed by selection rules. For electric dipole transitions between states of opposite parity, F can change either by $+1$, -1, or 0 (transitions between a state $F = 0$ and another $F' = 0$ are not allowed). M_F can change by $+1$, -1, or 0. The M_F selection rule depends further on the polarization of the laser light. SEE HYPERFINE STRUCTURE; NUCLEAR MOMENTS.

As an example, consider the optical pumping of a sodium -23 (^{23}Na) atom using σ^+ (circularly) polarized laser light. The quantization axis is chosen as the direction of propagation of the laser beam. For the ground state of ^{23}Na, $I = 3/2$, $J = 1/2$, and $F = 1, 2$. For the first excited state of opposite parity, $I = 3/2$, $J = 1/2$, and $F' = 1, 2$. These levels are shown in **Fig. 2**a. The allowed transitions ($\Delta M_F = 1$) from the $F = 1$ ground state to the $F' = 1, 2$ excited states are shown in Fig. 2b as solid arrows. Transitions from and to the $F = 2$ level have been omitted for clarity. The spontaneous decay paths from the excited states have been shown as broken arrows. This process of excitation and decay constitutes an optical pumping cycle. A cursory inspection of Fig. 2b reveals that transitions populating the $M_F = 1$ level far outnumber those depleting it. After relatively few such pumping cycles the population in the atomic ensemble is found predominantly in the higher M_F levels. Under these conditions the spins of most of the nuclei are oriented in the direction of the laser beam. Such a technique has been used successfully to produce several species of nuclei with essentially all of the nuclear spins oriented in the same direction. Nuclear orientation can also be achieved by using a linearly polarized laser beam.

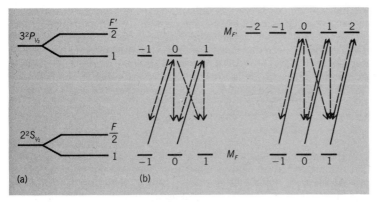

Fig. 2. Optical pumping of sodium-23. (a) Fine and hyperfine levels in the ground and first excited state of ^{23}Na. (b) An optical pumping cycle showing the allowed transitions, for excitation with a σ^+ polarized laser beam, from and to the $F = 1$ hyperfine level in the ground state. For the excitation, M_F can change by only $+1$, and for the decay, M_F can change by $+1$, -1, or 0.

The extent to which nuclei can be oriented in this manner is limited by several factors. Some of the factors limiting the efficiency of an optical pumping cycle are: how well the laser bandwidth is matched to the absorption profile of the transitions being pumped; the oscillator strengths of the atomic transitions; the length of the laser-atom interaction time; and excited and ground-state relaxation through atom-atom and atom-wall collisions and through precession in stray magnetic fields. The thrust of research in the field of laser-induced nuclear orientation has been in overcoming these difficulties.

Bibliography. C. E. Bemis, Jr., and H. K. Carter (eds.), *Lasers in Nuclear Physics*, Nuclear Science Research Conference Series, vol. 3, 1982; R. C. Fernow, Radiation damage in polarized target materials, *Nucl. Instrum. Meth.*, 148:311–316, 1978; D. E. Murnick and M. S. Feld, Applications of lasers to nuclear physics, *Annu. Rev. Nucl. Part. Sci.*, 29:411–454, 1979; G. G. Ohlsen et al. (eds.), *Polarization Phenomena in Nuclear Physics, 1980*, 1981; W. J. Thompson and T. B. Clegg, Physics with polarized nuclei, *Phys. Today*, 32(2):32–39, 1979.

ISOTOPE SHIFT
Peter M. Koch

A small difference between the different isotopes of an element in the transition energies corresponding to a given spectral line transition. For a spectral line transition between two energy levels a and b in an atom or ion with atomic number Z, the small difference $\Delta E_{ab} = E_{ab}(A') - E_{ab}(A)$ in the transition energy between isotopes with mass numbers A' and A is the isotope shift. It consists largely of the sum of two contributions, the mass shift (MS) and the field shift (FS), also called the volume shift. The mass shift is customarily divided into a normal mass shift (NMS) and a specific mass shift (SMS); each is proportional to the factor $(A' - A)/A'A$. The normal mass shift is a reduced mass correction that is easily calculated for all transitions. The specific mass shift is produced by the correlated motion of different pairs of atomic electrons and is, therefore, absent in one-electron systems.

It is generally difficult to calculate precisely the specific mass shift, which may be 30 times larger than the normal mass shift for some transitions. The field shift is produced by the change in the finite size and shape of the nuclear charge distribution when neutrons are added to the nucleus. Since electrons whose orbits penetrate the nucleus are influenced most, S-P and P-S transitions generally have the largest field shift.

For very light elements, $Z \lesssim 37$, the mass shift dominates the field shift. For $Z = 1$, the 0.13 nanometer shift in the red Balmer line led to the discovery of deuterium, the $A = 2$ isotope of hydrogen. For medium-heavy elements, $38 \lesssim Z \lesssim 57$, the mass shift and field shift contributions to the isotope shift are comparable. For heavier elements, $Z \gtrsim 58$, the field shift dominates the mass shift. A representative case is shown in the **illustration**.

When isotope shift data have been obtained for at least two pairs of isotopes of a given element, a graphical method introduced by W. H. King in 1963 can be used to evaluate quantitatively the separate contributions of the mass shift and the field shift. Experimentally determined field shifts can be used to test theoretical models of nuclear structure, shape, and multipole moments. Experimentally determined specific mass shifts can be used to test detailed theories of atomic structure and relativistic effects. *See* Nuclear moments; Nuclear structure.

Experimental techniques that have greatly increased both the amount and the precision of

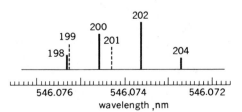

Some isotope shifts in the green line of mercury, $Z = 80$. In this heavy element the contribution of the field shift is much larger than that of the mass shift.

isotope shift data that can be obtained include on-line isotope separators for the study of isotopes with half-lives as short as a few seconds and spectroscopic methods employing high-resolution tunable lasers. Active development of isotope separation schemes based on these laser techniques has been undertaken. Isotope shift data have also been obtained for x-ray transitions of electrons in inner atomic shells and of muons in muonic atoms.

HYPERFINE STRUCTURE
Louis D. Roberts

A closely spaced structure of the spectrum lines forming a multiplet component in the spectrum of an atom or molecule, or of a liquid or solid. In the emission spectrum for an atom, when a multiplet component is examined at the highest resolution, this component may be seen to be resolved, or split, into a group of spectrum lines which are extremely close together. This hyperfine structure may be due to a nuclear isotope effect, to effects related to nuclear spin, or to both.

Isotope effect. The element zinc, for example, has three relatively abundant naturally occurring nuclear isotopes, ^{64}Zn, ^{66}Zn, and ^{68}Zn. The radius of a nucleus increases with the nuclear mass and, for a given element, the Coulomb interaction of the nucleus with the atomic s-electrons will be slightly weaker when the nuclear size is larger. This nuclear size effect causes a slight shift of certain of the spectrum lines, and this shift will be different for each isotope. For a mixture of ^{64}Zn, ^{66}Zn, and ^{68}Zn, certain of the multiplet components will thus consist of three closely spaced lines, one line for each isotope. A study of the isotope effect for an element leads, for example, to information about the dependence of the nuclear size on isotope mass, that is, on the number of neutrons in the isotope. *See Isotope shift*.

Structure due to nuclear spin. For the zinc isotopes discussed above, the nuclear spin $I = 0$, and these nuclei will be nonmagnetic and, in effect, have a spherical shape. If $I \neq 0$, however, two new nuclear properties may be observed. The nucleus may have a magnetic moment, and the shape of the nucleus may not be spherical but rather may be that of a prolate or oblate spheroid; that is, it may have a quadrupole moment. *See Spin*.

Atoms and molecules. If the electrons in an atom or a molecule have an angular momentum, the electron system may likewise have a magnetic moment. An electron quadrupole moment may also exist. The magnetic moment of the nucleus may interact with the magnetic moment of the electrons to produce a magnetic hyperfine structure. The quadrupole moments of the nucleus and of the electrons may couple to give an electric quadrupole hyperfine structure. In a simple example, the magnetic and the electric quadrupole hyperfine structure may be described by an energy operator H_{hfs} which has the form below, where $\bar{S}$ is an operator describing electron spin, $\bar{I}$

$$H_{\text{hfs}} = A\bar{I} \cdot \bar{S} + P(\bar{I}_z^2 - 3I(I+1))$$

and $\bar{I}_z$ are operators describing the nuclear spin and its z component, and I gives the magnitude of the nuclear spin. A and P are coupling constants which may take positive or negative values, and may range in magnitude from zero to a few hundred meters^{-1}. The term in A describes a magnetic, and the term in P a quadrupole hyperfine structure.

The measurement of a hyperfine structure spectrum for a gaseous atomic or molecular system can lead to information about the values for A and P. These values may be interpreted to obtain information about the nuclear magnetic and quadrupole moments, and about the atomic or molecular electron configuration.

Important methods for the measurement of hyperfine structure for gaseous systems may employ an interferometer, or use atomic beams, electron spin resonance, or nuclear spin resonance.

Liquid and solid systems. Hyperfine structure coupling may also occur and may be measured for liquid and solid systems. For liquids and solids, measurements are often made by electron spin or nuclear spin resonance methods. For solids, and for radioactive nuclei, one may, for example, also employ the Mössbauer effect or the angular correlation of nuclear gamma rays. *See Gamma rays; Mössbauer effect*.

For a diamagnetic solid, $A = 0$ in the equation above, and if the crystalline environment of an atom is cubic, $P = 0$ also. If this environment is not cubic, P may have a finite measurable value. If the solid is paramagnetic, ferromagnetic, or antiferromagnetic, A may be finite and measureable, and again P may or may not be zero depending on whether the atomic environment is cubic or not.

One may gain information about the nuclear moments and about electron bonding and magnetic structure from measurements of hyperfine structure for liquids and solids. Such measurements are extensively used, for example, in atomic and condensed matter physics, chemistry, and biology. SEE NUCLEAR MOMENTS.

Bibliography. A. Abragam, *The Principles of Nuclear Magnetism*, 1961, reprint 1983; L. C. Biedenharn and J. C. Louck, *Angular Momentum in Quantum Physics*, 1984; R. S. Raghavan and D. E. Murnick (eds.), *Hyperfine Interactions IV*, 1978; M. E. Rose, *Elementary Theory of Angular Momentum*, 1957.

NUCLEAR COLLISIONS AND REACTIONS

Scattering experiments (nuclei)	**104**
Coulomb excitation	**113**
Quasiatom	**114**
Nuclear molecule	**122**
Nuclear reaction	**127**
Deep inelastic collisions	**132**
Nuclear fusion	**133**
Thermonuclear reaction	**149**
Proton-proton chain	**149**
Carbon-nitrogen-oxygen cycles	**152**
Nuclear fission	**154**
Chain reaction	**162**
Spallation reaction	**163**
Anomalons	**164**

SCATTERING EXPERIMENTS (NUCLEI)
K. A. Erb

Experiments in which beams of particles such as electrons, nucleons, alpha particles and other atomic nuclei, and mesons are deflected by elastic collisions with atomic nuclei. Much is learned from such experiments about the nature of the scattered particle, the scattering center, and the forces acting between them. Scattering experiments, made possible by the construction of high-energy particle accelerators and the development of specialized techniques for detecting the scattered particles, are one of the main sources of information regarding the structure of matter. SEE PARTICLE ACCELERATOR; PARTICLE DETECTOR.

In the broad sense, any nuclear reaction is an example of scattering. However, this article treats elastic scattering only in the more restricted sense given below, and only insofar as it involves atomic nuclei. SEE COULOMB EXCITATION; GIANT NUCLEAR RESONANCES; NEUTRON SPECTROMETRY; NUCLEAR REACTION; NUCLEAR SPECTRA; NUCLEAR STRUCTURE.

Definitions of elastic scattering. The word "elastic" is used to indicate the absence of energy loss. If particle A collides with particle B of finite mass, there is a loss in the energy of A even if no energy has been transferred to the internal degrees of freedom of either A or B. Sometimes such a collision is referred to as inelastic, in order to distinguish its character from that of a collision with a particle having an infinite mass or its idealization, a fixed center of force. This terminology is not useful in the present context, because in the center-of-mass system of the two particles the sum of kinetic energies after the collision is the same as before. The distinction between elastic and inelastic scattering is made therefore on the basis of whether there are internal energy changes in the colliding particles. The collision is said to be inelastic even if the energy changes of the two particles compensate so as to leave the sum of the kinetic energies in the center-of-mass system unaltered. The treatment of inelastic scattering involves nuclear reaction theory because nuclear reactions markedly influence the scattering.

Cross sections. The results of scattering measurements and calculations are generally expressed in terms of differential cross sections, which furnish a quantitative measure of the probability that the incident particle is scattered through an angle, θ. Cross sections have units of area, and in nuclear physics the convenient measure of area is the barn and its subunits, the millibarn and microbarn (1 barn = 10^{-28} m^2).

Coulomb scattering by nuclei. The simplest type of elastic scattering experiment relevant to the study of nuclei involves the deflection of incident electrically charged particles by the Coulomb field of the target nucleus. Such experiments provided the first evidence for atomic nuclei. H. Geiger and E. Marsden observed in 1909 that low-energy alpha particles could be scattered through large angles in collisions with gold and silver targets, and Ernest Rutherford showed in 1911 that these results could be understood if the scattering center consisted of a positively charged region (the nucleus) considerably smaller in size than that occupied by an individual target atom. The yield of scattered alpha particles as a function of angle, called the angular distribution, was measured in detail in a subsequent (1913) experiment by Geiger and Marsden, and the results were entirely consistent with the Rutherford expression for the differential cross section given by Eq. (1), where Z_1 and Z_2 are the atomic numbers of the target and

$$\frac{d\sigma}{d\Omega_{point}}(\theta) = [Z_1 Z_2 e^2/16\pi\epsilon_0 E]^2 \sin^{-4}\theta/2 \qquad (1)$$

projectile, e is the electronic charge, ϵ_0 is the permittivity of free space, E is the center-of-mass energy of the alpha particle projectile, and θ is the center-of-mass scattering, or observation angle.

Equation (1) is appropriate only when the classical distance of closest approach, $d = Z_1 Z_2 e^2/4\pi\epsilon_0 E$, is greater than the combined size of the colliding pair; that is, only if the two charge distributions never overlap during the course of their interaction. The 1913 experiments involved collisions of 7.68-MeV alpha particles (1 MeV = 1.6×10^{-13} joule) with gold targets, and for this situation $d = 2.96 \times 10^{-14}$ m (29.6 fermis or femtometers in conventional nuclear physics notation, with 1 fm = 10^{-15} m). The detailed agreement between the Rutherford prediction of Eq. (1) and the 7.68-MeV data thus implied that the nuclear charge is contained within a region smaller than 30 fm (1.2×10^{-12} in.) in radius. Subsequent experiments involving higher-energy projectiles have determined the actual half-density radius of the gold nucleus to be approximately 7 fm (3 × 10^{-13} in.).

Although nuclear charge radii can be measured by systematically increasing the energy of the incident alpha particle until Eq. (1) no longer describes the angular distributions, specifically nuclear interactions come into play when the colliding nuclei overlap. In that situation, nuclear size is no longer the only factor being probed. This complexity is avoided when high-energy electrons are used as projectiles, since electrons are not subject to the nuclear forces. At very low energies the elastic scattering of electrons by nuclei conforms to the predictions of Eq. (1), but much more information is revealed at higher energies where the electrons can penetrate into the target nucleus. In the latter case and neglecting magnetic effects for the moment, the finite extent of the nucleus requires that Eq. (1) be modified by the presence of a nuclear form factor, $F(q^2)$, as in Eq. (2), in which $\vec{q}$ is the momentum transfer in the collision. The significance of $F(q^2)$ can be

$$\frac{d\sigma}{d\Omega}(\theta) = \frac{d\sigma}{d\Omega_{point}}(\theta) \, [F(\vec{q}^2)]^2 \tag{2}$$

appreciated by noting that at nonrelativistic energies, and assuming the Born approximation to be valid, $F(q^2)$ is the Fourier transform of the nuclear charge distribution.

If sufficiently complete measurements over a wide range of momentum transfer are available, the form factor can be determined in detail, and the nuclear charge density, $\rho_{ch}(r)$, obtained directly through the inverse transform. In practice, a functional form is usually assumed for $\rho_{ch}(r)$, and its parameters are adjusted to fit the electron scattering data. In this manner it is found that the experiments are consistent with nuclear charge distributions specified by Eq. (3). The param-

$$\rho_{ch}(r) = \rho_0[1 + e^{(r-c)/a}]^{-1} \tag{3}$$

eter c is the radius at which the density falls to one-half its central value. It assumes the typical value $c \approx 1.1 A^{1/3}$ fm, where A is the atomic number of the target nucleus. The quantity a (typically ≈ 0.55 fm) reflects the fact that nuclear surfaces are not sharply defined, so that the density falls off gradually with increasing radius. The so-called central charge density, ρ_0, is found to be approximately 1.1×10^{25} coulombs/m³, which corresponds to 0.07 proton per cubic femtometer (1.1×10^{39} protons per cubic inch). That ρ_0 is approximately the same for all but the lightest nuclei and $c \propto A^{1/3}$ implies that nuclear matter is nearly incompressible. The charge distributions of a variety of nuclei, as determined by electron scattering measurements, are shown in **Fig. 1**.

Electron scattering measurements involving large angular momentum transfer are sensitive to the distribution of magnetism as well as charge, and Eq. (2) then must be generalized to reflect the contributions from the two distributions. In addition, very precise measurements show that Eq. (3) is only approximately correct and the form of $\rho_{ch}(r)$ varies slightly from nucleus to nucleus as a consequence of variations in the structure of the nuclei in question.

Electron-nucleon scattering. Scattering of electrons by hydrogen gives information regarding the electron-proton interaction. From measurements of the variation of the differential cross section, it has been found necessary to postulate that both the proton charge and its intrinsic magnetic moment are distributed through a finite volume. Existing work favors the assumption of similarity of shape of these distributions. Energies in excess of 200 MeV are required for the detection of these effects, and the anomalous magnetic moment of the proton plays an important part at high energies. The experiments make it probable that the charge density has a root-mean-square radius of approximately 8×10^{-14} cm (3×10^{-14} in.). Measurements on the scattering of electrons by deuterium at large angles and higher energies lead to the conclusion that the magnetic moment of the neutron is not concentrated at this point. If it is assumed that neutron and proton magnetic moments are distributed through nearly the same volumes, a good representation of the scattering measurements is obtained.

Analyses of scattering data at various angles and energies indicate that there is no net charge density within the volume occupied by the neutron. This result is in agreement with measurements of the neutron-electron interaction made by scattering very slow neutrons from atomic electrons and atomic nuclei. While there is an interaction equivalent to a potential energy of approximately -3900 eV through a distance of e^2/mc^2, where m is the electron mass and c is the speed of light, amounting to approximately 2.8×10^{-13} cm (1.1×10^{-13} in.), it is accounted

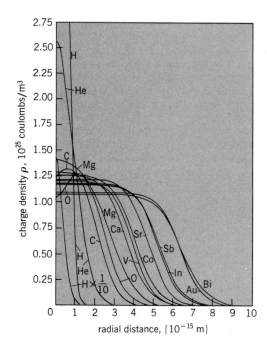

Fig. 1. Summary of the nuclear charge distributions found for various nuclei by electron scattering methods. (*After R. Hofstadter, Nuclear and nucleon scattering of high-energy electrons, Annu. Rev. Nucl. Sci., 7:231–316, 1957*)

for qualitatively as a consequence of what E. Schrödinger called the *Zitterbewegung* (tremblatory motion) expected for the neutron.

Electron scattering by nucleons has also provided valuable information concerning subnuclear processes. *See* ELEMENTARY PARTICLE; QUARKS.

Low-energy np scattering. Since the neutron is electrically neutral, an understanding of how it scatters in an interaction with a proton requires, even for low collision energies, a knowledge of the properties of the nuclear forces. Conversely, the scattering experiments probe these properties. The *np* force is responsible for the binding together of a neutron, *n*, a proton, *p*, to form a deuteron, and the detailed connection between *np* scattering cross sections and the properties of the deuteron was perceived in the early 1930s. For example, the assumption that the spatial extension, or range, of the *np* force is relatively small makes it possible to estimate the magnitude of the force, given the measured binding energy of the deuteron, and Eugene Wigner used this information in 1933 to calculate *np* scattering cross sections. The measured yield was found to be larger than would be expected if the forces between free neutrons and free protons were the same as those in the deuteron. To explain this difference, it was postulated by Wigner in 1935 that the *np* interaction is spin-dependent; that is, it depends on the relative orientation of the spins of the interacting particles. The proton and neutron are known to have a spin of ½; that is, their intrinsic angular momenta are known to be $(h/2\pi)/2 = \hbar/2$, where h is Planck's constant. According to quantum mechanics, when two spins s_1, s_2 combine vectorially, only the values given by Eq. (4) are possible for the resultant s. For the *np* system, therefore, the resultant spins

$$s = s_1 + s_2, s_1 + s_2 - 1, \ldots, |s_1 - s_2| \qquad (4)$$

are 0 or 1. In the first case one speaks of a singlet, and in the second of a triplet. *See* SPIN.

The singlet state behaves much like a round and perfectly smooth object which has the same appearance no matter how it is viewed, corresponding to only one possibility of forming a state with $s = 0$. The state with $s = 1$, on the other hand, can have three distinct spin orientations. Measurement of the projection of s on an axis fixed in space can give only the three values (1,0,−1), again in units $\hbar$. When protons with random spin directions collide with neutrons also having random spin directions, the triplet state is formed three times as often as the singlet. The deuteron, however, is in a triplet state. Thus the hypothesis of spin dependence can account for the difference between the forces in the deuteron and those in np scattering.

The neutron-hydrogen scattering experiments on which these conclusions were based were performed with slow neutrons having energies of a few electronvolts or less. The general quantum-mechanical theory of scattering is much simplified in this case. Because of the small range of nuclear forces, only the collisions with zero orbital angular momentum ($L\hbar = 0$) play a role, collisions with higher orbital angular momenta missing the region within which nuclear interactions take place. States with $L = 0$ (called S states) have the property of spherical symmetry, and nuclear forces matter in this case only inasmuch as they modify the spherically symmetric part of the wave functions.

The long-wavelength or low-energy scattering cross section can be described completely and simply by a quantity called the scattering length. For interparticle distances r greater than the range of nuclear forces, the spherically symmetric part of the wave function has the form given by expression (5), where a and C are constants. The constant a is called the scattering

$$C[1 + (a/r)] \tag{5}$$

length. It has the following meaning. If $R(r)$ denotes the wave function describing the np relative motion, the product $rR(r)$, when plotted against r, is represented by a straight line which cuts the axis of r at a distance a from the origin of coordinates. If the intersection is to the left (right) of the origin, a is counted as positive (negative). The two conditions are illustrated in **Figs. 2** and **3**. The possibilities $a > 0$ and $a < 0$ are sometimes referred to as those of the virtual and real level, respectively. Sometimes the opposite convention regarding the sign of a is used. The convention adhered to here provides the simplest connection with phase shifts.

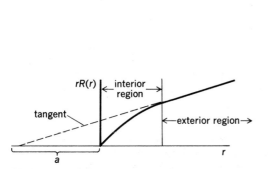

Fig. 2. Scattering length in the case of a virtual level.

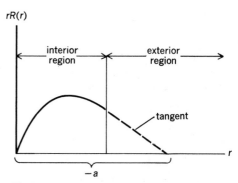

Fig. 3. Scattering length in the case of a real level.

The scattering cross section for a state with well-defined a is $4\pi a^2$, and thus low-energy np scattering is described by Eq. (6), where σ_{np} is the total cross section and a_s and a_t are the

$$\sigma_{np} = 4\pi(1/4\, a_s^2 + 3/4\, a_t^2) \tag{6}$$

singlet and triplet scattering lengths, respectively. Comparison of the scattering yields with the deuteron data provides the values $a_s = 24$ fm and $a_t = 5.4$ fm.

There are several additional means by which np scattering can be calculated. For example, with the techniques outlined below ("Potential scattering"), effective interaction potentials can be found which simulate the np nuclear forces to the extent that calculations using the potentials reproduce the measured cross sections. Two potentials corresponding to singlet and triplet $L = 0$ scattering are required for a description of low-energy np data, as is illustrated in **Fig. 4**. The term "potential well" is often used to describe these potential energy surfaces, because the system can be trapped in the region of space occupied by the potential somewhat similarly to the

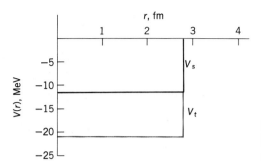

Fig. 4. Examples of singlet (V_s) and triplet (V_t) potential wells suitable for low-energy np scattering. The potentials are not uniquely determined; many other equally satisfactory possibilities exist.

way in which water is trapped in a well. For nucleon-nucleon scattering, it is frequently useful to express the potential energy in the form $V(r) = V_0 f(r/b)$, where f is a function which determines the shape of the well. The constants V_0 and b are usually referred to as the depth and range parameters. The scattering length determines approximately the product $V_0 b^2$ for a potential well of assigned shape. The variation of the cross section with energy E through an energy range of a few MeV can be used for the determination of b.

Low-energy pp scattering. The pp and pn interactions are believed to be closely equal. Gregory Breit proposed in 1936 the hypothesis of charge independence of nuclear forces, which supposes that nuclear forces, acting in addition to the electrostatic (Coulomb) repulsion, in the pp, pn, and nn cases are equal to each other. Because of the limits of experimental accuracy and uncertainties in the theoretical interpretation, the hypothesis is not established with perfect accuracy, but it is believed to hold within a few percent for the depth parameter V_0 if the range parameter b is specified as the same in the three cases. The concept of charge independence also has important implications for the structure of nuclei. SEE ISOBARIC SPIN.

Phase shifts. Scattering can be treated by means of phase shifts, which will be illustrated for two spinless particles. The wave function of relative motion will be considered first for a state of definite orbital angular momentum $L\hbar$, with L representing an integer. Outside the range R of nuclear forces, the wave function may be represented by Eq. (7), where k is 2π times the recip-

$$\psi_L = Y_{LM}(\theta,\phi)\mathcal{F}_L(kr)/(kr) \qquad (7)$$
$$(r > R)$$

rocal of the wavelength, the so-called wave number, θ and σ are the colatitude and azimuthal angles of a polar coordinate system, and Y_{LM} is the spherical harmonic of order L and azimuthal quantum number M. The form of ψ_L is determined by the Schrödinger wave equation which restricts $\mathcal{F}_L$ by the differential equation, Eq. (8), it being supposed that there is no Coulomb field.

$$\{d^2/dr^2 + [k^2 - L(L+1)/r^2]\}\mathcal{F}_L = 0 \qquad (8)$$
$$(r > R)$$

In the absence of nuclear forces, $\mathcal{F}_L$ satisfies the same equation at all distances and, aside from a constant factor, has its asymptotic form determined by the boundary conditions at $r = 0$ as in notation (9). In the presence of nuclear interactions, the asymptotic form is given by notation (10),

$$\mathcal{F}_L \sim \sin(kr - L\pi/2) \text{ as } r \to \infty \qquad (9)$$
$$\mathcal{F}_L \sim \sin(kr - L\pi/2 + \delta_L) \qquad (10)$$

where δ_L is a constant, called the phase shift, which determines the scattering. SEE SPHERICAL HARMONICS.

When elastic scattering is the only process that can take place, as for low- and medium-energy nucleon-nucleon scattering, the phase shifts are real numbers. Above a few hundred MeV, nucleon-nucleon scattering becomes strongly inelastic because of pion production, however, and the phase parameters become complex numbers reflecting the loss of flux from the elastic channel.

Nucleon-nucleus and nucleus-nucleus interactions are characterized by the presence of many nonelastic scattering channels, except at very low energies.

The wave packet representing the scattered particle consists of a superposition of all possible partial waves, $\mathscr{F}_L$, and calculation of the scattering cross section requires a knowledge of all phase shifts. In practice, however, only partial waves with orbital angular momenta corresponding to impact parameters within the range of the nuclear forces suffer nuclear phase shifts different from zero. The direct method of varying phase shifts to reproduce experimental data is most useful, therefore, when only a few L values satisfy this condition, as is the case for low- and intermediate-energy nucleon-nucleon scattering and for low-energy nucleon-nucleus and nucleus-nucleus collisions.

For charged spinless particles, such as two alpha particles, the phase shifts caused by specifically nuclear forces add to the asymptotic phase of the functions $\mathscr{F}_L$ for the Coulomb case, which differs from the non-Coulomb case only through the replacement of $kr - L\pi/2$ by $kr - L\pi/2 - \eta \ln(2kr) + \arg \Gamma(L + 1 + i\eta)$, where $\eta = Z_1 Z_2 e^2 / 4\pi\epsilon_0 \hbar v$, $Z_1 e$ and $Z_2 e$ represent the charges on the colliding particles, e represents the electronic charge, and v represents the relative velocity.

As in the np scattering case discussed above, the presence of spin adds additional complications. For example, if two particles with spin ½ collide, it is necessary, in general, to introduce phase shifts for each state with definite total angular momentum $J\hbar$.

Potential scattering. The direct determination of phase shifts through comparison with scattering data becomes difficult in situations where large numbers of partial waves, $\mathscr{F}_L$, are modified by the nuclear scattering center. The usual procedure in this case is to simulate the influence of the nuclear interactions by means of potentials, $V(r)$. For the nonrelativistic scattering of two spinless, uncharged particles, Eq. (8) is supplemented by the Schrödinger equation for $r < R$, Eq. (11), in which m is the reduced mass of the colliding pair. Equations (8) and (11) are then solved

$$\{d^2/dr^2 + [k^2 - L(L + 1)/r^2 - 2mV(r)/\hbar^2]\}\mathscr{F}_L(r) = 0 \quad (11)$$
$$(r < R)$$

subject to appropriate boundary conditions to determine the $\mathscr{F}_L(r)$ and hence δ_L for each L. A functional form is generally assumed for $V(r)$, and its parameters are varied and Eqs. (8) and (11) solved iteratively until the measured cross sections, polarizations, and so forth are reproduced.

From the viewpoint of microscopic models of nucleon-nucleon interactions, it appears highly improbable that a description of nucleon-nucleon scattering in terms of two-body energy-independent local potentials can have fundamental significance. Nevertheless, in a limited energy range, it is practical and customary to represent scattering by means of such a potential. As noted above, the simple potentials illustrated in Fig. 4 suffice for the description of low-energy np scattering. At somewhat higher energies, but below the threshold of meson production, a real potential may still be used, but different potentials are required for triplet-even, triplet-odd, singlet-even, and singlet-odd states to account for the (even, odd) parity dependence of the interactions. SEE PARITY.

Intermediate energy np and pp scattering. In the intermediate-energy region (10–440 MeV) the analysis of experimental material is more difficult than at low energies because of the necessity of employing many phase shifts and coupling constants. Analysis in terms of phase parameters involves fewer assumptions than that in terms of potentials. Except for approximations connected with the infrared catastrophe and related small inaccuracies in relativistic treatment of Coulomb scattering, it is based on very generally accepted assumptions, such as the validity of time reversal and parity symmetries for strong interactions. With infinite experimental accuracy, it should be possible to extract all the phase parameters from measurements of the differential cross section, the polarization spin correlation coefficients, and "triple scattering" quantities describing spin orientation which, for unpolarized incident beams and unpolarized targets, require three successive scatterings.

The analysis is usually carried out by assuming that for sufficiently high L and J the phase parameters may be represented by means of the one-pion exchange approximation. The value of the pion-nucleon coupling constant g^2 is often varied in an attempt to improve the fit to experimental data, and values for best fits are compared with those from pion physics. Reasonable agreement usually results.

The consistency of values of g^2 from pp data with those from np measurements indicates approximate validity of charge independence at the larger distances.

Potentials to be used in a nonrelativistic Schrödinger equation and capable of representing pp and np scattering have been devised either on a purely phenomenological or semiphenomenological basis. The former way provides a more accurate representation of the data. Nonrelativistic local potentials required from 0 to 310 MeV are different according to whether the state is even or odd, singlet or triplet. It is necessary to use central, tensor, spin-orbit, and quadratic spin-orbit parts of the potential. Most of the accurately adjusted potentials employ hard cores within which the potential is infinite. The spin-orbit potential suggested by pp scattering data indicated the probable participation of vector-meson exchange in nucleon-nucleon scattering, anticipating the discoveries of the ω- and ρ-mesons, as well as fictitious mesons, the latter partly intended as a representation of simultaneous two-pion exchange. At short distances, the potentials are often modified in order to improve agreement with experiment. Superposition of the single-boson potentials combined with the short-range modifications of the resulting potential gives the so-called one-boson exchange (OBE) potentials, These provide a fair but not excellent reproduction of phenomenological fit phase shifts. *See* QUANTUM FIELD THEORY.

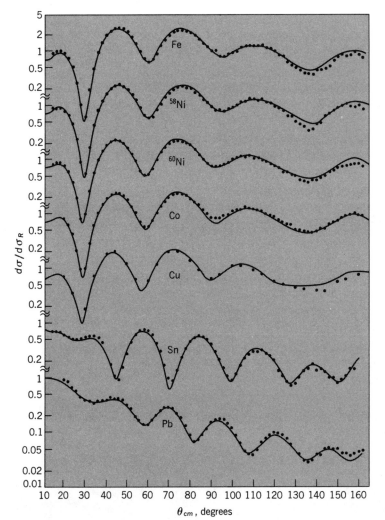

Fig. 5. Measured (points) and calculated (curves) differential cross sections for the scattering of 30-MeV protons from a variety of target nuclei, as a function of scattering angle in the center-of-mass system θ_{cm}. The theoretical calculations made use of optical potential wells. The vertical scale represents the differential cross section $d\sigma$ divided by the Rutherford differential cross section $d\sigma_R$. (After G. R. Satchler, Optical model for 30 MeV proton scattering, Nucl. Phys., A92:273–305, 1966)

Nucleon-nucleus scattering. Nucleon-nucleus scattering at low and intermediate energies ($E < 300$ MeV) is of interest primarily in connection with nuclear structure studies. The scattering wave functions, $\mathscr{F}_L(r)$, are required for quantitative interpretation of the inelastic scattering and reaction experiments which have been the main source of information regarding the properties of nuclear excited states. These scattering functions are obtained as the solutions of Eq. (11) when elastic data are reproduced by using potential models.

Nucleon-nucleus scattering experiments can be accounted for by a potential-well model with a complex potential (optical model) making use of a spin-orbit interaction term. The details of angular distributions of the cross section and polarization are reproduced remarkably well, and experiments favor some potential well shapes over others. Wells thus determined are wider than those obtained from electron-nucleus scattering experiments, and similar functional shapes work in both cases. The potential energy represented by the wells is added to the electrostatic potential energy in the calculations. The electrostatic potential energy is approximated by a central potential corresponding to the average distribution of nuclear charge. The quantitative success of the potential-well approach to the scattering of 30-MeV protons by a variety of nuclei is illustrated in **Fig. 5**.

Optical model potential fits to nucleon-nucleus data are in general agreement with the data in a wide energy range from several MeV to about 300 MeV, but the parameters of the potential have to be varied progressively. The spin-orbit potential found to represent the scattering data at the lower energies has the same value of its ratio to $(dV_{ct})/(rdr)$, where V_{ct} is the central potential, as it has in the shell theory of nuclear structure. At higher energies (300 MeV), it has been found that the spin-orbit potential has to be used with a smaller strength than in shell theory. The real part of the central potential has to be used with a smaller strength than in shell theory. The real part of the central potential decreases with energy and becomes almost zero at 300 MeV. Data at 1 GeV on proton scattering from carbon can be accounted for on the optical model by means of an imaginary central and real spin-orbit potential.

The real part of the volume potential $V(\vec{r}_p)$ can be calculated by folding the interaction potential between the projectile and a target nucleon, $v_{tp}(\vec{r})$, with the nuclear density of the target as in Eq. (12). Effective, rather than free, nucleon-nucleon interactions are used for $v(\vec{r}_{tp})$ because

$$V(\vec{r}_p) = \int \rho(\vec{r}_t) v(\vec{r}_{tp}) d\vec{r}_t \tag{12}$$

the target nucleon is embedded in a nucleus. The short range of $v(\vec{r}_{tp})$ implies that the resulting central potential has a spatial distribution closely similar to, but somewhat more rounded than, that of the target nuclear density, as is illustrated in **Fig. 6**. The imaginary part of the optical potential is phenomenologically determined.

Nucleus-nucleus scattering. Composite nuclei ranging from deuterium through uranium have been accelerated to energies as high as several GeV per constituent nucleon. Elastic scattering is an important process only toward the lower end of this energy range, perhaps only below a few tens of MeV per nucleon; at the higher energies, the principal interest is in nonelastic collisions, which are expected to provide information concerning properties of nuclear matter under unusual conditions (like high temperature and high density).

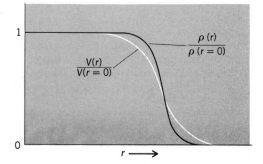

Fig. 6. Comparison between the shape of a nuclear density distribution, $\rho(r)$, and a potential, $V(r)$, obtained from it by folding in an interaction potential, $v(r)$, with a finite but short range as in Eq. (12). (*After G. R. Satchler, Introduction to Nuclear Reactions, John Wiley and Sons, 1980*)

Nucleus-nucleus scattering at low and intermediate energies is characterized by strong absorption associated with the relatively short mean free path for nuclei in nuclear matter. Only peripheral or glancing collisions are likely to lead to direct elastic scattering; more head-on collisions lead to more complicated processes such as compound nucleus formation which absorb flux from the incident beam. Since the de Broglie wavelengths of the incident nuclei are typically comparable to or smaller than the size of the target nuclei, the strong absorption for head-on collisions frequently causes the scattering to be diffractive in nature, and the elastic scattering angular distributions to resemble those observed in the scattering of light by small opaque obstacles. A typical example, corresponding to optical Fraunhöfer diffraction, is illustrated in **Fig. 7**.

Good fits to nucleus-nucleus scattering data are obtained by using potential wells. For collisions involving relatively light nuclei, the real potentials may be derived by folding an effective nucleon-nucleon interaction with the densities of the two colliding nuclei, in a slight generalization of Eq. (12). For more massive nuclei, analogies with the collisions of liquid droplets, provide guidance in determining the potentials. In many cases, purely phenomenological potentials provide the best fits to the data. There is considerable ambiguity in the potential parameters, because the scattering depends primarily on the values of the potentials at distances corresponding to glancing collisions, and is relatively insensitive to the values at smaller distances. The imaginary potentials are deeper than those used for nucleon-nucleus scattering because of the strong absorption property. Except for collisions involving light nuclei, no very strong evidence for the influence of spin-orbit potentials on elastic scattering yields has been found, but there is evidence suggesting that these may become important for more massive nuclei at energies in excess of 10 MeV per nucleon.

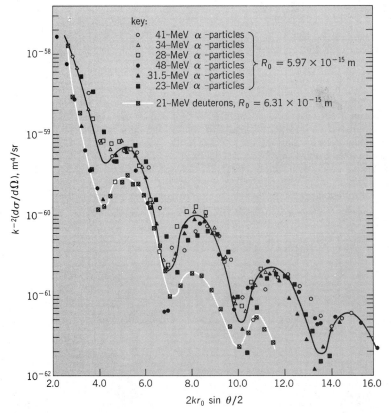

Fig. 7. Angular distribution for the elastic scattering of deuterons and alpha particles from magnesium, showing the typical Fraunhöfer-like diffraction pattern. (*After J. S. Blair et al., Diffraction analysis of elastic and inelastic scattering by magnesium, Nucl. Phys., 17:641–654, 1960*)

The optical model potential is not a potential in the ordinary sense. When inserted in the wave equation (11), it gives agreement with experiment by simulating the complicated interactions between the two colliding many-body systems in terms of a prescribed functional form. Possible variations in this form, associated with the varying proximity of the two nuclei during the collision are generally neglected. Alternative methods, such as the time-dependent Hartree-Fock approach, are being developed to treat nucleus-nucleus interactions from a more fundamental perspective.

Nucleus-nucleus scattering sometimes exhibits anomalous behavior which cannot be reproduced with potential models and which is believed to reflect the transient formation of moleculelike configurations. These configurations, or resonances, signal their presence by producing a rapid variation in the energy dependence of the scattering yields and by modifying the behavior of the angular distribution data. SEE NUCLEAR MOLECULE.

Meson scattering. The scattering of π mesons (pions) by nucleons has been studied intensively as an example of a strong interaction between a boson and a baryon, and particularly because of the connection with nucleon-nucleon interactions. Prominent resonances occur in pion-nucleon scattering yields, the best known of which appears for pion bombarding energies of about 200 MeV. Charge independence in pion-nucleon interactions is confirmed by the scattering measurements and their phase-shift analyses.

The scattering of pions by nuclei has been studied over the energy range extending from below 20 MeV to beyond 60 GeV. At low energies the experiments probe nucleon-nucleon correlations within nuclei. Attempts to relate the scattering behavior to the properties of the nuclear wave functions have been partially successful. Comparison of π^+ with π^- scattering yields provides information concerning differences between proton and neutron distributions within nuclei. Future study of pion-nucleus interactions at high energies is expected to reveal properties of the strong interactions, modified by the nuclear movement, at short distances.

The study of kaon-nucleus interactions, still in its infancy, is currently of interest primarily in connection with the production of nuclei in which, for example, a neutron is replaced by a lambda particle (Λ), and the investigation of Λ-nucleon and related interactions. SEE MESON.

Bibliography. R. C. Barrett and D. F. Jackson, *Nuclear Sizes and Structure*, 1977; E. H. S. Burhop (ed.), *High-Energy Physics*, vol. 1, 1967; M. Goldberger and K. M. Watson, *Collision Theory*, 1964, reprint 1975; G. R. Satchler, *Introduction to Nuclear Reactions*, 1980.

COULOMB EXCITATION
PAUL H. STELSON

A process in which two atomic nuclei that approach each other undergo transitions to excited states that are caused by the long-range Coulomb forces acting between the nuclei. The process can occur even when the nuclei do not come sufficiently close to allow the strong short-range nuclear forces to act.

That nuclei can be excited by this process is not so important in itself. What has proved to be enormously fruitful is the fact that experimental results based on this process have provided accurate values for electrical or electromagnetic properties of nuclei. To take an easily visualized example, Coulomb excitation measurements were instrumental in showing that many nuclear species are not spherical but are, instead, shaped like a football. SEE NUCLEAR STRUCTURE.

Experiments. The field of experimental Coulomb excitation began in the mid-1950s. Early experiments involved bombarding targets with beams of protons or alpha particles obtained from accelerators such as Van de Graaffs, cyclotrons, or linacs. Nuclei are made of protons and neutrons; the positively charged protons cause an intense electrostatic (Coulomb) field to surround nuclei. Positively charged bombarding particles, such as protons or alpha particles, must have a large kinetic energy in order to overcome the strong Coulomb repulsion of nuclei and thereby reach the close distances required for the attractive nuclear forces to act.

Coulomb excitation was discovered when protons of much too small an energy to overcome this Coulomb repulsion of the target nuclei were observed to still cause nuclear excitation. The occurrence of nuclear excitation was confirmed by detecting the gamma rays emitted from the target when the nuclei decayed back down to the ground state. See Alpha particles; Particle accelerator; Proton; Van de Graaff generator.

Although the first Coulomb excitation measurements were carried out with beams of protons or α-particles, it was soon realized that the use of beams of more highly charged projectiles would dramatically increase the probabilities for Coulomb excitation. Experiments have been done with beams of ^{16}O, ^{20}Ne, ^{40}Ar, ^{84}Kr, and ^{136}Xe ions. Coulomb excitation with these heavy ions is characterized by the term "multiple excitation." Many more excited states of nuclei are appreciably populated with these heavy-ion beams.

Theory and interpretations. The theory of pure Coulomb excitation is well understood, but it is mathematically complicated. The modern approach is to use computer programs to generate specific theoretical results applicable to a particular experimental situation.

In order to interpret Coulomb excitation results precisely, it is important that the influence of nuclear forces be negligible so that a "pure" Coulomb excitation situation exists. Of course, at higher bombarding energies, where nuclear forces are an important part of the reaction, Coulomb excitation is still present; and, in fact, the combination of Coulomb and nuclear forces produces an interesting reaction situation which has been the subject of much research. See Nuclear reaction; Scattering experiments (nuclei).

Measurement and applications. Coulomb excitation can be measured either by the direct analysis of the energy spectrum of the scattered projectiles or by the analysis of the gamma rays subsequently emitted by the excited nuclei. Both methods are used extensively. The chief advantage of a direct measurement of the spectrum of the scattered projectiles is the good accuracy with which excitation probabilities can be determined. A high degree of accuracy is valuable in determining such nuclear properties as static quadrupole moments of excited states, the existence and magnitude of hexadecapole or E4 moments, and the possibility of centrifugal stretching of nuclei.

The chief advantages offered by the detection of gamma rays following Coulomb excitation are: (1) The excellent energy resolution (approximately 2 keV) of Ge(Li) gamma ray detectors can be used to study the excited nuclear states; (2) good statistical accuracy (high counting rates) is achieved in much shorter times on the accelerator; and (3) more extensive information on the properties of the nuclear states can be extracted from gamma ray measurements. From measurements of angular distributions of the gamma rays, spin-parity values can be assigned to excited states, and information about the strength of magnetic dipole transitions in nuclei can be gained. See Gamma rays; Nuclear spectra.

Heavy-ion collisions. One of the important frontiers of nuclear physics research is the use of heavy-ion projectiles; producing collisions between large complex nuclei allows interesting questions to be asked about the behavior and structure of nuclei. In such collisions, the Coulomb interaction is always important. Thus it is likely that the Coulomb excitation process will continue to play a vital role in nuclear physics research.

QUASIATOM
Jack S. Greenberg

A transient electronic structure, found in atom-atom or ion-atom collisions, which approximates the characteristics associated with a stable atom whose atomic number equals the combined charge of the colliding nuclei. Such short-lived atomic species can be formed in energetic collisions which overcome the Coulomb repulsion between nuclei and carry the two nuclear charge centers well within the electron clouds. At the small internuclear separations, where the electrons cannot distinguish between the two nuclear centers, they act as if they are bound by a nearly monopolar field generated by a single charge center composed of the two nuclei. Thus, during this short period, the energy level structure resembles that of a united atom with an atomic number $Z = Z_1 + Z_2$, where Z_1 and Z_2 are the individual atomic numbers of the colliding atoms.

Quasiatom formation. The formation of quasiatoms in ion-atom collisions relies on a disparity between the nuclear collision velocity and the velocities of the orbiting electrons. The Coulomb potential produced by the colliding nuclei must vary sufficiently slowly compared with the orbiting time of the electrons so that the electrons adjust adiabatically to their changing environment at each stage of the collision as the internuclear separation changes with time. This condition is closely realized for collisions involving heavy systems at bombarding energies which bring the nuclear surfaces into contact against the Coulomb repulsion. The velocities of the inner-shell electrons are for such collisions approximately 10 times larger than the relative velocity of the two nuclei. **Figure 1** illustrates how the binding energies of some of the most strongly bound states are expected to evolve as a function of time and internuclear separation. Asymptotically, the quantum numbers are associated with the united quasiatom at one extreme, and with the separated atoms at the very large internuclear separations where the individual atoms do not interact. In the intermediate region, the electrons evolve through a series of quasimolecular states in the two-center field. The independent particle states shown in Fig. 1 are calculated at each internuclear separation by freezing the nuclear motion at the specific separation to obtain the two-center potential. Their detailed evolution forms the basis for understanding a number of observed collision phenomena in atom-atom scattering.

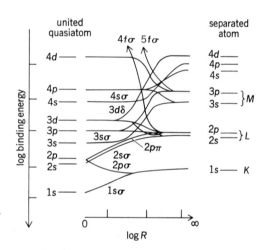

Fig. 1. Schematic representation of evolving quasimolecular states as a function of internuclear separation R. Both the separated atom and united quasiatomic limits are also shown.

Electron promotion. An important characteristic property of the transient molecular orbitals, introduced by F. Hund and R. Mullikan, is a promotion or demotion of the principal quantum number and binding energy in passing from the separated atom to the united atom limits. In their evolution with changing internuclear separation, molecular orbitals of very different principal quantum number in the separated atom limit may cross or approach very closely in binding energy. The decreased binding energy or the close proximity of molecular orbitals at finite internuclear separations can lead to excitation induced by a departure from complete adiabaticity due to radial variation or rotation of the internuclear axis with time. Thus, upon separation, the electrons may not necessarily return to their original atomic states, having crossed into available empty orbits, and the atom may then be left in an excited configuration leading to x-ray emission or Auger electron emission after the ions separate. This mechanism of electron excitation by promotion led to a revival of interest in the molecular formulation of atomic collisions. The molecular model explains the large energy transfer even for slow ions, the large cross sections corresponding to geometric atomic dimensions which are 14 orders of magnitude larger than expected for Coulomb excitation, the threshold behavior of these cross sections, and the discovery of resonance behavior in the production of K, L and M x-rays when levels of the target and projectile match.

Quasimolecular x-ray spectra. While the studies of characteristic x-ray production provide this indirect information on quasimolecular phenomena, more direct signatures for quasimolecule and quasiatom formation are supplied by other radiations from heavy-ion collisions. The x-ray spectra consist not only of characteristic lines, associated with the isolated atomic or ionic species after the collision, but also of x-ray continua, which are not characteristic of the separated atoms but may originate from the collision complex. Both atomic and molecular mechanisms can be responsible for the emission of the continuum x-rays, depending largely on the ratio of the ion and orbiting electron velocities.

For example, radiative electron capture (REC) of target electrons into projectile vacancies is associated with atomic processes and high collision velocities, while x-rays emitted during radiative transitions between molecular orbitals (MO x-rays) represent the other extreme. REC transitions are between atomic states, and the emission spectrum is determined predominantly by the momentum change occurring during the transfer of the electron. The spectrum, therefore, reflects the momentum distribution of the electrons in the initial target atomic state. On the other hand, since MO x-ray emission originates from transitions between the molecular orbitals, in principle, it can yield information on the structure of the collision complex and on the time evolution of these states. As such, it provides the most direct demonstration of electronic quasimolecule formation.

The study of quasimolecular x-rays has been extended to M, L, and K radiations in a variety of collision systems ranging from carbon + carbon (C + C) to lead + lead (Pb + Pb). Examples of the continuum K radiations from nickel + nickel (Ni + Ni) and niobium + niobium (Nb + Nb) collisions are shown in **Fig. 2**. It has been demonstrated that these continuum radiations exhibit a characteristic anisotropic emission pattern which depends on the photon frequency. This feature has been utilized to develop a spectroscopy of quasimolecular and quasiatomic states. Techniques have also been developed to selectively identify L- and K-type quasimolecular transitions in the continuum spectra by exploring cascade relationships between these radiations and characteristic x-rays. Due to gamma-ray background radiations from nuclear excitations during the collision, the exceptionally nuclear-stable Pb + Pb collision system (Z_{total} = 164) has been the heaviest quasi-atom explored via quasimolecular x-rays. Information on the quasimolecular energy level structure beyond this combined nuclear charge has been extracted from studies of the K-vacancy excitation

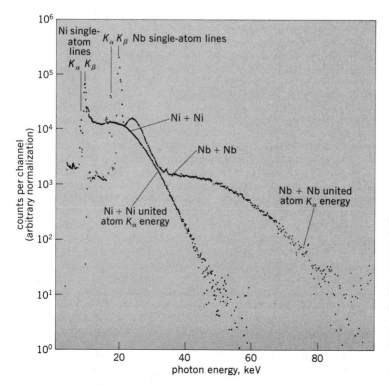

Fig. 2. Quasimolecular K x-ray spectra from Ni + Ni and Nb + Nb collisions. Incoming ion energies are 64.8 MeV for Ni + Ni and 75.9 MeV for Nb + Nb, and x-rays are observed at angle of 90° to beam direction. Extension of spectra beyond the united atom energies is produced by collision broadening. The characteristic K_α and K_β single-atom lines are also shown.

probability as a function of internuclear separation in the lead + uranium (Pb + U with Z_{total} = 174) and lead + curium (Pb + Cm with Z_{total} = 178) collision systems.

Delta-electron spectra. Delta electrons (high-energy electrons) ejected during the ionization process also provide an important experimental approach to study united atom characteristics, particularly in superheavy collision systems. The detection of these electrons provides information which is not available in vacancy production studies via x-ray deexcitation alone: through measurement of the electron energies, the energy transfer is clearly defined. The delta-electron spectrum reflects the momentum components of the inner-shell wave functions and thus is a major source of information on the quasiatom structure. For example, the mere presence of high-energy electrons in the delta-electron spectrum emitted from superheavy collision systems demonstrates that the bound electrons possess high momenta which, in turn, reflect the relativistic shrinkage of the atomic space produced by the large binding potentials in superheavy collision systems. SEE DELTA ELECTRONS.

Superheavy atoms. In addition to its utilization to explain the more traditional phenomena encountered in atomic collisions, the quasiatom provides unique access to a new domain in atomic physics. It was suggested in 1969 that the quasiatom could serve as a vehicle for investigating superheavy atoms, considerably beyond the heaviest stable systems available. Radioactive instability and the considerable difficulty of synthesizing nuclei with $Z > 108$ set the upper limit on the atomic number that can be investigated with stable atoms. As a practical limit, fermium ($Z = 100$), with a $1s$ binding energy of approximately 142 keV, has been the highest-Z stable atomic system studied. The data derived from stable atoms, therefore, only cover a small fraction of the manifold of discrete states between $+m_0c^2$ and $-m_0c^2$ which make up the solutions to the Dirac equation. (The quantity $m_0c^2 = 511$ keV is the energy equivalent of the rest mass m_0 of the electron; c is the speed of light.) In principle, quasiatoms can extend the investigations to include the whole range of bound states in the energy gap of $2m_0c^2$, and the continuum states beyond, by providing access to systems with Z values that can exceed $1/\alpha \simeq 137$, where α is the fine-structure constant.

Binding energies and wave functions. The interest in investigating this region of superheavy atomic species is motivated by the expectation that these systems will exhibit new phenomena associated with the quantum electrodynamics of strong fields. Unusual features are associated with the binding energy and the spatial behavior of atomic states which reflect the dominance of the relativistic effects produced by the strong electric fields encountered in quasiatoms such as uranium + uranium. For example, while for low-Z atoms the binding energies scale with Z^2, calculations show that the binding energy of the $1s$ state displays a faster increase with Z as relativistic effects set in at larger atomic numbers; this increase becomes very rapid as $Z\alpha = 1$ is passed. In the early development of relativistic quantum mechanics, $Z\alpha = 1$ appeared to present an apparent boundary which reflected the well-known singularity in the energy of $s_{1/2}$ and $p_{1/2}$ atomic states associated with a point nucleus of this charge. Although this ambiguity is removed by considering nuclear charge centers of finite dimensions, the onset of the rapid increase of the binding energy for the $1s$ state beyond $Z\alpha = 1$ still marks the remnants of the effect of the wave-function collapse about the nucleus due to relativistic effects.

The binding energy of the $1s$ electron exceeds its own rest mass at $Z \simeq 150$, and continues to increase even more precipitously toward $-m_0c^2$ as the charge is increased even further (**Fig. 3a**). The relativistic effects which are responsible for this rapid growth of binding energy with Z also lead to further severe modifications of the energy level structure, so that it no longer bears much resemblance to a non-relativistic hydrogenlike structure. For example, the fine-structure splitting of the $2p_{3/2}$–$2p_{1/2}$ states becomes very pronounced, achieving values in excess of m_0c^2. Another prominent new feature is the interchange of the $2p_{1/2}$ and $2s_{1/2}$ energy levels and the appearance of a large energy splitting between them where in light atoms a near-degeneracy exists. The latter property particularly reflects the relativistic spinor character of the wave functions. The wave functions also exhibit the dominant influence of relativistic effects in other properties. Very striking is the prediction that for $Z \simeq 170$ the electron density at the nucleus calculated relativistically exceeds the corresponding value calculated nonrelativistically, by more than three orders of magnitude.

Instability of the vacuum. A feature of special interest occurs when the coulombic binding of the electron increases beyond a critical value so that the lowest bound state (the $1s$ state)

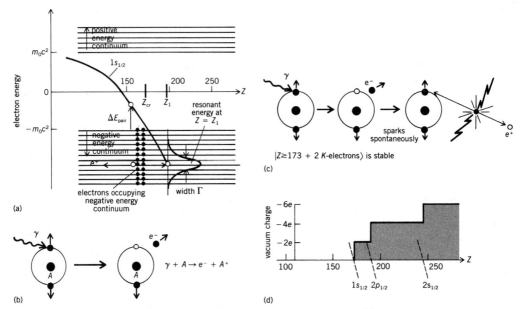

Fig. 3. Phenomena associated with strong electric fields in superheavy atoms. (a) Energy spectrum of electrons as a function of atomic number Z, showing diving of the 1s atomic state with increasing Z and its transformation into a decaying resonance state with the binding energy exceeding $2m_0c^2$. (b) Photoionization for $Z < Z_{cr} = 173$, leading to stable ionized atom. (c) Outcome of photoionization for $Z > Z_{cr}$, for which permanent ionization is not the stable state because the ionized 1s state is unstable to spontaneous positron emission. (d) Increase in the negative charge of the vacuum as successive vacant bound states join the negative-energy continuum.

becomes degenerate with the negative-energy continuum states, below $-m_0c^2$. The required critical charge Z_{cr} is predicted to be 173 for spherical nuclei of normal nuclear density. Under these conditions a fundamental instability in the electron-positron vacuum is initiated which can be described as follows.

Theory demonstrates that for finite nuclei there are no difficulties in tracing the energy levels of states, such as the 1s state in Fig. 3a, up to the value Z_{cr} where they merge with the negative-energy continuum. At this threshold, theory predicts that the bound state ceases to exist as it joins the negative-energy continuum but, instead, develops into a resonance shared over the continuum states. The width of this resonance, schematically shown in Fig. 3a, grows with increasing penetration into the continuum as Z increases beyond Z_{cr}. The appearance of this resonance marks a decaying state and, thus, an instability. The physical process that follows is suggested by considering the smallest energy required to create an electron-positron pair in the vicinity of such an overcritically charged bare nucleus. Below the critical charge, energy must be fed into the atomic system to create pairs. Figure 3a shows that the energy required, ΔE_{pair}, is the difference between $2m_0c^2$ and the binding energy of the 1s state. It decreases as Z increases and, eventually, at the critical charge it changes sign. Therefore, this boundary marks the threshold where it becomes energetically favorable to create an electron-positron pair, provided the 1s state is unoccupied by an electron. The spontaneous emission of such a pair is not forbidden by any conservation law, and thus a free positron escapes and an electron remains bound under supercritical binding conditions ($>2m_0c^2$).

There is, therefore, an essential difference between hydrogenlike atoms with $Z < Z_{cr}$ and with $Z > Z_{cr}$. Ionization of an electron from the former leads to a stable ionized atom (Fig. 3b), while permanent ionization of the inner-shell electrons of the superheavy atom with $Z > Z_{cr}$ is not a stable state (Fig. 3c). After the K electrons have been stripped off, a bare nucleus with $Z > 173$ will always spontaneously surround itself with two electrons. In the process, positrons are emitted with a rather well-defined kinetic energy centered at a value equal to the excess of the binding energy beyond $2m_0c^2$. This two-electron state becomes the lowest energetically stable state. Calculations show that for $Z \simeq 180$ it forms on a time scale of 10^{-19} s, corresponding to the width of the resonance state of approximately 10 keV. As the central charge is increased arbitrarily—so that successive bound states join the negative-energy continuum—successive phase transitions occur in which the vacuum increases its negative charge (Fig. 3d). The vacuum thus breaks down in supercritical fields and the new ground state is necessarily charged. It is stabilized by the Pauli principle, which forbids the occupation of each spatial state by more than two electrons.

Calculations of the interaction of the electron with the radiation field do not alter these predictions. Spontaneous positron emission is predicted to be, therefore, a unique signature for establishing the change in the ground state of quantum electrodynamics as the supercritical field threshold is passed. It is the search for this process that has motivated much of the experimental interest in ion-atom collisions of superheavy systems since stable nuclei with $Z > 173$ are not provided by nature. SEE ANTIMATTER; ELECTRON-POSITRON PAIR PRODUCTION; POSITRON; QUANTUM ELECTRODYNAMICS; QUANTUM FIELD THEORY.

Positron creation in quasiatoms. **Figure 4** illustrates how the formation of superheavy quasimolecules can simulate the conditions for observing spontaneous positron emission. If a vacancy is produced in the $1s\sigma$ molecular orbital by ionization at a suitable prior time, then spontaneous positron emission can occur during the fraction of the collision time when the $1s\sigma$ binding exceeds $2m_0c^2$ at the quasiatomic limit. However, in contrast to the stable atom situation referred to above, in dynamical systems such as quasiatoms other positron creation mechanisms,

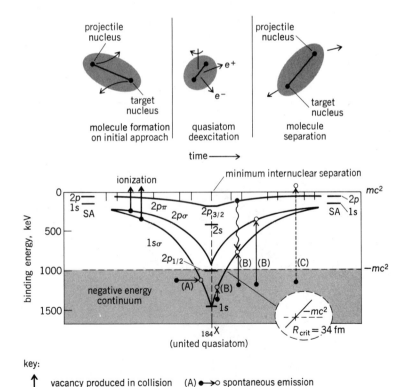

Fig. 4. Positron production mechanisms in heavy-ion collisions, shown on plot of energies of bound states as a function of time for U + U system.

not necessarily involving supercritically bound states, complicate the observation of spontaneous positron emission. In addition to the spontaneous process which proceeds without an external energy source, the time dependence of the electric field produced by the moving nuclear charges can also induce transitions of electrons from the negative-energy continuum both to unbound states and to unoccupied bound states even above the critical binding energy. The energy required for each of these two subcritical pair creation processes, induced emission into bound states and direct induced emission into continuum states, is supplied by the nuclear motion. Although they contribute backgrounds in the detection of spontaneous positron emission, the two effects are of considerable interest in themselves since they also reflect new aspects of the theory when strong electromagnetic fields are involved and when perturbation theory is not applicable.

Observation of dynamic positron creation. In 1976 a series of measurements were initiated to investigate positron creation in heavy-ion collisions. To selectively study both the subcritical and supercritical positron production processes discussed above, the studies included lead + lead, lead + uranium, uranium + uranium, and uranium + curium (Pb + Pb, Pb + U, U + U, and U + Cm) collision systems. The first generation of experiments developed information on the dynamic positron creation mechanisms (B) and (C) in Fig. 4. They established that in superheavy collision systems, positron production is observed considerably in excess of the intensity expected from nuclear background processes. They also showed that the magnitude of the cross sections, their dependence on the minimum internuclear separation, and, in particular, the almost exponential increase of the positron creation probability with increasing total nuclear charge ($Z_1 + Z_2$) cannot be accounted for without invoking the quasiatomic picture of ($Z_1 + Z_2$) acting in unison to produce very strong electric fields.

The general features predicted by the theory developed by W. Greiner and coworkers have been verified. The most striking of these first observations is the rapid increase of positron production with increasing Z_{total} ($Z_1 + Z_2$). Expressed as a power law proportional to ($Z_1 + Z_2$)^n, n takes on a value of about 21 (**Fig. 5**). This rapid increase with Z_{total} finds no other known analog in nature and singularly reflects the relativistic phenomena associated with very strong electric fields and their strongly bound atomic states.

Results of uncertainty principle. The positron spectra from the superheavy collision systems are emitted, predominantly, as continuous distributions in energy, with no apparent structure. Their mean energy and distribution in energy can be understood as roughly reflecting the time spent in the emission processes by utilizing the Heisenberg uncertainty principle, $\Delta E \cdot \Delta t > \hbar$, where ΔE represents the energy spread, Δt is the time during which emission takes place, and $\hbar$ is Planck's constant divided by 2π. For example, in a Coulomb scattering such as

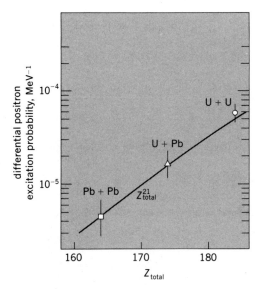

Fig. 5. Measured differential positron excitation probability as a function of the combined atomic number Z_{total} of the target and projectile. The distance of minimum approach and the relative ion velocity have been kept constant at 30 femtometers and 3.3×10^7 m/s respectively, and a positron energy, $E_{e+} = 478^{+54}_{-53}$ keV, has been selected near the peak of the positron spectrum.

illustrated in Fig. 4, the time spent by the *1s* electron in a state of supercritical binding is approximately 2×10^{-21} s. An application of the uncertainty principle in this case implies that even for the spontaneous emission process a broad spectral distribution with a width greater than 300 keV is expected and not the sharp line predicted for a static atom with $Z = 184$ (U + U). In fact, detailed theoretical calculations show that for experiments only involving sub-Coulomb energy scattering, where nuclear surfaces do not touch, the positron emission features do not show any distinguishing signatures that can identify the transition to supercritical binding. The calculations show that except for difficult-to-detect differences in intensity, the spectra emitted from systems with subcritical and supercritical binding are expected to be continuous and very similar, and this similarity is not expected to change appreciably as the scattering angle is changed. These predictions have been borne out by experiments.

Positron peaks in supercritical systems. However, subsequent experiments have revealed that very narrow positron peaks are produced in supercritical collision systems under very specific bombarding conditions. Three such examples, involving systems with supercritical binding, are shown in **Fig. 6**. The continua are produced primarily by the dynamic processes dis-

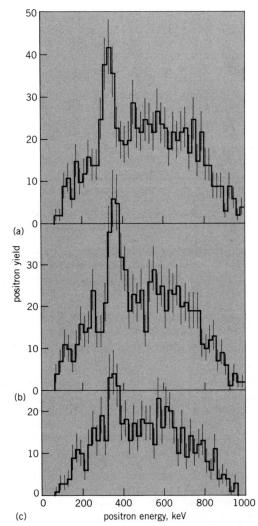

Fig. 6. Positron energy spectra for three collision systems with supercritical binding of the *1s* electron. These spectra were obtained at bombarding energies near the Coulomb barrier and for selected kinematic scattering conditions. (*a*) Uranium + curium (U + Cm with $Z_{total} = 188$). (*b*) Thorium + curium (Th + Cm with $Z_{total} = 186$). (*c*) Uranium + uranium (U + U with $Z_{total} = 184$).

cussed above. In each case the peak appears for a narrow interval of bombarding energies corresponding to the nuclei marginally touching, that is, as nuclear reactions set in. However, for each case a variety of experimental evidence rejects the possibility that the peaks originate from nuclear processes such as internal pair conversion of a gamma-ray transition in a nucleus excited in the collision.

From the arguments cited above using the uncertainty principle, the narrow widths of the peaks are inconsistent with the possibility that they are produced by spontaneous positron emission in a Coulomb scattering event which takes place in approximately 2×10^{-21} s. To explain the observed narrow width, the parent state must live for approximately 10^{-19} s. This relatively long-lived state is also required to prolong supercritical binding in order to allow the spontaneous positron creation amplitude to build up relative to the dynamic background. Therefore, to account for the peaks in the context of spontaneous positron emission, it has been suggested that metastable, giant nuclear systems are being formed at bombarding energies near the Coulomb barrier with mass and atomic numbers as large as $A = 486$ and $Z = 188$, respectively. A number of observations conform to this assumption. However, the similarity in the peak energies in Fig. 6 presents a contradiction with simple predictions by theory and precludes any straightforward association of the peaks with sponetaneous positron emission. The search for this process has been pursued with further experiments. More generally, the unexpected appearance of low-energy, narrow peaks in the positron spectra, which have not been observed in nature previously, could imply that a previously undetected source is being observed requiring an unusual explanation. Such a source could be the spontaneous positron emission process being sought, or possibly a heretofore undetected particle. SEE NUCLEAR MOLECULE; SUPERCRITICAL FIELDS.

Bibliography. D. A. Bromley (ed.), *Treatise on Heavy-Ion Science*, vol. 5., 1985; B. Crasemann (ed.), *X-ray and Atomic Inner-Shell Physics*, 1975; W. Greiner, *Quantum Electrodynamics of Strong Fields*, 1985; J. Rafelski, L. P. Fulcher, and A. Klein, Fermions and bosons interacting in arbitrary strong external fields, *Phys. Rep.*, 38C:227–361, 1978; J. Reinhardt and W. Greiner, Quantum electrodynamics of strong fields, *Rep. Prog. Phys.*, 40:219–295, 1977.

NUCLEAR MOLECULE
D. ALLAN BROMLEY

A quasistable entity of nuclear dimensions formed in nuclear collisions and comprising two or more discrete nuclei that retain their identities and are bound together by strong nuclear forces. Whereas the stable molecules of chemistry and biology consist of atoms bound through various electronic mechanisms, nuclear molecules do not form in nature except possibly in the hearts of giant stars; this simply reflects the fact that all nuclei carry positive electrical charges, and that under all natural conditions the long-range electrostatic repulsion prevents nuclear components from coming within the grasp of the short-range attractive nuclear force which could provide molecular binding. But in energetic collisions this electrostatic repulsion can be overcome.

Nuclear molecules were first suggested, in 1960, by D. A. Bromley, J. A. Kuehner, and E. Almqvist to explain very surprising experimental results obtained in the first studies on collisions between carbon nuclei carried out under conditions of high precision. In the original discovery, three sharp resonances appeared in the region of the Coulomb barrier corresponding to states in the ^{24}Mg compound nucleus near 20 MeV of excitation. These were completely unexpected and indicated a new mode of behavior for the 24-body ^{24}Mg system. Although such phenomena were known only in this ^{24}Mg system for almost 10 years, they have subsequently been shown to be ubiquitous features of both nuclear structure and nuclear dynamics. The exact mechanisms leading to the appearance of these molecular configurations are not yet known, however.

Thus far, attention has been focused on dinuclear molecular systems; the search for polynuclear configurations could provide important new insight into the behavior of many-body systems, generally.

^{12}C + ^{12}C system. In 1975 interest in molecular phenomena was renewed by discovery by Y. Abe that the original theoretical models proposed to explain the 1960 discovery were much richer in terms of predicted phenomena than had been realized. Experimental studies rapidly increased the number of resonances in the ^{12}C + ^{12}C system from the original 3 to a total of 39 (**Fig. 1***a*), established the spins and parities of many of them, and further demonstrated that the

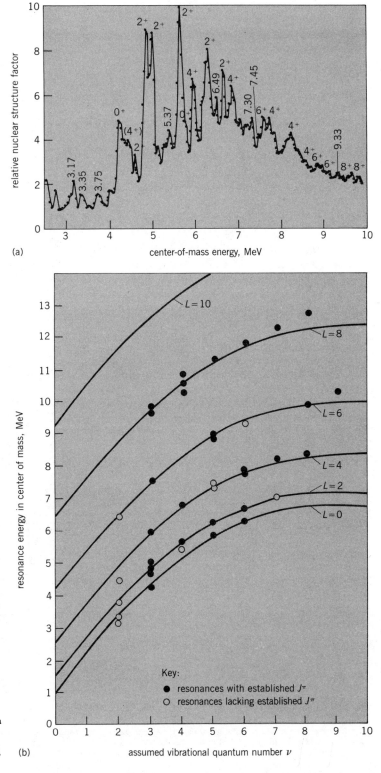

Fig. 1. Resonances in the $^{12}C + ^{12}C$ system. (a) Total cross section for $^{12}C + ^{12}C$ interactions as a function of the center-of-mass energy and expressed as a nuclear structure factor to remove penetrability effects. Numbers with superscript indicate spin (J) and parity (π). (b) Resonances correlated in terms of a U(4) symmetry and the corresponding analytic vibration-rotation spectrum. Solid curves are obtained from the equation in the text with $D = 0.34$ MeV; $a = 1.44$ MeV; $b = 0.08$ MeV; $c = 0.0757$ MeV.

wave functions of the states in ^{24}Mg that corresponded to the resonances had a marked molecular structure in which the ^{12}C nuclei retained their identity.

Models of molecular phenomena. Bromley, Kuehner, and Almqvist had noted that the effective potential, composed of the sum of nuclear, Coulomb, and centrifugal components, involved in the ^{12}C + ^{12}C collisions closely resembled a familiar Morse molecular one. In 1981 F. Iachello demonstrated that in such a potential it was possible, by using group theoretical techniques, to obtain an analytic expression for the allowed quantum states of the form of the equation below, where v and L are the vibrational and rotational quantum numbers and D, a, b, and c are

$$E(v,L) = -D + a(v + \tfrac{1}{2}) - b(v + \tfrac{1}{2})^2 + cL(L + 1)$$

constants. The governing group here is U(4). This simple expression provides a remarkably successful and complete description of all the experimental data (Fig. 1b).

The fact that this model predicts molecular dissociation at around 7 MeV suggested that the O$^+$ state in ^{12}C at 7.66 MeV might play an essential role in the molecular interaction rather

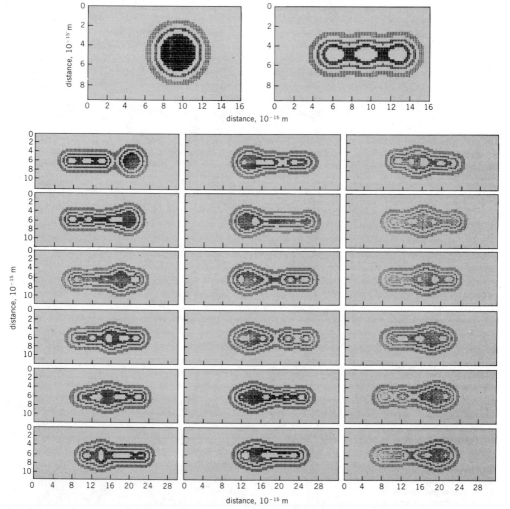

Fig. 2. Density contours calculated by using a constrained, time-dependent Hartree-Fock (TDHF) model. (*a*) Ground state of ^{12}C. (*b*) First excited O$^+$ state of ^{12}C. (*c*) Time sequence for the interaction of two carbon nuclei, one in each of these configurations.

than the 2^+ at 4.43 MeV which had been assumed to be critical to all earlier models. By using a constrained time-dependent Hartree-Fock (TDHF) approach, the density distribution for ^{12}C was first calculated in the ground (**Fig. 2**a) and excited (Fig. 2b) 0^+ states; the latter shows a linear, three-alpha-particle structure suggested several years previously. Assuming that one of the ^{12}C nuclei is inelastically excited to this state in an early stage of the collision, the TDHF model allows the collision to be followed in microscopic detail. At appropriately chosen relative energies, the excited carbon oscillates through the unexcited one, with a period of 2.3×10^{-21} s in the case shown in Fig. 2c. This is still a very crude model, but is suggestive of the kind of simple relative motions that are present in the molecular configurations.

In the ^{12}C + ^{16}O system, measurements at energies at, and below, the Coulomb barrier show that simple dinuclear molecular configurations are present in the form of rotational bands, just as in ^{12}C + ^{12}C, but at higher energies, although the resonances persist, their structure is substantially more complex. Below the barrier, a U(4) group description is again extremely successful.

Phenomena in heavier systems. Sharp, long-lived resonant states appear in the scattering of ^{28}Si on ^{28}Si up to the highest energies studied, 75 MeV of excitation in the ^{56}Ni compound system. Measurements of the angular distributions of the scattered nuclei (**Fig. 3**) show that these

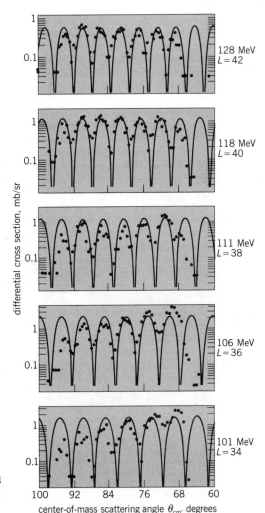

Fig. 3. Angular distributions of ^{28}Si nuclei elastically scattered from ^{28}Si at indicated energies of resonances. The solid curves are obtained by squaring the Legendre polynomial P_L of the indicated order L, giving the angular momentum of the resonance in units of $\hbar$ (Planck's constant divided by 2π). 1 mb = 10^{-31} m^2.

resonances have very high angular momentum; measurements have extended to $44\hbar$ (where $\hbar$ is Planck's constant divided by 2π), the highest angular momentum established in any quantum system. This is particularly interesting inasmuch as simple calculations suggest that at $50\hbar$ the ^{56}Ni nucleus should be torn apart by centrifugal forces. Resonance structure also appears in the ^{24}Mg + ^{24}Mg system, but not in ^{28}Si + ^{30}Si and ^{30}Si + ^{30}Si. Potential energy contours have been calculated for the compound nuclei involved in all these systems, at the angular momenta involved, assuming a generalized Nilsson shell model. It is suggestive that those systems which show a potential minimum at large prolate deformation appear to be those that show pronounced resonance structure. A quantitative theory that would allow calculation of the molecular phenomena from first principles, however, is still far from available.

Positron spectra have been measured from collisions of very heavy nuclei (uranium + uranium, uranium + thorium, uranium + curium) at energies in the region of the Coulomb barrier. A completely unexpected feature of these spectra is a sharp line at about 320 keV. If this is to arise from spontaneous positron creation in the supercritical Coulomb field created in the vicinity of the colliding ions, its width requires that some mechanism exist to maintain this supercritical field for much longer than would be the case in a simple Coulomb scattering. It has been suggested that this would occur if the interacting ions formed a molecular configuration and preliminary calculations show that this would be most probable in collisions in which the ends of the long prolate axes of these nuclei just came into contact. This configuration has not been established, but all available evidence points to the existence of superheavy nuclear molecular states in these interactions. *See Positron; Quasiatom; Supercritical fields.*

In all heavy-ion interactions yet studied, it has been the case that when adequate experimental energy resolution has been available to resolve sharp resonance structure corresponding to molecular phenomena, such structure has appeared. It will be an important challenge for higher-energy, high-precision nuclear accelerators to establish how high in excitation such structures can persist.

Radioactive decay. Indirect evidence supporting very asymmetric molecular configurations in heavy nuclei comes from the discovery of spontaneous radioactive decay of radium nuclei in which ^{14}C, and perhaps ^{16}O, nuclei are emitted. The probability of such decay is approximately 6×10^{-10} of that involving emission of an alpha particle, and the ratio of the probabilities is very closely equal to that calculated for the penetrability of the two products through the Coulomb barriers involved. This suggests that there is a substantial probability for both an alpha particle and ^{14}C to preexist in a molecular configuration. *See Radioactivity.*

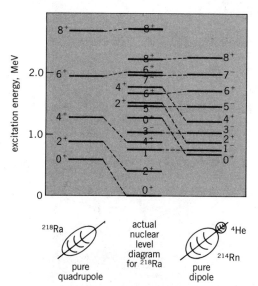

Fig. 4. Mixing of rotational bands based on pure quadrupole and dipole configurations in low-lying states of ^{218}Ra.

Dipole nuclear collectivity. That alpha particles do indeed preexist in these heavy nuclei has been suggested frequently on the basis of observed alpha-particle radioactivity. Rather striking confirmatory evidence has been obtained from measurements on the structure and electromagnetic deexcitation of bound states in radium nuclei. For example, in ^{218}Ra (**Fig. 4**) the presence of low-lying negative parity states and of very strong electric dipole transitions involving them provides strong evidence for the presence of a molecular dipole configuration involving an alpha particle and a ^{214}Rn nucleus. States having this configuration would be expected to mix with the normal quadrupole states except in the case of negative parity ones for which no low-lying quadrupole partner exists. Similar strong dipole effects provide evidence for the existence of alpha-particle molecular configurations in nuclei as light as ^{18}O where the molecular participants are ^{4}He and ^{14}C. *See* NUCLEAR STRUCTURE; SCATTERING EXPERIMENTS (NUCLEI).

Bibliography. D. A. Bromley, Nuclear molecules, *Sci. Amer.*, 239(6):58–69, 1978; N. Cindro (ed.), *Nuclear Molecular Phenomena*, 1982; K. A. Erb and D. A. Bromley, Nuclear molecular resonances in heavy-ion collisions, *Phys. Today*, 32(1):34–42, 1979; J. S. Lilley and M. A. Nagarajan (eds.), *Nuclear Cluster Phenomena*, 1984.

NUCLEAR REACTION
DENNIS G. KOVAR

A process that occurs as a result of interactions between atomic nuclei when the interacting particles approach each other to within distances of the order of nuclear dimensions ($\simeq 10^{-12}$ cm). While nuclear reactions occur in nature, understanding of them and use of them as tools have taken place primarily in the controlled laboratory environment. In the usual experimental situation, nuclear reactions are initiated by bombarding one of the interacting particles, the stationary target nucleus, with nuclear projectiles of some type, and the reaction products and their behaviors are studied. The study of nuclear reactions is the largest area of nuclear and subnuclear (or particle) physics; the threshold for producing pions has historically been taken to be the energy boundary between the two fields.

Types of nuclear interaction. As a generalized nuclear process, consider a collision in which an incident particle strikes a previously stationary particle, to produce an unspecified number of final products. If the final products are the same as the two initial particles, the process is called scattering. The scattering is said to be elastic or inelastic, depending on whether some of the kinetic energy of the incident particle is used to raise either of the particles to an excited state. If the product particles are different from the initial pair, the process is referred to as a reaction.

The most common type of nuclear reaction, and the one which has been most extensively studied, involves the production of two final products. Such reactions can be observed, for example, when deuterons with a kinetic energy of a few million electron-volts (MeV) are allowed to strike a carbon nucleus of mass 12. Protons, neutrons, deuterons, and alpha particles are observed to be emitted, and reactions (1)–(4) are responsible. In these equations the nuclei are indicated by

$$^{2}_{1}H + ^{12}_{6}C \rightarrow ^{2}_{1}H + ^{12}_{6}C \qquad (1)$$

$$^{2}_{1}H + ^{12}_{6}C \rightarrow ^{1}_{1}H + ^{13}_{6}C \qquad (2)$$

$$^{2}_{1}H + ^{12}_{6}C \rightarrow ^{1}_{0}n + ^{13}_{7}N \qquad (3)$$

$$^{2}_{1}H + ^{12}_{6}C \rightarrow ^{4}_{2}He + ^{10}_{5}B \qquad (4)$$

the usual chemical symbols; the subscripts indicate the atomic number (nuclear charge) of the nucleus, and the superscripts the mass number of the particular isotope. These reactions are conventionally written in the compact notation $^{12}C(d,d)^{12}C$, $^{12}C(d,p)^{13}C$, $^{12}C(d,n)^{13}N$, and $^{12}C(d,\alpha)^{10}B$, where d represents deuteron, p proton, n neutron, and α alpha particle. In each of these cases the reaction results in the production of an emitted light particle and a heavy residual nucleus. The (d,d) process denotes the elastic scattering as well as the ^{12}C nucleus to one of its excited states. The other three reactions are examples of nuclear transmutation or disintegration where the residual nuclei may also be formed in their ground states or one of their many excited states. The processes producing the residual nucleus in different excited states are considered to be the different reaction channels of the particular reaction. If the residual nucleus is formed in an excited state, it will subsequently emit this excitation energy in the form of gamma rays, or, in special cases, electrons. The residual nucleus may also be a radioactive species, as in the case of ^{13}N formed in the ^{12}C (d,n) reaction. In this case the residual nucleus will undergo further transformation in accordance with its characteristic radioactive decay scheme. SEE RADIOACTIVITY.

Nuclear cross section. In general one is interested in the probability of occurrence of the various reactions as a function of the bombarding energy of the incident particle. The measure of probability for a nuclear reaction is its cross section. Consider a reaction initiated by a beam of particles incident on a region which contains N atoms per unit area (uniformly distributed), and where I particles per second striking the area result in R reactions of a particular type per second. The fraction of the area bombarded which is effective in producing the reaction products is R/I. If this is divided by the number of nuclei per unit area, the effective area or cross section $\sigma = R/IN$. This is referred to as the total cross section for the specific reaction, since it involves all the occurrences of the reaction. The dimensions are those of an area, and total cross sections are expressed either square centimeters or barns (1 barn = 10^{-24} cm^2). The differential cross section refers to the probability that a particular reaction product will be observed at a given angle with respect to the beam direction. Its dimensions are those of an area per unit solid angle (for example, barns per steradian).

Requirements for a reaction. Whether a specific reaction occurs and with what cross section it is observed depend upon a number of factors, some of which are not always completely understood. However, there are some necessary conditions which must be fulfilled if a reaction is to proceed.

Coulomb barrier. For a reaction to occur, the two interacting particles must approach each other to within the order of nuclear dimensions ($\simeq 10^{-12}$ cm). With the exception of the uncharged neutron, all incident particles must therefore have sufficient kinetic energy to overcome the electrostatic (Coulomb) repulsion produced by the intense electrostatic field of the nuclear charge. The kinetic energy must be comparable to or greater than the so-called Coulomb barrier, whose magnitude is approximately give by the expression $E_{\text{Coul}} \approx Z_1 Z_2 / (A_1^{1/3} + A_2^{1/3})$MeV, where Z and A respectively refer to the nuclear charge and mass number of the interacting particles 1 and 2. It can be seen that while, for the lightest targets, protons with kinetic energies of a few hundred thousand electronvolts (keV) are sufficient to initiate reactions, energies of many hundred millions of electronvolts are required to initiate reactions between heavier nuclei. In order to provide energetic charged particles, to be used as projectiles in reaction studies, particle accelerators of various kinds (such as Van de Graaff generators, cyclotrons, and linear accelerators) have been developed, making possible studies of nuclear reactions induced by projectiles as light as protons and as heavy as ^{208}Pb. SEE PARTICLE ACCELERATOR.

Since neutrons are uncharged, they are not repelled by the electrostatic field of the target nucleus, and neutron energies of only a fraction of an electronvolt are sufficient to initiate some reactions. Neutrons for reaction studies can be obtained from nuclear reactors or from various nuclear reactions which produce neutrons as reaction products. There are two other means of producing nuclear reactions which do not fall into the general definition given above. Both electromagnetic radiation and high-energy electrons are capable of disintegrating nuclei under special conditions. However, both interact much less strongly with nuclei than nucleons or other nuclei, through the electromagnetic and weak nuclear forces, respectively, rather than the strong nuclear force responsible for nuclear interactions. SEE NEUTRON.

Q value. For a nuclear reaction to occur, there must be sufficient kinetic energy available to bring about the transmutation of the original nuclear species into the final reaction products. The sum of the kinetic energies of the reaction products may be greater than, equal to, or less than the sum of the kinetic energies before the reaction. The difference in the sums is the Q value for that particular reaction. It can be shown that the Q value is also equal to the difference in the masses (rest energies) of the reaction products and the masses of the initial nuclei. Reactions with a positive Q value are called exoergic or exothermic reactions, while those with a negative Q value are called endoergic or endothermic reactions.

In reactions (1)–(4), where the residual nuclei are formed in their ground states, the Q values are ^{12}C(d,d)^{12}C, Q = 0.0 MeV; ^{12}C(d,p)^{13}C, Q = 2.72 MeV; ^{12}C(d,n)^{13}N, Q = −0.28 MeV; and ^{12}C(d,α)^{10}B, Q = −1.34 MeV. For reactions with a negative Q value, a definite minimum kinetic energy is necessary for the reaction to take place. While there is no threshold energy for reactions with positive Q values, the cross section for the reactions induced by charged particles is very small unless the energies are sufficient to overcome the Coulomb barrier. A nuclear reaction and its inverse are reversible in the sense that the Q values are the same but have the opposite sign (for example, the Q value for the ^{10}B(α,d)^{12}C reaction is $+1.39$ MeV).

Conservation laws. It has been found experimentally that certain physical quantities must be the same both before and after the reaction. The quantities conserved are electric charge, number of nucleons, energy, linear momentum, angular momentum, and in most cases parity. Except for high-energy reactions involving the production of mesons, the conservation of charge and number of nucleons allow one to infer that the numbers of protons and neutrons are always conserved. The conservation of the number of nucleons indicates that the statistics governing the system are the same before, during, and after the reaction. Fermi-Dirac statistics are obeyed if the total number is odd, and Bose-Einstein if the number is even. The conservation laws taken together serve to strongly restrict the reactions that can take place, and the conservation of angular momentum and parity in particular allow one to establish spins and parities of states excited in various reactions. SEE PARITY; SYMMETRY LAWS.

Reaction mechanism. What happens when a projectile collides with a target nucleus is a complicated many-body problem which is still not completely understood. Progress made in the last decades has been in the development of various reaction models which have been extremely successful in describing certain classes or types of nuclear reaction processes. In general, all reactions can be classified according to the time scale on which they occur, and the degree to which the kinetic energy of the incident particle is converted into internal excitation of the final products. A large fraction of the reactions observed has properties consistent with those predicted by two reaction mechanisms which represent the extremes in this general classification. These are the mechanisms of compound nucleus formation and direct interaction.

Compound nucleus formation. As originally proposed by N. Bohr, the process is envisioned to take place in two distinct steps. In the first step the incident particle is captured by (or fuses with) the target nucleus, forming an intermediate or compound nucleus which lives a long time ($\simeq 10^{-16}$ s) compared to the approximately 10^{-22} s it takes the incident particle to travel past the target. During this time the kinetic energy of the incident particle is shared among all the nucleons, and all memory of the incident particle and target is lost. The compound nucleus is always formed in a highly excited unstable state, is assumed to approach themodynamic equilibrium involving all or most of the available degrees of freedom, and will decay, as ths second step, into different reaction products, or through so-called exit channels. In most cases, the decay can be understood as a statistical evaporation of nucleons or light particles. In the examples of reactions (1)–(4), the compound nucleus formed is ^{14}N, and four possible exit channels are indicated (**Fig. 1**). In reactions involving heavier targets (for example, $A \simeq 200$), one of the exit channels may be the fission channel where the compound nucleus splits into two large fragments. SEE NUCLEAR FISSION.

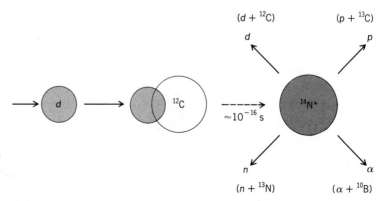

Fig. 1. Formation of the compound nucleus ^{14}N after capture of the deuteron by ^{12}C. Four exit channels are indicated.

The essential feature of the compound nucleus formation or fusion reaction is that the probability for a specific reaction depends on two independent probabilities: the probability for forming the compound nucleus, and the probability for decaying into that specific exit channel. While certain features of various interactions cannot be completely explained within the framework of the compound nucleus hypothesis, it appears that the mechanism is responsible for a large fraction of reactions occurring in almost all projectile-target interactions. Fusion reactions have been extremely useful in several kinds of spectroscopic studies. Particularly notable have been the resonance studies performed with light particles, such as neutrons, protons, deuterons, and alpha particles, on light target nuclei, and the gamma-ray studies of reactions induced by heavy projectiles, such as ^{16}O and ^{32}S, on target nuclei spanning the periodic table. These studies have provided an enormous amount of information regarding the excitation energies and spins of levels in nuclei. S*ee* N*uclear spectra*.

Direct interactions. Some reactions have properties which are in striking conflict with the predictions of the compound nucleus hypothesis. Many of these are consistent with the picture of a mechanism where no long-lived intermediate system is formed, but rather a fast mechanism where the incident particle, or some portion of it, interacts with the surface, or some nucleons on the surface, of the target nucleus. Models for direct processes make use of a concept of a homogeneous lump of nuclear matter with specific modes of excitation, which acts to scatter the incident particle through forces described, in the simplest cases, by an ordinary spherically symmetric potential. In the process of scattering, some of the kinetic energy may be used to excite the target, giving rise to an inelastic process, and nucleons may be exchanged, giving rise to a transfer process. In general, however, direct reactions are assumed to involve only a very small number of the available degrees of freedom.

Most direct reactions are of the transfer type where one or more nucleons are transferred to or from the incident particle as it passes the target, leaving the two final partners either in their ground states or in one of their many excited states. Such transfer reactions are generally referred to as stripping or pick-up reactions, depending on whether the incident particle has lost or acquired nucleons in the reaction. The (d,p) reaction is an example of a stripping reaction, where the incident deuteron is envisioned as being stripped of its neutron as it passes the target nucleus, and the proton continues along its way (**Fig. 2**).

The properties of the target nucleus determine the details of the reaction, fixing the energy and angular momentum with which the neutron must enter it. The energy of the outgoing proton is determined by how much of the deuteron's energy is taken into the target nucleus by the

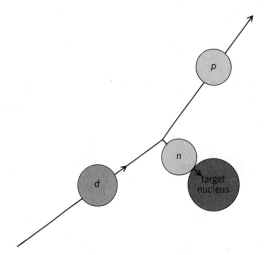

Fig. 2. A (d,p) **transfer reaction.**

neutron, and indeed serves to identify the final state populated by the Q value of the reaction as a whole. The angular distribution of the differential cross sections will, at appropriate bombarding energies, not be smooth but rather will show a distinct pattern of maxima and minima which are indicative of the spin and parity of the final state. The cross section for populating a specific final state in the target nucleus depends on the nuclear structure of the nuclei involved. This sensitivity has been used in studies of single-nucleon, two-nucleon, and four-nucleon transfer reactions, such as (d,p), (p,d), $(^3\text{He},d)$, (t,p), and $(^7\text{Li},d)$, to establish the validity and usefulness of the shell-model description of nuclei. Multinucleon transfer reactions with heavier projectiles have been powerful tools for reaching nuclei inaccessible by other means and in producing new isotopes.

Inelastic scattering is also a direct reaction whose angular distribution can provide information about the spin and parity of the excited state. Whereas the states preferentially populated in transfer reactions are those of specific single-particle or shell-model structure, the states preferentially excited in inelastic scattering are collective in nature. The states are most easily understood in the framework of macroscopic descriptions in which they are considered to be oscillations in shape about a spherical mean (vibrations) or to be the rotations of a statically deformed shape. The cross section for inelastic scattering is related to the shape or deformation of the target nucleus in its various collective excitations. Inelastic excitation can be caused by both a nuclear interaction and an electromagnetic interaction known as Coulomb excitation, where the target nucleus interacts with the rapidly changing electric field caused by the passage of the charged particle. Coulomb excitation is an important process at low bombarding energies in interactions involving heavy projectiles. Studies in inelastic scattering to low-lying states of nuclei across the periodic table have provided graphic demonstrations of the collective aspects of nuclei.

Elastic scattering is the direct interaction which leaves the interacting particles unchanged. For charged particles at low bombarding energies, the elastic scattering is well described in terms of the inverse-square force law between two electrically charged bodies. In this case, the process is known as Rutherford scattering. At higher energies the particles come into the range of the nuclear force, and the elastic scattering deviates from the inverse-square behavior. SEE NUCLEAR STRUCTURE; SCATTERING EXPERIMENTS (NUCLEI).

More complex reaction mechanisms. Processes intermediate between direct and compound nucleus formation do occur. The best example of such a process is the so-called preequilibrium emission, where light particles are emitted before the kinetic energy has been shared among all the nucleons in the compound nucleus. Another example is seen in the interaction of two massive nuclei, such as ^{84}Kr + ^{209}Bi, where the probability for formation of the compound nucleus is found to be very small. The experimental observations indicate that the nuclei interact for a short time and then separate in what appears to be a direct reaction. Although the interaction times for these so-called strongly damped or deep inelastic collisions are very small compared to that for compound nucleus formations, they are sufficiently long for considerable mass transfer and loss of relative kinetic energy to occur.

Nuclear reaction studies. In most instances the study of nuclear reactions is directed toward the long-range goal of obtaining information about the properties and structure of nuclei. Such studies usually proceed in two stages. In the first stage the attention focuses on the mechanism of the reaction and on establishing the dependence on the nuclei involved. The specific models proposed continue to be modified and improved, as they are confronted with more experimental results, until their predictions become reliable. At that point the second stage is entered, in which the focus of the effort is on extraction of information about nuclei.

There are other studies which are focused on reaction cross-section behaviors for other purposes. Examples of this are the neutron-capture reactions on heavy target nuclei that fission, and the $^3\text{H}(d,n)^4\text{He}$ reaction, important in thermonuclear processes, which continue to be studied because of their application as energy sources. Studies of the interactions of light nuclei at low energies are intensely studied because of their implications for astrophysics and cosmology. SEE NUCLEAR FUSION; THERMONUCLEAR REACTION.

Bibliography. R. Bass, *Nuclear Reactions with Heavy Ions*, 1980; J. Cerny (ed.), *Nuclear Spectroscopy and Reactions*, 1974; W. M. Gibson, *The Physics of Nuclear Reactions*, 1980; P. E. Hodgson, *Nuclear Reactions and Nuclear Spectroscopy*, 1971; G. R. Satchler, *Introduction to Nuclear Reactions*, 1980.

DEEP INELASTIC COLLISIONS
John R. Huizenga

Collisions between heavy ions in which the two nuclei interact strongly while their nuclear surfaces overlap. Deep inelastic collisions are characterized by features that are intermediate between those of comparatively simple quasielastic, few-nucleon transfer reactions and those of the highly complex compound-nucleus reactions. These deep inelastic or damped collisions often occur for heavy-ion reactions at center-of-mass energies less than 5 MeV per nucleon above the Coulomb barrier. During the brief encounter of the two nuclei, large amounts of kinetic energy of radial and orbital motion can be dissipated. The lifetime of the dinuclear complex (analogous to a chemical molecule) corresponds to the time required for the intermediate system to make a partial rotation (10^{-22} to 5×10^{-21} s). On separation, the final total kinetic energies of the two reaction fragments can be well below those corresponding to the Coulomb repulsion of spheres, indicating that the fragments are highly deformed in the exit channel, as is known to be the case for fission fragments. *See Nuclear fission*.

Characteristic features. The principal properties of damped or deep inelastic reactions at bombarding energies of a few MeV per nucleon above the interaction barrier are well established. Among these characteristic features are the following: (1) The reactions are binary with two massive reaction fragments in the exit channel. Subsequently, the excited primary fragments, which for large-energy losses have nearly the same temperature, decay via fission or the emission of light particles and gamma rays. (2) Large amounts of kinetic energy of radial and orbital motion can be dissipated in a reaction. (3) Fragment mass and charge distributions are bimodal and are centered close to the masses and charges of projectile and target nuclei. The second moments of these distributions are strongly correlated with kinetic energy dissipation. However, the first moment in $A(Z)$ is nearly independent of energy loss. (4) The angular distributions of the reaction fragments change in a characteristic way with the charge product $Z_P \cdot Z_T$ of the projectile-target combination and the bombarding energy.

Reaction fragment distributions. Experimental studies of the first and second moments of the mass, charge, and isobaric and isotopic distributions of damped reaction fragments and the variation of the moments with total kinetic energy loss and scattering angle have yielded new insight into the microscopic reaction mechanism for heavy ions. Evidence for the magnitude of the charge and mass flow is manifested by the second moments of these distributions which increase with increasing energy loss (where energy loss is a direct measure of the reaction time). Studies of the second moment (variance) in the charge distribution for fixed mass asymmetry (isobaric distribution) have shown that the charge equilibration degree of freedom, like the excitation energy, is relaxed in less than 2×10^{-22} s. In contrast to the short relaxation times for the above two intrinsic degrees of freedom, the mass asymmetry degree of freedom relaxes far more slowly.

Light-particle emission. Studies of light-particle emission from deep inelastic reactions are important because these particles can, in principle, be emitted at any stage of the reaction and hence carry signatures of the dominating mechanisms at work at the different reaction stages. In damped reactions with bombarding energies up to 5 MeV per nucleon above the Coulomb barrier, where hundreds of MeV of energy can be dissipated, neutron spectra are consistent with evaporation from the two fully accelerated heavy fragments. For kinetic energy losses greater than 50 MeV, analysis of the neutron spectra indicates that the two fragments have essentially the same temperature as a function of energy loss. In addition, analysis of the differential cross sections of emitted neutrons leads to the conclusion that the excitation energy is divided between the fragments in proportion to their masses. These interesting and important results are, at first, somewhat puzzling when interpreted in terms of simple nucleon exchange models, where the lighter fragment is predicted to have the higher temperature. However, such an initial temperature gradient over the dinuclear complex can be expected to decay through subsequent interactions. The nucleon exchange mechanism itself serves to produce a strong temperature-equalizing effect over the two components of the dinuclear complex with a temperature gradient. The dynamical competition between generation of a temperature difference and its decay can only be estimated from realistic model calculations.

Interpretation. The ultimate goal of experimental and theoretical investigations of damped or deep inelastic reactions between complex nuclei is to achieve an understanding of the reaction properties in terms of the underlying microscopic mechanisms. However, the development of microscopic quantal reaction theories has not progressed far enough to yield a consistent explanation of the observed reaction features. In this situation, a phenomenological interpretation appears to be useful, where certain reaction patterns are analyzed in terms of macroscopic or microscopic concepts, making it possible to study the dependence of model parameters on important reaction conditions. Comparisons of a variety of experimental results with dynamical transport calculations based on one-body nucleon exchange give strong evidence that stochastic exchange of independent nucleons accounts for the dominant part of the dissipative and fluctuative phenomena observed in damped reactions. However, the exact fraction of the energy loss that is due to nucleon exchange and the fraction due to other, more minor processes, including collective modes of energy transformation, remain to be determined. SEE NUCLEAR REACTION; NUCLEAR STRUCTURE.

Bibliography. M. Lefort and C. Ngô, Deep inelastic reactions with heavy ions: A probe for nuclear macrophysics studies, *Ann. Phys.*, 3:5–114, 1978; J. Randrup, Theory of transfer-induced transport in nuclear collisions, *Nucl. Phys.*, 327:490–516, 1979; W. U. Schröder and J. R. Huizenga, Damped heavy-ion collisions, *Annu. Rev. Nucl. Sci.*, 27:465–547, 1977; W. U. Schröder and J. R. Huizenga, Damped nuclear reactions, *Treatise on Heavy Ion Science*, vol. 2, pp. 113–726, 1984.

NUCLEAR FUSION
Richard F. Post

One of the primary nuclear reactions, the name usually designating an energy-releasing rearrangement collision which can occur between various isotopes of low atomic number. SEE NUCLEAR REACTION.

Interest in the nuclear fusion reaction arises from the expectation that it may someday be used to produce useful power, from its role in energy generation in stars, and from its use in the fusion bomb. Since a primary fusion fuel, deuterium, occurs naturally and is therefore obtainable in virtually inexhaustible supply (by separation of heavy hydrogen from water, 1 atom of deuterium occurring per 6500 atoms of hydrogen), solution of the fusion power problem would permanently solve the problem of the present rapid depletion of chemically valuable fossil fuels. As a power source, the lack of radioactive waste products from the fusion reaction is another argument in its favor as opposed to the fission of uranium.

In a nuclear fusion reaction the close collision of two energy-rich nuclei results in a mutual rearrangement of their nucleons (protons and neutrons) to produce two or more reaction products, together with a release of energy. The energy usually appears in the form of kinetic energy of the reaction products, although when energetically allowed, part may be taken up as energy of an excited state of a product nucleus. In contrast to neutron-produced nuclear reactions, colliding nuclei, because they are positively charged, require a substantial initial relative kinetic energy to overcome their mutual electrostatic repulsion so that reaction can occur. This required relative energy increases with the nuclear charge Z, so that reactions between low-Z nuclei are the easiest to produce. The best known of these are the reactions between the heavy isotopes of hydrogen, deuterium and tritium.

Fusion reactions were discovered in the 1920s when low-Z elements were used as targets and bombarded by beams of energetic protons or deuterons. But the nuclear energy released in such bombardments is always microscopic compared with the energy of the impinging beam. This is because most of the energy of the beam particle is dissipated uselessly by ionization and interparticle collisions in the target; only a small fraction of the impinging particles actually produce reactions.

Nuclear fusion reactions can be self-sustaining, however, if they are carried out at a very high temperature. That is to say, if the fusion fuel exists in the form of a very hot ionized gas of stripped nuclei and free electrons termed a plasma, the agitation energy of the nuclei can overcome their mutual repulsion, causing reactions to occur. This is the mechanism of energy generation in the stars and in the fusion bomb. It is also the method envisaged for the controlled generation of fusion energy.

PROPERTIES OF FUSION REACTIONS

The cross sections (effective collisional areas) for many of the simple nuclear fusion reactions have been measured with high precision. It is found that the cross sections generally show broad maxima as a function of energy and have peak values in the general range of 0.01 barn (1 barn = 10^{-28} m^2) to a maximum value of 5 barns, for the deuterium-tritium (D-T) reaction. The energy releases of these reactions can be readily calculated from the mass differences between the initial and final nuclei or determined by direct measurement.

Simple reactions. Some of the important simple fusion reactions, their reaction products, and their energy releases in MeV are given by reactions (1).

$$\begin{array}{rl}
\text{D} + \text{D} \to & {}^3\text{He} + n + 3.25 \text{ MeV} \\
\text{D} + \text{D} \to & \text{T} + p + 4.0 \text{ MeV} \\
\text{T} + \text{D} \to & {}^4\text{He} + n + 17.6 \text{ MeV} \\
{}^3\text{He} + \text{D} \to & {}^4\text{He} + p + 18.3 \text{ MeV} \\
{}^6\text{Li} + \text{D} \to & 2\,{}^4\text{He} \quad\quad + 22.4 \text{ MeV} \\
{}^7\text{Li} + p \to & 2\,{}^4\text{He} \quad\quad + 17.3 \text{ MeV}
\end{array} \quad (1)$$

If it is remembered that the energy release in the chemical reaction in which hydrogen and oxygen combine to produce a water molecule is about 1 eV per reaction, it will be seen that, gram for gram, fusion fuel releases more than 1,000,000 times as much energy as typical chemical fuels.

The two alternative D-D reactions listed occur with about equal probability for the same relative particle energies. Note that the heavy reaction products, tritium and helium-3, may also react, with the release of a large amount of energy. Thus it is possible to visualize a reaction chain in which six deuterons are converted to two helium-4 nuclei, two protons, and two neutrons, with an overall energy release of 43 MeV—about 10^5 kilowatthours (kWh) of energy per gram of deuterium. This energy release is several times that released per gram in the fission of uranium, and several million times that released per gram by the combustion of gasoline.

Cross sections. Figure 1 shows the measured values of cross sections as a function of bombarding energy up to 100 keV for the total D-D reaction (both D-D,n and D-D,p), the D-T reaction, and the D-^{3}He reaction. The most striking characteristic of these curves is their extremely rapid falloff with energy as bombarding energies drop to a few kilovolts. This effect arises from the mutual electrostatic repulsion of the nuclei, which prevents them from approaching closely if their relative energy is small. *See Nuclear structure.*

The fact that reactions can occur at all at these energies is attributable to the finite range of nuclear interaction forces. In effect, the boundary of the nucleus is not precisely defined by its classical diameter. The role of quantum mechanical effects in nuclear fusion reactions has been treated by G. Gamow and others. It is predicted that the cross sections should obey an exponential law at low energies. This is well borne out in energy regions reasonably far removed from resonances (for example, below about 30 keV for the D-T reaction). Over a wide energy range at low energies, the data for the D-D reaction can be accurately fitted by a Gamow curve, the result for the cross section being given by Eq. (2), where the bombarding energy W is in keV.

$$\sigma_{D\text{-}D} = \frac{288}{W} e^{-45.8 W^{-1/2}} \times 10^{-28} \text{ m}^2 \quad (2)$$

The extreme energy dependence of this expression can be appreciated by the fact that, between 1 and 10 keV, the predicted cross section varies by about 13 powers of 10, that is, from 2×10^{-46} to 1.5×10^{-33} m^2.

Energy division. The kinematics of the fusion reaction requires that two or more reaction products result. This is because both mass energy and momentum balance must be preserved. When there are only two reaction products (which is the case in all of the important reactions), the division of energy between the reaction products is uniquely determined, the lion's share always going to the lighter particle. The energy division (disregarding the initial bombarding energy) is as in reaction (3). If reaction (3) holds, with the As representing the atomic masses of

$$A_1 + A_2 \to A_1' + A_2' + Q \quad (3)$$

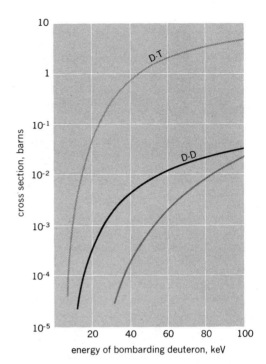

Fig. 1. Cross sections versus bombarding energy for three simple fusion reactions. (*After R. F. Post, Fusion power, Sci. Amer., 197(6):73–84, December 1957*)

the particles and Q the total energy released, then Eqs. (4) are valid, where $W(A'_1)$ and $W(A'_2)$ are

$$W(A'_1) + W(A'_2) = Q \qquad W(A'_1) = Q\left(\frac{A'_2}{A'_1 + A'_2}\right) \qquad W(A'_2) = Q\left(\frac{A'_1}{A'_1 + A'_2}\right) \qquad (4)$$

the kinetic energies of the reaction products.

Thus in the D-T reaction, for example, A'_1, the mass of the alpha particle, is four times A'_2, the mass of the neutron, so that the neutron carries off four-fifths of the reaction energy, or 14 MeV.

Reaction rates. When nuclear fusion reactions occur in a high-temperature plasma, the reaction rate per unit volume depends on the particle density n of the reacting fuel particles and on an average of their mutual reaction cross sections σ and relative velocity v over the particle velocity distributions. *See* THERMONUCLEAR REACTION.

For dissimilar reacting nuclei (such as D and T), the reaction rate is given by Eq. (5).

$$R_{12} = n_1 n_2 \langle \sigma v \rangle_{12} \quad \text{reactions/(m}^3)(\text{s}) \qquad (5)$$

For similar reacting nuclei (for example, D and D), the reaction rate is given by Eq. (6).

$$R_{11} = 1/2 n^2 \langle \sigma v \rangle \qquad (6)$$

Note that both expressions vary as the square of the total particle density (for a given fuel composition).

If the particle velocity distributions are known, $\langle \sigma v \rangle$ can be determined as a function of energy by numerical integration, using the known reaction cross sections. It is customary to assume a maxwellian particle velocity distribution, toward which all others tend in equilibrium. The values of $\langle \sigma v \rangle$ for the D-D and D-T reactions are shown in **Fig. 2**. In this plot the kinetic temperature is given in units of keV; 1 keV kinetic temperature = 1.16×10^7 K. Just as in the case of the cross sections themselves, the most striking feature of these curves is their extremely rapid falloff with temperature at low temperatures. For example, although at 100 keV for all reactions $\langle \sigma v \rangle$ is only weakly dependent on temperature, at 1 keV it varies as $T^{6.3}$ and at 0.1 keV as

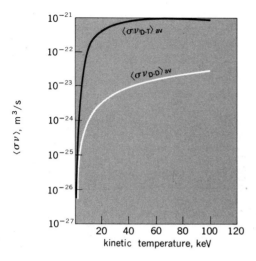

Fig. 2. Plot of the values of $\langle \sigma v \rangle$ versus kinetic temperature for the D-D and D-T reactions.

$T^{1/3}$! Also, at the lowest temperatures it can be shown that only the particles in the "tail" of the distribution, which have energies large compared with the average, will make appreciable contributions to the reaction rate, the energy dependence of σ being so extreme.

Critical temperatures. The nuclear fusion reaction can obviously be self-sustaining only if the rate of loss of energy from the reacting fuel is not greater than the rate of energy generation by fusion reactions. The simplest consequence of this fact is that there will exist critical or ideal ignition temperatures below which a reaction could not sustain itself, even under idealized conditions. In a fusion reactor, ideal or minimum critical temperatures are determined by the unavoidable escape of radiation from the plasma. A minimum value for the radiation emitted from any plasma is that emitted by a pure hydrogenic plasma in the form of x-rays or bremsstrahlung. Thus plasmas composed only of isotopes of hydrogen and their one-for-one accompanying electrons might be expected to possess the lowest ideal ignition temperatures. This is indeed the case: It can be shown by comparison of the nuclear energy release rates with the radiation losses that the critical temperature for the D-T reaction is about 4×10^7 K. For the D-D reaction it is about 10 times higher. Since both radiation rate and nuclear power vary with the square of the particle density, these critical temperatures are independent of density over the density ranges of interest. The concept of the critical temperature is a highly idealized one, however, since in any real cases additional losses must be expected to occur which will modify the situation, increasing the required temperature.

FUSION POWER

Worldwide interest in nuclear fusion arises from its promise as an inexhaustible and environmentally attractive source of energy for the future. Fusion power plants do not yet exist, but research into the physics and technology that will be required to construct such systems has been under way in many countries since the 1950s. As a result of this research, there is a growing consensus that fusion power is an achievable goal within the context of the present approaches. Consequently, the emphasis in fusion research is shifting toward resolving issues that are posed by economic constraints. That is, it is shifting toward optimization and simplification of the specific approaches to fusion power that have been developed in the fusion quest.

The two key problems in achieving net power from a fusion system are, first, to heat the fusion fuel charge to its required high kinetic temperature, and second, to confine the heated fuel for a long enough time for the fusion energy released to exceed the energy required to heat the fuel to its fusion combustion temperature, including all relevant losses and inefficiencies.

Pellet fusion, also called inertial confinement fusion, aims at the same objective, but by an entirely different route. Here the idea is to rapidly compress and heat a tiny pellet of fusion fuel (in the frozen state), carrying out the entire operation so quickly that a net fusion energy release can take place before the pellet flies apart. At present the main technical effort in pellet fusion is centered on the use of arrays of high-powered lasers to accomplish the necessary compression and heating. However, a substantial effort is also being devoted to achieving pellet fusion by the use of intense and highly focused beams of protons. The use of heavy ion beams as the "driver" for pellet fusion is also being seriously considered.

Magnetic confinement. The possibility of magnetically confining a fusion plasma derives from the nature of the plasma state of matter. A plasma may be viewed as an electrically conducting gas that exerts an outward pressure, or as a collection of free positive and negative charges. The pressure exerted by the plasma gas as it attempts to expand can be resisted by the electromagnetic stress field associated with a strong magnetic field; the individual charged particles can at the same time be guided by a properly shaped magnetic field that forces these particles to execute orbits that remain within the vacuum chamber surrounding the plasma and that do not contact the walls.

The difficulty with magnetic confinement arises from the fact that the plasma, in order to be confined, must possess self-generated particle flows and electrical currents (diamagnetic currents), whose interaction with the externally generated fields produces the confining forces. These particle flows and electrical currents may give rise to instabilities in the plasma behavior. Adequate stability of the confined plasma is a prime requirement for effective magnetic confinement; otherwise, particles can escape prematurely, before having a sufficient probability to fuse. Thus, finding means for suppressing the inherent tendency for confined plasma to become unstable has been one of the central goals of nuclear fusion from its inception.

There have been many types of magnetic confinement systems proposed since the initiation of fusion research. Three generic types appear to have the most promise: toroidal confinement systems (of which the tokamak is the most prominent member), mirror and tandem mirror systems, and field-reversed systems.

Tokamak. The tokamak (**Fig. 3**) is a closed or toroidal (doughnut-shaped) confinement system. It uses confining fields that represent a combination of a strong toroidal field (that is, field lines directed the long way around the toroid) with a weaker poloidal field (field lines circling the

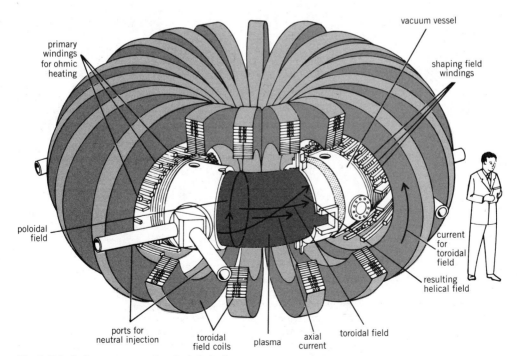

Fig. 3. Principal components of a tokamak device.

short way around the torus). The toroidal field is generated by external coils that encircle the chamber; the poloidal field is usually generated by a strong toroidally directed electric current that is induced to flow in the plasma by transformer action. The field line pattern that results from the super-position of the toroidal and poloidal fields is helical. In the tokamak, the circulating current not only provides the main confining force through the generation of the poloidal magnetic field but also performs the important function of initial heating (ohmic heating) of the plasma.

Closely related to the tokamak is the stellarator. In the stellarator the poloidal field is generated by an array of longitudinal conductors laid around the chamber with a helical pitch, creating in this way a composite field that resembles that in the tokamak. By contrast with the tokamak, however, confinement in the stellarator is not dependent upon induced currents flowing in the plasma, no transformer being required (except possibly for startup).

The strong toroidal field in the tokamak has the main function of stabilizing the plasma against "kinking" magnetohydrodynamic instability modes. The necessity of having a strong toroidal external field for this purpose has the disadvantage of limiting the magnetic efficiency factor, beta, of the tokamak. Beta, in magnetic fusion, is the ratio of the energy density of the plasma to that of the externally applied field. High beta is desirable from an economic standpoint to maximize the utilization of the externally generated magnetic field, thus minimizing the capital cost of the magnet system relative to the fusion power output.

Owing to the closed nature of its confining field (field lines do not leave the confinement

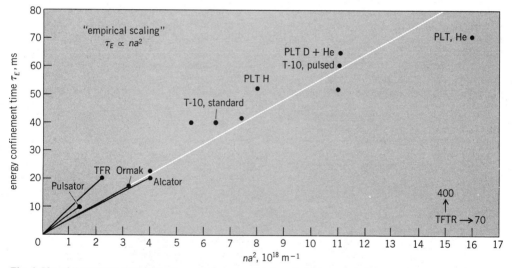

Fig. 4. Plot of "empirical scaling law" results for several tokamak devices, including results projected for the large tokamak fusion test reactor.

chamber), the tokamak has the important advantage that its plasma confinement time increases as the square of the radius, a, of the plasma column (actually, empirically, approximately as na^2, where n is the plasma density; **Fig. 4**). This property is a consequence of the circumstance that, as long as gross stability is maintained, plasma particles can escape only by diffusion across the field. Such diffusion-limited losses proceed on a time scale which increases with the square of the characteristic distances involved (here the radius of the plasma column). Thus, in toroidal systems such as the tokamak adequate confinement times can in principle always be achieved by scaling up the dimensions. Though large (several feet), the plasma diameters projected from present experiments (**Fig. 5**) to be needed for a tokamak fusion power plant appear to be acceptable.

Mirror machine and tandem mirror. The mirror machine (**Fig. 6**) is an open-ended system (field lines leave at the ends) in which a hot plasma is held trapped by repeatedly reflecting

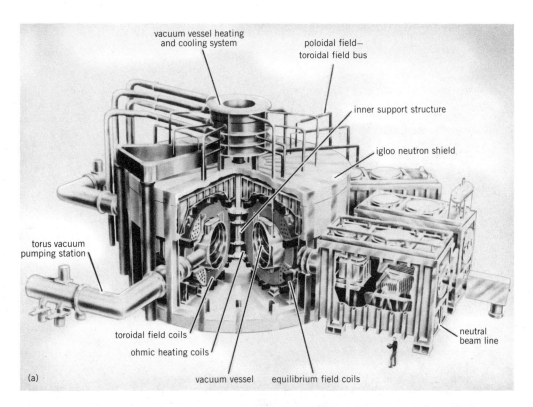
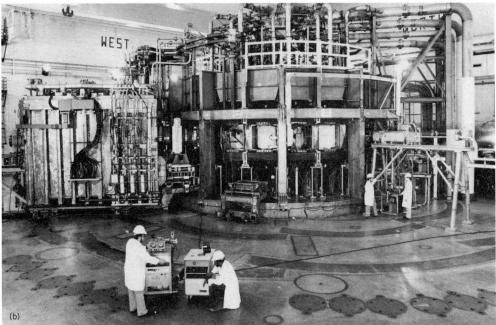

Fig. 5. Tokamak fusion test reactor, the largest magnetic confinement fusion research device in the United States. (a) Cutaway view of components. (b) Fully constructed reactor.

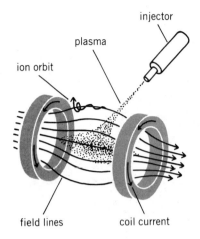

Fig. 6. Schematic illustration of mirror machine using simple mirrors. (*After R. F. Post and F. L. Ribe, Fusion reactors as future energy sources, Science, 186:397–407, 1974*)

from magnetic mirrors (regions of intensified magnetic field at each end of the confinement chamber).

An important attribute of open systems such as the mirror machine is that their fields can be shaped so as to create a magnetic well, that is, a confining field that has a nonzero minimum surrounded by closed contours of increasing magnetic intensity. An example of such a magnetic well field is the mirror field produced by a baseball coil (a coil winding resembling the seam on a baseball; **Fig. 7**). When confined in a magnetic well, a plasma is positively constrained from exhibiting any form of gross instability, even up to plasma pressures comparable to those exerted by the confining field, that is, up to plasma beta values of order unity.

Plasma confinement in mirror systems differs intrinsically from that in tokamaks or other closed systems. In the latter there is no need to confine particles as far as their motion along the field lines is concerned. By contrast, in a mirror machine longitudinal confinement is essential to prevent the immediate loss of particles out of the open ends. This confinement is provided by the repelling force exerted on charged particles as they spiral along field lines that are converging, that is, as they move toward regions of increasing field strength. Particles which spiral with sufficiently steep helical pitch angles (like a coil spring) will be repelled strongly enough to be reflected (trapped) by the mirrors. It follows that this type of mirror system cannot confine an isotropic plasma, one with uniformly distributed pitch angles. Only particles whose pitch angles are

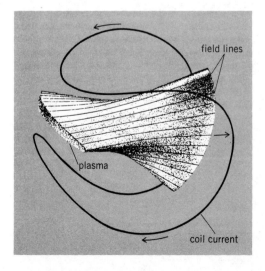

Fig. 7. "Baseball" coil configuration, producing a magnetic well mirror field.

sufficiently large to lie between the loss cones defined by the strength of the mirror fields relative to the field intensity between the mirrors will be confined.

Mirror systems typically rely on the injection of beams of energetic neutral atoms, directed transversely into the confinement region (Fig. 6), to maintain the plasma temperature and density. These energetic atoms, when ionized by collisions with already-trapped plasma particles, become newly trapped ions and electrons that maintain the plasma in competition with particle leakage through the mirrors.

An important disadvantage of conventional mirror systems for fusion power purposes is that the leakage of particles through the mirrors, arising as it does from collisions that deflect the ions into pitch angles lying within the mirror loss cone, results in a rate of loss of plasma energy that is comparable to the rate of energy generation from fusion reactions in the plasma. In this circumstance the energy gain factor Q—the ratio of fusion power to that required to maintain the plasma temperature—is at best not much larger than 1, implying an economically unacceptably large fraction of recirculated power needed to maintain the plasma against end losses. This deficiency in the original mirror concept has stimulated the development of the tandem mirror and field-reversed mirror concepts, discussed below.

An inherent characteristic of mirror systems is the positive ambipolar potential of the plasma that is a natural consequence of the confinement. Since the collision frequency for electrons is higher than that for ions (mainly because of the electrons' higher velocities), other factors being equal they would tend to diffuse into the loss cone and be lost much more rapidly than would the ions. But such a differential loss rate would automatically result in the buildup of a net positive charge in the confined plasma, thus driving it to a positive electrical potential with respect to its surroundings. In a steady state this potential is just that needed to reduce the electron loss rate (now only of the most energetic electrons, those on the "tail" of their maxwellian distribution) to equality with the ions.

In the tandem mirror (**Figs. 8** and **9**), the ambipolar potential is put to use to confine a

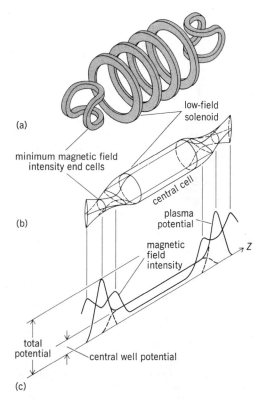

Fig. 8. Tandem mirror system. (a) Coils. (b) Configuration. (c) Variations of magnetic field intensity and plasma potential.

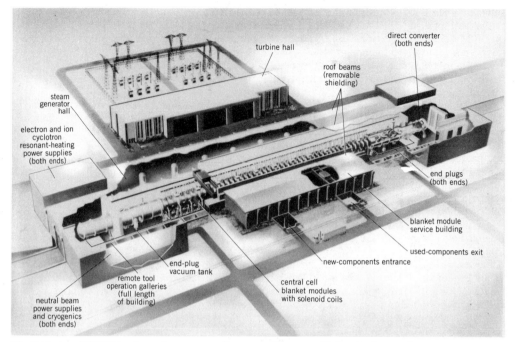

Fig. 9. Design of MARS tandem mirror reactor. (*Lawrence Livermore National Laboratory*)

fusion plasma, resulting in greatly improved confinement relative to that possible with a simple mirror cell. In the most simple embodiment of a tandem mirror (as in Fig. 8) a large-volume central mirror cell in which the fusion plasma is to be confined is stoppered at each end by two small-volume mirror cells having a high–ion-temperature plasma of higher density than that of the central cell plasma. The ambipolar potential of these plugs, being relatively more positive than that of the central cell, serves to confine (axially) the ions of the central cell plasma; its electrons are confined by the overall positive potential of the system.

Subsequent to the invention of the initial version of the tandem mirror, an important improvement, the thermal barrier, was introduced. The thermal barrier is a short region of negative-going potential introduced between the positive potential peak in the plugs and the constant potential of the central cell. The function of thermal barrier is to isolate the electrons of the plug from those in the central cell. Thus isolated, they can be heated (by microwave power, for example) to a higher temperature than that of the central cell electrons. This heating then gives rise to increased plugging potentials without the necessity for increasing the plasma density in the plugs, thus resulting in practical and economic gains. The thermal barrier itself is created by localized microwave heating aimed and tuned in such a way as to create a localized population of high-energy, magnetically trapped electrons. This superthermal trapped electron population, in displacing colder electrons, gives rise to a local depression of the potential, that is, to the thermal barrier.

Field-reversed systems. Magnetically confined plasmas exhibit the property of diamagnetism. That is, the confinement process necessarily involves the existence of internal electric currents that, in interaction with the magnetic field, give rise to the confining force. These internal currents act to reduce the strength of an externally applied confining field (Lenz's law), relative to the vacuum value it would have in the absence of the plasma. If these naturally arising diamagnetic currents could be employed to provide the major part of the confining force, thereby enhancing the efficiency of use of the external field, the resultant economic and practical benefits would be substantial.

The earliest attempted embodiment of this general concept was the toroidal pinch, the progenitor of the tokamak. Because it relied solely on the self-constricting effect of a current-carrying fluid to provide confinement, the toroidal pinch was not successful: its self-constricted plasma column was subject to rapidly growing kinking instabilities of hydromagnetic origin. However, with growth in the understanding of the magnetic confinement process, ways to circumvent such problems in pinchlike systems having substantial self-confinement effects have been proposed and studied. These approaches, generically describable as field-reversed systems, have taken four distinguishable forms: field-reversing particle rings; the reversed-field pinch; field-reversed mirror systems; and the self-field tokamak or spheromak.

1. Field-reversing particle rings. These plasma entities employ a ring of circulating high-energy charged particles. The particles execute large orbits in an externally applied mirror field, thereby generating a diamagnetic current that is sufficiently strong to reverse the direction of the field within the ring. In this way a poloidal field with closed lines is produced within which a fusion plasma is trapped. An early form of this device was called the Astron.

2. Reversed-field pinch. This is a toroidal pinch in which the pinch (poloidal) field has superposed on it a toroidal field, as in the tokamak. In the reversed-field pinch, however, the externally applied toroidal field is weaker, so that the internally generated toroidal field, arising from diamagnetic currents, is actually reversed relative to the external toroidal field. Theory shows that under the proper circumstances this particular configuration possesses an inherent gross stability in that it represents a minimum energy state relative to other adjacent equilibria.

3. Field-reversed mirror. Theory predicts and experiments (in devices called field-reversed theta pinches) have confirmed that an elongated field-reversed plasma configuration having only poloidal field components can exist within a mirror field. In principle, such a configuration could be sustained by tangentially injected neutral beams that would drive ion diamagnetic currents. The closed field-line pattern of such a state (**Fig. 10**) inhibits the escape of plasma ions, which now must diffuse across the field-reversed region before they reach the open field lines of the mirror field. Another possibility that has been considered is a repetitively pulsed system in which similar field-reversed entities would be created, translated, and compressed, thereby achieving fusion conditions in a high-density, high-beta plasma.

4. Spheromaks. The same theory that predicted the existence of a grossly stable, field-reversed state in the toroidal reversed-field pinch predicts that a spherical-shaped, field-reversed plasma entity having both poloidal and toroidal field components can exist immersed in an externally generated and properly shaped poloidal field. Since much of the confining force arises from the self-generated poloidal field and since, in contrast to the tokamak, there is no need for conductors down the axis to produce the toroidal field component (which is entirely self-generated), the spheromak offers many potential advantages: high beta, compactness, and so forth. As yet unresolved experimentally are issues such as the degree of control of particle and energy transport under fusion conditions in spheromaks.

All of the field-reversed magnetic confinement systems described above have as their primary objective the more efficient use of confining magnetic fields for fusion purposes. Particularly as it is exemplified by the field-reversed mirror, the plasma beta value, as determined relative to the externally generated magnetic field, is very high. This circumstance could lead to compact fusion power systems with improved economics, that is, lowered capital costs for the magnet coils and their support structure, a major cost item in fusion systems. For example, the field-reversed mirror would have a fusion power density with the D-T reaction that is estimated to be more than 500 times that of a conventional tokamak. Alternatively, the greatly increased magnetic efficiency could in principle allow the use of so-called advanced fusion fuel cycles, involving the D-^{3}He or other reactions. These reactions have special advantages in terms of reduced neutron fluxes and in terms of their suitability for direct conversion of the reaction energy to electricity. The understanding of the confinement qualities of these several configurations under fusion-relevant plasma conditions is as yet insufficient to assure the practical success of any of them, although each one of them has been shown in the laboratory to be stable or stabilizable against gross instability modes.

Pellet fusion. The basic idea behind pellet fusion (**Fig. 11**) is the rapid implosion of a high-density fusion fuel pellet to produce a heated core that will fuse before it can fly apart. As usually conceived, the implosion would result from the rapid heating and subsequent ablation of the surface of the pellet, giving rise to an inward-acting reaction force that compresses the core.

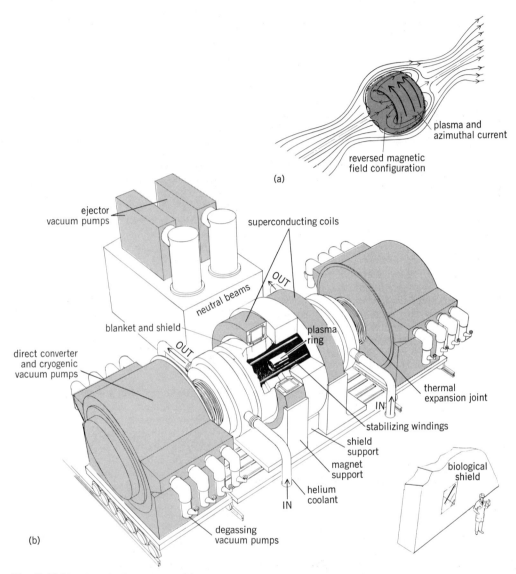

Fig. 10. Field-reversed mirror system. (*a*) Configuration, showing field lines and diamagnetic currents. (*b*) 20-MW demonstration power plant. (*Lawrence Livermore National Laboratory*)

But for this process to yield a net energy, that is, for it to achieve the required $n\tau$ value, very large compression factors, of order 10,000, are required. That this should be the case can be seen from simple considerations: As matter is compressed spherically, its density (n in the $n\tau$ product) increases as the cube of the radial compression factor. But the confinement time τ, here measured by the time of flight of an average particle out of the compressed core, decreases only as the first power of the radius. Thus $n\tau$ increases with the square of the radial compression factor. However, it is necessary to use tiny pellets to make this approach to fusion technically accessible from the standpoint of engineering limits on the amount of energy deliverable from the pellet drivers (lasers or particle accelerators), and on the amount of fusion energy released from the pellet that can be

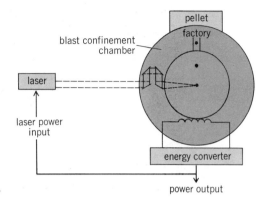

Fig. 11. Conceptual laser pellet fusion power plant.

absorbed in surrounding structures. These limitations, taken together with the Lawson requirement, dictate the need for very large density compression factors, in turn leading to a requirement for very high compression forces [of order 10^{12} atm (10^{17} pascals), ten times greater than those existing at the center of the Sun].

The problems thereby posed are twofold. First, very high pulsed powers of hundreds of terawatts must be focused down to millimeter (0.04 in.) dimensions and delivered in times of nanoseconds or less. The implied peak power densities are thus of order 10^{16} W/cm^2, made necessary by the requirement that an almost instantaneous ablation of the outer surface of the pellet should occur before heat can flow into the interior of the pellet; premature heat flow would prevent reaching the required compression factors. This problem can in principle be solved by using a sufficiently large array of high-power lasers, focusing their beams so as to uniformly illuminate the surface of the pellet (**Figs. 12, 13,** and **14**). Alternatively, converging beams of electrons or ions from particle accelerators would be employed.

Fig. 12. Output section of 4 of the 10 Nova laser-beam lines. Two 30-ft-high (9-m), 200-ft-long (60-m) support frames, designed for alignment and seismic stability, support the complete Nova amplifier chains. (*Lawrence Livermore National Laboratory*)

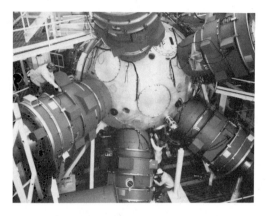

Fig. 13. Nova target chamber, a 15-ft-diameter (4.6-m), 29,000-lb (13,000-kg) aluminum vacuum vessel. Mounted inside the housings surrounding the chamber are 32-in.-diameter (80-cm) fused silica lenses, and also arrays of potassium dihydrogen phosphate (KDP) crystals to convert the 1.05-mm infrared laser light to 0.525-mm (green) or 0.35-mm (near-ultraviolet) light for more effective target irradiation. (*Lawrence Livermore National Laboratory*)

Second, if the compression process is not carried out with high uniformity, it will go askew; small errors in uniformity can lead to a major reduction in the achievable compression. Sophisticated aiming and timing techniques must be used in the driver, and any instabilities must not be allowed to spoil the symmetry of the compression or lead to undue mixing or preheating.

High temperatures and substantial $n\tau$ values, comparable to those achieved in magnetic

Fig. 14. Design of Nova inertial-confinement fusion research facility at Lawrence Livermore National Laboratory. Laboratory building containing neodymium:glass laser system is separated into four clean-room areas: master oscillator room on lower level where initial pulse light is generated; laser bay containing 10 main laser amplifier chains; optical switchyard in which the 29-in.-diameter (74-cm) beams are diagnosed and reflected into the target room; and target area which houses the 15-ft-diameter (4.6-m) aluminum target chamber. (*Lawrence Livermore National Laboratory*)

Representative inertial confinement fusion laser facilities

Laser and location	Type	Number of beams	Total optical aperture, cm²	Wavelength, μm	Power in 10^{-10} s pulse, 10^{12} watts	Energy in 10^{-9} s pulse, 10^3 joules
Nova; Lawrence Livermore National Laboratory, United States	Neodymium:glass	10	43,000	1.05 0.53 0.35	80–120 30–60 20–40	80–120 50–80 40–70
Gekko XII; Osaka University, Osaka, Japan	Neodymium:glass	12	11,500	1.05	40	20
Omega; University of Rochester/Laboratory for Laser Energetics, United States	Neodymium:glass	24	7,000	1.05 0.35	12 —	4 2
Vulcan; Rutherford-Appleton Laboratory, United Kingdom	Neodymium:glass	12	1,140	1.05 0.53	4.0 2.0	1.0 0.4
Pharos III; Naval Research Laboratory, United States	Neodymium:glass	3	940	1.05 0.53	— —	1.5 ~1.0
Chroma; KMS Fusion, Inc., Ann Arbor, Michigan, United States	Neodymium:glass	2	630	1.05 0.53	2 1	0.7 0.3
Antares; Los Alamos National Scientific Laboratory, United States	Carbon dioxide	24	21,600	10.6	—	40
Asterix III; Max Planck Institute for Quantum Optics, West Germany	Iodine	1	230	1.3 0.65	— —	0.1–0.2 (3×10^{-10} s) 0.04–0.08

fusion, have been attained in pellet fusion experiments. The **table** lists some of the parameters of equipment used in laser pellet experiments. Continued progress can be expected, as new and larger facilities come on line, but increase in performance by some orders of magnitude is needed before breakeven can be attained.

POWER PLANTS

Preliminary studies have been made of the forms that fusion power plants might take, following some of the approaches outlined above. These studies cannot of course be definitive, but they have helped to indicate the sizes, capital costs, and special engineering problems that are likely to characterize fusion power plants, insofar as they can now be visualized.

The types of power plants that have been studied encompass both pulsed and steady-state fusion systems, operating in either a driven mode or one in which plasma ignition would be achieved. A driven fusion power system is one in which the plasma temperature is maintained primarily by a continuous input of energy—for example, by the injection of high-intensity beams of energetic neutral fuel atoms—thereby maintaining the required kinetic temperature and density of the plasma. Electrical power needed to produce the beams would be obtained by recirculating a portion of the electrical output of the plant. A positive power balance, that is, net output power ($Q > 1$), would be possible here only when the recirculated power is less than the electrical output as recovered from the plasma, including not only the energy content of the reaction products but that of the unreacted part (electrons and heated fuel ions); economical power would likely be possible only if the recirculated power fraction were to be much smaller than the power produced (that is, Q much greater than 1).

By contrast, a fusion system operated in an ignition mode is one where the energy deposited directly within the plasma by charged reaction products (3.5-MeV alpha particles in the case of the D-T reaction, that is, only 20% of the total fusion energy release) is sufficient to maintain the plasma temperature, including the requirement for heating up new cold fuel particles introduced to maintain the fuel plasma density as fusion "combustion" proceeds. The ignition mode is therefore more demanding with respect to confinement time than is a driven mode, but in principle could be technically simpler, since it does not impose as strict requirements on the efficiencies of the energy recovery and plasma heating systems.

By exploiting the increase in confinement time associated with an increase in plasma radius and extrapolating from the performance of present devices, it appears that large tokamaks could operate in an ignition mode. Conventional and tandem mirror systems, requiring as they do the maintenance of plasma in the mirror cells, would need to be operated in a driven mode. However, in the case of the tandem, it seems that it will be possible to achieve ignition in the central cell plasma, with attendant simplifications and other advantages. Laser pellet systems, although the pellet itself would be expected to ignite, necessarily require recirculated power to initiate the burn. To achieve high net power relative to recirculated power would seem to imply the need for pellet energy gains (Q) of 100 or more.

Considering the magnetic confinement approach, systems studies have led to some important conclusions: Fusion power plants based on the tokamak principle in its conventional form will be relatively large both in size and in power output; electrical power outputs of 500 to some thousands of megawatts are likely to be typical. Power plants utilizing the tandem idea might be somewhat smaller in physical size than conventional tokamaks, and possibly also capable of somewhat lower plant electrical outputs than tokamaks, still satisfying economic requirements. Power plants which are based on field reversal technology, such as the field-reversed mirror, might be the smallest of all, exhibiting the highest fusion power density while being both compact in size and permitting (at least in the demonstration phase) electrical power outputs as low as tens of megawatts.

Another general result of the design studies is to show the importance of choice of materials and heat transfer characteristics of the inner wall of the containment chamber. The flux of 14-MeV neutrons through this wall coming from D-T reactions in the plasma will cause localized heating, radiation damage, and induced radioactivity. Thus the design of this portion of any D-T fusion power plant can be expected to be of critical importance. The materials chosen need to be picked not only for their resistance to radiation damage but also for minimum activation (that is, minimal yield and short half-life for the neutron-induced radioactivity). It does appear that first-wall materials having the desired characteristics can be developed.

Another critical factor is that of the generation of the confining magnetic field. Here the criterion is to achieve the required field (which may be very high for some approaches) at the least capital cost and for the least expenditure of energy. Fortunately, the development of practical high-current-density, high-field superconductors appears to provide an almost ideal solution to this problem.

Bibliography. G. Casini (ed.), *Engineering Aspects of Nuclear Fusion Reactors*, 1982; F. Chen, *Introduction to Plasma Physics and Controlled Fusion*, 2d ed., 1984; R. A. Gross, *Fusion Energy*, 1984; R. F. Post, Controlled fusion research and high temperature plasmas, *Annu. Rev. Nucl. Sci.*, 20:509–558, 1970; J. Raeder et al., *Controlled Nuclear Fusion*, 1986; G. Schmidt, *Physics of High Temperature Plasmas*, 2d ed., 1979; W. M. Stacey, *Fusion and Technology*, 1984; E. Teller (ed.), *Fusion*, vol. 1, parts A and B, 1981.

THERMONUCLEAR REACTION
Richard F. Post

A nuclear fusion reaction which occurs between various nuclei of the light elements when they are constituents of a gas at very high temperatures. Thermonuclear reactions, the source of energy generation in the Sun and the stable stars, are utilized in the fusion bomb. *See* Nuclear fusion.

Thermonuclear reactions occur most readily between isotopes of hydrogen (deuterium and tritium) and less readily among a few other nuclei of higher atomic number. At the temperatures and densities required to produce an appreciable rate of thermonuclear reactions, all matter is completely ionized; that is, it exists only in the plasma state. Thermonuclear fusion reactions may then occur within such an ionized gas when the agitation energy of the stripped nuclei is sufficient to overcome their mutual electrostatic repulsions, allowing the colliding nuclei to approach each other closely enough to react. For this reason, reactions tend to occur much more readily between energy-rich nuclei of low atomic number (small charge) and particularly between those nuclei of the hot gas which have the greatest relative kinetic energy. This latter fact leads to the result that, at the lower fringe of temperatures where thermonuclear reactions may take place, the rate of reactions varies exceedingly rapidly with temperature.

The reaction rate may be calculated as follows: Consider a hot gas composed of a mixture of two energy-rich nuclei, for example, tritons and deuterons. The rate of reactions will be proportional to the rate of mutual collisions between the nuclei. This will in turn be proportional to the product of their individual particle densities. It will also be proportional to their mutual reaction cross section σ and relative velocity v. Thus Eq. (1) gives the rate of reaction. The quantity $\langle\sigma v\rangle_{12}$

$$R_{12} = n_1 n_2 \langle \sigma v \rangle_{12} \text{ reactions/(cm}^3\text{)(s)} \tag{1}$$

indicates an average value of σ and v obtained by integration of these quantities over the velocity distribution of the nuclei (usually assumed to be maxwellian). Since the total density $n = n_1 + n_2$, then if the relative proportions of n_1 and n_2 are maintained, R_{12} varies as the square of the total particle density.

The thermonuclear energy release per unit volume is proportional to the reaction rate and the energy release per reaction, as in Eq. (2). If this energy release, on the average, exceeds the

$$P_{12} = R_{12} W_{12} \text{ ergs/(cm}^3\text{) (s)} \tag{2}$$

energy losses from the system, the reaction can become self-perpetuating. *See* Carbon-nitrogen-oxygen cycle; Nuclear reaction; Proton-proton chain.

Bibliography. S. Glasstone and R. H. Lovberg, *Controlled Thermonuclear Reactions*, 1960, reprint 1975; R. A. Gross, *Fusion Energy*, 1984; K. Miyamoto, *Plasma Physics for Nuclear Fusion*, 1980; G. Schmidt, *Physics of High Temperature Plasmas*, 2d ed., 1979; W. M. Stacey, *Fusion and Technology*, 1984.

PROTON-PROTON CHAIN
Georgeanne R. Caughlan

A group of nuclear reactions involving fusion of light nuclei that converts hydrogen into helium. It is believed to be the principal source of energy in main sequence stars of a little more than a solar mass and of less massive stars. Completion of a chain results in the consumption of four protons (hydrogen-1 nuclei, designated ^{1}H), and the production of a helium (^{4}He) nucleus plus two positrons (e^+) and two neutrinos (ν). The two positrons are annihilated along with two electrons (e^-), and the total energy release is 26.73 MeV. Approximately 0.58 MeV is released as neutrino energy and is not available as thermal energy in a star. The chain can be thought of as the conversion of four hydrogen atoms into a helium atom plus energy in the form of photons or neutrinos, or the kinetic energy of particles. The energy $E = 26.73$ MeV arises from the mass difference between four hydrogen atoms and the helium atom, and is calculated from the Einstein mass-energy equation $E = \Delta m c^2$, where Δm is the mass difference and c^2 is the square of the velocity of light. Because hydrogen is the fuel consumed in the process, it is referred to as hydrogen burning by means of the proton–proton chain.

Reactions. The first reaction involves the fusion of two protons to form a nucleus of heavy hydrogen (the deuteron, ^{2}H) with the release of a positron and a neutrino. In a relatively rare alternate reaction, one proton may capture an electron and fuse with a second proton to form the deuteron with release of a monoenergetic neutrino. This is followed by the fusion of a proton and a deuteron to form a helium nucleus of mass 3 (^{3}He) with the release of a 5.494-MeV photon (γ). The electromagnetic energy of the photon is promptly converted into thermal energy. Most often, the chain is completed by fusion of two ^{3}He nuclei to form a ^{4}He nucleus with the release of two protons. These reactions are represented by reactions (1)–(4) in **Tables 1** and **2**. In Table 1, the energy release, Q, of each reaction is indicated following the reaction. The energy convertible into thermal heat, Q (thermal), is also given, as is the maximum neutrino energy E_ν^{max}. The average neutrino energy loss is given by $Q - Q$(thermal).

If there is sufficient ^{4}He already present at the site of hydrogen burning, it is possible to complete the proton–proton chain by reactions (5)–(7) in Tables 1 and 2, or at higher hydrogen-burning temperatures by reactions (8)–(10). Reaction (6) in Tables 1 and 2 represents the capture of an electron by the beryllium-7 nucleus to produce a lithium-7 nucleus and a neutrino. Boron-8 is unstable to β^+ decay and is shown releasing a positron (e^+) and a neutrino in its decay to an excited state in ^{8}Be (the asterisk on ^{8}Be* shows that it is an excited state rather than the ground state of that nucleus). The ^{8}Be* then decays [reaction (10)] to two alpha particles (^{4}He nuclei) releasing 3.03 MeV, which represents the 2.94-MeV excitation energy plus the 0.092 MeV by which the mass of two alpha particles differs from that of the ground state of ^{8}Be.

Reaction rates and mean lifetimes τ in the Sun for the reactions in the proton–proton chain, calculated from the reaction rate equations of M. J. Harris and colleagues using the conditions of temperature, density, and mass fractions of hydrogen and helium from a standard model of the Sun developed by J. N. Bahcall and colleagues, are shown in Table 2. In this model, the principal site of hydrogen burning is just outside the central core. At a distance of 0.0511 solar radius from the center, chosen to lie in the most active part of that site, the mass fraction of ^{1}H is calculated to be $X(^1\text{H}) = 0.483$, that of ^{3}He is $X(^3\text{He}) = 2.70 \times 10^{-5}$, and that of ^{4}He is $X(^4\text{He}) = 0.500$. The remainder $X(A, \geq 12) = 0.017$ is in the form of heavy elements. The moles per gram of electrons is $Y(e^-) = 0.738$; this value is approximately the number of electrons per nucleon (in this context, nucleons include neutrons bound in nuclei plus free and bound protons). The density at this distance is $\rho = 112$ g/cm^3 (112 times the density of water), and the temperature is 14×10^6 K (25×10^6 °F). The thermonuclear reaction rates R, calculated for this temperature, are shown in Table 2. Small corrections for screening of the Coulomb field between nuclei by electrons and for partial electron degeneracy have been neglected.

Table 1. Energies involved in the proton–proton chain reaction

Reaction	Q, MeV	Q (thermal), MeV	E_ν^{max}, MeV	
^{1}H + ^{1}H → ^{2}H + e^+ + ν	1.442	1.192	0.420	(1)
or				
^{1}H + e^- + ^{1}H → ^{2}H + ν	1.442	0.001	1.441	(2)
^{2}H + ^{1}H → ^{3}He + γ	5.494	5.494		(3)
^{3}He + ^{3}He → ^{4}He + 2^{1}H	12.859	12.859		(4)
or				
^{3}He + ^{4}He → ^{7}Be + γ	1.586	1.586		(5)
^{7}Be + e^- → ^{7}Li + ν	0.862	0.050	0.862 (89.5%)	(6)
			0.384 (10.5%)	
^{7}Li + ^{1}H → 2^{4}He	17.347	17.347		(7)
or				
^{7}Be + ^{1}H → ^{8}B + γ	0.137	0.137		(8)
^{8}B → ^{8}Be* + e^+ + ν	15.04	7.41	14.06	(9)
^{8}Be* → 2^{4}He	3.03	3.03		(10)

Table 2. Reaction rates and mean lifetimes in the Sun

Reaction	R^*	$\lambda^\dagger$, s^{-1}	$\tau^\ddagger$, y	%§ termination	
$^1H + {}^1H \rightarrow {}^2H + e^+ + \nu$	5.90×10^{-20}	3.17×10^{-18}	1.00×10^{10}	99.75	(1)
or					
$^1H + e^- + {}^1H \rightarrow {}^2H + \nu$	2.10×10^{-24}	9.33×10^{-21}	3.40×10^{12}	0.25	(2)
$^2H + {}^1H \rightarrow {}^3He + \gamma$	9.63×10^{-3}	5.18×10^{-1}	6.12×10^{-8}	100.	(3)
$^3He + {}^3He \rightarrow {}^4He + 2{}^1H$	7.74×10^{-11}	7.76×10^{-14}	4.08×10^5	86.	(4)
or					
$^3He + {}^4He \rightarrow {}^7Be + \gamma$	7.61×10^{-16}	1.06×10^{-14}	2.98×10^6		(5)
$^7Be + e^- \rightarrow {}^7Li + \nu$	1.52×10^{-9}	1.26×10^{-7}	2.52×10^{-1}	14.	(6)
$^7Li + {}^1H \rightarrow 2{}^4He$	7.73×10^{-6}	4.16×10^{-4}	7.63×10^{-5}		(7)
or					
$^7Be + {}^1H \rightarrow {}^8B + \gamma$	1.71×10^{-12}	9.19×10^{-11}	3.45×10^2		(8)
$^8B \rightarrow {}^8Be^* + e^+ + \nu$		9.01×10^{-1}	3.52×10^{-8}	0.015	(9)
$^8Be^* \rightarrow 2{}^4He$		2.37×10^{21}	1.34×10^{-29}		(10)

*The units of the thermonuclear reaction rate R are reactions per second per mole/cm^3 for reactions involving two interacting nuclei and in reactions per second per (mole/cm^3)^2 for three interacting particles. (After M. J. Harris et al., Thermonuclear reaction rates, III, *Annu. Rev. Astron. Astrophys.*, 1983.)
†The quantity λ is the destruction rate per second for the first nucleus in column one.
‡The mean lifetime of the first nucleus in the first column is $\tau = 1/(3.1558 \times 10^7 \lambda)$ in years.
§After J. N. Bahcall et al., Standard solar models and uncertainties in predicted capture rates of solar neutrinos, *Rev. Mod. Phys.*, 54:767–799, 1982.

The quantity λ in Table 2 is calculated from $\rho R X({}^1H)/1.0078$ for two-body reactions induced by ^1H, from $\rho R X({}^3He)/3.0160$ for reactions induced by ^3He, and from $\rho R X({}^4He)/4.0026$ for reactions induced by ^4He. In the three-body reaction $^1H + e^- + {}^1H \rightarrow {}^2H + \nu$, $\lambda = \rho R Y(e^-) X({}^1H)/1.0078$. Proper allowance has been made for interactions involving identical particles.

Using the rates of the nuclear reactions in the chain under conditions believed to be prevalent in the hottest part of the core of the Sun, Bahcall and colleagues calculated the probable percentages of the different modes of completion of the p-p chain. They determined that about 86% of the reaction chains in the Sun are completed through the $^3He + {}^3He$ reaction, and 14% through the $^3He + {}^4He$ reaction. This 14% divides up with only 0.015% through $^7Be + {}^1H$, and the remainder, approximately 14%, through $^7Be + e^-$.

Neutrino emission. The electromagnetic photon energies and the particle kinetic energies produced in these nuclear reactions are promptly converted into thermal energy in the central region of a star. This energy is transported rather slowly to the surface of the star by radiative transfer or by convection. In the Sun, for example, several millions of years are required for the energy to reach the surface. In contrast, the neutrinos that are produced interact only weakly with matter and thus travel directly out of the Sun at the velocity of light in about 2 s. Although the neutrinos carry only a small percentage of the energy produced in the nuclear reactions, their penetrability provides a means of determining what is happening at the center of the Sun now, rather than millions of years ago. R. Davis and his collaborators at the Brookhaven National Laboratory devised an ingenious experiment to measure the influx of these neutrinos reaching the Earth from the Sun. Their detector is ^{37}Cl, the heavy isotope of chlorine. This detector is not sensitive to the abundant low-energy neutrinos from reaction (1) in the tables, but is particularly sensitive to the relatively high-energy neutrinos of the ^8B decay, so the analysis of the importance of the infrequent branch of the chain which produces ^8B is of particular interest. Scientists are puzzled by the fact that the flux of neutrinos expected on the basis of laboratory reaction-rate measurements and theories of the present structure and evolutionary state of the Sun is several times larger than that found in Davis's very careful measurements. Astrophysicists are reexamining all aspects of their theories of stellar structure and conditions of the core of the Sun in their attempts to solve the solar neutrino problem. Nuclear and elementary particle physicists are studying the properties of the electron-type neutrinos emitted by the Sun, particularly the remote possibility that they may transform into undetectable muon and tauon neutrinos on their journey from the Sun to the Earth. An experiment has been designed using ^{71}Ga, the heavy isotope of gallium.

This isotope is sensitive to the abundant low-energy neutrinos from the Sun and should provide a definitive solution to the solar neutrino problem. SEE CARBON-NITROGEN-OXYGEN CYCLES; NEUTRINO; NUCLEAR FUSION; NUCLEAR REACTION.

Bibliography. J. Audouze and S. Vauclair, *Introduction to Nuclear Astrophysics*, 1980; J. N. Bahcall et al., Standard solar models and the uncertainties in predicted capture rates of solar neutrinos, *Rev. Mod. Phys.*, 54:767–799, 1982; C. A. Barnes, D. D. Clayton, and D. N. Schramm (eds.), *Essays in Nuclear Astrophysics*, 1982; M. J. Harris et al., Thermonuclear reaction rates, III, *Annu. Rev. Astron. Astrophys.*, 21:165–176, 1983.

CARBON-NITROGEN-OXYGEN CYCLES
GEORGEANNE R. CAUGHLAN

A group of nuclear reactions that involve the interaction of protons (nuclei of hydrogen atoms, designated by ^{1}H) with carbon, nitrogen, and oxygen nuclei. The cycle involving only isotopes of carbon and nitrogen is well known as the carbon-nitrogen (CN) cycle. These cycles are thought to be the main source of energy in main-sequence stars with mass 20% or more in excess of that of the Sun. Completion of any one of the cycles results in consumption of four protons (4 ^{1}H) and the production of a helium (^{4}He) nucleus plus two positrons (e^+) and two neutrinos (ν). The two positrons are annihilated with two electrons (e^-), and the total energy release is 26.73 MeV. Approximately 1.7 MeV is released as neutrino energy and is not available as thermal energy in the star. The energy $E = 26.73$ MeV arises from the mass difference between four hydrogen atoms and the helium atom, and is calculated from the Einstein mass-energy equation $E = \Delta mc^2$, where Δm is the mass difference and c^2 is the square of the velocity of light. Completion of a chain can be thought of as conversion of four hydrogen atoms into a helium atom. Because the nuclear fuel that is consumed in these processes is hydrogen, they are referred to as hydrogen-burning processes by means of the carbon-nitrogen-oxygen (CNO) cycles.

Carbon-nitrogen cycle. The original carbon-nitrogen cycle was suggested independently by H. A. Bethe and C. F. von Weiszäcker in 1938 as the source of energy in stars. In the first reaction of the carbon-nitrogen cycle, a carbon nucleus of mass 12 captures a proton, forming a nitrogen nucleus of mass 13 and releasing a photon of energy, 1.943 MeV. This may be written: ^{12}C + ^{1}H → ^{13}N + γ, or ^{12}C(p,γ). ^{13}N is an unstable nucleus that decays by emitting a positron (e^+) and a neutrino (ν). In reaction form, ^{13}N → ^{13}C + e^+ + ν, or ^{13}N$(e^+\nu)^{13}$C. The cycle continues through ^{14}N, ^{15}O, and ^{15}N by the reactions and decays shown in ^{12}C$(p,\gamma)^{13}$N$(e^+\nu)$-^{13}C$(p,\gamma)^{14}$N$(p,\gamma)^{15}$O$(e^+\nu)^{15}$N$(p,\alpha)^{12}$C. These reactions form the second cycle from the right in the **illustration**. The reaction ^{15}N$(p,\alpha)^{12}$C represents emission of an alpha particle (^{4}He nucleus) when ^{15}N captures a proton and this cycle returns to ^{12}C. Because of the cycling, the total number of carbon, nitrogen, and oxygen nuclei remains constant, so these nuclei act as catalysts in the production of a helium nucleus plus two positrons and two neutrinos with the release of energy. The positrons that are created annihilate with free electrons rapidly after creation, so the energy used in their creation is returned to the energy fund. Synthesis of some of the rare carbon, nitro-

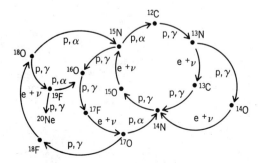

Carbon-nitrogen-oxygen cycles. (*After J. Audouze, ed., CNO Isotopes in Astrophysics, D. Reidel, Dordrecht, 1977*)

Table 1. Energies involved in the carbon-nitrogen cycle

Reaction	Q, MeV	Q (thermal), MeV	E_ν^{max}, MeV
$^{12}C + {}^1H \rightarrow {}^{13}N + \gamma$	1.943	1.943	
$^{13}N \rightarrow {}^{13}C + e^+ + \nu$	2.221	1.51	1.199
$^{13}C + {}^1H \rightarrow {}^{14}N + \gamma$	7.551	7.551	
$^{14}N + {}^1H \rightarrow {}^{15}O + \gamma$	7.297	7.297	
$^{15}O \rightarrow {}^{15}N + e^+ + \nu$	2.753	1.75	1.731
$^{15}N + {}^1H \rightarrow {}^{12}C + {}^4He$	4.966	4.966	

gen, and oxygen isotopes is accomplished through hydrogen burning by means of the carbon-nitrogen-oxygen cycles.

Table 1 shows the energy release, Q, and the average release of thermal energy, Q (thermal), in each reaction of the cycle, as well as the maximum neutrino energy E_ν^{max} in the two reactions in which a neutrino is produced.

A standard model for the Sun has been developed by J. N. Bahcall and colleagues to determine the flux of solar neutrinos that should be coming to the Earth. In this model, the site of maximum hydrogen burning is just outside the central core of the Sun. At a distance of 0.0511 solar radius from the center, chosen to lie in the most active part of that site for hydrogen burning in the proton-proton chains, the temperature is 14×10^6 K (25×10^6 °F), the density is $\rho = 112$ g/cm^3 (112 times the density of water), and the mass fraction of hydrogen is $X(^1H) = 0.483$. The reaction rates, R, calculated for the carbon-nitrogen cycle reactions at that temperature, are shown in **Table 2**. Small corrections for screening of the Coulomb field between nuclei by electrons have been neglected. Taking into account the mass fraction of hydrogen and the density at the site, the number of reactions per nucleus per second is given by $\lambda = \rho R X(^1H)/1.0078$. By inverting λ, the mean lifetimes, τ, of the carbon, nitrogen, and oxygen nuclei are obtained; these are shown in the last column of Table 2 in years. The fraction of energy from CNO cycles in the Sun is calculated to be only 1.5%; the remaining 98.5% of the energy is from hydrogen burning in the proton-proton chains.

Table 2. Reaction rates and mean lifetimes for carbon-nitrogen cycle in the Sun

Reaction	R, s^{-1} per (mole/cm^3)*	λ, s^{-1}	τ, years
$^{12}C + {}^1H \rightarrow {}^{13}N + \gamma$	8.19×10^{-17}	4.40×10^{-15}	7.20×10^6
$^{13}N \rightarrow {}^{13}C + e^+ + \nu$		1.16×10^{-3}	2.73×10^{-5}
$^{13}C + {}^1H \rightarrow {}^{14}N + \gamma$	2.83×10^{-16}	1.52×10^{-14}	2.08×10^6
$^{14}N + {}^1H \rightarrow {}^{15}O + \gamma$	3.05×10^{-19}	1.64×10^{-17}	1.93×10^9
$^{15}O \rightarrow {}^{15}N + e^+ + \nu$		5.68×10^{-3}	5.58×10^{-6}
$^{15}N + {}^1H \rightarrow {}^{12}C + {}^4He$	7.27×10^{-15}	3.91×10^{-13}	8.11×10^4

*From M. J. Harris et. al., Thermonuclear reaction rates, III, Annu. Rev. Astron. Astrophys., 21:165–176, 1983.

CNO bicycle. Nuclear research in the 1950s led to the addition of a second cycle to the processes of hydrogen burning by the CNO cycles. Laboratory research has shown that one in 880 proton captures by ^{15}N results in the formation of an oxygen nucleus of mass 16, leading to reactions which may cycle back to ^{14}N by $^{15}N(p,\gamma)^{16}O(p,\gamma)^{17}F(e^+\nu)^{17}O(p,\alpha)^{14}N$. The pair of cycles shown earlier, forming the main cycle and the first cycle to the left of it in the illustration, is called the carbon-nitrogen-oxygen bicycle.

Other CNO cycles. In research since 1960, it has become apparent that many possible branches among the nuclei must be included in any analysis of hydrogen burning by carbon, nitrogen, and oxygen nuclei. For example, if the unstable nucleus ^{13}N manages to capture a proton before it decays in its mean lifetime of 862 s, a third cycle can occur through ^{13}N$(p,\gamma)^{14}$O$(e^+\nu)^{14}$N. This cycle is displayed in the illustration by the branch on the right-hand side of the main carbon-nitrogen cycle and is known as the fast or the hot carbon-nitrogen cycle. The other possible branches leading to additional cycles shown in the diagram are due to competition between (p,γ) and (p,α) reactions. The added reactions are ^{17}O$(p,\alpha)^{18}$F$(e^+\nu)^{18}$O$(p,\alpha)^{15}$N and ^{18}O$(p,\alpha)^{19}$F$(p,\alpha)^{16}$O. The reaction ^{19}F$(p,\gamma)^{20}$Ne shown in the diagram leads out of the carbon, nitrogen, and oxygen nuclei and hence away from CNO cycles.

There are two additional branches that may occur if the unstable fluorine nuclei (^{17}F and ^{18}F) capture protons before they can decay. SEE NUCLEAR FUSION; NUCLEAR REACTION; PROTON-PROTON CHAIN.
Bibliography. J. Audouze (ed.), *CNO Isotopes in Astrophysics*, 1977; J. Audouze and S. Vauclair, *Introduction to Nuclear Astrophysics*, 1980; J. N. Bahcall et al. Standard solar models and the uncertainties in predicted capture rates of solar neutrinos, *Rev. Mod. Phys.*, 54:767–799, 1982; E. M. Burbidge et al., Synthesis of elements in stars, *Rev. Mod. Phys.*, 29:547–650, 1957; M. J. Harris et al., Thermonuclear reaction rates, III, *Annu. Rev. Astron. Astrophys.*, 21:165-176, 1983.

NUCLEAR FISSION
JOHN R. HUIZENGA

An extremely complex nuclear reaction representing a cataclysmic division of an atomic nucleus into two nuclei of comparable mass. This rearrangement or division of a heavy nucleus may take place naturally (spontaneous fission) or under bombardment with neutrons, charged particles, gamma rays, or other carriers of energy (induced fission). Although nuclei with mass number A of approximately 100 or greater are energetically unstable against division into two lighter nuclei, the fission process has a small probability of occurring, except with the very heavy elements. Even for these elements, in which the energy release is of the order of 200,000,000 eV, the lifetimes against spontaneous fission are reasonably long. SEE NUCLEAR REACTION.

Liquid-drop model. The stability of a nucleus against fission is most readily interpreted when the nucleus is viewed as being analogous to an incompressible and charged liquid drop with a surface tension. Such a droplet is stable against small deformations when the dimensionless fissility parameter X in Eq. (1) is less than unity, where the charge is in esu, the volume is in

$$X = \frac{(\text{charge})^2}{10 \times \text{volume} \times \text{surface tension}} \qquad (1)$$

cm^3, and the surface tension is in ergs/cm^2. The fissility parameter is given approximately, in terms of the charge number Z and mass number A, by the relation $X = Z^2/50\,A$.

Long-range Coulomb forces between the protons act to disrupt the nucleus, whereas short-range nuclear forces, idealized as a surface tension, act to stabilize it. The degree of stability is then the result of a delicate balance between the relatively weak electromagnetic forces and the strong nuclear forces. Although each of these forces results in potentials of several hundred megaelectronvolts, the height of a typical barrier against fission for a heavy nucleus, because they are of opposite sign but do not quite cancel, is only 5 or 6 MeV. Investigators have used this charged liquid-drop model with great success in describing the general features of nuclear fission and also in reproducing the total nuclear binding energies. SEE NUCLEAR BINDING ENERGY; NUCLEAR STRUCTURE.

Shell corrections. The general dependence of the potential energy on the fission coordinate representing nuclear elongation or deformation for a heavy nucleus such as ^{240}Pu is shown in **Fig. 1**. The expanded scale used in this figure shows the large decrease in energy of about 200 MeV as the fragments separate to infinity. It is known that ^{240}Pu is deformed in its ground state, which is represented by the lowest minimum of -1813 MeV near zero deformation. This energy represents the total nuclear binding energy when the zero of potential energy is the energy of the individual nucleons at a separation of infinity. The second minimum to the right of zero deformation illustrates structure introduced in the fission barrier by shell corrections, that is, corrections

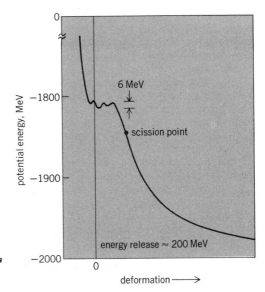

Fig. 1. Plot of the potential energy in MeV as a function of deformation for the nucleus ^{240}Pu. (*After M. Bolsteli et al., New calculations of fission barriers for heavy and superheavy nuclei, Phys. Rev., 5C:1050–1077, 1972*)

dependent upon microscopic behavior of the individual nucleons, to the liquid-drop mass. Although shell corrections introduce small wiggles in the potential-energy surface as a function of deformation, the gross features of the surface are reproduced by the liquid-drop model. Since the typical fission barrier is only a few million electron volts, the magnitude of the shell correction need only be small for irregularities to be introduced into the barrier. This structure is schematically illustrated for a heavy nucleus by the double-humped fission barrier in **Fig. 2**, which represents the region to the right of zero deformation in Fig. 1 on an expanded scale. The fission barrier has two maxima and a rather deep minimum in between. For comparison, the single-humped liquid-drop barrier is also schematically illustrated. The transition in the shape of the nucleus as a function of deformation is schematically represented in the upper part of the figure.

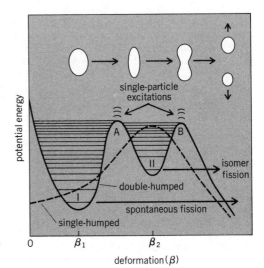

Fig. 2. Schematic plots of single-humped fission barrier of liquid-drop model and double-humped barrier introduced by shell corrections. (*After J. R. Huizenga, Nuclear fission revisited, Science, 168:1405–1413, 1970*)

Double-humped barrier. The developments which led to the proposal of a double-humped fission barrier were triggered by the experimental discovery of spontaneously fissionable isomers by S. M. Polikanov and colleagues in the Soviet Union and by V. M. Strutinsky's pioneering theoretical work on the binding energy of nuclei as a function of both nucleon number and nuclear shape. The double-humped character of the nuclear potential energy as a function of deformation arises, within the framework of the Strutinsky shell-correction method, from the superposition of a macroscopic smooth liquid-drop energy and a shell-correction energy obtained from a microscopic single-particle model. Oscillations occurring in this shell correction as a function of deformation lead to two minima in the potential energy, shown in Fig. 2, the normal ground-state minimum at a deformation of β_1 and a second minimum at a deformation of β_2. States in these wells are designated class I, and class II states, respectively. Spontaneous fission of the ground state and isomeric state arises from the lowest-energy class I and class II states, respectively. SEE NUCLEAR ISOMERISM.

The calculation of the potential-energy curve illustrated in Fig. 1 may be summarized as follows. The smooth potential energy obtained from a macroscopic (liquid-drop) model is added to a fluctuating potential energy representing the shell corrections, and to the energy associated with the pairing of like nucleons (pairing energy), derived from a non-self-consistent microscopic model. The calculation of these corrections requires several steps, namely, (1) specification of the geometrical shape of the nucleus, (2) generation of a single-particle potential related to its shape, (3) solution of the Schrödinger equation, and (4) calculation from these single-particle energies of the shell and pairing energies.

The oscillatory character of the shell corrections as a function of deformation is caused by variations in the single-particle level density in the vicinity of the Fermi energy. For example, the single-particle levels of a pure harmonic oscillator potential arrange themselves in bunches of highly degenerate shells at any deformation for which the ratio of the major and minor axes of the spheroidal equipotential surfaces is equal to the ratio of two small integers. Nuclei with a filled shell, that is, with a level density at the Fermi energy that is smaller than the average, will then have an increased binding energy compared to the average, because the nucleons occupy deeper and more bound states; conversely, a large level density is associated with a decreased binding energy. It is precisely this oscillatory behavior in the shell correction that is responsible for spherical or deformed ground states and for the secondary minima in fission barriers, as illustrated in Fig. 2.

More detailed theoretical calculations based on this macroscopic-microscopic method have revealed additional features of the fission barrier. In these calculations the potential energy is regarded as a function of several different modes of deformation. The outer barrier B (Fig. 2) is reduced in energy for shapes with pronounced left-right asymmetry (pear shapes), whereas the inner barrier A and deformations in the vicinity of the second minimum are stable against such mass asymmetric degrees of freedom. Similar calculations of potential-energy landscapes reveal the stability of the second minimum against gamma deformations, in which the two small axes of the spheroidal nucleus become unequal, that is, the spheroid becomes an ellipsoid.

Experimental consequences. The observable consequences of the double-humped barrier have been reported in numerous experimental studies. In the actinide region more than 30 spontaneously fissionable isomers have been discovered between uranium and berkelium, with half-lives ranging from 10^{-11} to 10^{-2} s. These decay rates are faster by 20 to 30 orders of magnitude than the fission half-lives of the ground states, because of the increased barrier tunneling probability (see Fig. 2). Several cases in which excited states in the second minimum decay by fission are also known. Normally these states decay within the well by gamma decay; however, if there is a hindrance in gamma decay due to spin, the state (known as a spin isomer) may undergo fission instead.

Qualitatively, the fission isomers are most stable in the vicinity of neutron numbers 146 to 148, a value in good agreement with macroscopic-microscopic theory. For elements above berkelium the half-lives become too short to be observable with available techniques; and for elements below uranium, the prominent decay is through barrier A into the first well, followed by gamma decay. It is difficult to detect this competing gamma decay of the ground state in the second well (called a shape isomeric state), but identification of the gamma branch of the 200-ns ^{238}U shape isomer has been reported. SEE RADIOACTIVITY.

Direct evidence of the second minimum in the potential-energy surface of the even-even nucleus ^{240}Pu has been obtained through observations of the E2 transitions within the rotational band built on the isomeric 0+ level. The rotational constant (which characterizes the spacing of the levels and is expected to be inversely proportional to the effective moment of inertia of the nucleus) found for this band is less than one-half that for the ground state and confirms that the shape isomers have a deformation β_2 much larger than the equilibrium ground-state deformation β_1. From yields and angular distributions of fission fragments from the isomeric ground state and low-lying excited states some information has been derived on the quantum numbers of specific single-particle states of the deformed nucleus. (Nilsson single-particle states) in the region of the second minimum.

At excitation energies in the vicinity of the two barrier tops, measurements of the subthreshold neutron fission cross sections of several nuclei have revealed groups of fissioning resonance states with wide energy intervals between each group where no fission occurs. Such a spectrum is illustrated in **Fig. 3**a, where the subthreshold fission cross section of ^{240}Pu is shown for neutron energies between 500 and 3000 eV. As shown in Fig. 3b, between the fissioning resonance states there are many other resonance states, known from data on the total neutron cross sections, which have negligible fission cross sections. Such structure is explainable in terms of the double-humped fission barrier and is ascribed to the coupling between the compound states of normal density in the first well to the much less dense states in the second well. This picture requires resonances of only one spin to appear within each intermediate structure group illustrated in Fig. 3a. In an experiment using polarized neutrons on a polarized ^{237}Np target, it was found

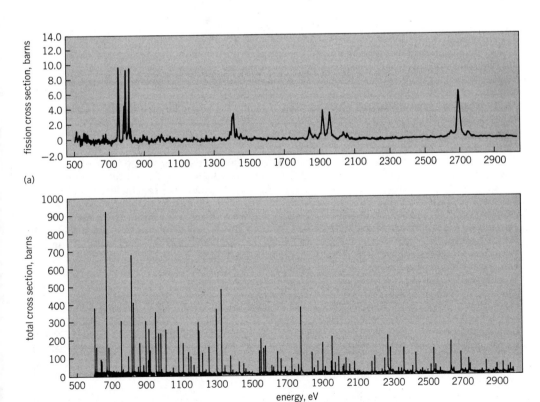

Fig. 3. Grouping of fission resonances demonstrated by (a) the neutron fission cross section of ^{240}Pu and (b) the total neutron cross section. (*After V. M. Strutinsky and H. C. Pauli, Shell-structure effects in the fissioning nucleus, Proceedings of the 2d IAEA Symposium on Physics and Chemistry of Fission, Vienna, pp. 155–177, 1969*)

that all nine fine-structure resonances of the 40-eV group have the same spin and parity: $I = 3+$. Evidence has also been obtained for vibrational states in the second well from neutron (n,f) and deuteron stripping (d,pf) reactions at energies below the barrier tops (f indicates fission of the nucleus). SEE NEUTRON SPECTROMETRY.

A. Bohr suggested that the angular distributions of the fission fragments are explainable in terms of the transition-state theory, which describes a process in terms of the states present at the barrier deformation. The theory predicts that the cross section will have a steplike behavior for energies near the fission barrier, and that the angular distribution will be determined by the quantum numbers associated with each of the specific fission channels. The theoretical angular distribution of fission fragments is based on two assumptions. First, the two fission fragments are assumed to separate along the direction of the nuclear symmetry axis so that the angle θ between the direction of motion of the fission fragments and the direction of motion of the incident bombarding particle represents the angle between the body-fixed axis (the long axis of the spheroidal nucleus) and the space-fixed axis (some specified direction in the laboratory, in this case the direction of motion of the incident particle). Second, it is assumed that the transition from the saddle point (corresponding to the top of the barrier) to scission (the division of the nucleus into two fragments) is so fast that Coriolis forces do not change the value of K (where K is the projection of the total angular momentum I on the nuclear symmetry axis) established at the saddle point.

In several cases, low-energy photofission and neutron fission experiments have shown evidence of a double-humped barrier. In the case of two barriers, the question arises as to which of the two barriers A or B is responsible for the structure in the angular distributions. For light actinide nuclei like thorium, the indication is that barrier B is the higher one, whereas for the heavier actinide nuclei, the inner barrier A is the higher one. The heights of the two barriers themselves are most reliably determined by investigating the probability of induced fission over a range of several megaelectronvolts in the threshold region. Many direct reactions have been used for this purpose, for example, (d,pt), (t,pf), and $(^3\text{He}, df)$. There is reasonably good agreement between the experimental and theoretical barriers. The theoretical barriers are calculated with realistic single-particle potentials and include the shell corrections.

Fission probability. The cross section for particle-induced fission $\sigma(y,f)$ represents the cross section for a projectile y to react with a nucleus and produce fission, as shown by Eq. (2).

$$\sigma(y, f) = \sigma_R(y) \, (\Gamma_f/\Gamma_t) \qquad (2)$$

The quantities $\sigma_R(y)$, Γ_f, and Γ_t are the total reation cross sections for the incident particle y, the fission width, and the total level width, respectively where $\Gamma_t = \Gamma_f + \Gamma_n + \Gamma_\gamma + \cdots$ is the sum of all partial-level widths. All the quantities in Eq. (2) are energy-dependent. Each of the partial widths for fission, neutron emission, radiation, and so on, is defined in terms of a mean lifetime τ for that particular process, for example, $\Gamma_f = \hbar/\tau_f$. Here $\hbar$, the action quantum, is Planck's constant divided by 2π and is numerically equal to 1.0546×10^{-34} J s = 0.66×10^{-15} eV s. The fission width can also be defined in terms of the energy separation D of successive levels in the compound nucleus and the number of open channels in the fission transition nucleus (paths whereby the nucleus can cross the barrier on the way to fission), as given by expression (3), where I is the

$$\Gamma_f(I) = \frac{D(I)}{2\pi} \sum_i N_{fi} \qquad (3)$$

angular momentum and i is an index labeling the open channels N_{fi}. The contribution of each fission channel to the fission width depends upon the barrier transmission coefficient, which, for a two-humped barrier (see Fig. 2), is strongly energy-dependent. This results in an energy-dependent fission cross section which is very different from the total cross section shown in Fig. 3 for ^{240}Pu.

When the incoming neutron has low energy, the likelihood of reaction is substantial only when the energy of the neutron is such as to form the compound nucleus in one or another of its resonance levels (see Fig. 3b). The requisite sharpness of the "tuning" of the energy is specified by the total level width Γ. The nuclei ^{233}U, ^{235}U, and ^{239}Pu have a very large cross section to take up a slow neutron and undergo fission (see **table**) because both their absorption cross section and

Cross sections for neutrons of thermal energy to produce fission or undergo capture in the principal nuclear species, and neutron yields from these nuclei*

Nucleus	Cross section for fission, σ_f, 10^{-24} cm^2	σ_f plus cross section for radiative capture, σ_r	Ratio, $1 + \alpha$	Number of neutrons released per fission, v	Number of neutrons released per slow neutron captured, $\eta = v/(1 + \alpha)$
^{233}U	525 ± 2	573 ± 2	1.093 ± 0.003	2.50 ± 0.01	2.29 ± 0.01
^{235}U	577 ± 1	678 ± 2	1.175 ± 0.002	2.43 + 0.01	2.08 ± 0.01
^{239}Pu	741 ± 4	1015 ± 4	1.370 ± 0.006	2.89 ± 0.01	2.12 ± 0.01
^{238}U	0	2.73 ± 0.04			0
Natural uranium	4.2	7.6	1.83	2.43 ± 0.01	1.33

*Data from *Brookhaven National Laboratory 325*, 2d ed., suppl. no. 2, vol. 3, 1965. The data presented are the recommended or least-squares values published in this reference for 0.0253-eV neutrons. All cross sections are in units of barns (1 barn = 10^{-24} cm^2 = 10^{-28} m^2).

their probability for decay by fission are large. The probability for fission decay is high because the binding energy of the incident neutron is sufficient to raise the energy of the compound nucleus above the fission barrier. The very large, slow neutron fission cross sections of these isotopes make them important fissile materials in a chain reactor. SEE CHAIN REACTION.

Scission. The scission configuration is defined in terms of the properties of the intermediate nucleus just prior to division into two fragments. In heavy nuclei the scission deformation is much larger than the saddle deformation at the barrier, and it is important to consider the dynamics of the descent from saddle to scission. One of the important questions in the passage from saddle to scission is the extent to which this process is adiabatic with respect to the particle degrees of freedom. As the nuclear shape changes, it is of interest to investigators to know the probability for the nucleons to remain in the lowest-energy orbitals. If the collective motion toward scission is very slow, the single-particle degrees of freedom continually readjust to each new deformation as the distortion proceeds. In this case, the adiabatic model is a good approximation, and the decrease in potential energy from saddle to scission appears in collective degrees of freedom at scission, primarily as kinetic energy associated with the relative motion of the nascent fragments.

On the other hand, if the collective motion between saddle and scission is so rapid that equilibrium is not attained, there will be a transfer of collective energy into nucleonic excitation energy. Such a nonadiabatic model, in which collective energy is transferred to single-particle degrees of freedom during the descent from saddle to scission, is usually referred to as the statistical theory of fission.

The experimental evidence indicates that the saddle to scission time is somewhat intermediate between these two extreme models. The dynamic descent of a heavy nucleus from saddle to scission depends upon the nuclear viscosity. A viscous nucleus is expected to have a smaller translational kinetic energy at scission and a more elongated scission configuration. Experimentally, the final translational kinetic energy of the fragments at infinity, which is related to the scission shape, is measured. Hence, in principle, it is possible to estimate the nuclear viscosity coefficient by comparing the calculated dependence upon viscosity of fission-fragment kinetic energies with experimental values. The viscosity of nuclei is an important nuclear parameter which also plays an important role in collisions of very heavy ions.

The mass distribution from the fission of heavy nuclei is predominantly asymmetric. For example, division into two fragments of equal mass is about 600 times less probable than division into the most probable choice of fragments when ^{235}U is irradiated with thermal neutrons. When the energy of the neutrons is increased, symmetric fission (**Fig. 4**) becomes more probable. In general, heavy nuclei fission asymmetrically to give a heavy fragment of approximately constant mean mass number 139 and a corresponding variable-mass light fragment (see **Fig. 5**). These experimental results have been difficult to explain theoretically. Calculations of potential-energy

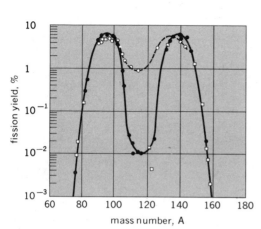

Fig. 4. Mass distribution of fission fragments formed by neutron-induced fission of $^{235}U + n = ^{236}U$ when neutrons have thermal energy, solid curve (*Plutonium Project Report, Rev. Mod. Phys.*, 18:539, 1964), and 14-MeV energy, broken curve (based on R. W. Spence, Brookhaven National Laboratory, AEC-BNL (C-9), 1949). Quantity plotted is 100 × (number of fission decay chains formed with given mass)/(number of fissions).

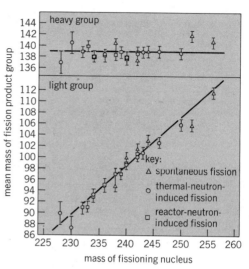

Fig. 5. Average masses of the light- and heavy-fission product groups as a function of the masses of the fissioning nucleus. Energy spectrum of reactor neutrons is that associated with fission. (After K. F. Flynn et al., Distribution of mass in the spontaneous fission of ^{256}Fm, *Phys. Rev.*, 5C:1725–1729, 1972)

surfaces show that the second barrier (B in Fig. 2) is reduced in energy by up to 2 or 3 MeV, if octuple deformations (pear shapes) are included. Hence, the theoretical calculations show that mass asymmetry is favored at the outer barrier, although direct experimental evidence supporting the asymmetric shape of the second barrier is very limited. It is not known whether the mass asymmetric energy valley extends from the saddle to scission; and the effect of dynamics on mass asymmetry in the descent from saddle to scission has not been determined. Experimentally, as the mass of the fissioning nucleus approaches $A \approx 260$, the mass distribution approaches symmetry. This result is qualitatively in agreement with theory.

A nucleus at the scission configuration is highly elongated and has considerable deformation energy. The influence of nuclear shells on the scission shape introduces structure into the kinetic energy and neutron-emission yield as a function of fragment mass. The experimental kinetic energies for the neutron-induced fission of ^{233}U, ^{235}U, and ^{239}Pu have a pronounced dip as symmetry is approached, as shown in **Fig. 6**. (This dip is slightly exaggerated in the figure because the data have not been corrected for fission fragment scattering.) The variation in the neutron yield as a function of fragment mass for these same nuclei (**Fig. 7**) has a "saw-toothed" shape which is asymmetric about the mass of the symmetric fission fragment. Both these phenomena are reasonably well accounted for by the inclusion of closed-shell structure into the scission configuration.

A number of light charged particles (for example, isotopes of hydrogen, helium, and lithium) have been observed to occur, with low probability, in fission. These particles are believed to be emitted very near the time of scission. Available evidence also indicates that neutrons are emitted at or near scission with considerable frequency.

Postscission phenomena. After the fragments are separated at scission, they are further accelerated as the result of the large Coulomb repulsion. The initially deformed fragments collapse to their equilibrium shapes, and the excited primary fragments lose energy by evaporating neutrons. After neutron emission, the fragments lose the remainder of their energy by gamma radia-

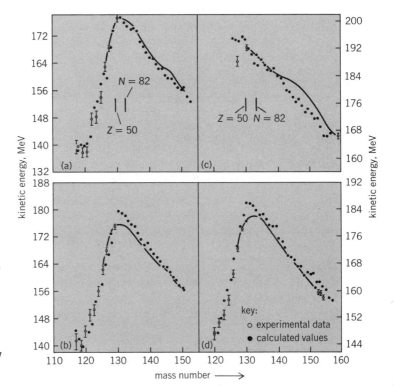

Fig. 6. Average total kinetic energy of fission fragments as a function of heavy fragment mass for fission of (a) ^{235}U, (b) ^{233}U, (c) ^{252}Cf, and (d) ^{239}Pu. Curves indicate experimental data. (*After J. C. D. Milton and J. S. Fraser, Time-of-flight fission studies on ^{233}U, ^{235}U and ^{239}Pu, Can. J. Phys., 40:1626–1663, 1962*)

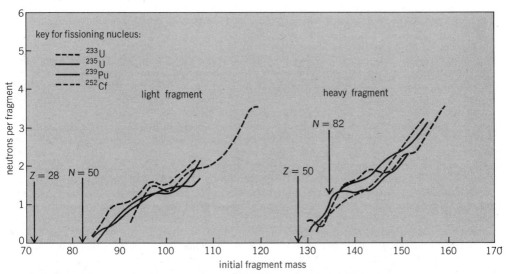

Fig. 7. Neutron yields as a function of fragment mass for four types of fission as determined from mass-yield data. Approximate initial fragment masses corresponding to various neutron and proton "magic numbers" N and Z are indicated. (*After J. Terrell, Neutron yields from individual fission fragments, Phys. Rev., 127:880–904, 1962*)

tion, with a life-time of about 10^{-11} s. The kinetic energy and neutron yield as a function of mass are shown in Figs. 6 and 7. The variation of neutron yield with fragment mass is directly related to the fragment excitation energy. Minimum neutron yields are observed for nuclei near closed shells because of the resistance to deformation of nuclei with closed shells. Maximum neutron yields occur for fragments that are "soft" toward nuclear deformation. Hence, at the scission configuration, the fraction of the deformation energy stored in each fragment depends on the shell structure of the individual fragments. After scission, this deformation energy is converted to excitation energy, and hence, the neutron yield is directly correlated with the fragment shell structure. This conclusion is further supported by the correlation between the neutron yield and the final kinetic energy. Closed shells result in a larger Coulomb energy at scission for fragments that have a smaller deformation energy and a smaller number of evaporated neutrons.

After the emission of the prompt neutrons and gamma rays, the resulting fission products are unstable against beta decay. For example, in the case of thermal neutron fission of ^{235}U, each fragment undergoes on the average about three beta decays before it settles down to a stable nucleus. For selected fission products (for example, ^{87}Br and ^{137}I) beta decay leaves the daughter nucleus with excitation energy exceeding its neutron binding energy. The resulting delayed neutrons amount, for thermal neutron fission of ^{235}U, to about 0.7% of all the neutrons given off in fission. Though small in number, they are quite important in stabilizing nuclear chain reactions against sudden minor fluctuations in reactivity. SEE NEUTRON.

Bibliography. K. Depta et al., Bimodal fission in ^{258}Fm, *Mod. Phys. Lett.*, A1:377–381, 1986; E. K. Hulet et al., Bimodal Symmetric fission observed in the heaviest elements, *Phys. Rev. Lett.*, 56:313–316, 1986; *Proceedings of the 3d IAEA Symposium on Physics and Chemistry of Fission*, Rochester, New York, 1973; R. Vandenbosch and J. R. Huizenga, *Nuclear Fission*, 1973.

CHAIN REACTION
NORMAN C. RASMUSSEN

A succession of generation after generation of acts of division (called fission) of certain heavy nuclei. The fission process releases about 200 MeV (3.2×10^{-4} erg = 3.2×10^{-11} joules) in the form of energetic particles including two or three neutrons. Some of the neutrons from one generation are captured by fissile species (^{233}U, ^{235}U, ^{239}Pu) to cause the fissions of the next generations. The process is employed in nuclear reactors and nuclear explosive devices. SEE NUCLEAR FISSION.

The ratio of the number of fissions in one generation to the number in the previous generation is the multiplication factor k. The value of k can range from less than 1 to less than 2, and depends upon the type and amount of fissile material, the rate of neutron absorption in nonfissile material, the rate at which neutrons leak out of the system, and the average energy of the neutrons in the system. When $k = 1$, the fission rate remains constant and the system is said to be critical. When $k > 1$, the system is supercritical and the fission rate increases.

A typical water-cooled power reactor contains an array of uranium rods (about 3% ^{235}U) surrounded by water. The uranium in the form of UO_2 is sealed into zirconium alloy tubes. The water removes the heat and also slows down (moderates) the neutrons by elastic collision with hydrogen nuclei. The slow neutrons have a much higher probability of causing fission in ^{235}U than faster (more energetic) neutrons do. In a fast reactor, no light nuclei are present in the system and the average neutron velocity is much higher. In such systems it is possible to use the excess neutrons to convert ^{238}U to ^{239}U. Then ^{239}U undergoes radioactive decay into ^{239}Pu, which is a fissile material capable of sustaining the chain reaction. If more than one ^{239}Pu atom is provided for each ^{235}U consumed, the system is said to breed (that is, make more fissile fuel than it consumes). In the breeder reactor, the isotope ^{238}U (which makes up 99.3% of natural uranium) becomes the fuel. This increases the energy yield from uranium deposits by more than a factor of 60 over a typical water-moderated reactor, which mostly employs the isotope ^{235}U as fuel.

A majority of power reactors use water as both the moderator and the coolant. However, a limited number of reactors use heavy water instead of light water. The advantage of this system is that it is possible to use natural uranium as a fuel so that no uranium enrichment is needed. Some other power reactors are gas-cooled by either helium or carbon dioxide and are moderated with graphite.

SPALLATION REACTION
David K. Scott

A nuclear reaction in which the interacting nuclei are disintegrated into a large number of the constituent protons, neutrons, and other light particles. Compared to more conventional reactions involving the simple transfer of nucleons between the colliding nuclei, spallation is a violent process occurring at high incident energy. An extreme case is shown in the **illustration**, the interaction of an argon nucleus (18 protons and 22 neutrons) with a lead nucleus (82 protons and 126 neutrons) in a streamer chamber at an incident energy of 72,000 MeV. The emitted charged particles appear as tracks of electrical discharge in a gas. In this example the nuclei are apparently completely shattered into individual nucleons. Spallation can also be induced by lighter particles, such as high-energy protons; however, these reactions are usually less violent and produce only a few particles (typically 3 to 10). Although the most abundant final products are protons, neutrons, and alpha particles, the spallation of larger fragments such as lithium and carbon also occurs. The short, thick tracks in the illustration are probably created by the emission of these larger fragments of nuclear matter.

Collision of an argon nucleus with a lead nucleus at an incident energy of 72,000 MeV in a streamer chamber. The nucleus is incident from the left. (*From L. S. Schroeder, Streamer chambers—their use for nuclear science experiments, Nucl. Instrum. Meth., 162:395-404, 1979*)

The detailed mechanism whereby the incident energy is communicated to the nuclei in the spallation process is not well understood, but there may well be analogies with the breakup of other forms of matter. Like the nucleus, a substance such as bouncing putty will deform into new shapes under a gentle collision, but will shatter like a piece of glass when subjected to a hard, rapid blow. There is evidence that the breakup of nuclei also sets in suddenly at energies of a few tens of MeV per colliding nucleon. The process may be used to measure the tensile strength of nuclear matter. It is suggested that the spallation debris of a nuclear collision could contain information about high density and shock waves in nuclei. The spallation reaction also plays an important role in determining the abundance of elements in the universe. In particular, ^{6}Li, ^{9}Be, ^{10}B, and ^{11}B are created from spallation of interstellar gas by galactic cosmic rays, which leads to collisions of hydrogen with abundant nuclei such as carbon, oxygen, and silicon. See Nuclear Reaction.

Bibliography. H. Reeves, W. A. Fowler, and F. Hoyle, Galactic cosmic ray origin of Li, Be and B in stars, *Nature*, 226:727–729, 1970; D. K. Scott, Towards relativistic heavy ion collisions, *Prog. Particle Nucl. Phys.*, 4:5–93, 1980.

ANOMALONS
Harry H. Heckman

High-energy nuclear fragments associated with the fragmentation of projectile (beam) nuclei at relativistic energies that exhibit anomalously short interaction mean free paths. Because the mean free paths of nuclei in matter are inversely proportional to their reaction cross sections, that is, nuclear sizes, the shortening of the mean free paths of projectile fragments implies the existence of nuclear entities—anomalons—that possess nuclear dimensions beyond those known for conventional nuclei.

Experimental evidence for anomalons was first noted in cosmic-ray experiments in the early 1950s, with subsequent evidence coming from experiments carried out at high-energy heavy-ion accelerators in the United States and Soviet Union. However, conclusive proof that anomalons exist as a physical entity has yet to be fully established. Still, the possibility that anomalons may exist has far-reaching consequences for theoretical notions involving the basic composition of nuclear matter.

The salient properties of anomalons have been derived from experiments carried out by using various nuclear beams, up to and including iron, at beam energies in excess of 2 GeV per nucleon. Anomalons constitute about 3–6% of all particle fragments, the remaining particle fragments being normal nuclei. Anomalons exhibit an anomalously short mean free path of about 2 cm (0.8 in.) in nuclear emulsion and plastic track detectors, shorter by factors of 2–10 than those observed and theoretically expected for beam nuclei of the same charge. [For comparison, the mean free paths of helium and iron nuclei in nuclear emulsion are 22 and 7.3 cm (8.7 and 2.9 in.), respectively.] No decay of anomalons via charged-particle emission has been observed. Their lifetime appears to be greater than 10^{-11} s. The anomalon effect persists, and is possibly enhanced, in subsequent collisions of the projectile fragments.

Plausible theoretical speculations on anomalons have ventured into notions of quark structures in nuclei based on quantum chromodynamics, a theory that has had great success in high-energy physics. Nonquark approaches involve the formation of bubble and quasimolecular nuclei and clusters of π^- mesons and neutrons (pineuts) bound to nuclear fragments. SEE NUCLEAR REACTION; NUCLEAR STRUCTURE; QUANTUM CHROMODYNAMICS.

Bibliography. E. M. Friedlander et al., Anomalous reaction mean free paths of nuclear projectile fragments from heavy ion collisions at 2A GeV, *Phys. Rev.*, C27:1489–1520, 1983; A. L. Robinson, A nuclear puzzle emerges at Berkeley, *Science*, 210: 174–175, 1980; B. M. Schwarzchild, New evidence for anomalously large nuclear fragments, *Phys. Today*, 35(4): 17–19, April 1982.

ENERGETIC PARTICLES AND RADIATION

Nuclear radiation	166
Alpha particles	167
Beta particles	170
Gamma rays	173
Compton effect	176
Electron-positron pair production	182
Charged particle beams	183
Electron capture	191
Delta electrons	192

NUCLEAR RADIATION
Dennis G. Kovar

All particles and radiations emanating from an atomic nucleus due to radioactive decay and nuclear reactions. Thus the criterion for nuclear radiations is that a nuclear process is involved in their production.

The term was originally used to denote the ionizing radiations observed from naturally occurring radioactive materials. These radiations were alpha rays (energetic helium nuclei), beta rays (negative electrons), and gamma rays (electromagnetic radiation with wavelength much shorter than visible light). SEE ALPHA PARTICLES; BETA PARTICLES; GAMMA RAYS.

Nuclear radiations have traditionally been considered to be of three types based on the manner in which they interact with matter as they pass through it. These are the charged heavy particles with masses comparable to that of the nuclear mass (for example, protons, alpha particles, and heavier nuclei), electrons (both negatively and positively charged), and electromagnetic radiation. For all of these, the interactions with matter are considered to be primarily electromagnetic. (The neutron, which is also a nuclear radiation, behaves quite differently.) The behavior of mesons and other particles is intermediate between that of the electron and heavy charged particles. SEE CHARGED PARTICLE BEAMS.

A striking difference in the absorption of the three types of radiations is that only heavy charged particles have a range. That is, a monoenergetic beam of heavy charged particles, in passing through a certain amount of matter, will lose energy without changing the number of particles in the beam. Ultimately, they will be stopped after crossing practically the same thickness of absorber. The minimum amount of absorber that stops a particle is its range. The greatest part of the energy loss results from collisions with atomic electrons, causing the electrons to be excited or freed. The energy loss per unit path length is the specific energy loss, and its average value is the stopping power of the absorbing substance.

For electromagnetic radiation (gamma rays) and neutrons, on the other hand, the absorption is exponential; that is, the intensity decreases in such a way that the equation below is valid,

$$-\frac{dI}{I} = \mu \, dx$$

where I is the intensity of the primary radiation, μ is the absorption coefficient, and dx is the thickness traversed. The difference in behavior reflects the fact that charged particles are not removed from the beam by individual interactions, whereas gamma radiation photons (and neutrons) are. Three main types of phenomena involved in the interaction of electromagnetic radiation with matter (namely, photoelectric absorption, Compton scattering, and electron-positron production) are responsible for this behavior.

Electrons exhibit a more complex behavior. They radiate electromagnetic energy easily because they have a large charge-to-mass ratio and hence are subject to violent acceleration under the action of the electric forces. Moreover, they undergo scattering to such an extent that they follow irregular paths. SEE ELECTRON.

Whereas in the case of the heavy charged particles, electrons, or gamma rays the energy loss is primarily the result of electromagnetic effects, neutrons are slowed down by nuclear collisions. These may be inelastic collisions, in which a nucleus is left in an excited state, or elastic collisions, in which the colliding nucleus acquires part of the energy (of the order of 1 MeV) to excite the collision partner. With less kinetic energy, only elastic scattering can slow down the neutron, a process which is effective down to thermal energies (approximately 1/40 keV). At this stage the collision, on the average, has no further effect on the energy of the neutron. SEE NEUTRON.

As noted previously, the other nuclear radiations such as mesons have behaviors which are intermediate between that of heavy charged particles and electrons. Another radioactive decay product is the neutrino; because of its small interaction with matter, it is not ordinarily considered to be a nuclear radiation. SEE MESON; NEUTRINO; NUCLEAR REACTION.

ALPHA PARTICLES
Robley D. Evans

Helium nuclei which have been ejected at high velocity from atomic nuclei as products of radioactive decay or as products of induced nuclear reactions. Helium nuclei which have been accelerated to high velocities for use as bombarding particles in nuclear reactions also may be called alpha particles. The helium nucleus has a charge of $+2e$, that is, twice the magnitude of the charge e of a proton or of an electron, and a mass of 4.00015 atomic mass units. The velocity of ejection of the alpha particle varies from one radioactive substance to another, but is usually in the domain of $4-8 \times 10^{10}$ in./s ($1-2 \times 10^9$ cm/s). Many radioactive substances emit alpha particles in two or more discrete energy groups, usually in the domain of $4-6$ MeV, but occasionally as low as 2 MeV or as high as 10 MeV. The other radiations emitted by radioactive atoms are beta particles (negative or positive electrons) and gamma rays (high-frequency electromagnetic radiation). For the theory of radioactive decay SEE Radioactivity. For alpha-particle detection SEE Particle Detector. See also Beta particles; Gamma rays; Nuclear reaction; Nuclear spectra.

As an alpha particle passes through matter, it ionizes many of the atoms along its path, thereby gradually loses its kinetic energy, and is stopped (absorbed). For example, a 5-MeV alpha particle is stopped after traversing 1.4 in. (3.5 cm) of air at 100 kilopascals (1 atm) pressure and 59°F (15°C) and engaging in enough collisions with atomic electrons along its path to produce about 150,000 ion pairs.

Average charge. The charge on the alpha particle is $2e$ when it is emitted from its parent nucleus. While passing through the electron cloud of the emitting atom and the other atoms which lie along its paths, the alpha particle will capture one or two electrons and thus become a singly ionized or neutral helium atom. After this capture the swiftly moving atom will be ionized quickly by collisions with other atoms. There is therefore a rapid exchange of electrons between the moving alpha particle and its absorbing medium. About 1000 exchanges occur along the path of a single alpha particle, and the process of capture and loss of electrons becomes most rapid as the alpha-particle velocity declines near the end of its range. The net average charge on an alpha particle is therefore less than $2e$, and decreases as its velocity V decreases, reaching $1.99e$ at $V = 6.4 \times 10^{10}$ in./s (1.6×10^9 cm/s), $1.5e$ at $V = 2.2 \times 10^{10}$ in./s (0.56×10^9 cm/s), $1.0e$ at $V = 1.3 \times 10^{10}$ in./s (0.33×10^9 cm/s), $0.5e$ at $V = 0.6 \times 10^{10}$ in./s (0.16×10^9 cm/s), and zero at the end of the range.

Ionization by alpha particles. The theory of the stopping of alpha particles by matter deals with the kinetic energy lost by the moving charge, not with the ionization produced in the absorbing medium. The actual number of ion pairs produced by a given transfer of kinetic energy depends in a complicated way upon the nature and purity of the absorber. Present knowledge in this area is almost entirely empirical.

Delta electrons. Along the path of the alpha particle in any absorber, the magnitude of the energy transfer to individual atomic electrons can vary from an amount just sufficient to produce ionization or excitation (about 14 eV in air) up to about 2000 eV. Those electrons which receive a relatively large energy transfer (above about 100 eV) are called delta electrons. An appreciable fraction, roughly one-half, of the energy lost by alpha rays appears as delta electrons. These delta electrons lose their energy by further ionizing collisions with other atoms in the absorber. The total ionization in the absorber is the sum of the primary ionization produced by collisions of the alpha particle with atomic electrons and the secondary ionization produced by the delta electrons. See Delta electrons.

Ion pairs. In each ionizing event an originally neutral atom or molecule is divided into a free electron and a residual positive ion. Depending upon the nature of the nearby atoms, the liberated electron may remain free or it may become attached to a neutral atom to form a negative ion. The term ion pair means the residual positive ion and its negative counterpart, regardless of whether the electron is free or attached.

Specific ionization. Along the path of an alpha particle in normal air, some 50,000–150,000 ion pairs/in. (2000–6000 ion pairs/mm) are produced, depending upon the velocity of the alpha particle at the point under consideration. The number of ion pairs per unit path length is

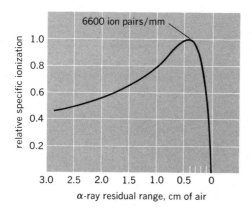

Fig. 1. Specific ionization of a single alpha particle in air at 59°F (15°C) and 29.92 in. (760 mm) Hg (101.3 kPa). 1 cm = 0.4 in.

called the specific ionization. **Figure 1** shows the specific ionization produced by an alpha particle in air for various values of the residual range, that is, at various distances measured back from the end of its range. The maximum specific ionization occurs at a residual alpha-particle range of 0.16 in. (0.40 cm) of air, where the average charge on the alpha particle has fallen to about $1.5e$. With the further decline in the effective charge, the specific ionization, which is proportional to the square of the charge, declines nonlinearly toward zero as the velocity decreases.

The specific ionization, or more exactly the linear energy transfer, also depends upon the velocity V of the alpha particle, and is proportional to $1/V^2$. This dependence is seen in the left side of Fig. 1.

Energy loss per ion pair. The energy loss per ion pair depends on the chemical nature, density, and purity of the absorber. In air an average of about 33.5 eV is lost by the alpha particle per ion pair formed. More than half of this energy transfer results in excitation without ionization, as can be seen from the fact that the ionization potentials of oxygen (O_2) and nitrogen (N_2) are only 13.6 eV and 14.5 eV, or less than half the average energy lost per ionization.

Straggling of alpha particles. Identical alpha particles, all having the same initial velocity, do not all have the same range. The observed ranges of individual particles from any monoenergetic source show a substantially normal (or gaussian) distribution about the mean range. The standard deviation of this distribution is of the order of 1% for 5-MeV alpha particles in any absorber. This distribution of ranges is due to statistical fluctuations in the number and magnitude of the individual collisions between the alpha particle and the atomic electrons, and is known as range straggling.

Because of range straggling, the average specific ionization in a beam of well-collimated alpha particles as a function of distance from the source is significantly different from the specific ionization curve for a single alpha particle shown in Fig. 1. The average specific ionization curve, called the Bragg curve after W. H. Bragg, who first examined it systematically, is shown in **Fig. 2**. The Bragg ionization curve is seen to be broader at the peak and to have a pronounced "tail" near the end of the range due to straggling. The extrapolated ionization range, R_i in Fig. 2, is slightly greater than the true mean range R of the distribution of alpha-particle ranges.

Range versus energy. The mean range R of a monoenergetic group of alpha particles cannot be derived wholly from theory but must rest on the results of many careful measurements plus interpolation and extrapolation based on theory. **Figure 3** shows the mean range of alpha particles in air for alpha-particle energies up to 17 MeV.

Geiger-Briggs rule. It is often convenient to use a rule of thumb to obtain relative ranges in a particular material if the range is known for one value of the energy. H. Geiger first pointed out that the range R of an alpha particle is roughly proportional to the cube of its initial velocity V. A more exact relationship, for R greater than 2 in. (5 cm) of air, and an improvement of Geiger's rule of G. H. Briggs is $R = $ constant $\times V^{3.26}$ (Geiger-Briggs rule).

Range in other absorbers. Dry air has long been the standard reference absorber for alpha-particle range-versus-energy relationships, such as that shown in Fig. 3. An approximate

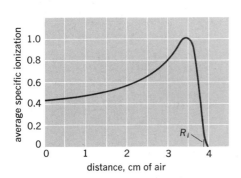

Fig. 2. The Bragg ionization curve for polonium alpha particles (5.305 MeV) in air at 59°F (15°C) and 29.92 in. (760 mm) Hg (101.3 kPa). 1 cm = 0.4 in. (*After R. Naidu, Ann. Phys., ser. 11, 1:72–122, 1934*)

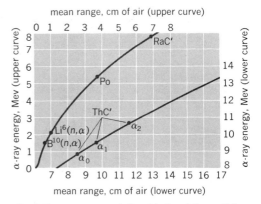

Fig. 3. Range-energy relationship for alpha particles in dry air at 59°F (15°C) and 29.92 in. (760 mm) Hg (101.3 kPa). 1 cm = 0.4 in. (*After R. D. Evans, The Atomic Nucleus, McGraw-Hill, 1955*)

relationship for the relative ranges of alpha particles of the same energy in various absorbers rests on early observations by Bragg and R. Kleeman, who found that that relative stopping power per atom is approximately proportional to the square root of the atomic weight A, and that in mixtures and compounds of several elements each atom acts independently of all the others. Then the range R of a given alpha particle in an absorber is related to its range R_0 in a reference material by Eq. (1), where N is the number of atoms per cubic centimeter and S is the relative atomic

$$\frac{R}{R_0} = \frac{N_0}{N} \cdot \frac{1}{S} \tag{1}$$

stopping power of atoms of the absorber compared with atoms of the reference material. For elementary materials $N = (\rho/A)K$, where ρ is the density and K is Avogadro's number, and $S = \sqrt{A/A_0}$. For mixtures Eqs. (2) and (3) hold, where $n_1, n_2, \ldots$ are the atomic fractions of the

$$N = \frac{\rho K}{(n_1 A_1 + n_2 A_2 + n_3 A_3 + \cdots)} \tag{2}$$

$$S = n_1 \sqrt{A_1/A_0} + n_2 \sqrt{A_2/A_0} + n_3 \sqrt{A_3/A_0} + \cdots \tag{3}$$

elements whose atomic weights are $A_1, A_2, \ldots$.

Then the Bragg-Kleeman rule, for elementary absorbers, is given by Eq. (4), and for mix-

$$\frac{R}{R_0} = \frac{N_0}{N} \cdot \frac{1}{S} = \frac{(\rho_0/A_0)}{(\rho/A)} \cdot \frac{\sqrt{A_0}}{\sqrt{A}} = \frac{\rho_0 \sqrt{A}}{\rho \sqrt{A_0}} \tag{4}$$

tures the effective value of $\sqrt{A}$ is to be taken as in Eq. (5).

$$\sqrt{A} = \frac{n_1 A_1 + n_2 A_2 + n_3 A_3 + \cdots}{n_1 \sqrt{A_1} + n_2 \sqrt{A_2} + n_3 \sqrt{A_3} + \cdots} \tag{5}$$

For air, $\sqrt{A_0} = 3.82$, $\rho_0 = 1.226 \times 10^{-3}$ g/cm^3, and the Bragg-Kleeman rule then takes its most common form, shown in Eq. (6). In aluminum the range of alpha particles in the domain of

$$R = 3.2 \times 10^{-4} \frac{\sqrt{A}}{\rho} R_{\text{air}} \tag{6}$$

2–10 MeV is then about 1/1600 of the range in air. In general, the range of ordinary alpha particles in various solids is only a few thousandths of an inch (tens of micrometers). Thus the range of the 5.15-MeV alpha particles from plutonium-239 is 1.47 in. (3.68 cm) in standard air, about 35 μm in

Initial kinetic energies of principal alpha-particle groups	
Source	Energy, MeV
Americium-241	5.486 ± 0.001
Plutonium-239	5.156 ± 0.001
Plutonium-238	5.4988 ± 0.0008
Uranium-238	4.200 ± 0.005
Uranium-235	4.396 ± 0.005
Thorium-232	4.011 ± 0.005
Radium-226	4.782 ± 0.001
Polonium-210	5.3049 ± 0.0006

water, 25 μm in compact bone, 21 μm in Ilford track emulsion, 23 μm in aluminum, and 27 μm in silicon.

The Bragg-Kleeman rule is usually good to ± 15%. A closer estimate can be obtained from the detailed theory of energy loss, but accurate range-versus-energy relationships for all absorbers must be based on direct experiments.

The inital kinetic energies of the principal alpha-particle group emitted by some common and standard alpha particle sources are shown in the **table**.

BETA PARTICLES
Gunnar Backstrom

Fast, charged particles emitted from certain radioactive nuclei. These particles, which are all identical except for the sign of their charge, are classified as positrons (+) and negatrons (−). The latter class is identical with atomic electrons. The kinetic energies range from zero up to 3–5 MeV. For the theory of beta decay SEE RADIOACTIVITY. SEE ALSO ALPHA PARTICLES; ELECTRON; POSITRON.

Interaction with matter. Beta particles, often designated β particles or β rays, interact strongly with matter: The particles are generally completely stopped in passing through 0.4 in. (1 cm) of solid material. If various thicknesses of some material are placed across a beam of beta particles, the number of penetrating particles is found to decrease gradually to zero as the thickness increases (**Fig. 1**). There is a definite thickness of absorber which is just sufficient to stop all the beta particles. This quantity, which is called the range of the beta particle, is usually expressed as the product ρx (g/cm^2) of density and thickness, since then this "range" will be approximately independent of the material (**Fig. 2**).

The passage of beta particles through matter is macroscopically observable as heat, due to the kinetic energy dissipated. This radiation also may promote certain chemical reactions and cause structural changes in materials, for instance, the discoloring of glass.

The track of a fast electron, as observed in a photographic emulsion, does not follow a straight line until the particle is stopped, but zigzags in a random way because of collisions with the much heavier nuclei. A beta particle loses its energy by excitation and ionization of atoms and at higher energy also by emitting bremsstrahlung, or brake radiation, when scattered. The first class of energy loss may be pictured as an inelastic collision with a bound electron, which thus reaches a higher energy state. The energy absorbed may or nay not be sufficient to liberate the electron. When a beta particle passes near a nucleus, the electrostatic attraction deflects its path. When so accelerated, the electron emits one or more bremsstrahlung quanta. This type of energy loss increases strongly with the atomic number of the absorber and becomes important in heavy elements at energies of a few MeV.

Apart from the stopping process, which proceeds similarly, negatrons and positrons behave differently in matter. Usually after it has been stopped, the positron combines with an atomic negatron and both vanish (annihilate). The two electron masses appear in the form of the electro-

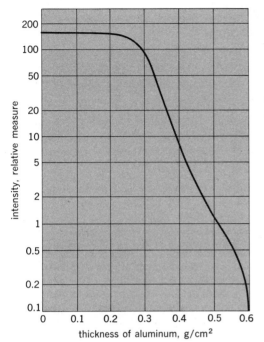

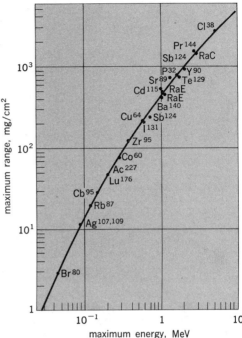

Fig. 1. The intensity of β-particles from indium-116 on passage through aluminum. (*After J. M. Cork, Radioactivity and Nuclear Physics, 3d ed., Van Nostrand, 1957*)

Fig. 2. Dependence of the range ρx of beta particles on their energy. (*After K. Siegbahn, ed., Beta- and Gamma-Ray Spectroscopy, North Holland, 1955*)

magnetic energy of two photons, each of energy 0.511 MeV, which are emitted in opposite directions in order to conserve momentum. Before annihilation the positron-negatron pairs form bound systems, called positronium atoms. SEE POSITRONIUM.

Detection of beta particles. Beta detectors may function so as to produce visible tracks of the particles, as in photographic emulsions and cloud chambers, or may give a signal at the advent of each individual particle. A detector may also indicate the total number of particles incident in a definite time. All detectors work on the principle of observing the interaction with matter, in particular ionization and excitation. For further information SEE PARTICLE DETECTOR.

Beta-particle spectrometers. These are used for measuring the energy or momentum distribution of beta particles. The energy of an electron is in principle inferred from the way it is influenced by a magnetic field. If a charged particle is moving perpendicularly to the flux of a uniform magnetic field, its orbit is bent into a circle of radius ρ. The momentum is found to be proportional to the product $B\rho$ of magnetic flux density and radius.

A simple and generally used spectrometer is the semicircular type, shown schematically in **Fig. 3**. From the electrons that leave the source in all directions, a thin bundle (represented by an angular spread 2ϕ in Fig. 3) is selected by the entrance slit E to improve resolution. Particles of a definite energy move in congruent circles and are brought to a focus after going through an angle of 180°. Particles of different energies come to a focus at different distances. The particles may be detected by means of a photographic plate along the focal line x, which permits the simultaneous recording of a large energy range. The angular spread 2ϕ allowed by the finite slit width introduces a spread Δx in the position at which the particle strikes the photographic plate. Alternatively a Geiger counter may be located at a fixed distance from the source and the field strength varied to focus one energy after another on a slit in front of the counter. Whereas this

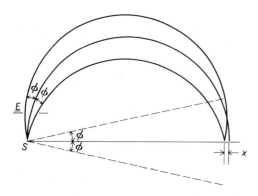

Fig. 3. Principle of the semicircular spectrometer. The magnetic field lines are perpendicular to the paper. The broken lines are diameters of the circular orbits.

instrument can focus a bundle of electrons divergent in the plane of the figure, it does nothing to collect particles which originally make an angle with this plane.

Focusing in both directions is achieved by using a cylindrically symmetric field with a density decreasing radially as $(\rho)^{-1/2}$ (**Fig. 4**). This spectrometer needs a fixed counter position.

An important figure of merit for a spectrometer is its resolving power, usually given by the width at half-height of a peak due to electrons of a definite energy. Although both instruments described yield relative peak widths down to 0.02%, the latter type realizes the same resolving power with appreciably higher transmission; that is, it accepts a larger percentage of the electrons emitted. Most of the precise work in beta spectroscopy is done with such double-focusing spectrometers.

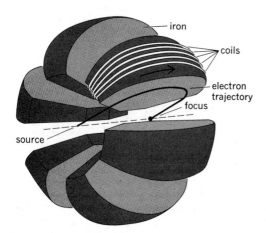

Fig. 4. The orange spectrometer, with the very high transmission. (*After K. Siegbahn, ed., Beta- and Gamma-Ray Spectroscopy, North Holland, 1955*)

There exist many types of spectrometers which make use of the focusing properties of magnetic lenses. They have found application especially in cases where high transmission is more important than high resolution.

An instrument with exceptionally high transmission is the so-called orange spectrometer (Fig. 4), which is a number of modified double-focusing spectrometers employing a common source and a common detector.

Bibliography. H. Behrens, *Electron Radial Wave Functions and Nuclear Beta-Decay*, 1982; R. D. Evans, *The Atomic Nucleus*, 1955, reprint 1982; K. S. Krane, *Introductory Nuclear Physics*, 1987; H. J. Lipkin, *Beta Decay for Pedestrians*, 1962; M. Morita, *Beta Decay and Muon Capture*, 1973; E. Segre, *Nuclei and Particles*, 2d ed., 1977.

GAMMA RAYS
J. W. OLNESS

Electromagnetic radiation emitted from excited atomic nuclei as an integral part of the process whereby the nucleus rearranges itself into a state of lower excitation (that is, energy content). For the theory of gamma emission SEE NUCLEAR STRUCTURE; RADIOACTIVITY.

Nature of gamma rays. The gamma ray is an electromagnetic radiation pulse—a photon—of very short wavelength. The electric (**E**) and magnetic (**H**) fields associated with the individual radiations oscillate in planes mutually perpendicular to each other and also the direction of propagation with a frequency v which characterizes the energy of the radiation. The **E** and **H** fields exhibit various specified phase-and-amplitude relations, which define the character of the radiation as either electric (EL) or magnetic (ML). The second term in the designation indicates the order of the radiation as 2^L-pole, where the orders are monopole (2^0), dipole (2^1), quadrupole (2^2), and so on. The most common radiations are dipole and quadrupole. Gamma rays range in energy from a few keV to 100 MeV, although most radiations are in the range 50–6000 keV. As such, they lie at the very upper high-frequency end of the family of electromagnetic radiations, which include also light rays and x-rays. SEE MULTIPOLE RADIATION.

Wave-particle duality. The dual nature of gamma rays is well understood in terms of the wavelike and particlelike behavior of the radiations. For a gamma ray of intrinsic frequency v, the wavelength is $\gamma = c/v$, where c is the velocity of light; energy is $E = hv$, where h is Planck's constant. The photon has no rest mass or electric charge but, following the concept of mass-energy equivalence set forth by Einstein, has associated with it a momentum given by $p = hv/c = E/c$.

Origin. One of the most frequently utilized sources of nuclear gamma rays is ^{60}Co (that is, the cobalt isotope of $N = 33$ neutrons, $Z = 27$ protons, and thus of atomic mass number $A = N + Z = 60$). As shown in **Fig. 1**, the decay process begins when ^{60}Co (in its ground state, or state of lowest possible excitation) decays to ^{60}Ni ($N = 32$, $Z = 28$) by the emission of a β^- particle. More than 99% of these decays lead to the 2506-keV level of ^{60}Ni; this level subsequently deexcites by an 1173-keV gamma transition to the 1332-keV level, which in turn emits a 1332-keV gamma ray leading to the ^{60}Ni ground state.

The gamma rays from ^{60}Ni carry information not only on the relative excitation of the ^{60}Ni levels, but also on the quantum-mechanical nature of the individual levels involved in the gamma decay. From the standpoint of nuclear physics, the levels of a given nucleus can be described most simply in terms of their excitation energies ($\mathbf{E}_x$) relative to the ground state, and in terms of

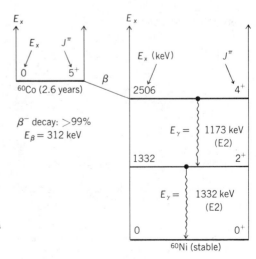

Fig. 1. Energy-level diagram illustrating the gamma decay of levels of ^{60}Ni resulting from beta decay of ^{60}Co.

the total angular momentum (**J**) and parity (π) quantum numbers given as $\mathbf{J}^\pi$. For a gamma-ray transition from initial state i to final state f, one obtains $E_x^i - E_x^f = E_\gamma'$, where E_γ' is the measured gamma energy after small (second-order) corrections for nuclear recoil and relativistic effects. Nuclear selection rules restrict the multipole character of the radiation according to the change in the quantum numbers J^π of the initial and final states. In Fig. 1, for example, the transitions must be electric quadrupole (E2), since they connect states of similar parity ($\pi = +$) by radiation of order $L \geq J_i - J_f = 2$. SEE NUCLEAR ISOMERISM; NUCLEAR SPECTRA; PARITY; SELECTION RULES; SPIN.

Use as nuclear labels. Various nuclear species exhibit distinctly different nuclear configurations: the excited states, and thus the gamma rays which they produce, are also different. Precise measurements of the gamme-ray energies resulting from nuclear decays may therefore be used to identify the gamma-emitting nucleus, that is, not only the atomic number Z but also the specific isotope that is designated by A. This has ramifications for nuclear research and also for a wide variety of more practical applications.

The two most widely used detectors for such studies are the NaI(Tl) detector and the Ge(Li) detector. **Figure 2** shows typical gamma spectra measured for sources of ^{60}Co and ^{54}Mn. Full-

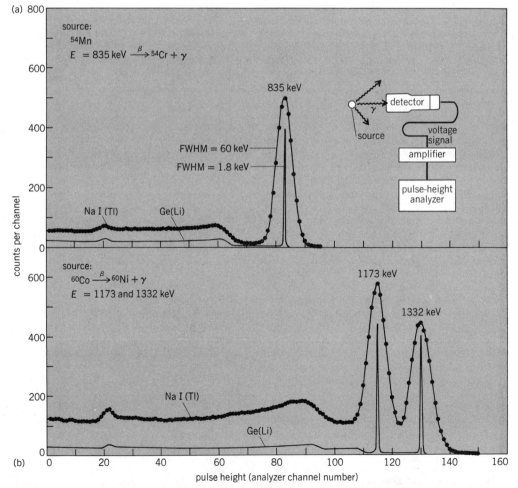

Fig. 2. Gamma-ray spectra from radioactive sources as measured with both NaI(Tl) and Ge(Li) detectors. Inset shows the components of detector apparatus. (*a*) ^{54}Mn source. (*b*) ^{60}Co sources.

energy peaks are labeled by the gamma-ray energy, given in keV. The figure of merit for these detectors, defined for a given gamma energy as the full-width-at-half-maximum (FWHM) for the full energy peak, is indicated. Although the more efficient NaI(Tl) detector can clearly distinguish the ^{60}Co and ^{54}Mn gamma rays, it is evident that the Ge(Li) detector, having a line width of only 1.8 keV, is more appropriate for complex nuclei, or for studies involving a greater number of source components. SEE GAMMA-RAY DETECTORS.

Applications to nuclear research. One of the most useful studies of the nucleus involves the bombardment of target nuclei by energetic nuclear projectiles, to form final nuclei in various excited states. For example, ^{48}Ca bombarded by ^{16}O makes ^{60}Ni quite strongly via the ^{48}Ca(^{16}O,4n)^{60}Ni reaction, as well as numerous other final species. Ge(Li) measurements of the decay gamma rays are routinely used to identify the various final nuclei according to their characteristic gamma rays, that is, the 1332- and 1173-keV gamma rays of ^{60}Ni, for example.

Precise measurements of the gamma energies, together with intensity and time-coincidence measurements, are then used to establish the sequence of gamma-ray decay, and thus construct from experimental evidence the nuclear level scheme. Angular correlation and linear polarization measurements determine the radiation character (as M1, E1, E2, or M2 or mixed) and thus the spin-parity of the nuclear levels. These studies provide a very useful tool for investigations of nuclear structure and classification of nuclear level schemes.

Practical applications. In these applications, the presence of gamma rays is used to detect the location or presence of radioactive atoms which have been deliberately introduced into the sample. In irradiation studies, for example, the sample is activated by placing it in the neutron flux from a reactor. The resultant gamma rays are identified according to isotope by Ge(Li) spectroscopy, and thus the composition of the original sample can be inferred. Such studies have been used to identify trace elements found as impurities in industrial production, or in ecological studies of the environment, such as minute quantities of tin or arsenic in plant and animal tissue.

In tracer studies, a small quantity of radioactive atoms is introduced into fluid systems (such as the human blood stream), and the flow rate and diffusion can be mapped out by following the radioactivity. Local concentrations, as in tumors, can also be determined.

Doppler shift. If $E_{\gamma 0} = h\nu_0$ is the gamma ray energy emitted by a nucleus at rest, then the energy $E_\gamma = h\nu$ emitted from a nucleus moving with velocity v at angle θ (with respect to the direction of motion) is given by Eq. (1) where c is the velocity of light. In terms of the frequency

$$E_\gamma = E_{\gamma 0}\left(1 + \frac{v}{c}\cos\theta\right) \quad (1)$$

v, this expression is entirely analogous to the well-known Doppler shift of sound waves. Experimental measurements of the Doppler shift are used to determine the velocity of the nucleus and, more importantly, to shed light on the lifetime of the nuclear gamma-emitting state. A major advantage of this technique is that the same nuclear reaction which produces the excited nuclear states can also be employed to impart a large velocity to the nucleus.

For example, the velocity of ^{60}Ni nuclei produced via the ^{48}Ca(^{16}O,4n)^{60}Ni reaction at $E(^{16}$O$) = 50$ MeV is $v/c = 0.00204$, and instead of $E_\gamma = 1332$ keV one should observe $E_\gamma = 1359$ keV. The extent of the shift is clearly within the resolving power of the Ge(Li) detector, which may therefore be used to measure E_γ and thus infer v. In most nuclear reactions, v is a known function of time t [that is, $v = v(t)$], and one therefore obtains a distribution of E_γ's whose precise shape may be related to the lifetime of the nuclear state.

Doppler-shift measurements of gamma rays from recoil nuclei produced in nuclear reactions have been routinely used since the mid-1960s to measure nuclear lifetimes of 10^{-9} to 10^{-14} s—a range previously considered inaccessible to study.

Interaction with matter. For the three types of interaction with matter which together are responsible for the observable absorption of gamma rays, namely, Compton scattering, the photoelectric effect, and pair production, SEE COMPTON EFFECT; ELECTRON-POSITRON PAIR PRODUCTION.

The energy of a photon may be absorbed totally or partially in interaction with matter; in the latter case the frequency of the photon is reduced and its direction of motion is changed. Photons are thus absorbed not gradually, but in discrete events, and one interaction is sufficient to remove a photon from a collimated beam of gamma rays. The intensity I of a beam decreases

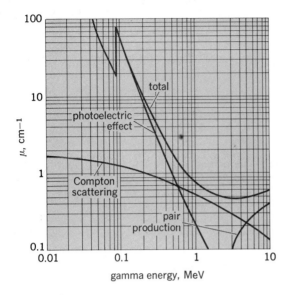

Fig. 3. Graphic representation of partial and total attenuation coefficients for lead as a function of gamma energy. (*National Bureau of Standards*)

exponentially, as in Eq. (2), where x is the path length, I_0 is the initial intensity, and μ is the

$$I = I_0 e^{-\mu x} \qquad (2)$$

linear attenuation coefficient, which is characteristic of the material and the gamma energy.

The dependence of the attenuation coefficient on gamma-ray energy is shown in **Fig. 3** for a lead absorber. For different absorbers, the attenuation is generally greater for the more dense materials. Most attenuation coefficients are tabulated as mass attenuation coefficients μ/ρ where ρ is the material or elemental density.

Bibliography. J. Cerny (ed.), *Nuclear Spectroscopy and Reactions*, pt. C, 1974; K. Debertin and W. B. Mann, *Gamma and X-ray Spectrometry Techniques and Applications*, 1983; R. D. Evans, *The Atomic Nucleus*, 1955, reprint 1982; K. S. Krane, *Introductory Nuclear Physics*, 1987; E. Segré, *Nuclei and Particles*, 2d ed., 1977; K. Siegbahn (ed.), *Alpha-, Beta-, and Gamma-Ray Spectroscopy*, 1965.

COMPTON EFFECT
EASTMAN N. HATCH

The increase in wavelength of electromagnetic radiation, observed mainly in the x-ray and gamma-ray region, on being scattered by material objects. This increase in wavelength, $\lambda_2 - \lambda_1 = \Delta\lambda$, of the scattered radiation, which is caused by the interaction of the radiation with the weakly bound electrons in the matter in which the scattering takes place, is given to good approximation by Eq. (1). Here λ_1 is the wavelength of the incident radiation, λ_2 is the wavelength

$$\lambda_2 - \lambda_1 = \Delta\lambda = (h/m_0 c)(1 - \cos \phi) \qquad (1)$$

of the radiation scattered at the angle ϕ, h is Planck's constant, m_0 is the rest mass the electron, c is the speed of light, and ϕ is the angle that the direction of the scattered radiation makes with the direction of the incident radiation.

The Compton effect illustrates one of the most fundamental interactions between radiation and matter and displays in a very graphic way the true quantum nature of electromagnetic radiation. Together with the laws of atomic spectra, the photoelectric effect, and pair production, the

Compton effect has provided the experimental basis for the quantum theory of electromagnetic radiation.

The Compton effect represents a great departure from earlier ideas concerning electromagnetic radiation. According to the original theory for the scattering of electromagnetic radiation by electrons in matter, which was developed by J. J. Thomson about 1900, the scattered radiation should have exactly the same wavelength as the incident radiation. This theory considers the incident radiation to have an oscillating electric field and shows that an electron would be forced by this electric field to oscillate with the same frequency as the field. The theory of electromagnetic radiation developed in the latter part of the nineteenth century by James Clerk Maxwell predicts that a point charge, such as an electron, when oscillating with a given frequency, will itself emit in all directions waves of electromagnetic radiation of exactly the same frequency. Therefore an increase in the wavelength, corresponding to a decrease in the frequency, of the scattered radiation is not to be expected if the scattering of x-rays takes place according to Thomson's theory.

In a series of experiments, beginning in 1922, A. H. Compton confirmed the earlier conclusions of other scientists that the wavelengths of scattered x-rays increase, depending on the angle of scattering, a result in direct conflict with Thomson's theory. This discovery, along with its subsequent explanation by Compton, is regarded as one of the most significant contributions in physics. Compton showed that a beam of x-rays is composed of individual particles, called photons, each of which carries the energy $h\nu$ and also the linear momentum $h\nu/c$ in the direction of the beam, where h is Planck's constant, ν is the frequency of the radiation, and c is the speed of light. Moreover, the photon can impart energy and linear momentum to an individual electron in an elastic collision with the electron.

Experimental results. Using a crystal spectrometer, Compton made careful measurements of the wavelength spectrum of molybdenum K x-rays after they had been scattered at different angles from graphite. A diagram of the experimental apparatus for these measurements is shown in **Fig. 1**a. The spectrum of wavelengths of the molybdenum K x-rays after scattering from graphite at various angles is shown in Fig. 1b. While the incident radiation before scattering consists mainly of a fairly narrow range of wavelengths (that is, the molybdenum K line), the spectrum observed after scattering consists of two peaks. One of these peaks P has essentially the same wavelength as the molybdenum K line, but the second peak M has a longer wavelength. The wavelength of this second peak depends on the scattering angle, and is longer for larger scattering angles (Fig. 1b).

The dependence of the wavelength shift $\lambda_2 - \lambda_1 = \Delta\lambda$ on scattering angle ϕ is plotted in Fig. 1c. The range of wavelengths due to the inhomogeneity in the scattering angle required by these experiments is also shown for the angles 45, 90, and 135°, and labeled $\delta\lambda$ in the figure. The first peak in Fig. 1b, which has the same wavelength as the incident radiation, is called the unmodified line. The longer-wavelength peak is called the modified line and is clearly due to a different type of scattering than Thomson predicted, since the wavelengths of the x-rays are increased in the scattering process. This is the type of scattering which yields the longer wavelengths in Compton scattering.

Theoretical explanation. Compton's explanation of the observed wavelength shift in x-ray scattering is based on developments which took place early in the twentieth century. Max Planck provided an explanation for the observed intensity distribution with wavelength of electromagnetic radiation from a blackbody by introducing the idea that energy could be emitted or absorbed in the blackbody only in discrete amounts equal to $h\nu$, where h is Planck's constant and ν is the frequency of the radiation. Planck's discovery, which marked the birth of quantum theory, was followed by Albert Einstein's explanation for photoemission. When light falls upon a surface, energetic electrons are emitted, and the energy of the electrons is found to be independent of the intensity but dependent on the frequency of the incident light which liberates the electrons. Einstein supplied the solution for this puzzle by postulating that the radiation field, that is, the beam of light, consists of particles called photons, each having energy $h\nu$, where h and ν are as Planck defined them. The photoelectric effect takes place when a photon is absorbed by an electron, so that the photon disappears and the electron assumes all of the energy $h\nu$ of the photon, less the energy required to bind the electron in its medium.

Compton's explanation of the wavelength shift assumes that, since x-rays are electromag-

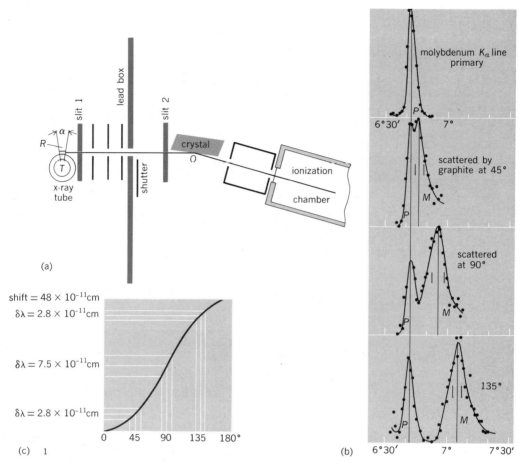

Fig. 1. Compton's experiment. (a) Characteristic K-lines from molybdenum target T of x-ray tube (shown in cross section) fall on graphite scatterer R. Scattered radiation is passed through slits 1 and 2, and analyzed spectrally by slow rotation of the calcite crystal and ionization chamber around pivot O. Longer wavelengths are observed by Bragg diffraction for larger angles of the calcite crystal with respect to the scattered beam defined by slits 1 and 2. Spread, or inhomogeneity, in scattering angle φ due to width of scatterer R is denoted by α. (b) Resulting spectra at three scattering angles. The ordinate is the intensity of the beam detected with ionization chamber and the abscissa is the angle that diffracting planes of calcite crystal make with scattered beam defined by slits 1 and 2. Larger angles of the calcite crystal spectrometer correspond to longer wavelengths in the spectrum of scattered radiation. (c) Graph showing dependence of shift $\lambda_2 - \lambda_1$ on φ. Effect of inhomogeneity α, about 0.31 radian, in scattering angle φ is shown in the graph. This spread gives rise to apparent spread δλ in wavelength shift due to thickness of scatterer R, as shown on the ordinate of the graph.

netic radiation, the picture given by Einstein of the quantum nature of the radiation field in the photoelectric effect describes the incident beam of x-rays. The scattering of the x-rays which gives rise to the peak of the modified line in the spectrum of the scattered x-rays (Fig. 1b) is due to photons or quanta of electromagnetic radiation scattering like material particles in collision with free or loosely bound electrons in the scattering material. Further, a photon, in addition to the energy $h\nu$, has a linear momentum $h\nu/c$ in the direction of travel of the beam. This momentum corresponds to that possessed by a massless particle having energy $h\nu$ and moving with the speed of light c, a fact which follows from Einstein's theory of special relativity.

In other words, the x-ray scattering process, which gives rise to the increase in wavelength, is a process in which an x-ray photon of energy $h\nu$ and linear momentum $h\nu/c$ scatters as a mechanical particle would in colliding eleastically with an electron which is at rest. In this type of collision, the x-ray particle transmits some of its energy and linear momentum to the electron which recoils. Therefore the x-ray photon after the collision has less energy $h\nu_2$ than before the collision. Since $\nu_2 = c/\lambda_2$ for a wave traveling with speed c, the wavelength λ_2 for the photon after the collision will be longer than λ_1, the wavelength it had before the scattering.

An important property of the elastic collision between two mechanical particles is that the energy and linear momentum will be conserved in the process. These two principles were used by Compton to calculate the increase in wavelength of the x-rays after scattering through a definite angle.

Scattering process. **Figure 2** shows the Compton scattering process. The incident photon with linear momentum p_1 and wavelength λ_1 collides with the electron, which is initially at rest. After the collision, the photon scatters off at an angle ϕ with respect to the incident direction.

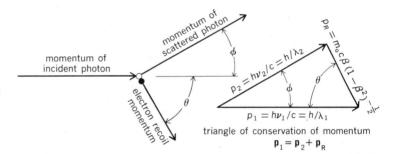

Fig. 2. Diagram for derivation of wavelength shift, electron recoil energy, and angular correlation between scattering angle ϕ and electron recoil angle θ.

The electron recoils and moves off at an angle θ with respect to the incident direction. A triangle resulting from the conservation of linear momentum of the particles involved in the collision is also shown in Fig. 2. This triangle shows the angular relationship which must be satisfied by ϕ and θ.

The principle of conservation of energy gives Eq. (2), where $h\nu_1$ is the energy of the inci-

$$h\nu_1 = h\nu_2 + m_0 c^2 \left[\frac{1}{\sqrt{1 - \beta^2}} - 1 \right] \tag{2}$$

dent photon, $h\nu_2$ is the energy of the scattered photon, and the second term on the right-hand side of the equation is the relativistic form of the kinetic energy of the recoiling electron.

In applying the principles of conservation of linear momentum, two additional equations are obtained. Conservation of the horizontal component of linear momentum (Fig. 2) gives Eq. (3), while conservation of the vertical component yields Eq. (4). In these equations V is the velocity of the recoiling electron and β is V/c.

$$\frac{h\nu_1}{c} = \frac{h\nu_2}{c} \cos\phi + \frac{m_0 V}{\sqrt{1 - \beta^2}} \cos\theta \quad (3) \qquad 0 = \frac{h\nu_2}{c} \sin\phi - \frac{m_0 V}{\sqrt{1 - \beta^2}} \sin\theta \quad (4)$$

After making the substitution $\nu = c/\lambda$, squaring the equations, and combining the results, Eq. (1) is obtained. The quantity in Eq. (1), $h/m_0 c = 24.26 \times 10^{-11}$ cm, is called the Compton wavelength of a free electron, and is the wavelength of a photon having energy $h\nu = m_0 c^2$, the rest energy of the electron. The Compton wavelength is the shift in the wavelength of a photon which is scattered through 90°, as seen in Eq. (1).

Two important physical results follow from Eq. (1). First, the wavelength shift $\Delta\lambda$ is independent of the wavelength of the incident photons and is, therefore, independent of the photon energy. Second, the shift $\Delta\lambda$ is independent of the type of material of the scatterer.

From Eq. (1) it follows that the energy of the scattered photon can be expressed as shown in Eq. (5). It then follows from Eqs. (2) and (5) that the kinetic energy E_R of the recoil electron can be expressed as in Eq. (6).

$$h\nu_2 = \frac{h\nu_1}{1 + \frac{h\nu_1}{m_0 c^2}(1 - \cos \phi)} \tag{5}$$

$$E_R = h\nu_1 - h\nu_2 = h\nu_1 \frac{\frac{h\nu_1}{m_0 c^2}(1 - \cos \phi)}{1 + \frac{h\nu_1}{m_0 c^2}(1 - \cos \phi)} \tag{6}$$

From the geometry of the momentum triangle of Fig. 2, Eq. (7) can be obtained. This is the

$$\cot \theta = -\left(1 + \frac{h\nu_1}{m_0 c^2}\right) \tan \frac{\phi}{2} \tag{7}$$

required relationship for the angular correlation between the scattered photon and the recoil electron.

Compton-Debye effect. Peter Debye knew of Compton's published measurements of the wavelength shift of scattered x-rays, and independently developed the same theoretical explanation as Compton. His results were published at about the same time as Compton's; in Europe the effect has been known as the Compton-Debye effect.

Experimental verification. Verification of Compton's ideas appeared soon after his theoretical explanation for the shift. The recoil electrons, which are predicted by Compton's scattering theory, were detected independently by C. T. R. Wilson and Walther Bothe in Wilson cloud chambers. The momenta of the recoil electrons were measured by magnetic deflection and found to agree with Compton's theory by A. A. Bless. Bothe and Hans Geiger demonstrated the simultaneity of the appearance of the scattered photon and recoil electron. Then Compton and A. W. Simon, using a cloud chamber with partitions, were able to show that the predicted correlation in Eq. (7) between the direction of the scattered photon and that of the recoil electron is satisfied in individual Compton scattering processes. Later experiments with improved techniques showed the correctness of Compton's theory with greater accuracy. SEE CLOUD CHAMBER.

The results of all these experiments led to the definite conclusion that electromagnetic radiation scattering from an electron, instead of spreading out in all directions around the scattering center as waves are expected to do, takes a definite direction in each individual process as a particle would.

Intensity distribution. Thomson's theory for the scattering of electromagnetic radiation predicts that the relative intensity of scattered radiation is symmetrical about a 90° scattering angle and proportional to the factor $1 + \cos^2 \phi$, where ϕ is the scattering angle. The angular distribution of scattered electromagnetic radiation as given by Thomson's theory is written as Eq. (8). Here I_ϕ/I_0 is the ratio of the radiation scattered at an angle ϕ to the incident radiation intensity,

$$I_\phi/I_0 = \frac{ne^4}{2r^2 m_0^2 c^4}(1 + \cos^2 \phi) \tag{8}$$

n is the effective number of independently scattering electrons, e is the charge on the electron, and r is the distance from the scatterer.

Thomson's theory did not take into account the effect of the magnetic vector of the radiation. Including the effect of the magnetic vector would have provided the electron with a recoil in the scattering of radiation from the effect known as radiation pressure. Consideration of recoil of the electron from radiation pressure would have led to the prediction of an increase in the wavelength of the scattered radiation from the Doppler shift. Paul Dirac, Gregory Breit, and others took into account the effect of the relativistic Doppler shift of the photon in scattering from the electron, and found that the intensity distribution from the Thomson theory was to be multiplied by the Breit-Dirac recoil factor R, given by Eq. (9).

$$R = \left[1 + \frac{h\nu_1}{m_0 c^2}(1 - \cos \phi)\right]^{3-} \tag{9}$$

In addition to their charge and mass, electrons also have a quantized spin or angular momentum and an associated magnetic moment. Using Dirac's relativistic quantum mechanics and taking into account the interaction of the spin and magnetic moment of the electron with the electromagnetic radiation, O. Klein and Y. Nishina obtained a further factor which modifies the angular distribution of scattered x-rays. This factor can be written as notation (10). Combining

$$1 + \frac{\left(\frac{h\nu_1}{m_0 c^2}\right)^2 (1 - \cos \phi)^2}{(1 + \cos^2 \phi)\left[1 + \frac{h\nu_1}{m_0 c^2}(1 - \cos \phi)\right]} \tag{10}$$

this factor and the Breit-Dirac recoil factor with the Thomson formula for the scattered x-ray angular distribution, the Klein-Nishina formula is obtained. This is written as Eq. (11). The quan-

$$I_\phi/I_0 = \frac{ne^4}{2r^2 m_0^2 c^4} \cdot \frac{1 + \cos^2 \phi}{\left[1 + \frac{h\nu_1}{m_0 c^2}(1 - \cos \phi)\right]^3} \cdot \left\{1 + \frac{\left(\frac{h\nu_1}{m_0 c^2}\right)^2 (1 - \cos \phi)^2}{(1 + \cos^2 \phi)\left[1 + \frac{h\nu_1}{m_0 c^2}(1 - \cos \phi)\right]}\right\} \tag{11}$$

tities on the left-hand side of the formula represent the same quantities as in the Thomson formula, Eq. (8). For the limiting case of low-energy x-rays ($h\nu_1 \to 0$), the Klein-Nishina formula reduces to the Thomson scattering law. But at photon energies above a few hundred kilovolts, the angular distribution departs significantly from the $1 + \cos^2 \phi$ distribution of the Thomson law.

Figure 3 shows the predicted results of the x-ray scattering angular distribution, comparing the Klein-Nishina formula with the Thomson formula at various energies. The full lines in Fig. 3 are the theoretical predictions, and the points are experimental measurements.

If the modified line of Fig. 1b is due to scattering of an x-ray photon from an electron which recoils, the presence of the unmodified line in the spectrum of scattered radiation must be due to some other type of scattering. The fact that the wavelength changes in Compton scattering, giving rise to the modified line of Fig. 1b, is due to the decrease in energy of the scattered photon in imparting recoil energy to the electron. The unmodified line is due to photons that scatter from electrons which are too tightly held in the atom to recoil from the impact of the photon. That is, these interactions correspond to photon-electron interactions for which the conservation of momentum and energy would require the free-electron recoil energy to be comparable to, or less

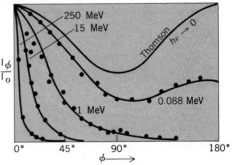

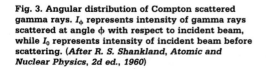

Fig. 3. Angular distribution of Compton scattered gamma rays. I_ϕ represents intensity of gamma rays scattered at angle ϕ with respect to incident beam, while I_0 represents intensity of incident beam before scattering. (*After R. S. Shankland, Atomic and Nuclear Physics, 2d ed., 1960*)

than, the electron binding energy in the atom. The momentum change accompanying scattering is communicated in such cases to the atom as a whole. Because of the atom's larger mass when compared to that of an electron, the photon transmits only a small contribution of linear momentum to the atom. Therefore the photon itself loses only a negligible amount of energy. The unmodified line is more prominent in scattering from atoms of higher atomic number because of the greater number of tightly bound electrons in these atoms. The theory giving the ratio of intensities of modified and unmodified scattering has been developed by I. Waller and D. R. Hartree.

There are many ways in which photons can interact with matter. In these interactions, usually photons either disappear completely or are scattered out of the initial beam of photons, so that the intensity of the beam is diminished as it moves through the scattering material. Of the many possible modes of interaction of photons with matter, there are four which predominate: Compton scattering (with a change in wavelength of the photon giving rise to the modified line in the scattered spectrum), elastic scattering (with no change in wavelength giving rise to the unmodified line), photoelectric effect or photoemission (when the visible and ultraviolet regions are most often involved), and electron-positron pair production (for photon energies above 1 MeV).

The Klein-Nishina formula, Eq. (11), can be integrated over the angle ϕ to obtain a formula which gives the relative intensity of photons scattered by the Compton effect compared with the initial intensity of the incident beam in the scattering material. This formula is related to the probability that Compton scattering will occur to a photon of a given energy $h\nu$. Formulas have been obtained giving the probability that the other three processes can occur. It has been demonstrated that, while all these processes can happen to some extent at most energies (except for electron-positron pair production which has a threshold at 1 MeV below which it cannot occur), the Compton effect is the predominant interaction for a large range of atomic numbers for photon energies between 1 and approximately 10 MeV. SEE GAMMA RAYS.

Bibliography. A. Beiser, *Concepts of Modern Physics*, 4th ed., 1987; A. H. Compton, The scattering of x-rays as particles, *Amer. J. Phys.*, 29:817–820, 1961; R. D. Evans, The Compton effect, *Handbuch der Physik*, vol. 34, pp. 218–298, 1958; J. H. Hubbell et al., Pair, triplet and total atomic cross sections and mass attenuation coefficients for 1 MeV–100 GeV photons in elements $Z = 1$ to 100, *J. Phys. Chem. Ref. Data*, 9(4):1023–47, 1980; K. S. Krane, *Modern Physics*, 1983; R. H. Stuewer, *The Compton Effect: Turning Point in Physics*, 1975; B. G. Williams (ed.), *Compton Scattering*, 1976.

ELECTRON-POSITRON PAIR PRODUCTION
GUNNAR BACKSTROM

A process in which a negative electron (negatron) and a positive electron (positron) are simultaneously created in the vicinity of a nucleus or an elementary particle. In external pair production, an electromagnetic wave (photon) is absorbed and creates an electron pair. The absorption of high-energy gamma rays is due mainly to this effect (see **illus**.). Internal pair production is not associated with observable electromagnetic radiation and may occur when an excited nucleus releases some of its internal energy.

Pair production is of considerable theoretical interest, not only as an exmaple of the materialization of energy, but also as a striking confirmation of the relativistic quantum theory proposed

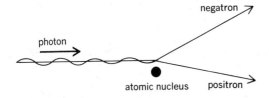

External pair (negatron-positron) production.

by P. A. M. Dirac. This theory has made possible quantitative predictions of production probability, differential electron distribution, and kinetic energy partition. The results are in satisfactory agreement with experimental findings.

External pair production can take place only if the energy of the photon exceeds $2mc^2$ (m = electron mass, c = velocity of light) or 1.02 MeV, which is the energy required for production of an electron pair at rest. The energy excess, $h\nu - 2mc^2$ (ν is the frequency of the photon, h is Planck's constant), appears as kinetic energy of the created particles. The sharing of this energy between the electrons takes place in a random way, such that the positron, for example, may assume any energy between zero and $h\nu - 2mc^2$ with about the same probability. Because of the electrostatic repulsion of the nucleus, the positron actually obtains a higher energy on the average than the negatron.

Conservation laws require that the momentum of the initial photon be transferred to the product particles. Simple calculations show that this can be fulfilled only if a third particle or system of particles takes part in the process. It may be a nucleus, as is usually the case, but in principle any charged particle may restore the momentum balance. For a given energy division between the electrons, the nucleus may recoil in any direction; consequently, the direction in which the electrons are emitted is not fixed, but is randomly distributed. As a consequence of its large mass, the nucleus receives only a vanishingly small part of the initial photon energy. For a discussion of conservation laws SEE NUCLEAR REACTION.

Internal pairs are often emitted from radioactive substances. After radioactive decay, the daughter nucleus may be left with excess energy. Although this energy is usually released as electromagnetic radiation, pair production may compete when the energy exceeds $2mc^2$, the probability increasing with higher energy release. The angular correlation of the pairs and the production probability also depend on the multipole order of the transition. SEE GAMMA RAYS; MULTIPOLE RADIATION; POSITRON; QUANTUM FIELD THEORY.

Bibliography. A. E. Beiser, *Concepts of Modern Physics*, 4th ed., 1987; M. L. Burns and A. K. Harding (eds.), *Positron-Electron Pairs in Astrophysics*, 1983; R. D. Evans, *The Atomic Nucleus*, 1955, reprint 1982.

CHARGED PARTICLE BEAMS
HANS BICHSEL

Unidirectional streams of charged particles traveling at high velocities. Charged particles can be accelerated to high velocities by electromagnetic fields. They are then able to travel through matter (termed an absorber), interacting with it, losing energy, and causing various effects important in many applications. The velocities under consideration in this article exceed 100,000 m/s (about 60 mi/s, or 200,000 mi/h), equivalent to an energy of 100 eV for a proton, and can approach the speed of light c (about 3×10^8 m/s, or 6.7×10^8 mi/h). Examples of charged particles are electrons, positrons, protons, antiprotons, alpha particles, and any ions (atoms with one or several electrons removed or added). In addition, some particles are produced artificially and may be short-lived (pions, muons).

Excluded from consideration are particles of, for example, cosmic dust (micrometeorites), which are clumps of thousands or millions of atoms. SEE ELEMENTARY PARTICLE; PARTICLE ACCELERATOR.

Particle properties. Fast charged particles are described in terms of the following properties (values for some particles are given in **Table 1**):

Charge, z, in multiples of the electron charge $e = 1.6022 \times 10^{-19}$ coulomb. At small velocities the charge may be less than the charge of the nucleus because electrons may be present in some of the atomic shells.

Rest mass, M; usually the energy equivalent Mc^2 (in MeV) is given.
Rest mass, m, of electron; $mc^2 = 0.51104$ MeV.
Mass in atomic mass units, u; $A_m = Mc^2/931.481$ MeV.
Kinetic energy, T, in MeV.
Velocity, v, in cm/s or m/s.
$\beta = v/c$; $c = 299{,}792{,}458$ m/s = speed of light.

Table 1. Properties of charged particles*

Ion	z	Lifetime, ns	Mass 10^{-24} g	Mass u	Mass MeV
Electron†	−1	Stable	0.910956	0.548593	511.004
Muon	1	2198.3	0.188357	0.113432	105.6598
Pion	1	26.04	0.248823	0.149846	139.578
Kaon	1	12.35	0.880322	0.530147	493.82
Sigma⁺	1	0.081	2.120318	1.276895	1189.40
Sigma⁻	−1	0.164	2.134436	1.285398	1197.32

Ion	z	Mass excess,‡ MeV	10^{-24} g	u	MeV
1N	0	8.0714	1.674920	1.0086652	939.553
1H	1	7.2890	1.672614	1.0072766	938.259
2H	1	13.1359	3.343569	2.0135536	1875.587
3H	1	14.9500	5.007334	3.0155011	2808.883
3He	2	14.9313	5.006390	3.0149325	2808.353
4He	2	2.4248	6.644626	4.0015059	3727.328
6Li	3	14.0884	9.985570	6.0134789	5601.443
7Li	3	14.9073	11.647561	7.0143581	6533.743
7Be	4	15.7689	11.648186	7.0147345	6534.093
9Be	4	11.3505	14.961372	9.0099911	8392.637
10B	5	12.0552	16.622243	10.0101958	9324.309
11B	5	8.6677	18.276741	11.0065623	10252.406
12C	6	0	19.920910	11.9967084	11174.708
13C	6	3.1246	21.587011	13.0000629	12109.314
14C	6	3.0198	23.247356	13.9999504	13040.691
14N	7	2.8637	23.246166	13.9992342	13040.024
15N	7	.1004	24.901771	14.9962676	13968.741
16O	8	−4.7365	26.552769	15.9905263	14894.875
17O	8	−.8077	28.220304	16.9947441	15830.285
18O	8	−.7824	29.880881	17.9947713	16761.791
19F	9	−1.4860	31.539247	18.9934674	17692.058
20Ne	10	−7.0415	33.188963	19.9869546	18617.472
21Ne	10	−5.7299	34.851833	20.9883627	19550.265
22Ne	10	−8.0249	36.508273	21.9858989	20479.451

*From *American Institute of Physics Handbook*, 3d ed., McGraw-Hill, 1972.
†Electron masses to be divided by 1000.
‡Mass excess given for neutral atoms; it is used to calculate nuclear reaction Q values.

For absorbers, the information needed is:
Atomic number, Z (number of protons in nucleus).
Average atomic weight, A (usually in g/mole, but numerically equal to mass in u).
Absorber thickness, x, in cm or g/cm^2.
Physical state (solid, liquid, gas, plasma).

Relation of velocity and energy. For small velocities ($\beta < 0.2$), the approximation $T = \frac{1}{2}Mc^2\beta^2$ is accurate to 3%. The expressions $\beta^2 = \zeta(\zeta + 2)/(\zeta + 1)^2$, and $T = Mc^2[(1/\sqrt{1 - \beta^2}) - 1]$, with $\zeta = T/Mc^2$, are correct for all velocities. For $\zeta > 100$, the expression $\beta^2 = 1 - (1/\kappa^2)$, with $\kappa = \zeta + 1$, is more suitable to provide accurate values.

Range. If a parallel beam of monoenergetic particles (that is, a beam in which all particles have exactly the same kinetic energy T) enters an absorber with a flat and smooth surface, it is found that all the particles (with the exception of electrons) travel along almost straight lines, slow down at approximately the same rate, and stop at approximately the same distance x from the surface (**Fig. 1**). The average distance traveled by the particles is called the mean range $R(T)$.

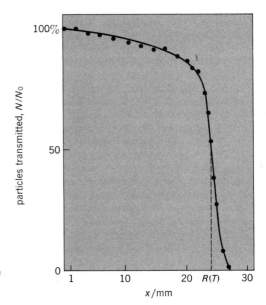

Fig. 1. Transmission curve for protons with kinetic energy $T = 144$ MeV through copper. (*After L. Koschmieder, Zur Energiebestimmung von Protonen aus Reichweitmessungen, Z. Naturforsch.*, 19a:1414–1416, 1964)

Energy loss and straggling. If the same beam travels through a thin absorber, it will emerge from it with a reduced average energy $\langle T_1 \rangle$ (**Table 2**). The difference between T and $\langle T_1 \rangle$ is called the average energy loss $\Delta = T - \langle T_1 \rangle$. Owing to the randomness of the number of collisions experienced by each particle, the range and the reduced energy fluctuate around the average values. This fluctuation is called straggling of energy loss or of range. (The shape of the curve in Fig. 1 for $x > 22$ mm is determined by range straggling.) SEE ALPHA PARTICLES.

Charge state. At high velocities, an ion usually has the full charge ze of the nucleus. As soon as v drops below the velocity $u_K \cong zc/137$ of the K-shell electrons, electrons in the absorber will be attracted into K-shell orbits, thus reducing the total charge of the ion to a value $z^* = z - 1$ or $z - 2$. As the velocity drops further, more and more electrons will be attached to the ion; but since some of these electrons will be lost or gained as successive collisions take place, z^* must be considered as an average value which changes with v (see **Fig. 2**).

Interactions. In traveling through matter, charged particles interact with nuclei, producing nuclear reactions and elastic and inelastic collisions with the electrons (electronic collisions) and with entire atoms of the absorber (atomic collisions). Usually, in its travel through matter a charged particle makes few or no nuclear reactions or inelastic nuclear collisions, but many elec-

Table 2. Calculated reduced average energy at various depths in copper of protons with $T = 144$ MeV*	
x/mm	T_1/MeV
1	141
5	127
10	107
15	85
20	57

*From *American Institute of Physics Handbook*, 3d ed., McGraw-Hill, 1972.

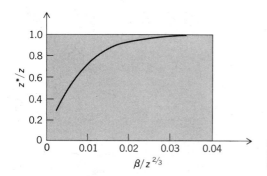

Fig. 2. Average charge z^* of a particle of velocity $v = \beta c$. (After *American Institute of Physics Handbook*, 2d ed., McGraw-Hill, 1963)

tronic and atomic collisions. The average distance between successive collisions is called the mean free path, λ. In solids, it is of the order of 10 cm (4 in.) for nuclear reactions. It ranges from the diameter of the atoms (about 10^{-10} m) to about 10^{-7} m for electronic collisions. The mean free path, λ, depends on the properties of the particle and, most importantly, on its velocity.

Nuclear interactions. In nuclear reactions [for example, $Be(d,n)B$] the incident particle is removed from the beam. Therefore, a reduction in the fluence (number of particles in the beam per cm^2) will be observed. (The decrease in the number of particles in Fig. 1 for x less than about 22 mm (0.88 in.) is due to nuclear interactions.) Such an attenuation can be described by Eq. (1),

$$\frac{N}{N_0} = e^{-x\Sigma} = e^{-x/\lambda} \qquad (1)$$

where $e = 2.71828\ldots$, $N_0 =$ initial particle fluence at $x = 0$, $N =$ particle fluence at x, and $\Sigma =$ probability for an interaction to take place per centimeter of absorber. The mean free path is $\lambda = 1/\Sigma$; Σ is equal to $n\sigma$, where $\sigma =$ cross section per atom for nuclear reactions in square centimeters; and $n =$ number of atoms per cubic centimeter of absorber material. SEE NUCLEAR REACTION; SCATTERING EXPERIMENTS (NUCLEI).

A rough estimate of the cross section for particles with $T/A_m > 10$ MeV can be obtained from Eq. (2). For thin absorbers, $N/N_0 \cong 1 - x/\lambda + \cdots$.

$$\sigma = 5 \times 10^{-26} \text{ cm}^2 \, (A_m^{1/3} + A^{1/3})^2 \qquad (2)$$

An important nuclear interaction is Coulomb (or Rutherford) scattering: because both the incident particles and the nuclei of the absorber atoms have electric charge with values ze and Ze respectively, a change in the direction of motion of the particles will take place during the passage of the particles near the nuclei. The total cross section for this process is only slightly less than the cross-sectional area of the total atom. Usually, though, the angular deflection is much less than 0.01°, and only multiple scattering, the compounding of collisions with many atoms, will cause noticeable total deflections. If an observation of a very fine beam of particles were made along the direction of motion, the scattering events would be seen as small lateral displacements in random directions, and the final lateral displacement would be their vector sum. Although very few particles experience no deflection, the most probable location of the particles is still on the original line of the beam.

Bremsstrahlung. If a charged particle is accelerated, it can emit photons called bremsstrahlung. This process is of great importance for electrons as well as for heavy ions with $T \gg Mc^2$. It is used extensively for the production of x-rays in radiology. Electrons circulating in storage rings emit large numbers of photons with energies (100–1000 eV) not readily available from other sources.

Atomic collisions. At low velocities it may be convenient to consider separately collisions in which most of the energy loss is given as kinetic energy to a target atom. Usually, electronic excitation, electron rearrangements, and possibly ionization accompany this process. The term

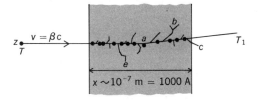

Fig. 3. Energy loss by heavy charged particle. No details shown for energy losses of secondary electrons (delta rays). Rutherford scattering at point a. At point b, delta ray experiences collision that results in tertiary delta ray. At c, delta ray escapes from absorber.

key:

⊤ ionization event
•— excitation event

"nuclear collision" is used by some scientists. No simple quantitative description of atomic collisions is available.

Electronic collisions. The interaction and energy transfer (see **Fig. 3**) between the charged particle and the electrons are caused by the Coulomb force. In general, except in tenuous plasmas, electrons are bound. In gases, all electrons are bound to individual atoms or molecules in well-defined orbits. For these isolated molecules (henceforth, atoms will be included with molecules), electrons can be moved into other bound orbits (excitation) requiring a well-defined energy ϵ_2. Another possibility is the complete removal of the electron from the atom (ionization) requiring an energy $\epsilon \geq I$, where I is the ionization energy for the particular electron. The secondary electron, which is called a delta ray, will have kinetic energy $K = \epsilon - I$. In both processes, the charged particle will lose energy; the energy loss is ϵ_e or ϵ, respectively. Also, it will be deflected very slightly. However, the change in direction is so small that it does not show in Fig. 3; the larger deflection caused by Rutherford scattering at point a is visible. SEE DELTA ELECTRONS.

In liquids and solids, only the inner electrons are associated with a specific nucleus (in aluminum metal the K- and L-shell electrons). Excitation and ionization processes for these electrons are very similar to those in free molecules. The outer electrons are either associated with several neighboring nuclei (nonconducting materials) or, in metals (in Al, the three M-shell electrons), form a plasma-like cloud. Collective or plasma excitations ($\epsilon \cong 20$ eV) take place with high probability, but direct ionization ($\epsilon \gg 20$ eV) also occurs. Because of the requirements of momentum conservation, the maximum energy loss which can occur is given by $\epsilon_M \approx 2mv^2 = 2mc^2\beta^2$ for particles heavier than electrons.

The probability for energy losses ϵ by the incident particle is described by the energy loss spectrum $w(\epsilon)$. Theoretical values have been calculated by H. Bethe. An energy loss spectrum for heavy charged particles in adenine ($C_5N_5H_5$) is shown in **Fig. 4**.

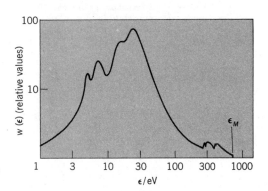

Fig. 4. Schematic single-collision spectrum $w(\epsilon)$ for heavy charged particles in adenine ($C_5N_5H_5$).

The structure between 3 and 30 eV relates to the outer electrons. Similar structures have been observed for many solids (including metals). Excitation and ionization of the K-shell of C (above 280 eV) and N (above 400 eV) cause further structure. The average energy loss per collision is defined by Eq. (3).

$$\langle \epsilon \rangle = \int \epsilon \, w(\epsilon) \, d\epsilon \Big/ \int w(\epsilon) \, d\epsilon \tag{3}$$

Statistics of energy loss. The total energy loss δ of a particle traveling through matter is the sum of the energy losses ϵ_i in each collision: $\delta = \Sigma \epsilon_i = \epsilon_1 + \epsilon_2 + \epsilon_3 + \epsilon_4 + \cdots \epsilon_v$, where v collisions have occurred, each with a probability given by $w(\epsilon)$ (see Fig. 4). If a large number of particles are observed, they will experience on the average $q = \langle v \rangle$ collisions (q is not an integer), and an average energy loss $\Delta = \langle \delta \rangle = q \langle \epsilon \rangle$, as long as collisions are uncorrelated with another. The number of collisions is distributed according to a Poisson distribution; the fraction $P(v)$ of particles having experienced exactly v collisions is given by Eq. (4). The distribu-

$$P(v) = \frac{q^v}{v!} e^{-q} \tag{4}$$

tion function for energy losses is called a straggling function $f(\Delta, x)$. Examples are given in **Figs. 5** and **6**. For fairly thick absorbers, $f(\Delta, x)$ is approximately a gaussian of width proportional to $\sqrt{x}$.

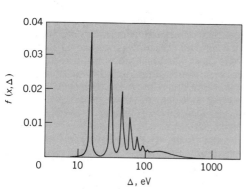

Fig. 5. Calculated straggling curve $f(x, \Delta)$ for 20-MeV protons incident on Al absorber of thickness 5.8×10^{-8} m. Spikes represent multiples of the "plasma loss" at 15 eV. (*After H. Bichsel and R. Saxon, Comparison of calculational methods for straggling in thin absorbers, Phys. Rev., A11:1286–1296, 1975*)

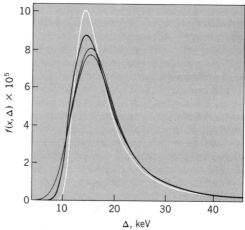

Fig. 6. Calculated straggling curve $f(x, \Delta)$ for 20-MeV protons incident on Al absorber of thickness 3.71×10^{-6} m. Various theoretical calculations are presented. (*After H. Bichsel and R. Saxon, Comparison of calculational methods for straggling in thin absorbers, Phys. Rev., A11:1286–1296, 1975*)

In many applications, the details of energy loss are not important, and a knowledge of the mean or average energy loss Δ is sufficient. If the total collision cross section per atom $w_t = \int w(\epsilon) d\epsilon$ is known (w_t in cm^2), the mean number of collisions is given by $q = xnw_t$, and the mean energy loss by Eq. (5). The quantity stopping power S thus is defined by Eq. (6), in MeV/cm.

$$\Delta = xn \langle \epsilon \rangle w_t \equiv xS \tag{5}$$

$$S \equiv n \langle \epsilon \rangle w_t \tag{6}$$

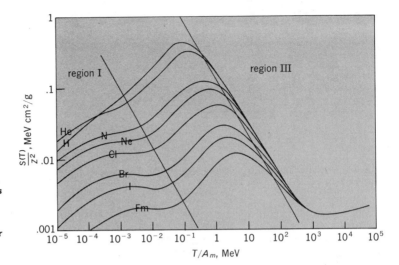

Fig. 7. S(T) for heavy ions in Al. (After L. C. Northcliffe and R. F. Schilling, Range and stopping power tables for heavy ions, Nucl. Data Tables, A7:233, 1970)

Since knowledge of w_t and $\langle \epsilon \rangle$ is not extensive, S is frequently determined in experimental measurements in which a beam passes through an absorber, and S is calculated from Eq. (7)

$$S = \lim_{x \to 0} \frac{\Delta}{x} = \lim_{x \to 0} \frac{T - \langle T_1 \rangle}{x} \tag{7}$$

Equation (5) is valid only if x is much smaller than the mean range, $R(T)$; otherwise, S varies significantly as the particle loses energy in the absorber. If x is not small the following procedure can be used to obtain $\langle T_1 \rangle$, provided that $R(T)$ has been tabulated: Find $R(T)$ in the range table, calculate $y = R(T) - x$, and then find the energy T_1 corresponding to y in the range table.

Stopping power. The stopping power is a function of the velocity v of the incident particle, its effective charge z^*, and the absorber material. **Figure 7** shows $S(T)$ for heavy ions of various elements in aluminum. S is expressed in MeV cm^2/g. This can be converted to MeV/cm by the formula S(MeV/cm) = ρS(MeV cm^2/g), where ρ is the density of absorber material in g/cm^3. Atomic collisions dominate in region I. For region III, S can be calculated by using the theoretical Bethe expression, Eq. (8), where I_A = average excitation energy of absorber (approximately

$$S = \frac{.30708}{\beta^2} \frac{Z}{A} \left[\ln \frac{2mc^2\beta^2}{I_A(1-\beta^2)} - \beta^2 - \frac{C}{\beta^2} - d(\beta) \right] \cdot (z^*)^2 \left[1 + G(z^*,\beta) \right] \text{ MeV cm}^2/\text{g} \tag{8}$$

$Z \times 10^{-5}$ MeV); C = shell correction constant; $d(\beta)$ = density correction, important for $T > Mc^2$; and $G(z^*,\beta)$ = correction due to the second Born approximation, important only for $\beta^2 < 0.01$.

A simpler approximate expression valid to about 10% for $z \leq 10$ in the same region is given by Eq. (9).

$$S = \frac{2.6 z^2}{\beta^{1.66} Z^{.25}} \text{ MeV cm}^2/\text{g} \qquad 0.1 < \beta < 0.88 \tag{9}$$

Range. A good approximation to the mean range $R(T)$ can be calculated from S using Eq. (10).

$$R(T) = \int_0^T \frac{d\tau}{S(\tau)} \tag{10}$$

Usually, a numerical integration is performed to obtain R from a table of S or the Bethe

Table 3. Range of validity of range formula

Range of kinetic energy T	Range of β	r	s
$10 \leq T < 90$ MeV	$.145 < \beta < .4$	$.116$ g/cm^2	1.84
$90 < T < 400$ MeV	$.4 < \beta < .7$	$.275$ g/cm^2	2.33
$400 < T < 1000$ MeV	$.7 < \beta < .88$	$.532$ g/cm^2	3.34

formula. An approximation formula is Eq. (11). The range of validity for this range formula is given in **Table 3**.

$$R = r \cdot \beta^s \cdot \frac{Z^{.25}}{z^2} \quad z \leq 10 \tag{11}$$

Channeling. In absorbers consisting of a single crystal, it has been found that the energy loss will be reduced if the direction of the particle beam coincides with certain preferred alignments of the crystal. It is believed that the particles travel through "open spaces" in the crystal, thus suffering a succession of collisions with relatively small energy losses and angular deflections, and tending to stay in the preferred direction (channel). Thus, even if the total number of collisions q were unchanged, the average energy loss per collision (ϵ) would be reduced, and the total energy loss Δ would be less.

Ionization. The secondary electrons of energies K produced in ionizing electronic collisions will travel through the absorber and will also suffer various collisions, producing further energetic electrons, and so on. This process continues until the electrons have energy $K < I$. It has been found experimentally that the total average number j of ions produced in this way (in the distance x) is proportional to the energy Δ lost by the particle: $j = \Delta/\omega$.

The constant of proportionality ω introduced in this definition has values between 20 eV and 45 eV for gases, about 3.6 eV for silicon, and 2.96 eV for germanium. It is almost, but not exactly, independent of particle energy and type. If particles lose all their energy in the material, the total ionization J is related to the kinetic energy T: $J = T/W$. The relation between W and ω is: $T/W = \int dT/\omega$. $W = \omega$ only if ω is exactly independent of energy.

The ionization j along the path of a single particle increases with the distance traveled, approximately at the same rate as S. For a beam of particles, straggling also influences the total ionization at a given location. The function obtained from the combination (actually the convolution) of both effects is the Bragg curve.

Electrons. Although the interactions discussed earlier all occur for electrons, there are some major differences between electron beams and beams of heavier particles. In general, the path of an electron will be a zigzag. Angular deflections in the collisions will frequently be large. Electron beams therefore tend to spread out laterally, and the number of primary electrons in the beam at a depth x in the absorber decreases rapidly.

Since it is not possible to distinguish individual electrons, it is customary in a collision between two electrons to consider the one emerging with the higher energy as the primary electron. The maximum energy loss in a collision therefore is $mv^2/4$ (for $T \ll mc^2$). The stopping power expression therefore is somewhat different for electrons. SEE BETA PARTICLES.

Biological effects. In general, for the same dose (the energy deposited per gram along the beam line) heavy charged particles will produce, because of their higher local ionization, larger biological effects than electrons (which frequently are produced by x-rays).

Observation. The most direct method of observing a beam of charged particles is to observe the electric current that they form (any flow of electric charges is an electric current). In all accelerators (such as cyclotrons, Van de Graaff generators, linear accelerators, and x-ray tubes) the beam current is measured as a primary monitor of the proper operation of the machine. It is not possible to identify the type of particles with current measurements (except for their electric charge). SEE PARTICLE DETECTOR.

Devices using ionization. If an electric field **E** is applied to an absorber irradiated with charged particles, the ions and electrons produced will travel in the direction **E,** and the resulting ionization current can be measured (electronic amplification usually is needed). If an oscilloscope is available, the ionization J associated with a single particle can be observed, and the energy (or energy loss) of the particle can be calculated from $T = JW$. Semiconductor detectors (chiefly silicon and germanium) are used extensively for this purpose, but gas-filled ionization chambers have also been used. Cloud chambers and bubble chambers use this principle, but individual ions or clumps of ionization are observed visually. Proportional counters, spark chambers, and Geiger-Müller tubes also operate on the same principle; but in the latter two only the presence of a particle is indicated, and J is not related to T. SEE BUBBLE CHAMBER; CLOUD CHAMBER; IONIZATION CHAMBER; SPARK CHAMBER.

Devices using excitation. The excited state of energy ϵ_e produced in excitation can decay with the emission of light (luminescence or scintillation). Early observations of radioactivity were made with this method (using ZnS screens and visual observation, usually with microscopes), and the method is used extensively with luminescent dials (for example, on wristwatches). The light emitted usually is detected and amplified with photomultipliers. Again, the energy T can be measured. Scintillators used are NaI(Tl), CsI, anthracene, stilbene, and various solids and liquids. SEE LIQUID SCINTILLATION DETECTOR; SCINTILLATION COUNTER.

A more indirect use of excitations and ionizations is in "chemical" devices (such as photographic emulsions, $FeSO_4$ solutions, thermoluminescence).

Applications. Electron beams are used in the preservation of food. In medicine, electron beams are used extensively to produce x-rays for both diagnostic and therapeutic (cancer irradiation) purposes. Also, in radiation therapy, deuteron beams incident on Be and ^{3}H targets are used to produce beams of fast neutrons, which in turn produce fast protons, alpha particles, and carbon, nitrogen, and oxygen ions in the irradiated tissue. Energetic pion (~ 100 MeV), proton (~200 MeV), alpha (~ 1000 MeV), and heavier ion beams can possibly be used for cancer therapy. The existence of a Bragg peak for these particles promises improvements in the dose distribution within the human body.

The well-defined range of heavy ions permits their implantation at given depths in solids (this is useful in the production of integrated circuits). Radiation damage studies are performed with charged particles in relation to development work for nuclear fission and fusion reactors.

Charged particle beams are used in many methods of chemical and solid-state analysis. Nuclear activation analysis can be performed with heavy ions. Isotopes can be produced with fast charged ions. SEE SUPERTRANSURANICS; TRANSURANIUM ELEMENTS.

Bibliography. F. H. Attix, *Introduction to Radiological Physics and Radiation Dosimetry*, 1986; H. Bichsel, in *American Institute of Physics Handbook*, 3d ed., 1972; R. D. Evans, *The Atomic Nucleus*, 1955; M. Inokuti, *Rev. Mod. Phys.*, 43:297–347, 1971; G. F. Knoll, *Radiation Detection and Measurement*, 1979; M. Mladjenovic, *Radioisotope and Radiation Physics*, 1973; C. G. Orton (ed.), *Radiation Dosimetry: Physical and Biological Aspects*, 1986.

ELECTRON CAPTURE
MCALLISTER H. HULL

The process in which an atom or ion passing through a material medium either loses or gains one or more orbital electrons. In the passage of charged particles (defined here as nuclei having more or less than Z atomic electrons, where Z is the atomic number) through matter, the capture (and loss) of electrons is an important process in the slowing down of the particles and therefore has a strong influence on their range. Thus a neutral hydrogen atom loses only about half as much energy per centimeter as the positively charged proton in passing through matter consisting of light elements.

For the ordinary charged particles (alpha particles and protons) the capture process is important only at low energies, when the particle velocity is of the order of electron velocities in the stopping material, and thus is important at the end of the range. For fission fragments, however, which initially have a large excess of positive charges, electron capture occurs immediately and continues throughout the slowing-down process. This fact causes the energy-loss mechanisms at the latter part of the range to be different for fission fragments and protons or alpha particles. The heavy ions (nuclei of oxygen, argon, and so forth with all atomic electrons stripped away) now available are intermediate in mass between fission fragments and the light particles and have higher velocity than fission fragments. Thus their energy loss is relatively unaffected by electron capture in the early part of the range, but in later stages this process has important consequences. SEE NUCLEAR FISSION.

The nuclear capture of electrons (K capture) occurs by a process quite different from atomic capture and is in fact a consequence of the general beta interaction. This general interaction includes β^- decay (the oldest known beta transformation and hence the name), β^+ decay (or positron decay), and K capture, the latter so called because the electron captured by the nucleus is taken from the K shell (the shell nearest the nucleus) of atomic electrons. The probability of occurrence of electron capture by the nucleus obviously depends on the amount of time the electrons spend at the nucleus, that is, on the size of the electron wave function at the nuclear center. Since to a very good approximation only electrons with zero orbital angular momentum have a wave function that is finite at the center, capture is not expected from any but the K-shell. However, a second-order process can occur, in which (to speak pictorially and thus somewhat imprecisely) an s electron (from the K shell) is captured with the simultaneous transition of a p electron (from the L shell) to the K shell with the emission of gamma radiation. This differs from ordinary x-radiation following K capture by the fact that energy is not conserved in the transition (it must, of course, be conserved in the whole process). Since K electrons spend more time in the nucleus for large-Z nuclei than for small, K capture is more probable im heavier nuclei. The second-order process (called L capture) is even more strongly Z-dependent and actually controls the shape of the gamma-ray spectrum for very heavy nuclei. SEE RADIOACTIVITY.

What has been concluded so far depends on the atomic (that is, electromagnetic) interactions. But the process itself is a result of the beta interaction, which is between the electron field and the nucleon field. This weak interaction (so called because the processes involving the interaction take place in times that are long on the nuclear time scale), which couples electrons (or positrons) with nuclei, gamma rays, and neutrinos, has been the subject of increased study because it has been shown to demonstrate nonconservation of parity. The electron and positron are identical except for electromagnetic interactions; thus a nucleus which is energetically capable of K capture will also usually be capable of positron emission, and the two processes do indeed compete in several nuclei. SEE PARITY; QUANTUM FIELD THEORY.

DELTA ELECTRONS
D. ALLAN BROMLEY

Energetic electrons ejected from atoms in matter by the passage of ionizing particles. In every primary ionizing collision between a charged particle and an atom, one or more electrons are ejected. Delta electrons are, by definition, that small fraction of these emitted electrons having energies which are large compared to the ionization potential. The name is a traditional one—comparable to alpha particles, for energetic helium nuclei, and beta particles, for energetic electrons emitted in radioactive decays. SEE ALPHA PARTICLES; BETA PARTICLES; CHARGED PARTICLE BEAMS.

The energy W of the delta electron emitted at angle θ in the laboratory is given by Eq. (1),

$$W = 2mv^2 \cos^2 \theta \qquad (1)$$

where v is the velocity of the incident charged particle, and m is the electron mass. The electrons in the forward direction are the most energetic and have an energy near $4(m/M)E$, where M and E are the mass and energy of the incident charged particle. The cross section for the ejection of a delta electron in the energy range between W and given by Eq. (2), where e is the electron

$$d\sigma = \frac{2\pi e^4 z^2}{mv^2} \cdot \frac{dW}{W^2} \qquad (2)$$

charge and ez is the charge of the incident particle, and the cross section for finding an electron

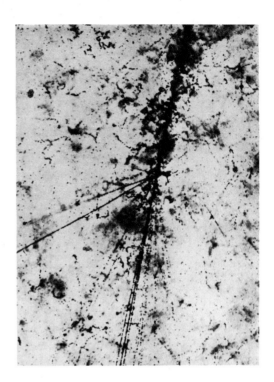

Nuclear emulsion showing iron nucleus ($Z = 26$) in cosmic radiation entering from upper right and producing large numbers of delta electrons, giving its track a heavy, "hairy" appearance.

between θ and $\theta + d\theta$ is given by Eq. (3). The number of electrons emitted per centimeter of path

$$d\sigma = \frac{2\pi e^4 z^2}{m^2 v^4} \cdot \frac{\sin \theta \, d\theta}{\cos^3 \theta} \tag{3}$$

length is obtained by multiplying these cross sections by the number of electrons per cubic centimeter in the stopping material.

Delta electrons are responsible for the "hairy" appearance of charged particle tracks when they are observed in cloud chambers or in photographic emulsions (see **illus**.). In studies of superhigh-energy particles in cosmic radiation and from the highest-energy accelerators, observation of the number of delta electrons per centimeter of path length has been shown to lead to a reliable determination of the charge of the energetic particle.

ELEMENTARY PARTICLES

Elementary particle	196
Antimatter	212
Lepton	214
Electron	215
Positron	217
Neutrino	217
Quarks	220
Photon	228
Gluons	228
Intermediate vector boson	232
Graviton	236
Magnetic monopoles	237
Tachyon	239

ELEMENTARY PARTICLE
Charles J. Goebel

A particle which cannot be described as compound, in the present state of knowledge. The elementary particles are thus the fundamental constituents of matter.

The known elementary particles are listed in **Table 1**. The graviton, the quantum of the gravitational field, has been omitted from Table 1 since it plays no role in high-energy particle physics: it is firmly predicted by theory, but the prospect of direct observation is exceedingly remote. The heavy vector bosons $W\pm$ and Z^0 were observed in 1983; their properties had been deduced from the weak interactions, for which they are responsible. Gluons and quarks are never seen as free particles; this phenomenon is known as confinement. Hadrons, strongly interacting particles (**Table 2**), are compounds of quarks and gluons; essentially, mesons are composed of a quark-antiquark pair, $q\bar{q}$, and baryons are three quarks, qqq, bound together by the exchange of gluons. *See* Baryon; Gluons; Graviton: Hadron; Intermediate vector boson; Meson; Quarks.

The gauge bosons, γ, g, $W^\pm$, and Z^0, are the quanta of gauge fields. Their fundamental couplings are indicated in Table 1. (The couplings have each been stated in one form, but by the

Table 1. The elementary particles[a]

Gauge bosons $J^P_C = 1^-_-$ Self-conjugate except $\overline{W^+} = W^-$.

Name	Symbol	Charge[b]	Mass, GeV	Couplings
Photon	γ	0	0	$A \Rightarrow \gamma A$
Gluon[c]	g	0	0	$A \Rightarrow g A'$
Weak bosons				
Charged	$W^\pm$	± 1	85	$U \Rightarrow W^+ D$
Neutral	Z^0	0	95	$A \Rightarrow Z^0 A$

Fermions $J = \tfrac{1}{2}$ All have distinct antiparticles, except perhaps the neutrinos.

Name	Charge[b]	Symbol	Mass, GeV	Symbol	Mass, GeV	Symbol	Mass, GeV
Leptons							
Neutrinos	0	ν_e	$<10^{-7}$	ν_μ	$<.0006$	ν_τ	$<.5$
Charged leptons[d]	-1	e	.0005	μ	.106[e]	τ	1.78[e]
Quarks[c]							
Up type	$\tfrac{2}{3}$	u	.005	c	1.4	t	≈ 40
Down type	$-\tfrac{1}{3}$	d	.01	s	.15	b	4.8

[a]The graviton, with $J^P_C = 2^+_+$, has been omitted, since it plays no role in high-energy particle physics.
[b]In units of the proton charge.
[c]The gluon is a color SU_3 octet {8}; each quark is a color triplet {3}. These colored particles are confined constituents of hadrons; they do not appear as free particles.
[d]Any further charged leptons have mass greater than 15 GeV.
[e]The μ and τ leptons are unstable, with the following mean life and principal decay modes (branching ratios in %):

$\mu \quad \tau_\mu = 2.2 \times 10^6$ s $\qquad e\bar{\nu}_e\nu_\mu$ 100

$\tau \quad \tau_\tau = 3 \times 10^{-13}$ s $\qquad \mu\bar{\nu}_\mu\nu_\tau$ 18
$\qquad\qquad\qquad\qquad\qquad e\bar{\nu}_e\nu_\tau$ 17
$\qquad\qquad\qquad\qquad\quad$ (hadrons)$^-\nu_\tau$ 65

principle of line reversal there may be other equivalent forms; for instance, $U \Rightarrow W^+D$ means also $W^- U \Rightarrow D$, $W^- \Rightarrow \bar{U}D$, and so forth.) There are also couplings between three or four gauge bosons. The photon γ, the quantum of the electromagnetic field, is coupled to charge: that is, the coupling $A \Rightarrow \gamma A$, the amplitude of a process in which a γ is emitted by the elementary particle A, is proportional to Q_A, the charge of A. Similarly, the gluon g, the quantum of the gluon (or color) field, is coupled to color, discussed below. Of the elementary particles, only the quarks and gluons carry color. The different kinds of quarks, called flavors of quark, all have the same color (that is, they are all color SU_3 triplets) and hence are coupled equally to the gluon. This flavor independence of the gluon coupling results in the flavor SU_N symmetries of the hadrons and their strong interactions, discussed below. The coupling $A \Rightarrow gA'$ vanishes unless A and A' belong to the same color multiplet, that is, unless A and A' are both gluons or both the same flavor of quark. Thus, flavor is conserved in strong interactions. The term flavor may be used in a broader sense in which, for example, ν_e and e are two lepton flavors. In this broader sense, the flavor of a quark or lepton changes when it emits or absorbs a $W^{\pm}$, $U \Rightarrow W^+D$, as discussed below. The exchange of a $W^{\pm}$ between two particles, changing their flavors, is a weak interaction. The exchange of a Z^0 is a neutral-current weak interaction. *See* COLOR; FLAVOR; LEPTON.

Experimental evidence for confinement is found, for example, in inelastic electron-proton scattering at high energy, in particular, in deep inelastic scattering, in which the electron loses a sizable fraction of its energy. The observed cross section shows that the charge of the proton is carried by pointlike (radius less than 10^{-1} femtometer) particles of small mass. But no such particles are seen in the final state of this process, or indeed of any other high-energy collision. What is seen is a narrow shower of hadrons. The interpretation is that the electron scatters off one of the quarks in the proton and gives it a large energy and momentum, the quark responding as though it were a free particle of mass much less than 100 MeV (consistent with the masses of the u and d quarks in Table 1). Later, through the production of quark-antiquark pairs, the energy and momentum of the struck quark is divided up among a number of hadrons, mostly pions, a process called hadronization or fragmentation of the quark, which is to be distinguished from the decay of a free particle. The string model of confinement, discussed below, gives a concrete picture of this process. The resulting shower of hadrons whose total momentum vector is roughly that of the original quark is called a hadronic jet (like a jet of water which breaks up into a spray of droplets). Such jets are the closest available phenomenon to the actual observation of a quark as a free particle.

General properties. In relativistic quantum mechanics, a particle is a system of definite mass and spin. Thus an O_2 molecule, say, in a definite energy level is just as much a particle as is an electron or a pi meson; the concept of particle has nothing to do with elementarity or structure. An unstable particle has a complex mass m, whose imaginary part Im m is equal to $-\frac{1}{2}\Gamma$. Here $\Gamma = \hbar/\tau$, where τ is the mean life of the particle and $\hbar$ is Planck's constant h divided by 2π. The spin, the intrinsic part of the angular momentum of a particle (the part which does not vanish in the rest frame), is a half integer, in units of $\hbar$. A particle is also characterized by other quantum numbers, such as parity and lepton number, which are eigenvalues of further symmetry operators that commute with mass and spin. *See* SPIN; SYMMETRY LAWS.

Associated fields. An individual particle is never unique; that is, any number of identical copies can exist as well. This means that a field whose quantum is the particle can be defined. There is thus no distinction between kinds of fields and kinds of particles: one can speak interchangeably of the electromagnetic field or of the photon particle, and of the electron or of the electron field. Each particle is either a boson or a fermion, according to whether its spin is an integral or a half-odd-integral multiple of $\hbar$. In the macroscopic world, where classical physics holds, fermions can only appear as particles, but bosons can appear either as particles or as fields if many identical bosons are put into the same mode. *See* QUANTUM FIELD THEORY.

Antiparticles. To each kind of particle there corresponds an antiparticle, or conjugate particle, which has the same mass and spin, also has the same space parity and charge parity (quantum numbers which have the values + or − and are conserved multiplicatively), belongs to the conjugate representation (multiplet) of internal symmetry (for example, an antiquark belongs, to an anti-triplet $\{\bar{3}\}$ of color SU_3), and has opposite values of charge, I_3, strangeness, and so forth (quantum numbers which are conserved additively). For instance, the antielectron is the positron. Particles for which the antiparticle is the same as the particle are called self-conjugate; examples

Table 2. The hadrons (strongly interacting particles)

	Hadronic quantum numbers†	Symbol (mass, MeV)	J_C^P	Width, MeV	Decay products and branching ratios, %
Mesons					
		$\pi(138)$	0_-^-	Hadronically stable	
		$\rho(775)$	1_-^-	160	$\pi\pi$ 100
		$\delta(980)$	0_+^+	50	$\eta\pi$, $K\bar{K}$
		$A_1(1100-1300)$	1_+^+	300	$\rho\pi$ 100?
	$I = 1$	$B(1230)$	1_+^+	130	$\omega\pi$ 100
		$A_2(1315)$	2_+^+	100	$\rho\pi$ 70, $\eta\pi$ 15, $\omega\pi\pi$ 10, $K\bar{K}$ 5
		$\rho'(1600)$	1_-^-	300	4π 85, $\pi\pi$ 15
		$A_3(1660)$	2_-^-	200	$f\pi$ 60, $\rho\pi$ 30
		$g(1690)$	3_-^-	200	$\pi\pi$ 25, $\omega\pi$, $A_2\pi$, $\rho\rho$, ...
		$\eta(549)$	0_-^-	Hadronically stable	
		$\omega(782)$	1_-^-	10	$\pi\pi\pi$ 90, $\pi\gamma$ 9, $\pi\pi$ 1
		$\eta'(958)$	0_-^-	0.3	$\eta\pi\pi$ 66, $\rho^0\gamma$ 30, $\omega\gamma$ 3, $\gamma\gamma$ 2
		$S^*(980)$	0_+^+	40	$K\bar{K}$, $\pi\pi$
		$\phi(1020)$	1_-^-	4	K^+K^- 49, K_LK_S 35, $\rho\pi$ 15, $\eta\gamma$ 2
		$H(1190)$	1_+^+	300	$\rho\pi$
		$f(1275)$	2_+^+	180	$\pi\pi$ 83, $K\bar{K}$, $\pi\pi\pi\pi$ 3
		$\zeta(1275)$	0_-^-	70	$\eta\pi\pi$
		$D(1285)$	1_+^+	30	$\eta\pi\pi$ 50, $\rho\pi\pi$ 40, $K\bar{K}$ 10
		$\epsilon(1300)$	0_+^+	200–400	$\pi\pi$ 90, $K\bar{K}$ 10
		$E(1420)$	1_+^+	50	$K^*\bar{K} + \bar{K}^*K$, $\delta\pi$?
		$\iota(1440)$	0_-^-	50	$\delta\pi$
		$f'(1515)$	2_+^+	70	$K\bar{K}$ 100?
		$\theta(1650)$	2_+^+	200	$\eta\eta$
		$\omega(1665)$	3_-^-	165	$\rho\pi$, $B\pi$
		$h(2040)$	4_+^+	150	$\pi\pi$, $K\bar{K}$
		$r(2510)$	6_+^+	250	$\pi\pi$
	$I = 0$	$\eta_c(2980)$	0_-^-	<20	$\eta\pi\pi$, $K\bar{K}\pi$, 4π, $K\bar{K}\pi\pi$, ...
		$\psi(3097) = \psi/J$	1_-^-	0.06	mesons 86, $\mu\bar{\mu}$ 7, $e\bar{e}$ 7
		$\chi(3415) = \chi_0$	0_+^+		$\pi\pi$, $K\bar{K}$, 4π, ..., $J\gamma$?
		$\chi(3510) = \chi_1$	1_+^+		$J\gamma$ 30, 4π, ...
		$\chi(3556) = \chi_2$	2_+^+		$J\gamma$ 15, $\pi\pi$, $K\bar{K}$, 4π, ...
		$\eta(3590) = \eta_c'$	0_-^-		$\eta\pi\pi$, $K\bar{K}\pi$, 4π, $K\bar{K}\pi\pi$, ...
		$\psi(3686) = \psi'$	1_-^-	0.2	$J\pi\pi\pi$ 50, $\chi(3413)\gamma$ 7, $\chi(3508)\gamma$ 7, $\chi(3554)\gamma$ 7, $J\eta$ 4, $\mu\bar{\mu}$ 1, $e\bar{e}$ 1
		$\psi(3770) = \psi''$	1_-^-	25	$D\bar{D}$ 100
		$\psi(4030)$	1_-^-	50	hadrons
		$\psi(4160)$	1_-^-	80	hadrons
		$\psi(4415)$	1_-^-	~40	hadrons
		$\Upsilon(9460) = \Upsilon$	1_-^-	0.04	$\mu^+\mu^-$ 3, e^+e^-, hadrons
		$\chi_b(9875) = \chi_{b0}$	0_+^+		
		$\chi_b(9896) = \chi_{b1}$	1_+^+		
		$\chi_b(9916) = \chi_{b2}$	2_+^+		
		$\Upsilon(10024) = \Upsilon'$	1_-^-	0.02	$\mu^+\mu^-$ 2, e^+e^- 2, hadrons
		$\chi_b(10233) = \chi_{b0}'$	0_+^+		
		$\chi_b(10253) = \chi_{b1}'$	1_+^+		
		$\chi_b(10271) = \chi_{b2}'$	2_+^+		
		$\Upsilon(10355) = \Upsilon''$	1_-^-	0.01	$\mu^+\mu^-$, e^+e^-, hadrons
		$\Upsilon(10580) = \Upsilon'''$	1_-^-	15	$B\bar{B}$ 100
		$K(495)$	0^-	Hadronically stable	
		$K^*(892) = K^*$	1^-	50	$K\pi$ 100
		$Q_1(1280)$	1^+	120	$K\rho$ 40, $\kappa\pi$ 30
	$I = \frac{1}{2}$ $s = +1$	$Q_2(1400)$	1^+	150	$K^*\pi$ 95
		$K^*(1435)$	2^+	100	$K\pi$ 49, $K^*\pi$ 27, $K^*\pi\pi$ 11, $K\rho$ 7, $K\omega$ 4, $K\eta$ 3
		$\kappa(1500)$	0^+	250	$K\pi$
		$L(1770)$	2^-	200	$K^*(1435)\pi$, $K^*\pi$, Kf
		$K^*(1780)$	3^-	130	$K\pi\pi(I\rho, K^*\pi)K\pi$ 20
	$I = \frac{1}{2}$ $c = +1$	$D(1866)$	0^-	Hadronically stable	
		$D(2008) = D^*$	1^-	<2	$D\pi$, $D\gamma$
	$I = 0$ $s = +1, c = +1$	$F(1970) = F$	0^-?	Hadronically stable	$\phi\pi$, $\phi\pi\pi\pi$
		$F(2140) = F^*$	1^-?		$F\gamma$
	$I = \frac{1}{2}$ $b = +1$	$B(5275) = B$	0^-	Hadronically stable	$D/\bar{\nu}$ or $D^*/\bar{\nu}$ [$I = e$, μ or τ] 80

ELEMENTARY PARTICLES

Table 2. The hadrons (strongly interacting particles) [cont.]

Hadronic quantum numbers[†]		Symbol (mass, MeV)	J_C^P	Width, MeV	Decay products and branching, %				
Baryons		$N(939) = N$	$\frac{1}{2}^+$	Hadronically stable					
					$N\pi$	$\Delta\pi$	$N\rho$	$N\epsilon$	$N\eta$
		$N(1450)$	$\frac{1}{2}^+$	200	50–65,	23,	7,	7,	18
		$N(1520)$	$\frac{3}{2}^-$	125	55,	23,	19,	<5	
		$N(1535)$	$\frac{1}{2}^-$	150	40,	1,	3,	2,	55
		$N(1650)$	$\frac{1}{2}^-$	150	60,	4–15,	7–21,	<10,	ΛK10, ΣK2–7
		$N(1680)$	$\frac{5}{2}^-$	155	40,	50,	5		
		$N(1680)$	$\frac{5}{2}^+$	130	60,	18,	13,	22	
$I = \frac{1}{2}$		$N(1700)$	$\frac{3}{2}^-$	120	10,	15–40,	<5,	<40,	4
		$N(1710)$	$\frac{1}{2}^+$	120	20,	10–20,	40–65,	15–40,	2–20, ΣK10
		$N(1810)$	$\frac{3}{2}^+$	200	17,	20,	45–70,	20	
		$N(1990)$	$\frac{7}{2}^+$	250	5,				3
		$N(2190)$	$\frac{7}{2}^-$	250	15,				2
		$N(2200)$	$\frac{9}{2}^-$	250	10,				2
		$N(2220)$	$\frac{9}{2}^+$	300	20,				1
		$N(2650)$	$11\frac{1}{2}^-$	400	5				
		$N(3030)$	?	400					
		$\Delta(1232) = \Delta$	$\frac{3}{2}^+$	115	100				
		$\Delta(1620)$	$\frac{1}{2}^-$	140	32,	40,	<50		
		$\Delta(1640)$	$\frac{3}{2}^+$	250	20,	30–45,	<10		
		$\Delta(1700)$	$\frac{3}{2}^-$	200	15,	<50,	40		
		$\Delta(1910)$	$\frac{1}{2}^+$	220	20–25,	<40,			ΣK2–20
$I = \frac{3}{2}$		$\Delta(1920)$	$\frac{5}{2}^-$	200	4–12				
		$\Delta(1920)$	$\frac{5}{2}^+$	250	15,	10–30,	60		
		$\Delta(1950)$	$\frac{7}{2}^+$	240	40,	30,	20		
		$\Delta(2420)$	$11\frac{1}{2}^+$	300	10				
		$\Delta(2850)$	$15\frac{1}{2}^+$?	400					
		$\Delta(3230)$	$19\frac{1}{2}$?	440					
		$\Lambda(1116) = \Lambda$	$\frac{1}{2}^+\{8\}$	Hadronically stable					
					$N\overline{K}$	$\Sigma\pi$			
		$\Lambda'(1405)$	$\frac{1}{2}^-\{1\}$	40		100			
		$\Lambda'(1520)$	$\frac{3}{2}^-\{1\}$	16	46,	42,	$\Lambda\pi\pi$10		
		$\Lambda(1670)$	$\frac{1}{2}^-\{8\}$	40	20,	20–60,	$\Lambda\eta$ 15–35		
		$\Lambda(1690)$	$\frac{3}{2}^-\{8\}$	60	25,	20–40,	$\Lambda\pi\pi$25, $\Sigma\pi\pi$20		
	$I = 0$	$\Lambda(1800)$	$\frac{1}{2}^-\{8\}$	300	25–40				
		$\Lambda(1815)$	$\frac{1}{2}^+\{8\}$	80	60,	10,	$\Sigma^*\pi$5–10		
		$\Lambda(1830)$	$\frac{5}{2}^-\{8\}$	95	<10,	35–75,	$\Sigma^*\pi > 15$		
		$\Lambda(1870)$	$\frac{3}{2}^+\{8\}$	100	15–40,	3–10			
		$\Lambda'(2100)$	$\frac{7}{2}^-\{1\}$	250	30,	5			
$s = -1$		$\Lambda(2110)$	$\frac{5}{2}^+$	200	5–25,	<40,	$N\overline{K}^*(892)$ 20–60		
		$\Lambda(2350)$	$\frac{9}{2}^+$	120	12,	10			
		$\Sigma(1193) = \Sigma$	$\frac{1}{2}^+\{8\}$	Hadronically stable					
					$N\overline{K}$	$\Lambda\pi$	$\Sigma\pi$		
		$\Sigma(1385) = \Sigma^*$	$\frac{3}{2}^+\{10\}$	35		88,	12		
		$\Sigma(1660)$	$\frac{1}{2}^+\{8\}$	100	<30				
		$\Sigma(1670)$	$\frac{3}{2}^-\{8\}$	50	10,	<20,	20–60		
		$\Sigma(1750)$	$\frac{1}{2}^-\{8\}$	75	10–40,	5–20,	<8,	$\Sigma\eta$15–55	
	$I = 1$	$\Sigma(1765)$	$\frac{5}{2}^-\{8\}$	120	41,	14,	1,	$\Lambda(1520)\pi$19, $\Sigma^*\pi$9	
		$\Sigma(1915)$	$\frac{5}{2}^+\{8\}$	100	10	15			
		$\Sigma(1940)$	$\frac{3}{2}^-$	220	<20				
		$\Sigma(2030)$	$\frac{7}{2}^+\{10\}$	180	20,	20,	5–10,	$\Lambda(1520)\pi$15, $\Sigma^*\pi$10	
		$\Sigma(2250)$	$\frac{7}{2}^-$?	100	<10				
		$\Sigma(2455)$	?	120					
		$\Sigma(2595)$	?	200					
		$\Xi(1318) = \Xi$	$\frac{1}{2}^+\{8\}$	Hadronically stable					
$s = -2$		$\Xi(1530)$	$\frac{3}{2}^+\{10\}$	10	$\Xi\pi$100				
$I = \frac{1}{2}$		$\Xi(1820)$	$\frac{3}{2}^-$?$\{8\}$	20	$\Lambda\overline{K}$45, $\Xi(1530)\pi$45, $\Sigma\overline{K}$10				
		$\Xi(2030)$	?	16	$\Sigma\overline{K}$80, $\Lambda\overline{K}$20				
$s = -3$, $I = 0$		$\Omega^-(1672)$	$\frac{3}{2}^+$	Hadronically stable					
$c = +1$	$I = 0$	$\Lambda_c^+(2280)$	$\frac{1}{2}^+$	Hadronically stable					
	$I = 1$	$\Sigma_c(2440)$	$\frac{1}{2}^+$		$\Lambda_c^+\pi$				

[†]The hadronic quantum numbers are *i*-spin magnitude I, the third component of *i*-spin I_3, strangeness s, charm c, bottomness b, and so forth. Each entry in the table represents an *i*-spin multiplet of $2I + 1$ states with $I_3 = -I, -I + 1, \ldots, I$. The values of s, c, b, and so on are specified in the table only if nonzero.

are the photon γ and the neutral pion π^0. The equality of masses implies the equality of lifetimes of particle and antiparticle. Thus the positron is stable; however, in the presence of ordinary matter it soon annihilates with an electron, and thus is not a component of ordinary matter. SEE ANTIPROTON; PARITY; POSITRON.

The conjugation operator, which transforms a particle to an equal-mass antiparticle, is a symmetry operator. A self-conjugate particle is an eigenstate of the operator, and so the eigenvalue of the latter is a quantum number of the particle. If weak interactions are neglected, the conjugation operator can be taken to be C, the charge conjugation operator. The γ and π^0 have the C values $-$ and $+$ respectively. When weak interactions are taken into account, C is not a symmetry operator, but CP, the product of the charge conjugation and space inversion operators, is. Usually the effect of the weak interactions is negligible: for instance, they mix into the π^0 unmeasurably small amounts of channels with $C = -$ and $P = +$. But there are exceptions. Neutrinos interact only through the weak interactions; the effect of C on a neutrino is to make a particle which is uncoupled from the weak interactions, which means that it is unobservable. The operation which produces the observed antineutrino is CP (neglecting violation of invariance under time reversal T). Another example is the K^0: the K^0 is strange and therefore not self-conjugate; that is, $\overline{K^0} \neq K^0$ (the K^0 has quark content $d\bar{s}$ so the $\overline{K^0}$ is $\bar{s}d$). Here, weak interactions cannot be neglected, because in their absence the K^0 and $\overline{K^0}$ are degenerate: the weak interactions determine the eigenstates according to degenerate perturbation theory. The mass eigenstates if nondegenerate must be CP eigenstates (neglecting T violation). They are nondegenerate because the states to which they can decay through the weak interaction are different: a $J = 0$ $\pi\pi$ system has $CP = +$; a $J = 0$ $\pi\pi\pi$ system has $CP = -$. The mass eigenstate, that is, particle, called K_S (short-lived neutral K) decays almost always into $\pi\pi$; the particle called K_L (long-lived neutral K) decays into $\pi\pi\pi$ and also $\pi e \nu$ and $\pi \mu \nu$, about equally. SEE TIME-REVERSAL INVARIANCE.

Strictly, the conjugation operator is CPT, that is, the product of charge conjugation, space inversion, and time reversal; this is because CPT is always a symmetry operator according to the CPT theorem of W. Pauli and G. Lüders, and experiment shows that none of the factors of CPT are. The violation of the CP symmetry (or equivalently violation of T, according to the CPT theorem) is observed only in K_L decay: the K_L decays into $\pi\pi$ with a small branching ratio, 0.3%; therefore it is not exactly a $CP = -$ eigenstate. CP violation is also seen in the unequalness of the branching ratios of $K_L \rightarrow \pi^+ l^- \bar{\nu}$ and $\pi^- l^+ \nu (l = e$ or $\mu)$; they differ by 0.3%.

Interactions. The interactions of particles are responsible for their scattering and transformations (decays and reactions). Because of interactions, an isolated particle will decay into other particles if this does not violate the selection rules of the symmetry group of the interaction responsible for the decay. Two particles passing near each other may transform, perhaps into the same particles but with changed momenta (elastic scattering) or into other particles (inelastic scattering). The rates or cross sections of these transformations, and so also the interactions responsible for them, fall into three groups: strong (typical decay rates of 10^{21}–10^{23} s^{-1}), electromagnetic (10^{16}–10^{19} s^{-1}), and weak ($<10^{15}$ s^{-1}). Strong interactions occur only between hadrons and have the largest symmetry group. Electromagnetic interactions result from the coupling of charge to the electromagnetic field. They are the best-understood interactions. Weak interactions are usually unobservable in competition with strong or electromagnetic interactions. They are observable only when they do something which those much stronger interactions cannot do (forbidden by the selection rules); for instance, by changing flavors they can make a particle decay which would otherwise be stable, and by making parity-violating transition amplitudes they can produce an otherwise absent asymmetry in the angular distribution of a reaction. SEE FUNDAMENTAL INTERACTIONS; STRONG NUCLEAR INTERACTIONS; WEAK NUCLEAR INTERACTIONS.

Stability. Most particles are unstable and decay into smaller-mass particles. The only particles which appear to be stable are the massless particles (graviton, photon), the neutrinos (possibly massless), the electron, the proton, and the ground states of stable nuclei, atoms, and molecules. It is speculated that some or all of the neutrinos may be massive and unstable, and that the proton (and therefore all nuclei) may be unstable. The present view is that the only massive particles which are strictly stable are the electron and the lightest neutrino(s). The electron is the lightest charged particle; its decay would be into neutral particles and could not conserve charge. Likewise, the lightest neutrino is the lightest fermion; its decay would be into bosons and could not conserve angular momentum. SEE ELECTRON; NEUTRINO; PROTON.

The unstable elementary particles must be studied within a short time of their creation, which occurs in the collision of a fast (high-energy) particle with another particle. Such fast particles exist in nature, namely the cosmic rays, but their flux is small; thus most elementary particle research is based on high-energy particle accelerators. SEE NUCLEAR REACTION; PARTICLE ACCELERATOR; PARTICLE DETECTOR.

Hadrons can be divided into the quasistable (or semistable) and the unstable. The quasistable hadrons are simply those that are too light to decay into other hadrons by way of the strong interactions, such decays being restricted by the requirement that isotopic spin I, strangeness s, charm c, and any other flavors, be conserved. The quasistable hadrons that decay through weak interactions have long mean lives—greater than 10^{10} times the characteristic time of strong interactions, $\hbar/m_\pi c^2 = 0.5 \times 10^{-23}$ s where m_π is the mass of the π meson, or pion, and c is the speed of light. Three hadrons, π^0, η, and Σ^0, can decay by way of the electromagnetic interaction, which conserves the flavor quantum number I_3 but not I. These three have mean lives of the order of $10^5 \times \hbar/m_\pi c^2$. **Figure 1** shows masses and primary decay modes of the quasistable baryons.

The important practical distinctions in the experimental study of interactions are among (1) the stable massive particles (electrons and nuclei), which can be used as target particles as well as in beams; (2) the particles with mean lives greater than 10^{-8} s (γ, ν, η, $\mu^\pm$, $\pi^\pm$, K_L, $K^\pm$), which can only be used in beams as incident particles; (3) the quasistable hadrons with means lives of the order of 10^{-10} s ($\Xi^{0,-}$, Λ, $\Sigma^\pm$, K_S, Ω^-), which have only a small, but usable, chance of interacting in matter before decaying: and (4) the remaining hadrons, which have a vanishingly small chance of reinteracting except when produced within a nucleus.

Resonances. The unstable hadrons are also called particle resonances or excited hadrons. Their lifetimes, of the order of $\hbar/m_\pi c^2$, are much too short to be observed directly. Instead they

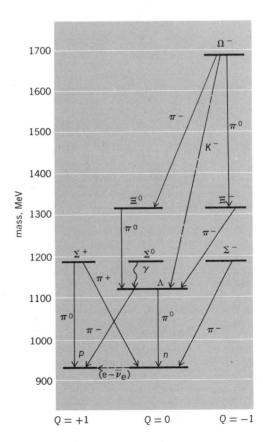

Fig. 1. Decay modes of quasistable baryons.

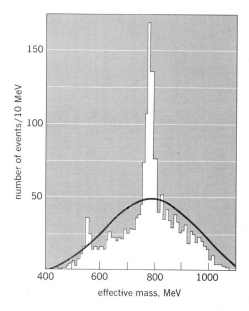

Fig. 2. Observation of the η meson (m = 549 MeV) and the ω meson (m = 783 MeV) as resonances in the reaction $\pi^+ p \to \pi^+ p \pi^+ \pi^- \pi^0$. Effective mass is the total relativistic energy of three of the emerging pions in their center-of-mass coordinate system. Peak in curve indicates that collision can create a short-lived particle of corresponding mass, which decays into three pions.

appear, through the uncertainty principle, as spreads in the masses of the particles—that is, in their widths—just as in the case of nuclear resonances (**Fig. 2**).

The first unstable hadron observed was the "3,3" resonance [in current notation Δ(1232)] in pion-nucleon scattering. That process, $\pi N \to \Delta \to \pi N$, is an example of making a resonance by formation. In principle, any of the decay channels of a resonance can be used as an initial channel to form the resonance, for example, $\omega \pi \pi \to A_2 \to \rho \pi$. But practically, the initial channel must be composed of just two particles, at least one of which is stable so that it can be used as a target, and the other sufficiently stable so that it can be used as a beam particle. In many cases the resonance decays only weakly, or not at all, into such channels; then the resonance can only be made in production along with other particles. An example of this is the production of the ρ^0 meson in the reaction of $\pi^- p \to n \rho^0 \to n \pi^+ \pi^-$. Production is a less effective way of observing a resonance than is formation. The only resonances which can be made by formation are the neutral $J_C^P = 1^-$ mesons (where J_C^P is a combined symbol for spin J, parity P, and charge parity C), which can be formed in $e^+ e^-$ collisions, the ordinary (nonstrange, noncharmed, and so forth) baryons, which can be formed in πN collisions, and the singly-strange ($s = -1$) baryons, which can be formed in $\overline{K} N$ collisions.

Multiplets. A characteristic of the hadrons is that they are grouped into isotopic-spin (*i*-spin) multiplets (for example, n, p; π^-, π^0, π^+); the masses of the particles in each multiplet differ by only a few MeV. Each multiplet can be assigned a certain magnitude I of the isotopic-spin vector **I** (sometimes denoted by **T**). Just as the spin-projection states of a particle with spin S form a multiplet with $2S + 1$ members, the charge-projection states of a hadron with isotopic spin I form a multiplet with $2I + 1$ members. The charge Q (in units of the proton charge e) of each member of a multiplet is related to its value of I_3, the third component of isotopic spin, by $Q = I_3 + \frac{1}{2} Y$, where Y is an integer characteristic of the multiplet, termed hypercharge. SEE ISOBARIC SPIN.

According to the quark model, *i*-spin symmetry results from the smallness of the mass difference of the lightest kinds (flavors) of quarks, the u and the d, together with the flavor independence of the glue force which binds quarks together to form hadrons. The fundamental *i*-spin doublet is (u, d), with $I_3 = (\frac{1}{2}, -\frac{1}{2})$; all heavier quarks are *i*-spin singlets, with $I = I_3 = 0$. Since their charges are $Q_u = \frac{2}{3}$, $Q_d = -\frac{1}{3}$ (in units of the proton charge) the charge of a hadron is $Q = \frac{2}{3} N_u - \frac{1}{3} N_d + Q' = I_3 + \frac{1}{6} (N_u + N_d) + Q'$, where N_u is the net number of u quarks (number of u minus number of $\bar{u}$) in the hadron, N_d is the net number of d quarks, and Q' is the charge carried by heavier quarks (if any) in the hadron. Only the term I_3 in this formula for Q varies in a given multiplet, and so the formula agrees with $Q = I_3 + \frac{1}{2}Y$. For "ordinary" hadrons (nonstrange, noncharmed, nonbottom, and so forth), which are made of the "ordinary" quarks u and d, the hypercharge Y is seen to be 0 for a meson ($q\bar{q}$) and 1 for a baryon (qqq).

Supermultiplets. The hadrons exhibit a further grouping into supermultiplets of i-spin multiplets, the masses of the multiplets in a supermultiplet differing by a few hundred MeV. These supermultiplets appear to be the consequence of a symmetry that is less exact than i-spin symmetry. This symmetry is with respect to a group of transformations SU_3 and is called flavor SU_3 to distinguish it from color SU_3 symmetry. The adjoint representation of the group SU_3 is an octet, consisting of i-spin multiplets with $I = 0, \frac{1}{2}, \frac{1}{2}$, and 1 (hence $1 + 2 + 2 + 3 = 8$ members); the lightest mesons (η, K, $\bar{K}$, π, all with $J^P = 0^-$) and baryons (Λ, N, Ξ, Σ, all with $J^P = \frac{1}{2}^+$) in fact form just such octets.

The vector ($J^P = 1^-$) mesons ω, ϕ, K^*, $\bar{K}^*$, and ρ form another octet plus a singlet, and the $J^P = \frac{3}{2}^+$ baryon resonances $\Delta(1232)$, $\Sigma(1385)$, $\Xi(1530)$, and $\Omega(1672)$ form a deket ($4 + 3 + 2 + 1 = 10$), which is a representation of SU_3.

According to the quark model, this SU_3 symmetry and the pattern of charges in the SU_3 multiplets results from the existence of a third kind (flavor) of quark, the s (strange) quark, with charge the same as the d quark, namely $-\frac{1}{3}$, together with the flavor independence of the glue force: that is, all three quarks u, d, and s have the same interaction with the glue field. The resulting flavor SU_3 symmetry is broken by the relatively large mass of the s, approximately 150 MeV. The three quarks make up the fundamental triplet, $\{3\}$, representation of SU_3; it follows from the vector addition rules that mesons, with the composition $q\bar{q}$, occur only as singlets or octets, $\{1\}$ or $\{8\}$, and baryons, qqq, occur only as singlets, octets, or dekets, $\{1\}$, $\{8\}$, or $\{10\}$. This is consistent with the nonobservation of so-called exotic hadrons which other SU_3 multiplets would contain, such as doubly charged mesons or positive strangeness baryons.

Hadrons are known which contain yet more massive quarks, the c and the b (see Table 1). The resulting symmetry is badly broken, and the supermultiplets hardly recognizable. The hadronic quantum numbers I_3, strangeness s, charm c, bottomness b, and so forth, listed in Table 2, specify the net flavor content of a hadron according to the scheme $I_3 = \frac{1}{2}(N_u - N_d)$, $s = -N_s$, $c = N_c$, $b = -N_b$, and so forth, where, as before, N_u is the net number of u quarks, and so on.

Nomenclature. Nomenclature is closely tied to knowledge. When i-spin symmetry became apparent, the term nucleon came into use to mean the multiplet whose two members are the proton and the neutron. Similarly, all other i-spin multiplets have been given names, for example, pi meson π and sigma hyperon Σ. The members of a multiplet are distinguished by writing charge as a superscript, for example, π^+ and Σ^0. (But p and n are usually written instead of N^+ and N^0.) Antiparticles are denoted by putting a bar over the symbol of the corresponding particle: antiproton $\bar{p}$, antisigma-plus [= anti(sigma-minus)] hyperon $\bar{\Sigma}^+$ ($= \Sigma^-$), and so forth. (But the antielectron $\bar{e}$, the positron, is usually written e^+; likewise $\bar{\mu} = \mu^+$.) The equality $\bar{\pi} = \pi$ is valid in the sense that conjugation only permutes the members of the multiplet: $\bar{\pi}^+ = \pi^-$, and so forth.

With the discovery of large numbers of unstable hadrons, it has become impractical to give each hadron an individual name; thus the custom has arisen of calling it by the name of a similar hadron of lower mass, the symbol (sometimes asterisked) being followed by its mass (approximate, in MeV). Baryons are named after the lowest-mass baryon with the same I and s, for example, $\Lambda(1520)$ and $\Xi(1530)$.

Baryons composed only of ordinary and strange quarks can only have six combinations of I and s, so there are just six names: N, Δ, Λ, Σ, Ξ, Ω. Further, the only flavor SU_3 multiplets which these baryons form are $\{1\}$: Λ, $\{8\}$: N, Λ, Σ, Ξ, and $\{10\}$: Δ, Σ, Ξ, Ω. Therefore an N belongs to an $\{8\}$, and a Δ or an Ω belongs to a $\{10\}$, uniquely, but a Λ belongs to either a $\{1\}$ or an $\{8\}$, and a Σ or a Ξ belongs to an $\{8\}$ or a $\{10\}$. In fact, a Λ is never a pure $\{1\}$ or $\{8\}$, because the mass of the s quark mixes a Λ in a $\{1\}$ with a Λ of the same J^P in an $\{8\}$; this mixing also occurs for Σ or Ξ hyperons. If the masses of the members of the multiplets concerned are known, one can estimate the relative amount of $\{1\}$ and $\{8\}$ in a given Λ. In Table 2 the dominant SU_3 multiplet is given. The singly charmed analogs of the strange baryons Λ and Σ (the baryons in which the s quark is replaced by a c quark) are called Λ_c and Σ_c respectively.

Most established mesons have their own proper names. But, in the manner of baryons, some are given the same name as the lowest-mass meson of the same I and s (and C, if applicable). Irregularly, D is the name of two unrelated mesons, as is B. The lightest charmonium ($c\bar{c}$) pseudoscalar is named η_c, in the manner of the Λ_c, though not very logically, because the η is not close to 100% $s\bar{s}$. (Following the same scheme, it would have been logical to call the $D(1866)$ "K_c," and the $B(5200)$ "K_b.")

In contrast to the i-spin multiplets, there are no accepted names or symbols for the supermultiplets. One must denote them by their $J^P_{(C)}$ and their rank in mass. For instance, the "lowest (or ground state) $J^P = \frac{1}{2}^+$" specifies the baryon supermultiplet consisting of N, Λ, Σ, Ξ, Λ_c, Σ_c, and so forth.

Regge recurrences. It is known from the interactions of hadrons that they are not point particles, and so it could be expected that hadrons would have rotationally excited states, as do molecules, for instance. Such excited states would be a sequence of hadrons with increasing spins (J_0, $J_0 + 1$, . . .) and masses, but with the same values of other quantum numbers (except for parities: P, C, and G, which would alternate in sign). For historical reasons, such hadrons are called Regge recurrences. The relation between their spin and mass is termed a Regge trajectory. For a rigid body, this would be of the form $m = m_0 + J^2/2\mathcal{I}$ ($\mathcal{I}$ = moment of inertia), but for hadrons, it is found empirically to be the form $m^2(J) = b(J - \alpha_0)$, where $b \cong 1$ GeV^2. For example, the $I = 1$ mesons ρ, A_2, and g are recurrences with $J^P(G) = 1^-(+)$, $2^+(-)$, and $3^-(+)$ respectively. There is no $0^-(-)$ member of this trajectory; according to the empirical mass-spin relation, it would have had an imaginary mass.

Because of the exchange character of many forces, it is expected that alternate Regge recurrences will actually fall on separate trajectories (if the two trajectories happen to coincide, as in the case of the mesons ρ, A_2, and g, one says that there is exchange degeneracy). For instance, the $I = \frac{1}{2}$ baryons $N(939)$ and $N(1680)$, with $J^P = \frac{1}{2}^+$ and $\frac{5}{2}^+$ respectively, are recurrences on a trajectory on which a $\frac{3}{2}^-$ baryon would have a mass 1370 MeV; in fact, there is such a baryon but with a different mass, namely 1520. Thus this baryon, together with a recurrence, the $N(2190)$, $\frac{7}{2}^-$, lies on a different trajectory. SEE REGGE POLE.

Quantum chromodynamics. It appears that the "glue" field which binds quarks together to make hadrons is a Yang-Mills (that is, a non-Abelian) gauge field of an SU$_3$ symmetry group, color SU$_3$. The quanta of the field are called gluons, and its quantum theory is called quantum chromodynamics (QCD). The gluon field resembles the electromagnetic field, but has an internal symmetry index (octet index) which runs over eight values; that is, there are really eight fields, corresponding to the eight parameters needed to specify an SU$_3$ transformation. (Technically, the field transforms under the group in the same manner as the generators, that is, as the adjoint representation.) Just as the electromagnetic field is coupled to (that is, photons are emitted and absorbed by) the density and current of a conserved quantity, charge, the gluon field is coupled to color. This is a shorthand way of saying the following: The octet of gluon fields is coupled to an octet of color charges; the matrix element of the color charge a of a particle belonging to the representation (multiplet) $\mathcal{R}$ of color SU$_3$ (all fields and particles belong to color SU$_3$ multiplets because color SU$_3$ is a good symmetry) which makes a transition between color states A and B (A and B label members of the multiplet $\mathcal{R}$) is given in Eq. (1), where $(T_a)^\mathcal{R}$ is the matrix

$$g(T_a)^\mathcal{R}_{BA} \qquad (1)$$

of the generator T_a of SU$_3$ in the representation $\mathcal{R}$, and g is a constant which is the same for all particles. The important thing is that the coupling of the gluon to a particle is fixed by the color of the particle (that is, what member of what color multiplet) and just one universal coupling constant g, analogous to the electronic unit of charge e. (The analogy breaks down in quantum theory, as discussed below; the quantity g is no longer constant but it is still universal.) SEE GAUGE THEORY.

A color singlet particle does not couple to gluons because $(T_a)^1 = 0$; such colorless particles include the photon, leptons, and any collections of colored particles (quarks and gluons) which are vector-coupled to form a color singlet. Gluons are not colorless, and therefore they are coupled to themselves. This situation is very different from electromagnetism, where the photon does not carry charge. The consequence of this self-coupling of massless particles is a severe infrared (small momentum transfer or large distance) divergence of perturbation theory. In particular, the interaction between two colored particles through the gluon field, which in lowest order is an inverse-square Coulomb force, proportional to g^2/r^2 (where r is the distance between the particles), becomes stronger than this inverse-square force at larger r. A way of describing this is to say that the coupling constant g is effectively larger at larger r; this defines the so-called running coupling constant $g(r)$. According to the first-order radiative correction, $g(r)$ becomes infinite at a certain distance r_c. As r is raised toward r_c, $g(r)$ rises, and the perturbation series becomes less reliable; all that can be said is that the interaction is very different from Coulomb for $r \gtrsim r_c$.

This situation is very different from that in quantum electrodynamics (QED). There, the interaction between, say, two μ^+ leptons becomes precisely of the inverse-square form at larger distance, with coefficient equal to the product of the leptons' charges, e^2. In dimensionless form, this is $e^2/\hbar c \approx 1/137$, the dimensionless parameter α of QED. At smaller distance the interaction becomes stronger (vacuum polarization), but this is a very small effect at distances larger than the Compton wavelength of the lightest charged particle (the electron). In QCD the lightest colored particle (the gluon) is massless, and so vacuum polarization, the deviation from the inverse-square law, occurs at all r. There is also a sign difference: in QED, vacuum polarization shields a charge—just as in any polarizable medium—resulting in an increase in the apparent charge as one gets closer, where the shielding is less effective; in QCD, vacuum polarization—the effect of the gluon self-coupling—is antishielding, resulting in an apparent decrease in the color strength as one gets closer.

There is no way to define a coupling constant g^2 in QCD analogous to e^2 in QED. Instead of such a dimensionless parameter, QCD has the scale parameter r_c. At small r, or equivalently at large momentum transfer, the running coupling becomes small (asymptotic freedom); consequently, perturbation theory becomes reliable. Calculations in perturbative QCD, such as for the cross section for gluon bremsstahlung or the gluonic radiative corrections to the cross section for $e^+e^- \rightarrow$ hadrons, are in good agreement with experiment. This is an important reason for the present view that QCD is the correct theory of the hadron glue. SEE QUANTUM CHROMODYNAMICS; QUANTUM ELECTRODYNAMICS.

String model of confinement. A specific form for the gluonic force between two colored particles, at large r, namely that it falls to a nonzero constant value λ, of the order of $\hbar c r_c^{-2}$, is suggested by a model, the superconductor analogy. This force is confining. (Evidence for this form of the gluonic force also comes from a nonperturbative method of calculating QCD, lattice QCD.) A superconductor excludes a magnetic field from its interior, below a penetration depth D, except in the form of flux bundles, which have a diameter of order D, and one unit of flux, namely $\Phi_0 = 2\pi/Q_{sc}$, where Q_{sc} is the charge of the carriers of the supercurrent. (For a real superconductor, the carriers are Cooper pairs of electrons, so $Q_{sc} = 2e$.) Superconductivity is destroyed along the core of the bundle. If a magnetic monopole were available and put into a superconductor, the magnetic flux leaving the monopole would gather into a fluxbundle (in a real superconductor, two bundles). That is, closer to the monopole than D, the magnetic field would be roughly that of an isolated monopole; farther away, it would be in the form of a bundle of diameter D. A flux bundle has a certain energy per unit length, of order $\Phi_0^2 D^2 \equiv \lambda$; hence if a pair of monopoles of equal but opposite magnetic charge were put into the superconductor, the lowest energy state would be one in which the flux bundle went straight from one to the other. This means that the monopoles are confined by a long-distance force of constant magnitude λ. (The energy per unit length of the bundle is the same as the tension it exerts, just as the energy per unit area of a surface is its surface tension.) At distances much less than D, the force is of the usual inverse-square form. SEE MAGNETIC MONOPOLES.

The conjecture is that the vacuum is like a superconductor with respect to color, with the interchange, however, of electric and magnetic quantities. That is, the vacuum acts like a color magnetic superconductor which confines color flux into bundles which have a diameter of order r_c and an energy per unit length equal to λ of order $\hbar c r_c^{-2}$. The color flux bundles run between colored particles; they can also form closed loops. These flux bundles are often idealized as having vanishing diameter and are then called strings. This idealization is obviously good only if the flux bundles are long compared to r_c, and if their local radius of curvature is always much larger than r_c.

The motion, according to classical mechanics, of such a string with a particle (quark) at each end, which has the lowest energy for a given angular momentum J, is as usual a rigid rotation. The string is straight, rotating in a plane; centrifugal force balances the tension of the string. At large J, the masses of the quarks at the ends are irrelevant; most of the energy and angular momentum of the system is in the string. The result in this limit is that $E^2 = 2\pi\lambda J$, where E is the total rest energy, that is, mass. Such a linear relation between E^2 and J is just what is found experimentally for the relation between the mass and spin of hadrons of the lowest mass for given spin (the leading Regge trajectory), aside from a constant term, which can be viewed as a quantum effect. The value of λ is 0.185 GeV^2/$\hbar c$ (1.5 × 10^4 newtons or 17 tons weight). An idealization of the hadron S matrix, the Veneziano or dual model, which embodied the feature of linear Regge trajectories, turned out to have a spectrum of hadrons which was just the spectrum of a string (with nothing at its ends); this was the first appearance of strings in the theory of hadrons. In view of the idealization of the string model and the neglect of the quark masses, it is a mystery that the leading Regge trajectories are so accurately linear.

The string (flux bundle) model also gives a picture for the process of fragmentation: If two quarks, tied with a string, move apart at high velocity, the string is stretched. When it is longer than L, where $\lambda L = 2m_q$, it is unstable to the process of breaking with the creation of a pair of quarks of mass m_q to cap the new string ends (similar to breaking a bar magnet, and creating a new pair of north-south poles). The energy required for the pair creation comes from the loss of a length L of string (that is, L is the length of the created gap); hence the formula above for L. If the new lengths of string are still being stretched enough, they too will break, and so on. Thus the process of hadronization of quarks is really the fragmentation of strings. This process in various reactions is shown schematically in **Fig. 3**.

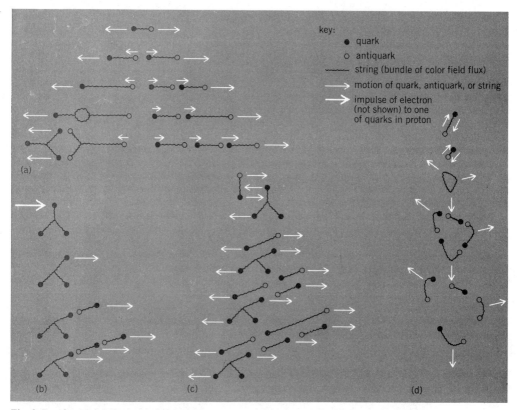

Fig. 3. Development of hadronic jets in high-energy collisions according to the string model. (a) Annihilation of e^+e^- into hadrons. First picture is just after the virtual photon has materialized into a quark pair. (b) Hard ep collision (deep inelastic scattering). (c) Soft πp collision. String exchange occurs between first and second pictures. (d) Zweig's rule-violating hadronic decay of a heavy quarkonium $J_C^P = 1$-meson.

Electron-positron annihilation. Electron-positron annihilation at high energies is observed in colliding-beam machines. To lowest order in the fine-structure constant, $\alpha = e^2/\hbar c \approx 1/137$, there are two reactions. One is $e^+e^- \to \gamma\gamma$, which is completely described by quantum electrodynamics and will not be discussed further. The other is $e^+e^- \to \gamma^* \to x\bar{x}$, where γ^* is a virtual photon. If x is a lepton, the cross section at a total center-of-mass energy $\sqrt{s}$ well above threshold (that is, $\sqrt{s} \gg 2m_x$) is independent of the lepton's mass m_x and proportional to the square of its charge, Q_x. If x is a quark, its coupling to the gluon field makes a small correction to the cross section and also later "fragments" the quarks into hadrons (Fig. 3a). Thus the ratio R of the cross sections for $e^+e^- \to$ hadrons and $e^+e^- \to \mu^+\mu^-$ is given by Eq. (2), where the sum is

$$R \equiv \sigma(\text{hadrons})/\sigma(\mu\bar{\mu}) = 3\sum_f Q_f^2/e^2 \quad (2)$$

over all the flavors of quarks whose mass is less than the beam energy and the factor of three comes from the sum over the color. For example, at $\sqrt{s} = 8$ GeV, the flavors are u, d, s, and c, and so $R = 3[(2/3)^2 + (1/3)^2 + (1/3)^2 + (2/3)^2] = 10/3$. The reaction $e^+e^- \to$ hadrons has the great advantage in studying hadron properties that the hadronic final state is created as a single high-energy $q\bar{q}$ pair, whereas in ep, pp, πp, and other reactions more quarks are involved.

Equation (2) for R is not valid near the threshold for making a quark pair, for instance in the case of the c quark near $s = 2m_c = 2.8$ GeV. (The zero-point energy of the quarks, due to confinement, moves the effective threshold energy a few hundred MeV higher.) This is because there is little if any energy for fragmentation. The cross section is strongly modulated, with peaks at $c\bar{c}$ resonant states with $J_C^P = 1^-$, the quantum numbers of the γ^*. Below threshold for decaying into the lightest hadrons which contain a c or $\bar{c}$, that is, $\sqrt{s} < 2m_{D(1866)}$, these states, namely $\Psi(3097)$ and $\Psi(3686)$, can decay hadronically only through annihilation of the $c\bar{c}$ pair into gluons. The selection rules require that the number of gluons is at least three (Fig. 3d), so that the "fragmentation" of these gluons does not make the back-to-back double jets seen in $e^+e^- \to \gamma^* \to q\bar{q} \to$ hadrons. This contrast in the distribution of hadrons between events seen on-resonance and off-resonance is more marked at higher energy, namely at the Υ mesons. At still higher energy, the hadronic decays of the expected $t\bar{t}$ mesons (mass ≈ 80 GeV) should show clearly three jets, from the fragmentation of the three gluons. *See* J PARTICLE; UPSILON PARTICLES.

The process $e^+e^- \to \gamma^* \to x\bar{x}$ can also proceed through the Z^0: $e^+e^- \to Z^{0*} \to x\bar{x}$. At low energy this additional amplitude is negligible, but at higher energy it is larger, relative to the γ^* amplitude, and at the mass of the Z^0, $\sqrt{s} = m_{Z^0} = 90$ GeV, it dominates, forming a real Z^0 boson. Already at $\sqrt{s} \approx 35$ GeV, the interference of the parity violating Z^{0*} amplitude with the γ^* amplitude makes an easily observable forward-backward asymmetry in the angular distribution of the x and $\bar{x}$.

High-energy hadron-hadron collisions. According to the quark model (QCD), the dominant process in high-energy hadron-hadron collisions is the exchange of a gluon when the two hadrons overlap. This leaves each hadron in a color octet state (by the vector-coupling rule, singlet + octet = octet), and hence they are tied (confined) to one another. Described in more detail, the effect of the exchange of a gluon between a quark in one hadron and a quark in the other hadron is to exchange the strings which were attached to the quarks, so that there are now two strings between the hadrons (Fig. 3c). This color transfer interaction between the hadrons is usually soft, transferring little momentum. The strings stretch as the particles continue on their way and then fragment, just as in deep inelastic ep scattering (Fig. 3b) or in e^+e^- annihilation into hadrons (Fig. 3a). Because in the present case there are two fragmenting strings, twice as great a density of produced hadrons is expected, compared to e^+e^- annihilation and this is roughly so. The model also agrees with the rough constancy of the cross section with energy.

Hadron structure. A generally good semiquantitative understanding of the properites of hadrons follows from assuming that the force between quarks in a hadron is described by a potential of Coulomb form at small r, less than $r_c \approx 1$ femtometer, and linear in r, namely λr, at large r. The description of hadrons composed of heavy quarks (the $c\bar{c}$ and $b\bar{b}$ systems) is particularly good, because their motion in their low-lying bound states is nonrelativistic, so the eigenfunction of a Schrödinger equation is a good approximation to the wave function. (For light quarks, which move relativistically in their bound state, the approximation of the interaction by a static potential is not very good.)

Mesons. The simplest hadrons are the mesons, composed of a quark and an antiquark (called quarkonium, after positronium). The colors of the quark and the antiquark, a color triplet and antitriplet respectively, must be vector-coupled to a singlet. Since the spins of the q and $\bar{q}$ can be vector-coupled to either singlet or triplet, the possible states are (in spectroscopic notation) ^1L, with $J = L$ and $P = (-)^{L+1}$, $C = (-)^L$ and ^3L, with $J = L - 1, L,$ or $L + 1$, $C = (-)^{L+1}$. (The values of charge parity C apply only when the quark and antiquark are conjugates of one another, that is, of conjugated flavors.) Thus there are sequences of states with $J_C^P = 0_+^-, 1_-^+, \ldots$, $0_+^+, 1_-^-, \ldots, 1_-^-, 2_+^+, \ldots$, and $1_+^+, 2_-^-, \ldots$. The state 0_+^- and the sequence $0_-^+, 1_+^-, \ldots$ are missing; in fact no such mesons have been found.

The lowest states of a given quark pair are expected to be the S states, $^1S_0$ and $^3S_1$, with $J_C^P = 0_+^-$ and 1_-^- respectively. They would be degenerate in mass if it were not for the spin-spin interaction which is the color analog of the interaction of magnetic moments through the magnetic field, that is, the hyperfine interaction. This makes the pseudoscalar, 0_+^-, lighter than the vector, 1_-^-.

In each spin-space state, the four mesons composed of light quarks, $u\bar{d}$, $u\bar{u}$, $d\bar{d}$, and $d\bar{u}$, constitute an i-spin triplet $[u\bar{d}, (u\bar{u} - d\bar{d})/\sqrt{2}, d\bar{u}]$ and a singlet $[(u\bar{u} + d\bar{d})/\sqrt{2}]$. These mesons would be expected to have about the same mass (of course the $u\bar{d}$ and $d\bar{u}$ have exactly the same mass, being mutual charge conjugates) to the accuracy to which the u and d quarks have the same mass, namely about 5 MeV. The four mesons $u\bar{s}$, $d\bar{s}$, $s\bar{d}$, and $s\bar{u}$ constitute an i-spin doublet $[u\bar{s}, d\bar{s}]$ and its conjugate $[s\bar{u}, s\bar{d}]$, with strangeness $+1$ and -1 respectively. These four mesons have about the same mass, which is higher than the mass of the first four, nonstrange, mesons by about 150 MeV, the mass of the s quark. The meson with composition $s\bar{s}$, an i-spin singlet, has a mass which exceeds the strange mesons again by about m_s. A set of mesons with the just-described properties is called an ideal nonet. The lowest 1^- mesons form such a nonet, since $m_\rho \approx m_\omega$ (775 $\approx$ 782) and $m_{K^*} - m_\omega \approx m_\phi - m_{K^*}$ (110 $\approx$ 128). Also the $s\bar{s}$ composition of the ϕ is shown by the large branching ratio of its decay into $K\bar{K}$, 84% (despite the small phase space available), since that decay proceeds merely by the breaking of the string between the s and $\bar{s}$ with creation of a pair of light quarks, whereas a decay such as $\phi \to \pi\rho$ requires the annihilation of the $s\bar{s}$ and the creation of two pairs of light quarks. The observation that the latter reaction is weaker than the former is known as Zweig's rule.

The ideal nonet can be extended to an ideal 16-plet by letting the quarks be u, d, s, or c. The largeness of m_c causes the wave function of the $c\bar{c}$ ground state, the ψ, to be concentrated at small distance (this is an elementary property of the nonrelativistic Schrödinger equation) where the potential is negative, and consequently the mass of the ψ exceeds the rest mass $2m_c$ by noticeably less than in the other mesons where at least one of the quarks is light. A consequence is that the ψ is lighter than the lightest pair of mesons containing c quarks, that is, $m_\psi < 2m_D$, so that all of its hadronic decays must violate Zweig's rule (Fig. 3d), resulting in the quite small total width of the ψ.

There is a nonet of pseudoscalar, 0_+^-, mesons below 1 GeV, but it is far from ideal: the lowest i-spin singlet $\eta(549)$ is much heavier than the triplet $\pi(138)$. On the evidence of their masses, the π, K, $\bar{K}$, and η form roughly an SU_3 octet, having the form $(u\bar{u} + d\bar{d} - 2s\bar{s})/\sqrt{6}$, so that $m_\eta \approx (4m_K - m_\pi)/3$. This is because there is a large interaction matrix element in the 0^- state between the quark pair channel and the two gluon channel: $q\bar{q} \leftrightarrow gg$: since gluons are flavorless, this can happen only in the flavor singlet state, which is thus shifted away in mass from the octet state. (The largeness of m_c, m_b, and so forth, means that $c\bar{c}$, $b\bar{b}$, and so forth mix very little with the glue.) By contrast, in the 1^- state, selection rules force the glue state to have at least three gluons: $q\bar{q} \leftrightarrow ggg$; the "glue" mixing effect is therefore weaker. The foregoing explanation of the low-lying pseudoscalar mesons has a glaring defect, however: the effect of mixing the $q\bar{q}$ channel with "glue" should drive down the lowest i-spin singlet state below the π mass (according to the rule that two energy levels are repelled when coupled by an interaction); this is in fact observed in the 2_+^+ mesons, where $m_f = 1275 < 1315 = m_{A_2}$. The nonobservation of an isosinglet 0^- meson lighter than the π has been explained in sophisticated ways, but not so far in any simple convincing way.

Baryons. Baryons, according to the quark model, are composed of three quarks, qqq, in a color singlet; this is antisymmetric in the color indices of the quarks, so the baryon wave function is symmetric in the remaining coordinates: space, spin, and flavor. The spins of the three quarks can be added (vector-coupled) to give $S = 3/2$, symmetric, or $S = 1/2$, of mixed symmetry. The flavors of the quarks can be vector-coupled to a total flavor state (multiplet) which is symmetric, mixed, or antisymmetric. For instance, for three flavors (the quark is then a flavor SU$_3$ triplet, {3}), these three total flavor states are {10}, {8}, and {1} respectively. It is useful to couple the spin and flavor into states of overall symmetry (these are states of the group SU$_{2N}$, where N is the number of flavors), because if the spin dependence of the quark-quark interaction is neglected, all members of one such state have the same spatial wave function and the same mass.

The ground-state spatial wave function is symmetric and has vanishing angular momentum, $L = 0$, and even parity. Hence its states have $J = S$, and are a $3/2^+$ {10} and a $1/2^+$ {8}. The spin-spin (hyperfine) interaction raises the mass of the $3/2^+$ and lowers the mass of the $1/2^+$. The states of {8} containing 0, 1, or 2 s quarks are i-spin multiplets with $I = 1/2$, 0 and 1, and $1/2$ respectively. The observed lightest $1/2^+$ baryons are $N(939)$, $\Lambda(1116)$, $\Sigma(1193)$, and $\Xi(1318)$. Although the Λ and Σ have the same number of s quarks, their masses are different. The following is the explanation for this mass difference: The two light quarks in the Λ form an isosinglet, and hence (because the spin-flavor wave function is symmetric) a spin singlet, and so there is no spin-spin interaction between the strange quark and the light quarks; the entire spin-spin interaction is between the light quarks. This makes the hyperfine energy greater in magnitude (and hence lowers the mass more) in the Λ than in the Σ, because the (color) magnetic moment is inversely proportional to the quark mass. Because the lowest members of the $1/2^+$ octet are all hadronically stable, they are long-lived enough that their magnetic moments can be measured; these are all consistent with the quark model.

For any reasonable quark-quark interaction, the first excited orbital state of qqq is 1^-, of mixed symmetry. The spin-spin and spin-orbit interactions mix states of the same J^P and i-spin. (Thus, for instance, there are three $1/2^-$ Λ states which are mixed.) The calculation of the mixing shows a remarkable agreement with the observed masses and couplings of the low-lying negative-parity baryons. (This is not a calculation from first principles: several parameters, representing the matrix elements of the spin-dependent interactions, are chosen to best fit the data. But the number of data compared to the number of parameters is very large.) The only notable discrepancy is that the lowest $1/2^-$ Λ, $\Lambda(1405)$, is predicted about 100 MeV too high: presumably its mass has been depressed by coupling to $\overline{K}N$ continuum states lying higher in energy.

Calculations of the second excited orbital level, using a harmonic oscillator model, yield a large number of positive-parity states, far more than are experimentally observed. The discrepancy has been explained by calculations showing that most of these baryons are weakly coupled to pseudoscalar mesons, so that they cannot be seen in the usual way, using incident π or K beams to form them, or looking for their decays into π, K, or η plus a baryon.

Unconventional hadrons. In addition to the "conventional" hadrons with the structure $q\bar{q}$ or qqq, there are other assortments of quarks and/or gluons which can form a color singlet. It is not known whether such hadrons can be long-lived enough to be seen as a resonance peak. If such an object were observed, there would be two possible ways of knowing that it was not a "conventional" hadron: one is that the spectrum of "conventional" hadrons are well enough understood that an interloper would be noticeable; the other would be that the unconventional hadron was an exotic, that is, with quantum numbers which were impossible for a conventional hadron. The most discussed unconventional hadrons are "glueballs" and baryonium.

A glueball is made only of gluons. (In the string picture, it is an endless string, a loop.) It is a flavorless meson, therefore self-conjugate, and with $I = 0$. There are two candidates for glueballs which seem to be interlopers, the $\iota(1440)$, with $J_C^P = 0_+^-$ and the $\theta(1640)$ with $J_C^P = 2_+^+$. Neither, however, has the expected decay branching ratios for a glueball. Since these mesons are not exotics, they can have $q\bar{q}$ channels mixed in, which will change the branching ratios. Conversely, this would mean that the conventional 0_+^- and 2_+^+ mesons have pure gluon channels mixed in; this is certainly not ruled out. The lightest exotic glueball is a three-gluon system with $J_C^P = 1_+^-$: no such meson has been seen.

Baryonium is the name given to a meson with the structure $qq\overline{qq}$ for the reason that by merely making a quark pair (in the string model, breaking one string) it can decay (if heavy enough) into a baryon-antibaryon pair. (The name is rather misleading since the baryon pair has nothing to do with the basic structure of the meson, in contrast to the meaning of positronium or quarkonium.) It is not known whether it can more easily decay into a pair of mesons, without having even to make one pair of quarks. An example of an unmistakably exotic baryonium would be a doubly charged meson.

Weak interaction. From the quark point of view, the so-called charged-current weak interactions are interactions which are flavor-changing; here flavor is meant in a broad sense which includes leptons. In the older language, these interactions violate I_3 conservation ($I_3 = \pm \frac{1}{2}$ for the u and d quarks respectively, 0 for all others), and/or s (strangeness), and so forth. All experimental results to date are consistent with the weak reactions being point four-Fermi interactions, that is, occurring when the four fermions are at one point in space-time. But such an interaction is not renormalizable and thus not acceptable as an elementary interaction. The properties of the known charged-current four-Fermi interactions are consistent with being the result of the exchange of a massive charged vector (spin 1) particle, called the $W^\pm$ (W^+ and W^- are antiparticles of one another), which is emitted and absorbed by the so-called charged current. It turns out that this particle can be part of a renormalized field theory, which is a spontaneously broken gauge theory (by the Higgs mechanism). This is the electro-weak theory of S. Weinberg, S. Glashow, and A. Salam. It has four gauge fields, whose quanta are the W^+ and W^-, a massive neutral vector particle Z^0, and the photon. The Z^0 is coupled to all particles with strengths controlled by a parameter, the single Weinberg (or weak) angle θ_W, whose value is determined by observation of the neutral-current weak interactions which exchange of the Z^0 yields. These were unknown before this theory suggested their existence. They have been observed in $v e$ scattering, high-energy $e^+ e^-$ scattering, deep inelastic ep scattering, and the low-energy e-nucleus interaction in the atom. In many of these, the Z^0 exchange is observable only because its coupling is partly axial vector, and so interferes with photon exchange to produce a parity-violating effect.
See Neutral currents; Symmetry breaking; Weinberg-Salam model.

The property of the Z^0 coupling, that it is diagonal (that is, the particle which emits a Z^0 remains unchanged, as with the photon), is deduced from the nonobservation of decays such as $K^+ \to \pi^+ \nu \bar{\nu}$ and $\mu \to e e \bar{e}$, which show the absence of a $Z^0 d \bar{s}$ coupling, or a $Z^0 e \bar{\mu}$ coupling respectively. Another place where a $Z^0 d \bar{s}$ coupling would be noticeable is in the $K^0 - \bar{K}^0$ system: exchange of a Z^0 would produce the transition $d\bar{s} \leftrightarrow s\bar{d}$, that is, $K^0 \leftrightarrow \bar{K}^0$, which would make a much larger difference in the masses of K_S and K_L than is observed. In fact, this mass difference is even smaller than what is produced by the second-order effect of two exchanges of a $W^\pm$: $d\bar{s} \leftrightarrow u\bar{u} \leftrightarrow s\bar{d}$: this discrepancy led to the prediction of the charmed quark c, since if properly coupled, its contribution in $d\bar{s} \leftrightarrow c\bar{c} \leftrightarrow s\bar{d}$ can cancel against the previous one.

An important property of the weak interaction is universality. This was first observed in the fact that the interactions responsible for the beta decays $\mu \to e \bar{\nu}_e \nu_\mu$ and $n \to p e \bar{\nu}_e$ are nearly equal; that is, the couplings of the $W^\pm$, $W\nu_\mu \bar{\mu}$ and Wud, are nearly equal. (Likewise it is found that the couplings $W\mu \bar{\nu}_\mu$ and $We\bar{\nu}_e$ are equal.) This points to a deep similarity of leptons and quarks. Despite the fact that the strangeness-changing-coupling $Wu\bar{s}$ is about ¼ the size of the previously mentioned couplings, universality is maintained by Cabibbo's scheme: the couplings $We\nu_e$ and Wud_C are exactly equal (and like $We\nu_\mu$, the coupling Wus_C vanishes). Here, d_C and s_C are a mixture of d and s, given by Eqs. (3), where θ_c is called the Cabibbo angle; in words, the doublets (d_C, s_C) and (d, s) are related by an SU_2 transformation, the Cabibbo rotation. With the discovery of the b quark (and expectation of the t) the scheme of Eqs. (3) has been extended to

$$d_C = \cos \theta_C d + \sin \theta_C s \qquad\qquad s_C = -\sin \theta_C d + \cos \theta_C s \qquad (3)$$

all three $Q = -\frac{1}{3}$ quarks: that is, an SU_3 rotation relates (d_C, s_C, b_C) and (d, s, b). The precise statement of universality is that the couplings of the $W^\pm$ are of the form $\overline{WUD}$, that is $U \Rightarrow W + D$, where the pair (U, D) stands for any of the pairs (ν_e, e), (ν_μ, μ), (ν, τ), (u, d_C), (c, s_C), and (t, b_C). These pairs are called weak i-spin doublets. (The coupling is V-A, that is, to the left-hand helicity component only.) It is at present unknown why the members of the weak i-spin doublets are not the mass eigenstates; a related mystery is the scheme of values of the quarks and leptons.

The weak bosons W^+, W^- ($= \overline{W^+}$), and Z^0 were observed in 1983 at the CERN (European Center for Nuclear Research) Super Proton Synchrotron (SPS) proton-antiproton ($p\bar{p}$) collider, a facility built for that purpose. The $W^\pm$, and similarly the Z^0, are made by the annihilation of a quark pair in a $p\bar{p}$ collision, $q\bar{q} \to W \to$ decay products, where the q and $\bar{q}$ are contained in an incident proton p and antiproton $\bar{p}$ respectively. The predictions of the standard electroweak

Table 3. Predicted properties of the weak bosons

	Mass, GeV	Width, GeV	Decay products and branching ratios, %
W^+	85	2.7	$\nu_e e^+, \nu_\mu \mu^+, \nu_\tau \tau^+$ 8, 8, 8
			$u\bar{d}, c\bar{s}, t\bar{b}$ 25, 25, <25†
W^-	(same as W^+, conjugated)		
Z^0	95	2.5	$e^+e^-, \mu^+\mu^-, \tau^+\tau^-$ 3, 3, 3
			$\nu_e \bar{\nu}_e, \nu_\mu \bar{\nu}_\mu, \nu_\tau \bar{\nu}_\tau$ 6, 6, 6
			$u\bar{u}, c\bar{c}, t\bar{t}$ 11, 11, <11†
			$d\bar{d}, s\bar{s}, b\bar{b}$ 14, 14, 14

†Depends on mass of t quark.

model for the weak bosons are listed in **Table 3**, where the value of $\sin^2 \theta_W$ is taken to be 0.23, and it is assumed that the only elementary particles whose mass is less than about 50 GeV are those listed in Table 1 (that is, the three generations of quarks and leptons). The observed properties of the W and Z are consistent with these predictions but are not yet well determined. The decay modes $e^\pm \nu_e$ and $\mu^\pm \nu_\mu$ of the W^+, and $e^+ e^-$ and $\mu^+ \mu^-$ of the Z^0, are easy to observe, despite the rarity of the W or Z production (about 1 per 10^6 $p\bar{p}$ collisions) because no other process makes many leptons with momenta as high as 40 GeV, at a large angle to the incident p and $\bar{p}$ beams. Because there are three times as many kinds of quarks as leptons (each flavor of quark comes in three colors), most decays of the W and Z are into pairs of quarks, which fragment into 40-GeV jets; but these decays are hard to observe because of the background of similar jets made by the scattering process $q\bar{q} \to q'\bar{q}'$. Particularly interesting decay modes are those involving new particles, such as $Z^0 \to t\bar{t}$ or $W^+ \to t\bar{b}$, where t is the top quark, $Z^0 \to E\bar{E}$, where E is a new heavy charged lepton, or $W \to H\gamma$, where H is a Higgs particle. These will occur if the particles involved are not too massive, but will be hard to observe because of the background of events yielding similar final states. The Z will be made cleanly by $e^+ e^-$ annihilation at the Stanford Linear Collider (SLC) and Large Electron Positron (LEP) collider, at CERN, which will enable its decay modes to be easily seen. Of course, the decays $Z \to \nu_\ell \bar{\nu}_\ell$, where ν_ℓ is a neutrino, cannot be seen directly, but at least the total branching ratio into these modes can be deduced by observation of the total width of the Z. With the assumption of universality, this will determine the number of kinds of neutral leptons (neutrinos; ν_e, ν_μ, and ν_π are presently known) whose mass is not greater than about 45 GeV.

Grand unified theories. As described above, the electroweak theory, starting from the observation that both the electromagnetic and weak interactions result from the exchange of vector (spin 1) bosons, has unified these interactions into a spontaneously broken gauge theory. Similarly, the observation that the strong (hadronic) interactions are also due to the exchange of vector bosons (gluons) suggests that all these vector bosons (the photon, the three weak bosons, and the eight gluons) are quanta of the components of the gauge field of a large symmetry group, SU_5 or larger. Such theories are called grand unified theories (GUTs). The large symmetry group of the grand unified theory must be spontaneously broken, making all the gauge bosons massive except the gluon octet and the photon, leaving $SU_3 \times U_1$ (color $\times$ electromagnetism) as the apparent gauge symmetry of the world. If the large symmetry group is semisimple (as in SU_5), the gauge field of the grand unified theory has only one coupling constant; this reduces the number of freely choosable parameters in the theory, and so increases its predictive power. The observed coupling constants g_{Strong} ($= g$), g_{Weak}, and g_{EM} ($= e$) are unequal. However, these couplings vary with momentum transfer (running coupling constants, as in QCD). With increasing momentum transfer, g and g_{Weak} decrease whereas e increases, and so they become more nearly equal. By a remarkable coincidence, all three become equal (within errors) at the same momentum transfer (this means that the value of θ_W is correctly given), namely 10^{14} GeV, the grand unification energy. In grand unified theories, this energy is the mass of superheavy leptoquark gauge bosons, analogous to the $W^{\pm}$ and Z^0 bosons of the electroweak subtheory. At momentum transfers larger than this, the mass of the leptoquark bosons is irrelevant and the large symmetry is unbroken; for smaller momentum transfers, the symmetry is broken and the three couplings g, g_{Weak}, and e become different, the more so the lower the momentum transfer.

In these theories, the leptons and quarks occur together in multiplets of the large symmetry group (this is how universality of the weak boson coupling comes about naturally). The couplings of the leptoquark gauge bosons turn leptons into quarks or vice versa (this is the reason for the name leptoquark), or quarks into antiquarks. The exchange of a leptoquark boson can therefore result in the transformation $qqq \to l\bar{q}q$, for example, $p \to e^+\pi^0$. This baryon and lepton-number-violating interaction is a much weaker interaction than the analogous ordinary weak interaction, because leptoquark bosons are much heavier than weak bosons. The predicted ratio of the lifetime of the proton to that of a hyperon is the order of the fourth power of the ratio of the boson masses, which yields the estimate of Eq. (4). A real calculation, assuming the simplest grand unified

$$\tau_p \approx (m_{l-q}/m_W)^4 \tau_\Lambda$$
$$\approx (10^{14} \text{ GeV}/10^2 \text{ GeV})^4 \, 10^{-10} \text{ s} \approx 10^{38} \text{ s} \tag{4}$$

theory, minimal SU$_5$, gives the estimate $10^{-37\pm 2}$ s^{-1} for the partial decay rate of proton into $\pi^0 e^+$, where the factor $10^{\pm 2}$ reflects the uncertainty of the grand unification mass (leptoquark boson mass) and the structure of the proton. Experimental upper limits on partial decay rates seem to rule out the simplest grand unification theory, but there are more elaborate models which will yield proton decay rates not in conflict with observation. Although baryon decay has not been observed, it is possible that baryon creation has been, in the sense that the most reasonable way of accounting for the nonvanishing density of baryons in the universe is to suppose that there was creation of baryons in the past, early in the big bang when the temperature was of the order of the grand unification mass. SEE GRAND UNIFICATION THEORIES.

Bibliography. M. K. Gaillard, Toward a unified picture of elementary particle interactions, *Amer. Sci.*, 70:506–514, 1982; F. Halzen and A. D. Martin, *Quarks and Leptons: An Introductory Course in Modern Particle Physics*, 1984; A. W. Hendry and D. B. Lichtenberg, The quark model, *Rep. Prog. Phys.*, 41: 1707–1780, 1978; I. S. Hughes, *Elementary Particles*, 2d ed., 1985; Y. Nambu, The confinement of quarks, *Sci. Amer.*, 235(5):48–60, 1976; G. K. O'Neill and D. Cheng, *Elementary Particle Physics*, 1979; Particle Data Group: Review of particle properties, *Rev. Mod. Phys.*, 56(2), part 2, April 1984; C. Quigg, Elementary particles and forces, *Sci. Amer.*, 252(4):84–95, April 1985; R. F. Schwitters, Fundamental particles with charm, *Sci. Amer.*, 237(4):56–70, 1977.

ANTIMATTER
JOSEPH LACH

A substance containing atoms that are charge conjugates of atoms found in ordinary matter. Since physicists have demonstrated the existence of the positron, the antiproton, and the antineutron, which are, respectively, the antiparticles (charge-conjugate particles) of the electron, the proton, and the neutron, it is clear that at least in principle it is possible to form the charge conjugate of an atom. Such an atom would be formed in analogy with an ordinary atom, with each particle replaced by its charge conjugate. Such an atom would constitute antimatter. In the laboratory it has been possible to form antideuterons, the nuclei of antihydrogen of mass 2, composed of an antineutron and an antiproton, and an isotope of antihelium of mass 3, composed of an antineutron and two antiprotons. All of the charge conjugates of the known elementary particles that have lifetimes greater than 10^{-10} have been produced by high-energy particle accelerators. SEE ANTINEUTRON; ANTIPROTON; POSITRON.

Dirac theory. The existence of antiparticles was predicted by P. A. M. Dirac in 1928. He attempted to formulate a quantum-mechanical equation that would describe particles with kinetic energies comparable with the energy E contained in their rest mass m. According to the Einstein equation $E = mc^2$, where c is the speed of light, Dirac sought a relation that would be linear in the two basic quantum-mechanical operators analogous to linear momentum p and total energy E (kinetic plus rest mass energy) given by Eq. (1), where $\hbar$ is Planck's constant divided by 2π and

$$\vec{P} \Rightarrow \frac{\hbar}{i}\vec{\nabla} \qquad E \Rightarrow \hbar\frac{\nabla}{\nabla t} \qquad (1)$$

i is the square root of -1. This led him to postulate what is now known as the Dirac equation. For a free particle (subject to no external fields such as the electromagnetic field), this can be written as Eq. (2), where ψ is the wave function describing the quantum-mechanical state of the

$$i\hbar\frac{2\psi}{2t} = -i\hbar\sum_{k=1}^{3}\alpha_k\frac{2\psi}{2x_k} + \beta mc^2\psi \qquad (2)$$

particle. The Dirac equation contains derivatives of ψ with respect to time t, and three spatial coordinates, x_k, $k = 1, 2, 3$. Dirac noted that unless the constants α_k and β were 4×4 matrices, the equation did not have any solutions. These 4×4 matrices implied that ψ also had four components. Dirac interpreted one pair of components as describing the two possible spin states of a spin-½ particle (the electron with spin up or down) and the other pair describing the two spin states of another spin-½ particle called the antielectron. The two particles would have opposite charges, but the same mass, spin, and so forth.

The Dirac equation predicted that particles could have an intrinsic spin as well as the existence of antiparticles. Thus with one bold insight, two of the most fundamental ingredients of modern particle theory were predicted. The creation of a particle and its antiparticle from the vacuum was interpreted by Dirac as lifting its anitparticle from a negative to a positive energy state. The state with all negative energy levels filled is equated to the vacuum, and the filled sea of negative energy states does not produce any physical effects. One cannot produce a Dirac particle from the vacuum without also producing its antiparticle; the antiparticle is the negative energy image or hole left in the negative energy sea. The more modern view utilizes a mathematical theory in which particles can be created and absorbed. This is a field theory and is analogous to the Maxwell theory of the electromagnetic field, which is a field theory of photons. *See* Electron-Positron Pair Production; Quantum Field Theory; Spin.

Discovery of antiparticles. The antielectron or positron was discovered by C. D. Anderson in 1932. In studying the interaction of cosmic rays, he observed that a high-energy photon was converted into a pair of particles, a positive electron and a negative electron. The kinetic energy of the photon was converted into the mass of the electron, positron pair. When coming to rest in ordinary matter, a positron will interact with an electron, releasing their combined energy (both kinetic and rest mass) as photons. This process is called annihilation. This can be viewed as the particle recombining with its negative energy image. The hole in the sea is filled, and both particle and antiparticle are destroyed.

The antiproton was discoverd in 1955 by E. Segré and his coworkers using a high-energy particle accelerator at the Lawrence Berkeley Laboratory. High-energy protons were directed toward a metal target, where some of the kinetic energy of the incident proton was converted to the mass of new particles. Among these new particles the antiproton was identified. Although an isolated antiproton is stable, when brought into contact with matter it annihilates to produce π mesons and other particles. All of these produced particles transform within microseconds into electrons, neutrinos (and their antiparticles), and photons. An atom of matter and its counterpart of antimatter, if brought in contact, would annihilate in a similar manner.

Antimatter in cosmology. Antimatter out of contact with ordinary matter would be stable, and there has been speculation about the presence in the cosmos of antiworlds, composed of antimatter. H. Alfvén studied extensively the possible role of antimatter in cosmology. Substantial concentrations of antimatter would have been detected within the solar system; however, astronomical observations cannot rule out large isolated concentrations of antimatter beyond the solar system.

The big bang theory of the creation of the cosmos provides a natural explanation for the observed apparent excess of matter over antimatter. In this theory the cosmos begins as an almost point-like concentration of energy. At times very close to creation, this fireball contains all known particles, and its expansion procedes in accordance with the known laws of particle physics and thermodynamics. If it is assumed that the big bang fireball originally contained as much matter as antimatter, then the very slight known difference in the decay properties of matter and antimatter will perturb this equality. This difference results from the fact that the combined symmetry operation of charge conjugation and parity (known as CP) is not exact. This gives rise to a very slight excess of matter over antimatter. The fireball cools as it expands, and the matter and antimatter annihilate. The very slight matter excess now becomes significant. All of this would have happened in a time interval in the order of 10^{-38} s. SEE ELEMENTARY PARTICLE; SYMMETRY LAWS.

LEPTON
Martin L. Perl

An elementary particle having no internal constituents which interacts through the electromagnetic, weak, and gravitational forces, but does not interact through the strong (nuclear) force. Leptons are very small, less than 10^{-18} m in size. This is less than 1/1000 the size of a nucleus and less than 10^{-8} the size of an atom. Indeed, existing measurements are consistent with leptons being point particles.

These properties of the lepton family of particles are to be contrasted with the properties of the hadron family of particles. Hadrons such as the π-meson and the proton are believed to be composed of internal constituents called quarks; and hadrons interact through the strong force as well as through the other forces. Hadrons have sizes about 10^{-15} m, which is at least 1000 times the size of a lepton. Thus leptons are much simpler than hadrons in both their structure and their behavior; hence leptons are considered to be more elementary than hadrons. At present the lepton family and the quark family are thought to lie at the same level of elementariness. SEE HADRON; QUARKS.

The **table** lists the properties of the six known leptons. There are three known charged leptons: the electron (e), muon (μ), and tau (τ). Associated with each charged lepton is a neutral lepton called a neutrino. SEE ELECTRON; NEUTRINO.

Properties of leptons

	Electron	Muon	Tau
Charged lepton name	Electron	Muon	Tau
Charged lepton symbol	$e^\pm$	$\mu^\pm$	$\tau^\pm$
Dates of discovery	1890s	Late 1930s	1974–1975
Charged lepton mass, MeV	0.51	105.7	~1785
Charged lepton lifetime, s	Stable	2.2×10^{-6}	$(3.4 \pm 0.5) \times 10^{-13}$
Associated neutrino symbols	$\nu_e, \bar{\nu}_e$	$\nu_\mu, \bar{\nu}_\mu$	$\nu_\tau, \bar{\nu}_\tau$
Associated neutrino mass, MeV	<0.00006 (may be 0)	<0.57 (may be 0)	<150 (may be 0)

Lepton conservation. The association between charged leptons and their neutrinos comes about through an empirical law called lepton conservation. Take as an example the electron (e^-). There are only two ways in which an e^- can be destroyed: it can be annihilated by combining it with an $e+$ (positron); or it can be changed into a ν_e. But an e^- cannot be changed into any other lepton or hadron such as an $e+$, μ^-, μ^+, τ^-, τ^+, π^-. And an e^- cannot be changed into a muon-associated neutrino (ν_μ, $\bar{\nu}_\mu$) or into a tau-associated neutrino (ν_τ, $\bar{\nu}_\tau$). Thus there is a unique property of the electron (e^-) which is preserved or conserved in all reactions; the only other particle carrying this unique property is the electron-associated neutrino (ν_e).

Similar lepton conservation laws hold for the μ and τ. The only other particle carrying the unique property of the μ^- is the muon-associated neutrino (ν_μ). The only other particle carrying the unique property of the τ^- is the tau-associated neutrino (ν_τ). There is at present no explanation for these observed lepton conservation laws. SEE SYMMETRY LAWS.

Decay. The electron is a stable particle; that is, it never decays. The muon and tau decay with average lifetimes listed in the table. The way in which they decay illustrates the lepton conservation laws. The muon decays in only one way, as given in reaction (1), using the μ^- as

$$\mu^- \rightarrow \nu_\mu + e^- + \bar{\nu}_e \qquad (1)$$

the example. The tau, being heavier, decays more rapidly than the muon and can decay in many ways. Some examples are given by (2). In all these examples of τ^- decay, the τ^- is converted to

$$\tau^- \to \nu_\tau + e^- + \bar{\nu}_e$$
$$\tau^- \to \nu_\tau + \mu^- + \bar{\nu}_\mu \quad (2)$$
$$\tau^- \to \nu_\tau + \pi^-$$
$$\tau^- \to \nu_\tau + \pi^- + \pi^+ + \pi^-$$

a ν_τ, with other leptons or hadrons also being produced.

Masses. The masses of the leptons (see table) are given in the energy equivalent unit megaelectron volts (MeV); 1 MeV is the energy gained by an electron that moves through a voltage of 10^6 V, and is equal to 1.602×10^{-13} joule. The masses of the electron, muon, and their neutrinos are smaller than the masses of any of the hadrons. However, the discovery of the tau, whose mass is larger than that of many hadrons, destroyed the concept that leptons had to be very small-mass particles.

All measurements of the masses of the neutrinos are upper limits on the masses; that is, it is possible that some or all of the neutrinos have zero mass. Indeed, most theoretical descriptions of lepton behavior assume that all neutrinos are massless.

Production. Leptons can be produced in various ways. One way is through the decay of a hadron. For example, a muon and its associated antineutrino are produced when a pion decays, as in reaction (3). Another way to produce leptons is through the annihilation of an electron with

$$\pi^- \to \mu^- + \bar{\nu}_\mu \quad (3)$$

a positron; for example, muons can be produced by reaction (4), and taus can be produced by reaction (5). It was through reaction (5) that the tau was discovered. SEE ELEMENTARY PARTICLE.

$$e^+ + e^- \to \mu^+ + \mu^- \quad (4) \qquad e^+ + e^- \to \tau^+ + \tau^- \quad (5)$$

Bibliography. F. Halzen and A. D. Martin, *Quarks and Leptons: An Introductory Course in Modern Particle Physics*, 1984; L. B. Okun, *Leptons and Quarks*, 1983; M. L. Perl, The tau lepton, *Annu. Rev. Nucl. Partic. Sci.*, vol. 30, 1980; M. L. Perl and W. T. Kirk, Heavy leptons, *Sci. Amer.*, 283:50–57, 1978.

ELECTRON
Charles J. Goebel

An elementary particle which is the negatively charged constituent of ordinary matter. The electron is the lightest known particle which possesses an electric charge. Its rest mass is $m_e \cong 9.1 \times 10^{-31}$ kg (2.0×10^{-30} lbm), about 1/1836 of the mass of the proton or neutron, which are, respectively, the positively charged and neutral constituents of ordinary matter. Discovered in 1895 by J. J. Thomson in the form of cathode rays, the electron was the first elementary particle to be identified. SEE ELEMENTARY PARTICLE; NUCLEAR STRUCTURE.

Charge. The charge of the electron is $-e \cong -4.8 \times 10^{-10}$ electrostatic unit (esu). The sign of the electron's charge is negative by convention, and that of the equally charged proton is positive. This is a somewhat unfortunate convention, because the flow of electrons in a metallic conductor is thus opposite to the conventional direction of the current.

The most accurate direct measurement of e is the celebrated oil drop experiment of R. A. Millikan in 1909, in which the charges of droplets of oil in air are measured by finding the electric field which balances each drop against its weight. The weight of each drop is determined by observing its rate of free fall through the air, and using Stokes' formula for the viscous drag on a slowly moving sphere. The charges thus measured are integral multiples of e.

Electrons and matter. Electrons are emitted in radioactivity (as beta rays) and in many other decay processes; for instance, the ultimate decay products of all mesons are electrons, neutrinos, and photons, the meson's charge being carried away by the electrons. The electron itself is completely stable, according to all available evidence. Electrons contribute the bulk to ordinary matter; the volume of an atom is nearly all occupied by the cloud of electrons surrounding the nucleus, which occupies only about 10^{-13} of the atom's volume. The chemical properties of ordinary matter are determined by the electron cloud. SEE BETA PARTICLES; RADIOACTIVITY.

The electron obeys the Fermi-Dirac statistics, and for this reason is often called a fermion. One of the primary attributes of matter, impenetrability, results from the fact that the electron, being a fermion, obeys the Pauli exclusion principle; the world would be completely different if the lightest charged particle were a boson, that is, a particle that obeys Bose-Einstein statistics.

Spin. Every elementary particle possesses an intrinsic angular momentum called its spin. The spin of the electron has the magnitude $\frac{1}{2}\hbar$, where $\hbar$ is Planck's constant h divided by 2π. An electron thus has two spin states: spin up and spin down. To describe this, the nonrelativistic wave function is a two-component function, that is, a vector in a two-dimensional spin-space; the two linearly independent vectors represent the two possible spin states. In 1928 P. A. M. Dirac derived the corresponding relativistic wave equation (Dirac equation). Here, the electron wave function must have four components; correspondingly, for a wave of given momentum, there are four internal states. In addition to the two-valued spin coordinate, there is an energy coordinate; that is, for a momentum p, the energy can be $\pm\sqrt{(m_e c^2)^2 + (pc)^2}$, where c is the velocity of light. SEE SPIN.

Positron. The negative energy states were at first an embarrassment, for they extend downward indefinitely; an electron would cascade indefinitely downward in energy, radiating photons. Electrons obey the exclusion principle, however, and therefore this conclusion can be avoided by assuming that in empty space all the negative energy states are already occupied, and so exclude any more electrons. A new process is possible now; such a negative energy electron, by absorbing energy, can go to a positive energy state. This leaves behind a hole in the sea of negative-energy electrons; the hole has a positive energy, because it represents a missing negative energy. In fact, such a hole has all the properties of an electron except that it appears to have a positive charge (because it represents a missing negative charge). This particle is the positron, first discovered in 1932 by C. D. Anderson in a cloud-chamber study of cosmic radiation. SEE POSITRON.

The electron (sometimes called a negatron) and the positron are on an equal footing; if one started with a Dirac wave equation for the positron, identifying electrons with holes in the negative-energy positron states, one would get an equivalent theory. The apparent dissymmetry inherent in the construction of the hole theory disappears from the results when the total charge, energy, and momentum of empty space is defined to be zero; actually, the dissymmetry can be avoided at all stages in the formalism of quantum field theory. SEE ANTIMATTER; QUANTUM FIELD THEORY; SYMMETRY LAWS.

Magnetic moment. The electron has magnetic properties by virtue of its orbital motion about the nucleus of its atom and its rotation (spin) about its own axis. The intrinsic magnetic moment of the electron is predicted by the Dirac equation to be the value (Bohr magneton) shown in the equation below. The actual moment μ differs from μ_D by a small amount (anomalous mag-

$$\mu_D = \frac{e\hbar}{2m_e}$$

netic moment) due to electromagnetic radiative corrections: $\mu = 1.0011\mu_D$. This theoretical value, calculated using renormalized quantum field theory, agrees with the experimental value. SEE MAGNETON; QUANTUM ELECTRODYNAMICS.

Other leptons. The electron is the lightest of a family of elementary particles, the leptons. The other known charged leptons are the muon and the tau. These three particles differ only (as far as known) in mass; they have the same spin, charge, strong interactions (namely none), and weak interactions. In a weak interaction a charged lepton is either unchanged (neutral weak current reaction) or changes (charged weak current reaction) into an uncharged lepton, that is, a neutrino. In the latter case, each charged lepton (electron, muon, or tau) is seen to change only into the corresponding neutrino (ν_e, ν_μ, or ν_τ). SEE LEPTON; NEUTRINO; WEAK NUCLEAR INTERACTIONS. For additional information SEE COMPTON EFFECT; ELECTRON CAPTURE; PARTICLE ACCELERATOR.

POSITRON
Charles J. Goebel and Jack Greenberg
C. J. Goebel wrote the first three paragraphs.

An elementary particle with mass equal to that of the electron, and positive charge equal in magnitude to the electron's negative charge. The positron is thus the antiparticle (charge-conjugate particle) to the electron. Its existence was predicted by P. A. M. Dirac. It was first observed by C. D. Anderson in 1932. The positron has the same spin and statistics as the electron. Positrons, like electrons, appear as decay products of many heavier particles; electron-positron pairs are produced by high-energy photons in matter. SEE ANTIMATTER; ELECTRON; ELECTRON-POSITRON PAIR PRODUCTION; ELEMENTARY PARTICLE.

A positron is, in itself, stable, but cannot exist indefinitely in the presence of matter, for it will ultimately collide with an electron. The two particles will be annihilated as a result of this collision, and photons will be created. However, a positron can first become bound to an electron to form a short-lived "atom" termed positronium. SEE POSITRONIUM.

The virtual production of electron-positron pairs by an electromagnetic field produces a polarization of the vacuum. This results in effects such as the scattering of light by light and modification of the electrostatic Coulomb field at short distances. SEE QUANTUM ELECTRODYNAMICS.

Quantum field theory predicts the occurrence of a fundamental positron creation process in the presence of strong, static electric fields. For electric fields that exceed the critical value of approximately 1.3 keV/femtometer, the neutral vacuum of quantum electrodynamics decays by positron emission into a new, lowest-energy state which is charged. The source of this vacuum instability and positron emission can be traced to the behavior of the electron when its binding energy in the electric field exceeds the energy equivalent of twice its rest mass, $2m_0c^2$, where m_0 is the rest mass and c is the speed of light. Such strong binding can be achieved in a superheavy atom with an atomic number $Z > 173$. For a bare nucleus with $Z > 173$, it becomes energetically favorable to transform the electron binding energy of larger than $2m_0c^2$ into simultaneously creating an electron bound to the nucleus and a positron that escapes from the nucleus.

This process of spontaneous positron emission has not been observed since atoms with $Z > 173$ are not available in nature. However, with the introduction of heavy-ion accelerators, it has become possible to simulate such an atom for a short period in a high-energy collision between two stable heavy atoms such as uranium. When the two nuclear charges, each with $Z = 92$, are in close proximity at the turning point of the collision, well within the innermost orbits of the electrons, they act together as a common source of nuclear charge with $Z = 184$ to form a quasiatom with many of the properties of an ordinary atom with the same atomic number. The quasiatom maintains a supercritical electric field intensity for a period of approximately 2×10^{-21} s, if the nuclei move on a Rutherford trajectory, or even longer if a giant nuclear complex is formed in the collision. Experiments have utilized a variety of such collision systems with total Z ranging from 180 to 188 to search for spontaneous positron emission. A number of these experiments reproduce the salient features expected for this process. However, some inconsistencies with the predictions of the theory have yet to be resolved before spontaneous positron emission is established experimentally. SEE NUCLEAR MOLECULE; QUASIATOM; SUPERCRITICAL FIELDS.

Bibliography. D. A. Bromley (ed.), *Heavy Ion Science*, vol. 5, 1985; J. W. Humberstone and M. R. C. McDowell (eds.), *Positron Scattering in Gases*, 1983; J. M. Jauch and F. Rohrlich, *The Theory of Photons and Electrons*, rev. ed., 1975; A. Mills and K. F. Canter, *Positron Studies of Solids, Surfaces and Atoms*, 1985.

NEUTRINO
Charles Baltay

An elementary particle designated by the Greek symbol ν, with zero rest mass and zero electric charge. It is classified as a lepton (the other classes of elementary particles being the hadrons and gauge bosons. The elementary particles are known to interact with each other by three basic types of forces: strong nuclear, electroweak (consisting of the electromagnetic and the weak interactions), and gravitational, in order of decreasing strength. All particles with nonzero electric charge, as well as the photon, have electromagnetic interactions, and the mesons and baryons have strong interactions. The neutrino is the only known particle that has only weak interactions. Thus, the neutrino is a unique tool in the study of weak forces, since the interactions are free of the effects of the strong interactions and the electromagnetic interactions, which are many orders of magnitude stronger than the weak interactions. For this reason, ν interactions have been the subject of active study at large particle accelerators. SEE ELEMENTARY PARTICLE; FUNDAMENTAL INTERACTIONS; LEPTON; WEAK NUCLEAR INTERACTIONS.

Basic properties. The existence of the neutrino was postulated in 1930 to explain the apparent nonconservation of energy in beta-decay process $n \rightarrow p + e^- + \bar{\nu}$. It was not unitl 1953 that the existence of the neutrino was experimentally verified, by observation of the interactions caused by free neutrinos. Quantitative studes of ν interactions, undertaken in 1961 by utilizing neutrinos produced by the Brookhaven National Laboratory alternating gradient synchrotron resulted in the descovery that there are two distinct neutrinos—the electron neutrino ν_e, associated with beta decay, and the muon neutrino, associated with pion decay, $\pi^+ \rightarrow \mu^+ + \nu\mu$. From the absence of neutrinoless double beta decay, it can be inferred that ν_e is not identical to its antiparticle $\bar{\nu}_e$. It is thus believed that there are four distinct neutrinos, namely, ν_e, ν_μ, $\bar{\nu}_e$, and $\bar{\nu}_\mu$. SEE RADIOACTIVITY.

In the late 1970s a new heavy charged lepton, the tau ($\tau^\pm$), was discovered to exist in addition to the previously known charged leptons, the electron ($e^\pm$) and the muon ($\mu^\pm$). There is some theoretical speculation that two new additional types of neutrinos, the tau-neutrino (ν_τ) and the anti-tau-neutrino ($\bar{\nu}_\tau$), which are associated with this new heavy lepton, exist in addition to the four types of neutrinos mentioned above, and experimental searches for these new types of neutrinos have been undertaken.

Considerations of angular momentum conservation in pion decay establish the spin of the neutrino to be ½ in units of $h/2\pi$, where h is Planck's constant. The masses of the neutrinos are experimentally consistent with zero; the best upper limits are $M_{\nu_e} \leq 60$ eV and $M_{\nu_\mu} \leq 1.2$ MeV. The experimental upper limit on the neutrino charge is less than 10^5 times the charge of the electron. It is therefore generally assumed that the rest mass and electric charge of the neutrinos are identically zero. However, there has been some experimental indication and theoretical speculation that the mass of at least some of the neutrinos is not identically zero but has some finite, although very small, value. This possiblity has fundamental cosmological implications since astrophysicists believe that there are a very large number of neutrinos in the universe, and if the neutrinos had a nonzero mass they would constitute a significant fraction of the total mass of the universe. Nonzero-mass neutrinos might also lead to some interesting experimental consequences, generally referred to as neutrino oscillations, in which the different types of neutrinos could transmute into each other. SEE SPIN.

According to the rules of quantum mechanics, a massless spin-½ particle such as the neutrino can have its spin lined up along its direction of motion or opposite to its direction of motion; these are called the helicity states, right-handed and left-handed respectively. Thus, in general, such a particle has four components—the particle and its antiparticle, right-handed and left-handed. It was, however, predicted by the two-component theory of the neutrino, and shown by experiment, that only two of the four components exist for the neutrinos—left-handed neutrinos and right-handed antineutrinos. SEE HELICITY.

Study of interactions. The interactions of neutrinos with electrons, protons, and neutrons have been studied in a number of experiments at large proton accelerator laboratories: the 12-GeV zero-gradient synchrotron at the Argonne National Laboratory (Argonne, Illinois); the 28-GeV proton synchrotron (the PS) and the 400-GeV proton synchrotron (the SPS) at CERN (Geneva, Switzerland); the 30-GeV alternating-gradient synchrotron at Brookhaven National Laboratory (Upton, New York); the 70 = GeV proton synchrotron at Serpukhov (Soviet Union); the 400-GeV proton synchrotron and the 1000-GeV superconducting proton synchrotron (the teratron) at the Fermi National Accelerator Laboratory or Femilab (Batavia, Illinois). At all these laboratories, the synchrotrons are used to accelerate on the order of 10^{12} protons/s to their peak energy. The proton beam then is focused onto a small metal target in which, among other things, π and K mesons are produced. Magnetic focusing lenses are used to form a beam of these mesons, which are then allowed to drift in a decay path, where their decays ($\pi \to \mu + \nu_\mu$, or $K \to \mu + \nu_\mu$, for example) are the source of the desired neutrinos. The decay path is followed by a thick shield (made out of iron or earth) that will absorb the μs, and πs and Ks that did not decay, but allow the neutrinos to pass through. SEE MESON.

The neutrino detector, in which the neutrino interactions occur, is located beyond the shield. Typically, 10^9 neutrinos traverse the detector per second. The probability of interactions for the neutrinos is so small that typically less than 1 of these 10^9 neutrinos interact in the detector. Two kinds of neutrino detectors have been used: large bubble chambers, containing 10,000–40,000 liters of liquid hydrogen, deuterium, or some other liquid; and spark chamber detectors, with aluminum, iron, or lead plates, with total masses in the range 10–100 tons (9–90 metric tons). The interactions of neutrinos in these detectors are studied by observation of tracks left by charged particles produced in the interactions. SEE BUBBLE CHAMBER; PARTICLE DETECTOR; SPARK CHAMBER.

The interactions of neutrons proceed via the weak interactions. The weak interactions can be classified in two types: the charged current interactions, in which the exchanged current or particle (the $W^\pm$) carries one unit of electric charge, and thus, for example, an incoming neutral lepton such as the ν_μ is changed into a charged lepton, the μ^-; and the neutral current interactions, where the exchanged current or particle (the Z^0) carries no electric charge, and thus an incident neutral ν_μ remains an outgoing ν_μ. SEE INTERMEDIATE VECTOR BOSON; NEUTRAL CURRENTS.

Charged current interactions. A large number of experiments have been carried out on charged current neutrino interactions such as the quasielastic process, $\nu_\mu + n \to \mu^- + p$, single-pion production $\nu_\mu + p \to \mu^- + p + \pi^+$, and the inclusive inelastic process, $\nu_\mu + N \to \mu^- +$ hadrons, as well as similar reactions with incident $\bar{\nu}_\mu$. The most striking result is related to the study of the internal structure of the proton and the neutron. According to the currently favored quark-parton model, the proton and the neutron are not the simple fundamental particles they were originally believed to be but are composites made up of more fundamental constituents called quarks or partons. The scattering of neutrinos by neutrons or protons can be used to study this internal structure, just as the scattering of alpha particles (Rutherford scattering) was used to investigate the internal structure of the atom. The neutrino scattering data were thus very important in establishing the basic correctness of the quark-parton model. A more complete theory of the forces between the quarks that bind them together in the proton and the neutron is called quantum chromodynamics (QCD). A detailed analysis of the neutrino scattering data has both provided a test of the validity of the QCD theory and yielded measurements of the distributions of the different kinds of quarks inside the proton and the neutron. SEE QUANTUM CHROMODYNAMICS; QUARKS.

Neutral current interactions. Since their discovery in 1973, the neutral current interactions of neutrinos have been studied extensively in a large variety of processes such as neutrino-electron scattering, $\nu + e^- \to \nu + e^-$, neutrino-proton scattering, $\nu + p \to \nu + p$, single-ion production, $\nu + p \to \nu + p + \pi^0$, and inclusive neutrino scattering, $\nu + p \to \nu +$ any number of hadrons. The extensive amount of data collected on these processes can all be understood in terms of the Weinberg-Salam model of the weak and the electromagnetic interactions. This model unifies the electromagnetic and the weak forces into a single force, the electroweak force, reducing the number of basic forces or interactions of nature from four (strong, electromagnetic, weak, and gravitational) to three. SEE WEINBERG-SALAM MODEL.

Charm production by neutrinos. One difficulty encountered in the early days of the Weinberg-Salam model was that the neutral current interactions predicted by the model and observed in neutrino interactions, as discussed above, also predicted certain K-meson decays such as $K^0 \to \mu^+ + \mu^-$, which were not observed experimentally. This difficulty was resolved by S. L. Glashow, J. Iliopoulos, and L. Maiani, who postulated a new quantum number (or property) of hadrons that they called charm, such that the effects of this new quantum number canceled out the unwanted neutral current contributions to K decays. This new quantum number also implied the existence of a new and until then experimentally unobserved class of particles, the so-called charmed particles. The first experimental evidence for the existence of this new class of particles came from neutrino interactions. Experiments at the Fermi National Accelerator lab observed events of the type $\nu_\mu + N \to \mu^+ + \mu^- +$ hadrons and $\nu_\mu + N \to \mu^- + e^+ + K^0 +$ hadrons, where the μ^+ or the e^+ and the K^0 were the decay products of the new charmed particles. An experiment at Brookhaven National Laboratory found an event of the reaction $\nu_\mu + p \to \mu^- + \Lambda^0 + \pi^+ + \pi^+ + \pi^+ + \pi^-$, where the Λ^0 and the four πs were interpreted to be the decay products of a charmed particle. Conclusive evidence for the existence of charmed particles was obtained in $e^+ + e^-$ collisions at the Stanford Linear Accelerator Center in California and in photoproduction experiments at Fermilab. Subsequently the production of four different charmed particles has been seen in neutrino reactions: the $D(1.86)$, the $D^*(2.01)$, the $\Lambda_c(2.26)$ and the $\Sigma_c(2.43)$, where the numbers in parentheses are the masses of the particles in units of GeV. The properties of these particles as produced in neutrino interactions have been found to be consistent with the predictions of the model of Glashow, Iliopoulus, and Maiani. SEE CHARM.

Bibliography. V. Barger and D. Cline (eds.), *Neutrino Mass and Low Energy Weak Interactions*, 1984; F. Boehm and P. Vogel, Low energy neutrino physics and neutrino mass, *Annu. Rev. Nucl. Part. Sci.*, 34:125, 1984; D. Cline and W. F. Fry, Neutrino scattering and new particle production, *Annu. Rev. Nucl. Part. Sci.*, 27:209–278, 1977; T. Ferbel (ed.), *Techniques and Concepts of High Energy Physics III*, 1985; E. Fiorini (ed.), *Neutrino Physics and Astrophysics*, 1982; H. E. Fisk and F. Sciulli, Charged current neutrino interactions, *Annu. Rev. Nucl. Part. Sci.*, 32:499–573; 1982; P. A. Huug and J. J. Sakurai, *Struct. Neutral Currents*, 31:375–438, 1981; J. Tran Thanh Van (ed.), *Proceedings of the 4th Moriond Workshop on Massive Neutrinos*, 1984.

QUARKS
LAWRENCE W. JONES AND O. W. GREENBERG
O. W. Greenberg wrote the section Theory.

The basic constituent particles, of which "elementary" particles are now believed to be composed. Theoretical models built on the quark concept have been very successful in explaining and predicting many phenomena in particle physics. However, the experimental observation of free quarks remains ambiguous.

Search for fundamental constituents. Physics research for almost two centuries has been probing progressively deeper into the structure of matter in order to seek at every stage the constituents of each previously "fundamental" entity. Thus a sequence proceeding from crystals through molecules, atoms, and nuclei to nucleons and mesons has been revealed. The energies required to dissociate each entity increase from thermal energies to gigaelectronvolts in proceeding from crystals to mesons. It is therefore only a logical extrapolation of past patterns to expect that hadrons—mesons and baryons—might be dissociated into more fundamental constituents if subjected to a sufficiently high energy. SEE HADRON; NUCLEON.

Evidence supporting the quark model. There were at least three factors which made a quark model plausible in the 1960s. First, the classic electron-proton elastic scattering experiments demonstrated that a proton has a finite form factor. Nonrelativistically, this is equivalent to a finite radial extent of the electric charge and magnetic moment distributions. It was plausible that the charge cloud which constitutes a proton is a probability distribution of some smaller, perhaps pointlike constituents, just as the charge cloud of an atom was learned to be a probability distribution of point electrons.

Second, the evolution in the late 1950s and early 1960s of hadron spectroscopy revealed an order and symmetry among the states of hadronic matter that could be interpreted in terms of representations of the SU(3) symmetry group. This in turn was interpreted by M. Gell-Mann, and independently by G. Zweig, as a consequence of the grouping of elementary constitutents of fractional electric charge (christened quarks by Gell-Mann) in pairs and triplets to form the observed hadrons. The general features of the quark model of hadrons have withstood the tests of time, and many of the static properties of hadrons are consistent with predictions of this model. SEE SYMMETRY LAWS.

Third, the deep inelastic scattering of electrons on protons revealed form factors corresponding to pointlike constituents of the proton. This is altogether consistent with the interpretation of the finite proton elastic form factor suggested above, in analogy to atomic electrons. These proton constituents were named partons, although from the beginning it was recognized that partons and quarks might be merely different manifestations of the same entities. The parton notion of proton structure was also invoked to explain the nature of secondary particle (pion) distributions in high-energy proton-proton collisions.

Physicists now believe that the proton and neutron are not fundamental constituents of matter, but that they are made of quarks, very much as the nuclei of ^{3}H and ^{3}He are made of protons and neutrons and as the molecules of NO_2 and N_2O are made of oxygen and nitrogen atoms.

Kinds of quarks. Until 1974 only three flavors of quarks were known; two of very nearly equal mass, of which the proton, neutron, and pi mesons are composed, and a third, more massive quark which is a constituent of strange particles such as the K mesons and hyperons such as Λ^0. The names attached to these quarks are the up quark (u), the down quark (d), and the strange quark (s). Baryons are presumed to be composed of three quarks, such as the proton (uud), neutron (udd), $\Lambda^0(uds)$, and $\Xi^-(dss)$. Mesons are composed of a quark-antiquark pair, such as the $\pi^+(u\bar{d})$, $\pi^-(\bar{u}d)$, $K^+(u\bar{s})$, and $K^-(\bar{u}s)$. Antiparticles such as the antiprotons are formed by the antiquarks of those forming the particle, for example, the antiproton $\bar{p}$ ($\bar{u}\bar{u}\bar{d}$). S EE B ARYON ; M ESON .

The quantum numbers of quarks are simply added to give the quantum numbers of the elementary particle which they form on combination. The natural unit of electric charge of a quark is $+2/3$ or $-1/3$ of the charge on a proton (1.6×10^{-19} coulomb), and the baryon number of each quark is $+1/3$; the charge, baryon number, and so forth, of each antiquark is just the negative of that for each quark.

The properties of quarks, to the extent that they are now understood, are presented in **Table 1**, where additional quarks which are discussed below are also included.

Table 1. Properties of quarks

Flavor	u	d	c	s	t*	b
Mass, (GeV/c^2)†	0.39	0.39	1.55	0.51	(30–50)	4.72
Electric charge	+2/3	−1/3	+2/3	−1/3	(+2/3)	−1/3
Baryon number	1/3	1/3	1/3	1/3	(1/3)	1/3
Spin in units of ($\hbar$)	1/2	1/2	1/2	1/2	(1/2)	1/2
Isotopic spin	+1/2	−1/2	0	0	(0)	0
Strangeness	0	0	0	−1	(0)	0
Charm	0	0	+1	0	(0)	0

*There are some reports of evidence for the t quark. Listed are its predicted properties.
†Masses are uncertain. Values listed here are half the mass of the lowest-lying quark-antiquark vector meson.

The manner in which the u, d, and s quarks and their antiquarks may combine to form families of mesons is indicated in **Fig. 1** for one such multiplet, a meson pseudoscalar nonet. On this plot, "hypercharge" is a quantum number related to the quark strangeness and baryon number, whereas "isospin" (here the third component of isotopic spin) is a quantum number related to the u-d quark difference. S EE H YPERCHARGE ; I SOBARIC SPIN .

In the strong interactions of elementary particles, quark-antiquark pairs may be created if sufficient energy is present, but a quark does not transform into another quark. The weak interaction, however, does permit quark transformations to occur, so that a d quark becomes a u quark in the radioactive decay of a neutron, and elementary particles containing an s quark (strange particles) decay to nonstrange particles only through the weak interaction. S EE S TRANGE PARTICLES ; W EAK NUCLEAR INTERACTIONS .

A particularly interesting set of particles is the group of vector mesons composed of a quark-antiquark pair. An example is the ϕ meson, composed of an $s\bar{s}$ quark pair. It may decay to

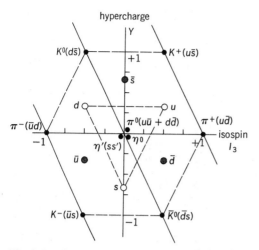

Fig. 1. Pseudoscalar meson nonet and the u, d, and s quarks and antiquarks as represented in isospin (I_3)–hypercharge space. The solid lines sloping downward to the right are lines of constant electric charge.

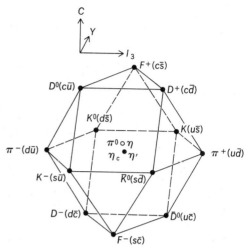

Fig. 2. Pseudoscalar mesons composed of u, d, s, and c quarks. The three orthogonal axes are the z component of isospin I_3, hypercharge Y, and charm C. (After C. Quigg, Lectures on charmed particles, Fermi National Accelerator Laboratory, Fermilab-Conf-78/37-THY, April 1978)

a K^+K^- meson pair, but is inhibited from decaying to a $\pi^+\pi^-$, as the latter contain no s or $\bar{s}$ quarks. The meson may also decay electromagnetically to a $\mu^+\mu^-$, or e^+e^- lepton pair; conversely it may be produced by a photon (which coincidentally has the same quantum numbers) or by the collision of an electron and positron.

Such quark-antiquark systems are very analogous to the positronium atom, composed of a positron, e^+, and an electron, e^-. The force binding this atom together is the electromagnetic Coulomb attraction, whereas the force binding the quark-antiquark system is the strong interaction. In both cases, the energy levels of excited states of the system are directly related to the details of the force holding the particles together. SEE POSITRONIUM.

Charm. In 1974 two very different experiments, at the Brookhaven National Laboratory and the Stanford Linear Accelerator Center, found evidence for unusual elementary particles with lifetimes about 1000 times longer than typical mesons which decay through the strong interaction. These particles (labeled the J or ψ) were the first evidence for the existence of a new quark, the c (charm) quark. In the Brookhaven experiment of S. C. C. Ting and his collaborators, a high-intensity proton beam of 30 GeV was directed onto a beryllium target, and the invariant mass of produced e^+e^- pairs studied for evidence of the new particle. B. Richter with a group from Stanford University and the University of California found a sharp resonance in the cross section for e^+e^- interactions in an electron-positron colliding-beam storage ring. Ting labeled his particle the J; Richter independently called it the ψ.

The J/ψ meson is understood to be the vector meson composed of the $c\bar{c}$ quark pair, and its decay into mesons is strongly inhibited, so that its lifetime is about 1000 times typical heavy, unstable mesons. This new quark had been predicted earlier by S. Glashow. Subsequently mesons and baryons containing the c quark were found and studied. The two-dimensional diagram of Fig. 1 may be elaborated by adding a third dimension corresponding to the charm quantum number, with integral values of charge, and so forth, defining a polyhedron whose vertices correspond to identified and to predicted charmed mesons. Such a diagram is reproduced in **Fig. 2** for the pseudoscalar mesons, including those of Fig. 1. SEE CHARM; J PARTICLE.

Upsilon particle. At the 400-GeV proton synchrotron of the Fermi National Accelerator Laboratory near Chicago, L. M. Lederman and his colleagues studied the production of $\mu^+\mu^-$ meson pairs in an experiment which in many ways resembled a scaled-up version of Ting's Brookhaven experiment. In 1977 the group reported a new resonance at about 9.4 GeV; subsequent detailed study by this and other groups confirmed that this is the lowest-lying state of a new quark system. Experiments at the electron-positron storage ring at the Deutsches Elektronen Synchrotron (DESY) laboratory near Hamburg strongly indicated that the new quark has a charge of $-\frac{1}{3}$. The quarks which make up Lederman's Υ (upsilon) particle have been called b for "bottom" (or "beauty"). There is a new quantum number associated with the b, and there are mesons which contain a b quark and an ordinary u or d antiquark. These mesons, referred to generically as B mesons, have been produced and studied at the e^+e^- colliding beam facilities at Stanford, Cornell, and DESY. A four-dimensional extension of Fig. 2 can be imagined which includes this added hierarchy of mesons, such as $b\bar{u}$, $b\bar{s}$, and $c\bar{b}$. SEE UPSILON PARTICLE.

Physicists expect that there is a heavier quark which has the same relationship to the b quark that the charm quark has to the strange quark. In 1984 a CERN group reported evidence that a state with a mass of 30 to 50 GeV had been seen and that it might be a meson containing this heavier t quark, where t may stand for "top" or "truth". The t quark is listed in the table, together with its presumed properties. Careful searches for the "toponium" state $t\bar{t}$ have been negative, up to total energies of 45 GeV (corresponding to a t quark mass of 22.5 GeV).

Color. The most natural spin assignment for quarks is ½, such that their intrinsic angular momentum is $\hbar/2$ (where $\hbar$ is Planck's constant h divided by 2π), exactly as for the electron and muon. This results in spins of 0,1, or higher integers (for mesons composed of quark-antiquark pairs with orbital angular momentum about each other), and in spins for three-quark baryons (protons and so forth) of ½, ³⁄₂, and so forth. A problem arose when the structure of observed baryons required two or, in some cases, three quarks of the same flavor in the same quantum state, a situation which is forbidden for spin-½ particles by the Pauli exclusion principle. In order to accommodate this contradiction, the concept of color was introduced, and it was proposed that each quark could be red, green, or blue. This color quantum number then breaks the degeneracy and allows up to three quarks of the same flavor to occupy a single quantum state. Since the original proposal, confirmation of the color concept has been obtained from experiments with electron-positron storage rings, and the theory of quantum chromodynamics (analogous to quantum electrodynamics) has been developed. SEE COLOR; QUANTUM CHROMODYNAMICS.

Searches for free quarks. One outstanding mystery remains: why are quarks not observed as free objects in high-energy collisions? Their fractional electric charge should render them easily detectable because the ionization, and hence the signal, produced by a charged particle of very high energy passing through a detector is proportional to the square of its electric charge. A quark would then give a signal of $\frac{1}{9}$ or $\frac{4}{9}$ that from an electron, proton, meson, or any other known charged particle.

One possible explanation for the failure of experiments to detect quarks would be that their rest mass is so great that searches at particle accelerators have failed to reach the threshold for their production. Alternatively (or in addition) they might be produced so rarely that the quark production cross section might be very low.

Sensitive searches have been made at all high-energy particle accelerators and with cosmic rays. In 1969 positive evidence for quarks was reported from cosmic rays. Subsequent, more sensitive cosmic-ray experiments have, however, failed to confirm those observations.

In an experiment at the intersecting storage rings (ISR) of the European Organization for Nuclear Research (CERN), several "telescopes" of scintillation counters, Cerenkov detectors, and multiwire proportional chambers were assembled to search for particles of fractional charge emerging from the proton-proton (p-p) intersection region. These p-p collisions at 52 GeV in the center of mass (equivalent to a proton of 1500 GeV striking a proton at rest) would be energetically above threshold for the production of quarks in pairs of up to 23 GeV rest mass. No quarks were seen among the products of 1.2×10^{10} proton-proton interactions, corresponding to an upper-limit quark production cross section of 4×10^{-35} cm², or less than one quark per billion proton-proton interactions.

Two other experiments were carried out at the synchrotron of the Fermi National Accelerator Laboratory (FNAL) near Chicago. Here protons of up to 300 GeV struck a beryllium target, and beams of secondary particles were detected emerging from it. In the first experiment, a telescope of scintillation counters was used to search for particles of fractional electric charge. No quarks were found among 1.4×10^9 negative particles (mesons) corresponding to an upper-limit

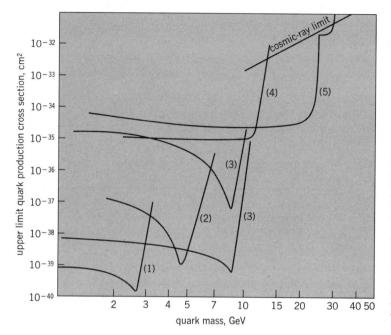

Fig. 3. Upper-limit cross section for production of quarks of charge-⅓ e as a function of quark mass from experiments at various particle accelerators and cosmic rays. The different experiments are indicated by numbers: (1) CERN 28-GeV proton synchrotron; (2) Serpukov 70-GeV proton synchrotron; (3, 4) Fermi National Accelerator Laboratory synchrotron; (5) CERN intersecting storage rings. (*After L. W. Jones, A review of modern quark search experiments, Rev. Mod. Phys.*, 49:717–752, 1977)

production cross section of about 10^{-35} cm^2. With 300-GeV protons incident, this search was sensitive to quarks of up to 12 GeV rest mass. The second search was made with similar beams over a somewhat longer period of time and with additional detectors. No quarks were detected with an upper-limit production cross section of 10^{-39} cm^2 for 9 GeV/c^2 rest mass quarks of ⅓ e charge and 10^{-38} cm^2 for 11 GeV/c^2 quarks of −⅔ e charge. SEE PARTICLE ACCELERATOR.

These searches, together with accelerator and cosmic-ray searches reported earlier, are summarized in **Fig. 3** for quarks of charge −⅓. The area of production cross section–quark mass to the lower right of the curves represents still possible quark production.

Search for quarks in stable matter. It is also possible to seek quarks in stable matter by exploring the net electric charge on small samples, such as oil droplets or metal pellets. The charge of the electron e was first determined in 1911 by R. A. Millikan in his oil drop experiments, and variations of that technique have been used to search for net electric charges of ±⅓ e. Whereas a number of experimenters have reported negative results of searches for fractional charges, W. M. Fairbank and colleagues have reported positive evidence for quarks. In their experiment, Fairbank's group suspended superconducting niobium pellets of about 100 micrograms in a magnetic potential well, and then caused them to oscillate under an applied electric field. Several of the pellets that they have studied have given evidence for fractional electric charge. A peculiar aspect of their evidence for free quarks is that the pellets seem to gain or lose their quarks somewhat capriciously. It is also disturbing that this is the only group to claim evidence for quarks, in spite of considerable efforts and comparable sensitivity elsewhere. The Fairbank result is difficult but not impossible to reconcile with the otherwise negative evidence. Most physicists would prefer to await an independent confirmation of this result before accepting the idea of free quarks. The prevailing view of theoretical physicists is that quarks may not exist as free particles. The quantum chromodynamics theory incorporates a force between quarks which grows with their separation so that they cannot be isolated.

Understanding of quarks. Quarks are now firmly established as fundamental constituents of matter. Theoretical physicists have evolved a theoretical structure, quantum chromodynamics (discussed in more detail below), to understand and predict the behavior of elementary particles, based on the interactions of their constituent quarks. With the understanding of quarks, it may be possible to evolve a grand unified theory combining the forces of electricity and magnetism, the strong and the weak interactions, all into one coherent theoretical framework, and a major theoretical effort in this direction has been undertaken. SEE FUNDAMENTAL INTERACTIONS; GRAND UNIFICATION THEORIES.

Unresolved questions. In spite of the dramatic progress in the understanding of elementary particles which has accompanied the quark concept and the corresponding revolution in the concept of matter at its most fundamental level, at least four major questions remain.

First, it has not been determined whether or not there are free quarks, and whether Fairbank's observations are correct, or there is an error in his experiment. Second, the number of flavors (kinds) of quarks has not been determined. The predicted t quark has not been definitely observed, although there is now some evidence for it. It is not known whether there are a large number of quarks, or whether the flavor spectrum stops at five or six. Third, it is not known whether quarks are truly fundamental entities, or whether quarks in turn have their own internal substructure. Finally, the quarks may not be stable for infinite times, but may undergo spontaneous decay to leptons. Grand unified theories predict that such decays would lead to the instability of all matter with a lifetime of at least 10^{30} years. Experimental searches have been unsuccessful in detecting evidence for such decays and have set limits on the lifetime of the proton of at least 10^{32} years.

Theory. The theory of the properties and interactions of quarks will now be discussed in detail.

Local gauge theories. The theory of the interactions of quarks is expressed, at a fundamental level, in terms of local gauge theories: theories which are invariant under independent symmetry transformations at each space-time point. Such a theory is characterized by its symmetry group, coupling constants, matter fields, and whether or not the symmetry is broken spontaneously. (Spontaneously broken symmetry is described below.) Local gauge theories first entered elementary particle physics in quantum electrodynamics (QED), the theory of the interactions of the electron-positron field with the Maxwell field of electromagnetism. For quantum electrodynamics, the symmetry group is U(1), the group of complex numbers of modulus one which can be identified with rotations of the complex plane; the coupling constant is the electric charge, the matter field is the electron-positron field, and the group is unbroken. SEE SYMMETRY LAWS.

Local gauge theories were generalized to "non-abelian" theories by C. N. Yang and R. L. Mills. Non-abelian theories have a non-abelian group, that is a group in which, in general, $g_2 g_1 \neq g_1 g_2$, where the g's are group elements. Typical non-abelian groups are the SU(N) and SO(N) groups. SU(N) is the group of N by N unitary (U) matrices which have determinant one (or are "special", S). Unitary matrices preserve the length in a complex space: they have the property $UU^\dagger = I$, where I is the unit matrix, and $U^\dagger$ is the complex conjugate transpose of U. SO(N) is the group of N by N orthogonal matrices. Orthogonal matrices preserve the length in a real space: they have the property $OO^t = I$, where O^t is the transpose of the matrix O. The group of rotations in three-dimensional space is the group SO(3). SEE UNITARY SYMMETRY.

The groups which are relevant for gauge theories are "Lie" groups, named after S. Lie. These groups are continuous, like the group of rotations in three-dimensional space, and have further smoothness properties, which are also shared by SO(3). An arbitrary element of a Lie group near the unit or identity can be written as the identity plus a sum of a certain number of "generators" multiplied by real numbers. Each generator corresponds to a small or "infinitesimal" transformation away from the identity. For SO(3), the generators can be chosen to be the deviations from the identity associated with small rotations around the x, y, and z axes.

In a local gauge theory there must a gauge field, whose associated gauge particle is a spin-1 boson, corresponding to each generator. For U(1), which has one generator, there is one gauge field. In quantum electrodynamics, this is the electromagnetic vector potential. For SU(N) and SO(N) there are $N^2 - 1$ and $\frac{1}{2}N(N - 1)$ generators and gauge fields, respectively. Gauge fields are the mediators of the interactions between the matter fields. When the local gauge symmetry is unbroken, the particles associated with the gauge fields are massless, corresponding to a long-range interaction.

For groups such as U(1), SU(N), or SO(N), the interactions are determined by a single constant. This fact, and the fact that the form of the interaction is completely fixed by the group and the matter fields, makes such gauge theories tightly constrained. There is a charge associated with each generator of a gauge theory. For quantum electrodynamics the charge is the electric charge of the electron. A crucial difference between non-abelian and abelian gauge theories is that the gauge fields themselves carry the charges in non-abelian theories, while in abelian theories the gauge fields, for example, the electromagnetic potential in quantum electrodynamics, are neutral (that is, do not carry the charges). Thus the gauge fields in a non-abelian theory can interact by exchanging other gauge fields, while the gauge field (or fields) in an abelian theory only interact with each other indirectly via the matter fields.

Spontaneous symmetry breaking is the phenomenon in which the symmetry is broken by the vacuum state, although the lagrangian and equations of motion of the theory obey the symmetry. P. W. Higgs, T. W. B. Kibble, and others showed that spontaneous symmetry breaking in a local gauge theory gives rise to a mass for the gauge particles associated with the generators for which the symmetry is broken. The interactions mediated by massive gauge particles are short-ranged. SEE GAUGE THEORY; SYMMETRY BREAKING.

Electroweak interactions of quarks. The electromagnetic interactions of quarks can be described by using the electromagnetic current given by Eq. (1), where the γ_μ are the Lorentz

$$j_\mu = \sum_i (2/3 \; \bar{u}_i \gamma_\mu u_i - 1/3 \bar{d}_i \gamma_\mu d_i) \tag{1}$$

covariant forms of the Dirac matrices, and the sum runs over the three presently known generations of quarks listed in **Table 2**. The u_i are the so-called up quarks with charge $2/3 e$ (u, c, and t), and the d_i are the so-called down quarks with charge $-1/3 e$ (d, s, and b). The six quark species are called "flavors." This form of the electromagnetic current says that the electromagnetic interaction does not change the flavor of the quark, and that quarks interact electromagnetically with a strength proportional to their electric charge. Electromagnetism is mediated by the photon, a neutral massless gauge particle.

The weak interactions of quarks can be described by using three weak currents: charge raising and lowering ones, and a neutral current. The prototype of charged-current weak interactions is beta decay, exemplified by the decay of the neutron, $n \rightarrow pe^-\bar{\nu}_e$. The charge raising current has the form of Eq. (2), where d'_i stands for a superposition or mixture of the three down

$$j_\mu^+ = \sum_i \bar{u}_i \gamma_\mu d'_i \tag{2}$$

quarks. N. Cabibbo introduced this mixing between the d and s quarks to take account of the fact that strangeness-conserving decays, such as neutron decay, go at a faster rate (when phase space factors are removed) than the analogous strangeness-violating decays, such as lambda decay, $\Lambda \rightarrow pe^-\bar{\nu}_e$. The theory with only three flavors of quarks leads to "strangeness-changing neutral-current weak decays," such as $K \rightarrow \mu\bar{\mu}$, which experimental data show occur at an extraordinarily low rate. To avoid these unseen decays, Glashow, J. Iliopoulos, and L. Maiani suggested that there should be a fourth flavor of quark, the charm quark c, whose presence in intermediate states would cancel the unobserved decays, provided the mass of the c quark was not too different from the mass of the u quark. The charm theory predicted new hadronic states containing the c quark. As discussed above, the discovery of the narrow J/ψ resonance at mass

Table 2. Three generations of quarks and leptons

	Generation:	1	2	3
Quarks*		u_α	c_α	(t_α)
		d_α	s_α	b_α
Leptons		ν_e	ν_μ	ν_τ
		e^-	μ^-	τ^-

*The index α runs over the three quark colors. The three quarks with charge $2/3 e$ (u, c, and t) are called the three up quarks u_i; and the three quarks with charge $-1/3 e$ (d, s, and b) are called the three down quarks d_j.

3.1 GeV/c^2 gave dramatic confirmation of the theory, and the further narrow T resonance discovered at mass 9.4 GeV/c^2 gave evidence for the b, the fifth flavor of quark. There is tentative evidence that t, the sixth flavor of quark, has been found in W decays with mass in the range 30–50 GeV/c^2.

Until the work of Glashow, S. Weinberg, and A. Salam, who contributed theoretical developments which gave a partial unification of electromagnetism and weak interactions, the theory of weak interactions suffered from divergences in calculations beyond lowest order in the Fermi constant, $G_F \sim 10^{-5} M_P^{-2}$, where M_P is the proton mass. This partial unification used a local gauge theory based on the group SU(2) × U(1), which has four generators, three from the group SU(2) and one from U(1). The gauge fields associated with two of the SU(2) generators are identified as the mediators of the charge-changing weak interactions mentioned above. The electromagnetic vector potential is associated with a linear combination (or "mixture") of two of the remaining generators. A fourth gauge field Z^0 associated with the orthogonal mixture of the last two generators is required by the theory. Spontaneous symmetry breaking simultaneously produces the mixing of the generators, and gives masses to the $W^\pm$ and Z^0 fields, as required by the short range of the weak interactions. Since the Z^0 is neutral, this partially unified theory predicts the existence of neutral-current weak processes, such as $\nu_\mu p \to \nu_\mu p$. Such processes were found in neutrino experiments at CERN, giving striking confirmation of the electroweak theory. SEE WEINBERG-SALAM MODEL.

When the electroweak theory was first proposed in essentially final form, it was not clear that it avoided divergences in higher-order calculations. An important theoretical step, taken by G. 't Hooft before the detection of neutral currents, showed that such divergences are absent, so that the theory is well-defined. Technically, the theory was shown to be renormalizable; that is, all divergences in higher-order calculations can be absorbed in redefinitions of a small number of physical constants, such as masses and coupling constants.

The CP violation observed in the $K^0 - \bar{K}^0$ system is a puzzling phenomenon. M. Kobayashi and T. Maskawa showed that at least six flavors of quarks are needed for quark mixing to lead to violation of CP. As mentioned above, the sixth flavor, t, which would complete the third generation, may have been found.

Quantum chromodynamics. The apparently contradictory low-energy (massive, strongly bound, indeed permanently confined) and high-energy (almost massless, almost free) behaviors of quarks can be reconciled by using quantum chromodynamics (QCD), which is a local gauge theory based on the group SU(3), as the theory of the strong interactions. The eight gauge fields of quantum chromodynamics are called gluons. SEE GLUONS; QUANTUM CHROMODYNAMICS.

The high-energy properties of quarks follow from the asymptotic freedom of quantum chromodynamics. The interquark potential has the form of Eq. (3). Because of quantum corrections,

$$V(r) = g_{\text{eff}}^2(r)/r \tag{3}$$

the effective or running coupling constant g_{eff} of a quantum field theory depends on the distance scale r at which the coupling constant is measured. A theory is asymptotically free if g_{eff} goes to zero as r goes to zero so that perturbation theory is valid for small r. For large r, the potential grows linearly without bound, as given in Eq. (4) where σ is a constant, so that single quarks can

$$V(r) \to \sigma r \tag{4}$$

never be free. This is the phenomenon of permanent quark confinement. Since confinement occurs when g_{eff} is large, nonperturbative methods are necessary to demonstrate confinement. The best evidence for confinement comes from the Monte Carlo numerical approximation to lattice gauge theory on finite lattices.

Standard model. The local gauge theory based on the group SU(3) × SU(2) × U(1), in which all fundamental interactions have a gauge theory origin, and SU(3) is unbroken, while SU(2) × U(1) is spontaneously broken to the U(1) of electromagnetism, serves as a standard model for all elementary particle physics at energies below about 100 GeV.

Beyond the standard model. As noted above, a major direction of theoretical research beyond the standard model is the search for a grand unified theory in which quantum chromodynamics and the electroweak interactions, and possibly gravity, are part of a larger scheme with a single coupling constant; however, the striking prediction of grand unified theories that the proton should decay has not been confirmed. The two other areas of such research are the exploration of supersymmetric theories in which all fermions have bosonic partners, and vice versa; and the study of models in which quarks, leptons, and other "elementary" particles are composites of still more basic particles. SEE PROTON; SUPERSYMMETRY.

Bibliography. S. Coleman, The 1979 Nobel prize in physics, *Science*, 206:1290–1292, 1979; E. D. Commins and P. H. Bucksbaum, *Weak Interactions of Leptons and Quarks*, 1983; R. P. Feynman, Structure of the proton, *Science*, 183:601–610, 1974; H. Fritzsch, *Quarks, the Stuff of Matter*, 1983; S. L. Glashow, Quarks with color and flavor, *Sci. Amer.*, 233(4):38–50, 1975; O. W. Greenberg, Resource Letter Q–1, Quarks, *Amer. J. Phys.*, 50(12): 1075–1089, 1982; F. Halzen and A. D. Martin, *Quarks and Leptons: An Introductory Course in Modern Particle Physics*, 1984; L. W. Jones, A review of quark search experiments, *Rev. Mod. Phys.*, 49(4)717–752, 1977; L. M. Lederman, The upsilon particle, *Sci. Amer.*, 239(4):72, 1978; Y. Nambu, The confinement of quarks, *Sci. Amer.*, 235(5):48–60, 1976; R. F. Schwitters, Fundamental particles with charm, *Sci. Amer.*, 237(4):56–70, 1977.

PHOTON
MURRAY SARGENT III

A quantum of a single mode (that is, single wavelength, direction, and polarization) of the electromagnetic field. There are also two other definitions of photon in use, not entirely consistent with the first definition or each other: an elementary light particle or "fuzzy ball," and an informal unit of light energy. The fuzzy-ball definition emphasizes a particle character of light suggested, for example, by momentum exhibited in the Compton effect and light levitation phenomena. Although this definition is often justified by the random arrivals of counts in photoelectron detection, light waves incident on a quantum-mechanical detector yield the same behavior. More critically, the fuzzy-ball picture lacks a rigorous foundation and is not required for the explanation of any fundamental phenomenon. As an informal unit of energy, the photon equals $h\nu$, where h is Planck's constant ($= 6.626 \times 10^{-34}$ joule-second), and ν is the frequency of the light in hertz.

The definition as a single-mode light quantum has rigorous foundation in quantum electrodynamics, and contradicts the fuzzy-ball definition in that, according to Fourier analysis, light of a single wavelength must be spread out. Other theories, typified by "neoclassical" theory, attempt to explain the interaction between light and matter by quantizing only the matter's response, that is, without using the photon. However, quantum electrodynamics remains the only theory capable of quantitatively explaining spontaneous emission, the Lamb shift, and the anomalous magnetic moment of the electron. SEE COMPTON EFFECT; QUANTUM ELECTRODYNAMICS.

GLUONS
C. QUIGG

The hypothetical force particles which are believed to bind quarks into "elementary" particles. Although theoretical models in which the strong interactions of quarks are mediated by gluons have been successful in predicting, interpreting, and understanding many phenomena in particle physics, free gluons remain undetected in experiments (as do free quarks). According to prevailing opinion, an individual gluon cannot be isolated.

Color. In 1961 M. Gell-Mann and Y. Ne'eman independently suggested that the strong (nuclear) interaction respected the unitary symmetry SU(3) and that the strongly interacting particles called hadrons could be classified according to the patterns prescribed by SU(3). The family groups or supermultiplets that emerged were confined to a few of the simplest possibilities permitted under SU(3) symmetry. Mesons, the hadrons with integral spin in units of $\hbar$ (Planck's constant h divided by 2π), occur only in families with 1 or 8 members. The baryons, which possess half-integral spin, fit into groups with 1, 8, or 10 members. Gell-Mann and G. Zweig separately showed in 1963 that this circumstance could be explained by the hypothesis that hadrons were composites of fundamental constituents that have come to be called quarks. In this quark model of hadrons, a meson is composed of one quark and one antiquark, and a baryon is composed of three quarks. All the hadrons then known could be built out of three different varieties (or flavors) of quarks, denoted up, down, and strange. To account for the observed pattern of mesons and baryons, quarks must be spin-½ particles. S*ee* U*nitary symmetry*.

Although these rules reproduce the properties of the observed hadron states, they lead to a theoretical inconsistency. The characteristics of the unstable hadron resonance known as Δ^{2+} (1232 MeV/c^2), which decays into a proton and a positively charged pi meson, require that it be composed of three quarks in a configuration which is symmetric under the interchange of any pair of quarks. However, according to the Pauli exclusion principle (which first emerged in the description of atomic structure), identical spin-½ particles cannot occupy the same quantum state. The quark model could be brought into agreement with the Pauli principle, without compromising any of its successes, if a new attribute were ascribed to the quarks which would make the three up quarks distinguishable. For fanciful reasons, this new attribute is now known as color, though it has no connection with the color of visible light. Quarks are said to come in three colors, most frequently given the arbitrary labels red, blue, and green. A Δ^{2+} resonance composed of one red up quark, one blue up quark, and one green up quark will then have the observed properties and be consistent with the laws of quantum mechanics. In this picture the antiparticle of a red up quark is an anti-red anti-up quark, so that the mesons are described as colorless quark-antiquark pairs.

Support for the idea that each quark flavor comes in three distinguishable colors has come from the lifetime of the neutral pi meson, and from the rate at which strongly interacting particles are produced in electron-positron annihilations. Theoretical predictions for these observables are sensitive to the number of distinct quark species, and thus to the number of colors.

The fundamental particles that do not experience strong interactions are the leptons, which like the quarks are spin-½ particles that are structureless at the current limits of resolution. The most familiar examples of leptons are the electron, the muon, and the neutrinos. Each lepton flavor comes in but a single species, which is to say that leptons are colorless. It is therefore appealing to regard color as the strong-interaction analog of the electric charge. Like electric charge, color cannot be created or destroyed in any of the known interactions; it is said to be conserved. S*ee* C*olor*; L*epton*.

Gauge symmetry. The existence of a conserved quantity is quite generally a consequence of a continuous group of symmetry transformations which leave the laws of physics unchanged in form. For example, the conservation of energy follows from the fact that physical laws depend upon the time interval between occurrences, and not upon an absolute time measured on some master clock. Translation in time (that is, the resetting of clocks) is a symmetry of the equations of physics. Symmetries relating to internal properties of particles, like electric charge, are known as gauge symmetries. In the case of conservation of the color charge, a natural choice is the unitary group SU(3), now applied to color rather than flavor. S*ee* S*ymmetry laws*.

Yang-Mills theory. It frequently happens that the symmetries respected by a phenomenon are recognized before a complete theory has been developed. The question thus arises as to whether a complete theory of nuclear forces could be deduced from a knowledge of the symmetry. If the equations of physics are required to be invariant in form under local symmetry transformations which may be different at every point in space and time, the interactions related to the symmetry are completely fixed. The manner in which this could be accomplished for any continuous symmetry was indicated by C. N. Yang and R. L. Mills in 1954. Nuclear forces had long been known not to distinguish between the proton and neutron. From the point of view of nuclear forces, the designations force among protons and neutrons could be derived by imposing local isospin invariance (which is to say that the convention could be chosen independently at every point of space and time). In general, the requirement of local gauge invariance implies that the interaction must occur through the exchange of massless spin-1 bosons. One species of force particle corresponds to each conserved quantity. This made the Yang-Mills theory unacceptable as a description of nuclear forces. It predicted that nuclear forces were mediated by three massless "gauge bosons," whereas the short range (on the order of 10^{-15} m) over which nuclear forces are observed to act demands that the force particles be massive, as proposed by H. Yukawa. S*ee* G*auge theory*; I*sobaric spin*; Q*uantum field theory*.

Quantum chromodynamics. Applying similar reasoning to the idea that a local color gauge symmetry should prescribe the strong interaction among quarks leads to the gauge theory of strong interactions which has been called quantum chromodynamics (QCD). The mediators of the strong interaction are eight massless vector bosons, which are named gluons because they make up the "glue" which binds quarks together. It is hoped that the infinite range of the forces mediated by the gluons may help to explain why free quarks have not been isolated. The gluons themselves carry color. Hence, strong interactions among gluons will also occur through the exchange of gluons. It is therefore believed that gluons, as well as quarks, may be permanently confined. According to this view, only colorless objects may exist in isolation. SEE QUANTUM CHROMODYNAMICS.

Experimental evidence. No evidence has been reported for isolated or free gluons. As indicated above, the current interpretation of quantum chromodynamics is that free gluons cannot exist. Therefore it is necessary to devise indirect means to test the idea that gluons exist with all the desired attributes.

Inelastic electron-proton scattering. Early support for the existence of an electrically neutral glue within the proton came from 1968 experiments on inelastic electron-proton scattering carried out at the Stanford Linear Accelerator Center (SLAC). These experiments indicated that the electrons were not scattered electromagnetically from the proton as a whole, but from individual pointlike charged objects subsequently identified with the quarks. They also showed that only about half of the energy of a rapidly moving proton is carried by its charged constituents. The remainder must then be borne by neutral constituents which do not interact electromagnetically. This role would naturally be played by the gluons.

Charmonium lifetimes. Further evidence for the utility of the gluon concept was provided by the unusually long lifetime of the strongly decaying charmonium state J/ψ. In quantum electrodynamics the atom composed of an electron and an antielectron (positron) is known as positronium. Positronium occurs in two forms: orthopositronium, in which the electron and positron spins are aligned, and parapositronium, in which the spins are opposite. The electron and positron may annihilate into photons. The spinless parapositronium state may decay into two photons (**Fig. 1a**), but the spin-1 orthopositronium state must decay into three photons (Fig. 1b). The difficulty of radiating an additional photon is reflected in the fact that orthopositronium lives 1120 times longer than parapositronium. In similar fashion, charmonium, the strong-interaction "atom" composed of a charmed quark and a charmed antiquark, decays by the annihilation of the quark and antiquark into gluons. The gluons materialize into the observed hadrons through the action of the confinement mechanism, with unit probability. For the pseudoscalar parachromonium level, designated η_c, the semifinal state is composed of two gluons (**Fig. 2a**). The vector

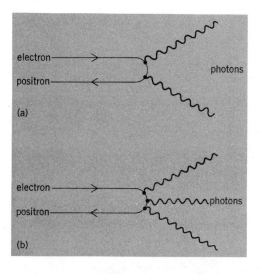

Fig. 1. Positronium decay. (a) Decay of parapositronium into two photons. (b) Decay of orthopositronium into three photons.

ELEMENTARY PARTICLES 231

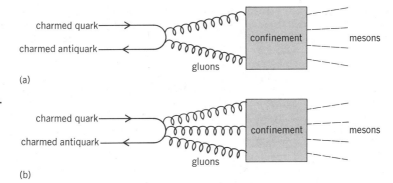

Fig. 2. Charmonium decay. (a) Decay of paracharmonium through a two-gluon semifinal state. (b) Decay of orthocharmonium through a three-gluon semifinal state.

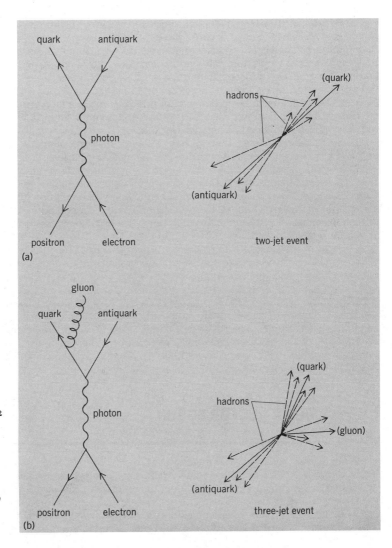

Fig. 3. Mechanisms for hadron production in electron-positron annihilation. (a) Two-jet event produced by the mechanism electron + positron → quark + antiquark. (b) Three-jet event produced by the mechanism electron + positron → quark + antiquark + gluon. The observed hadrons are represented by broken lines.

particle J/ψ, which corresponds to orthocharmonium, must decay into three gluons (Fig. 2b). The remarkably long lifetime of J/ψ and the large ratio of J/ψ to η_c lifetimes (approximately 500) argue for the aptness of the analogy. Decays of the still heavier quarkonium state upsilon also support this. See CHARM; J PARTICLE; POSITRONIUM.

Three-jet pattern. The most unambiguous evidence for the existence of gluons was reported in 1979 by a number of experimental groups working at the high-energy electron-positron storage ring PETRA at the Deutsches Elektron-Synchrotron (DESY) in Hamburg. It had earlier been established in experiments at SLAC and DESY that the dominant mechanism for hadron production in electron-positron annihilations is electron + positron → quark + antiquark, with the quarks materializing into hadrons. This interpretation explains the rate of particle production and the characteristic angular distribution of the sprays or jets of hadrons that emerge from the collisions (**Fig. 3**a). If quantum chromodynamics is correct, one of the outgoing quarks may occasionally radiate an energetic gluon, just as a fast electron may radiate an energetic photon. When this happens, the hadrons may be expected to emerge in a three-jet pattern (Fig. 3b). A number of examples of this behavior have been reported in electron-positron annihilations at center-of-mass energies exceeding 24 GeV. By far the most satisfactory interpretation of these findings is that gluon radiation is being observed indirectly.

Other implications. The existence of gluons with the properties implied by quantum chromodynamics has additional consequences. The interactions of quarks and gluons specified by quantum chromodynamics suggest the existence of a number of new species of hadrons. The most important of these from the gluon perspective are quarkless mesons, composed entirely of gluons. The simplest of these glueballs, as they are sometimes called, may be searched for in the radiative decay J/ψ → photon + 2 gluons, or in the decays of heavier quarkonium states. Several glueball candidates have been observed in this way, most notably the iota with mass 1440 MeV/c^2. Further experimental work will be required to determine whether the iota or other candidates must be identified as quarkless states. See MESON.

It is to be expected that further experimental tests of the properties which have been imputed to gluons will be forthcoming. The angular distribution of hadron jets from vector states of heavy quarkonia may exhibit a three-lobed pattern that reflects the three-gluon semifinal state. Detailed characteristics of the three-jet events in electron + positron → quark + antiquark + gluon → hadrons may reveal which jet of hadrons has emerged from the gluon, and test the spin-1 nature of the gluon. Other tests of the gluon spin are to be had from analysis of the pattern of scaling violations in inelastic lepton scattering, and from comparisons of the hadronic decay rates of various quarkonium states.

Proof of the existence of gluons with the canonical properties would vindicate the idea of color gauge symmetry and provide strong encouragement for quantum chromodynamics and, by derivation, for grand unified theories of the strong, weak, and electromagnetic interactions. See ELEMENTARY PARTICLE; FUNDAMENTAL INTERACTIONS; GRAND UNIFICATION THEORIES; QUARKS.

Bibliography. N. Calder, *The Key to the Universe*, 1977; M. Creutz, *Quarks, Gluons and Lattices*, 1985; S. L. Glashow, Quarks with color and flavor, *Sci. Amer.*, 233(4):38–50, October 1975; K. Ishikawa, Glueballs, *Sci. Amer.*, 247(5):142–156, 1982; T. B. W. Kirk and H. D. I. Abarbanel (eds.), *Proceedings of the 1979 International Symposium on Lepton and Photon Interactions at High Energies*, Fermilab, Batavia, Illinois, 1980; Y. Nambu, The confinement of quarks, *Sci. Amer.*, 235(5):48–60, November 1976.

INTERMEDIATE VECTOR BOSON
JAMES W. ROHLF

The fundamental particles that transmit the weak force. (An example of a weak interaction process is nuclear beta decay.) These elementary particles, which are also called W and Z particles, were discovered in 1983 in very high-energy proton-antiproton collisions. It is through the exchange of W and Z bosons that two particles interact weakly, just as it is through the exchange of photons that two charged particles interact electromagnetically. The intermediate vector bosons were postulated to exist in the 1960s; however, their large masses prevented their production and study at accelerators until 1983. Their discovery was a key step toward unification of the weak and electromagnetic interactions. SEE ELEMENTARY PARTICLE; FUNDAMENTAL INTERACTIONS; WEAK NUCLEAR INTERACTIONS; WEINBERG-SALAM MODEL.

Production and detection. The W and Z particles are predicted by electroweak theory to be very massive, roughly 100 times the mass of a proton. Therefore, the experiment to search for the W and the Z demanded collisions of elementary particles at the highest available center-of-mass energy. Such very high center-of-mass energies capable of producing the massive W and Z particles were achieved with collisions of protons and antiprotons at the laboratory of the European Organization for Nuclear Research (CERN) near Geneva, Switzerland. This was accomplished through conversion of the existing 4-mi-circumference (6-km) superproton syncrotron (SPS) into a proton-antiproton ($p\bar{p}$) collider. The antiprotons were created by collision of an intense beam of protons with a stationary nuclear target, and injected into a specially constructed ring called an antiproton accumulator where their momentum spread was narrowed by using an innovative technique called stochastic cooling. Dense bunches of antiprotons could then be injected into the SPS along with protons circulating in the opposite direction. After injection into the SPS, the protons and antiprotons were accelerated together to an energy of 270 GeV each, and stored for several hours at this energy. This made a center-of-mass energy of 540 GeV per $p\bar{p}$ collision, which was about a factor of 10 higher than previously achieved with any other accelerator. SEE PARTICLE ACCELERATOR.

The $p\bar{p}$ collisions were monitored in two underground experimental areas (UA1 and UA2). Experiment UA1 consisted of a large, 2000-ton magnetic spectrometer completely surrounding the $p\bar{p}$ collision region. The inner portion of the detector contained a precision device for charged-particle tracking. Immediately surrounding the central detector were devices (calorimeters) for measuring the energies of particles, both charged and neutral. The UA2 detector was a smaller, 200-ton nonmagnetic detector. SEE PARTICLE DETECTOR.

Discovery of the W particle. The intermediate vector bosons have very distinctive signatures in the UA1 detector. The W particle (which comes in two electric charges, W^+ and W^-) is identified through its decay into positron (e^+) and neutrino (ν), $W^+ \rightarrow e^+\nu$, or electron (e^-) and antineutrino ($\bar{\nu}$), $W^- \rightarrow e^-\bar{\nu}$. The trajectory of the electron is measured by the trail of ionization that it leaves in the central detector. The energy of the electron is measured by the calorimeters. The details of the energy deposition of the electron in the calorimeters distinguish it from heavier, strongly interacting particles, such as pions and protons. The neutrino has such a small interaction probability that it leaves no direct traces in the detector. However, strong evidence for the presence of a neutrino comes from an apparent lack of momentum conservation.

Analysis of data recorded in November and December 1982 from about 10^9 $p\bar{p}$ interactions revealed six events with an isolated electron or positron of very large transverse momentum (momentum perpendicular to the colliding beam axis) together with large missing transverse momentum (**Fig. 1**a). The UA1 group could find no explanation for these events other than the production of new massive charged particles (W^+ and W^-), which decayed into an electron (or positron) and neutrino. Confirmation of the UA1 result was soon announced by the UA2 group.

Discovery of the Z particle. Another run of the CERN $p\bar{p}$ collider in the spring of 1983 obtained nearly 10^{10} $p\bar{p}$ collisions, and confirmed the existence of the charged boson W with about 10 times the original statistics (Fig. 1b). Also, for the first time, four events were found in the UA1 data that contained both a high-energy electron and a high-energy positron (**Fig. 2**). All four events have a common value of invariant mass (about 95 GeV/c^2 where c is the speed of light) within experimental resolution. Events were also detected that contained a positive muon (μ^+) and a negative muon (μ^-) with invariant mass of about 95 GeV/c^2.

These data are clear evidence for the existence of a new massive particle that decays into e^+e^- and $\mu^+\mu^-$. The only known interpretation of this particle which fits the measured properties is that it is the neutral intermediate vector boson Z^0. Confirmation of this result came later from the UA2 group.

Properties of the W and Z particles. Striking features of both the charged W and the Z^0 particles are their large masses. The charged boson (W^+ and W^-) mass is measured to be 80.9 ± 1.5 GeV/c^2, and the neutral boson (Z^0) mass is measured to be 95.6 ± 1.5 GeV/c^2 from the UA1 experiment. (For comparison, the proton has a mass of about 1 GeV/c^2.) There is an additional systematic error on these mass values of 3%, mainly due to calibration of the calorimeter.

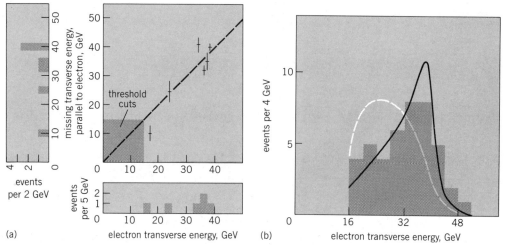

Fig. 1. Observation of the *W* particle in the UA1 data. (a) Scatter plot of electron (or positron) transverse energy versus the transverse energy imbalance (due to a neutrino) in the whole event, from a sample of six events. Threshold cuts select events with large electron transverse energy and missing transverse energy (*after G. Arnison et al., Experimental observation of isolated large transverse energy electrons with associated missing energy at $\sqrt{s}$ = 540 GeV, Phys. Lett., 122B:103–116, 1983*). (b) Electron transverse energy distribution from a large sample of 43 events. The solid curve is the expected distribution for the decay, $W \to e\nu$, while the broken curve is the distribution expected for the decay of a different particle "X" into an electron and two neutrinos (*after G. Arnison et al., Further evidence for charged intermediate vector bosons at the SPS collider, Phys. Lett., 129B:273–282, 1983*).

The *W* and Z^0 mass values reported by the UA2 experiment are in agreement with the UA1 results. Prior to the discovery of the *W* and the *Z*, particle theorists had met with some success in the unification of the weak and electromagnetic interactions. The electroweak theory as it is understood today is due largely to the work of S. Glashow, S. Weinberg, and A. Salam. Based on low-energy neutrino scattering data, which in this theory involves the exchange of virtual *W* and *Z* particles, theorists made predictions for the *W* and *Z* masses. The actual measured values are in agreement (within errors) with predictions. The discovery of the *W* and the *Z* particles at the predicted masses is an essential confirmation of the electroweak theory.

Only a few intermediate vector bosons are produced from 10^9 proton-antiproton collisions at a center-of-mass energy of 540 GeV. This small production probability per $p\bar{p}$ collision is understood to be due to the fact that the bosons are produced by a single quark-antiquark annihilation. The other production characteristics of the intermediate vector bosons, such as longitudinal and transverse momentum distributions (with respect to the $p\bar{p}$ colliding beam axis), all support this theoretical picture. SEE QUARKS.

The decay modes of the *W* and *Z* are well predicted. The simple decays $W^+ \to e^+\nu$, $W^- \to e^-\bar{\nu}$, $Z^0 \to e^+e^-$, and $Z^0 \to \mu^+\mu^-$ are all spectacular signatures of the intermediate vector bosons. However, these leptonic decays are expected to be only a few percent of the total number of *W* and *Z* decays, and there will be much experimental activity to find the other decay modes. The *W* particle also has an identifying feature in its decay, which is due to its production through quark-antiquark annihilation. The quark and antiquark carry intrinsic angular momentum (spin) that is conserved in the interaction. This means that the *W* is polarized at production; its spin has a definite orientation. This intrinsic angular momentum is conserved when the *W* decays, yielding a preferred direction for the decay electron or positron (**Fig. 3**). This distribution is characteristic of a weak interaction process, and it strongly favors the assignment of the spin of the *W* to be 1, as expected from theory. (Spin-1, or more generally integer-spin, particles are called bosons; another example is the photon.)

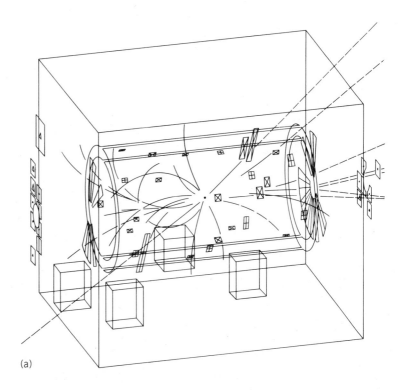

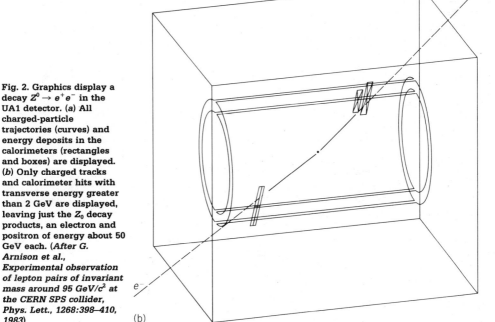

Fig. 2. Graphics display a decay $Z^0 \rightarrow e^+e^-$ in the UA1 detector. (a) All charged-particle trajectories (curves) and energy deposits in the calorimeters (rectangles and boxes) are displayed. (b) Only charged tracks and calorimeter hits with transverse energy greater than 2 GeV are displayed, leaving just the Z_0 decay products, an electron and positron of energy about 50 GeV each. (*After G. Arnison et al., Experimental observation of lepton pairs of invariant mass around 95 GeV/c^2 at the CERN SPS collider, Phys. Lett., 1268:398–410, 1983*)

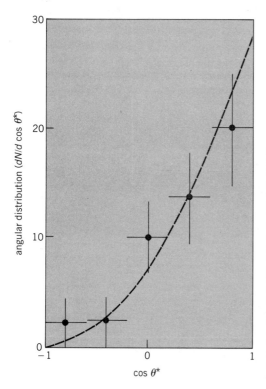

Fig. 3. Angular distribution of the emission angle θ^* of the e^- with respect to the proton direction (or the e^+ with respect to the antiproton direction) from W^+ and W^- events (in the rest frame of the W). The experimental data are given by the crosses. The curve is the expectation for a spin-1 particle produced by simple quark-antiquark annihilation followed by the weak interaction decay, $W^+ \to e^+\nu$ or $W^- \to e^-\bar{\nu}$. (After G. Arnison et al., *Further evidence for charged intermediate vector bosons at the SPS collider, Phys. Lett., 129B:273–282, 1983*)

Bibliography. G. Arnison et al., Experimental observation of isolated large transverse energy electrons with associated missing energy at $\sqrt{s}$ = 540 GeV, *Phys. Lett.*, 122B:103–116, 1983; G. Arnison et al., Experimental observation of lepton pairs of invariant mass around 95 GeV/c^2 at the CERN SPS collider, *Phys. Lett.*, 126B:398–410, 1983; M. Banner et al., Observation of single isolated electrons of high transverse momentum in events with missing transverse energy at the CERN $p\bar{p}$ collider, *Phys. Lett.*, 122B:476–485, 1983; C. Rubbia et al., Producing massive neutral intermediate vector bosons with existing accelerators, *Proceedings of the International Neutrino Conference*, Aachen, West Germany, pp. 683–687, 1976.

GRAVITON
Charles J. Goebel

A theoretically deduced particle postulated as the quantum of the gravitational field. According to Einstein's theory of general relativity, accelerated masses (or other distributions of energy) should emit gravitational waves, just as accelerated charges emit electromagnetic waves. And according to quantum field theory, such a radiation field should be quantized; that is, its energy should appear in discrete quanta, called gravitons, just as the energy of light appears in discrete quanta, namely photons.

The properties of the graviton follow from the properties of the classical gravitational field. That is, its rest mass and charge are zero (like the photon); it has spin 2 in units of $h/2\pi$, where h is Planck's constant, and is therefore a boson, which is a particle that obeys Bose-Einstein statistics. Because of its vanishing rest mass, its spin is restricted to be parallel to its motion, so that a graviton has only two independent spin states (again like a photon). *See* BOSE-EINSTEIN STATISTICS.

Attempts to observe gravitational radiation are discussed in the article on gravitation. Such observations are difficult, because matter is very weakly coupled to the gravitational field (the gravitational force between an electron and a proton is only 10^{-39} times the electrical force between them), so that the rate of emission and absorption of gravitational radiation is very small. The observation of quanta of the gravitational field, gravitons, is impossible according to present knowledge. As far as known, the only physical situations in which the quantization of gravitation plays any appreciable role are the evaporation of black holes (light ones particularly), and the very high-temperature phase early in the history of the universe, namely within 10^{-33} s after the big bang. SEE ELEMENTARY PARTICLE.

MAGNETIC MONOPOLES
ALFRED S. GOLDHABER

Magnetic charges; hypothetical entities designed to put magnetism on an equal footing with electricity. In this article, gaussian units are used throughout.

Electric fields are produced by electric charges according to the formula $\mathbf{E}(\mathbf{r}) = q\mathbf{r}/r^3$, where q is a point charge at the origin of coordinates, $\mathbf{r}$ is the position of observation at distance r from q, and $\mathbf{E}$ is the resulting field at $\mathbf{r}$. They exert forces on point charges $\mathbf{F} = q\mathbf{E}$, where $\mathbf{F}$ is the force, q is the charge, and $\mathbf{E}$ is the electric field from all other charges. A magnetic field about a wire carrying electric current is a well-known example. In turn, magnetic fields exert forces on charges according to $\mathbf{F} = q(\mathbf{v}/c) \times \mathbf{B}$, where $\mathbf{v}$ is the velocity, $\mathbf{B}$ is the magnetic field, and c is the speed of light. This force only acts on a moving charge and is perpendicular to the direction of motion and to the magnetic field. It is natural to speculate that there may be magnetic charges which act as sources of magnetic fields in the same way that electric charges produce electric fields. Two such magnetic charges would obey the famous rules that like charges repel and unlike charges attract each other.

The name monopole or one-pole for a magnetic charge is meant as a contrast to dipole or two-pole. A magnetic dipole is a system producing a magnetic field which looks like that of two equal but opposite magnetic charges close together, when the field is measured at a large distance from the dipole. Such a field can be produced by a loop of wire carrying an electric current. In this case, it is obviously impossible to tear the dipole apart and get two magnetic monopoles.

Parity violation. If there were only magnetic charges in the world, and no electric charges, one could simply interchange the names electric and magnetic and proceed to describe all electromagnetic phenomena in the same way one does now. Clearly, the interesting new results must come from the interaction of magnetic monopoles with electric charges. Consider an electric charge q in the presence of a magnetic monopole m. If both particles are at rest, there is no force between them. If the electric particle moves, it suffers a force perpendicular to its velocity and to the line of centers from q to m. Imagine that m is situated at the center of a clock face, and q, which is moving upward, is located at 3 o'clock. Then the resultant force will be into the clock if the product qm is positive, and out of the clock if qm is negative. If q were located at 9 o'clock, but still moving upward, the resultant force would be reversed from the previous case. Thus, the simultaneous presence of electric and magnetic charges would permit an absolute distinction between left and right. In other words, the laws of motion of charges would no longer be symmetric under reflection; they would violate parity. However, the laws would still by symmetric under a "generalized reflection," in which all magnetic charges present are reversed in sign when spatial coordinates are inverted. This is significant because a magnetically neutral system, even if it contained monopoles, could still obey reflection symmetry. Similar considerations apply to symmetry with respect to the direction of flow of time. Again, a generalized time-reversal symmetry is obeyed even if monopoles exist, provided there are also particles with exactly the same properties except for opposite magnetic charge. Monopoles cannot explain the well-known parity violation effects in weak interactions, but they might be connected with the much feebler violation of time-reversal symmetry found in the decays of neutral K mesons. SEE PARITY; SYMMETRY LAWS; TIME REVERSAL INVARIANCE.

Charge quantization. The quantity $S = qm/c$ has the dimensions of angular momentum. In fact, S is the magnitude of the angular momentum stored in the electromagnetic field when a point charge and a monopole are simultaneously present. The axis of this angular momentum is the line from charge to monopole. During the motion of a charge past a stationary monopole, the angular momentum $\mathbf{L} = \mathbf{r} \times \mu\mathbf{v}$ of the charge is not conserved (here $\mathbf{r}$ is the position and μ the mass of the charged particle). This is clear because $\mathbf{L}$ is perpendicular to the plane defined by the position and velocity, but the magnetic force tilts $\mathbf{v}$ out of the plane it is in at a given moment, so that $\mathbf{L}$ is also tilted. There does exist a total angular momentum which is conserved, $\mathbf{J} = \mathbf{L} + \mathbf{S}$, the sum of the angular momentum of the electric charge and that stored in the electromagnetic field. In quantum theory, angular momentum around a given axis is measured in multiples of $\hbar/2$, where $\hbar$ is Planck's constant divided by 2π. Application of this principle to the angular momentum S yields the Dirac quantization condition $qm/c = n\hbar/2$, where n is any integer. This result was obtained in 1931 by P. A. M. Dirac on the basis of quantum-mechanical analysis. As Dirac pointed out, the condition implies that the charge of any monopole must be a multiple of $m_0 = \hbar c/2e$, where e is the smallest known charge, that of an electron. Reciprocally, the existence of even one monopole m in the universe implies that electric charge must come in multiples of $q_0 = \hbar c/2m$. Thus, the existence of monopoles might be said to explain the quantization of electric charge. This argument is somewhat less attractive now than when Dirac first proposed it, since many properties of elementary particles besides charge are known to be quantized. Furthermore, in theories where the electromagnetic field belongs to a set of fields related by symmetry transformations in some abstract space (as happens in so-called non-abelian gauge field theories), it is often possible to deduce the quantization rules from the symmetry, analogous to the deduction of angular momentum quantization from rotation symmetry. Some such theories actually predict "composite" magnetic monopoles, made by coupling electromagnetic and other fields together. *See* GAUGE THEORY.

Properties of Dirac monopoles. The monopole obeying the condition $m = nm_0$ is called a Dirac monopole. The minimum magnetic charge is $m_0 = \hbar c/2e = e/2\alpha \approx 69e$, where α is the fine-structure constant $e^2/\hbar c$. Several characteristics would make a Dirac monopole easy to identify. Because of its great charge, a rapidly moving monopole would ionize atoms far more effectively than an electron, thus producing a unique, very dense track. The end of the track would show a decrease in ionization density, in contrast to that of an electric particle. This happens because the electric force on a monopole is proportional to its velocity. Monopoles would be expected to be bound in ferromagnetic materials, with binding to the bulk material that could approach the kiloelectronvolt level, and even stronger binding to small clusters of atoms. In other solids, electron-volt binding might occur. If these estimates were correct, then monopoles incident on the Earth (or the Moon) could be trapped in surface material for geologic times. Even such trapped monopoles could be accelerated to very high energies by applied magnetic fields, making them easily liberated and detected. Alternatively, material containing a monopole and passed through a wire coil would cause a transient rise in voltage across the coil, by Faraday's law of magnetic induction. No other entity could duplicate this effect.

Grand unification models. There has been much discussion of grand unification models, describing electromagnetic, weak, and strong forces in a single framework. The two most striking predictions of many such models are that protons should be able to disintegrate, albeit quite rarely, and that very heavy magnetic monopoles should exist in nature. In the most simple such model, minimal SU(5), the monopole mass would be comparable with that of an ice cube 0.001 in. (25 μm) on a side, but the diameter of the monopole core would be 10^{25} times smaller than that of the ice cube. This implies a core mass density so immense that a cubic inch of closely packed monopoles would weigh more than the visible universe. A heavy pole would be likely to act as a catalyst for proton disintegration, so that the observation of such events could be evidence for monopoles as well as for grand unification models. *See* GRAND UNIFICATION THEORIES; PROTON.

Monopole searches. The special features of monopoles have all been used in one or more of the many searches in magnetic or other materials, with particle accelerators, and in cosmic radiation. These searches have established stringent limits on monopole abundance, but no clear evidence for the existence of isolated poles. There have been two very different, unconfirmed experimental results which could be interpreted as monopole candidate events. A balloon-borne cosmic-ray detector in 1975 recorded a track with very large ionization which might have been due to a pole carrying two Dirac units of magnetic charge. However, there are other possible explanations of the event. In 1982 a magnetically shielded superconducting coil registered a pulse consistent with passage of a single unit of magnetic charge. Such an event has not recurred despite a more than tenfold increase in accumulated exposure of equivalent detectors. Although experimental searches are intensifying, knowledge about magnetic poles (other than abundance limits) remains purely theoretical.

Bibliography. B. Cabrera, First results from a superconductive detector for moving magnetic monopoles, *Phys. Rev. Lett.*, 48:1378–1381, 1982; R. A. Carrigan, Jr. and W. P. Trower, Superheavy magnetic monopoles, *Sci. Amer.*, 46:106–118, April 1982; R. A. Carrigan, Jr., and W. P. Trower (eds.), *Magnetic Monopoles*, 1983; D. E. Groom, In search of the supermassive magnetic monopole, *Phys. Rep.*, 140:324–373, 1986; J. L. Stone (ed.), *Monopole '83*, 1984.

TACHYON
ROLAND H. GOOD, JR.

A hypothetical faster-than-light particle consistent with the special theory of relativity. According to this theory, a free particle has an energy E and a momentum $\mathbf{p}$ which form a Lorentz four-vector. The length of this vector is a scalar, having the same value in all inertial reference frames. One writes Eq. (1), where c is the speed of light and the parameter m^2 is a property of the particle,

$$E^2 - c^2 p^2 = m^2 c^4 \tag{1}$$

independent of its momentum and energy. Three cases may be considered: m^2 may be positive, zero, or negative. The case $m^2 > 0$ applies for atoms, nuclei, and the macroscopic objects of everyday experience. The positive root m is called the restmass. If $m^2 = 0$, the particle is called massless. A few of these are known: the electron neutrino, the muon neutrino, the tau neutrino, the photon, and the graviton. The third case, $m^2 < 0$, was studied originally by S. Tanaka and by O. M. P. Bilaniuk, V. K. Deshpande, and E. C. G. Sudarshan. Further contributions were made by G. Feinberg, who gave the name tachyons (after a Greek word for swift) to the particles with $m^2 < 0$. Whether such particles exist is an interesting speculation, but, there has been no experimental evidence for them.

In general, the particle speed is given by Eq. (2). If $m^2 > 0$, Eq. (1) implies $E > cp$ and

$$v = \frac{cp}{E} c \tag{2}$$

Eq. (2) gives $v < c$. If $m^2 = 0$, then $E = cp$ and $v = c$. In case $m^2 < 0$, one finds $E < cp$ and $v > c$. Tachyons exist only at faster-than-light speeds.

The quantities p or E can be eliminated from Eqs. (1) and (2) to get expressions for E and p in terms of the speed. In case $m^2 > 0$, the familiar results are given by Eqs. (3), where the

$$E = \frac{mc^2}{\sqrt{1 - \frac{v^2}{c^2}}} \qquad p = \frac{mv}{\sqrt{1 - \frac{v^2}{c^2}}} \tag{3}$$

radical signs imply the positive roots. The way to make the analogous formulas for tachyons is to introduce the positive number μ such that $m^2 = -\mu^2$ and Eq. (4) holds. Then Eqs. (2) and (4)

$$E^2 - c^2 p^2 = -\mu^2 c^4 \tag{4}$$

give Eqs. (5). It is seen that for ordinary particles as v increases, E increases, but to speed them

$$E = \frac{\mu c^2}{\sqrt{\frac{v^2}{c^2} - 1}} \qquad p = \frac{\mu v}{\sqrt{\frac{v^2}{c^2} - 1}} \tag{5}$$

up to $v = c$ would involve an infinite amount of energy. In contrast for tachyons, as v decreases, E increases, but to slow them down to $v = c$ would involve an infinite amount of energy.

For an ordinary free particle with $m^2 > 0$, there is always a special reference frame, called the rest frame, in which $\mathbf{p} = 0$. From Eq. (1) it is seen that the energy there has the minimum value mc^2. No such special frame exists for massless particles. For tachyons also the situation is quite different. For a free tachyon a special frame can be found in which $E = 0$ and, according to Eq. (4), p has the minimum value μc. Since the tachyons may exist at zero energy, there is no energy obstacle in creating them in elementary particle reactions.

According to electromagnetic theory, a charged particle moving at a speed greater than the speed of light in a medium emits light, the Cerenkov radiation. If charged tachyons existed, they would spontaneously radiate light even in a vacuum. *See* Cerenkov radiation.

Attempts to detect tachyons have been made by looking for the Cerenkov radiation (T. Alväger and M. N. Kreisler) and by analyzing for negative m^2 values in elementary particle reactions (C. Baltay, R. Linsker, N. K. Yeh, and G. Feinberg). The conclusions were negative, and present indications are that this type of particle does not exist. *See* Elementary particle.

Bibliography. O. M. P. Bilaniuk, V. K. Deshpande, and E. C. G. Sudarshan, Meta relativity, *Amer. J. Phys.*, 30:718–723, 1962; G. Feinberg, Particles that go faster than light, *Sci. Amer.*, February 1970.

7

HADRONS

Hadron	242
Baryon	242
Nucleon	256
Proton	257
Neutron	260
Antiproton	263
Antineutron	263
Strange particles	264
Hyperon	264
Delta resonance	265
Meson	265
J particle	277
Upsilon particles	280

HADRON
A. K. Mann

The generic name of a class of particles which interact strongly with one another. Examples of hadrons are protons, neutrons, the π, K, and D mesons, and their antiparticles. Protons and neutrons, which are the constituents of ordinary nuclei, are members of a hadronic subclass called baryons, as are strange and charmed baryons, for example, Λ_s^0 and Λ_c^0. Baryons have half integral spin, obey Fermi-Dirac statistics, and are known as fermions. Mesons, the other subclass of hadrons, have zero or integral spin, obey Bose-Einstein statistics, and are known as bosons. The electric charges of baryons and mesons are either zero or ± 1 times the charge on the electron. Masses of the known mesons and baryons cover a wide range, extending from the pi-meson, with a mass approximately one-seventh that of the proton, to the upsilon-meson, with a mass about 10 times the proton mass. The spectrum of meson and baryon masses is not fully understood. SEE BARYON; MESON; NEUTRON; PROTON.

It has been believed that the net number of baryons in the universe is a conserved quantity. In making this count, baryons and antibaryons are arbitrarily assigned baryon numbers $+1$ and -1, respectively, so that production or annihilation of a baryon-antibaryon pair in a given reaction has no effect on the net baryon number. (Mesons are assigned baryon number zero.) In this scheme, the least massive baryon, the proton, would be stable, and all other baryons unstable. However, it has been suggested, as a result of some theoretical attempts to unify the strong, weak, and electromagnetic interactions, that protons may be unstable with a lifetime many orders of magnitude greater than the age of the universe; experimental searches for such an instability have been undertaken. SEE FUNDAMENTAL INTERACTIONS; SYMMETRY LAWS.

In addition to baryon number, hadrons and their antiparticles are assigned quantum numbers to represent other properties of hadronic matter. Among these are isobaric (i-) spin (related to electric charge), strangeness, charm, and quite probably others, all of which denote collateral families within the main family of hadrons. Strong interactions conserve i-spin, and hence hadrons may form i-spin multiplets whose members differ in mass by only a very small fraction of the proton mass, for example, the proton-neutron doublet. Hadrons with nonzero values of the strangeness and charm quantum numbers, as well as hadrons without a decay mode that conserves i-spin, are quasistable; their decays take place through the weak interaction, which does not conserve strangeness or charm, or through the electromagnetic interaction, which does not conserve i-spin. The quasistable hadrons have a lifetime many orders of magnitude longer than that of nonstrange, noncharmed hadrons, which decay with the conservation of i-spin; because of their very short lifetimes (less than 10^{-20} s), the latter are often called elementary particle-particle scattering resonances. SEE CHARM; ISOBARIC SPIN; STRANGE PARTICLES; WEAK NUCLEAR INTERACTIONS.

Much of the data relating to the properties of hadrons (both baryons and mesons) can be interpreted as if hadrons consist of more elementary constituents known as quarks and gluons. This conception of hadrons has widespread appeal in elementary particle physics. SEE ELEMENTARY PARTICLE; GLUONS; QUARKS.

BARYON
RICHARD H. DALITZ

The generic name for any hadronic particle with baryon number $B = +1$. By far the most common baryons are the proton and neutron, the two states of the nucleon doublet $N = (p,n)$, whose intrinsic properties are listed in **Table 1**. The baryon number of any particular state may be deduced from its production or decay processes, or both, since the total baryon number is conserved (with possible rare exceptions discussed below) and $B = 0$ holds for all mesons and leptons. SEE LEPTON; MESON; NEUTRON; NUCLEON; PROTON.

The scientific view of the hadrons changed greatly during the 1970s. It is now generally accepted that they are composite, consisting of spin-$\frac{1}{2}$ quarks (q), corresponding antiquarks ($\bar{q}$), and some number of gluons, the last being the quanta of the intermediate field which binds the quarks and antiquarks to form hadrons. $B = +\frac{1}{3}$ holds for a quark q, $B = -\frac{1}{3}$ for an antiquark

Table 1. Known stable and semistable baryons and their properties

Baryon	Mass, MeV	Spin-parity	Strangeness (s)	Charm (C)	Lifetime, s	Dominant decay modes	Magnetic moment, n.m.*
p	938.280 ± 0.003	½⁺	0	0	>10³⁹	—	2.7928
n	939.573 ± 0.003	½⁺	0	0	898 ± 16	$p\bar{\nu}_e e^-$	−1.9130
Λ	1115.60 ± 0.05	½⁺	−1	0	2.63 ± 0.02 × 10⁻¹⁰	$p\pi^-$ (64%) $n\pi^0$ (36%) $p\bar{\nu}_e e^-$ (0.084 ± 0.002%)	−0.613 ± 0.004
Σ^+	1189.36 ± 0.06	½⁺	−1	0	8.0 ± 0.04 × 10⁻¹¹	$p\pi^0$ (52%) $n\pi^+$ (48%) $p\gamma$ (0.12 ± 0.01%)	2.38 ± 0.02
Σ^0	1192.46 ± 0.08	½⁺	−1	0	6 ± 1 × 10⁻²⁰	$\Lambda\gamma$	
Σ^-	1197.34 ± 0.05	½⁺	−1	0	1.48 ± 0.01 × 10⁻¹⁰	$n\pi^-$ $n\bar{\nu}_e e^-$ (0.11 ± 0.005%)	−1.10 ± 0.05
Ξ^0	1314.9 ± 0.6	½⁺	−2	0	2.9 ± 0.1 × 10⁻¹⁰	$\Lambda\pi^0$	−1.25 ± 0.02
Ξ^-	1321.3 ± 0.15	½⁺	−2	0	1.64 ± 0.02 × 10⁻¹⁰	$\Lambda\pi^-$ $\Lambda\bar{\nu}_e e^-$ (0.028 ± 0.012%)	−1.85 ± 0.75
Ω^-	1672.5 ± 0.3	(3/2⁺?)	−3	0	0.82 ± 0.03 × 10⁻¹⁰	ΛK^- (69%) $\Xi^0\pi^-$ (23%) $\Xi^-\pi^0$ (8%)	
Λ_c	2282 ± 3	(½⁺?)	0	1	2.3 ± 0.8 × 10⁻¹³	$pK^-\pi^+$ (2.2 ± 1%)	

*The abbreviation n.m. denotes the unit $e\hbar/2M_p c$ (nuclear magneton).

$\bar{q}$, while $B = 0$ holds for a gluon. Thus, a baryon consists of three ("valence") quarks, together with some number of quark-antiquark ($q\bar{q}$) pairs (called the quark-antiquark sea) and of gluons. The known quarks are listed in **Table 2**. They must be assigned fractional charge values, relative to the proton charge. *See* Gluons; Hadron; Quarks.

Color and quantum chromodynamics. This quark theory of the hadrons has been proposed in a quite specific form, known as quantum chromodynamics (QCD). It is a gauge theory based on a symmetry hypothesized for the hadronic interactions of the quarks, which says that these interactions are invariant with respect to an SU(3)$_C$ group of unitary transformations with modulus unity acting in an abstract three-dimensional space known (whimsically) as color space. Each quark type then has three color states, usually labeled by the suffixes r = red, g = green, and b = blue, corresponding to the three axes of this space. The gauge particle of this symmetry theory is the gluon, a neutral vector particle coupled universally with the currents of color, just as the photon, the gauge particle of quantum electrodynamics (QED), is coupled universally with the electromagnetic current. However, whereas the photon has no charge, the gluon has ($3^2 - 1$) = 8 color components, so that it is a color octet. Consequently, there is a gluon contribution to the color currents, and so the gluon field must interact with itself, introducing a nonlinearity into quantum chromodynamics which has no parallel in quantum electrodynamics. This nonlinearity has important implications for quantum chromodynamics, leading to its asymptotic freedom, the

Table 2. Properties of established quarks*

Quark type	u (up)	d (down)	s (strange)	c (charmed)	b (bottom)		
Charge (Q/e_p)	2/3	−1/3	−1/3	2/3	−1/3	·	·
Mass, GeV	≈0.01	≈0.01	≈0.5	≈1.5	≈4.7	·	·
Flavor	$I_3 = +\tfrac{1}{2}$	$I_3 = -\tfrac{1}{2}$	$s = -1$	$C = +1$	$b = +1$	·	·

*To each quark, there exists an antiquark with the opposite flavor values and with opposite intrinsic parity.

property that the coupling of gluon to the color current approaches zero at short distances, which is essential for even qualitative agreement between quantum chromodynamics predictions and the empirical data on high-energy collision processes. *See* Color; Gauge theory; Quantum chromodynamics; Quantum electrodynamics; Symmetry laws.

An important element in quantum chromodynamics is the confinement dogma, the assertion that only color singlet states have finite energy. This assertion implies that neither a quark nor a gluon can exist in a free state, since the former is a color triplet and the latter a color octet. Zero mass is expected for a gauge particle (as is the case for the photon), hence for the gluon, and a mass of order 10 MeV is given for the (u,d) quarks in Table 2, but these values refer to the masses effective within hadronic states. In accord with this dogma, no observations of free gluons or quarks have yet been confirmed. Many theoreticians anticipate that this dogma will be deducible from quantum chromodynamics itself but, despite some favorable indications for this expectation, no rigorous proof that the dogma follows from quantum chromodynamics has been given.

Quantitative predictions of the properties of baryonic states are currently made using a simplified quark-quark (q-q) potential with the following features: (1) an attractive long-range potential, increasing with separation to ensure confinement, and (2) a spin-dependent potential representing one-gluon exchange, effective at small separation, where the regime of asymptotic freedom holds and perturbation theory is valid. Such predictions have had a great deal of success.

For a three-quark system, there is only one color-singlet (that is, scalar) wave function available, namely that given by Eq. (1), where the vector $\mathbf{q} = (q_r, q_g, q_b)$ refers to color space. This

$$\Phi_{\text{col}}(1,2,3) = \mathbf{q}(1) \cdot \mathbf{q}(2) \times \mathbf{q}(3) \tag{1}$$

is antisymmetric (A) with respect to permutation of the quark labels (1,2,3). The complete wave function for this system has the general form of Eq. (2), and must also have this permutation

$$\psi(1,2,3) = \Sigma \Phi_{\text{col}}(1,2,3)\, \psi_{\text{space}}(1,2,3)\, \chi_{\text{spin}}(1,2,3)\, \Phi_{\text{flavor}}(1,2,3) \tag{2}$$

symmetry A, from general considerations (the Pauli spin-statistics theorem). These remarks imply that the space × spin × flavor wave function for a baryonic state must be permutation symmetric (S), as if the quarks obeyed Bose statistics, for those variables. Indeed, the achievement of this result was precisely the historical purpose behind the introduction of the color degree of freedom. Within this article, mesons are regarded as bound antiquark-quark ($\bar{q}$-q) systems; their color wave functions have the singlet form $\bar{\mathbf{q}}(1) \cdot \mathbf{q}(2)$.

Nucleons and isospin. The quark content of the nucleons is given by Eqs. (3). The

$$p = (uud) \qquad n = (udd) \tag{3}$$

nucleons are ground states, so that their space wave functions ψ_{space} have orbital angular momentum $L = 0$ and are expected to be nodeless; this requires that these ψ_{space} wave functions have permutation symmetry S. The three quark spins sum to $S = \frac{1}{2}$, a spin state which has mixed (M) permutation symmetry. Antisymmetry for the wave function Eq. (2) then requires that the flavor wave functions (uud) and (udd) have mixed symmetry. (The product of two factors with M symmetry can be resolved into three terms, one with S, one with A, and one with M symmetry. Symbolically, M ⊗ M = S + A + M, whereas M ⊗ S = M and M ⊗ A = M.)

The u and d quarks are approximately degenerate in mass, their masses being small. For equal masses, their (kinetic + mass) energies would be the same. Their coupling with the gluon field g has the simple form of Eq. (4), where L_{int} is the lagrangian of the interaction, the $g_{\mu\alpha}$

$$L_{\text{int}} = f\Sigma g_{\mu\alpha}\,(\bar{u}\gamma_\mu \lambda_\alpha u + \bar{d}\gamma_\mu \lambda_\alpha d) \tag{4}$$

denote the components of the gluon field, the sum being over $\mu = 0, 1, 2, 3$ for time and space, and $\alpha = 1, 2, \ldots, 8$ for the components of octet color, the $\{\gamma_\mu\}$ are the matrices which appear in the Dirac equation, and the $\{\lambda_\alpha\}$ are the infinitesimal operators of the SU(3)$_C$ group. In this case, the total energy remains unchanged for any SU(2) transformation in the space (u,d). This invariance is the origin of the property known as charge independence for hadronic interactions. By analogy with the well-known group SU(2)$_\sigma$ for Pauli spin, this group is labeled SU(2)$_\tau$, its eigenvalues being known as the isospin I. The states (p,n) are then the $I_3 = (+\frac{1}{2}, -\frac{1}{2})$ components of an isospin doublet N. Isospin is well known in the classification of nuclear states; these occur as charge multiplets, a set of corresponding states in the nuclei with mass number A and the charge values $Z = I + A/2, I - 1 + A/2, \ldots, -I + A/2$, in which the nuclear interactions are the same. Thus, these $2I + 1$ states are said to have isospin component $I_3 = I, I - 1, \ldots, -I$, their energy being independent of I_3. The situation for systems of (u,d) quarks runs completely parallel with this.

This isospin symmetry is violated by any mass difference between the u and d quarks. It is also violated by the electromagnetic interactions, since the charge of the state depends on I_3 and the electromagnetic field couples with charge. The latter contribute further to the mass difference $(m(u) - m(d))$ and also give rise to electromagnetic interactions between the constituent quarks within the hadron. These effects contribute terms in the energy which are not invariant under the $SU(2)_\tau$ transformations, that is, terms which violate the isospin symmetry. Such charge-dependent effects are well known for the nuclear case, and the situation for (u,d) quarks is quite similar. *See* ISOBARIC SPIN.

Baryon octet. The replacement of a d quark in the nucleon by an s quark produces a baryon state with spin parity $\frac{1}{2}^+$ and strangeness number $s = -1$, the latter being given by $[n(\bar{s}) - n(s)]$, where $n(q)$ denotes the number of quarks of type q in the system considered. The states thus reached have the flavor structures of Eq. (5), the other factors in their wave functions Eq. (2)

$$(\Sigma^+, \Sigma^0, \Sigma^-) = (uus, (ud + du)s/\sqrt{2}, dds) \quad (5a) \qquad \Lambda = (ud - du)s/\sqrt{2} \quad (5b)$$

being identical with those for the nucleons; thus the isotriplet Σ and isosinglet Λ states are obtained. If a u quark and a d quark are each replaced by an s quark in Eq. (3), the isodoublet Ξ states of Eq. (6) are obtained. The s quark has $I = 0$, being unaffected by the $SU(2)_\tau$ transforma-

$$(\Xi^0, \Xi^-) = (uss, dss) \quad (6)$$

tions in the (u,d) space. The flavor wave function (sss) is necessarily symmetric and cannot occur with total spin $S = \frac{1}{2}$. Baryonic states with $s \neq 0$ are collectively termed hyperons. *See* HYPERON; STRANGE PARTICLES.

These eight baryon states $(p, n, \Sigma^+, \Sigma^0, \Sigma^-, \Lambda, \Xi^0, \Xi^-)$ all have the spin parity $\frac{1}{2}^+$ and the same internal wave functions. They are arrayed in Fig. 1, where the symmetry of their relationship is made evident; for this, it is helpful to use the quantum number $Y = (B + s)$, named hypercharge. In the approximation that the s quark has the same mass as the (u,d) quarks, the quark energies (kinetic + mass) are equal and the quark-gluon coupling has the form of Eq. (7),

$$L_{int.} = f\Sigma g_{\mu a} (\bar{u}\gamma_\mu \lambda_a u + \bar{d}\gamma_\mu \lambda_a d + \bar{s}\gamma_\mu \lambda_a s) \quad (7)$$

which is invariant under all $SU(3)_f$ transformations in the flavor space spanned by (u,d,s). The invariance of this expression for the pair-wise interchanges $(d \rightleftharpoons s)$, $(s \rightleftharpoons u)$, and $(u \rightleftharpoons d)$ is the origin of the threefold axes of symmetry shown in **Fig. 1**.

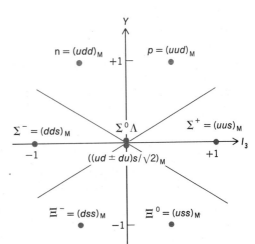

Fig. 1. Baryon octet states, arrayed with respect to I_3 as ordinate and $Y = (B + s)$ as abscissa. The charge number Q is given by $Q = I_3 + Y/2$. There are three axes of symmetry.

Mass differences. The mass difference $\delta m = (m(s) - m(u,d))$ is quite large, and so SU(3)$_f$ symmetry is much more strongly violated than SU(2)$_\tau$ symmetry. This is most apparent in the baryon mass values, which vary widely over the octet; the leading variation is that proportional to the strangeness s, which counts the s-quark content of each baryon. The approximately 75-MeV difference between the mean Σ mass $\bar{m}(\Sigma)$ and $m(\Lambda)$ has a more subtle origin, but is well accounted for on the basis of the quark-quark (qq) potential from quantum chromodynamics, described above. The small mass differences within each isospin multiplet are believed due to the intrinsic (u,d) mass difference and to electromagnetic effects. Assuming that these baryon states all have the same internal wave function, it follows that baryon octet states with the same charge Q have the same electromagnetic mass-shift $\delta m(Q)$; for example, $p = (uud)$ and $\Sigma^+ = (uus)$ have the same electromagnetic structure, since d and s have the same charge. From this observation, S. Coleman and S. Glashow pointed out in 1961 the remarkable SU(3)$_f$ relation of Eq. (8) connect-

$$m(\Xi^-) - m(\Xi^0) = m(\Sigma^-) - m(\Sigma^+) + m(p) - m(n) \tag{8}$$

ing the mass differences within the N, Σ, and Ξ multiplets, which is rather well satisfied by the empirical masses. These remarks about electromagnetic structure also imply the magnetic moment equalities of Eqs. (9), which are also in rough accord with the data.

$$\mu(\Sigma^+) = \mu(p) \qquad \qquad \mu(\Sigma^-) = \mu(\Xi^-) \qquad \qquad \mu(\Xi^0) = \mu(n) \tag{9}$$

Production and reaction processes. The production and reaction processes observed for these baryons are governed by flavor selection rules. For hadronic processes (but not for weak processes) these selection rules are simply Eq. (10) for each flavor $f = B, Q, s, c, \ldots$; that is,

$$\Delta n(f) = 0 \tag{10}$$

each type of quark is independent of the others, and quarks do not change type. Quantum chromodynamics provides a ready understanding of these rules; for example, each term of Eq. (7) preserves quark flavor. The isospin symmetry SU(2)$_\tau$ implies some additional constraints going beyond Eq. (10), especially in relating quantitatively various reactions involving baryons within the same charge multiplets. Otherwise, which reactions occur or do not occur can be accounted for in terms of Eq. (10) and the empirical baryon and meson mass values. *See Selection rules*.

The Λ and Σ hyperons are formed by closely related interactions. Their formation in pion-proton collisions of sufficiently high energy, or in K^- absorption by protons, given by reactions such as (11) has been especially well studied. These reactions illustrate the selection rules for B,

$$\pi^- + p \rightarrow \begin{cases} K^0 + \Lambda \\ K^0 + \Sigma^0 \\ K^+ + \Sigma^- \end{cases} \qquad K^- + p \rightarrow \begin{cases} \pi^- + \Sigma^+ \\ \pi^0 + \Sigma^0 \\ \pi^+ + \Sigma^- \\ \pi^0 + \Lambda \end{cases} \tag{11}$$

Q, and s; the (K^0, K^+) mesons have $s = +1$, while the ($K^-, \overline{K}^0$) mesons are their antiparticles, and have $s = -1$. It is illuminating to consider these reactions in terms of their constituent quarks, as shown in **Fig. 2**, where time progresses from left to right, and a quark line with backward-directed arrows represents an antiquark going forward in time. The simplest mechanisms for the reactions are equivalent to the exchange of $\bar{q}q$ (= mesons) or qqq (= baryon) systems between the meson and baryon. They generally involve intermediate $\bar{q}q$ creation and annihilation processes, but that of Fig. 2c involves only quark rearrangement. The simplest Ξ-production processes are given by reactions (12), where the $\Delta s = -2$ transition $N \rightarrow \Xi$ is balanced by the $\Delta s = +2$ transition $\overline{K} \rightarrow K$.

$$K^- + p \rightarrow \begin{cases} \Xi^- + K^+ \\ \Xi^0 + K^0 \end{cases} \tag{12}$$

Baryon-baryon interactions. The baryon-baryon interactions are of particular interest. Between nucleons, this interaction gives rise to the existence of atomic nuclei, and it has been particularly well studied, both empirically and theoretically, for the NN system. For large separations (greater than 0.8×10^{-15} m), the NN force is due to the exchange of pions and of other known mesons with masses less than about 1 GeV; for small separations (less than 0.4×10^{-15} m), a strong short-range repulsion is observed, possibly arising from the suppressive effects of the Pauli principle for quarks when the quark structures of the two nucleons overlap. At low energies,

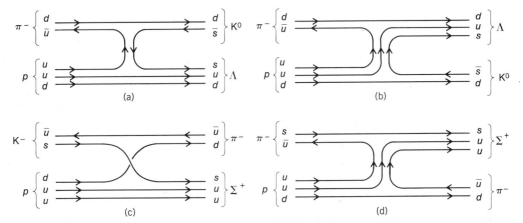

Fig. 2. Quark line graphs to illustrate meson-baryon reaction process. (a) $\pi^- + p \to K^0 + \Lambda$. (b) $\pi^- + p \to \Lambda + K^0$. (c) $K^- + p \to \pi^- + \Sigma^+$. (d) $\pi^- + p \to \Sigma^+ + \pi^-$. Time increases from left to right for these figures.

the outstanding feature of the NN interaction is its strong noncentral tensor component, which is due to one-pion exchange and is a direct consequence of the pseudoscalar nature of the pion. It also has a strong spin-orbit interaction, observed in NN interactions at higher energy and of much importance for the shell structure of nuclei.

Attractive ΛN forces of nuclear strength [with potentials $v(\Lambda N)$ of order 60% of $v(NN)$] are known to exist from the observation of Λ hypernuclei, each consisting of a Λ hyperon bound to an ordinary nucleus and denoted by $_\Lambda Z^A$. The $_\Lambda Z^A$ (ground state) systems remain stable until the Λ hyperon decays, since the conservation laws of Eq. (10) do not allow the deexcitation of the ΛN system by any hadronic process. Λ-proton elastic scattering has also been studied in the low-energy regime. Observations on excited Λ-hypernuclear states suggest that the ΛN spin-orbit interaction is relatively small. *See Hypernuclei.*

The Σ hyperons also interact strongly with nucleons, as shown by the studies of $\Sigma^\pm$-proton elastic scattering at low energies. However, Σ hyperons also react strongly with nucleons of suitable charge, transforming into a Λ hyperon with release of about 75 MeV of kinetic energy; for example, reactions (13) are well known. These absorptive reactions preclude the existence of long-

$$\Sigma^- + p \to \Lambda + n \quad (13a) \qquad \Sigma^+ + n \to \Lambda + p \quad (13b)$$

lived Σ hypernuclei, except for several special cases such as $n\Sigma^-$ or $nn\Sigma^-$ for which the selection rules of Eq. (10) allow no hadronic deexcitation process. There is no evidence for bound states of these last systems, but there is evidence suggesting that a class of exceptional Σ-hypernuclear levels exist with lifetimes long on the nuclear scale.

The Ξ hyperons also interact strongly with nucleons; the little evidence available is consistent with theoretical expectation that processes (14) should be strong, occurring through a quark

$$\left.\begin{array}{c}\Xi^- + p \\ \Xi^0 + n\end{array}\right\} \to \Lambda + \Lambda \qquad (14)$$

rearrangement process. A striking example of this process was the observation of reaction (15) for

$$\Xi^- + C^{12} \to {}_{\Lambda\Lambda}\text{He}^6 + \text{Li}^7 \qquad (15)$$

a Ξ^- stopping in nuclear emulsion. From the total binding energy for the two Λ hyperons in $_{\Lambda\Lambda}\text{He}^6$, deduced from the energy released in its weak decay to $_\Lambda\text{He}^5 p\pi^-$, the $^1S_0$ $\Lambda\Lambda$ interaction is known to have strength comparable with that for the $^1S_0$ ΛN interaction.

Baryon beta decay. The neutron beta-decay process (16) has been known for some

$$n \to p + e^- + \bar{\nu}_e \qquad (16)$$

decades. It was known quite early that the beta transition $n \to p$ involves two terms of opposite parity, vector (Fermi) and axial-vector (Gamow-Teller), with coupling amplitudes G_V and G_A. The empirical ratio $G_A/G_V = -1.253 \pm 0.007$ agrees well with the calculations of S. Adler (1965) and W. Weisberger (1966) based on the current algebra hypothesis (this reflects the property of chiral symmetry for the hadronic interactions). G_V is slightly smaller than the value G known for muon beta decay, $G_V/G = 0.97 \pm 0.01$. The related process $\mu^- p \to n\nu_\mu$ is also well known, and its inverse, the neutrino-induced reaction $\nu_\mu n \to p\mu^-$, has been much studied for high-energy ν_μ beams (up to about 200 GeV) obtained from the decay $\pi^+ \to \mu^+ \nu_\mu$ of π^+ beams from high-energy proton accelerators. SEE NEUTRINO.

Analogous beta-decay processes are known for most of the hyperons in the baryon octet, but their rates are about an order of magnitude smaller than expected if their beta-decay amplitudes were the same as for the $n \to p$ transition. The Λ beta-decay process (17) is well known,

$$\Lambda \to p + e^- + \bar{\nu}_e \qquad (17)$$

with a branching ratio of $8.4 \pm 0.2 \times 10^{-4}$. The transition $\Lambda \to p$ involves $\Delta s = +1$, whereas the transition $n \to p$ of Eq. (16) involves $\Delta s = 0$. For Σ hyperons, both types of transitions are known; $\Delta s = 0$ holds for the decay modes (18) and $\Delta s = +1$ holds for the decay (19). The

$$\Sigma^- \to \Lambda + e^- + \bar{\nu}_e \qquad\qquad \Sigma^+ \to \Lambda + e^+ + \nu_e \qquad (18)$$

$$\Sigma^- \to n + e^- + \bar{\nu}_e \qquad (19)$$

conceivable decay mode $\Sigma^+ \to n e^+ \nu_e$ is not observed, its branching ratio being less than 5×10^{-6}. For Ξ hyperons, the $\Delta s = +1$ decay mode $\Xi^- \to \Lambda e^- \bar{\nu}_e$ is known, but there is no evidence (branching ratio less than 3×10^{-3}) for the energetically favorable $\Delta s = +2$ mode $\Xi^- \to n e^- \bar{\nu}_e$. The only $\Delta s = +1$ decay mode for Ξ^0 would be $\Xi^0 \to \Sigma^+ e^- \bar{\nu}_e$, which has not yet been observed (branching ratio less than 10^{-3}).

Parity and charge-conjugation violation. All of these beta-decay processes violate both parity conservation and charge-conjugation symmetry. Parity violation in beta decay was first observed in 1957, as an asymmetry in the angular distribution of beta electrons emitted from polarized ^{60}Co nuclei. Later, this was demonstrated also for the beta decay of free neutrons, and subsequently also for the beta decay (17) of polarized Λ particles. SEE PARITY.

Cabibbo theory. A remarkable synthesis of all the data on baryon beta-decay processes has been achieved, by the work of many theoreticians. This is based on the notion of a weak $\Delta Q = +1$ current $J_\lambda^{\text{wk}+}$ which includes both vector and axial vector terms and which couples with itself. It consists of two parts, one for hadrons (meaning quarks) and the other for leptons, and so leads to the interaction amplitude (20). $(J_{\text{lept}}^{\text{wk}+})_\lambda$ consists of a sum of terms $(e\nu_e)$, (μ,ν_μ), (τ,ν_τ),

$$G\Sigma(J_{\text{hadr}}^{\text{wk}+} + J_{\text{lept}}^{\text{wk}+})_\lambda \dagger (J_{\text{hadr}}^{\text{wk}+} + J_{\text{lept}}^{\text{wk}+})_\lambda \qquad (20)$$

and so forth, for each lepton and its associated neutrino. The purely leptonic amplitude, which describes the muon decay $\mu^- \to \nu_\mu (e^- \bar{\nu}_e)$ for example, is then given by the product of $J_{\text{lept}}^{\text{wk}+}$ with its conjugate, as it occurs in expression (20), and has the universal amplitude G. The baryon beta-decay amplitudes stem from the hadron-lepton cross terms in Eq. (20). The crucial step forward was taken in 1963 by N. Cabibbo, who proposed that $J_{\text{hadr}}^{\text{wk}+}$ for the quark flavors then known should be based on the single quark doublet $(u, d\cos\theta_C + s\sin\theta_C)$, thus linking the $\Delta s = 0$ and $\Delta s = +1$ transitions. This leads to expression (21) for the semileptonic weak transition processes, the

$$G\{\Sigma_\lambda (J_{\text{lept}}^{\text{wk}+}(e^+\nu_e) + \cdots)_\lambda \dagger (\cos\theta_C J_{\text{hadr}}^{\text{wk}+}(d \to u)_\lambda$$
$$+ \sin\theta_C J_{\text{hadr}}^{\text{wk}+}(s \to u)_\lambda) + \text{hermitian conjugate}\} \qquad (21)$$

$\Delta s = +1$ transition $s \to u$ having an intrinsic amplitude weaker by the factor $\tan\theta_C$ than that for the $\Delta s = 0$ transition $d \to u$. Each $J_{\text{hadr}}^{\text{wk}+}$ has two terms $(V_\lambda + A_\lambda)$, as was known for neutron beta decay. Equation (21) prescribes the relationship between the values for A_λ for the various baryonic transitions in terms of two numbers, one being (G_A/G_V) for $n \to p$ and the other being θ_C. These currents $(J_{\text{hadr}}^{\text{wk}+})_\lambda$ for single quark transitions are automatically members of an octet, the property Cabibbo assumed; also the transitions induced by these currents necessarily have $\Delta Q = 1$ for $\Delta s = +1$, or $\Delta Q = -1$ for $\Delta s = -1$, so that the interaction (21) forbids $\Sigma^+ \to n e^+ \nu_e$ while allowing $\Sigma^- \to n e^- \bar{\nu}_e$. The value appropriate for the Cabibbo angle θ_C is given consistently by a wide variety of data, as $\theta_C = 0.223 \pm 0.005$ radians. From Eq. (21), the value predicted for G_V/G is $\cos\theta_C = 0.97$, which is consistent with the observed value given above.

$\Delta Q = 0$ hadron transitions are also possible, based on neutral currents $(J_{hadr}^{wk0})_\lambda$, but the data on mesonic transitions have shown clearly that there are no such transitions for $\Delta s = +1$, so they need not be considered for baryon decays. $\Delta Q = 0$, $\Delta s = 0$ transitions can and do occur, but are most accessible in the related inelastic neutrino scattering processes, of the type $\nu N \to \nu' +$ (hadrons). In 1970, S. Glashow, J. Iliopoulos, and L. Maiani proposed a specific mechanism to eliminate all $\Delta Q = 0$, $\Delta s = +1$ transitions, involving a new quark c, with the proposal that its contribution to $(J_{hadr}^{wk})_\lambda$ should be based on a second quark doublet, $(c, -d \sin\theta_C + s \cos\theta_C)$. Assuming the same strength for the neutral currents constructed from each of these two doublets, the two neutral currents of the type $J_{hadr}^{wk0}(d \to s)$ then formed cancel exactly. The dominant weak transition for the c quark is then the $\Delta Q = -1$ transition $c \to s$, with amplitude $G \cos\theta_C$.

Nonleptonic baryon decay. These are the dominant modes of hyperon decay, as shown in Table 1, typically having lifetimes of the order of 10^{-10} s. The decay $\Lambda \to p\pi^-$ has been especially well studied and shows a strong violation of parity conservation, as a large forward-backward asymmetry (coefficient $\alpha_\Lambda = 0.64$) in the pion distribution relative to the initial Λ-spin direction; this allows efficient measurements of Λ-spin polarization to be made for Λ-production processes. The same holds for the decay $\Sigma^+ \to p\pi^0$, where $\alpha_{\Sigma^+}^0 = -0.98$. The modes $\Sigma^\pm \to n\pi^\pm$ are also well known, with asymmetries $\alpha(\Sigma^+) = -\alpha(\Sigma^-) = 0.07 \pm 0.01$. The Ξ-hyperon decays are $\Xi^- \to \Lambda\pi^-$ with $\alpha(\Xi^-) = -0.435 \pm 0.015$, and $\Xi^0 \to \Lambda\pi^0$ with $\alpha(\Xi^0) = -0.41 \pm 0.02$. All of these decays are weak, with $\Delta s = +1$.

The decay $\Sigma^0 \to \Lambda\gamma$ is not a weak process but an electromagnetic transition, allowed ($\Delta s = 0$) by the hadronic selection rules, with measured lifetime about 10^{-19} s. The decay $\Sigma^+ \to p\gamma$ is well known but involves $\Delta s = +1$, so that it is due to the joint action of weak and electromagnetic interactions. No $\Delta s = +2$ decay, such as $\Xi^0 \to p\pi^-$ or $\Xi^- \to p\pi^-\pi^-$, has been detected.

The observed weak nonleptonic decay processes are all implied by the terms $(J_{hadr}^{wk})_\lambda \dagger \times (J_{hadr}^{wk})_\lambda$ of Eq. (21), including both the $\Delta Q = +1$ and the (not specified here) $\Delta Q = 0$ components. For example, the Λ decay results from the transition $s \to u(\bar{u}d)$, but this is not simply related with the final state observed. Empirically, there is a $\Delta I = \frac{1}{2}$ rule, for $\Delta s = +1$ transitions, which means that the final state $B\pi$ is dominated by the isospins $|(I \pm \frac{1}{2})|$, where I is the initial baryon isospin. These nonleptonic hadronic terms of Eq. (21) also give rise to nonmesonic weak decay interactions; reaction (22a) represents a weak contribution to the NN forces, and its tiny effects have been identified in the study of nuclear processes, while reaction (22b) is well known

$$NN \to NN \quad (22a) \qquad\qquad \Lambda N \to NN \quad (22b)$$

from the energetic nonmesonic decays observed for Λ hypernuclei, for example, $_\Lambda\text{He}^5 \to n\text{He}^4$ with approximately 173 MeV energy release. *See* Weak nuclear interactions.

3/2$^+$ baryon decuplet. The three-quark state (uuu) with $S = \frac{3}{2}$ is flavor symmetric and spin symmetric. Antisymmetry for the full wave function [Eq. (2)] then requires the space wave function to have symmetry S, as holds for the nucleon states [Eq. (3)]. If the potentials $v(qq)$ are spin and flavor independent, this state will have the same $L = 0$ space wave function as the baryon octet, and it is therefore intimately connected with the latter. There are four such non-strange states, denoted by Δ and having isospin $I = \frac{3}{2}$, whose quark structures are shown in Eqs. (23). The replacement of one, two, or three of these d quarks by an s quark leads to an

$$\begin{aligned}\Delta^{++} &= (uuu) & \Delta^+ &= (uud)_S \\ \Delta^0 &= (udd)_S & \Delta^- &= (ddd)\end{aligned} \quad (23)$$

isotriplet Σ^*, an isodoublet Ξ^*, and an isosinglet Ω, respectively, forming the decuplet depicted in **Fig. 3**.

These Δ states are identified with the strong $I = \frac{3}{2}$, J^P (spin parity) $= \frac{3}{2}^+$ pion-nucleon resonances first observed in 1953, with mass about 1232 MeV. They are particle unstable, the breakup $\Delta \to N\pi$ being allowed by all the hadronic selection rules; their lifetimes are approximately 0.5×10^{-23} s. The Σ^* states have mass 1382 MeV and decay dominantly to $\Lambda\pi$. They were first established in 1960 from observations on $\Lambda\pi$ correlations in reaction (24), showing the

$$K^- + p \to \begin{Bmatrix} \Sigma^{*+} + \pi^- \\ \Sigma^{*-} + \pi^+ \end{Bmatrix} \to \Lambda + \pi^+ + \pi^- \quad (24)$$

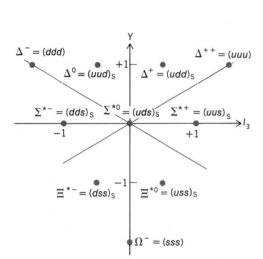

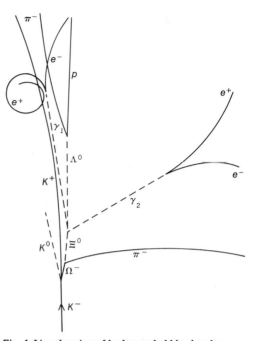

Fig. 3. The baryon decuplet states, arranged with respect to I_3 as ordinate and $Y = (B + s)$ as abscissa. The charge number Q is given by $Q = I_3 + Y/2$. There are three axes of symmetry.

Fig. 4. Line drawing of hydrogen bubble chamber event involving production and decay of Ω^- particle. An incident K^- meson (momentum 5000 MeV/c) interacts with a proton giving Eq. (7). The K^0 meson is not observed but its presence is deduced from analysis of energy and momentum balance. $\Omega^- \to \Xi^0 \pi^-$ decay occurs, the (invisible) Ξ^0 travels some distance and then decays to neutral system $\Lambda \pi^0$; the Λ particle is detected from its decay $\Lambda \to p\pi^-$, the π^0 from the electron pair conversions of the two photons γ_1 and γ_2 resulting from the $\pi^0 \to \gamma\gamma$ decay. Broken lines denote unobserved paths of neutral particles, deduced from measurements on visible particle tracks.

occurrence of the intermediate states shown there. The Ξ^* states have mass 1532 MeV and decay to $\Xi\pi$, as observed in 1962 for reactions (25). The isosinglet Ω^- was found in 1964, in reaction (26) depicted in **Fig. 4**, and has mass 1672 MeV. The Ω^- particle is semistable; that is, it decays

$$K^- + p \to \begin{cases} \Xi^{*-} + K^+ \to \Xi^- + \pi^0 + K^+ \\ \Xi^{*0} + K^0 \to \Xi^- + \pi^+ + K^0 \end{cases} \quad (25) \qquad K^- + p \to \Omega^- + K^+ + K^0 \quad (26)$$

through the weak interactions, with lifetime on the order of 10^{-10} s; hadronic decay is forbidden, since its mass lies below the lowest threshold $m(\Xi^0 K^-) = 1808$ MeV for known hadrons with the same net strangeness.

It is this occurrence of the $^2 8$ and $^4 10$ multiplets (where the notation is $^{2S+1}\alpha$, α specifying the SU(3)$_f$ multiplicity) in association which indicates (**Table 3**) that their (spin × flavor) wave function has symmetry S. Without color, the spin-statistics theorem would indicate that their Ψ_{space} wave function had symmetry A. Since these multiplets are ground states, it would be very difficult to achieve this, if it is indeed even possible. It was these baryonic resonance observations which provided the empirical basis for the introduction of a three-dimensional color space for quarks.

With quantum chromodynamics, the mass difference $[m(\Delta) - m(N)] = 293$ MeV is attrib-

Table 3. Permutation symmetries possible for spin × SU(3)$_f$ wave functions for three quarks

Permutation symmetry	SU(3) multiplets		
	Singlet	Octet	Decuplet
S		$S = \frac{1}{2}$	$S = \frac{3}{2}$
M	$S = \frac{1}{2}$	$S = \frac{1}{2}$ and $\frac{3}{2}$	$S = \frac{1}{2}$
A	$S = \frac{3}{2}$	$S = \frac{1}{2}$	

uted to the spin-spin component (Fermi-Breit term) of the one-gluon-exchange potential. Quantum chromodynamics also predicts approximately equal mass spacing between the four charge multiplets of the decuplet; the spacings observed from Δ to Ω are 150, 150, and 140 MeV, in turn. The existence of a semistable $\frac{3}{2}^+$ Ω^- hyperon with mass 1685 MeV had been predicted by M. Gell-Mann from SU(3)$_f$ symmetry in 1963, so the discovery of such a curious particle in 1964 quickly led to a general acceptance of SU(3)$_f$ symmetry.

Higher excited baryonic states. Many further particle-unstable baryon states, with lifetimes in the range 10^{-22} to 10^{-23} s, have become established. The $s = 0$ states have been explored most thoroughly, by the measurement of the angular and polarization angular distributions for pion-nucleon elastic and charge-exchange scattering and their partial-wave analysis to give phase shifts as a function of spin-parity and mass. Similarly, the $s = -1$ states have been studied, mostly using elastic scattering and reaction data for K^- on nucleons. Further evidence on the $s = -1$ states also comes from the study of final states in reactions $\overline{K}N \rightarrow (\Lambda$ or $\Sigma) +$ pions, just as the $\Sigma(1382)$ resonance (which lies below the $\overline{K}N$ threshold) was deduced from data on reaction (24). The $s = -2$ states can be studied only from the final state effects, for example, in $\overline{K}N \rightarrow K\Xi^*$, just as $\Xi(1530)$ was deduced from reaction (25), so that relatively little is known concerning Ξ^* states. The most prominent baryonic resonances established, up to mass values of order 2500 MeV, are listed in **Table 4**.

All of the resonances now established are consistent with the limitations of the three-quark model. For example, no resonances are established for $s = +1$; there is one candidate, for the

Table 4. Known prominent and well-established excited baryonic states*

N states	J_P	Δ states	J_P	Λ states	J_P	Σ states	J_P	Ξ states	J_P
N(1450)	$\frac{1}{2}^+$	$\Delta(1232)$	$\frac{3}{2}^+$	$\Lambda(1405)$	$\frac{1}{2}^-$	$\Sigma(1385)$	$\frac{3}{2}^+$	$\Xi(1530)$	$\frac{3}{2}^+$
N(1520)	$\frac{3}{2}^-$	$\Delta(1620)$	$\frac{1}{2}^-$	$\Lambda(1520)$	$\frac{3}{2}^-$	$\Sigma(1660)$	$\frac{1}{2}^+$	$\Xi(1815)$	$\frac{3}{2}^-$
N(1535)	$\frac{1}{2}^-$	$\Delta(1640)$	$\frac{3}{2}^+$	$\Lambda(1670)$	$\frac{1}{2}^-$	$\Sigma(1670)$	$\frac{3}{2}^-$	$\Xi(1930)$	?
N(1650)	$\frac{1}{2}^-$	$\Delta(1700)$	$\frac{3}{2}^-$	$\Lambda(1690)$	$\frac{3}{2}^-$	$\Sigma(1750)$	$\frac{1}{2}^-$	$\Xi(2030)$	?
N(1670)	$\frac{3}{2}^-$	$\Delta(1900)$	$\frac{1}{2}^-$	$\Lambda(1800)$	$\frac{3}{2}^+$	$\Sigma(1765)$	$\frac{5}{2}^-$		
N(1680)	$\frac{5}{2}^-$	$\Delta(1910)$	$\frac{1}{2}^+$	$\Lambda(1815)$	$\frac{5}{2}^+$	$\Sigma(1915)$	$\frac{5}{2}^+$		
N(1680)	$\frac{5}{2}^+$	$\Delta(1920)$	$\frac{5}{2}^-$	$\Lambda(1830)$	$\frac{5}{2}^-$	$\Sigma(1940)$	$\frac{3}{2}^-$		
N(1700)	$\frac{3}{2}^-$	$\Delta(1920)$	$\frac{5}{2}^-$	$\Lambda(1870)$	$\frac{3}{2}^+$	$\Sigma(2030)$	$\frac{7}{2}^+$		
N(1710)	$\frac{1}{2}^+$	$\Delta(1950)$	$\frac{7}{2}^+$	$\Lambda(2100)$	$\frac{7}{2}^-$	$\Sigma(2250)$	?		
N(1740)	$\frac{3}{2}^+$	$\Delta(1960)$	$\frac{3}{2}^+$	$\Lambda(2110)$	$\frac{5}{2}^+$	$\Sigma(2455)$	?		
N(1830)	$\frac{3}{2}^-$	$\Delta(2010)$	$\frac{3}{2}^-$	$\Lambda(2350)$	$\frac{9}{2}^+$	$\Sigma(2595)$	?		
N(1880)	$\frac{3}{2}^+$	$\Delta(2420)$	$\frac{11}{2}^+$						
N(1990)	$\frac{7}{2}^+$	$\Delta(2850)$	$\frac{15}{2}^+$						
N(2190)	$\frac{7}{2}^-$	$\Delta(3230)$	?						
N(2200)	$\frac{9}{2}^-$								
N(2220)	$\frac{9}{2}^+$								
N(2650)	$\frac{11}{2}^-$								
N(3030)	?								

*There should be further columns for excited Ω states, made of three s quarks, and for charmed baryon states Λ_c and Σ_c, consisting of one c quark and two u or d quarks. Only $\Omega(1672)$ and $\Lambda_c(2282)$, both semistable, are known. There is some evidence for an unstable excited state $\Sigma_c(2500)$.

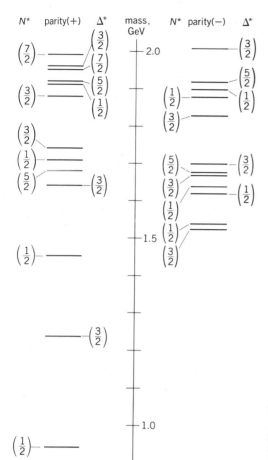

Fig. 5. Spectrum of known nucleonic excited states up to mass 2.0 GeV, with their spin values, those with positive parity being on the left half of the figure, those with negative parity being on the right half. N^* states are those with $I = \frac{1}{2}$, Δ^* states those with $I = \frac{3}{2}$.

$I = 0$, KN, $\frac{1}{2}^-$ partial wave, deduced from elastic scattering and charge-exchange data, but despite many years of work the resonance interpretation is still not uniquely established for it. No resonances with $I = 2$, $s = -1$ are established for the $\pi\Sigma$ channels, nor with $I = \frac{3}{2}$, $s = -2$ for the $\pi\Xi$ channels, and there are indeed no qqq configurations which have these quantum numbers, so none is expected.

The density of excited baryon states per unit energy interval increases rapidly with increasing mass, as illustrated for the N^* and Δ^* states in **Fig. 5**. The validity of SU(3)$_f$ symmetry requires that there should exist Λ^*, Σ^*, and Ξ^* states for each N^*, and Σ^*, Ξ^*, and Ω^* states for each Δ^*, most of which have not yet been detected. Further exploration is going on, especially for high mass values, and there is every reason to expect the density of states to continue to increase in the mass region above 2500 MeV. The higher states already known indicate at least the following four mechanisms of excitation:

1. Rotational excitation of low-lying multiplets, such as the ground state multiplets with symmetry S, comprising the $\frac{1}{2}^+$ octet and the $\frac{3}{2}^+$ decuplet. This is well illustrated by the sequence of prominent Δ^* states, for which **Fig. 6** shows a plot of (mass)^2 versus spin J. The established points lie on a straight line, with intervals $\Delta J = 2$, which is generally known as a Regge (rotational) trajectory. SEE REGGE POLE.

2. Vibrational (radial) excitation of the ground state multiplets of a given type. In Fig. 5 the $\frac{1}{2}^+$ state $N(1450)$ and $\frac{3}{2}^+$ state $\Delta(1640)$ constitute a repetition of the ground state multiplets with symmetry S. The precise nature of this radial excitation is still under investigation.

3. Excitations involving internal orbital angular momentum L with parity P. These are natural excitations to expect if baryons are composed of three quarks. Table 3 lists the SU(3)-spin combinations which go together for a particular permutation symmetry of the space wave function. The observed states then have parity P and form submultiplets for each J obtained by the vector addition of L and S. It is convenient to define internal coordinates, $\boldsymbol{\rho}$ for the vector separation between two quarks and $\boldsymbol{\lambda}$ for the vector separation of the third quark from their center of mass, and to specify excited configurations by the quantum numbers $(n_\rho l_\rho;\ n_\lambda l_\lambda)$ for these internal degrees of freedom. Then $P = (-1)^{l_\rho + l_\lambda}$ and $\boldsymbol{L} = \boldsymbol{l}_\rho + \boldsymbol{l}_\lambda$.

The ground state multiplets correspond to $(n_\rho l_\rho;\ n_\lambda l_\lambda) = (00;\ 00)$, which has symmetry S, assuming harmonic oscillator orbitals.

The first excited configuration is then a linear superposition of (11; 00) and (00; 11); its space wave function therefore has $L^P = 1^-$, with symmetry M. From Table 3, this comprises the multiplets $^2 10$, $^2 8$, and $^2 1$, together with $^4 8$, and so accounts for all the N^* and Δ^* states of negative parity lying below 1.8 GeV on Fig. 5. This supermultiplet also includes two SU(3)$_f$ singlets, $\Lambda(1520)$ with $3/2^-$ and $\Lambda(1405)$ with $1/2^-$. The various SU(3)$_f$ multiplets are not pure; in general, all states with the same s, I, J, and P in the same band $N = n_\rho + n_\lambda$ will mix strongly, and there can also be mixing between the bands. With quantum chromodynamics, these mixings are prescribed, and considerable success has been achieved, especially by N. Isgur and G. Karl, in accounting for the data in this way.

The second excited configurations have even parity, consisting of the configurations (22; 00), (20; 00), (11; 11), (00; 20), and (00; 22). These contain space wave functions with permutation symmetry S for $L = 2$ and 0, M for $L = 2$ and 0, and A for $L = 1$. The first group (S) includes the Regge excitations ($L = 2$) of the ground state and the radial vibrations ($L = 0$), as well as a number of SU(3)$_f$ multiplets with other values of J. All of the N^* and Δ^* states required by these multiplets have been found to exist. Some of the states required by the SU(3)$_f$ multiplets of the second group (M) are established, sufficient to suggest strongly that these M supermultiplets exist. There is good reason to believe that states of the supermultiplet A will be difficult to produce and to detect, and there are no candidates for it yet.

The third excited configurations will have $P = -1$ and will be very numerous. The ob-

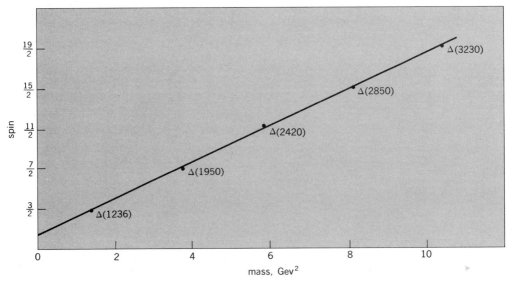

Fig. 6. (Mass)^2 for prominent $I = 3/2$ nucleonic excited states plotted against their spin J. The uppermost state has been placed on the assumption that its spin-parity value continues the sequence $(3/2+)$, $(7/2+)$, $(11/2+)$, $(15/2+)$, ... The excellent straight-line fit is interpreted as a Regge trajectory for a sequence of rotational levels.

served states include the $7/2^-$ Regge recurrence $N(2190)$ expected from the $3/2^-$ state $N(1520)$. A supermultiplet of symmetry S and $L = 1$ is also expected, and the five states above 1800 MeV, on the right half of Fig. 5, do have the (I,J,P) combinations to fill out this supermultiplet. No other assignment is available for the $5/2^-$ $\Delta(1920)$ state. It is unlikely that all the multiplets of the $N = 3$ band will ever be identified empirically, but at least no states have yet been found which are inconsistent with this model.

This qqq model for baryons is far from complete. It provides a good mnemonic for baryon spectroscopy, and even works well quantitatively, but it does not include any account of the quark sea or of gluonic components, even for the ground states. Other kinds of excitation are also possible; for example, if $s = +1$ resonances are established, they might correspond to multiquark states with the structure $(\bar{q}qqqq)$. There might also exist excited (hybrid) states where the excitations are gluonic. However, there is no data calling for such excitations at present.

Charmed baryons and beyond. Further baryon states can be formed by replacing one or more of the u, d and s quarks of the states discussed above by a c quark. If the s and c quarks both had the same mass as the (u,d) quarks, the states formed would correspond to an SU(4)$_f$ symmetry. This is emphasized by **Fig. 7**, which shows the extensions of the $1/2^+$ baryon octet (Fig. 7a) and the $3/2^+$ baryon decuplet (Fig. 7b) obtained in this way. The lowest plane of each part of the figure consists of the charmless baryon states; those in Fig. 7a are in accord with the octet of Fig. 1, and those in Fig. 7b with the decuplet of Fig. 3. If SU(4)$_f$ symmetry were exact, the states of the array in Fig. 7a or Fig. 7b would all have the same internal structure and would be the substates of a representation, $^2 20_M$ for Fig. 7a or $^4 20_S$ for Fig. 7b, of the SU(4)$_f$ group. These are two different 20-dimensional representations of this group; the suffix denotes the permutation symmetry of the flavor wave functions forming the basis of each SU(4)$_f$ representation. Figure 7a consists of four equal hexagons of side l put together to form a symmetric solid, whose outer surface consists of these hexagons plus four equilateral triangles of side l; Fig. 7b consists of four equilateral triangles of side $3l$, put together to form a pyramid.

In reality, the c-quark mass is so large that little quantitative detail of SU(4)$_f$ symmetry can survive in the physical situation, and the structure of Fig. 7 has value mainly for general comprehension and for the counting of states. This is clear from the quantum chromodynamics calculation of Isgur and his colleagues; the limitation of the space wave function to one particular symmetry and one particular band N is often a poor approximation. Their lowest calculated levels are given in **Table 5**; the states are labeled as for SU(3)$_f$, the suffix c meaning that the s quark has been replaced by a c quark (which increases the charge by $+1$). With these mass values, only

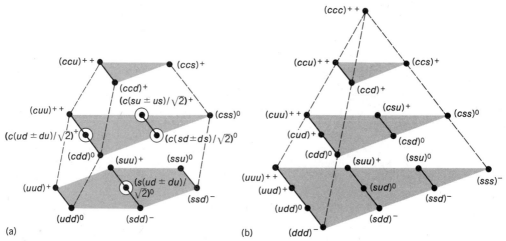

Fig. 7. Arrays of the states of baryons made from three quarks of the types u, d, s, and c, with no internal orbital angular momentum. The quark content and charge of each state is specified. The vertical axis specifies the number of c quarks. (a) $1/2^+$ baryons. (b) $3/2^+$ baryons. The former have flavor symmetry M; the latter have symmetry S.

Table 5. Mass, spin, and parity values calculated from quantum chromodynamics for the low-lying C = +1 charmed baryonic states

$J^P =$	$\frac{1}{2}^+$	$\frac{3}{2}^+$	$\frac{1}{2}^-$	$\frac{3}{2}^-$
	$\Lambda_c(2260)$	$\Sigma_c(2510)$	$\Lambda_c(2510)$	$\Lambda_c(2590)$
	$\Sigma_c(2440)$			

the isosinglet $\Lambda_c^+(2260)$ will be semistable, and its weak decay will be due to the dominant c-quark transition $c \to s$, as remarked above.

The observed state $\Lambda_c^+(2282)$ has a number of decay modes (Table 1), the strongest being to $pK^-\pi^+$, with branching ratio 2.2 ± 1.0%. Its isospin has not been demonstrated to be $I = 0$, but it is reasonable to identify it with the lowest calculated state on Table 5. There is also some evidence for a Σ_c state of mass about 2500 MeV, decaying hadronically to $\pi\Lambda_c$. This may be identified with $\Sigma_c(2510)$, for this is the charmed analog to $\Sigma(1385)$, the decay $\Sigma_c \to \pi\Lambda_c$ being parallel with the decay $\Sigma \to \pi\Lambda$, but the spin-parity for the observed state $\Sigma_c(\sim 2500)$ is not yet known empirically. SEE CHARM.

The existence of a fifth quark b was demonstrated in 1978 in work with the PETRA electron-positron storage ring at the DESY laboratory, Hamburg, and theoreticians anticipate the existence of a sixth quark, to be named t, whose mass must exceed 22 GeV in order to account for the lack of evidence for this quark even at the top energy of PETRA. These quarks will add two further additive quantum numbers with conservation laws of the form of Eq. (10). Of course, there could well exist still further, heavier quarks at present unsuspected. Baryon states built of these heavier quarks, together with the lighter quarks, will certainly exist and will be observed in the course of time; no candidates to be identified with Λ_b have yet been brought forward. SEE PARTICLE ACCELERATOR; UPSILON PARTICLES.

Baryonic spectroscopy. It is clear that the baryonic levels present a spectroscopic situation not unlike, although more complicated than, that already explored for atomic nuclei and their excited states. There are electromagnetic and mesonic transitions between the baryonic levels, whose systematics are just beginning to be studied. For example, the two mesonic transitions (27a) and (27b) are well known, and the radiative transition (28) has become established.

$$\Sigma(1765) \to \pi\Lambda(1520) \quad (27a) \qquad \Sigma(2030) \to \pi\Sigma(1385) \quad (27b) \qquad \Lambda(1520) \to \gamma\Lambda \quad (28)$$

Knowledge of this baryon spectroscopy has been developing rapidly as has also theoretical understanding of it. SEE NUCLEAR SPECTRA.

Antibaryons. For every baryonic state mentioned above, there will exist an antibaryon state with opposite flavor quantum numbers, in particular with $B = -1$. The antiproton $\bar{p}$ was first identified in 1954, and antiproton beams have been available at high-energy proton accelerators for many years. In 1981 the Super Proton Synchrotron (SPS) at CERN (Geneva) was adapted to become the CERN Collider, a proton-antiproton storage ring, with which $\bar{p}$-p collisions have been studied up to 540 GeV center-of-mass energy, with spectacularly fruitful results, especially the production and decay of the long-sought weak bosons $W^\pm$ and Z^0 and the clear observation of quark and gluon jets produced in these collisions. SEE INTERMEDIATE VECTOR BOSON; WEINBERG-SALAM MODEL.

Most of the expected antibaryon states have been detected and studied, in some detail, for example using $\bar{p}$ beams for reactions of the general type (29), where B_1 and B_2 denote any two

$$\bar{p} + p \to \bar{B}_1 + B_2 + \text{mesons} \quad (29)$$

baryonic states (stable or unstable) and the mesons are such that the reaction conserves charge, strangeness, charm, and so forth. However, the reaction rates fall rapidly with increasing mass, strangeness, and charm for the final baryons. Electron-positron annihilation reactions can be studied efficiently using storage rings, and these suffer much less from this disadvantage, mainly because the background events suffer a similar fall in rate; for example, at the storage rings SPEAR and PEP at SLAC (Stanford University), CESR at Cornell University, and DORIS and PETRA at DESY (Hamburg), the energy available is well above the $\Lambda_c\overline{\Lambda}_c$ threshold. The study of reactions of the type (30) with SPEAR has given much of the present knowledge of $\Lambda_c(2282)$.

$$e^+ + e^- \to \overline{\Lambda}_c + (\Lambda_c \text{ or } \Sigma_c) + \text{mesons} \qquad (30)$$

Antibaryon decay processes correspond directly to those for the corresponding baryons. They have the same lifetimes, within measurement accuracy, but some differences are to be expected when higher accuracy can be achieved. Examples of the decay processes which have been observed are given in Eq. (31).

$$\overline{\Lambda} \to \overline{p}\pi^+ \qquad \overline{\Sigma}^+ \to \overline{n}\pi^+ \qquad \overline{\Lambda}_c \to \overline{p}K^+\pi^- \qquad (31)$$

Interaction with matter. When negatively charged particles with lifetimes longer than about 10^{-12} s come to rest in condensed matter, they are first captured into outer atomic orbits in the constituent elements and then cascade down to inner orbits by the ejection of Auger electrons and the emission of photons. Finally, they undergo absorption by the atomic nucleus, through reactions which release energy and cause nuclear breakup, giving rise to a nuclear star. SEE HADRONIC ATOM.

Σ^- stops in (nuclear) photographic emulsion often (about 75% of the time) do not lead to the emission of charged particles. The visible stars are small, with visible energy 10–20 MeV, and frequently (about 10% of Σ^- stops) emit a light Λ hypernucleus. These observations are consistent with the reaction mechanism of Eq. (13a), releasing two neutral particles of energy 75 MeV, the capture occurring from the nuclear periphery. Ξ^--capture stars result from absorptive reaction (14), with even less energy released, and frequently lead to the emission of two Λ hypernuclei. Antiproton stops in emulsion lead to large stars because the annihilation reaction $\bar{p}N \to$ (mesons) releases about 1880 MeV, with the emission of many mesons (mostly pions).

Proton stability. In grand unified theories (GUT), which attempt to account for both quarks and leptons, together with their strong, electromagnetic, and weak interactions, transitions $q \to l$ generally exist, at some level, since such theories assign leptons and quarks to common multiplets. This violation of baryon conservation opens the possibility that the lightest baryon, the proton, might not be absolutely stable but undergoes decay processes such as $p \to e^+\pi^0$ at a very low rate. Indeed, cosmology appears to require the existence of nucleon decay processes in order to account for the baryon-antibaryon asymmetry of the universe. Empirically, the proton is known to have a half-life exceeding about 10^{39} s. More precisely, its partial decay rate to $e^+\pi^0$ has been measured to be less than 2×10^{-40} s^{-1}, with 95% probability. The simplest GUT [known as the SU(5) theory] predicts a total proton decay rate of about $3 \times 10^{-37\pm 2}$ s^{-1}, and so does not appear acceptable, as things now stand. Other GUTs may predict smaller decay rates, even zero. Since proton decay offers the possibility of discriminating between various GUTs, the detection of proton decay (and neutron decay involving baryon nonconservation), or at least the improvement of the present empirical limits on its rate, is an important subject of investigation. SEE ELEMENTARY PARTICLE; FUNDAMENTAL INTERACTIONS; GRAND UNIFICATION THEORIES.

Bibliography. A. Garcia and P. Kielanowski, *The Beta Decay of Hyperons*, 1985; A. W. Hendry and D. B. Lichtenberg, The quark model, *Rep. Prog. Phys.*, 41: 1707–1780, 1978; N. Isgur and G. Karl, Hadron spectroscopy and quarks, *Phys. Today*, 36(11):36–42, 1983; Particle Data Group, Review of particle properties, *Phys. Lett.*, vol. 111B, April 1984.

NUCLEON
EMILIO SEGRÈ

A collective name for a proton or a neutron. Protons and neutrons are the main constituents of the nuclei of atoms and have considerable similarity between themselves. They have the same spin, the same statistics, approximately the same mass, and, through the process of beta decay, can transform into each other. Thus it is convenient to have a common term to designate them both.

Occasionally neutrons and protons are considered as two states of a single particle called the nucleon, the two states being distinguished by the special value of an internal variable which can then assume only two values and is called the third component of the isotopic spin.

The nucleon description sufficient for the purpose of nuclear physics is relatively simple. The fundamental properties of charge, mass, spin, statistics, magnetic moment, and a phenomenological potential between nucleons are sufficient data.

However, in order to investigate the properties of matter on a finer scale, increasingly more powerful tools are needed, which in practice often mean higher energy particles and a more elaborate theory. Attempts can then be made to find a structure in the nucleons. This is a goal of particle physics. For instance, it is possible in a certain sense to ascertain the spatial distribution of the electric charge and of the magnetization in a nucleon.

In interpreting experimental data of this type, the nucleon may be envisaged as a complex system in a state of dynamical equilibrium where a central core continuously emits or absorbs pions from a surrounding pionic cloud. The central core is a Dirac particle of intrinsic magnetic moment, either zero or equal to a nuclear magneton and neutral, or of protonic electric charge. This core is often called the "bare" nucleon. In such a picture the magnetic moment of the physical nucleon is attributed in part to the core and in part to the orbital contribution of the surrounding pions. In the same spirit the empirical potential between nucleons is analyzed through a mechanism of pion exchange, in which the pions appear as quanta of the field of the nuclear forces. However, since pions are insufficient to completely describe either the forces or the magnetic moments, it is necessary to invoke additional heavier particles called ρ and ω mesons. The structure becomes then increasingly complicated.

At the unusual dimensions involved in particle physics the observed phenomena depart more and more from the everyday experience based on objects directly accessible to the unaided senses. The apparatus required for study of particle physics becomes of stupendous complexity and magnitude; concomitantly, theory becomes increasingly abstract and mathematical. SEE ELEMENTARY PARTICLE; ISOBARIC SPIN; NEUTRON; PROTON.

Bibliography. F. Halzen and A. D. Martin, *Quarks and Leptons: An Introductory Course in Modern Particle Physics*, 1984; I. S. Hughes, *Elementary Particles*, 2d ed., 1985; G. K. O'Neill and D. Cheng, *Elementary Particle Physics: An Introduction*, 1979; E. Segrè, *Nuclei and Particles*, 2d ed., 1977.

PROTON
THOMAS H. FIELDS

A positively charged particle that is the nucleus of the lightest chemical element, hydrogen. The hydrogen atom consists of a proton as the nucleus, to which a single negatively charged electron is bound by an attractive electrical force (since opposite charges attract). The proton is about 1836 times heavier than the electron, so that the proton constitutes almost the entire mass of the hydrogen atom. Most of the interior of the atom is empty space, since the sizes of the proton and the electron are very small compared to the size of the atom.

For chemical elements heavier than hydrogen, the nucleus can be thought of as a tightly bound system of Z protons and N neutrons. An electrically neutral atom will then have Z electrons bound comparatively loosely in orbits outside the nucleus. SEE NEUTRON; NUCLEAR STRUCTURE.

Most of the overall properties of the proton can be experimentally determined in straightforward ways. Free protons can be easily obtained for experimental purposes by ionization (electron removal) of hydrogen atoms in an electrical gas discharge. In some cases, measurements on the proton are best carried out by using neutral hydrogen atoms. In magnetic resonance experiments, the magnetic dipole moment of the proton has been measured with high accuracy. The numerical values of some overall properties of the proton can be summarized as follows: charge, 1.602×10^{-19} coulomb; mass, 1.673×10^{-27} kg; spin, $(\frac{1}{2})\hbar$ (where $\hbar$ is Planck's constant h divided by 2π); magnetic dipole moment, 1.411×10^{-26} joule/tesla; radius, about 10^{-15} m. SEE NUCLEAR MOMENTS; SPIN.

Size and structure. In order to understand basic questions concerning the size and structure of the proton, it is instructive to contrast its properties with those of the electron. The behavior of the electron has also been studied in many precision experiments at both high and low energies, and all of the electron's properties have been found to be those expected of a spin-$\frac{1}{2}$ particle which is described by the Dirac equation of quantum mechanics. Such a Dirac particle has no internal size or structure. SEE ELECTRON.

By contrast, although it also has a spin of ½, the proton definitely does not behave as a geometrical point Dirac particle. Its magnetic moment, which is different from that for a Dirac particle, and its binding with neutrons into nuclei strongly suggest that the proton has some kind of internal structure, rather than being a point particle. That this is actually the case has been shown in many high-energy physics experiments. Experiments to explore the structure of the proton need to be carried out by using high-energy particles for two reasons. First, the uncertainty relationship between momentum and position, which is a fundamental principle of quantum mechanics, requires the use of high-momentum particles to observe small regions of space. Second, Einstein's equation of special relativity, $E = mc^2$, where E is energy, m is mass, and c is the speed of light, shows that to produce the additional mass which is needed to form excited states of the proton (described below) requires high-energy collisions.

Correspondingly, two different kinds of high-energy physics experiments have been used to study the internal structure of the proton. An example of the first type of energy is the scattering of high-energy electrons, above say 1 GeV, from a target of protons. The angular pattern and energy distribution of the scattered electrons give direct information about the size and structure of the proton. The second type of high-energy experiment involves the production and study of excited states of the proton, often called baryonic resonances. It has been found that the spectrum of higher-mass states which are produced in high-energy collisions follows a definite pattern. SEE BARYON.

In 1963, M. Gell-Mann and, independently, G. Zweig pointed out that this pattern is what would be expected if the proton were composed of three spin-½ particles, quarks, with two of the quarks (labeled u) each having a positive electric charge of magnitude equal to ⅔ of the electron's charge (e), and the other quark (labeled d) having a negative charge of magnitude of ⅓ e. Subsequently, the fractionally charged quark concept was developed much further, and has become central to understanding every aspect of the behavior and structure of the proton.

Quark structure. As a result of experimental and theoretical research in high-energy physics since the 1960s, a broad understanding of the internal structure and constituents of the proton has been achieved. The behavior of the proton is essentially that of a system of three quarks, with the charges specified above. The excited, higher-mass states of the proton have been shown to correspond in some detail to the theoretical expectations for such a three-quark system. This correspondence is not yet on a precise basis, because the equations of the quantum-mechanical theory of interactions between quarks, quantum chromodynamics (QCD), have not yet been solved rigorously for a bound system of three quarks.

Quantum chromodynamics has led to other radically new insights into the behavior of the quarks. Among the most fundamental and yet paradoxical properties of the quarks, according to quantum chromodynamics, is the existence of a new quantum number called color, which acts in such a way as to forbid the stable existence of a single, isolated quark. The requirement that isolated physical systems have zero net color in fact forbids the existence of any free particles which have fractional charge. This conclusion has been tested by many experiments which have searched unsuccessfully for such particles. SEE COLOR.

For the proton, the same principle leads to the conclusion that the proton is the lowest allowed energy state of three quarks. The proton cannot dissociate or decay into separated quarks, for that would lead to the existence of free quarks. In other words, the proton is absolutely stable with respect to the quantum chromodynamics interactions of quarks. SEE QUANTUM CHROMODYNAMICS; QUARKS.

Search for decay. For many years, it has been recognized not only that the stability of the proton is of great practical importance, but also that this stability is probably due to some fundamental conservation law in physics. The proton and the neutron are the building blocks of which all nuclei are formed, and the observed stability of the atoms in the universe is a direct consequence of the stability of the proton itself.

As described above, quantum chromodynamics gives a clear description of why the proton, as a three-quark system, cannot break apart into free quarks, but this leaves open the question as to whether the quarks themselves might decay. Such a question goes beyond the scope of quantum chromodynamics, but it can be answered by the unified electroweak theory of S. Weinberg, A. Salam, and S. Glashow. This theory gives a description of all known electromagnetic and weak (radioactive decay) phenomena, and has so far correctly predicted the results of all experimental measurements of these phenomena. The theory also gives a definite prediction that the proton is stable. SEE WEAK NUCLEAR INTERACTIONS; WEINBERG-SALAM MODEL.

However, an important class of fundamental theories, called grand unification theories (GUTs), makes the opposite prediction—that the proton will decay. The predicted lifetime of the proton is very long, about 10^{30} years or more—which is some 10^{20} times longer than the age of the universe—but this predicted rate of proton decay may be detectable in practical experiments. SEE GRAND UNIFICATION THEORIES.

The fundamental physics issue at stake in the search for proton decay is whether quarks, whose behavior is described in quantum chromodynamics, will ultimately be transformed into electrons or muons, which are described by a completely separate (electroweak) theory. If this process does occur, it will be unambiguous evidence for a single mechanism being at the root of all known forces in nature, excepting gravity. It will also permit the unification of all known particle types into one family. SEE FUNDAMENTAL INTERACTIONS.

To search for proton decay, experiments are being carried out which can detect and identify one or two proton decay events per year in a sample whose mass is of the order of 1000 tons (900 metric tons). To achieve such unprecedented sensitivity, the entire experiment is carried out in a deep underground laboratory. This shields the apparatus from most of the background events which would otherwise be caused by cosmic rays. The **illustration** shows a proton decay detector being constructed in a laboratory beneath the Alps. The detector is constructed of 900 metric tons (992 short tons) of thin iron plates, each 20 ft × 20 ft × 0.12 in. (6 m × 6 m × 3 mm). Particles resulting from proton decay within an iron nucleus will be detected as they traverse Geiger tubes and flash chambers which are sandwiched between the thin iron plates. Even in an underground laboratory, there is an unavoidable background caused by neutrinos, which are secondary products of cosmic rays. This neutrino background will set an ultimate limit on the sensitivity of proton decay experiments. SEE NEUTRINO; PARTICLE DETECTOR.

If the proton is observed to decay, this new interaction will also have profound consequences for understanding of cosmology. The very early times of the big bang (about 10^{-30} s) are characterized by energies so high that the same grand unified interaction which would allow

Proton decay experiment being assembled in an underground laboratory in the Frejus tunnel under the French Alps.

proton decay would also completely determine the subsequent evolution of the universe. This could then explain the remarkable astrophysical observation that the universe appears to contain only matter and not an equal amount of antimatter. The present standard theories of fundamental particle interactions, quantum chromodynamics and the electroweak theory, allow only processes in which an equal number of particles and antiparticles is created. These fundamental theories are therefore inconsistent with the apparent absence of antimatter in the universe. SEE ELEMENTARY PARTICLE.

Bibliography. M. Gaillard, Toward a unified picture of elementary particle interactions, *Amer. Sci.*, 70:506–514, 1982; H. Georgi, A unified theory of elementary particles and forces, *Sci. Amer.*, 244(4):48–63, 1981; L. Sulak, Waiting for the proton to decay, *Amer. Sci.*, 70:616–625, 1982; S. Weinberg, The decay of the proton, *Sci. Amer.*, 244(6):64–75, 1981; V. F. Weisskopf, The origin of the universe, *Amer. Sci.*, 71:473–480, 1983.

NEUTRON
ARTHUR H. SNELL

An elementary particle having approximately the same mass as the proton, but lacking a net electric charge. It is indispensable in the structure of the elements, and in the free state it is an important reactant in nuclear research and the propagating agent of fission chain reactions. Neutrons, in the form of highly condensed matter, constitute the substance of neutron stars.

Neutrons in nuclei. Neutrons and protons are the constituents of atomic nuclei. The number of protons in the nucleus determines the chemical nature of an atom, but without neutrons it would be impossible for two or more protons to exist stably together within nuclear dimensions, which are of the order of 10^{-13} cm. The protons, being positively charged, repel one another by virtue of their electrostatic interactions. The presence of neutrons weakens the electrostatic repulsion, without weakening the nuclear forces of cohesion. In light nuclei the resulting balanced, stable configurations contain protons and neutrons in almost equal numbers, but in heavier elements the neutrons outnumber the protons; in ^{238}U, for example, 146 neutrons are joined with 92 protons. Only one nucleus, ^{1}H, contains no neutrons. For a given number of protons, neutrons in several different numbers within a restricted range often yield nuclear stability—and hence the isotopes of an element. SEE ISOTOPE; NUCLEAR STRUCTURE; PROTON.

Sources of free neutrons. Free neutrons have to be generated from nuclei, and since they are bound therein by cohesive forces, an amount of energy equal to the binding energy must be expended to get them out. Usually the binding energy for each neutron amounts to 1–1.4 picojoules (6–8 MeV). Nuclear machines, such as cyclotrons and electrostatic generators, induce many nuclear reactions when their ion beams strike target material. Some of these reactions release neutrons, and these machines are sources of high neutron flux. If the accelerator is sharply pulsed, the time-of-flight method can be used for accurate energy resolution of the neutrons up to energies of about 0.3 pJ (2 MeV), the flight paths being up to 660 ft (200 m) long. SEE NEUTRON SPECTROMETRY; NUCLEAR BINDING ENERGY; NUCLEAR REACTION.

There are several kinds of portable neutron sources. Some consist of an intimate mixture of an alpha-emitting radionuclide with beryllium powder. Neutrons are released from the nuclear reaction ^{9}Be $(\alpha n)^{12}$C, which is the reaction by which the neutron was discovered in 1932. Intense sources that emit about 5×10^{10} n/s are now made, using mixtures of beryllium with, for example, ^{238}Pu (half-life 89 years) or ^{241}Am (half-life 458 years). Such a source generates several hundred watts of heat. Pure ^{252}Cf (half-life 2.65 years) needs no admixture of beryllium, because neutrons are emitted in its spontaneous fission; such a source is especially compact. The neutrons emitted by the foregoing sources have energies that extend up to 0.8–1 pJ (5–6 MeV). A source that gives neutrons of lower energy is the Sb-Be photoneutron source. Here the 1.70-MeV gamma rays of ^{124}Sb (half-life 60 days) slightly exceed the binding energy of neutrons in beryllium (1.67 MeV), so the neutrons have an energy of 1.70−1.67 = 0.03 MeV (4.6 femtojoules).

Neutrons are released in the act of fission, and nuclear reactors are unexcelled as intense neutron sources. The absorption of one neutron by a ^{235}U nucleus is required to induce fission, but 2.5 neutrons are on the average released; this regeneration makes possible the nuclear chain reaction. A powerful research reactor may generate neutrons in such abundance that 1 cm² (0.15 in.²) of a sample placed therein would be traversed by 10^{15} neutrons per second. A hole through the surrounding shield can yield a collimated beam having a unidirectional flux of 10^9 neutrons/(cm²)(s) or 6.5×10^9 neutrons/(in.²)(s). The explosion of a 10-kiloton (4×10^{13} J) nuclear bomb releases about 10^{30} neutrons in about 1 microsecond. SEE CHAIN REACTION; NUCLEAR FISSION.

Neutrons occur in cosmic rays, being liberated from atomic nuclei in the atmosphere by collisions of the high-energy primary or secondary charged particles. They do not themselves come from outer space.

Penetrating power. Neutrons resulting from nuclear reactions usually possess kinetic energies of the order of 0.2 pJ (1 MeV). Having no electric charge, they interact so slightly with atomic electrons in matter that energy loss by ionization and atomic excitation is essentially absent. Consequently they are vastly more penetrating than charged particles of the same energy. The main energy-loss mechanism occurs when they strike nuclei. As with billiard balls, the most efficient slowing-down occurs when the bodies that are struck in an elastic collision have the same mass as the moving bodies; hence the most efficient neutron moderator is hydrogen, followed by other light elements; deuterium, beryllium, and carbon.

The great penetrating power of neutrons imposes severe shielding problems for reactors and other nuclear machines, and it is necessary to provide walls, usually of concrete, several feet in thickness to protect personnel. The currently accepted health tolerance levels for an 8-h day correspond for fast neutrons to a flux of 20 neutrons/(cm²)(s) or 130 neutrons/(in.²)(s); for slow neutrons, 700/(cm²)(s) or 4500/(cm²)(s). On the other hand, fast neutrons are useful in some kinds of cancer therapy.

Detection of neutrons. In pulse counting, neutrons are allowed to produce exothermic (energy-releasing) nuclear reactions, the ionizing products of which are made to generate electrical impulses, in a proportional counter, ionization chamber, or scintillation counter, that can be amplified for individual counting. A proportional counter containing boron, either as a coating on the inner walls or as a filling gas (boron trifluoride), counts neutrons by virtue of the reaction $^{10}B(n,\alpha)^7Li + 0.451$ pJ (2.78 MeV). An ionization chamber coated internally with ^{235}U gives ionization pulses from the energy of fission fragments as they travel through the gas. A lithium iodide crystal (europium-activated) scintillates because of the energy released by the reaction $^6Li(n,\alpha)^3H + 0.765$ pJ (4.78 MeV). The light pulses (scintillations) are reflected onto a photomultiplier, which transforms them to electrical pulses. Capture gamma-rays emitted from strong neutron absorbers, such as cadmium, can similarly be registered by scintillation counting. Large andsensitive neutron detectors have been made by dissolving cadmium or boron salts in tanks containing scintillating liquids. SEE IONIZATION CHAMBER; LIQUID SCINTILLATION DETECTOR; PARTICLE DETECTOR; SCINTILLATION COUNTER.

In detection by activation, advantage is taken of the fact that many elements become radioactive under neutron irradiation. A sample is exposed, and its radioactive strength is subsequently measured by conventional counting equipment. Gold and indium foils are convenient and sensitive detectors of this kind. Their applications can be further specialized by taking advantage of resonance absorption. If, for example, gold foil is enclosed in cadmium, the cadmium will exclude thermal neutrons, and the gold will be activated mainly by neutrons with an energy of 0.78 attojoule (4.9 eV) because gold has a neutron capture resonance at that energy. Other elements can be similarly used for other selected energies. The converse also occurs; for example, if a thick plug of ^{57}Fe is placed in a beam of fast neutrons, it will preferentially transmit neutrons with an energy of 4 pJ (25 keV) because at that energy the neutrons interact only weakly with the ^{57}Fe nuclei.

Detection by recoil is particularly applicable to the counting of fast neutrons. A counter with hydrogenous walls or filling gas (for example, methane) gives pulses because the protons produce ionization when they recoil after being struck by the fast neutrons.

Intrinsic properties. Free neutrons are themselves radioactive, each transforming spontaneously into a proton, an electron (β^- particle), and an antineutrino. The energy release is 0.125 pJ (0.782 MeV) per event, and the half-life is 10.61 ± 0.16 min. This instability is a reflection of the fact that neutrons are slightly heavier than hydrogen atoms. The neutron's rest mass is 1.0086649 atomic mass units on the unified mass scale (1.67262×10^{-24} g), as compared with 1.0078250 atomic mass units for the hydrogen atom.

Neutrons are, individually, small magnets. This property permits the production of beams of polarized neutrons, that is, beams of neutrons whose magnetic dipoles are aligned predominantly parallel to one direction in space. The magnetic moment is -1.913042 nuclear magnetons. The magnetic structure has a finite size, being roughly exponential in intensity, with a root-mean-square radius of 0.9×10^{-13} cm. Neutrons spin with an angular momentum of ½ in units of $h/2\pi$, where h is Planck's constant. The negative sign attached to the magnetic moment indicates that the magnetic moment vector and the angular momentum vector are oppositely directed. SEE MAGNETON; NUCLEAR MOMENTS; SPIN.

Despite its overall neutrality, the neutron does have an internal distribution of electric charge, as has been revealed by scattering experiments. On a still finer scale, the neutron can also be presumed to have a quark structure in analogy of that of the proton. SEE QUANTUM CHROMODYNAMICS; QUARKS.

If the centers of the $+$ and $-$ charge distributions in the neutron should be slightly displaced from each other, the neutron would have an electric dipole moment. This possibility has a fundamental importance because it is linked through various interaction theories with the conservation of parity and with the symmetry of time reversal. (If time-reversal invariance holds, the neutron should have no electric dipole moment.) So far it has been found that if the separated charges are equal to $\pm e$ (the electronic charge), the distance between their centers must be less than 10^{-24} cm. This limit is not yet sufficiently small to give a conclusive choice between the various forms of theoretically possible interactions, but it is likely that the sensitivity of the experiments can be increased through the use of ultracold neutrons. SEE PARITY; SYMMETRY LAWS.

Ultracold neutrons. When neutrons are completely slowed down in matter, they have a maxwellian distribution in energy that corresponds to the temperature of the moderator with which they are in equilibrium. At room temperature their mean energy is about 0.004 aJ (0.025 eV), their mean velocity is about 2200 m/s (7300 ft/s), and their de Broglie wavelength is about 0.2 nm. (The approximate coincidence of this wavelength with the interatomic distances in solids is the basis for the science of neutron diffraction.) The maxwellian distribution has a tail extending to very low energies, and a few neutrons (about 10^{-11} of the main neutron flux) at this extreme have energies less than 5×10^{88} aJ (3×10^{-7} eV), and hence velocities of less than about 7 m/s (23 ft/s). The de Broglie wavelength of these ultracold neutrons is greater than 50 nm, which is so much larger than interatomic distances in solids that they interact with regions of a surface rather than with individual atoms, and as a result they are reflected from polished surfaces at all angles of incidence.

A typical source of ultracold neutrons consists of a "converter" in the reflector of a neutron reactor, together with an internally smooth, evacuated tube several centimeters in diameter that leads the neutrons out through the shield. The lead-out duct has three or four bends; the ultracold neutrons are preferentially reflected at these bends and are thus selected from the numerous faster neutrons. The neutrons can be further slowed either by sending them upward against gravity (they can rise only 2–3 m (6–9 ft), and the lead-out duct can be vertical if desired), or by means of a "neutron turbine," which is a paddle wheel whose curved blades move in the same direction as the neutrons, but with lower velocity. The neutrons can be polarized by passage through or reflection from a sheet of magnetized material.

The ultracold neutrons can be stored in "neutron bottles," of which there are two kinds. One is simply a vacuum vessel with a door that can be closed after a batch of neutrons has entered. Populations of about 100 neutrons have been retained in such vessels, but the storage times are considerably shorter than the half-life of the neutrons against their natural radioactive decay, and the nature of the extra loss mechanisms is not yet fully understood. The other kind of bottle is again a vacuum vessel, but it uses a multipolar magnetic field that contains the neutrons, because the field gradients act upon the neutrons' magnetic dipole moments so as to keep the neutrons away from the walls. Such a configuration is realized in a torus with hexapole windings around its major circumference, or a sphere with polar and equatorial windings carrying opposing currents.

Ultracold neutrons are important in basic physics and have applications in studies of surfaces and of the structure of inhomogeneities and magnetic domains in solids. SEE ANTINEUTRON; ELEMENTARY PARTICLE.

Bibliography. S. Cierjacks (ed.), *Neutron Sources*, 1982; T. von Egidy (ed.), *Fundamental Physics with Reactor Neutrons and Neutrinos*, Conf. Ser. no. 42, Institute of Physics, London, 1978; R.

Golub and J. M. Pendlebury, Ultra-cold neutrons, *Rep. Progr. Phys.*, 42:439, 1979; P. Schofield, *The Neutron and Its Applications 1982*, 1983; E. Sheldon (ed.), *Proceedings of International Conference on the Interactions of Neutrons with Nuclei*, vols. 1 and 2, Technical Information Center, Department of Energy, Oak Ridge, 1976.

ANTIPROTON
Michael E. Zeller

The antiparticle of the proton. Its existence was implied in 1928 by the relativistic wave equation of P. A. M. Dirac, and was experimentally observed by O. Chamberlain and colleagues in 1955. The mass and intrinsic angular momentum of the antiproton are equal to those of the proton, while the electrical properties, that is, charge and magnetic moment, are equal in magnitude but opposite in sign. SEE PROTON.

The proton belongs to a class of particles termed baryons, which in the composite quark model of elementary particles consist of three quarks. The antiproton is an antibaryon made of three antiquarks. This structure is in contrast to mesons, which are composed of quark-antiquark pairs. To an experimentally very high accuracy, the number of baryons is conserved in any particle interaction; the baryons are counted as positive and the antibaryons as negative. The creation of antiprotons by high-energy collisions of particles with matter must thus involve the associated production of protons or other baryons to conserve total baryon number. Likewise, antiprotons can annihilate protons or other baryons, producing primarily mesons and electromagnetic radiation. SEE BARYON; MESON; QUARKS; SYMMETRY LAWS.

In the annihilation process the rest mass energy of the proton and antiproton and the kinetic energy of these particles are converted to mass and kinetic energy of the annihilation products. By creating and accelerating antiprotons to high energies and colliding these with equally energetic protons moving in the opposite direction, physicists have produced some of the more massive elementary particles. The most notable example is the production of the W^+ and Z particles, which have approximately 100 times more mass than the proton, and whose existence and properties were postulated in the formulations of the unification of the weak and electromagnetic interactions. SEE FUNDAMENTAL INTERACTIONS; INTERMEDIATE VECTOR BOSONS; WEAK NUCLEAR INTERACTIONS.

The abundance of protons in cosmic rays is greater than 10,000 times larger than that of antiprotons. Since the total number of baryons is conserved in particle interactions, this implies an asymmetry in matter to antimatter in the universe, not only at present but at very early times. Such an asymmetry is not understood in the standard "big bang" model of the origin of the universe. In that model the initial state of the universe is composed of energy alone, with zero net baryon number. Current theories involving grand unification of elementary particle interaction hypothesize baryon nonconserving processes, but at a level below present experimental limits. SEE ANTIMATTER; ELEMENTARY PARTICLE; GRAND UNIFICATION THEORIES.

ANTINEUTRON
Richard Wilson

The antiparticle of the neutron. According to the equations describing particles, for each particle there also exists an antiparticle, in which many of the properties of the particle are inverted, the sign of the charge is opposite, and the sign of the magnetic moment (μ) is inverted. In a region free of magnetic field, the masses of the neutron and antineutron are the same, the spins ($\frac{1}{2}$) are the same, the statistics (according to the Fermi formulas) are the same, and the decay constants for β-decay are the same. In a magnetic field B, the energies differ by $2\ \mu B$ and the effective masses by $2\ \mu B/c^2$, where c is the speed of light.

Some antiparticles mix freely with their particles (K^0 and $\overline{K^0}$ mesons). A proton cannot mix with an antiproton without violating conservation of charge. A neutron can mix with an antineutron without violation of charge conservation, but it is known that the neutron does not do so very often, for otherwise matter would rapidly be annihilated. The concept of conservation of baryon number (the neutron has baryon number $+1$, the antineutron -1) was invented to describe this fact. SEE SYMMETRY LAWS.

Antineutrons have been identified only when they are formed in conjunction with another baryon, usually a neutron. Antineutrons are annihilated in matter, and this gives them an optical potential in nuclear matter that is different from that of neutrons, and hence a different total energy. The possible mixing of neutrons and antineutrons can be described by a small mixing mass term, and its reciprocal, a mixing time. The lifetime of nucleons in matter has been measured to be greater than 10^{31} years, which, by means of an optical model calculation, suggests a mixing time greater than 10^7 s. Direct searches for the conversion of free neutrons to antineutrons suggest a mixing time greater than 10^6 s. SEE ANTIMATTER; ELEMENTARY PARTICLE; NEUTRON; NEUTRON-ANTINEUTRON OSCILLATIONS.

STRANGE PARTICLES
HORACE D. TAFT

Particles possessing the attribute of strangeness, a quantum number associated with one of the several quarks which are thought to constitute their structure. The nomenclature arose from the fact that the production rates for strange particles, which are compatible with strong interaction times of the order of 10^{-23} s, appeared to be inconsistent with their relatively long lifetimes (in the range 10^{-8} to 10^{-10} s). This inconsistency was resolved by experimental observations that strange particles are invariably produced in pairs (associated production) of equal and opposite strangeness, while their decay channels are restricted to those in which the strangeness quantum number is not conserved. This nonconservation of strangeness greatly suppresses the decay rates relative to the production rates. SEE QUARKS.

Strange particles with baryon number 0 are designated K mesons; those with baryon number 1 are designated hyperons. When characterized by a new quantum number called the hypercharge Y (related to the previously used strangeness quantum number S by $Y = S + B$, where B is the baryon number), strange particles are seen as members of multiplets which include the smaller nonstrange particle multiplets of mesons and nucleons. Additional quantum numbers, designated as charm and beauty (and perhaps there are others), detected since 1974 are expected to expand this multiplet structure still further. Families of particles possessing both strangeness and one or more of these new quantum numbers are also predicted. For discussion of strange particle production processes, reactions, and decay properties SEE BARYON; MESON. SEE ALSO ELEMENTARY PARTICLE; SYMMETRY LAWS.

Bibliography. R. K. Adair and E. C. Fowler, *Strange Particles*, 1963; F. E. Close, *An Introduction to Quarks and Partons*, 1979; F. Halzen and A. D. Martin, *Quarks and Leptons: An Introductory Course in Modern Particle Physics*, 1984; I. S. Hughes, *Elementary Particles*, 2d ed., 1985.

HYPERON
RICHARD H. DALITZ

A collective name for any baryon with nonzero strangeness number s. The name hyperon has generally been limited to particles which are semistable, that is, which have long lifetimes relative to 10^{-22} s and which decay by photon emission or through weaker decay interactions. Hyperonic particles which are unstable (that is, with lifetimes shorter than 10^{-22} s) are referred to as excited hyperons. The known hyperons with spin $\frac{1}{2}\hbar$ (where $\hbar$ is Planck's constant divided by 2π) are Λ, Σ^-, Σ^0, and Σ^+, with $s = -1$, and Ξ^- and Ξ^0, with $s = -2$, together with the Ω^- particle, which has spin $\frac{3}{2}\hbar$ and $s = -3$. The corresponding antihyperons have baryon number $B = -1$, opposite strangeness s, and charge Q; they are all known empirically, except for $\overline{\Omega}^+$.

The first excited hyperon was reported in 1960. The state $\Sigma(1385)^+$ was observed as a $\pi\Lambda$ resonance in the final state of reaction (1). The symbol $\Sigma(m)$ or $\Lambda(m)$ indicates that the strangeness

$$K^- + p \rightarrow \pi^- + \Sigma(1385)^+ \rightarrow \pi^- + \pi^+ + \Lambda \qquad (1)$$

is $s = -1$ and that the isospin is $I = 1$ or 0, respectively. The superscript gives the charge and m is the mass in MeV; if no m is given, the symbol refers to the ground state, for example, Λ means $\Lambda(1115.5)$. SEE ISOBARIC SPIN.

Reaction (1) is an example of an excited hyperon production reaction. Formation reactions are also possible for most excited hyperons with $s = -1$. For example, the properties of $\Lambda(1520)$ are particularly well known from observations on its formation and decay, for K^- mesons incident on hydrogen, reaction (2). $\Sigma(1385)$ cannot be formed in this way, because its mass lies below the

$$K^- p \rightarrow \Lambda(1520) \rightarrow \begin{cases} K^- + p \\ \pi^\pm + \Sigma^\pm, \text{etc.} \end{cases} \quad (2)$$

$K^- p$ threshold energy, $m_K + m_p \simeq 1432$ MeV.

There is no deep distinction between hyperons and excited hyperons, beyond the phenomenological definition above. Indeed, the hyperon $\Omega(1672)^-$ and the excited hyperons $\Xi(1530)$ and $\Sigma(1385)$, together with the unstable nucleonic states $\Delta(1236)$ are known to form a unitary decuplet of states with spin $3/2\hbar$. SEE BARYON; ELEMENTARY PARTICLE; SYMMETRY LAWS; UNITARY SYMMETRY.

DELTA RESONANCE
CARL DOVER

A member of a class of subatomic particles called baryons, which exists in four electric charge states and has a total spin of $J = 3/2$. In the underlying quark model, the delta resonance (Δ) consists of three quarks whose intrinsic spins of $1/2$ are lined up in the same direction. The Δ is closely related to the more familiar nucleon constituents of atomic nuclei, the neutrons (n) and protons (p). SEE NUCLEON; QUARKS.

The Δ was first observed as a resonant interaction of a beam of pi mesons (π) with a proton target. The probability of a scattering interaction between the π and the proton is strongly dependent on energy, attaining a maximum at the Δ mass of 1236 MeV/c^2 (where c is the speed of light). The formation of the very short-lived Δ (with a lifetime on the order of 10^{-23} s) is followed immediately by its decay back into pion and nucleon. SEE MESON; SCATTERING EXPERIMENTS (NUCLEI).

The understanding of some nuclear phenomena requires explicit treatment of π or Δ degrees of freedom. For instance, the Δ modifies the polarizability of the nucleus, and thus contributes to the quenching of the transition strength observed in charge exchange reactions [that is, (p, n) reactions] on nuclei. The Δ also plays a role in the quenching of the β-decay process ($n \rightarrow p + e^- + \bar{\nu}_e$, where e^- denotes an electron and $\bar{\nu}_e$ an antineutrino) in the nucleus. High-momentum components are induced in the nucleus through the presence of Δ's. These are probed by a variety of transfer processes. The effect of the Δ is also seen in electromagnetically induced reactions [(γ,p), (γ,π^0), $(e,e'p)$, and so forth] involving the interaction of a real or virtual photon (γ) with the nucleus. SEE NUCLEAR REACTION; NUCLEAR STRUCTURE; RADIOACTIVITY.

The Δ thus plays an important role in a wide variety of nuclear phenomena, even under conditions of low energy and momentum transfer. The study of these phenomena reveals much about the presence of pions in nuclei, in addition to neutrons and protons. SEE BARYON; ELEMENTARY PARTICLE.

MESON
RICHARD H. DALITZ

The generic name for any hadronic particle with baryon number zero. Such particles were first envisaged in 1935 by H. Yukawa, who pointed out that the main features of nuclear forces would be explained if these forces were transmitted between nucleons through an intermediate field coupled with nucleons, provided that its quanta (nuclear force mesons) were massive [200 to 300 electron masses (m_e)] and could carry electric charge between the nucleons. SEE BARYON; HADRON; NUCLEAR STRUCTURE; QUANTUM FIELD THEORY.

Table 1. Semistable pseudoscalar mesons now known

Characteristics	$\pi^+(\pi^-)$		π^0		$K^+(K^-)$		K^0_S		K^0_L	
Mass M, MeV	139.567		134.963		493.67		497.67			
σ_M	±0.001		±0.004		±0.015		±0.13			
Lifetime T, s	2.603×10^{-8}		0.83×10^{-16}		1.237×10^{-8}		0.892×10^{-10}		5.18×10^{-8}	
σ_T	±0.002		±0.06		±0.003		±0.002		±0.04	
Major decay modes	$\mu^+\nu_\mu$	100%	$\gamma\gamma$	98.8%	$\mu^+\nu_\mu$	63.5%	$\pi^+\pi^-$	69%	$3\pi^0$	22%
	$e^+\nu_e$	1.23×10^{-4}	$\gamma e^+ e$	1.20%	$\pi^0\mu^+\nu_\mu$	3.2%	$\pi^0\pi^0$	31%	$\pi^+\pi^-\pi^0$	12%
	$\mu^+\nu_\mu\gamma$	1.24×10^{-4}	$2e^+2e^-$	3.2×10^{-5}	$\pi^0 e^+\nu_e$	4.8%			$\pi^\pm e^\mp\nu_e$	39%
					$\pi^+\pi^0$	21.2%			$\pi^\pm\mu^\mp\nu_\mu$	27%
					$\pi^+\pi^+\pi^-$	5.6%				
					$\pi^+\pi^0\pi^0$	1.7%				
Nonzero flavor number	$I_3 = +1(-1)$		$I_3 = 0$		$I_3 = +½(-½)$ $s = +1(-1)$		Mixture of two states with $(I_3, s) = \pm(-½, 1)$			

Discovery. In 1937 C. Anderson established the existence of positively and negatively charged μ mesons (now known as muons) of mass 105.6 MeV ($\simeq 206.5 m_e$) in cosmic radiation, but these were soon shown to have very weak coupling with nucleons, that is, to be nonhadronic and therefore not "mesons," as this name is used today. They are members of the lepton family, together with the electron and the neutrinos. The nuclear force mesons turned out to be the π mesons (pions), of mean mass about 138 MeV ($\simeq 273 m_e$), first identified in cosmic radiation by C. F. Powell in 1947. Pions with positive, negative, or zero charge are produced copiously in nuclear collisions of sufficiently high energy. Their properties have been studied intensively in the laboratory, using the powerful beams of charged pions now available from many types of high-energy particle accelerators. *See* Lepton; Particle accelerator.

Already in 1948, studies of the cosmic radiation indicated the existence of heavier mesons (masses about 495 MeV), now known as K mesons (kaons). Their detailed study developed much later, partly because their laboratory production required the development of multi-GeV proton accelerators and partly because their production is less copious than that for pions at all proton energies available to date (typically, several percent of pion production). As discussed below, there are four K mesons, K^+, K^-, and two neutral particles, K^0_S and K^0_L, and strong K^+, K^-, and K^0_L beams are obtainable from higher-energy ($\geq$30 GeV) proton accelerators. Since the $\pi^\pm$, $K^\pm$, and $K^0_{L,S}$ mesons have relatively long lifetimes (about 10^{-8} to 10^{-10} s), they will be referred to as semistable; their properties are listed in **Table 1**.

In 1961 the ω meson of mass 783 MeV was discovered through its rapid decay to three pions in a bubble chamber investigation at Berkeley. It is highly unstable, with lifetime 6.6×10^{-23} s, as corresponds to the natural width 10.1 MeV observed in its mass value. [From the uncertainty relation for energy and time, the lifetime τ and the natural width Γ of any state are related by $\tau(s) \times \Gamma(\text{MeV}) = 6.582 \times 10^{-22}$.] Such a short lifetime implies that the ω meson decays through hadronic interactions. As discussed below, many such highly unstable heavy mesons are now established, with lifetimes shorter than 10^{-22} s, decaying hadronically to lighter mesons, and more continue to be discovered, as higher mass ranges are explored and as the statistics of earlier experiments are increased. *See* Bubble chamber.

In 1974 the J/ψ meson of mass 3097 MeV was discovered at Brookhaven National Laboratory in proton experiments and at the electron-positron storage ring SPEAR at Stanford. Although it decays to mesons, its natural width is only 0.063 MeV (lifetime 10^{-20} s). Also, it decays electromagnetically, yielding an e^+e^- or $\mu^+\mu^-$ pair, at a rate comparable with those for mesonic final states. This new phenomenon gave the first indication of the existence of heavy new quarks. Hence its study has often been termed the "new physics." *See* J particle; Quarks; Upsilon particles.

Quark structure. Hadrons are now considered to be composite, consisting of spin-½ quarks (q), corresponding antiquarks ($\bar{q}$), and some number of gluons (g), the last being the quanta

Table 1. Semistable pseudoscalar mesons now known (cont.)

η	$D^+(D^-)$	$D^0(\bar{D}^0)$	$F^+(F^-)$	$B^+(B^-)$	$B^0(\bar{B}^0)$
548.8 ±0.6	1869.4 ±0.6	1864.7 ±0.6	1971 ±6	5270.8 ±3.0	5274.2 ±2.8
7.5×10^{-19} ±1.1	9×10^{-13} ±2	4.5×10^{-13} ±1	3×10^{-13} ±1	1.5×10^{-12} ±0.5	1.5×10^{-12} ±0.5
$3\pi^0$ 32%	$\bar{K}^0\pi^+$ 2%	$K^-\pi^+$ 2%	$\phi\pi^+$	all $e^+\nu$ 13%	all $e^+\nu$ 13%
$\pi^+\pi^-\pi^0$ 24%	$K^-\pi^+\pi^+$ 5%	$K^-\pi^+\pi^0$ 9%	$\phi\eta^+$	all $\mu^+\nu$ 12%	all $\mu^+\nu$ 12%
$\pi^+\pi^-\gamma$ 5%	$K^-\pi^+\pi^+\pi^0$ 3%	$K^-\pi^+\pi^+\pi^-$ 5%	$\eta\pi^+\pi^+\pi^-$	$\bar{D}^0\pi^+$ 4%	$\bar{D}^0\pi^+\pi^-$ 13%
$\gamma\gamma$ 39%	$\bar{K}^0\pi^+\pi^0$ 13%	all K^- 44%	Are seen	$D^{*-}\pi^+\pi^+$ 5%	$\bar{D}^{*-}\pi^+$ 3%
	$\bar{K}^0\pi^+\pi^+\pi^-$ 8%	all K^0_S 17%			
	all $e^+\nu$ 19%	all $e^+\nu$ 5%			
$I = 0$	$I_3 = +\frac{1}{2}(-\frac{1}{2})$	$I_3 = -\frac{1}{2}(+\frac{1}{2})$	$s = -1(+1)$	$I_3 = +\frac{1}{2}(-\frac{1}{2})$	$I_3 = -\frac{1}{2}(+\frac{1}{2})$
	$C = +1(-1)$	$C = +1(-1)$	$I = 0$	$b = +1(-1)$	$b = +1(-1)$
			$C = +1(-1)$		

of the intermediate field which binds the quarks and antiquarks to form hadrons. Baryon number $B = +\frac{1}{3}$ holds for a quark q, $B = -\frac{1}{3}$ for antiquark $\bar{q}$, while $B = 0$ holds for a gluon. In this view, the simplest possibility is that each meson is a quark-antiquark ($q\bar{q}$) pair bound together by the gluon field, and this model does account quite well for most of the known mesons and their properties. However, more complicated systems (for example, consisting of two quarks with two antiquarks) can be considered and may even be required by some of the present data. The known quarks are listed in **Table 2**. They must be assigned fractional charge values, relative to the proton charge. *See* GLUONS.

Table 2. Quarks now established*

Quark type:	u	d	s	c	b	t
Flavor name:	Charge, isospin I_3		Strange	Charm	Bottom	Top
Charge Q/e	2/3	−1/3	−1/3	2/3	−1/3	(2/3)
Mass, GeV	≈0.01	≈0.01	≈0.5	≈1.5	≈4.7	≈40 ± 10
Flavor†	$I_3 = \frac{1}{2}$	$I_3 = -\frac{1}{2}$	$s = -1$	$C = +1$	$b = +1$	$t = +1$

*To each quark, there exists an antiquark with the opposite flavor values and with opposite intrinsic parity.
†Only nonzero flavor values are entered on this table.

Color and quantum chromodynamics. An exactly conserved variable, whimsically known as color, may be attributed to the quarks. The facts about the baryons require the color space to be three-dimensional. This law of conservation of the color current is then equivalent to an invariance of all hadronic interactions through all the transformations of the special unitary symmetry group $SU(3)_c$ acting on this color space. Quantum chromodynamics is the gauge theory associated with this exact symmetry, just as quantum electrodynamics is the gauge theory associated with the law of conservation of electric charge. The quanta of its gauge field are known as the gluons. These are neutral vector particles, eight in number—a color octet—coupling directly with the color current, just as the photon, the quantum of the electromagnetic field, couples directly with the electric current. *See* COLOR; SYMMETRY LAWS.

Asymptotic freedom. The coupling constant for the gluon interaction with the color current depends on the momentum transfer. Although it is large for low-momentum transfer, since it is responsible for the strong interactions which bind quarks and antiquarks to form hadrons, it approaches zero asymptotically as the momentum transfer increases to infinity.

Color current. The net color current receives separate contributions from each quark type, as well as from the gluon field. Since each quark type is color triplet, each of these contributions has the same form, except for the label α (known as the quark flavor) which characterizes the quark type. Since the gluon couples with the total color current, the theory of quantum chromodynamics is uniquely defined, apart from the universal coupling constant. Further, the color currents do not change quark flavor, so that, for any hadronic reaction, Eq. (1) holds, where $n(q_\alpha)$ and

$$n(q_\alpha) - \bar{n}(\bar{q}_\alpha) = \text{constant} \qquad (1)$$

$\bar{n}(\bar{q}_\alpha)$ denote the numbers of quarks and of antiquarks, respectively, for each quark flavor α.

Confinement. The confinement dogma holds that only color singlet states are physically realizable. With it, neither a quark nor a gluon can exist in isolation. Thus, for any color singlet system of quarks and antiquarks, quantum chromodynamics must generate an appropriate confining potential. This has not yet been proved to be the case, although some arguments have been given which suggest that it may be so.

A mesonic system therefore consists of an equal number of quarks and antiquarks, with perhaps some gluons, all in a color singlet configuration. SEE QUANTUM CHROMODYNAMICS.

Mesons as quark-antiquark systems. Quarks q (and antiquarks $\bar{q}$) have spin ½, so that a $(q\bar{q})$ system may have net spin $S = 0$ or 1 (in units of $\hbar$, Planck's constant divided by 2π). The relative angular momentum within the $(q\bar{q})$ system must be quantized, with value $L\hbar$, where $L = 0, 1, 2 \ldots$ is a positive integer. The total angular momentum J is the vector sum of L and S, and so takes the value $J = L$ for $S = 0$, and $J = L, L \pm 1$ for $S = 1$. The notation used for these configurations is $^{2S+1}L_J$, the letters S, P, D, F, ... being used for $L = 0, 1, 2, 3 \ldots$, respectively. There will be a series of radial excitations for each configuration, so that it is also necessary to specify a principal quantum number n. SEE SPIN.

Dirac's theory of spin-½ particles indicates that an antiquark has intrinsic parity opposite that for the corresponding quark. Since orbital angular momentum L contributes parity $(-1)^L$, the net parity P is $(-1)^{L+1}$. Another operation of importance, known as charge conjugation, is the replacement of quark by antiquark, and vice versa, in the system. This is of interest only when the system contains an antiquark $\bar{q}_\alpha$ for every quark q_α it contains, so that the operation can reproduce the original system, albeit with coefficient ± 1, known as the charge conjugation parity C. This operation is clearly of interest only to systems with value zero for charge, baryon number, and all other flavors. For the $(q\bar{q})$ system with configuration $^{2S+1}L_J$, C has the value $(-1)^{L+S}$ and C is conserved through all hadronic interaction processes. The notation JPC will be used to characterize a multiplet which includes one state having zero values for all flavor quantum numbers, C being the value for this state. The photon has $C = -1$, which leads to the forbiddenness of some $(q\bar{q})$ radiative transitions, such as $^3S_1 \to \gamma^1P_1$ and $^1S_0 \to 3\gamma$. SEE PARITY.

Unitary flavor symmetries. It appears plausible that all the flavor dependence of mesonic states is due to the mass differences between different quark flavors. The quark mass values known are given in Table 2; these are effective mass values for quarks as constituents within hadrons.

The u and d quarks have essentially the same mass $\bar{m}$. The color current contributed by them is proportional to expression (2), as is also the contribution of their masses to the total

$$\sim (u^*u + d^*d) \qquad (2)$$

energy, in the limit where $m_u = m_d = \bar{m}$. Form (2) remains the same under all transformations of the SU(2) group, acting in the (u,d) flavor space, this being the group of all the complex linear transformations in a two-dimensional space which do not change the magnitude of vectors in that space. This is the well-known isospin group, isomorphic with the group of real rotations in three-dimensional space. It is this invariance which is responsible for the fact that hadronic particle states are observed to occur in isospin multiplets, characterized by a total isospin I, there being $(2I + 1)$ states with the same mass but different charge values, running from $(Q_0 + I)$ in unit steps to $(Q_0 - I)$, where the value of Q_0 depends on the system which is being considered. SEE ISOTOPIC SPIN.

Similarly, the color current for (u,d,s) quarks is proportional to expression (3), which remains

$$(u^*u + d^*d + s^*s) \qquad (3)$$

invariant under all transformations of the SU(3)$_f$ group, acting in the (u,d,s) flavor space, as indicated by the suffix f. However, the energy due to the quark masses takes form (4), neglecting the

$$\bar{m}(u^*u + d^*d + s^*s) + (m_s - \bar{m})(s^*s) \tag{4}$$

small difference between m_u and m_d, and this is invariant under SU(3)$_f$ only to the extent that the physical quantities considered may be insensitive to this quark mass difference $(m_s - \bar{m}) \simeq 500$ MeV. This is the SU(3) symmetry proposed for the hadronic interactions by M. Gell-Mann and by Y. Ne'eman in 1961. It works remarkably well in many contexts, but its approximate nature is evident from the above discussion.

This discussion can be extended to the space of (u,d,s,c) quarks, with a corresponding unitary flavor symmetry group SU(4)$_f$. In this case, there are mass differences of order $(m_c - \bar{m})$ $\simeq$1500 MeV, so that the symmetry is very badly broken. Nevertheless, some remnants of this symmetry remain, and it provides a useful basis for the classification of states and for discussing relationships between processes related by SU(4)$_f$ transformations, if used with caution.

The situation for vector mesons is illustrated on **Fig. 1**, where the mesonic states have been plotted in a symmetrical three-dimensional pattern. The charm C has values -1, 0, and $+1$ on the three planes shaded, going from bottom to top. The plane containing the four D-mesons holds all the mesons with zero strangeness; those to the right of it have $s = -1$, those to the left have $s = +1$. On each shaded plane there is a definite SU(3)$_f$ pattern; on the bottom and top planes, the patterns correspond to the 3 and 3* representations of SU(3)$_f$ symmetry, respectively. The isospin multiplets are marked by prominent heavy lines. On the central plane, the pattern consists of an SU(3)$_f$ octet together with two singlets. The latter mix strongly with the $I = 0$, $s = 0$ member of the SU(3)$_f$ octet, owing to the strong breaking of the SU(3)$_f$ and SU(4)$_f$ symmetries. The physical states ω, φ, and ψ at the center of the pattern are quite well approximated by $(\bar{u}u + \bar{d}d)/\sqrt{2}$, $(\bar{s}s)$, and $(\bar{c}c)$, as expected if symmetry breaking is dominated by quark mass-differences, rather than to the $C = 0$, $s = 0$, $I = 0$ states characteristic of an SU(4)$_f$ symmetry, and their masses range from 782 to 3097 MeV. *See* Unitary symmetry.

Quark and antiquark interactions. These are of several classes, including hadronic interactions, electromagnetic interactions, and weak interactions.

Hadronic interactions. At the fundamental level, these are due to the quark-gluon and gluon-gluon couplings discussed above. They are flavor-independent, apart from their dependence on the quark masses m_α. The quark-antiquark interactions have some close parallels with the electromagnetic interactions, although the former are much stronger; the quark-gluon coupling constant α_s has a value $\simeq 0.5$ for (momentum transfer)$^2 = 5(\text{GeV}/c)^2$, compared with $\alpha \simeq 1/137$ for the electromagnetic coupling constant.

At the phenomenological level, prominent contributions may still arise from more specific mechanisms. In this respect the pion plays an outstanding role, since it is the lightest meson, the next being the kaon, about 3.5 times heavier. For example, the long-range part of the nucleon-nucleon interaction is dominated by the one-pion-exchange mechanism. *See* Strong nuclear interactions.

Electromagnetic interactions. These are carried by the electromagnetic field which cou-

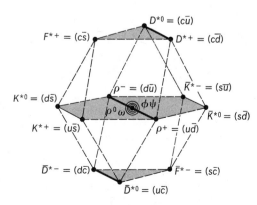

Fig. 1. The 16 vector mesons arrayed on a flavor SU(4)$_f$ plot.

ples with the electric current of the quarks. As can be seen from the quark charge values listed in Table 2, these interactions are flavor-dependent, violating even the SU(2) isospin symmetry.

Weak interactions. These have been unified with the electromagnetic interactions, forming the electroweak interaction theory developed by S. Glashow, A. Salam, and S. Weinberg. This unification required the existence of some very massive vector bosons and these were found in 1983 through experiments using the proton-antiproton collider at CERN, the $W^\pm$ bosons having mass 80.8 ± 2.7 GeV and the Z^0 boson having mass 92.9 ± 1.6 GeV, in excellent accord with expectation for the proposed electroweak theory known as SU(2) × U(1), or more popularly, as the "standard model." These bosons are not mesons since they have no strong nuclear interactions. Together with the massless photon, they carry the electroweak interactions between the currents with which they interact. However, for energies much lower than the boson masses, the weak interactions separate out from the electromagnetic interactions, being effectively pointlike and many orders of magnitude weaker than the latter. SEE INTERMEDIATE VECTOR BOSON; WEINBERG-SALAM MODEL.

Since the electroweak interaction is universal, the $W^\pm$-current interaction has essentially the same coupling constant e (the electron charge) as holds for the electromagnetic interaction. The reason for the weakness of weak decay processes is the large W mass M_W. For example, for the purely leptonic decay $\mu^+ \to \bar{\nu}_\mu e^+ \nu_e$, the interaction amplitude reduces to the form of expression (5), where J represents the current giving the transition indicated. This current × current

$$\frac{e^2}{M_W^2} J(\mu^+ \to \bar{\nu}_\mu) \cdot J(\text{Vac} \to e^+ \nu_e) \tag{5}$$

term has the general form proposed for all beta-decay interactions by E. Fermi in 1933, a proposal now well verified. The coefficient of expression (5) defines the weak coupling amplitude G_F, leading to relation (6), where the empirical value for G_F is also given, and c is the speed of light and

$$G_F = \frac{4\pi(e^2/\hbar c)}{M_W^2 \sqrt{2}} = \frac{1.02678(2) \times 10^{-5}}{M_P^2} \tag{6}$$

M_P is the proton mass. This equality is quite well satisfied by the observed W mass. The small value for G_F given in Eq. (6) causes the weak decay processes relevant for semistable mesons to have decay rates of order 10^{-12} relative to those for unstable mesons, whereas meson decay processes involving electromagnetic effects are suppressed typically by only a factor of about 10^{-3} per photon interaction involved, as in the case of $\pi_0 \to \gamma\gamma$ decay, even though the electromagnetic and weak interactions are now part of the one electroweak interaction.

The study of the weak decays of the semistable mesons is an important means of learning the precise constitution of the hadronic weak currents. It is known that the amplitude ratio $(c \to s)/(c \to d)$ is about 4; D-meson decay modes generally include a $\bar{K}$ meson (see Table 1). The b-quark transition $(b \to c)$ has amplitude only about one-twentieth that for $(c \to s)$; despite their larger energy release, B-mesons have lifetimes longer than those for D mesons. The $(b \to u)$ amplitude is less than 0.15 times the $(b \to c)$ amplitude; the known B-meson decay modes all include a D meson (see Table 1). This is an area of very active research. SEE FUNDAMENTAL INTERACTIONS; WEAK NUCLEAR INTERACTIONS.

Pseudoscalar mesons. These have q-$\bar{q}$ configuration $^1S_0$, corresponding to total spin zero and odd parity. In most cases, they are the lightest meson for their set of flavor quantum numbers, presumably because the spin dependence of the flavor-independent q-$\bar{q}$ potential has most attraction when $S = 0$. These mesons (Table 1) are semistable because the only decay processes energetically available to them involve violations of flavor conservation, which can occur only through the weak interactions. The exceptions to these remarks are the states with zero for all flavor quantum numbers, the π^0, η, η', and η_c; π^0 and η undergo electromagnetic decay, whereas η' and η_c have strong but suppressed hadronic decay modes.

Pions. The pions π^-, π^0, and π^+ form a charge triplet (isospin $I = 1$). The $\pi^\pm$ mesons decay dominantly by reactions (7). The muons in cosmic radiation or in beams from accelerators

$$\pi^+ \to \mu^+ + \nu_\mu \qquad \pi^- \to \mu^- + \bar{\nu}_\mu \tag{7}$$

originate mainly from these π-μ decay processes, following pion production in high-energy nuclear collisions. The muon emitted in $\pi^+(\pi^-)$ decay spins in a right-handed (left-handed) sense relative to its direction of emission, which illustrates directly the nonconservation of both parity and charge-conjugation parity in the π-μ decay process. Since its flavor quantum numbers are all zero, and it has charge-conjugation parity $C = +1$, the π^0 meson can decay to $\gamma\gamma$ electromagnetically. It therefore decays rather rapidly, with lifetime $\sim 10^{-16}$ s.

Kaons. The kaons form two charge doublets (isospin $I = \frac{1}{2}$), (K^+, K^0) and $(\bar{K}^0, K^-)$, each doublet being antiparticle to the other. The K^+ meson has many decay modes, the dominant being to $(\mu^+\nu_\mu)$ with branching ratio 63.5%. The K^- meson decays to the corresponding antiparticle systems, with the same branching ratios. It was recognized quite early that K^+ decay led to two systems, $\pi^+\pi^0$ and $\pi^+\pi^+\pi^-$, which have opposite parity. This discrepancy was resolved in 1957, when it was shown experimentally that parity conservation does not generally hold for the weak interactions, so that the existence of both pionic modes for K^+ decay was simply an illustration of this fact. Parity-nonconservation is demonstrated explicitly by the observation of longitudinal polarization for the μ^+ meson emitted in K^+ semileptonic decay modes.

Two neutral kaons are known empirically, both semistable (Table 1). The K_S^0 meson is short-lived and decays dominantly to $\pi\pi$ states. The K_L^0 meson has lifetime several times longer than $\pi^\pm$ and $K^\pm$, so that K_L^0 beams can also be prepared at accelerators and used for experments on K_L^0 properties. A series of ingenious experiments has established that (because of weak interaction effects) the K_L^0 meson is just a little heavier than the K_S^0 meson, the masses differing by $3.52 \pm 0.015 \times 10^{-12}$ MeV (the K_S^0 lifetime width is 7.4×10^{-12} MeV). To a good first approximation, these K_S^0 and K_L^0 states are the coherent superpositions (8) of the K_0 and $\bar{K}_0$ states, which have

$$\psi(K_S^0) = \psi_+ = \frac{\{\psi(K^0) - \psi(\bar{K}^0)\}}{\sqrt{2}} \quad (8a) \qquad \psi(K_L^0) = \psi_- = \frac{\{\psi(K^0) + \psi(\bar{K}^0)\}}{\sqrt{2}} \quad (8b)$$

definite values under the combined operation CP, $+1$ for K_S^0 and -1 for K_L^0. Charge conjugation C induces $K^0 \leftrightarrow \bar{K}^0$, and the parity P has the value -1 for both K^0 and $\bar{K}^0$. For $J = 0$, CP has the value $+1$ for the final nonleptonic state $\pi\pi$, and -1 for the final state 3π, assuming that the latter has only S-wave internal motions, as is compatible with the data on this decay mode. This suggested for a time that, although both charge-conjugation and space-reflection symmetries are separately violated in the weak interactions, the weak interactions might still be invariant under the combined operation CP.

In 1964 observation of the decay mode $K_L^0 \to \pi^+\pi^-$ was reported but with a low branching ratio, now known to be $2.03 \pm 0.05 \times 10^{-3}$, by J. Cronin, V. Fitch, and coworkers. This observation demonstrates directly the failure of CP invariance in the weak interactions, at the level of 10^{-3}. This failure of CP invariance was soon confirmed by the observation of the decay mode $K_L^0 \to \pi^0\pi^0$, with branching ratio $0.94 \pm 0.18 \times 10^{-3}$. It follows that the physical K_L^0 meson does not correspond to ψ_- in Eqs. (8), but to ψ_- with a small admixture of ψ_+, the latter with amplitude $\simeq 10^{-3}$; and vice versa for K_S^0.

Despite considerable exploration, no other evidence has been found for the failure of CP invariance, nor for the failure of time-reversal invariance T, which is considered equivalent in view of a rather general theorem that CPT invariance must always hold, in any other weak interaction process. These $K_L^0 \to \pi\pi$ phenomena enable an experiment to be prescribed for a laboratory far out of contact with the solar system, whose results will determine absolutely whether that laboratory is made of matter or antimatter. Also, the observation of K_S^0-K_L^0 interference effects in high-momentum neutral kaon beams has demonstrated the property of quantum-mechanical coherence over macroscopic distances, measured in tens of meters.

Eta mesons. The neutral $\eta(549)$ decays all require the intervention of the electromagnetic interaction. Even the mode $\eta \to \pi^+\pi^-\pi^0$ violates isospin conservation, and virtual electromagnetic effects are the most plausible cause of this. The neutral $\eta'(958)$ is the ninth member of this SU(3) nonet, lying far above all the other members in mass. Its dominant decay mode $\eta' \to \eta\pi\pi$ is hadronic but slow, the η' full width being only (0.29 ± 0.05) MeV; photon-emitting modes, $\rho^0\gamma$ and even $\gamma\gamma$, compete quite strongly with it, their branching fractions being 30% and 2%, respectively. An eta meson with structure $(c\bar{c})$ is now known, η_c (2981), whose decay modes include $\eta\pi\pi$, $2\pi^+2\pi^-$, and $K^+K^-\pi^+\pi^-$. Their relative rates are not well known, nor the total η_c width. A similar $(b\bar{b})$ meson η_b is predicted but not yet observed.

Table 3. Lightest vector mesons established for the u, d, s, c, and b quarks

Characteristics	ρ	$K^*(\bar{K}^*)$	ω	φ	J/ψ	$D^*(\bar{D}^*)$	$F^*(\bar{F}^*)$	Υ	$B^*(\bar{B}^*)$
Mass, MeV	769 ± 3	892.1 ± 0.4	782.6 ± 0.2	1019.5 ± 0.1	3096.9 ± 0.1	2009 ± 1	2118 ± 10	9460.0 ± 0.3	≈5325 ± 4
Width, MeV	154 ± 5	50 ± 1	9.9 ± 0.3	4.2 ± 0.1	0.063 ± 0.009	<2	?	0.044 ± 0.007	?
Nonzero flavor numbers	$I=1$	$s=+1(-1)$ $I=½$				$C=+1(-1)$ $I=½$	$C=+1(-1)$ $s=+1$		$b=+1(-1)$ $I=½$
Dominant decay modes	ππ	Kπ	$\pi^+\pi^-\pi^0$ 90% $\pi^0\gamma$ 9%	K^+K^- 49% $K^0_LK^0_S$ 35% $\pi^+\pi^-\pi^0$ 15%	e^+e^- 7% $\mu^+\mu^-$ 7% $2\pi^+2\pi^-\pi^0$ 4% + many hadronic modes	$D\pi$ ~80% $D\gamma$ ~20%	$F\pi$? $F\gamma$ seen	e^+e^- 3% $\mu^+\mu^-$ 3% $\tau^+\tau^-$ + very many hadronic modes	$B\gamma$ 100% 3%

Vector mesons. These mesons have JPC = (1 − −), consistent with the parameters of the lowest $(q\bar{q})$ state with the configuration $^3S_1$. The 16 states made from (u,d,s,c) quarks and $(\bar{u},\bar{d},\bar{s},\bar{c})$ antiquarks are arrayed on Fig. 1, and their masses, widths, and dominant decay processes are given in **Table 3**.

The "old" vector mesons form the nonet depicted on the central $C=0$ plane of Fig. 1, excluding the $(c\bar{c})$ state J/ψ. They were all discovered in production reactions, and identified by the analysis of their decay processes.

Transformation from photons. The existence of some of these mesons had already been conjectured before 1961, on the basis of knowledge of the electromagnetic structure of the proton and neutron. The neutral mesons ρ_0, ω, φ, and ψ all have the same quantum numbers as the photon and therefore couple directly with it; it is therefore natural that the ρ^0 and ω mesons, which are made of the same quark flavors as are the nucleons, should play an important role in the photon-nucleon interaction. A free photon cannot transform in vacuum to become a free vector meson V, of course, but a photon with sufficiently high energy can so transform if there is some particle present which can absorb the (small) momentum transfer then involved. Reaction (9) provides such a situation, one where the nucleons can even act coherently, taking up this momentum transfer while leaving the nucleus unexcited. Process (9) is then diffractive in nature, and its investigation has been important in the study of vector mesons.

$$\gamma + \text{nucleus} \rightarrow \text{nucleus} + V \tag{9}$$

Formation experiments. However, the major part of present knowledge of the vector mesons comes from formation experiments, through the annihilation processes between head-on electron and positron beams in a storage ring, process (10). (Here and on the figures, γ* denotes

$$e^+ + e^- \rightarrow \gamma^* \rightarrow \text{vector meson } V \rightarrow V\text{-meson decay products} \tag{10}$$

an intermediate electromagnetic field such that its energy is not equal to its momentum, as is required for a free photon. The technical term used to describe this field is virtual photon.) The inverse process, the decay of a vector meson, all of whose flavor quantum numbers are zero, into a lepton pair, is shown in **Fig. 2a**, while the decay of such a meson to a quark pair (seen subsequently as two jets of outgoing hadrons) through an intermediate virtual photon is shown in Fig.

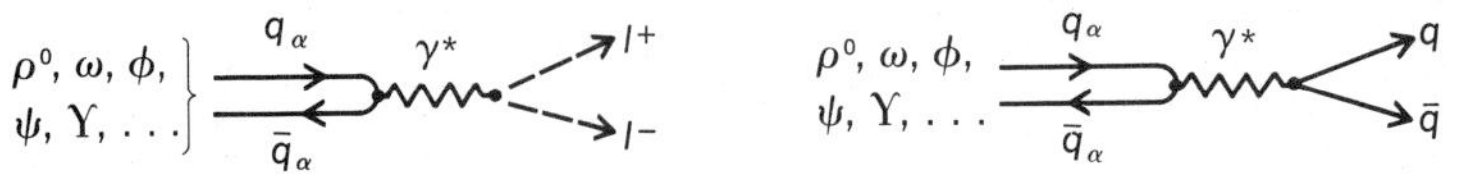

(a) $V^0 \rightarrow l^+l^-$ (b) $V^0 \rightarrow q\bar{q} \rightarrow$ hadrons

Fig. 2. Quark and lepton line figures showing annihilation of quark q_α with antiquark $\bar{q}_\alpha$ in a vector meson whose flavor quantum numbers are all zero, to give rise (a) to a lepton pair or (b) to a quark pair through an intermediate virtual photon γ*.

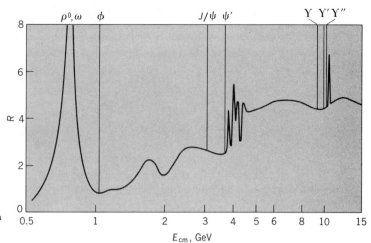

Fig. 3. Observed ratio $R = \sigma(e^+e^- \text{ hadrons})/\sigma(e^+e^- \to \mu^+\mu^-)$ plotted as a function of center-of-mass energy E_{cm}.

2b. (However, the major contribution to the latter decay mode arises from transitions due to intermediate gluons, discussed below.) The rates observed for e^+e^- production of hadrons are shown on **Fig. 3**, where the ratio $R = \sigma(e^+e^- \to \text{hadrons})/\sigma(e^+e^- \to \mu^+\mu^-)$, where σ stands for the indicated cross section, is plotted against the center-of-mass energy $E_{cm} = 2E_e$, the electron and positron having the same energy E_e, but opposite directions, in the storage ring. The excitation of the ρ^0 and ω mesons is a strong and broad resonance peak. The excitation of the ϕ meson is strong, but this state has a width of only 4 MeV, too narrow to be shown.

Decay processes. The ρ^0 decay is depicted in terms of quarks on **Fig. 4**a. From the $u\bar{u}$ part of the ρ^0 state, this decay requires the creation of a $d\bar{d}$ pair, and their interaction with $\bar{u}$ and u, respectively, leading to the configuration ($\pi^+ + \pi^-$); from the $d\bar{d}$ part, it requires the creation of a $u\bar{u}$ pair, leading to the same final state. The kinetic energy released is about 500 MeV, and the transition $\rho^0 \to \pi^+\pi^-$ is rapid, as the observed width $\Gamma(\rho^0) = 154$ MeV attests.

The ϕ decays are illustrated on Figs. 4b and c. Without the gluon coupling in Fig. 4b, the initial ϕ and final $\pi^+\rho^-$ systems would have no quarks in common. Such disconnected graphs

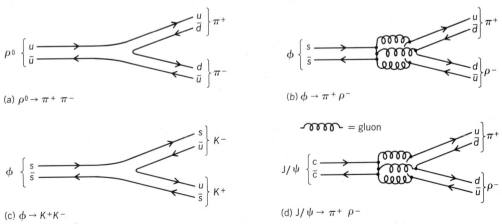

Fig. 4. Quark line figures to illustrate the decay processes of vector mesons. (a) Decay $\rho^0 \to \pi^+\pi^-$. (b) Decay $\phi = (s\bar{s}) \to \pi^+\rho^-$. (c) Decay $\phi \to K^+K^-$. (d) Decay $(c\bar{c}) = J/\psi \to \pi\rho$, which is observed to have branching fraction $1.2 \pm 0.1\%$.

are to be disregarded, as noted by S. Okubo, G. Zweig, and J. Iizuka (OZI rule), since the creation point of the final quarks is then not causally connected with the initial state. However, the gluons have a strong interaction with quarks, and they can provide the necessary connection between the initial and final quarks. But they do not make this decay transition rapid, owing to the property of asymptotic freedom for the gluon interactions. Annihilation between s and $\bar{s}$ quarks requires their very close approach. By the uncertainty principle, small spatial distance implies large momentum transfer p, and asymptotic freedom requires that the hadronic coupling constant $\alpha_s(p^2)$ approach zero logarithmically as p^2 increases to infinity. This $s\bar{s}$ annihilation also requires the formation of at least three gluons, so that the amplitude $(s\bar{s}) \to \pi^{\pm}\rho^{\pm}$ is proportional to α_s^3 and the asymptotic freedom of quantum chromodynamics is quite sufficient to make it rather small. In fact, even the transitions $\phi \to \pi^+\pi^-\pi^0$ observed (Table 3) are attributable to a small admixture of ω state with $(s\bar{s})$ in the physical ϕ state. This conclusion about the role of asymptotic freedom in heavy vector meson decay does not affect the amplitude for $\rho^0 \to \pi^+\pi^-$ of Fig. 4a, nor that for $(s\bar{s}) \to K^+K^-$ in Fig. 4c. The $\bar{d}d$ or $\bar{u}u$ pair creation required occurs through their exchange of gluons with the quarks initially present, but α_s is large for these gluons since they are dominantly of low momentum transfer.

The decay mode $\phi \to K\bar{K}$ (Fig. 4c) is the natural one for a state $(s\bar{s})$, involving the pair creation of only the lightest quarks. However, the K^+K^- threshold mass is at 987.3 MeV, whereas $M(\phi) = 1019.6$ MeV, so that the kinetic energy release is only 32 MeV. Since the decay $\phi \to K\bar{K}$ requires orbital angular momentum $l_{K\bar{K}} = 1$, there is a centrifugal barrier against this decay, and the small energy release means that its rate is very much suppressed relative to that for the otherwise parallel decay mode $\rho^0 \to \pi^+\pi^-$.

J/ψ family. With this background, the immediate interpretation of the narrow resonant states shown on Fig. 3 at around 3 to 4 GeV and around 10 GeV is that each cluster of states signals the threshold for a new quark flavor. The J/ψ(3097) is interpreted as the $^3S_1$ state for the charmed system $(c\bar{c})$ and ψ′ (3685) as its first excited $^3S_1$ state, that is, the first radial excitation of the J/ψ meson. The lowest state of the system $(c\bar{u})$ is the pseudoscalar $(^1S_0)$ meson $D^0(1863)$, so the threshold $D\bar{D}$ analogous to the $K\bar{K}$ threshold in the case of the ϕ meson lies at 3727 MeV. Since the mass values of the J/ψ and ψ′ mesons are below this $D\bar{D}$ threshold, they can decay only through the OZI-forbidden processes mediated by intermediate gluons, such as given in Fig. 4d or through the electromagnetic processes shown in Fig. 2a. The strong suppression of J/ψ decay to hadrons by the effect of asymptotic freedom is underlined by the fact that the decays given by Fig. 2a for the leptons $l = e$ and μ amount to 15% of the total J/ψ decay rate. The peaks occurring just above the ψ′ mass are due to the excitation of further states in the $(^3S_1, ^3D_1)$ $c\bar{c}$ system. These states have widths characteristic of normal hadronic processes, since they lie above the $D\bar{D}$ threshold.

Mesons in (u,d,s,c) system. All 16 of the vector mesons appropriate to the (u,d,s,c) system are now known empirically. Their known properties are given in Table 3, and they are displayed symmetrically on Fig. 1, although their masses range from 776 to 3097 MeV. All of the corresponding pseudoscalar mesons are now known and may be arrayed similarly. Those which are semistable are entered in Table 1; the D-meson doublet are the lightest mesons with $(s,C) = (0,+1)$, and the F-meson singlet the lightest meson with $(s,C) = (+1,+1)$. The others are the isosinglet mesons η′ (958) and η_c (2980), both of which can decay hadronically, through OZI-forbidden processes; their decays involve only two gluons, so that their hadronic decays are less suppressed than are the J/ψ hadronic decays.

Upsilon mesons. The situation for the Υ mesons around 10 GeV is quite similar to the above, and corresponds to the onset of processes involving the fifth quark b. Three narrow vector states, Υ(9460), Υ′(10023), and Υ″(10356), with widths ranging from 44 to 18 keV, are well established, as well as a fourth state Υ‴(10577) massive enough for decay to $B\bar{B}$ (Table 1) and with width (about 20 MeV) to match. Heavier Υ states have been reported, Υ^{iv}(10845), Υ^v(11020), and Υ^{vi}(11200), with widths on the order of 100 MeV. These states find a natural interpretation as a sequence of radially excited $L = 0$ states of the $b\bar{b}$ system.

Meson multiplets with positive parity. The $q\bar{q}$ model for meson multiplets suggests that the first excited multiplets should correspond to $L = 1$, the net parity then being $+$. Four such multiplets should exist: ^3L_J with $J = 0, 1,$ and 2, for $L = 1$, and ^1L_J with $J = L = 1$. The former three have $C = +1$, and there is quite detailed evidence for their existence as χ states in the ψ spectroscopy. The $^1P_1$ state has $C = -1$; for the $(c\bar{c})$ system, it is difficult to reach from the initial ψ' state, so that the present experiments would not be expected to find it. Positive parity states for the systems $(c\bar{u})$, $(c\bar{d})$, and $(C\bar{S})$ are also difficult to produce and there is no empirical knowledge of them yet.

In the narrower realm of (u,d,s) quarks, many resonances with positive parity have become established, requiring the existence of a nonet for each of these four P configurations (**Table 4**).

Table 4. "Old" mesonic states below about 1800 MeV assigned to $(q\bar{q})$ meson nonets on the basis of their (JPC) values*

		(I,s)			
State	(JPC)	(1,0)	(0,0)	(0,0)	(½,1)
$L = 0$	(0−+)	π(137)	η(549)	η'(958)	K(495)
	(1−−)	ρ(769)	ω(783)	φ(1020)	K*(892)
$L = 1$	(0++)	δ(983)	ε(1300)	S*(975)	K*(1350)
	(1++)	A_1(1275)	D(1283)	E(1418)	Q_1(1270)
	(1+−)	B(1234)	H(1190)	?	Q_2(1406)
	(2++)	A_2(1318)	f(1275)	f'(1525)	K*(1425)
$L = 0^*$	(0−+)	π(1300)	ι(1440)	?	?
	(1−−)	ρ(1590)	?	φ'(1685)	$K^{+\prime}$(1650)?
$L = 2$	(3−−)	g(1691)	ω(1690)	φ(1853)	K*(1780)
	(2−+)	A_3(1680)	?	?	L(1770)

*A question mark indicates either that the assignment given is controversial or that no meson has been found to have the required quantum numbers. $L = 0^*$ means the first radially excited $L = 0$ state. The lightest charmed meson D(1866) lies just above this mass range.

The (2++) nonet is well established, and the (1++) nonet has also become established, after long controversy, while the 1(+−) nonet is not yet complete.

The (0++) states offer serious problems. The states ε(1400) and K*(1350) can plausibly belong to a common nonet; similarly the states δ(983) and S*(975) can plausibly belong to a common nonet. However, all four of these states can belong to the same nonet only with difficulty. For example, $K\bar{K}$ is the dominant decay mode for S*(980) while ππ is dominant for ε(1300), just opposite to expectation for a nonet.

Corresponding states are known for the $(c\bar{c})$ and $(b\bar{b})$ systems, where they are denoted by χ_c and χ_b. The ^3P_J states $\chi_c(J)$ lie at masses 3415 MeV for $J = 0$, 3510 MeV for $J = 1$, and 3556 MeV for $J = 2$, in the same order as for the nonets $(J++)$ entered on Table 4. The $^1P_1$ χ_c state has not yet been observed. The three observed states are conveniently reached following ψ' formation, through the radiative transitions ψ' → $\gamma\chi_c(J)$. They can all decay hadronically, although at a suppressed rate; the dominant transition for the $J = 1$ and $J = 2$ states is $\chi_c → \gamma\psi/J$. The ^3P_J states $\chi_b(J)$ have mass values less well known, at about 9870 MeV for $J = 0$, 9895 MeV for $J = 1$, and 9915 MeV for $J = 2$; the $^1P_1$ χ_b state is not yet known. The observed χ_b states are most conveniently reached by radiative decay from the T′ (10023) state formed in e^+e^- collisions.

Higher excited mesonic states. Rotationally excited nonets are expected to exist, with mass increasing with increasing L. No nonets with $L > 2$ are completely established yet, but the $(I = 1, s = 0)$ state with $J = L + 1$ is known for all $L \leq 6$. The value (mass)^2 has been plotted against J on Fig. 6, and is well fitted by a linear dependence, passing through 0.5 at (mass)$^2 = 0$. This line is known as a Regge trajectory—a trajectory of rotationally excited states based on the ρ meson as its lowest member—hence the ρ trajectory.

The K* member of the $J = L + 1$ nonets is known for $L \leq 3$, and their (mass)^2 values are also plotted against J on **Fig. 5**. They are well fitted by a linear K* trajectory, parallel with the ρ trajectory. SEE REGGE POLE.

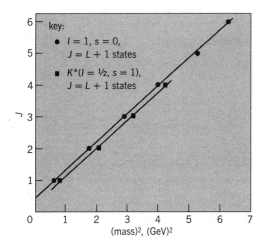

Fig. 5. Spin J plotted against (mass)^2 for the $I = 1$, $s = 0$ mesons and the $I = 1/2$, $s = 1$ mesons with $J = L + 1$, fitted by straight lines.

Radial excitations of these nonets also occur, denoted by $n^{(2S+1)}L_J$, where $n = 1$ denotes the lowest state. These may be sought most easily for the $JPC = (1 - -)$ nonets, since the e^+e^- annihilations excite selectively those states with the same quantum numbers as the photon. The states involving only (u,d,s) quarks will be relatively broad, since the decay channels available for the $n = 1$ states will necessarily be equally available for the radially excited $n \geq 2$ states. Three candidates for the $n = 2$ vector nonet, $\rho'(1590)$, $K^{*\prime}(1650)$, and $\phi'(1685)$, have been identified (Table 4).

Quarkonium is the name now used for the system $(\bar{q}q)$. For the c quark, its investigation has given a rather detailed understanding of the ψ and χ_c states and their γ spectroscopy. The known states (**Fig. 6**) are almost all of those predicted by this model; below the $D\bar{D}$ threshold, only the state $\eta_c{}'$ remains to be found. Simple calculations based on the flavor-independent $q\bar{q}$ potential of quantum chromodynamics (but retaining the flavor-dependent quark masses), consisting of a linear confining potential, and an attractive one-gluon-exchange potential, provide an acceptable account of the various branching fractions measured, including the net hadronic rate, for the ψ, ψ', and the three χ_c states.

For the b quark, a similar investigation is under way, both experimentally and theoretically. As mentioned above, six radial excitations of the Υ state are known, two of them lying below the $B\bar{B}$ threshold and therefore very narrow. The first radial excitations χ_b' of the χ_b states are known, with masses about 10233 MeV, 10254 MeV, and 10271 MeV for $J = 0$, 1, and 2, respectively. They are most conveniently produced using Υ''' formation in e^+e^- collisions in a storage ring, followed by the radiative transitions $\Upsilon''' \rightarrow \gamma\chi_b'(J)$. Thus the bottonium system offers a rich spectroscopy whose study is still in its early stages.

Other excitations. Higher-mass mesonic states can also result from the excitation of further $q\bar{q}$ pairs, or of the gluonic degrees of freedom, or both. The most simple quark-field excitation is that offered by the four-quark configuration $(\bar{q}^2q^2)$. Even with all internal orbital angular momenta limited to zero values, and considering only (u,d,s) quarks, this configuration would predict the existence of a very large number of positive parity mesonic multiplets, some of which are predicted to lie quite low in mass. It has been suggested, for example, that the $(0++)$ states, $\delta(983)$ and $S^*(975)$, which appeared so anomalous above may be states of this type. The most simple gluon field excitation leads to "hybrid states," such as the structure $(\bar{q}qg)$ where g denotes a gluon. Glueballs, systems consisting only of gluons, are also conceivable, with structure (g^n) and these can mix with states of the same JPC which consist of $q\bar{q}$ pairs with or without gluons. However, although a number of mesonic states have been analyzed with one of these further interpretations in mind, it is now clear that such an interpretation can only be convincing when the mesonic state considered is exotic, that is, when it has a JPC value not possible for quarkonium. SEE ELEMENTARY PARTICLE.

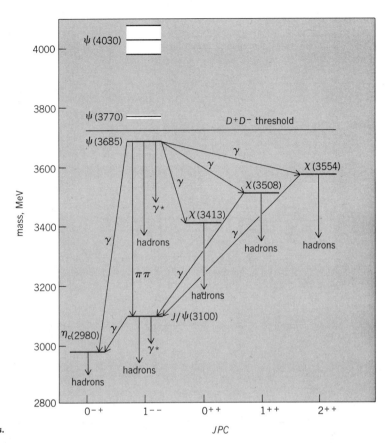

Fig. 6. Spectrum of the known ψ, χ, and η_c states below $D\bar{D}$ threshold showing their γ and ππ transitions, together with the nature (that is, hadronic or mediated by a virtual photon γ*) of their absorptive decay processes.

Bibliography. M. K. Gaillard, Toward a unified picture of elementary particle interactions, *Amer. Sci.*, 70:506–514, 1982; S. L. Glashow, Quarks with color and flavor, *Sci. Amer.*, 233(4):38–50, 1975; A. Hendry and D. B. Lichtenberg, The quark model, *Rep. Prog. Phys.*, 41:1707–1780, 1980; N. Isgur and C. Karl, Hadron spectroscopy and quarks, *Phys. Today*, 36(11):36–43, 1983; Particle Data Group, Review of particle properties, *Phys. Lett.*, vol. 111B, April 1984; G. t'Hooft, Gauge theories of the forces between elementary particles, *Sci. Amer.*, 242(6):90–116, 1980.

J PARTICLE
Samuel C. C. Ting

An elementary particle with an unusually long lifetime and large mass which does not fit into any of the schemes for classifying the large number of previously known particles. The discovery of J particles in proton-proton (p,p) and electron-positron (e^-,e^+) collisions created excitement in elementary particle physics in the mid-1970s.

Discovery of J particle. Prior to 1974, there had been much theoretical speculation on the existence of long-lived neutral (no electric charge) particles with superheavy masses larger than 10 GeV/c^2 (10^{10} electronvolts divided by the speed of light squared, or about 10 times the mass of the hydrogen atom). These were thought to play the role in weak interactions that photons play in electromagnetic interactions. There was however, no theoretical justification, and no predictions existed, for long-lived particles in the mass region 1–10 GeV/c^2. *See* Intermediate vector boson.

The J particles are rarely produced in p-p collisions. Statistically, they occur once after many millions of subnuclear reactions, in which most of the particles are "ordinary" elementary particles, such as kaon (K), pion (π), or proton (p). One searches for the J particle by detecting its e^+e^- decays. A two-particle spectrometer was used by a group from Massachusetts Institute of Technology (MIT) at the Brookhaven National Laboratory (BNL) to discover this particle. A successful experiment must have: (1) A very high-intensity incident proton beam to produce a sufficient amount of J particles for detection; the Alternating Gradient Synchrotron (AGS) accelerator at BNL provided a beam of 10^{12} 30-GeV protons per second for this experiment. (2) The ability, in a billionth of a second, to pick out the $J \to e^-e^+$ pairs amidst billions of other particles through the detection apparatus. SEE PARTICLE ACCELERATOR.

The detector is called a magnetic spectrometer. A positive particle and a negative particle each traversed one of two 70-ft-long (21-m) arms of the spectrometer. The e^+ and e^- were identified by the fact that a special counter, called a Cerenkov counter, measured their speed as being slightly greater than that of all other charged particles. Precisely measured magnetic fields bent them and measured their energy. Finally, as a redundant check, the particles plowed into high-intensity lead-glass and the e^+ and e^- immediately transformed their energy into light. When collected, this light "tagged" these particles as e^+ and e^-, and not heavier particles such as π, K, or p. The simultaneous arrival of an e^- and e^+ in the two arms indicated the creation of high-energy light quanta from nuclear interactions. The sudden increase in the number of e^+e^- pairs at a given energy (or mass) indicated the existence of a new particle. SEE CERENKOV RADIATION; PARTICLE DETECTOR.

The trajectory of electrons was measured by precision devices called multiwire proportional chambers. They consisted of 10,000 very fine gold-plated wires of 0.08-in. (2-mm) spacing, each with its own amplifier and recording system. The signals from the Cerenkov counters were collected by thin spherical and elliptical mirrors measuring about 3½ ft (1.1 m) in diameter. The counters were filled with gaseous hydrogen so that only energetic e^- (and e^+) would produce the light which is due to the Cerenkov effect. To measure the arrival time of e^+e^- pairs to one-billionth of a second, there were 100 elements of thin plastic scintillation counters, each less than 0.08 in. (2 mm) thick. SEE SCINTILLATION COUNTER.

By August 1974 the MIT group began to observe abundant numbers of e^+e^- pairs with a total combined mass of 3.112 GeV (see **illus.**), and line width Γ much smaller than the resolution of the detector: $\Gamma < 5$ MeV (5×10^6 eV). From August through October, many experimental checks on the detector were made. The most important of these checks consisted of changing the magnetic field of the detecting magnets. This moved the particle trajectories to different regions of the detector. Still the abundance of e^+e^- pairs did not change, indicating a real particle had been discovered. This new particle was called the J particle since J is the symbol used to denote electromagnetic current and spin in elementary particle physics. A joint announcement of the discovery was made together with a team from Stanford Linear Accelerator Center (where it was called the psi particle) in November 1974.

Properties of J particle. The $J \to e^-e^+$ production rate from proton-proton reactions at an incident energy of 30 GeV is 10^{-34} cm^2, or one part in 10^7 of "ordinary" particle yields. The yield increases by a factor of about 50 with a 300 GeV/c incident neutron beam. This increase of yield with energy is very similar to that of the K meson and antiproton productions. In the Brookhaven experiment, the yield of J decreased by almost a factor of 10 when the incident beam energy was reduced to 20 GeV/c. J particles have also been produced by bombardment of complex nuclear targets with high-energy photons. Production rates seem to be consistent with diffractive production like photoproduction of ordinary vector particles (the ρ, ω, and ϕ). Analyses of photoproduction data indicate the J is not an ordinary intermediate vector boson. Rather, it belongs to a strong-interaction family.

Most of the properties of J particles have been measured at various electron-positron storage rings. The data show that the measured line width of J is less than 2 MeV. By measuring $e^+e^- \to J \to e^-e^+$ decay rate, one obtains a total width of 63 ± 9 keV. The observed mass, m_J, varies from 3090 MeV/c^2 to 3112 MeV/c^2 from one laboratory to another, with an average of 3097 MeV. The spin (intrinsic angular momentum) is the same as that of the photon. Modes of decay into $\pi^+\pi^-$ and K^+K^- were not found. From this it has been concluded that the J particle does not belong to families of particles of about equal mass (like the π-mesons, π^+, π^0, π^-, or the two

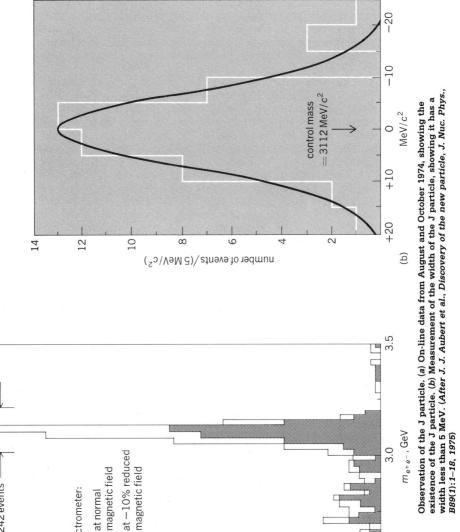

Observation of the J particle. (a) On-line data from August and October 1974, showing the existence of the J particle. (b) Measurement of the width of the J particle, showing it has a width less than 5 MeV. (After J. J. Aubert et al., *Discovery of the new particle*, J. Nuc. Phys., B89(1):1–18, 1975)

nucleons p, n), but that these particles are "single." Decays of J $\to \bar{p}p$, $\Lambda\bar{\Lambda}$, $n\pi^0$ have been found. SEE ISOBARIC SPIN.

There is a family of particles which, by their emission and absorption of monochromatic gamma rays, transform into each other and into the J particle. The energy spectrum of the J-particle family is very similar to the positronium spectrum. (Positronium is the simplest atom, consisting of an electron bound with a positron.) The similarity between the J-particle spectrum and the positronium spectrum, together with the fact that the lifetime of the J particle is about 1000 to 10,000 times longer than that of other known elementary particles, indicates that the J particle is a bound state of a new type of quark and antiquark. This idea was first proposed by S. Glashow, who called it the charm quark. The J particle therefore is a charmonium state. SEE CHARM; POSITRONIUM.

The existence of the charm quark has stimulated experiments to search for more quark states. Indeed, a new quark state (the upsilon) was found at a mass of 9.4 GeV by Leon Lederman and collaborators, again by measuring proton nuclei interactions. SEE UPSILON PARTICLE.

With the construction of a 46-GeV electron-positron colliding-beam accelerator (known as PETRA) at Deutsches Elektronen-Synchrotron (DESY), physicists have been searching for the sixth quark, the top quark. Although there were many theoretical predictions that the toponium of the (top quark, antitop quark) bound state should exist around 30–40 GeV, experimental results indicate that if the toponium state exists at all its mass is larger than 46 GeV. There are indications from the proton-antiproton collider at the European Organization for Nuclear Research (CERN) that toponium may exist with a mass between 70 and 80 GeV. SEE ELEMENTARY PARTICLE; QUARKS.

Bibliography. J. J. Aubert et al., *Phys. Rev. Lett.*, 33:1404 and 1624, 1974; E. J. Augustin et al., *Phys. Rev. Lett.*, 33:1406, 1974; W. Braunschweig et al., *Phys. Lett.*, 53B:393, 1974.

UPSILON PARTICLES
PAOLO FRANZINI

A family of elementary particles whose first three members were discovered in 1977. The upsilon mesons, Υ, are the heaviest known vector mesons, with masses greater than 10 times that of the proton. They are bound states of a heavy quark and its antiquark. The quarks which bind to form the upsilons carry a new quantum number called beauty or bottomness, and they are called b-quarks or b. The mass of the b-quark is around 5 GeV. The anti-b-quark or $\bar{b}$ carries antibeauty, and therefore the upsilons carry no beauty and are often called hidden beauty states. Direct proof of the existence of the b-quark was obtained at the Cornell collider CESR by observing the existence of B mesons which consist of a b-quark bound to a lighter quark. Twelve $b\bar{b}$-mesons have also been observed thus far. SEE QUARKS.

Heavy quarks. In addition to electric charge, all quarks carry a new nonscalar "charge," called color, which results in forces much stronger than the electric Coulomb force. There is, however, a major difference between the electric and the color forces. While the Coulomb force becomes weaker at large distances, color forces are believed to become stronger at large distances, resulting in the confinement and thus unobservability of free quarks. The theory of the color forces is known as quantum chromodynamics (QCD). SEE COLOR; QUANTUM CHROMODYNAMICS.

A bound system of a heavy quark and its antiquark ($q\bar{q}$ or quarkonium) has very small dimensions. As a result, the color force becomes relatively weak and the quarks move with low velocity. The quark-anti-quark system can therefore be described by a Schrödinger equation if the potential between the quarks is known. In a pure coulombic potential, a $q\bar{q}$ system becomes very similar to an hydrogen atom. A better analogy, however, is the case of positronium, an atomlike system consisting of an electron and a positron bound together by their mutual electric attraction. Just as for the case of positronium, which can decay into photons, the quanta of the electromagnetic field, so can the quarkonium state annihilate into gluons, the quanta of the field coupled to the color charge. According to older ideas, heavy quarkonium states should annihilate in times of the order of 10^{-24} s, while quantum chromodynamics suggests much longer lifetimes. SEE GLUONS; POSITRONIUM.

In 1974 the first heavy vector meson was discovered at the Brookhaven National Laboratory

AGS proton accelerator and at the Stanford positron-electron (e^+e^-) collider SPEAR. The mass of this new particle, called J/ψ, is around a 3 GeV and its width of the order of 60 keV, corresponding to a lifetime of about 10^{-19} s. The discovery of the J/ψ (a $c\bar{c}$ bound system, where c stands for the charm quark of mass around 1.5 GeV), was a triumph for quantum chromodynamics. SEE CHARM; J PARTICLE.

The upsilon mesons provide even better testing grounds for quantum chromodynamics. Because they contain the much heavier b quarks, the color force is slightly weaker than for the J/ψ case. Thus the upsilon annihilation rate was expected to be slower or, equivalently, its width to be narrower. This is indeed so: the width of the upsilon is only 40 keV while its mass is 9.4 GeV.

Potential models. Quantum chromodynamics has not yet been able to derive the exact form of the potential. It can, however, suggest its form. In particular, the strength of the potential is supposed to increase linearly at large distances and to obey a $1/r$ law at short distances, like the Coulomb potential. On this basis, the properties of the ψ and Υ families have been thoroughly calculated. The amount of experimental information is now so large that in fact the process can be reversed, and from the data the form of the potential has been extracted to great accuracy.

Family of states. Just as for the case of the hydrogen atom or of positronium, the $b\bar{b}$ pair can bind in many different ways, properly classified by the quantum numbers of the various states. The bound pair can have orbital angular momentum of 0, 1, 2, and so forth, the corresponding states being referred to as S-wave, P-wave, D-wave, and so forth. Each quark, in addition, has a spin angular momentum of ½. The two spins can combine to give a total spin of 0 or 1. Spin-0 states, as in all quantum systems, are called singlet states, and spin-1 states, triplet states. Orbital and spin angular momentum combine to give the total angular momentum of the state. Bound states are usually classified by the principal quantum number n, the total angular momentum J, and the total spin S.

The masses of all the known $b\bar{b}$ states are listed in the **table**. The known $b\bar{b}$ levels and their intertransitions are shown in the **illustration**, using the standard spectroscopic notation.

Masses of the known member of the upsilon family	
Meson name	Mass, GeV/c^2
Υ	9.45999 ±0.00012
Υ'	10.02345 ±0.00035
Υ''	10.3555 ±0.0005
Υ'''	10.5790 ±0.0008
Υ(5S)	10.845
Υ(6S)	11.02
$\chi_{b,J=2}$	9.915 ±0.005
$\chi_{b,J=1}$	9.894 ±0.006
$\chi_{b,J=0}$	9.875 ±0.005
$\chi'_{b,J=2}$	10.271 ±0.005
$\chi'_{b,J=1}$	10.254 ±0.005
$\chi'_{b,J=0}$	10.233 ±0.005

S-wave states. While for a pure coulombic potential there exists an infinite number of bound states, the quarkonium potential requires that there should be only three S-wave, spin-triplet, bound $b\bar{b}$ states. These are, in fact, the first three upsilon mesons discovered at Fermilab in 1977 and later confirmed at the DORIS positron-electron (e^+e^-) collider of the DESY laboratory in Hamburg and at the CESR collider. The Υ mass is now known to about 10 parts per million, an accuracy close to that for the electron mass.

The fourth upsilon, Υ''', was discovered at CESR in 1980. In contrast to the first three upsilons, which have widths ranging from 40 to 20 keV, the width of the Υ''' is about 20 MeV, almost a thousand times greater, indicating that it is slightly heavier than two B mesons into

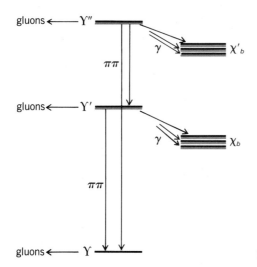

Known $b\bar{b}$ levels and their intertransitions.

which it decays. Direct evidence for the decay $Y''' \to B + \bar{B}$ was obtained by the CESR experiments, by observing the beta decay of the B mesons.

The fifth and sixth radial excitations of the triplet S-wave $b\bar{b}$ system were also observed at CESR in 1984. These mesons are usually called $Y(5S)$ and $Y(6S)$.

P-wave states. Six spin-triplet, P-wave bound $b\bar{b}$ states, called χ_b, are expected to exist. These states cannot be produced in e^+e^- annihilations because they have positive parity and charge conjugation. They can, however, be produced in a two-step process given by expressions (1) and (2). In this way the lowest spin-triplet P-wave state can be reached, after emission of a

$$e^+e^- \to Y \quad (1) \qquad Y \to \chi_b + \gamma \quad (2)$$

photon, from the Y', and the second P-wave state can be reached via decays of the Y''. Because the triplet P-wave states have total spin of 1 and also orbital angular momentum 1, the total angular momentum of these states can be, according to the rules of quantum mechanics, 0 or 1 or 2. The spin-orbit interaction shifts the mass of states with different total angular momentum, resulting in six different states: $\chi_{b,J=2}$, $\chi_{b,1}$, $\chi_{b,0}$, $\chi'_{b,2}$, $\chi'_{b,1}$, and $\chi'_{b,0}$. The three χ'_b were discovered in 1982 at CESR by observing the decays $Y'' \to \chi + \gamma$, $\chi \to Y' + \gamma$, and $\chi \to Y + \gamma$, and the three χ_b were discovered in the same manner at CESR in 1983. Later the χ_b states were also observed by other experiments at CESR and DORIS.

In addition to electromagnetic transitions between Y's and χ's, transitions between Y's, with emission of two pions, have also been observed. *See* ELEMENTARY PARTICLE; MESON.

Bibliography. P. Franzini and J. Lee-Franzini, Upsilon resonances, *Annu. Rev. Nucl. Part. Phys.*, 33:1–29, 1983; L. M. Lederman, The upsilon particle, *Sci. Amer.*, 239(4):72–80, 1978.

EXOTIC ATOMS

Positronium	284
Muonium	285
Pionium	286
Hadronic atom	287

POSITRONIUM
Vernon Hughes

The bound state of an electron and a positron. Positronium was discovered by studies of the so-called annihilation radiation from positrons stopped in gases. It is formed in a collision between a positron and a gas atom which results in the capture of an atomic electron by the positron. The positron is the antiparticle to the electron and hence has an inertial mass equal to that of the electron, a positive charge equal in magnitude to the charge of the electron, and a spin of $\hbar/2$, where $\hbar$ is Planck's constant h divided by 2π. SEE POSITRON.

Positronium is of particular interest because it is the two-body system to which quantum electrodynamics is applicable, and its study has served as an important confirmation of the theory of quantum electrodynamics. SEE QUANTUM ELECTRODYNAMICS.

No states of positronium other than the ground $n = 1$ state ($n = 1, 2, 3, \ldots$, being the principal quantum number) have been found. Studies of positron annihilation in solids and liquids indicate that a perturbed form of positronium exists under certain conditions.

Energy levels. The approximate energy levels of positronium can be calculated from the Schrödinger equation with the nonrelativistic hamiltonian, H_0, as shown in Eq. (1), where $p_1(p_2)$ is

$$H_0 = \frac{p_1^2}{2m} + \frac{p_2^2}{2m} - \frac{e^2}{r} \tag{1}$$

the electron (positron) linear momentum, m is the mass of the electron or positron, $-e$ is the charge of the electron, and r is the distance between the positron and the electron.

The energy levels of the bound states are given by Eq. (2), where the quantity r_{yp} is defined

$$W_n = -\frac{\pi^2 m e^4}{h^2 n^2} \equiv -\frac{r_{yp}}{n^2} \tag{2}$$

as the Rydberg constant for positronium. The binding energies W_n of positronium are one-half the corresponding binding energies of the hydrogen atom (if the proton-to-electron mass ratio is considered infinite). In particular, the ionization energy of positronium (the binding energy of the ground $n = 1$ state) is 6.8 eV.

Fine structure to the energy levels of positronium as given by Eq. (2) of the order $\alpha^2 r_{yp}$ (here $\alpha = e^2/\hbar c \cong 1/137$ is called the fine-structure constant) arises from relativistic effects, including the electron and positron spin magnetic moments, and from the interaction with the electromagnetic field, which causes electron-positron pair annihilation. Since the electron and positron intrinsic spin angular momenta are ½ in units of $\hbar$, the total spin angular momentum quantum number S of positronium can be either 0 (singlet state, parapositronium) or 1 (triplet state, orthopositronium). For each n value, positronium can exist in either a singlet or a triplet state. The orbital angular momentum quantum number L can assume the values $L = 0, 1, \ldots, n - 1$. In particular, the ground $n = 1$ state of positronium is split into two levels, $^1S_0$ and $^3S_1$, which are separated in energy by the amount given by Eq. (3). The term of order $\alpha^3 r_{yp}$ arises from virtual quantum elec-

$$W(^3S_1) - W(^1S_0) = \alpha^2 r_{yp}\left[\frac{7}{3} - \frac{2\alpha}{\pi}\left(\frac{16}{9} + \ln 2\right)\right] \tag{3}$$

trodynamic processes. This energy separation, often called the hyperfine structure of the ground state of positronium, corresponds to a frequency difference $\Delta\nu$ of 2.0337×10^5 MHz. SEE HYPERFINE STRUCTURE.

The dependence of the energy levels of positronium on an external magnetic field (Zeeman effect) can be determined from the hamiltonian term given in Eq. (4), in which subscript 1 refers

$$H_H = \mu_0 g_{s_1} \mathbf{s}_1 \cdot \mathbf{H} + \mu_0 g_{l_1} \mathbf{l}_1 \cdot \mathbf{H} + \mu_0 g_{s_2} \mathbf{s}_2 \cdot \mathbf{H} + \mu_0 g_{l_2} \mathbf{l}_2 \cdot \mathbf{H} \tag{4}$$

to the electron and subscript 2 to the positron, μ_0 is the Bohr magneton ($= e\hbar/2mc$), g_l is the orbital g value ($=1$), g_{s_1} is the electron spin g value $[=2(1 + \alpha/2\pi - 0.328\,\alpha^2/\pi^2)]$, $g_{s_2} = -g_{s_1}$, $\mathbf{l}_{1(2)}$ is the electron (positron) orbital angular momentum, $\mathbf{s}_{1(2)}$ is the electron (positron) spin angular

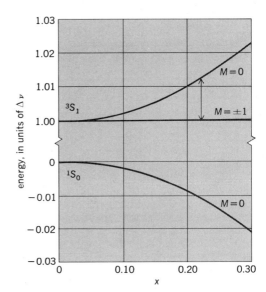

Zeeman energy levels of positronium in its ground $n = 1$ state. The quantity $\Delta\nu$ is the hyperfine structure separation between the $^3S_1$ and the $^1S_0$ states of positronium at zero static magnetic field. The M values designate the magnetic substates, and $x = 2g_{s1}\mu_0 H/(h\Delta\nu)$.

momentum, and **H** is the external magnetic field intensity. Positronium has no permanent magnetic moment, but there can be a magnetic moment induced by the external magnetic field, and hence an energy level can depend on H^2 or higher powers of H. The energy level diagram for the ground state of positronium in a magnetic field is shown in the **illustration**. From measurements of the frequency of the Zeeman transition $\Delta M = \pm 1$ between the magnetic sublevels of the $^3S_1$ state, the hyperfine structure interval $\Delta\nu$ has been determined and is given by Eq. (5). This ex-

$$\Delta\nu = (2.0333 \pm 0.0004) \times 10^5 \text{ MHz} \tag{5}$$

perimental value agrees with the theoretical value, and the agreement constitutes the principal test of the quantum electrodynamics of the two-body problem.

Decay. Positronium is an unstable atom and annihilates with the emission of photons. From its ground $^1S_0$ state a positronium atom at rest decays into two gamma rays, each having an energy of mc^2 (~510 keV) with a decay rate of 8.03×10^9 s^{-1}; from its ground $^3S_1$ state a positronium atom at rest decays into three gamma rays whose energies total $2mc^2$ with a decay rate of 7.21×10^6 s^{-1}.

Bibliography. H. J. Ache (ed.), *Positronium and Muonium Chemistry*, 1979; P. Hautojaervi (ed.), *Positrons in Solids*, 1979; J. W. Humberstone and M. R. C. McDowell (eds.), *Positron Scattering in Gases*, 1983; A. Mills and K. F. Canter, *Positron Studies of Solids*, 1985; A. Rich, Recent experimental advances in positronium research, *Rev. Mod. Phys.*, 53:127–166, 1981.

MUONIUM
Patrick O. Egan

An exotic atom, mu or (μ^+e^-), formed when a positively charged muon (μ^+) and an electron are bound by their mutual electrical attraction. It is a light, unstable isotope of hydrogen, with a muon replacing the proton. Muonium has a mass 0.11 times that of a hydrogen atom due to the lighter mass of the muon, and a mean lifetime of 2.2 microseconds, determined by the spontaneous decay of the muon ($\mu^+ \rightarrow e^+ \nu_e \bar{\nu}_\mu$).

Muonium is formed when beams of μ^+ produced in particle accelerators are stopped in certain nonmetallic targets. The μ^+ beams are generally spin-polarized; that is, the average spin angular momentum of the muons in the beam points in a definite spatial direction. The muon spin retains this spatial orientation after picking up an electron to form muonium in the reaction $\mu^+ + X \rightarrow Mu + X^+$, where X represents an atom of the target material. Since the positron in μ^+ decay is emitted preferentially along the muon spin direction, the muonium polarization can be monitored very simply by measuring the spatial distribution of the decay positron. This technique of polarization measurement allows extremely small amounts of muonium to be detected and studied. SEE SPIN.

Muonium was first observed by V. W. Hughes and coworkers in 1960. The characteristic Larmor precession of the muonium polarization in a magnetic field when positive muons were stopped in a target of argon gas indicated the formation of muonium. The observed Larmor frequency of 14 GHz/tesla is a unique signature of the triplet ($F = 1$) bound state of an electron and a muon.

Since muonium is a system consisting only of leptons, it serves as an ideal testing ground for the theory of quantum electrodynamics (QED), which describes the electromagnetic interaction between particles. Indeed, experimental measurement of the hyperfine structure interval of the ground-state muonium levels have shown that the quantum electrodynamic theory of this system is accurate to the level of one part per million. Such measurements also provide the best available values for the muon mass and magnetic moment. SEE LEPTON; QUANTUM ELECTRODYNAMICS.

Muonium chemistry and muonium spin rotation (MSR) are subfields which concentrate on study of muonium in matter. Chemical reaction rates for a wide variety of muonium reactions have been measured and compared with hydrogen rates. Formation, spin precession, and depolarization of muonium in gases, semiconductors, and insulators have been studied in some detail. Such experiments seek to understand the chemical and physical behavior of a light hydrogen isotope in matter, and to use the extremely sensitive muon polarization measurement technique to probe the structure of materials. SEE POSITRONIUM.

Bibliography. H. J. Ache (ed.), *Positronium and Muonium Chemistry*, 1979; R. H. Heffner and D. G. Fleming, Muon spin relaxation, *Phys. Today*, 37(12):38–46, 1984; V. W. Hughes, Muonium, *Phys. Today*, 20(12):29, 1967; V. Hughes and C. S. Wu (eds.), *Muon Physics*, 3 vols., 1975; D. C. Walker, *Muon and Muonium Chemistry*, 1984.

PIONIUM
PAUL A. SOUDER

An exotic atom, also called the pi-mu atom, which is similar in structure to the hydrogen atom but with the proton replaced by a pion and the electron replaced by a muon. Pionium is unique among atoms that have been observed in the laboratory in that all of its constituents are unstable particles not found in ordinary matter. The pion, a particle involved in nuclear forces, lives only 2.6×10^{-8} s, and the muon, a particle like the electron except that it is 207 times heavier, lives 2.2×10^{-6} s. The lifetime of this atom is thus determined by the pion lifetime. Due to the large mass of the muon, the atom is about 120 times smaller in radius than is hydrogen. SEE ELEMENTARY PARTICLE; LEPTON; MESON.

Most exotic atoms, such as muonium or positronium, contain only one short-lived particle and are formed by causing that particle to interact with an ordinary atom. This method is impossible for pionium since neither constituent is found in normal material. Instead, pionium is formed during the decay of a certain heavier particle called the neutral kaon. A kaon has many modes of decay, one of which results in the formation of a pion, a muon, and a neutrino ($K_L^0 \rightarrow \pi^+ \mu^- \nu$). For this decay, both the muon and pion occasionally have almost exactly the same speed and direction, resulting in the formation of a bound atomic system. Although this process is extremely rare, less than 1 per 1,000,000 kaon decays, pionium has been observed in the laboratory. This decay of a particle of radius 5×10^{-14} cm (2×10^{-14} in.) into an atom of radius 5×10^{-11} cm (2×10^{-11} in.) is in striking contrast to the usual scheme of things where atoms are broken into smaller constituents. SEE MUONIUM; POSITRONIUM.

Pionium is of interest to physicists for two reasons. First, the rate of formation provides information about the details of kaon decay. The observed formation rate is indeed compatible with present understanding of such decays. Another potentially interesting application of pionium involves atomic spectroscopy. Although pionium is usually formed in the ground state, it is presumably also produced in the excited 2S state. A measurement of the Lamb shift, which is the energy difference between the 2S and the 2P state, would serve to measure the pion radius. SEE WEAK NUCLEAR INTERACTIONS.

The nomenclature of exotic atoms is not well established, and the name pionium may also refer to the pion-electron atom. Although this atom is probably formed when pions are stopped in gases, there has been no interest in studying it in the laboratory.

Bibliography. R. Coombes et al., Detection of π-μ Coulomb bound states, *Phys. Rev. Lett.*, 37:249–252, 1976; I. L. Nemenov, Atomic decays of K_L^0 mesons, *Sov. J. Nucl. Phys.*, 16:67–69, 1973 (*Yad. Fiz.*, 16:125, 1973); Pionium, another new quasi-atom, *Sci. News*, 109:356–357, 1976.

HADRONIC ATOM
CLYDE E. WIEGAND AND BOGDAN POVH

A hydrogenlike system that consists of a strongly interacting particle (hadron) bound in the Coulomb field and in orbit around any ordinary nucleus. The kinds of hadronic atoms that have been made and the years in which they were first identified include pionic (1952), kaonic (1966), Σ^- hyperonic (1968), and antiprotonic (1970). They were made by stopping beams of negatively charged hadrons in suitable targets of various elements, for example, potassium, zinc, or lead. The lifetime of these atoms is of the order of 10^{-12} s, but this is long enough to identify them and study their characteristics by means of their x-ray spectra. They are available for study only in the beams of particle accelerators. Pionic atoms can be made by synchrocyclotrons and linear accelerators in the 500-MeV range. The others can be generated only at accelerators where the energies are greater than about 6 GeV. SEE ELEMENTARY PARTICLE; HADRON; PARTICLE ACCELERATOR.

The hadronic atoms are smaller in size than their electronic counterparts by the ratio of electron to hadron mass. For example, in pionic calcium, atomic number $Z = 20$, the Bohr radius of the ground state is about 10 fermis (1 fermi = 10^{-15} m), and in ordinary calcium it is about 2500 fermis. Thus the atomic electrons are practically not involved in the hadronic atoms, and the equations of the hydrogen atom are applicable. The close approach of the hadrons to their host nuclei suggests that hadron-nucleon and hadron-nucleus forces will be in evidence, and this is one of the motivations for studying these relatively new types of atoms.

X-ray emissions. Negative hadrons are captured into orbits of large principal quantum number n by the attraction of the positively charged nuclei. As the hadrons fall through successively smaller Bohr orbits (cascade), electrons are ejected from the cloud of atomic electrons (Auger effect). When a hadron has reached about the same radius as that of the electronic atom ground state, x-ray emission becomes the dominant method for the system to shed its excitation energy. X-rays, whose energy increases with each successive jump, are emitted until the hadron reaches the ground state ($n = 1$) of the hadronic atom or is absorbed by the nucleus in a strong interaction.

The precise measurement of the energies and shapes of the x-ray transitions not perturbed by the strong interaction is an excellent method to determine masses and magnetic moments of the hadrons. This is the case for the masses of negative pions, kaons, antiprotons, and Σ^-, as well as the magnetic moments of the last two.

The lines of the spectra of special interest are due to transitions between the lowest quantum levels because the hadrons are then closest to the nuclei and the nuclear forces perturb the orbits. The effects expected and observed are that some of the lines are slightly broadened (energy indefinite) and the average transition energy is different from that predicted solely on the basis of Coulomb effects.

The series of x-ray lines generally cuts off rather abruptly at $n > 1$. However, in light pionic atoms the ground state is reached. Kaons, Σ^-, and antiprotons ($\bar{p}$) can probably reach the ground state only when the nucleus is singly charged. In kaonic chlorine atoms, for example, the series ends at $n = 3$, and in kaonic lead the series ends at $n = 7$.

Experimentally, the hadrons are generated by a beam of protons incident on a metallic

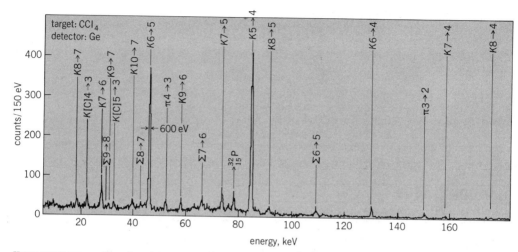

X-ray spectrum resulting from kaons stopped in carbon tetrachloride. Lines from Σ^- hyperonic atoms are seen along the kaonic x-rays. The pionic lines came from decay products. Nuclear gamma rays of the first excited state of phosphorus-32 are seen at 78 keV. (*Lawrence Berkeley Laboratory*)

target. A secondary beam is used to transport the particles to the target in which the atoms are to be made and studied. The arrival of a hadron is signaled by a set of scintillation counters as it is slowed down by passage through a moderator of carbon or beryllium. The thickness of the moderator is adjusted so that a maximum number of hadrons stops in the target under investigation. The x-ray detectors are semiconductors of silicon or germanium. Efficiencies for detecting an x-ray that comes from within the target are around 5×10^{-3}. The energy resolution of the detectors is of paramount importance. The **illustration** is an example of a kaonic x-ray spectrum of chlorine obtained by an ultrapure germanium detector. The lines are labeled according to their hadronic transitions. The intensity of the lines average about 0.3 x-ray per stopped kaon for the principal lines ($\Delta n = -1$). About 5,000,000 kaons were stopped to obtain the spectrum. In special cases where the resolution is of vital importance, as was the case in determination of the pionic mass, a crystal spectrometer is used. SEE PARTICLE DETECTOR.

Pionic atoms. The interpretation of the x-ray spectra of pionic atoms is complicated by the necessity for the pion to react with two nucleons rather than one nucleon alone, as in reaction (1), where N stands for either a proton or a neutron. The two-nucleon final state is required for the reaction to conserve momentum.

Through the use of the line width and energy shift data of all the measurements on many elements throughout the periodic table, a calculation was made to determine the parameters of the low-energy pinion-nucleus interaction. An important ingredient of this calculation is the short-range nucleon-nucleon correlation in nuclei, which determines the probability for reaction (1). The

$$\pi + N + N \rightarrow N + N + \text{kinetic energy} \qquad (1)$$

experimental data on the pion absorption in the nuclei can be fairly well accounted for in this phenomenological approach.

Kaonic atoms. Kaonic atoms were expected to yield valuable information concerning the surface of nuclei because a kaon reacts very strongly with either a single neutron or a single proton, as in reaction (2).

$$K^- + N \rightarrow \pi + \text{hyperon} \qquad (2)$$

On the basis of theory it had been predicted that more neutrons than protons would be found on the surfaces of nuclei. One of the first interpretations of the behavior of the series of x-ray lines of various elements ranging from $Z = 3$ through 92 suggested that the neutron dominance of nuclear surfaces was verified. However, it was later pointed out that a resonance between K^- and proton, $Y^*(0)1405$, would probably enhance the affinity of kaons for protons on the nuclear outskirts, and this could account for the increased capture rate observed experimentally. The data on the kaonic atoms could not be analyzed uniquely in terms of the difference of the proton and neutron distributions at the nuclear surface. The K^- nucleon resonances at the threshold complicate the parametrization of elementary interaction (2) so much that the present quality of the data is not sufficient to determine the elementary interaction on the one hand and the nuclear matter distribution on the other. SEE NUCLEAR STRUCTURE.

Σ^- hyperonic atoms. When kaons are absorbed by nucleons, about 20% of the hyperons produced are Σ^- particles. In the light elements most of the Σ^- hyperons are ejected from the nucleus in which they are generated. Some of them are captured by target nuclei; Σ^- hyperonic atoms are formed and emit characteristic x-rays, just as do the kaonic atoms. Weak x-ray lines due to the hyperonic atoms are found along with the kaonic x-ray lines (see illus.). The hyperonic lines are of special interest because they are actually doublets due to the magnetic moment of the Σ^-. X-ray lines of Σ^- atoms in some heavy elements have been measured, and the Σ^- magnetic moment was determined to be $\mu = -(1.097 \pm 0.044)$ nuclear magnetons with an additional systematic error of ± 0.04 nuclear magneton. Perhaps Σ^- will turn out to be a suitable probe of the nucleus.

Antiproton atoms. Antiproton atoms are the latest in the series of hadronic atoms to be observed. Their x-ray lines are doublets due to the magnetic moment of $\bar{p}$. The splitting has been measured and μ found to be -2.795 ± 0.019 nuclear magnetons, which agrees very well with the proton magnetic moment but has a sign opposite to that expected for antiparticles.

The main research effort involving antiproton atoms has been dedicated to the investigation of the x-ray spectra of the antiprotonic hydrogen. The transitions to the ground state depend directly on the elementary antiproton-proton interaction at the threshold. If this interaction turns out to be simple enough, the antiprotonic atoms will be a future tool for measuring the matter distribution of the nuclear surface. Another source of low-energy antiprotons—the Low Energy Antiproton Ring (LEAR), which makes precision measurements on antiprotonic atoms feasible—was put into operation at CERN near Geneva, Switzerland.

There are two more hadrons with lifetimes long enough to be candidates for hadronic atom formation: the negative xi (Ξ^-) and the negative omega (Ω^-), but even at the largest accelerators, these particles are too scarce for their atoms to be detected.

Bibliography. C. J. Batty, Exotic atoms, *Sov. J. Part. Nuc.*, 13:71–96, 1982; J. B. Warren (ed.), *Nuclear and Particle Physics at Intermediate Energies*, 1976.

FUNDAMENTAL INTERACTIONS AND FIELDS

Fundamental interactions	**292**
Quantum field theory	**297**
Feynman diagram	**300**
Quantum electrodynamics	**301**
Supercritical fields	**307**
Weak nuclear interactions	**315**
Strong nuclear interactions	**321**
Quantum chromodynamics	**323**
Gauge theory	**331**
Instanton	**333**
Weinberg-Salam model	**334**
Neutral currents	**335**
Renormalization	**337**
Grand unification theories	**340**
Neutron-antineutron oscillations	**342**
Supergravity	**343**

FUNDAMENTAL INTERACTIONS
Abdus Salam

Fundamental forces that act between elementary particles, of which all matter is assumed to be composed.

PROPERTIES OF INTERACTIONS

At present, four fundamental interactions are distinguished. The properties of each are summarized in the **table**.

Properties of the four fundamental interactions

Interaction	Range	Exchanged quanta
Gravitational	Long-range	Gravitons (g)
Electromagnetic	Long-range	Photons (γ)
Weak nuclear	Short-range $\approx 10^{-18}$ m	W^+, Z^0, W^-
Strong nuclear	Short-range $\approx 10^{-15}$ m	Gluons (G)

Gravitational interaction. This interaction manifests itself as a long-range force of attraction between all elementary particles. The force law between two particles of masses m_1 and m_2 separated by a distance r is well approximated by the newtonian expression $G_N(m_1 m_2/r^2)$, where G_N is the newtonian constant, equal to $6.6720 \pm 0.0041 \times 10^{-11}$ m$^3 \cdot$ kg$^{-1} \cdot$ s^{-2}. The dimensionless quantity $(G_N m_e m_p)/\hbar c$ is usually taken as the constant characterizing the gravitational interaction, where m_e and m_p are the electron and proton masses, $2\pi\hbar$ is Planck's constant, and c is the velocity of light.

Electromagnetic interaction. This interaction is responsible for the long-range force of repulsion of like, and attraction of unlike, electric charges. The dimensionless quantity characterizing the strength of electromagnetic interaction is the fine-structure constant, given by Eq. (1) in

$$\alpha = e^2/4\pi\epsilon_0 \hbar c = 1/(137.03604 \pm 0.00011) \tag{1}$$

SI units, where e is the electron charge and ϵ_0 is the permittivity of empty space. At comparable distances, the ratio of gravitational to electromagnetic interactions (as determined by the strength of respective forces between an electron and a proton) is given by the quantity $4\pi\epsilon_0 G_N m_e m_p/e^2$, which is approximately 4×10^{-37}.

In modern quantum field theory, the electromagnetic interaction and the forces of attraction or repulsion between charged particles are pictured as arising secondarily as a consequence of the primary process of emission of one or more photons (particles or quanta of light) emitted by an accelerating electric charge (in accordance with Maxwell's equations) and the subsequent reabsorption of these quanta by a second charged particle. The space-time diagram (introduced by R. P. Feynman) for one photon exchange is shown in **Fig. 1**. A similar picture may also be valid for the gravitational interaction (in accordance with the quantum version of A. Einstein's

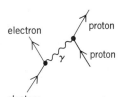

Fig. 1. Feynman diagram of electromagnetic interaction between an electron and a proton.

gravitational equations), but with exchanges of zero-rest-mass gravitons (g) rather than zero-rest-mass photons. (The existence of the graviton, however, has not yet been experimentally demonstrated.) SEE FEYNMAN DIAGRAM.

In accordance with this picture, the electromagnetic interaction (to one photon exchange approximation) is usually represented by reaction (2), where γ is the photon, emitted by the elec-

$$e + P \rightarrow (e + \gamma) + P \rightarrow e + (P + \gamma) \rightarrow e + P \qquad (2)$$

tron and reabsorbed by the proton. For this interaction, and also for the gravitational interaction represented by reaction (3), the nature of the participating particles (electron e and proton P) is

$$e + P \rightarrow (e + g) + P \rightarrow e + (P + g) \rightarrow e + P \qquad (3)$$

the same, before and after the interaction, and the exchanged quanta (γ or g) are electrically neutral. SEE GRAVITON; PHOTON; QUANTUM ELECTRODYNAMICS; QUANTUM FIELD THEORY.

Weak nuclear interactions. The third fundamental interaction is the weak nuclear interaction, which is responsible for the decay of a neutron into a proton, an electron, and an antineutrino. Its characteristic strength for low-energy phenomena is measured by the Fermi constant G_F, which is equal to $1.026 \times 10^{-5} m_p^{-2} \hbar^3/c$. Unlike electromagnetism and gravitation, weak interactions are short-range, with a force law of the type $e^{-M_W c r/\hbar}$, the range of the force ($\hbar/M_W c$) being of the order of 10^{-18} m. Until 1973, the only known weak interactions were those which changed the nature of the interacting particles (unlike electromagnetism and gravity). For example, consider reactions (4), where P is the proton, N is the neutron, μ^- is the negatively charged

$$P + e^- \xrightarrow{\text{weak}} N + \nu_e \quad \text{(equivalent to } \beta \text{ decay of the neutron } N \rightarrow P + e^- + \bar{\nu}_e\text{)} \qquad (4a)$$

$$P + \mu^- \xrightarrow{\text{weak}} N + \nu_\mu \quad \text{(muon capture by a proton with the emission of a neutrino)} \qquad (4b)$$

$$\mu^- + \nu_e \xrightarrow{\text{weak}} \nu_\mu + e^- \quad \text{(equivalent to muon decay } \mu^- \rightarrow e^- + \bar{\nu}_e + \nu_\mu\text{)} \qquad (4c)$$

muon, ν_e and ν_μ are the electronic and muonic neutrinos, and $\bar{\nu}_e$ and $\bar{\nu}_\mu$ are the corresponding antineutrinos. In reaction (4a), the weak interaction transforms a proton into a neutron and at the same time an electron into a neutrino.

An important question was finally answered in 1983: Is the weak interaction similar to electromagnetism in being mediated primarily by intermediate objects, the W^+ and W^- particles. If this is the case, then reactions (4a) and (4c), for example, would in detail be represented as reactions (5a) and (5b). (The superscript on each particle gives its electrical charge ($+$, 0, $-$) in

$$P^+ + e^- \rightarrow (N^0 + W^+) + e^- \rightarrow N^0 + (W^+ + e^-) \rightarrow N^0 + \nu^0 \qquad (5a)$$

$$\mu^- + \nu_e^0 \rightarrow (\nu_\mu^0 + W^-) + \nu_e^0 \rightarrow \nu_\mu^0 + (W^- + \nu_e^0) \rightarrow \nu_\mu^0 + e^- \qquad (5b)$$

units of the proton's charge.) The experimental answer (discovered at the CERN laboratory at Geneva) is that W^+ and W^- do exist, with a mass m_w of 80.9 ± 2.0 GeV/c^2. Each carries a spin of magnitude $\hbar$ just as does the photon (γ). The mass of these particles gives the range ($\hbar/m_W c = 10^{-18}$ m) of the weak interaction, and is also related to its strength G_F, as discussed below. SEE INTERMEDIATE VECTOR BOSON.

Another crucial discovery in weak interaction physics was the neutral current phenomenon in 1973, that is, the discovery of new types of weak interactions where (as in the case of electromagnetism or gravity) the nature of the interacting particles is not changed during the interaction, as in reactions (6).

$$\nu_\mu + e^- \xrightarrow{\text{weak}} \nu_\mu + e^- \qquad (6a) \qquad \qquad \nu_\mu + P \xrightarrow{\text{weak}} \nu_\mu + P \qquad (6b)$$

$$\nu_e + N \xrightarrow{\text{weak}} \nu_e + N \qquad (6c) \qquad \qquad e^- + P \xrightarrow{\text{weak}} e^- + P \qquad (6d)$$

The 1983 experiments at CERN also gave evidence for the existence of an intermediate particle z^0 which is believed to mediate such reactions. Thus reaction (6a), expressed in detail, is reaction (7). The mass m_Z of the z^0 has been found to be 93.0 ± 2.0 GeV/c^2. The magnitudes of

$$\nu_\mu + e^- \rightarrow (\nu_\mu + z^0) + e^- \rightarrow \nu_\mu + (z^0 + e^-) \rightarrow \nu_\mu + e^- \qquad (7)$$

the w^+, w^-, and z^0 masses had been predicted by the unified theory of electromagnetic and weak interactions (the electroweak interaction, discussed below), 16 years before the experiments which discovered them.

In contrast to gravitation, electromagnetism, and strong nuclear interactions, weak interactions violate left-right and particle-antiparticle symmetries. SEE PARITY; SYMMETRY LAWS; WEAK NUCLEAR INTERACTIONS.

Strong nuclear interaction. The fourth fundamental interaction is the strong nuclear interaction between protons and neutrons, which resembles the weak nuclear interaction in being short-range, although the range is of the order of 10^{-15} m rather than 10^{-18} m. Within this range of distances the strong force overshadows all other forces between protons and neutrons, with a characteristic strength parameter of the order of unity (compared with the electromagnetic strength parameter $\alpha \approx 1/137$).

Protons and neutrons are themselves believed to be made up of yet more fundamental entities, the up (u) and down (d) quarks ($P = uud$, $N = udd$). Each quark is assumed to be endowed with one of three color quantum numbers [conventionally labeled red (r), yellow (y), and blue (b)]. The strong nuclear force can be pictured as ultimately arising through an exchange of zero rest-mass color-carrying quanta of spin $\hbar$ called gluons (G) [analogous to photons in electromagnetism], which are exchanged between quarks (contained inside protons and neutrons), as in reaction (8). Since neutrinos, electrons, and muons (the so-called leptons) do not contain quarks,

$$\begin{aligned} \text{Quark} + \text{quark} &\rightarrow (\text{quark} + \text{gluon}) + \text{quark} \\ &\rightarrow \text{quark} + (\text{gluon} + \text{quark}) \\ &\rightarrow \text{quark} + \text{quark} \end{aligned} \quad (8)$$

their interactions among themselves or with protons and neutrinos do not exhibit the strong nuclear force. There is indirect experimental evidence for the existence of the gluons and of their spin being $\hbar$. SEE COLOR; GLUONS; LEPTON; QUANTUM CHROMODYNAMICS; QUARKS; STRONG NUCLEAR INTERACTIONS.

Gauge interactions. Three of the four fundamental interactions (electromagnetic, weak nuclear, and strong nuclear) appear to be mediated by intermediate quanta (photons γ; W^+, Z^0, and W^-; and gluons G, respectively), each carrying spin of magnitude $\hbar$. This is characteristic of the gauge interactions, whose general theory was given by H. Weyl, C. N. Yang, R. Mills, and R. Shaw. This class of interactions is further characterized by the fact that the force between any two particles (produced by the mediation of an intermediate gauge particle) is universal in the sense that its strength is (essentially) proportional to the product of the intrinsic charges (electric, or weak-nuclear, or strong-color) carried by the two interacting particles concerned.

The fourth interaction (the gravitational) can also be considered as a gauge interaction, with the intrinsic charge in this case being the mass; the gravitational force between any two particles is proportional to the product of their masses. The only difference between gravitation and the other three interactions is that the gravitational gauge quantum (the graviton) carries spin $2\hbar$ rather than $\hbar$. As discussed below, it is an open question whether all fundamental interactions are gauge interactions. SEE GAUGE THEORY.

UNIFICATION OF INTERACTIONS

Ever since the discovery and clear classification of these four interactions, particle physicists have attempted to unify these interactions as aspects of one basic interaction between all matter. The work of M. Faraday and J. C. Maxwell in the nineteenth century, which united the distinct forces of electricity and magnetism as aspects of a single interaction (the gauge interaction of electromagnetism), has served as a model for such unification ideas.

Gravitation and electromagnetism. The first attempt in this direction was made by Einstein who, having succeeded in understanding gravitation as a manifestation of the curvature of space-time, tried to comprehend electromagnetism as another geometrical manifestation of the properties of space-time, thus achieving a unification between these forces. In this attempt, to which he devoted all his later years, he is considered to have failed.

Electroweak interaction. A unification of weak and electromagnetic interactions, employing the gauge ideas discussed above, was suggested by S. Glashow and by A. Salam and J. C. Ward in 1959. This followed a parallel between these two interactions, pointed out by J. S. Schwinger in 1957. Assuming that (the then known) weak interactions (4) were mediated by exchanges of (the then hypothetical) W^+ and W^- particles, it could be shown from the empirical properties of weak interaction phenomena, that if the W's existed, they must carry an intrinsic spin of magnitude $\hbar$, just as does the photon, the gauge quantum of electromagnetism. If a bold unifying assumption was made that this magnitude of spins $\hbar$ for W^+, W^-, and the photon γ connotes a gauge character for a unified electroweak interaction, and that the intrinsic coupling strength of weak interactions is universally the same as that for electromagnetism (that is, $\alpha = 1/137$), then it could be shown that the masses of the W^+ and W^- particles must be in excess of the quantity given in Eq. (9).

$$\sqrt{\pi\alpha\hbar^3/\sqrt{2}G_F c} = 37.4 \text{ GeV}/c^2 \quad (9)$$

Following this initial attempt, Glashow (and independently later Salam and Ward) noted that such a unification hypothesis is incomplete, inasmuch as electromagnetism is a left-right symmetry-preserving interaction, in contrast to the weak interaction, which violates this symmetry. A gauge unification of such disparate interactions could be effected only if, additionally, new weak interactions represented by reactions (5) are also postulated to exist. Equivalently, there must exist a new electrically neutral intermediate weak-quantum Z^0 besides the (hypothetical) W^+ and W^-.

Spontaneous breaking and renormalization. There were two major problems with this unified electroweak gauge theory considered as a fundamental theory. Yang and Mills had shown that masslessness of gauge quanta is the hallmark of unbroken gauge theories. The origin of the masses of the weak interaction quanta W^+, W^-, and Z^0 (or equivalently the short-range of weak interactions), as contrasted with the masslessness of the photon (or equivalently the long-range character of electromagnetism), therefore required explanation. The second problem concerned the possibility of reliably calculating higher-order quantum effects with the new unified electroweak theory, on the lines of similar calculations for the "renormalized" theory of electromagnetism elaborated by S. Tomonaga, Schwinger, Feynman, and F. J. Dyson around 1949. The first problem was solved by S. Weinberg and Salam and the second by G. t'Hooft and by B. W. Lee and J. Zinn-Justin. SEE RENORMALIZATION.

Weinberg and Salam considered the possibility of the electroweak interaction being a "spontaneously broken" gauge theory. By introducing an additional self-interacting Higgs-Englert-Brout-Kibble particle into the theory, they were able to show that the W^+, W^-, and Z^0 would acquire well-defined masses through the so-called Higgs mechanism, these masses being given by Eqs. (10), where 37.4 GeV/c^2 is the combination of constants given by Eq. (9). Here θ_W is a

$$m_W = \frac{37.4 \text{ GeV}/c^2}{\sin \theta_W} \qquad m_Z = \frac{37.4 \text{ GeV}/c^2}{\sin \theta_W \cos \theta_W} \quad (10)$$

weak mixing parameter for electromagnetism and weak interactions. The constant $\sin^2 \theta_W$ can be determined from experiments which give the ratios of cross-sections of Z^0-mediated reactions (6) to the W^+ and W^--mediated reactions (4). The best available value, calculated from all low-energy experiments, is given by Eq. (11). SEE SYMMETRY BREAKING.

$$\sin^2 \theta_W = 0.218 \pm .010 \quad (11)$$

The predicted theoretical mass values of the W and Z particles deduced by substituting Eq. (11) into Eqs. (10) are in good accord with the experimental values found by the CERN 1983 experiments. The existence of the W and Z particles and this accord with regard to mass values give support to the basic correctness of the electroweak unification ideas, as well as to the gauge character of the electroweak interaction.

Prior to this direct evidence, indirect evidence for the existence of the characteristic reactions (6), predicted by the electroweak theory, had existed since 1973. The most crucial experiment in this respect, carried out at Stanford during 1978, exhibited interference effects between the photon (γ) and the Z^0 particle in the scattering of polarized electrons from protons (**Fig. 2**). These effects were established through observing the characteristic weak left-right symmetry violation in the reaction $e^- + P \rightarrow e^- + P$. The findings of this experiment provided indirect but quantitative confirmation of the predictions of the electroweak theory.

Higgs particle. The Weinberg-Salam electroweak theory contains an additional neutral particle (the Higgs) but does not predict its mass. A search for this particle will be seriously

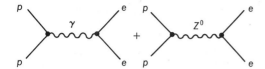

Fig. 2. Feynman diagrams corresponding to the Stanford (1978) experiment, which showed the interference effects between mediating photons and Z^0 exchanges in electron-proton scattering.

undertaken when the electron-positron accelerators LEP (Large Electron-Positron storage ring) at CERN and SLC (Stanford Linear Collider) come into commission. Meanwhile, there has been theoretical speculation on whether the Higgs particle is a composite object held together by a new, fundamental type of very strong interaction, the so-called technicolor interaction. This suggestion has the possible merit of eliminating the need for introducing a fundamental nongauge (self-) interaction among Higgs particles. SEE PARTICLE ACCELERATOR; WEINBERG-SALAM MODEL.

Electronuclear interaction. The gauge unification of weak and electromagnetic interactions, which started with the observation that the relevant mediating quanta (W^+, W^-, Z^0, and γ) possesses intrinsic spin $\hbar$, can be carried further to include strong nuclear interactions as well, if these strong interactions are also mediated through quanta (gluons) carrying spin $\hbar$. Since indirect experimental evidence already supports gluonic spin being $\hbar$, a complete gauge unification of all three forces (electromagnetic, weak-nuclear, and strong-nuclear) into a single electronuclear interaction seems plausible. Such a (so-called grand) unification necessarily means that the distinction between quarks on the one hand and neutrinos, electrons, and muons (leptons) on the other, must disappear at sufficiently high energies, with all interactions (weak, electromagnetic, and strong) clearly manifesting themselves then as facets of one universal gauge force with a primitive universal strength equal to $\alpha/\sin^2 \theta_w$. The fact that at low energies presently available, these interactions exhibit vastly different effective strengths is ascribed to differing renormalizations due to successive spontaneous symmetry breakings. A startling consequence of the eventual universality and the disappearance of distinction between quarks and leptons is the possibility, first discussed by J. C. Pati and Salam within their electronuclear model, of protons transforming into leptons and pions. Contrary to the older view, protons would therefore decay into leptons and pions and not live forever. A somewhat different model, elaborated later by H. Georgi and Glashow, predicts a lifetime of the order of 10^{29} years for the proton P, with decay principally through the mode $P \rightarrow e^+ + \pi^0$, where π^0 is the neutral pion and e^+ is the positron. Experiments carried out during 1983 to search for this mode of proton decay gave negative evidence for protons decaying at this rate, although other types of decay modes may have been observed in other experiments. The discovery of proton instability (with decays into leptons or antileptons) would be an epic discovery and a direct confirmation of the electronuclear (grand) unification. SEE GRAND UNIFICATION THEORIES; PROTON.

Consequences of symmetry breaking. Spontaneous symmetry breaking of gauge interactions has the characteristic that symmetry breaking is a phase phenomenon and disappears in a high-temperature environment. This implies that, at temperatures T in excess of 10^{15} K (T greater than $m_Z c^2/k$, where k is the Boltzmann constant), that is, up until 10^{-12} s after the outset of the big bang, there was no spontaneous breaking of the symmetry of electroweak interactions, and the W and the Z particles were massless, like the photons and the gluons. The onset of such phase transitions plays a crucial role in modern cosmological theories of the early universe, resolving some old dilemmas. For example, proton decay, and left-right and particle-antiparticle symmetry violations, provide a natural explanation for the fact that the present universe contains a preponderance of protons and neutrons rather than of their antiparticles. However, the existence of such phase transitions also poses some new dilemmas, such as the prediction of the existence of heavy magnetic monopoles (in the early universe), with abundances surviving into the present epoch, for which there is no experimental evidence. SEE ANTIMATTER; MAGNETIC MONOPOLES.

Prospects for including gravity. Research in unification theories of fundamental interactions is now concerned with uniting the gauge theories of gravity and of the electronuclear interactions. One promising approach is the extension of space-time to more than four dimensions, following ideas developed by T. Kaluza and O. Klein in the 1920s. Remarkably, the formal expression for Einstein's gravitational interaction in a space-time of dimensions higher than four, is equivalent to the standard Einstein theory of spin-$2\hbar$ gravitons in four dimensions plus a Yang-Mills theory of spin-$\hbar$ particles (that is, a theory describing the electronuclear type of gauge interactions) when the extra dimensions are contracted down to less than 10^{-35} m. No realistic model of such a compactified unified theory has emerged, though Einstein-like supersymmetric theories in 10-space and 1-time (a total of 11 dimensions) are the favored candidates. (Supersymmetry is the principle which treats gauge and Higgs particles on a par with quarks and leptons.) SEE SUPERGRAVITY; SUPERSYMMETRY.

Bibliography. M. K. Gaillard, Toward a unified picture of elementary particle interactions, *Amer. Sci.*, 70:506–514, 1982; H. Georgi and S. L. Glashow, Unified theory of elementary particle forces, *Phys. Today*, 33(9):30–39, September 1980; S. Glashow et al., Nobel Lectures in Physics, *Rev. Mod. Phys.*, 52:515–543, 1980; F. J. Hasert et al., Search for elastic muon-neutrino electron scattering, *Phys. Lett.*, 46R:121–124, 1973; J. C. Pati and A. Salam, Is baryon number conserved?, *Phys. Rev. Lett.*, 31:661–664, 1973; C. Prescott et al., Parity nonconservation in inelastic electron scattering, *Phys. Lett.*, 77B:347–352, 1978; C. Quigg, Elementary particles and forces, *Sci. Amer.*, 252(4):84–95, April 1985; C. Rubbia, *The Physics of the Proton-Anti-Proton Collider: Report to the International Europhysics Conference on High Energy Physics*, Rutherford Appleton Laboratory, Didcot (England), 1983.

QUANTUM FIELD THEORY
J. D. BJORKEN

Quantum theory of physical systems possessing an infinite number of degrees of freedom, such as the electromagnetic field, gravitational field, or wave fields in a medium. Its major applications lie in the attempted description of fundamental particles and their associated wave fields under circumstances in which both the effects of quantum mechanics and of special relativity are important. Quantum field theory is also used in nonrelativistic quantum theory for systems of many particles such as electrons in a metal, or for sound waves in liquids and solids. Here the discussion is mainly confined to the use of quantum field theory as a description of the properties of fundamental particles.

Yukawa force. The classical electromagnetic field is a dynamical system which must be subjected to the rules of quantum mechanics. The consequence of this requirement is the emergence of a particle (photon) interpretation from the classical Maxwell equations, manifesting the particle-wave duality of quantum mechanics in its most perfect form. The experimental success of this theory for electrodynamics has led to its imitation for the more complex interactions of other fundamental particles. For example, in 1935 H. Yukawa tried to interpret the strong force between protons and neutrons responsible for the existence of nuclei in the same way. This force is only effective when the protons and neutrons are within a distance, 10^{-13} cm, of each other. Yukawa supposed the nuclear force was due to the exchange of a particle, called a meson, between the nucleons, just as the electrostatic force between charged particles is interpreted in terms of exchange of a photon (see **illus**.). SEE ELEMENTARY PARTICLE; MESON.

The short range of the force can be understood as a consequence of a finite rest mass m for the meson. When the nucleon emits the meson, the energy cost is approximately mc^2. According to the uncertainty principle, if the energy of the system is observed to the accuracy ΔE, the

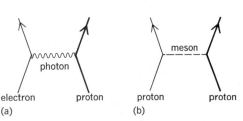

Analogy between (*a*) electromagnetic force via photon exchange and (*b*) nuclear force via meson exchange.

observation must be made over an interval of time Δt longer than $h/\Delta E$. Thus, an energy fluctuation of a magnitude of $\sim mc^2$ can only take place over a time interval $\Delta t < h/mc^2$, consistent with the uncertainty principle. In this time interval, the meson can travel no further than $c\Delta t < h/mc^2$, a distance identified with the range of the force. In this way Yukawa was led to predict the mass of the meson ($m\ 270 \simeq m_{\text{electron}}$) intermediate between the electron and proton mass. To describe this Yukawa meson, now known as π-meson, one writes an analog of the Maxwell field equations, described below, and then passes to the quantum description in a way similar to that described for the electromagnetic field. A satisfactory particle description is thereby obtained, again embodying the characteristic wave-particle duality of the quantum theory.

However, although the theory of isolated mesons is satisfactory, when interactions such as the couplings to the nucleons are introduced, chaos emerges. The field equations have not been solved. There are grave doubts as to whether solutions exist. There is, aside from the electromagnetic interactions, no small parameter in terms of which there might exist a series expansion. Finally, it is not even known what equation to write. For by now there are dozens of strongly interacting particles, none of which appear any more fundamental than any other. If field variables of all these particles appear in the basic dynamical equations, they are complicated indeed.

Faced with this dilemma, the theoretical physicist has concentrated his efforts on finding general properties of any relativistic quantum field theory, rather than detailed consequences of a specific set of equations. These efforts have been fruitful in three different areas: (1) the quantum theory of symmetries and their connection with conservation laws such as conservation of energy, momentum, and angular momentum; (2) the study of certain exact properties; in particular the mathematical property of analyticity of the functions used to describe collision processes; (3) the use of the diagrammatic method of R. P. Feynman, developed for quantum electrodynamics and appropriately modified for the study of more general interactions, to interpret qualitatively many of the features of collision processes between relativistic particles.

Axiomatic quantum field theory. For the study of general and presumably exact properties of a quantum field theory, one begins by assuming that it makes sense and that it satisfies the following axioms: (i) The solutions are consistent with the theory of special relativity and with basic quantum mechanical principles. (ii) The vacuum state (empty space) exists and is the state of lowest energy. (iii) The field variables are local; if the spacetime coordinates of two field variables cannot be connected by a signal traveling with a velocity less than or equal to the velocity of light, the fields are dynamically independent. This is the principle of microscopic causality. One also assumes that the dynamical equations do not depend explicitly upon absolute position in space or upon the orientation of the coordinates. These requirements lead, via the field theory formalism, to the existence of conservation laws of energy, momentum, and angular momentum, and are a deep and important result of the fundamentals of quantum field theory. Other more subtle symmetry properties are found as a consequence of the third, locality assumption. It is found, for example, that to every particle there must exist an antiparticle with opposite charge but precisely the same mass (in some cases, the particle and antiparticle can be identical). This prediction has been well verified experimentally, and the equality of mass of particle and antiparticle has been verified in one case to 1 part in 10^{17}. Furthermore, aggregates of identical particles of integer spin angular momentum 0, h, $2h$, . . . must have wave functions symmetric under interchange of any two particles, while aggregates of identical half-integer spin ($h/2$, $3h/2$, . . .) particles must have antisymmetric wave functions. Finally, the simultaneous operation (TCP) of space-inversion (P), time reversal (T), and replacement of particle by antiparticle (C) should be a symmetry of nature; that is, all dynamical equations are left unchanged in form by the TCP operation.

All these results are a consequence of "axioms" (i), (ii), and (iii), and form a cornerstone of axiomatic quantum field theory. In addition to these results, study of the analyticity properties of the functions describing collision processes has led to "dispersion relations," relations similar to the Kramers-Kronig relation in optics which relate the real part of the index of refraction at one frequency to an integral over the imaginary part at all frequencies. In the case of quantum field theory, the corresponding index of refraction is that associated with an elementary particle such as a π-meson propagating through matter and interacting with it by means of the strong nuclear force. The main input required to obtain these relations, both in the case of optics and the present case, is axiom (iii) of microscopic causality, that no signal is propagated with a velocity greater than that of light. SEE DISPERSION RELATIONS.

Feynman diagrams. For the special field theory of quantum electrodynamics, Feynman diagrams are extremely useful because the parameter e coupling the radiation field to matter is small. This means only the simplest diagrams are in practice important. For the strong interactions of elementary particles responsible for the nuclear force, there are no known small parameters, and complicated diagrams are just as important as simple ones, if indeed the concept may even be used under these circumstances. However, in practice it is found that for qualitative interpretation of many phenomena simple diagrams may be used to provide a partial understanding of the underlying dynamics; they remain a tool for the "theoretical laboratory," but their ultimate role in a correct theory is not known. *See* F*eynman diagram*.

Mathematical structure. As an illustration of the mathematical structure, the description of the Yukawa π-meson, which has zero spin angular momentum, is given below.

The first step in building a field theory for this meson is to construct a classical field equation analogous to the Maxwell equations. This is done by writing the relativistic energy-momentum relation $E^2 = \mathbf{p}^2 c^2 + m^2 c^4$, making the quantum-mechanical replacement $E \to i\hbar\,(\partial/\partial t)$, $\mathbf{p} \to -i\hbar\nabla$. One obtains in Eq. (1) the relativistic version of the Schrödinger equation, but treated as a

$$-\hbar^2 \frac{\partial^2}{\partial t^2}\, \phi(\mathbf{x},t) = (-\hbar^2 c^2 \nabla^2 + m^2 c^4)\, \phi(\mathbf{x},t) \tag{1}$$

classical equation. Upon Fourier transformation, one finds that Eq. (2) holds. For each Fourier

$$-\hbar^2 \frac{\partial^2}{\partial t^2}\, \widetilde{\phi}(\mathbf{p},t) = (\mathbf{p}^2 c^2 + m^2 c^4)\, \widetilde{\phi}(\mathbf{p},t) \tag{2}$$

component, this is the equation for a classical harmonic oscillator.

In passing to quantum mechanics, each $\widetilde{\phi}(\mathbf{p},t)$ becomes an operator, just as coordinate and momentum in the nonrelativistic quantum mechanics, which acts on an abstract wave function, or "state vector." Because the dynamics is that of an infinite assembly of independent harmonic oscillators, one for each momentum $\mathbf{p}$, the stationary states are labeled by the oscillator quantum numbers $n(\mathbf{p})$ for each and every $\mathbf{p}$, with the energy of such a state given by Eq. (3). For example,

$$E = \sum_{\mathbf{p}} E n(\mathbf{p}) = \sum_{\mathbf{p}} [n(\mathbf{p}) + 1/2]\sqrt{\mathbf{p}^2 c^2 + m^2 c^4} \tag{3}$$

the state with $n(\mathbf{p}) \neq 0$ and $n(\mathbf{p}') = 0$ for $\mathbf{p}' \neq \mathbf{p}$ represents a state of n mesons of momentum $|\mathbf{p}|$. The field operator $\widetilde{\phi}(\mathbf{p},t)$, the oscillator "coordinate," acting on a state containing $n(\mathbf{p})$ mesons changes $n(\mathbf{p})$ by ± 1, as in the nonrelativistic oscillator. It therefore "creates" or "destroys" single mesons of momentum $\mathbf{p}$. In this way the particle interpretation emerges from the quantum field formalism.

For the electron, as described by the wave equation of P. A. M. Dirac, a similar procedure may be carried out. The equation is considered a classical field equation, a Fourier transformation performed, and quantum conditions imposed upon the Fourier coefficients $\widetilde{\psi}(\mathbf{p},t)$, again treated as operators acting upon an abstract state vector or wave function. The only difference, aside from more complicated algebra, is that while meson operators for different momenta are taken to commute, as in Eq. (4) they must for electrons be taken to anticommute as in Eq. (5), to obtain a

$$\widetilde{\phi}(\mathbf{p},t)\widetilde{\phi}(\mathbf{p}'t) = \widetilde{\phi}(\mathbf{p}'t)\widetilde{\phi}(\mathbf{p},t) \quad (4) \qquad \widetilde{\psi}(\mathbf{p},t)\widetilde{\psi}(\mathbf{p}'t) = -\widetilde{\psi}(\mathbf{p}'t)\widetilde{\psi}(\mathbf{p},t) \quad (5)$$

satisfactory physical interpretation. This change of sign is related to the antisymmetry of electron many-particle wave functions, as compared with the symmetry of meson wave functions alluded to previously.

When interactions are introduced such as coupling of mesons to nucleons, the meson theory is equivalent to an infinite assembly of coupled nonlinear oscillators, treated quantum-mechanically. It is no wonder that progress has been slow in studying such a complex dynamical system.

Nonrelativistic applications. The Schrödinger equation, as well as the Dirac equation, can also be treated as a classical field equation and subjected to quantization. When this is done, a many-particle nonrelativistic theory of electrons is obtained. This theory, along with the Feynman diagram techniques, has been applied successfully in the study of properties of electrons in solids, notably in the theory of collective oscillations (plasma oscillations) of electrons in metals and in the presently accepted theory of superconductivity. Along with quantum electrodynamics, these studies are thus far the most successful applications of quantum field theory to physical phenomena. SEE SYMMETRY LAWS.

Bibliography. J. Bjorken and S. Drell, *Relativistic Quantum Fields*, 1965; N. Bogoliubov and D. Shirkov, *Introduction to the Theory of Quantized Fields*, 3d ed., 1980; L. D. Fadeev and A. A. Slavonov, *Gauge Fields: Introduction to Quantum Theory*, 1981; C. Itzykson and J. B. Zuber, *Quantum Field Theory*, 1980; F. Mandl and G. Shaw, *Quantum Field Theory*, 1985; L. H. Ryder, *Quantum Field Theory*, 1985; D. Thouless, *The Quantum Mechanics of Many-Body Systems*, 2d ed., 1972.

FEYNMAN DIAGRAM
PETER MOHR

A pictorial representation of elementary particles and their interactions. The diagrams show paths of particles in space and time as lines, and interactions between particles as points where the lines meet.

These diagrams were introduced by R. P. Feynman in 1949 in the context of quantum electrodynamics, the quantum field theory of electromagnetic interactions of charged particles. The **illustration** shows Feynman diagrams for electron-electron scattering. In each diagram, the straight lines represent space-time trajectories of noninteracting electrons, and the wavy lines represent photons, particles that transmit the electromagnetic interaction. External lines at the bottom of each diagram represent incoming particles (before the interactions), and lines at the top, outgoing particles (after the interactions). Interactions between photons and electrons occur at the vertices where photon lines meet electron lines. SEE ELECTRON; PHOTON.

Each Feynman diagram corresponds to the probability amplitude for the process depicted in the diagram. This amplitude can be calculated from the structure of the diagram according to specific rules. It is a product of wave functions for external particles, a propagation function for each internal electron or photon line, and a factor for each vertex that is proportional to the

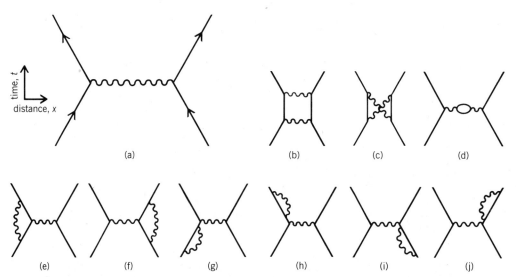

Feynman diagrams for electron-electron (Møller) scattering: (a) second-order diagram (two-vertices); (b–j) fourth-order diagrams.

strength of the interaction, summed over all possible space-time locations of the vertices. The propagation function, corresponding to a line joining two vertices, is the amplitude for the probability that a particle starting at one vertex will arrive at the other.

The set of all distinct Feynman diagrams with the same incoming and outgoing lines corresponds to the perturbation expansion of a matrix element of the scattering matrix in field theory. This correspondence can be used to formulate the rules for writing the amplitude associated with a particular diagram. The perturbation expansion and the associated Feynman diagrams are useful to the extent that the strength of the interaction is small, so that the lowest-order terms, or diagrams with the fewest vertices, give the main contribution to the matrix element. SEE SCATTERING MATRIX.

Since their introduction in quantum electrodynamics, Feynman diagrams have been widely applied in other field theories. They are employed in studies of electroweak interactions, certain situations in quantum chromodynamics, and in many-body theory in atomic, nuclear, plasma, and condensed matter physics. SEE ELEMENTARY PARTICLE; FUNDAMENTAL INTERACTIONS; QUANTUM CHROMODYNAMICS; QUANTUM ELECTRODYNAMICS; QUANTUM FIELD THEORY; WEAK NUCLEAR INTERACTIONS; WEINBERG-SALAM MODEL.

Bibliography. J. Bjorken and S. Drell, *Relativistic Quantum Fields*, 1965; J. Bjorken and S. Drell, *Relativistic Quantum Mechanics*, 1964; C. Itzykson and J. Zuber, *Quantum Field Theory*, 1980; F. Mandl and G. Shaw, *Quantum Field Theory*, 1985; J. Schwinger, *Selected Papers on Quantum Electrodynamics*, 1958; D. Thouless, *The Quantum Mechanics of Many-Body Systems*, 2d ed., 1972.

QUANTUM ELECTRODYNAMICS
J. D. BJORKEN AND HOWARD GROTCH
H. Grotch wrote the sections Feynman Diagrams, Renormalization, and Relation to Other Theories.

The study of the properties of electromagnetic radiation and its interaction with electrically charged matter, in particular, with atoms and their constituent electrons. The fundamental equations governing quantum electrodynamics actually are believed to encompass all of atomic physics, chemistry, properties of bulk matter, and classical electromagnetic theory. Almost all the phenomena readily perceived by the senses are believed to be ultimately understandable in terms of the laws of quantum electrodynamics. In the scope of its application and implications, and in the simplicity of the underlying assumptions, quantum electrodynamics is rivaled only by gravitational theory. In practice, the term quantum electrodynamics refers to those phenomena specific to the quantum nature of electromagnetic radiation. This includes the study of emission and absorption of light by atoms and the basic interactions of light with electrons and other fundamental particles. SEE QUANTUM FIELD THEORY.

Quantum electrodynamics was formulated by P.A.M. Dirac, W. Heisenberg, and W. Pauli shortly after the foundations of quantum mechanics were laid down in 1925. This formulation gave a satisfactory interpretation of the wave-particle duality of light, in which it is possible for light to manifest itself both in ways appropriate to a particle (photon) description or a wavelike description, in accordance with the experimental foundations of quantum mechanics. About the same time, Dirac discovered an equation describing the motion of electrons which incorporated both the requirements of quantum theory and of relativity. However, introduction of electromagnetic interaction into Dirac's equation, while successfully describing the magnetic properties of the electron and the existence of its oppositely charged counterpart, or antiparticle, the positron, led to mathematical disasters which inhibited the further development of the theory until after World War II. Stimulated by experiments using microwave techniques developed during World War II, S. Tomonaga, R. P. Feynman, J. S. Schwinger, F. J. Dyson, and many others found a way to bypass, but not solve, these difficulties. Since that time the theory has been confronted by very sophisticated experimental tests and has passed them satisfactorily. The accuracy of the comparisons of theory and experiment is well illustrated by the measurements of the magnetic moment of the electron (the strength of the elementary magnet associated with an electron), which gives Eqs. (1) and (2), where e = electron charge, h = Planck's constant, m_e = electron mass, and c = velocity of light.

Theory: $$\mu = \frac{eh}{4\pi m_e c}\left(1.001\ 159\ 652\ 460 \atop \pm\ 127\right) \quad (1)$$

Experiment: $$\mu = \frac{eh}{4\pi m_e c}\left(1.001\ 159\ 652\ 200 \atop \pm\ 40\right) \quad (2)$$

General applications. The equations of quantum electrodynamics have found many applications in the study of the delicate details of the structure of atoms (the so-called fine-structure and hyperfine structure of atoms). In particular, in the hydrogen atom the frequency of a certain spectral line (for the $2S_{1/2} \to 2P_{1/2}$ transition) is predicted, in the absence of quantum electrodynamic effects, to be almost zero. The effects of the radiation field modify this prediction. This shift in frequency was accurately measured by W. E. Lamb and computed by H. A. Bethe and others, and the theory and experiment agree to within 50 parts per million. Several similar modifications in the frequency of spectral lines in hydrogen have been accurately computed and compared with experiment.

As mentioned above, the magnitude of the magnetic moment of the electron and μ-meson (an elementary particle with properties similar to the electron, except for a larger mass) is also predicted by quantum electrodynamics. In all cases, measurements agree with theory to very high accuracy.

Quantum electrodynamics also provides an account of collision processes of electrons, positrons, and photons in matter. For example, an electron in passing through matter is accelerated by the Coulomb fields of the atoms and radiates light. The light (photons) in turn can materialize into matter, provided its frequency is high enough; this effect in fact was predicted by the equations of quantum electrodynamics. The matter produced is almost always an electron and a positron. These in turn, in passing through matter, are accelerated and radiate more light, which produces more electrons and positrons. In this way an avalanche of electrons, positrons, and photons is created; this is called a cascade shower. The details of the evolution of such showers can be computed from the equations of quantum electrodynamics and they agree with experiment, even for enormous showers containing 10,000,000 electrons created by very energetic cosmic rays entering the Earth's atmosphere. The collision processes involved in the cascade showers can also be studied under more controlled conditions by using particle accelerators. In this way, detailed tests probing the structure of the theory have been made employing very energetic beams of photons, electrons, positrons, and μ-mesons from the accelerators. Again, the agreement with the theoretical predictions has been consistently excellent. SEE ELECTRON-POSITRON PAIR PRODUCTION; PARTICLE ACCELERATOR.

Free electromagnetic field. The simplest electrodynamic system is that of radiation in free space, described classically by the Maxwell equations. To construct a quantum theory, one may first visualize the classical radiation confined in a box (such as a microwave cavity) and consider separately the normal modes, that is, components oscillating with a well-defined frequency. If for each normal mode (in musical terminology, fundamental, first harmonic, and so on) the amplitude of the oscillating electromagnetic field is given, the complete field may be constructed by adding together the contribution from each mode. In this way the dynamics of the electromagnetic field is reduced to the dynamics of each normal-mode amplitude, which in turn undergoes sinusoidal oscillation in time, just like a mass hung from a spring. In quantum theory one finds that any oscillator (such as the mass on a spring) can be found only in a discrete set of quantum states, with energy $E_n = h\nu(n + \frac{1}{2})$, where h is Planck's constant and ν is the oscillator frequency. For the radiation field, the same picture is taken over. The state above represents, in this case, n photons each of energy $h\nu$ and associated with the normal mode of frequency ν. The energy in the field is changed in discrete amounts corresponding to transitions between different oscillator states. Such changes in the state of the radiation field are associated physically with the addition or removal of photons from the box. The state of no photons present (the vacuum) is characterized by all the radiation oscillators in their lowest state $n = 0$. However, the energy of such a state

$$E = \sum_\nu \tfrac{1}{2} h\nu$$

where the sum is taken over all normal modes and is not zero, because of the existence of quantum fluctuations associated with the uncertainty principle. These quantum fluctuations of the radiation field have observable consequences. As an example, one may consider the energy contained between two neutral parallel conducting plates of area A, separated by a distance d. As d is decreased, fewer normal modes can fit inside the plates, and thus the vacuum energy is decreased. This leads to an attractive force between the plates $F = hc\,A/360\,d^4$, which, although quite small, has been measured and agrees with theory. Another example of the existence of quantum fluctuations is the Lamb shift of the spectral lines in hydrogen, mentioned above.

Free electron field. Because the electron is the lightest of all charged particles, it is most easily accelerated, and it most easily emits radiation. Most experimental studies of quantum electrodynamics have dealt with the electromagnetic properties of electrons. The first requirement in a theory of the electron is to have an equation of motion which incorporates both the theory of special relativity and quantum theory. Such an equation was found by Dirac. The wave function, or electron field, replacing the two-component Schrödinger-Pauli field $\psi(x)$ for a nonrelativistic electron, now has four components. The two new components are associated with negative energy states. Although this appears at first to be unacceptable physically, Dirac found a way around this problem, and the negative energy states were eventually associated with the electron's antiparticle, the positron. Aside from the negative-energy problem, which in time has become its most striking success, the Dirac equation immediately accounted for many of the observed properties of the electron. It predicts correctly an electron spin angular momentum of $h/4\pi$, and when coupling to external electromagnetic potentials are introduced, the correct interaction of electron spin with magnetic field (magnetic moment) is predicted. The relativistic corrections to the frequencies of hydrogen spectral lines are also correctly given from the solutions of the Dirac equation for an electron moving in the proton Coulomb field.

Interaction between radiation and electrons. The classical Maxwell equations contain in a well-defined way the current of charged particles as the source of radiation. This current is constructed from the electron field and in turn determines the nature of coupling of electrons to the radiation field. The resulting equations, just as their classical counterparts, have resisted exact solutions. However, the dynamics of the coupled electron-photon system can be determined in terms of a power series expansion in a dimensionless parameter, the fine-structure constant $\alpha = 2\pi e^2/hc \sim 1/137$, which is fortunately small (for a reason not totally understood). An intuitive physical picture of the meaning of this expansion was given by Feynman, and will now be discussed.

Feynman diagrams. In quantum electrodynamics the basic interactions may be described by means of space-time diagrams referred to as Feynman diagrams or graphs. With the help of these pictures a given process can be readily visualized and a set of rules can be utilized to compute the likelihood that this process will occur. The graphs or diagrams tell how to write expressions called amplitudes, and when the sum of appropriate amplitudes has been evaluated its absolute square directly determines the probability of the process.

These diagrams contain a number of features that will be discussed here qualitatively. Generally, the graph contains several solid or wavy lines which originate in the distant past and which characterize incident particles which will subsequently interact. As a result of this interaction the particles are mutually deflected (scattered), and then other solid or wavy lines emerge and move toward the distant future. The solid lines always represent charged particles such as electrons or positrons, while the wavy lines depict photons. The particles entering the graph from the past or emerging toward the future are free particles, which means that there is a unique relationship between their energy and momentum. As an example, **Fig. 1** illustrates an electron and a photon entering the diagram, interacting by some mechanism within the shaded interaction region, and then emerging. This process is known as Compton scattering. The lines originating in the past or emerging to the future are referred to as external lines in order to distinguish them from the internal lines discussed below. A convention is used in which time proceeds upward from the bottom toward the top of the page.

In the quantum electrodynamics of electrons e^- and positrons e^+ (antiparticles of the electron) it is necessary to find a graphical way of distinguishing e^- and e^+. In Fig. 1 the arrow on the initial and final solid line is drawn in the forward direction, which signifies that these particles are electrons. If a positron is incident from the past, it would be indicated with an arrow directed toward the past. Likewise, a positron emerging from the graph would also have an arrow drawn toward the past. Thus **Fig. 2** would represent the process of electron-positron scattering, while

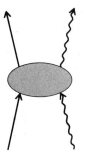

Fig. 1. Feynman diagram for Compton scattering.

Fig. 2. Feynman diagram for electron-positron scattering.

Fig. 3 is a process in which an incident electron and positron interact and subsequently disappear (annihilation) producing two photons in the final state.

Fundamental interaction. In order for something interesting to occur, namely a scattering or deflection, the solid and wavy lines representing the particles must somehow interact. In quantum electrodynamics the fundamental interaction, out of which all processes may be constructed, can be represented as in **Fig. 4**, where two solid and one wavy line meet or interact at a single space-time point x. The fundamental process of Fig. 4 cannot occur all by itself since it presents a situation in which an incident electron absorbs an incident photon to produce a final electron. This can be shown to violate the fundamental principle of energy-momentum conservation. Nevertheless, Fig. 4 does serve as a building block out of which more involved allowed processes can be constructed.

Perturbation theory. One of the most interesting and useful aspects of quantum electrodynamics is to be found in the statement that once the initial and final particles of a process have been selected, the most important or most probable diagrams are those which involve the fewest number of space-time interactions of the type shown in Fig. 4. That is, nature appears to select as most important the diagrams in which Fig. 4 occurs infrequently as part of a larger diagram. There is an important reason for this. The entire Feynman diagram procedure presents a method of solution of a complex problem by an approximation technique known as perturbation theory. Naturally, if an exact solution could be discovered, that would be preferable to perturbation theory, but thus far only the approximate methods of the diagrammatic approach have been successful. In this approach space-time points such as x are referred to as vertices and at each vertex energy and momentum must be conserved. Vertices are characterized by a certain strength or likelihood parameter, which in quantum electrodynamics is the electron charge e. If a graph has n vertices

Fig. 3. Feynman diagram for electron-positron annihilation.

Fig. 4. Electron-photon interaction vertex, representing the fundamental interaction out of which all processes may be constructed in quantum electrodynamics.

the mathematical expression (the amplitude) which corresponds to the graph will be proportional to e^n, and hence if n is large the amplitude for the process is as small as terms proportional to e^n. When such calculations are carried out, the fine-structure constant α is the expansion parameter, and, as a consequence of its smallness, complicated graphs are less important than simple ones. Therein lies the power and the success of perturbation theory.

Simple diagrams. For the process of Fig. 1, the simplest interaction which can occur is given by the graphs shown in **Fig. 5***a* and *b*. The difference is that in *a* the incident photon is

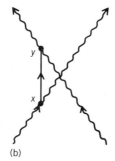
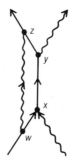

Fig. 5. Leading diagrams for Compton scattering. (*a*) Incident photon is absorbed at *x*; photon is emitted at *y*. (*b*) Incident photon is absorbed at *y*; photon is emitted at *x*.

Fig. 6. Feynman diagram that gives a radiative correction to Compton scattering.

absorbed at *x* while in *b* it is absorbed at *y*. Since each of these graphs has two space-time points at which interaction occurs, the amplitude has a strength proportional to e^2 or to α.

In Fig. 5 a new feature manifests itself in the appearance of a solid line connecting space-time points *x* and *y*. This is known as an internal line, and it represents an electron Green's function or propagator. This propagator provides a measure of the likelihood that an electron at space-time point *x* will traverse to space-time point *y* without any interactions at intermediate stages. Internal wavy lines also can occur in diagrams, and whenever they appear a photon Green's function or propagator occurs. For example, a photon propagator can be added to Fig. 5*a*, thus creating **Fig. 6**, a diagram which is proportional to α^2 and which contains one photon propagator and three electron propagators. This diagram gives one of the so-called radiative corrections to Compton scattering, and because of the extra two powers of *e* it should be less important than Fig. 5*a*. Although such radiative corrections may be less important, their contribution to a given process can be vital if the calculation is being carried out to considerable accuracy in order to make a comparison with precise experimental results.

Figure 6 also illustrates another feature of Feynman diagrams. There is a closed path connecting the space-time points *x*, *y*, *z*, and *w*. Whenever a single closed path, also called a closed loop, occurs, a complicated four-dimensional integration must be performed, and hence it is often arduous to evaluate the expression. The situation becomes even more difficult when there are multiple closed loops. Nevertheless, with the advent of high-speed computers, many integrals which were formidable several decades ago can now be worked out numerically with relative ease.

Renormalization. The appearance of closed loops creates additional problems beyond the technical ones of implementation of integration. The integrals that arise can be put in the form of integration of some four-momentum, and it often happens that some of these integrals diverge; that is, they are infinite. Three basic examples of where this occurs are illustrated in **Fig. 7**. These diagrams are generally a part of larger, more complex graphs, but wherever they appear an infinite integral is present. Figure 7*a* and *b* are respectively called the electron and photon self-energies, while Fig. 7*c* is referred to as a vertex correction.

In the early days of quantum electrodynamics these divergent integrals posed a considerable obstacle which had to be overcome in order to extract from the theory sensible conclusions.

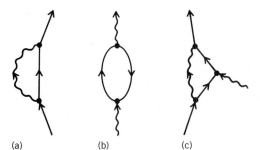

(a) (b) (c)

Fig. 7. Basic divergent closed loops. (a) Electron self-energy. (b) Photon self-energy. (c) Vertex correction.

The details of the procedure for rendering the results finite are known as the renormalization procedure. Only the basic idea of this procedure will be discussed.

When a theory such as quantum electrodynamics is constructed, there are certain basic parameters such as electron mass m_0 and electron charge e_0 which are a part of the basic description. In the absence of any interaction only m_0 appears, but when interaction is allowed to occur the interaction term is proportional to e_0. Now m_0 and e_0 are called the bare mass and the bare charge respectively (also, unrenormalized mass and charge). They are presumed to be the values these parameters would have if the interaction was reduced to zero. Thus they do not represent the values of the electron mass or charge which an experimentalist would determine in some measurement. When the Feynman diagrams are evaluated, they are expressed in terms of the parameters m_0 and e_0, and if there are divergent integrals they too are dependent on these bare parameters. Expressing results in terms of these quantities is clearly unsatisfactory, since all results should be expressed in terms of measured or "physical" values of the parameters. Only in that way is it possible to substitute numerical values and come up with an answer. Thus m and e, the measured values, are introduced instead, but it must be clearly understood that these measured values include all interactions, since the experimentalist cannot turn them off. Perhaps the most straightforward way to express answers in terms of the measured quantities is to replace m_0 and e_0 wherever they occur by writing them as m and e plus correction terms. This leads to some new terms (the corrections) to compute. It is these new terms, which are presumed to be divergent in perturbation theory, which now bring about a cancellation of the infinite results contained in the expressions of Fig. 7, when these are expressed as functions of m and e.

The entire procedure of renormalization of mass and charge is truly remarkable, and although not entirely satisfactory from a strict mathematical point of view, it leads to predictions of anomalous magnetic moments $(g - 2)/2$, Lamb shifts in simple atoms, and other effects which agree incredibly well with experiment. The high-precision tests of quantum electrodynamics provide excellent confirmation for both the validity of quantum electrodynamics and of the renormalization procedure employed to obtain finite results.

Relation to other theories. Quantum electrodynamics may be considered to be a model field theory in which the successes have been so great that attempts have been made to use it as a guide in constructing other theories. For many years, theorists struggled to develop a satisfactory framework to describe weak interactions, which are mainly responsible for instability of matter and for beta decay. It is now widely accepted that this theory has been unified with electrodynamics, the combined theory being called electroweak theory. The most satisfactory theory of this type is known as the Weinberg-Salam model. In this theory, known as a gauge theory, there are four bosons which mediate interactions between pointlike particles. The photon, which is massless, mediates the electromagnetic forces, while very massive particles discovered in 1982 and 1983, known as the W^+, W^-, and Z^0, mediate the weak force. The beauty of this theory is to be found in its renormalizability and in its calculability since it provides a framework which allows the use of perturbation theory. SEE GAUGE THEORY; INTERMEDIATE VECTOR BOSON; WEAK NUCLEAR INTERACTIONS; WEINBERG-SALAM MODEL.

In the 1970s another gauge theory was developed, again using some of the ideas of QED.

This theory, which is known as quantum chromodynamics (QCD), gives an underlying renormalizable theory of the constituents of strongly interacting particles such as protons, neutrons, pi mesons, and other particles collectively called hadrons. This theory contains quarks and gluons as fundamental pointlike entities and as basic building blocks. The quarks can interact with each other by exchanging gluons, and the gluons can also interact among themselves. In this theory the strength of interactions is often quite large, and hence the application of perturbation theory is limited to some special cases, thus making it exceedingly difficult to carry out calculations involving sums of Feynman graphs. Thus such questions as why free quarks have never been seen have not yet been answered. Nevertheless, this theory offers much promise. SEE GLUONS; QUANTUM CHROMODYNAMICS; QUARKS.

Bibliography. J. Bialyniccy-Birula, *Quantum Electrodynamics*, 1975; J. Bjorken and S. Drell, *Relativistic Quantum Mechanics*, 1964; S. N. Gupta, *Quantum Electrodynamics*, 1977; C. Itzykson and J.-B. Zuber, *Quantum Field Theory*, 1980; J. M. Jauch and F. Rohrlich, *The Theory of Photons and Electrons*, 2d ed., 1975; L. D. Landau et al., *Quantum Electrodynamics*, 2d ed., 1982; C. Quigg, *Gauge Theories of the Strong, Weak, and Electromagnetic Interactions*, 1983; J. Schwinger, *Selected Papers on Quantum Electrodynamics*, 1958.

SUPERCRITICAL FIELDS
WALTER GREINER

Static fields that are strong enough to cause the normal vacuum, which is devoid of real particles, to break down into a new vacuum in which real particles exist. This phenomenon has been observed for electric fields, and is predicted for other fields such as gravitational fields and the gluon field of quantum chromodynamics.

Concept of the vacuum. The entities of space, time, and matter can be said to form the basis of physics; and the concept of the vacuum, which is intimately connected with these entities, is therefore among the most fundamental issues in the scientific interpretation of the world. Newton's absolute space formed the basis for the hypothesis of the vacuum as an elastic medium, the so-called ether. The latter concept was developed in the early nineteenth century when the wave nature of light had been firmly established, in close analogy to the theory of elasticity. In Einstein's theory of relativity and gravity the absolute space and the ether were abandoned and replaced by a collection of inertial frames.

Quantum mechanics and quantum field theory laid the grounds for the present conception of the vacuum: a polarizable gas of virtual particles, fluctuating randomly. The concept of virtual particles not only expresses a philosophical notion but implies observable effects:

    1. The occurrence of spontaneous radiative emission from atoms and nuclei can be attributed to the action of the fluctuations of the virtual gas of photons.

    2. The zero-point motion of the virtual particles gives rise to such effects as the Casimir effect. (Two conducting, uncharged plates attract each other in a vacuum enviroment with a force that varies as the inverse fourth power of their separation.)

    3. The electrostatic polarizability of the virtual fluctuations can be measured in the Lamb shift and Delbrück scattering.

However, the most fascinating aspect of the vacuum of quantum field theory, which will be discussed below, is the possibility that it allows for the creation of real particles in strong, time-independent external fields. In this case the normal vacuum state is unstable and decays into a new vacuum that contains real particles. This, in itself, is a deep physical and philosophical insight. But it is more than an abstract problem: first, very strong electric fields are available for laboratory experiments; second, it can be shown that the quantum theory of interacting fields may be constructed from the vacuum-to-vacuum amplitude $W(J)$ of a quantized field in the presence of an arbitrary external source J. Effects that occur in strong external fields may, therefore, in some way be carried over to strongly coupled, interacting fields that form the basis of the strong and superstrong interactions. Extensive theoretical studies have led to new insights and clarification of the strong field problem. SEE QUANTUM ELECTRODYNAMICS; QUANTUM FIELD THEORY.

Vacuum decay in quantum electrodynamics. The decay of the vacuum in strong electrostatic fields is a phenomenon in quantum electrodynamics that can be studied only through low-energy heavy-ion collisions. The original motivation for developing the new concept of a charged vacuum arose in the late 1960s in connection with attempts to understand the atomic structure of superheavy nuclei expected to be produced by heavy-ion linear accelerators. *See Particle Accelerator.*

The best starting point for discussing this concept is to consider the binding energy of atomic electrons as the charge Z of a heavy nucleus is increased (**Fig. 1**). If the nucleus is assumed to be a point charge, the total energy E of the $1s_{1/2}$ level drops to 0 when $Z = 137$. This so-called $Z = 137$ catastrophe had been well known, but it was argued loosely that it disappears when the finite size of the nucleus is taken into account. However, in 1969 it was shown that the problem is not removed but merely postponed, and reappears around $Z = 173$. Any level $E(nj)$ can be traced down to a binding energy of twice the electronic rest mass if the nuclear charge is further increased. At the corresponding charge number, called Z_{cr}, the state dives into the negative-energy continuum of the Dirac equation (the so-called Dirac sea). The overcritical state acquires a width and is spread over the continuum (indicated by bars in Fig. 1, magnified by a factor of 10). Still, the electron charge distribution does remain localized. *See Antimatter.*

When Z exceeds 145, $E(1s_{1/2})$ is less than 0; that is, the binding energy exceeds the rest mass of the electron. Adding an electron therefore diminishes the mass of the atom. It would be energetically advantageous for an electron to be spontaneously created, thereby reducing the total energy. This is not possible because it would violate the conservation of charge and lepton number. Similarly, when Z exceeds Z_{cr} a K-shell electron is bound by more than twice its rest mass, so that it becomes energetically favorable to create an electron-positron pair. Now, however, the spontaneous appearance of such a pair is not forbidden by any conservation law. The electron becomes bound in the $1s_{1/2}$ orbital and the positron escapes. *See Positron.*

The overcritical vacuum state is therefore said to be charged. According to the hole theory, which is a consistent model for interpreting the field-theoretical (quantum-electrodynamical) calculations, the states of negative energy are occupied with electrons. This was postulated by Dirac to avoid the decay of electronic states with emission of an infinite amount of energy. In the undercritical situation a vacuum state $|0>$ can be defined without charges or currents by choosing the Fermi surface (up to which the levels are occupied) at $E_F = -m_e c^2$, below the lowest bound state (where m_e is the electron mass and c is the speed of light). The negative-energy continuum states occupied with electrons represent the model for this vacuum; its infinite charge

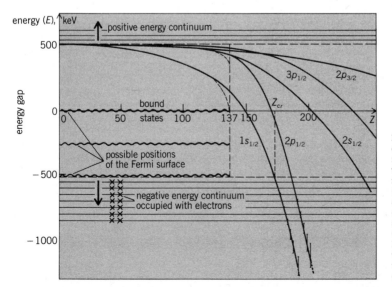

Fig. 1. Lowest bound states of the Dirac equation for nuclei with charge Z. While the energy levels in the field of a point charge (broken lines) for total angular momentum $j = \frac{1}{2}$ end at $Z = 137$, the solutions with extended Coulomb potential (solid lines) can be traced down to the negative comtinuum which is reached at critical charge Z_{cr}.

is renormalized to zero, and so it is a neutral vacuum. If now an empty atomic state dives into the negative continuum, it will be filled spontaneously with an electron from the Dirac sea with the simultaneous emission of a free positron moving to infinity. The remaining electron cloud of the supercritical atom is necessarily negatively charged. Thus, the vacuum becomes charged.

If the central charge is further increased to $Z = 184$ (the diving point of the $2p_{1/2}$ level), the vacuum acquires a charge of $-4e$; with increasing field strength, more electronic bound states join the negative continuum, and each time the vacuum undergoes a new phase transition and becomes more highly charged: the vacuum sparks in overcritical fields.

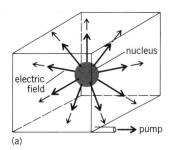 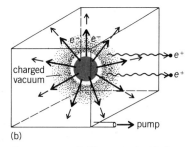

Fig. 2. Vacuum surrounding a heavy nucleus, represented by space inside box. The central nucleus serves as a source of a strong electric field. (a) Undercritical field: the vacuum is empty, that is, no particles (besides the central source) are in the box. (b) Overcritical field: space becomes charged through the emission of antiparticles. In principle, the vacuum is no longer empty under these conditions.

This situation is visualized in **Fig. 2**, where the vacuum is represented by a box surrounding the nucleus. If the electrons of the charged vacuum, represented in Fig. 2b by a diffuse cloud surrounding the nucleus, are somehow pumped out, new positrons (represented by e^+) will be emitted and the electronic cloud will reappear. The positrons, being in continuum states, can freely move around and are pumped out easily, so that a charged vacuum is again left behind.

Clearly, the charged vacuum is a new ground state of space and matter. The normal, undercritical, electrically neutral vacuum is no longer stable in overcritical fields: it decays spontaneously into the new stable but charged vacuum. Thus the standard definition of the vacuum, as a region of space without real particles, is no longer valid in very strong external fields. The vacuum is more accurately defined as the energetically deepest and most stable state that a region of space can have while being penetrated by certain fields.

Superheavy quasimolecules. Inasmuch as the formation of a superheavy atom of $Z > 173$ is very unlikely, a new idea is necessary to test these predictions experimentally. That idea, based on the concept of nuclear molecules, was put forward in 1969: a superheavy quasimolecule forms temporarily during the slow collision of two heavy ions (**Fig. 3**). It is sufficient to form the quasimolecule for a very short instant of time, comparable to the time scale for atomic processes to evolve in a heavy atom, which is typically of the order 10^{-18} to 10^{-20}. Suppose a uranium ion is shot at another uranium ion at an energy corresponding to their Coulomb barrier, and the two, moving slowly (compared to the K-shell electron velocity) on Rutherford hyperbolic trajectories, are close to each other (compared to the K-shell electron orbit radius). Then the atomic electrons move in the combined Coulomb potential of the two nuclei, thereby experiencing a field corresponding to their combined charge of 184 (Fig. 3). This happens because the ionic velocity (of the order of $c/10$) is much smaller than the orbital electron velocity (of the order of c), so that there is time for the electronic molecular orbits to be established, that is, to adjust to the varying distance between the charge centers, while the two ions are in the vicinity of each other. *See* QUASIATOM.

Dynamical heavy-ion collision processes. Several dynamical processes contribute to the ionization of the inner shells and to the production of positrons in undercritical as well as

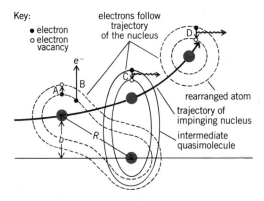

Fig. 3. Collision of two heavy ions in which a quasimolecule is formed. Distance between the nuclei is indicated by R, the impact parameter by b. Processes of type A (excitations of electrons into higher shells) and B (excitations of electrons into the upper continuum) empty the K shell. Processes C and D indicate molecular and atomic x-ray transitions, respectively.

overcritical systems. This is illustrated in **Fig. 4** for a system that becomes over-critical at small nuclear separation distances, less than a critical value R_{cr}. Processes A and B are electron excitation and ionization. Process C is the spontaneous filling of a previously produced vacancy when the level acquires a binding greater than $2m_ec^2$ and is the spontaneous decay of the vacuum described above. Because of the lack of full adiabaticity, energy can be drawn from the nuclear motion to lead to filling of the hole even at distances larger than R_{cr}. This effect (processes D and E) is called an induced transition, and its effect on positron production is twofold: it causes a washed-out threshold for the spontaneous positron production, and it greatly enhances the production cross section.

Process F is the direct pair production process. Whereas in ordinary pair production in a Coulomb scattering process a photon is exchanged between two hadrons only once, now there are multiple interactions with the joint Coulomb field of both nuclei. Because of the very strong field, the cross section for the pair production varies as $(Z_1 + Z_2)^{20}$, where Z_1 and Z_2 are the charges of the nuclei, which means that about 10 photons are exchanged. This behavior illustrates the nonperturbative character of this process, which (like the induced decay mechanism) overwhelms the spontaneous positron production process. The pair production process F can be interpreted as the shake-off of the vacuum polarization cloud. All these processes and predictions have been quantitatively confirmed by experiments. SEE ELECTRON-POSITRON PAIR PRODUCTION.

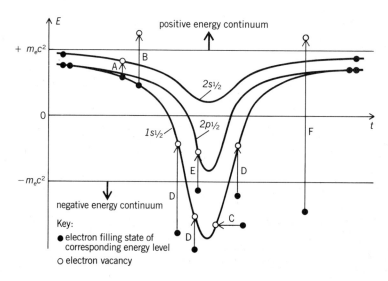

Fig. 4. Dynamical processes connected with positron production in overcritical heavy-ion collisions. Inner electron levels in the quasimolecule are shown as a function of time t. At the deepest point of the $1s$ level, the colliding nuclei are at the distance of closest approach. Processes are identified in text.

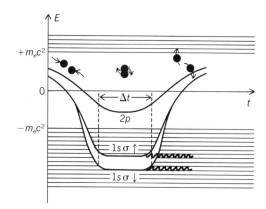

Fig. 5. Innermost shells of a superheavy molecule (atom) as a function of time t. Due to the sticking of the two nuclei, the superheavy atom lives for the time Δt, thus being able to emit positrons spontaneously.

Giant nuclear systems. The energy spectrum for positrons created in, for example, a uranium-curium collision consists of three components: the induced, the direct, and the spontaneous, which add up to a smooth spectrum. The presence of the spontaneous component leads only to 5–10% deviations for normal nuclear collisions along Rutherford trajectories. This situation raises the question as to whether there is any way to get a clear qualitative signature for spontaneous positron production, as opposed to detecting it through a quantitative comparison with theory. Suppose that the two colliding ions, when they come close to each other, stick together for a certain time Δt before separating again. This will in general require the use of bombarding energies slightly above the Coulomb barrier. Then the quasimolecular levels in the overcritical region get stretched out as shown in **Fig. 5**, which is to be contrasted with Fig. 4. During the sticking, the energies of the electronic states do not change, and this has two effects: the emission of positrons from any given state occurs with a fixed energy; and the induced production mechanisms do not contribute, whereas the spontaneous production (for overcritical states) continues to contribute.

The longer the sticking, the better is the static approximation. For Δt very long, a very sharp line should be observed in the positron spectrum with a width corresponding to the natural lifetime of the resonant positron-emitting state (approximately 3 keV for the uranium-uranium system). The observation of such a sharp line will indicate not only the spontaneous decay of the vacuum but also the formation of giant nuclear systems ($Z > 180$). (There should in general be two positron lines because of the Zeeman splitting of the $1s\sigma$ level, shown in Fig. 5, due to strong magnetic fields from the heavy ion currents.) *See* NUCLEAR MOLECULE.

Observation of spontaneous positron emission. The search for spontaneous positron emission in heavy-ion collisions began in 1976. Subsequent experiments have focused on studying positron spectra and on extending the investigations to collision systems with higher total nuclear charge. Of special interest are peak structures in the positron energy distribution. The most compelling evidence for these comes from experiments where coincidences between two scattered ions are used to define clearly events with two-body final states consistent with, or bordering on, elastic scattering.

The uranium-curium collision system has a total charge $Z = 188$. **Figure 6** shows positron spectra from uranium-238 and curium-248 colliding at an energy close to that of the Coulomb barrier. Particularly striking is the well-defined peak centered at an energy of about 320 keV. Analysis indicates that the intrinsic width of the peak is less than 20 keV. It is apparent that an explanation must be sought outside the scope of the theory based on Rutherford scattering alone, because this theory of dynamic positron creation does not allow for narrow peak structures in the positron spectrum. Deviations from this theory also have been demonstrated for uranium-uranium collisions in other experiments. All experiments indicate that there is a new source of positrons— a source that does not originate with the known dynamic mechanisms associated in a simple way with the time-varying electric field produced in Coulomb trajectories.

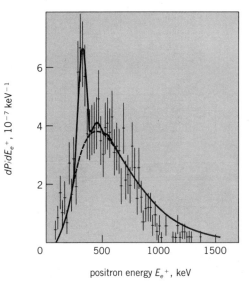

Fig. 6. Spectrum of emitted positrons in uranium-curium collisions with backward-scattered ions at a laboratory energy of 5.9 MeV per nucleon. dP/dE_e^+ is the probability that a positron will be produced within a unit energy interval. Crosses indicate experimental results. The solid curve indicates the theoretical prediction of the total positron yield (spontaneous plus dynamical positrons); the broken curve indicates the predicted yield of dynamical positrons only.

It has also been shown that these deviations from smooth positron spectra cannot be attributed to purely nuclear effects. Two prominent candidates for such effects are the internal pair conversion of a nuclear transition leading to a positron energy distribution that may be peaked; and the internal pair conversion process followed by the capture of the electron into empty atomic orbits, which leads to positron line spectra. These mechanisms were excluded in experiments which showed that both the x-ray spectra and the delta-electron spectra measured simultaneously with the positrons are smooth; that is, they show no structure. This proves that vacuum decay has been observed: The narrow positron peak does indeed represent spontaneous positron emission. The parent nuclear supercritical charge must exist for a long time compared to the collision times for scattering beneath the Coulomb barrier, as discussed above. SEE DELTA ELECTRONS.

Therefore, it has been suggested that the observation of spontaneous positron emission as a sharp line necessarily implies that, at bombarding energies close to that of the Coulomb barrier, metastable giant nuclear composite systems are formed with a rather long lifetime. Widths of 20 keV or less correspond to lifetimes for the nuclear molecular system longer than about 1000 times the Rutherford scattering collision time, during which the $1s\sigma$ state is overcritically bound. Such a lifetime could be supplied by the formation of a rather cold intermediate giant nuclear complex as the nuclei barely touch in overcoming the Coulomb barrier (**Fig. 7**). Due to their peculiar shape, the touching area of uranium nuclei in the position shown in Fig. 7b is particularly large, which could lead to a kind of nuclear cohesion.

Other field theories. The idea of overcriticality also has applications in other field theories, such as those of pion fields, gluon fields (quantum chromodynamics), and gravitational fields (general relativity).

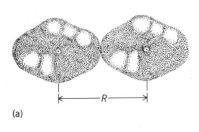
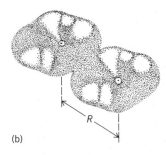

Fig. 7. Two uranium nuclei in typical positrons upon contact. (a) Small contact area. (b) Large contact area.

FUNDAMENTAL INTERACTIONS AND FIELDS

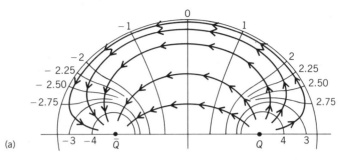

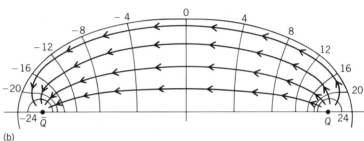

Fig. 8. Model of a heavy meson. Color-electric field lines (with arrows) and equipotential lines (with numbers to indicate relative values of potential) are shown within an ellipsoidal bag with a heavy quark Q and antiquark $\bar{Q}$ at the foci. (a) Small deformation. (b) Large deformation.

Gluon field. **Figure 8** shows a heavy meson with a heavy quark Q and antiquark $\bar{Q}$ located in the foci of an ellipsoidal bag. The color-electric or glue-electric field lines are also indicated; they do not penetrate the bag surface. The solutions of the Dirac equation for light quarks q in this field of force are shown in **Fig. 9**. In the spherical case (diagram 1 in Fig. 9b) the potential is zero, and the solutions with different charges degenerate. As the source charges Q and $\bar{Q}$ are pulled apart, the wave functions start to localize (diagrams 2 and 3). At a critical deformation of

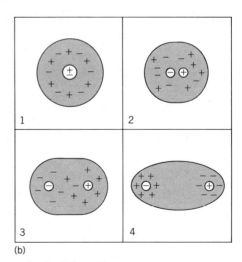

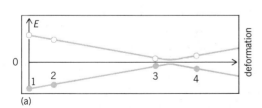

Fig. 9. Fission of a heavy meson. (a) Energy levels of Dirac equation for light quarks q as a function of meson deformation. In the vacuum state all negative energy modes are occupied (solid dots). (b) Diagrams of meson at successive stages of deformation, numbered corresponding to points in the energy level diagram. Heavy quarks Q and $\bar{Q}$ are indicated by + and − signs within circles. Other + and − signs indicate wave functions of light quarks q and $\bar{q}$. (After D. Vasak et al., Fission of bags by spontaneous quark-antiquark production in supercritical colour fields, Zeitschrift für Physik, C21:119–125, 1983)

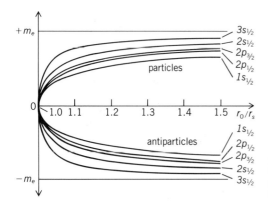

Fig. 10. Lowest bound states of the Dirac equation in the field of a gravitating mass shell, as function of the radius of the shell r_0 relative to the Schwarzschild radius r_s.

the bag, positive and negative energy states cross (Fig. 9a); that is, overrcriticality is reached and the color field is strong enough that the so-called perturbative vacuum inside the bag rearranges (diagram 4 in Fig. 9b) so that the wave functions are pulled to opposite sides and the color charges of the heavy quarks are completely shielded. Hence two new mesons of types $\bar{Q}q$ and $Q\bar{q}$ appear; the original meson fissions. This gives a vivid picture of how quark confinement works. SEE GLUONS; QUANTUM CHROMODYNAMICS; QUARKS.

Gravitational field. An example of superstrong gravitational fields is given by the solutions of the Dirac equation in the field of a gravitating mass shell of radius r_0. The spectrum shown in **Fig. 10** indicates that the energy gap vanishes completely when the radius of the shell equals the Schwarzschild radius r_s. This is where overcriticality is reached and pair production will set

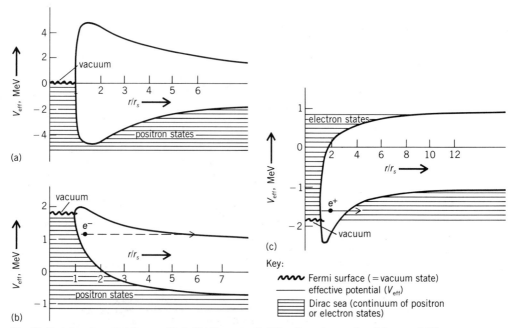

Fig. 11. Fermi surface and the so-called effective potential V_{eff} of an electron in a (a) neutral, (b) negatively charged, and (c) positively charged Reissner-Nordström geometry (the vicinity of a black hole).

in. This phenomenon is connected with the so-called Hawking radiation. In the case of a charged, rotating black hole, that is, for the solutions of the Dirac equation in a Reissner-Nordström geometry, the results can be best described by displaying the positive and negative energy levels together with the so-called effective potential of an electron (**Fig. 11**). Even for a neutral black hole (Fig. 11a) the gap between particle and antiparticle states is narrowed by the attractive gravitational interaction and vanishes for $r = r_s$. With the introduction of a charged center the symmetry between electron and positron states is broken: In the language of hole theory, the occupied negative continuum states are raised by a negatively charged center (Fig. 11b), and vice versa (Fig. 11c). If the Fermi surface is above $m_e c^2$, spontaneous pair creation sets in. A negatively charged black hole emits electrons to infinity while the positrons diminish the central charge, and similarly for a positively charged black hole. This process manifests the phase transition to a charged electron-positron vacuum. As a consequence, black holes can acquire only negligible charge. (More precisely, their charge-to-mass ratio, Q/M, is of the order of 10^{-42} e/m, where e/m is the charge-to-mass ratio of an electron.)

WEAK NUCLEAR INTERACTIONS
Eugene D. Commins

Fundamental interactions of nature that play a significant role in elementary-particle and nuclear physics, and are distinguished from other such interactions by special properties such as participation of all the fundamental fermions and failure to conserve parity. Of the four fundamental interactions of nature (gravitational, strong, electromagnetic, and weak), only the strong, electromagnetic, and weak forces are significant for elementary-particle and nuclear physics, given present understanding and foreseeable methods of observation. The weak force has very short range (less than 10^{-17} m) and is extremely feeble compared to strong and electromagnetic forces, but can be distinguished from these two by its special character. For example, according to the present view, all of matter consists of certain fundamental spin-½ constituents, the quarks and leptons, collectively called the fundamental fermions (**Table 1**). While only the quarks participate in strong interactions, and only the quarks and charged leptons e, μ, and τ participate in electromagnetic interactions, all of the fundamental fermions, including neutrinos, engage in weak interactions. Also, the strong and electromagnetic interactions respect spatial inversion symmetry (they conserve parity) and are also particle-antiparticle (charge conjugation) symmetric, whereas the weak interaction violates these two symmetries. SEE FUNDAMENTAL INTERACTIONS; LEPTON; PARITY; QUARKS; SYMMETRY LAWS.

Table 1. Fundamental constituents of matter*

		Generation†		
Constituent	Electric charge	1	2	3
Quarks	$+\frac{2}{3}\|e\|$ $-\frac{1}{3}\|e\|$	$\begin{pmatrix}u\\d\end{pmatrix}$	$\begin{pmatrix}c\\s\end{pmatrix}$	$\begin{pmatrix}t\\b\end{pmatrix}^{\ddagger}$
Leptons	0 $-\|e\|$	$\begin{pmatrix}\nu_e\\e^-\end{pmatrix}$	$\begin{pmatrix}\nu_\mu\\\mu^-\end{pmatrix}$	$\begin{pmatrix}\nu_\tau\\\tau^-\end{pmatrix}$

*For each quark and lepton there exists a corresponding antiparticle with the same mass but opposite charge.
†More quark and lepton generations may exist.
‡Experimental evidence for t is still lacking.

Weak interactions are classified as charged or neutral, depending on whether or not a particle participating in a weak reaction suffers a change of electric charge of one electronic unit. For example, the weak neutrino-nucleon scattering reaction of Eq. (1) is charged since the neu-

Table 2. Classification of weak processes, with some examples

Charged weak interactions
Purely leptonic — Muon decay
$\mu^- \to e^- \bar{\nu}_e \nu_\mu$
$\nu_\mu e^- \to \mu^- \nu_e$
$\nu_e e^- \to \nu_e e^-$
Semileptonic
$n \to p e^- \bar{\nu}_e$ — Neutron decay
$\nu_e n \to p e^-$ — Inverse beta decay
$\nu_\mu n \to p \mu^-$ — Pion decay
$\pi^+ \to \mu^+ \nu_\mu$ — Pion decay
$\mu^- p \to n \nu_\mu$ — Muon capture
$\Sigma^- \to n e^- \bar{\nu}_e$ — Hyperon decay
Purely hadronic
$K^+ \to \pi^+ \pi^0$ — Kaon decay
$K^- \to \pi^+ \pi^- \pi^0$ — Kaon decay
$\Lambda^0 \to p \pi^-$ — Hyperon decay

Neutral weak interactions
Neutrino-lepton scattering
$\nu_\mu + e \to \nu_\mu + e$
Neutrino-nucleon scattering
$\nu_\mu + N \to \nu_\mu + N$
Electron-nucleon scattering
$e + N \to e + N$ — Competes with electromagnetic interaction
Lepton-lepton scattering
$e^+ + e^- \to \mu^+ + \mu^-$
Hadron-hadron interaction
$N + N \to N + N$ — Competes with electromagnetic and strong interactions

$$\nu_\mu + n \to \mu^- + p \qquad (1)$$

trino ν_μ with zero charge transforms into a muon μ^- with negative charge, while the neutron n (zero charge) becomes a proton p (positive charge). (Here the neutron may be considered as being composed of two d quarks and one u quark, and the proton of one d quark and two u quarks. In Eq. (1) a d becomes a u.) On the other hand, the weak elastic scattering reaction of Eq. (2) is

$$\nu_\mu + p \to \nu_\mu + p \qquad (2)$$

neutral. Observed charged weak interactions include nuclear beta decay and electron capture, muon capture on nuclei, and the slow decays of unstable elementary particles such as the μ and τ leptons, π, K, charmed, and bottom mesons, and hyperons and charmed baryons (**Table 2**). Also, there are the charged neutrino-nucleon and neutrino-lepton scattering reactions. Neutral weak interactions were first observed in 1973, and include neutrino-nucleon and neutrino-lepton scattering as well as the electron-nucleon reaction of Eq. (3), which can also occur by electromag-

$$e^- + N \to e^- + N \qquad (3)$$

netic interaction. *See* BARYON; ELEMENTARY PARTICLE; HYPERON; MESON; NEUTRAL CURRENT.

In recent years the most important development in the study of weak interactions has been the creation by many workers, but principally by S. Glashow (1961), S. Weinberg (1967), and A. Salam (1968), of a successful theory based on the principles of local gauge invariance and spontaneous symmetry breaking. This theory proposes a single basis for the weak and electromagnetic interactions, and indeed, despite striking differences in the observed characteristics of strong, electromagnetic, and weak interactions, important theoretical ideas of a similar type suggest that all these interactions possess a common origin.

Early study. The development of understanding about weak interactions is inseparably linked with other major developments of twentieth-century physics: relativity, quantum mechanics and quantum field theory, and nuclear and elementary-particle physics in general. The study of weak interactions began with the discovery of radioactivity by H. Becquerel in 1896, and the recognition shortly thereafter that in one form of radioactivity the decaying nucleus emits "beta rays" (electrons). Nuclear beta decay was thus the first known weak process. In 1914 J. Chadwick observed that the electrons in beta decay are emitted with a continuous spectrum of energies. This result and subsequent observations in the 1920s led to a crisis, for since the energy available to the electron in beta decay is essentially the difference in rest energies of the initial and final nuclei, which is a definite quantity, it appeared as though the principle of conservation of energy was violated. In order to rescue that fundamental law as well as those of conservation of linear and angular momentum, also in jeopardy, W. Pauli proposed in 1930 and again in 1933 that a neutral particle of small or vanishing rest mass (later called the neutrino) is emitted along with the electron in nuclear beta decay and that it escapes observation because of its feeble interactions with surrounding matter. *See* BETA PARTICLES; NEUTRINO; RADIOACTIVITY.

Fermi theory. In 1934 Enrico Fermi proposed a theory of beta decay based on Pauli's neutrino hypothesis and constructed by analogy with quantum electrodynamics, which had been developed a few years before by P. A. M. Dirac. In quantum electrodynamics the interaction between an electron and the electromagnetic field is described by the interaction lagrangian density of Eq. (4), where $A^\lambda(\mathbf{x},t)$ is the four-vector potential of the electromagnetic field, and ej_λ is the

$$\mathcal{L}_{EM} = -ej_\lambda(\mathbf{x},t)\, A^\lambda(\mathbf{x},t) \tag{4}$$

electromagnetic current density of the electron, another four-vector. Fermi proposed that the beta-decay interaction is also described by the coupling of two four-vectors. In the simplest of nuclear beta decays, namely neutron beta decay, Eq. (5), a four-vector current density describes the neu-

$$n \to p + e^- + \bar{\nu}_e \tag{5}$$

tron-proton transformation: $\bar{\psi}_p \gamma^\lambda \psi_n$; and another four-vector current density describes the creation of e^- and $\bar{\nu}_e$: $\bar{\psi}_e \gamma^\lambda \psi_{\nu_e}$, where ψ_p, ψ_n, ψ_e, and ψ_{ν_e} are Dirac field operators and the γ^λ are 4 × 4 matrices appearing in Dirac's relativistic quantum theory. Thus Fermi obtained the beta-decay lagrangian density of Eq. (6), where the hermitian conjugate term (h.c.) was intended to account

$$\mathcal{L}_\beta = \frac{G}{\sqrt{2}}\, \bar{\psi}_p \gamma_\lambda \psi_n \cdot \bar{\psi}_e \gamma^\lambda \psi_{\nu_e} + \text{h.c.} \tag{6}$$

for nuclear β^+ decay and electron capture. G, the so-called weak interaction, or Fermi coupling constant, must be determined by experiment and is found to be given by Eqs. (7) or in units $\hbar = c = 1$ by Eq. (8), where m_p is the proton mass. *See* QUANTUM ELECTRODYNAMICS; QUANTUM FIELD THEORY.

$$\begin{aligned} G &= 1.43506 \pm 0.00026 \times 10^{-49}\ \text{erg}\cdot\text{cm}^3 \\ &= 1.43506 \pm 0.00026 \times 10^{-62}\ \text{J}\cdot\text{m}^3 \end{aligned} \tag{7}$$

$$G = 1.03 \times 10^{-5}\, m_p^{-2} \tag{8}$$

Fermi's theory gave a good account of many aspects of nuclear beta decay, especially when generalized somewhat to include other bilinear covariants than vector × vector, by G. Gamow and E. Teller in 1936. Also, it contained the essence of many future developments. However, it was recognized almost immediately (by W. Heisenberg in 1936, among others) that the Fermi theory cannot be fundamental, since when it is applied to high-energy processes such as neutrino-electron scattering, it leads to a failure of unitarity (nonconservation of probability). Also, the Fermi theory contains incurable divergences which occur in the calculation of higher-order corrections: the theory is not renormalizable. *See* RENORMALIZATION.

Parity violation. During the 25 years following Fermi's proposal, many weak processes were uncovered in addition to nuclear beta decay, with the discovery of new elementary particles and elucidation of their decay schemes. Gradually it became clear that these bear many similarities to nuclear beta decay and all are but different manifestations of a universal weak interaction. Perhaps the most dramatic achievement of this period was the important discovery in 1956 by T. D. Lee and C. N. Yang that parity is violated in the weak interaction. An example of parity violation is the decay of the pion, $\pi^+ \to \mu^+ \nu_\mu$, as seen in the pion rest frame (**Fig. 1**). In observed $\pi^+ \to \mu^+ \nu_\mu$ decay (Fig. 1a), the μ^+ spin is found experimentally to be opposite to its motion

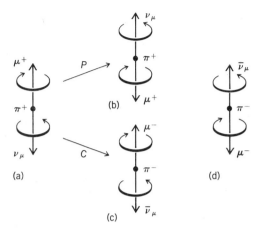

Fig. 1. Schematic diagram of weak decay of the pion, $\pi^+ \to \mu^+ \nu_\mu$, illustrating parity (P) and charge conjugation (C) violation in weak interactions. (a) Decay of π^+. (b) Result of P transformation not observed. (c) Result of C transformation not observed. (d) Result of CP transformation: observed $\pi^- \to \mu^- \bar{\nu}_\mu$ decay.

[helicity $h(\mu^+) = -1$]. Since the π^+ spin is zero and angular momentum is conserved, this implies $h(\nu_\mu) = -1$ as well. A parity (P) transformation results in reversal of μ^+ and ν_μ momenta but leaves spins invariant. Thus under P, $h(\mu^+)$ is reversed (Fig. 1b); but this is never observed, implying maximal P violation. SEE HELICITY.

A charge conjugation (C) transformation on $\pi^+ \to \mu^+ \nu_\mu$ decay changes the signs of all charges, but leaves spins and momenta unchanged yielding a μ^- with $h(\mu^-) = -1$ (Fig. 1c). Again, this is never observed; that is, the weak interactions violate charge conjugation maximally. On the other hand, a CP transformation on $\pi^+ \to \mu^+ + \nu_\mu$ decay results in observed $\pi^- \to \mu^- \bar{\nu}_\mu$ decay (Fig. 1d).

Feynman–Gell-Mann theory. The discovery of parity violation led, through a spurt in experimental and theoretical work, to the ultimate generalization and refinement of Fermi's theory, proposed by R. Feynman and M. Gell-Mann in 1958, and independently by R. Marshak and E. Sudarshan in the same year. In this formulation, the charged weak interactions are described by the lagrangian density of Eq. (9), where J_λ, the "universal" charged weak current, is a generaliza-

$$\mathcal{L}_w = \frac{G}{\sqrt{2}} J_\lambda^\dagger J^\lambda \qquad (9)$$

tion of the original four-vector currents proposed by Fermi. The new current J_λ contains not only a vector (V) but also an axial vector (A) portion. Thus in the lagrangian of Eq. (9) one obtains terms of the form $V \times V$ and $A \times A$ (which are scalars and do not change sign under spatial inversion) and also terms $V \times A$ and $A \times V$ (which are pseudoscalars and change sign under spatial inversion). The appearance of scalar and pseudoscalar terms in $\mathcal{L}_w$ yields parity-violating effects such as the emission of particles with definite helicity in weak decays, the asymmetric angular distribution of beta electrons emitted in the decay of polarized nuclei, and so forth.

The Feynman–Gell-Mann scheme provides an excellent account of the observed features of charged weak interactions, but like the Fermi theory it cannot be fundamental, since it is not renormalizable, and leads to absurdities at sufficiently high energies.

Weinberg-Salam model. The Fermi–Feynman–Gell-Mann scheme may be modified by assuming that the weak interaction does not occur at a single space-time point, but proceeds by exchange of an intermediate boson. This idea, originally suggested by J. Schwinger in 1957, arises naturally by analogy with electromagnetism, where the Coulomb force between two charged particles (say electron and proton) occurs by exchange of a photon, the zero-mass, spin-1 quantum of the electromagnetic field (**Fig. 2**a). However, in the case of the weak interaction, the intermediate bosons (again vector or spin-1 quanta) must be massive, since the weak interaction has short range. Moreover, there must exist at least three such bosons: a charged W^- and its charge-conjugate W^+ to transmit the charged weak interaction (Fig. 2b) and a neutral boson Z^0 for the

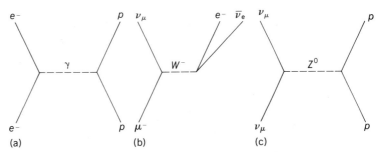

Fig. 2. Schematic (Feynman) diagrams of electromagnetic and weak interactions, occurring by exchange of intermediate bosons. In each diagram time flows upward. (a) Coulomb interaction (electromagnetic), occurring by photon exchange. (b) Example of charged weak interactions: muon decay ($\mu^- \to e^- \nu_\mu \bar{\nu}_e$), occurring by W^- exchange. (c) Example of neutral weak interaction: $\nu_\mu\, p$ scattering, occurring by Z^0 exchange.

neutral weak interaction (Fig. 2c). The problem was how to construct a renormalizable theory of weak interactions in which massive vector bosons are exchanged between fundamental fermions.

Local gauge invariance. The solution was indeed found in a theory uniting weak and electromagnetic interactions and based on a combination of subtle ideas. First is the notion that the theory must be invariant under local gauge transformations, that is, gauge transformations which can vary in an arbitrary manner from one space-time point to another. A theory with this property, first considered by Yang and R. L. Mills in 1954, represents a generalization of electrodynamics, and introduces the intermediate bosons, charged and neutral, in a natural way. At this stage the theory is renormalizable, but the bosons are massless, which cannot correspond to reality. SEE GAUGE THEORY.

Spontaneous symmetry breaking. The second important idea, which solves the problem of massless vector bosons, is spontaneous symmetry breaking. This means that one has a quantum field theory in which the lagrangian possesses a certain symmetry not shared by a particular state of the system (that is, the ground state). When this situation occurs in an ordinary nongauge field theory, it was shown by J. Goldstone in 1961 that there is a massless spin-0 excitation—the so-called Goldstone boson—corresponding to each degree of freedom in which the symmetry is broken. There is no experimental evidence for such bosons, so it would appear that spontaneous symmetry breaking cannot occur. However, the proof of the Goldstone theorem is based on two assumptions, namely, a positive metric in Hilbert space, and manifest covariance. In a gauge theory (for example electrodynamics, or the Yang-Mills theory) these assumptions cannot both be valid. Thus gauge theories evade the Goldstone theorem, as was first noted by P. Higgs in 1964. A most extraordinary result of these arguments is that the difficulties associated with massless vector bosons and Goldstone bosons neutralize one another: by a suitable transformation the Goldstone bosons disappear from the theory and the vector bosons acquire mass. Weinberg (1967) and independently Salam (1968) thus showed how to construct in a natural way a gauge theory of the Yang-Mills type combining weak and electromagnetic interactions. Essential earlier ideas had also been contributed by Glashow. Finally, as was proved by G. 't Hooft in 1971, the Weinberg-Salam theory is renormalizable even after the $W^\pm$ and Z^0 acquire mass. SEE SYMMETRY BREAKING.

Observation of neutral interactions. The Glashow-Weinberg-Salam model or standard model is constructed to conform to known and valid results of quantum electrodynamics and the older charged weak interactions scheme of Feynman and Gell-Mann. The startling new predictions arising from this theory are in the domain of neutral weak interactions, where experiments carried out between 1973 and 1980 confirmed the correctness of the theory in the "low-energy" domain (that is, for experiments where the energy transmitted by the intermediate boson is small compared to its rest energy). These included neutrino-electron scattering, neutrino-nucleon scattering, parity violation in atoms, and most notably scattering of polarized electrons on nucleons. The latter two cases involve interference between the weak and electromagnetic interactions.

Direct observations of intermediate bosons. The foregoing developments culminated in experiments performed in 1982 and 1983 at CERN (European Center for Nuclear Research) in which the intermediate bosons $W^{\pm}$ and Z^0 were directly produced and detected. For this purpose, high-energy counter-propagating beams of protons (p) and antiprotons ($\bar{p}$) were arranged to collide with one another. In some $p\bar{p}$ collisions, W's are created; these almost immediately undergo the decays of Eqs. (10) or (11). The W mass and electroweak coupling are identified from detection

$$W^{\pm} \rightarrow e^{\pm}\, \nu_e(\bar{\nu}_e) \qquad (10) \qquad\qquad W^{\pm} \rightarrow \mu^{\pm}\, \nu_\mu(\bar{\nu}_\mu) \qquad (11)$$

of the charged decay product ($e^{\pm}$ or $\mu^{\pm}$). Similarly, in some $p\bar{p}$ collisions, Z_0 bosons are produced, which decay to a pair of charged particles, as in Eqs. (12) or (13). The results of these experiments

$$Z^0 \rightarrow e^+ e^- \qquad (12) \qquad\qquad Z^0 \rightarrow \mu^+ \mu^- \qquad (13)$$

agree with expectations based on the standard model. Further detailed studies of $W\pm$ and Z^0 remain to be carried out at the CERN $p\bar{p}$ facility and at high-energy positron-electron (e^+e^-) colliders under construction at CERN and at the Stanford Linear Accelerator Center in the United States. SEE INTERMEDIATE VECTOR BOSON; WEINBERG-SALAM MODEL.

Present and future problems. Although the standard model must now be regarded as an accurate phenomenological description of electroweak interactions, many important questions remain unanswered.

Higgs bosons. An inevitable consequence of spontaneous symmetry breaking in the standard model is the appearence of certain hypothetical particles of zero spin, called Higgs bosons. Although these particles are required, no definite prescription is given for their mass. So far, there is no experimental evidence for their existence.

Fermion masses. The mechanism whereby the fundamental fermions acquire mass has not been determined. In the simplest version of the theory, they have zero mass, but additional mass-generating mechanisms can be invoked without spoiling gauge invariance. However, there is then no longer any particular reason to assume that the neutrino masses are zero or even equal. This leads to the important question of coupling between the various types of neutrinos, or as it is called, to neutrino oscillations. Experiments to detect finite neutrino mass and the existence of neutrino oscillations have been discussed and attempted, but no definite results have been obtained. This question has important implications for astrophysics and cosmology.

Unitary transformation of quark states. Although the fundamental quark states for strong interactions are as shown in Table 1, extensive experimental data on weak decays of mesons and baryons reveal the necessity of a unitary transformation of quark states, d, s, b, to a new basis suitable for weak interactions. This is described by the Kobayashi-Maskawa matrix, which contains three Cabibbo rotation angles, and possibly a complex phase δ related to CP violation. The physical origin of Cabibbo rotations is not understood.

CP violation. All experimental evidence is consistent with CPT (combined charge conjugation, parity, and time reversal) invariance of strong, electromagnetic, and weak interactions. However, it was discovered by J. Cronin, V. Fitch, and coworkers in 1964 that in certain weak decays of neutral K mesons, CP invariance is violated. In spite of much experimental work since then, in which the phenomenological parameters of CP violation have been determined ever more precisely, it is still not known whether CP violation originates in the strong, electromagnetic, or weak interaction, or possibly in a "superweak" interaction, or what relationship CP violation has to the standard model. The phenomenon may possibly be described by a nonzero complex phase δ in the Kobayashi-Maskawa matrix.

Cause of parity violation. There is as yet no satisfactory explanation for parity violation, perhaps the most striking experimental fact about the weak force.

Bibliography. G. Arnison et al., Experimental observation of isolated large transverse energy electrons associated with missing energy at $\sqrt{s}$ = 540 GeV, *Phys. Lett.*, 122B:103–116, 1983; G. Arnison et al., Further evidence for charged intermediate vector bosons at the SPS collider, *Phys. Lett.*, 129B:273–282, 1983; G. Arnison et al., Observation of the muonic decay of the charged intermediate vector boson, *Phys. Lett.*, 134B:469–476, 1984; E. D. Commins and P. H. Bucksbaum, *Weak Interactions of Leptons and Quarks*, 1983; H. Georgi, *Weak Interactions and Modern Particle Theory*, 1984; R. E. Marshak, Riazuddin, and C. P. Ryan, *Theory of Weak Interactions in Particle Physics*, 1969; L. B. Okun, *Weak Interactions of Elementary Particles*, 1965; A. Salam, Weak and electromagnetic interactions, *Nobel Symposium, Gothenburg*, pp. 367–377, 1968; S. Weinberg, *Phys. Rev. Lett.*, 19:1264–1266, 1967.

STRONG NUCLEAR INTERACTIONS
HERMAN FESHBACH

One of the fundamental physical interactions, which acts between a pair of hadrons. Hadrons include the nucleons, that is, neutrons and protons; the strange baryons, such as lambda (Λ) and sigma (Σ); the mesons, such as pion (π) and rho (ρ); and the strange meson, kaon (K). The nature of the interaction is determined principally through observation of the collision of a hadron pair. From this one learns that the interaction has a short range of about 10^{-15} m (10^{-13} in.) and is by far the dominant force within this range, being much larger than the electromagnetic interaction, which is next in magnitude. The strong interaction conserves parity and is time-reversal-invariant. SEE BARYON; HADRON; MESON; NUCLEON; PARITY; STRANGE PARTICLES; SYMMETRY LAWS.

Meson exchange. The interaction between the baryons, including the nucleons and the strange baryons, is thought to arise from the exchange of mesons. The interaction for relatively large distances between nucleons is generated by the exchange of single pions (**Fig. 1a**). At shorter separation distances the exchange of two-pion systems, such as the ρ (Fig. 1b), dominates. An important contribution is furnished by the exchange of two separate pions with (Fig. 1c) or without (Fig. 1d) the formation of an excited state of the nucleon. The interaction between the strange baryons, and between the strange baryons and the nucleons, also arises in part from the exchange of pions, but the exchange of kaons can be equally important (**Fig. 2**).

Range. The range of the interaction generated by the exchanges illustrated in Figs. 1 and 2 can be calculated as $\hbar/mc$, where m is the mass of the exchanged particles, $\hbar$ is Planck's constant divided by 2π, and c is the speed of light. Accordingly, the range of the interaction developed when a single pion is exchanged (Fig. 1a) is equal to 1.4×10^{-15} m (5.5×10^{-14} in.), while that due to Fig. 1c is 0.7×10^{-15} m (2.8×10^{-14} in.).

The internal structure and finite size of the baryons (they consist of three quarks enclosed in a "bag," the quarks interacting through the exchange of gluons) must be considered when the baryons are very close together, that is, for separation distances less than approximately 0.7×10^{-15} m (2.8×10^{-14} in.). The consequences for nuclear forces are being actively investigated. SEE GLUONS; QUARKS.

Nucleon-nucleon interaction. The nucleon-nucleon interaction is the most thoroughly investigated strong interaction. This strong interaction is found to be approximately charge-independent; that is, the interactions between two protons and between a neutron and proton are found to be equal when their configurations are identical. The strong interaction is found to be charge-symmetric; that is, the strong interaction between protons is identical to that between neutrons, again in similar states. SEE ISOBARIC SPIN.

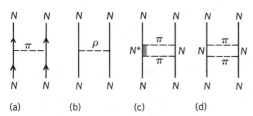

Fig. 1. Interaction between nucleons (a) from exchange of single pion, (b) from exchange of ρ-meson, a two-pion system, (c) from exchange of two separate pions with formation of excited state of nucleon, and (d) without formation of excited state.

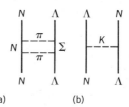

Fig. 2. Interaction between nucleon and strange (Λ) baryon from (a) exchange of pions and (b) exchange of kaon.

The nucleon-nucleon interaction is most precisely known when the nucleons are separated by more than 0.7×10^{-15} m (2.8×10^{-14} in.). For much smaller separation the interaction can be described empirically as strongly repulsive, this region being referred to as the hard core. For larger separations the interaction is attractive, falling off exponentially with distances in a manner dictated by the range of the interaction as given by the equation above.

Spin dependence. The nucleons have a spin and therefore their interaction is spin-dependent, that is, dependent on the spatial orientation of their spins. Because of this spin dependence, the simplest atomic nucleus, the deuteron, consisting of a neutron and proton, does not have a spherical shape. The spin dependence of the nucleon-nucleon interaction can be investigated by collisions between polarized nucleons, that is, nucleons whose spin have a definite spatial orientation. A resonance in the proton-proton collision has been discovered when energetic protons polarized in their direction of motion interact with protons polarized in either the same or opposite direction (**Fig. 3**). This resonance is thought to be a consequence of the possible excitation of one (or both) nucleons to an excited state (Fig. 1d). SEE SCATTERING EXPERIMENTS (NUCLEI); SPIN.

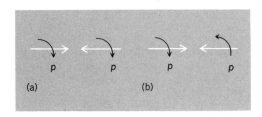

Fig. 3. Collision of two protons polarized in (a) same direction and (b) opposite direction.

Nuclear structure. The properties of nuclear forces are reflected in the structure of atomic nuclei. Qualitative aspects such as the short range and strength were first discovered in this manner. Another property is that of saturation; in other words, their general character is such that inside a nucleus each nucleon cannot interact with more than a few neighboring nucleons. The quantitative connection between nuclear forces and nuclear structure is not completely known and continues to be under active investigation. The fundamental goal is the determination of how the forces between a nucleon pair inside the nucleus are modified by the presence of the other nucleons in the nucleus. SEE NUCLEAR STRUCTURE.

Hypernuclei. The force between a nucleon and a Λ or Σ are being determined through the study of the properties of hypernuclei. The hypernuclei under current investigation consist of neutrons and protons and one Λ or Σ. The interaction between a Λ and a nucleon is not strong enough to form a stable two-body system like the deuteron. Another new feature is the existence of a substantial breakdown in charge symmetry, since it is found that the interaction between a Λ and a neutron is not the same as that between a Λ and a proton.

Exotic atoms. Further information about these systems is obtained from the exotic atoms, atoms formed by capturing negatively charged Σs in the attractive electric field of the nucleus. The radius of these Σ atomic orbits is much smaller than that of orbits of electrons in the electric field of the same nucleus, and as a consequence the Σ can interact via the strong interactions with the nucleons in the nucleus. This interaction is revealed by the properties of the x-rays emitted when the Σ changes its orbit.

Exotic atoms can also be formed by an atomic nucleus plus an antiproton. An antiproton has the same mass as the proton but carries a negative charge, whereas the proton carries a positive charge. Experiments involving the collision between antiprotons and protons promise to provide important insights into the strong interactions. SEE HADRONIC ATOM.

Interactions of mesons. Strong interactions exist between the pion and kaon with baryons and between pions and kaons. The interaction most thoroughly studied is that between pions and nucleons. It is found that when these interact they form a system which lives for about 6×10^{-22} s. It is referred to as the delta (Δ). The dynamics of the strong interaction of a pion

with a nucleus depends largely on the behavior of the Δ formed by the pion with one of the nucleons in the nucleus. SEE ELEMENTARY PARTICLE; FUNDAMENTAL INTERACTIONS.
Bibliography. C. J. Batty, Exotic atoms, *Sov. J. Particles Nucl.* 13:71–96, 1982; K. Hidaka and A. Yokosawa, *Survey in High Energy Physics 1*, 1980; B. Povh, *Annu. Rev. Nucl. Particle Sci.*, 28:1, 1978.

QUANTUM CHROMODYNAMICS
C. QUIGG

A theory of the strong ("nuclear") interactions among quarks, which are regarded as fundamental constituents of matter. Quantum chromodynamics (QCD) seeks to explain why quarks combine in certain configurations to form the observed patterns of subnuclear (or "elementary") particles, such as the proton and pi meson. According to this picture, the strong interactions among quarks are mediated by a set of force particles known as gluons. Strong interactions among gluons may lead to new structures that correspond to as yet undiscovered particles. The long-studied nuclear force which binds protons and neutrons together in atomic nuclei is regarded as a collective effect of the elementary interactions among constituents of the composite protons and neutrons. SEE NUCLEAR STRUCTURE.

It is believed that free quarks and gluons cannot be isolated, and it is hoped that quantum chromodynamics will provide a justification for this belief. By construction, quantum chromodynamics embodies many features abstracted from empirical observations. However, new mathematical inventions appear required before all the consequences of the theory can be reliably calculated. In part because of the shortcomings of known techniques for extracting predictions from the theory, quantum chromodynamics has not yet been subjected to rigorous experimental tests. Several qualitative predictions of quantum chromodynamics do seem to have been borne out. Part of the esthetic appeal of the theory is due to the fact that quantum chromodynamics is nearly identical in mathematical structure to quantum electrodynamics (QED) and to the unifed theory of weak and electromagnetic interactions put forward by S. Weinberg and A. Salam. This resemblance encourages the hope that a unified description of the strong, weak, and electromagnetic interactions may be at hand. SEE QUANTUM ELECTRODYNAMICS; WEAK NUCLEAR INTERACTIONS; WEINBERG-SALAM MODEL.

Gauge theories. At the heart of current theories of the fundamental interactions is the idea of gauge invariance, which draws its name from some early investigations by H. Weyl into a possible connection between scale changes and the equations of electrodynamics. Weyl attempted to deduce electromagnetism from a symmetry principle, the conjectured invariance of physical laws under a change of length scale chosen independently at every position of space and time. This specific undertaking ran afoul of quantum mechanics, but the general strategy and the name have survived. It is widely believed that gauge theories constructed to embody various symmetry principles represent the correct quantum-mechanical descriptions of the strong, weak, and electromagnetic interactions. SEE SYMMETRY LAWS.

Electromagnetism. The simplest example of a gauge theory is electromagnetism. It may be derived from a symmetry principle as follows. Quantum-mechanical observables do not depend upon the phase of the complex wave function which describes the state of a system. Therefore, the phase of a wave function can be rotated by an amount which is the same at all times and all places without affecting the physical consequences of the theory. The choice of phase is thus conventional, as opposed to observable. This is known as a global symmetry principle. It is natural to ask whether it should not be possible to choose this arbitrary convention independently at each point of space-time, again without affecting the physical consequences of the theory. It turns out to be possible to construct a quantum theory which is invariant under local (that is, position- and time-dependent) phase rotations that are proportional to the electric charge of the particles, but only if the theory contains an electromagnetic field with precisely the observed properties as summarized by Maxwell's equations. In the quantum theory, a massless spin-1 particle identified as the photon mediates the electromagnetic interaction. The interactions of matter with electromagnetism thus are essentially prescribed by the requirement of local phase invariance.

Local gauge invariance. Local phase rotations of the kind described above are the simplest examples of local gauge transformations. For a continuous symmetry, global gauge invariance implies the existence of a set of conserved currents. In the case of electromagnetism, it is the electric current which is conserved. A local gauge invariance requires in addition the existence of a massless gauge field corresponding to the set of phase transformations which forms the one-parameter unitary group U(1). The theory of electromagnetism was codified by James Clerk Maxwell more than 60 years before the local gauge invariance of its equations was discovered. However, it frequently happens in physics that the symmetries respected by a phenomenon are recognized before a complete theory has been developed. The question therefore arises as to whether the notion of local gauge invariance can be used to deduce the theory of nuclear forces.

Yang-Mills theory. This question was addressed in 1954 by C. N. Yang and R. L. Mills, and independently by R. Shaw. Early in the study of nuclear forces it was established that the nuclear interaction is charge-independent; it acts with the same strength between proton and proton, or proton and neutron, or neutron and neutron. This may be understood by saying that the proton and neutron represent two states of the same particle, called the nucleon. Just as an electron can be in a state with spin-up or spin-down, a nucleon can be in a state with the internal quantum number isospin-up (defined as the proton) or isospin-down (defined as the neutron). Charge independence then would reflect the invariance of the strong interactions under isospin rotations, characterized by the group SU(2). If isospin is regarded as a gauge group, local gauge invariance requires the existence of three massless spin-1 gauge particles, corresponding to the three generators of SU(2). The interactions of the gauge particles with nucleons are prescribed by the gauge principle. All of this is entirely parallel to the theory of electromagnetism. What distinguishes this SU(2) gauge theory from its U(1) counterpart is that the SU(2) gauge fields carry isospin and thus couple among themselves, whereas the photon is electrically neutral and does not interact with itself. Interacting gauge fields are an attribute of any theory based upon a non-abelian gauge group. SEE ISOBARIC SPIN.

Spontaneous symmetry breaking. Its mathematical properties notwithstanding, the Yang-Mills theory was unacceptable as a description of nuclear forces because, as H. Yukawa had shown, they are mediated by massive particles, whereas the gauge particles are required to be massless. Attempts in the early 1960s to generalize the Yang-Mills theory to the newly discovered SU(3) symmetry of the strong interactions encountered similar experimental objections. Gauge theories nevertheless continued to hold considerable appeal for theorists. Beginning in the late 1950s, a succession of gauge theories of the weak interactions appeared. At first these too foundered on the prediction of massless gauge bosons, but it was ultimately learned from the work of P. Higgs and others that spontaneous breakdown of the gauge symmetry would endow the gauge bosons with masses. All the elements were successfully combined in 1967 in the theory of weak and electromagnetic interactions proposed separately by Weinberg and Salam. When in 1971 G. 't Hooft and others demonstrated that spontaneously broken gauge theories were renormalizable, and hence calculable in the same sense as quantum electrodynamics, it stimulated experimental interest in the predictions of the Weinberg-Salam model, and renewed theoretical enthusiasm for gauge theories in general. Spontaneous symmetry breaking was not a cure for the shortcomings of gauge theories of the strong interactions. Instead, thanks to parallel developments described below, a new candidate emerged for the strong gauge group, and with it arose the idea of quantum chromodynamics. SEE GAUGE THEORY; RENORMALIZATION.

Color. It had been shown in 1963 by M. Gell-Mann and G. Zweig, working independently, that the observed pattern of strongly interacting particles, or hadrons, could be explained if the hadrons were composed of fundamental constituents called quarks. According to the quark model, a baryon such as the proton, which has half-integral spin in units of $\hbar$ (Planck's constant h divided by 2π) is made up of three quarks each with spin ½. An integral-spin meson, such as the pi meson, is made up of one quark and one antiquark. Three varieties (or flavors) of quarks, denoted up, down, and strange, could be combined to make all of the known hadrons, in precisely the families identified according to the eightfold way.

The idea that the strongly interacting particles are built up of quarks brought new order to hadron spectroscopy and suggested new relations among mesons and baryons. The constituent description brought with it a number of puzzles. These seemed at first to indicate that the quark model was nothing more than a convenient mnemonic recipe. In pursuing and resolving these puzzles, physicists have found a dynamical basis for the quark model which promises to give a complete description of the strong interactions.

An obvious question concerns the rules by which the hadrons are built up out of quarks. Mesons are composed of a quark and antiquark, while baryons are made of three quarks. What prevents two-quark or four-quark combinations? Within this innocent question lurks a serious problem of principle. The Pauli exclusion principle of quantum mechanics is the basis for understanding of the periodic table of the elements. It restricts the configurations of electrons within atoms and of protons and neutrons within nuclei. It should be a reliable guide to the spectrum of hadrons as well. According to the Pauli principle, identical spin-$\frac{1}{2}$ particles cannot occupy the same quantum state. As a consequence, the observed baryons such as Δ^{++} (uuu) and Ω^- (sss), which would be composed of three identical quarks in the same state, would seem to be forbidden configurations.

To comply with the Pauli principle, it is necessary to make the three otherwise identical quarks distinguishable by supposing that every flavor of quark exists in three varieties, fancifully labeled by the colors red, green, and blue. Then each baryon can be constructed as a "colorless" (or "white") state of a red quark, a green quark, and a blue quark. Similarly, a meson will be a colorless quark-antiquark combination. The rule for constructing hadrons may then be rephrased as the statement that only colorless states can be isolated. *See Color.*

A second issue is raised by the fact that free quarks have not been observed. This suggests that the interaction between quarks must be extraordinarily strong, and perhaps permanently confining. That free quarks are not seen is of course consistent with the idea that colored states cannot exist in isolation. On the other hand, the quark-parton model description of violent collisions rests on the assumption that quarks within hadrons may be regarded as essentially free, as explained below.

This paradoxical state of affairs may be visualized as follows. A hadron may be thought of as a bubble within which the constituent quarks are imprisoned. The quarks move freely within the bubble, but cannot escape from it. This picturesque representation yields an operational understanding of many aspects of hadron structure and interactions, but it falls far short of a dynamical explanation for the puzzling behavior of quarks. What remains to be understood is that quarks apparently interact only weakly when they are close together and yet cannot be pulled apart.

The quarks are the constituents of strongly interacting particles. The leptons, of which the electron and neutrino are the most common examples, are the fundamental particles which do not interact strongly. Each lepton flavor appears in only a single species. In other words, the leptons are colorless. In other respects, leptons resemble quarks: they are spin-$\frac{1}{2}$ particles which have no internal structure, at the current limits of resolution. Color may therefore be regarded as the strong-interaction analog of the electric charge. Color cannot be created or destroyed by any of the known interactions. Like electric charge, it is said to be conserved. *See Lepton.*

In the face of evidence that color could be regarded as the conserved charge of the strong interactions, it was natural to seek a gauge symmetry which would have color conservation as its consequence. An obvious candidate for the gauge symmetry group is the unitary group SU(3), now to be applied to color rather than flavor. The theory of strong interactions among quarks which is prescribed by local color gauge symmetry is known as quantum chromodynamics. The mediators of the strong interactions are eight massless spin-1 bosons, one for each generator of the symmetry group. These strong-force particles are named gluons because they make up the "glue" which binds quarks together into hadrons. Gluons also carry color, and hence have strong interactions among themselves.

Asymptotic freedom. The theoretical description of the strong interactions has historically been inhibited by the very strength of the interaction, which renders low order perturbative calculations untrustworthy. In 1973 a remarkable observation was reported by H. D. Politzer and by D. J. Gross and F. Wilczek. While studying the properties of Yang-Mills theories, they found that in many circumstances the effective strength of the interaction becomes increasingly feeble at short distances. For quantum chromodynamics, this implies that the interaction between quarks becomes weak at small separations. This discovery potentially has numerous important consequences, some of which will be described below. It raises the hope that some aspects of the strong interactions might be treated by using familiar computational techniques that are predicated upon the smallness of the interaction strength.

The physical basis for the change in the strength of the strong interaction as a function of the distance may be understood by examining first the corresponding question in electrodynamics. The electric charge carried by an object is customarily spoken of as a fixed and definite quantity, as indeed it is. However, if a charge is placed in surroundings in which other charges are free to move about, the effects of the charge may be modified. An example is a medium composed of many molecules, each of which has a positively charged end and a negatively charged end. In the absence of an intruding charged particle, the molecules are oriented randomly and the medium is electrically neutral not only in the large but locally as well—down to the submolecular scale. Introducing a test charge polarizes the medium (**Fig. 1**a). Charges of opposite sign are oriented toward the test charge, while those with like charge are repelled, as illustrated. This familiar screening effect means that the influence of the charge is diminished by the surrounding medium. Viewed from afar (but within the medium), the charge appears smaller in magnitude than its true or unscreened value. Only by inspecting the test charge at short range is it possible to feel its full effects.

A closely related phenomenon in quantum electrodynamics is known as vacuum polarization (Fig. 1b). The vacuum, or empty space, is normally thought of as the essence of nothingness. However, in quantum theory the vacuum is a complicated and seething medium in which "virtual" pairs of charged particles, most importantly electrons and positrons, have a fleeting existence. These ephemeral vacuum fluctuations are polarizable in the same way as the molecules of the above example, and serve to screen the charge at large distances. Consequently, in quantum electrodynamics it is also expected that the effective electric charge should increase at short distances, and indeed the consequences of this variation are observed in atomic spectra. The polarizability of the vacuum is related to the number of species of charged particles which take part in the screening. The behavior of the effective charge in quantum electrodynamics is opposite to that required in the realm of the strong interactions, where the interaction between quarks must diminish in strength at short distance.

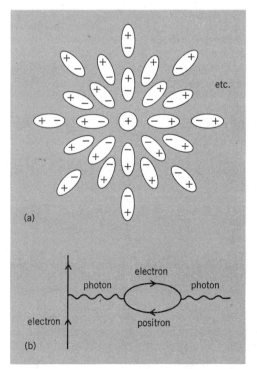

Fig. 1. Screening in electrodynamics. (a) Polarization of a dielectric medium by a test charge. (b) Feynman diagram contributing to vacuum polarization.

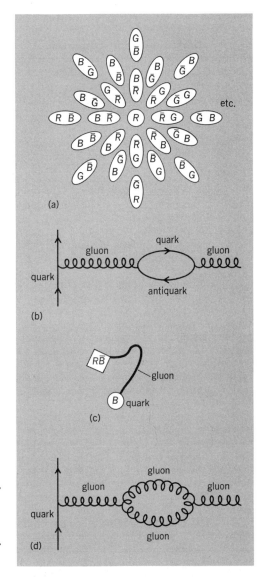

Fig. 2. Screening and antiscreening in quantum chromodynamics. (a) Screening of a colored quark by vacuum polarization. Color charges are denoted by R = red, $\bar{R}$ = anti-red, B = blue, $\bar{B}$ = anti-blue, G = green, $\bar{G}$ = anti-green. (b) Feynman diagram contributing to color polarization of the vacuum. (c) Delocalization of the color charge by gluon radiation. (d) Feynman diagram contributing to color antiscreening.

The situation in quantum chromodynamics is complicated by the fact that gluons carry the strong (color) charge, whereas photons are electrically neutral. This means that in addition to a color polarization phenomenon like the related charge screening of quantum electrodynamics (**Fig. 2**a and b), it is possible for the color charge of a quark to be shared with the gluon cloud (Fig. 2c and d). Because the color charge is spread out rather than localized, the effective color charge will tend to appear larger at long distances and smaller at short distances. The outcome of the competition between these opposing tendencies depends on the number of gluon species that can share the color charge and on the number of quark types that can screen the color charge. If the color gauge group is SU(3), the net effect is one of antiscreening, that is, of a smaller effective charge at short distances, provided the number of quark flavors is less than 17. Only five quark flavors are now known. Extremely close to a quark, the effective color charge becomes vanishingly small, so that nearby quarks behave as if they are noninteracting free particles. This is the origin of the term asymptotic freedom.

Experimental consequences. Asymptotic freedom has the immediate consequence, within the context of quantum chromodynamics, of providing a partial justification of the parton model description of violent scattering processes. The parton model was invented to explain features of electron-proton collisions in which many particles are created. Experiments first carried out at the Stanford Linear Accelerator Center in 1968 indicated that in these deeply inelastic collisions the electron "sees" the proton, not as an amorphous whole, but as a collection of structureless entities that have been identified as quarks. According to this picture, an electron scatters from a single parton, free from the influence of any other parton. This is reminiscent of electron scattering from nuclei, in which an electron may scatter from an individual proton or neutron as if it were a free particle. There is an important difference. Protons and neutrons are lightly bound in nuclei, and may be liberated in violent collisions. In contrast, free quarks have not been observed, so they must be regarded as very deeply bound within hadrons. Asymptotic freedom offers a resolution to the paradox of quasifree quarks that are permanently confined. At the short distances probed in deep inelastic scattering, the effective color charge is weak, so the strong interactions between quarks can largely be neglected. As quarks are separated, the effective color charge grows, so the strong interaction becomes more formidable. This characteristic may provide a mechanism for quark confinement, but calculational difficulties have prevented a verification of this conjecture.

In the regime of short distances probed in violent high-energy collisions, the strong interactions are sufficiently feeble that reaction rates may be calculated by using the diagrammatic methods developed for quantum electrodynamics. In some measure these calculations reproduce the simple quark model results as first approximations. This is the case, for example, in electron-positron annihilations into hadrons. The quark-antiquark production rate correctly anticipates both the structure of the dominant two-jet events and the approximate rate of hadron production. The strong-interaction corrections to this process include the process in which a gluon is radiated by one of the outgoing quarks. Like the quarks, the gluon materializes as a jet of hadrons. The resulting three-jet events are commonplace in the electron-positron annihilations studied at the PEP and PETRA storage rings.

The highest energies yet attained in collisions of the fundamental constituents are those reached in proton-antiproton interactions at the CERN Superproton Synchrotron (SPS) Collider, a storage ring in which 315-GeV protons collide head-on with counterrotating 315-GeV antiprotons. Collisions among quarks and gluons have been recorded at energies approaching 200 GeV. The hard scatterings of the partons (**Fig. 3**a) lead to striking jets of hadrons at large angles to the direction defined by the incident proton and antiproton beams. An example is shown in Fig. 3b. Events of this kind are observed at approximately the frequency suggested by quantum chromodynamics.

While diagrammatic methods are of great value in the study of strong interactions, several considerations prevent the resulting calculations from being as precise as those long familiar in quantum electrodynamics. The first is that at the energies currently accessible (or, in other words, at the distances currently probed), the strong interaction is still considerably stronger than electromagnetism. As a consequence the "higher-order" corrections to simple processes are in general more significant than in quantum electrodynamics. They are also somewhat more involved to calculate because quantum chromodynamics has three kinds of charge (red, green, and blue) in place of just one. This is not an issue of principle, but only of human endurance, so the precision of calculations will certainly be improved.

Although quantum chromodynamics provides support for the spirit of the parton model, it also exposes the incompleteness of the parton model description. Because quantum chromodynamics is an interacting field theory of quarks and gluons, probes of different wavelengths, which are analogous to microscopes of different resolving power, may map out different structures within the proton. This is indicated in **Fig. 4**, which shows, under increasing magnification, the virtual dissociation of quarks into quarks and gluons. There is evidence from high-energy muon-nucleon and neutrino-nucleon scattering of changes in the perceived structure of the proton of the kind suggested by quantum chromodynamics. Whether these changes are quantitively described by the theory remains to be seen.

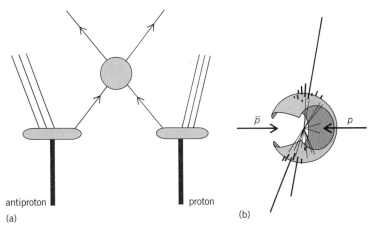

Fig. 3. High-energy proton-antiproton collisions. (a) Parton-model description of a violent collision between constituents of a proton and an antiproton. The active particles may be quarks, antiquarks, or gluons. (b) Configuration of a typical hard-scattering event observed at the CERN collider. Directions and energies of the produced charged particles are indicated by lines with lengths proportional to energies. The two-jet structure is apparent.

Quarkonium. The light hadrons are sufficiently large that the forces between the quarks within them are quite formidable. A fundamental description of the spectroscopy of light hadrons therefore awaits solution of the confinement problem. It is, however, possible to imagine special situations in which hadron spectroscopy is completely tractable by using available theoretical techniques. T. Appelquist and H. D. Politzer suggested late in 1974 that the bound system of an extremely massive quark with its antiquark would be so small that the strong force would be extremely feeble. In this case, the binding between quark and antiquark is mediated by the exchange of a single massless gluon, and the spectrum of bound states resembles that of an exotic atom composed of an electron and an antielectron (positron) bound electromagnetically in a Coulomb potential generated by the exchange of a massless photon (**Fig. 5***a*). Since the electron-positron atom is known as positronium, the heavy quark-antiquark atom has been called quarkon-

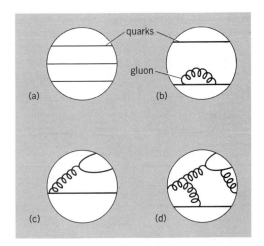

Fig. 4. Structure of the proton as observed with probes of increasing resolution. (a) Quarks appear as simple and noninteracting objects under "low magnification." (b–d) Quarks reveal increasing complexity because of their interactions with gluons as resolving power is improved.

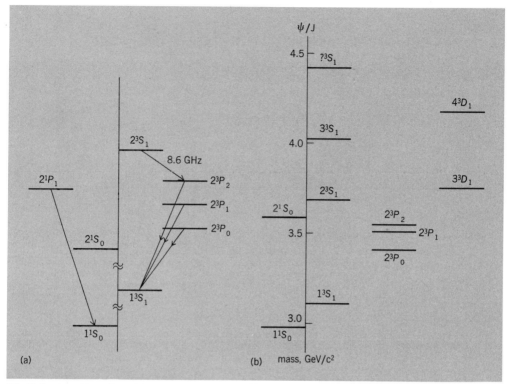

Fig. 5. Spectra of bound particle-antiparticle systems. (a) Positronium. (b) ψ/J family, known as charmonium.

ium. Two families of heavy quark-antiquark bound states, the ψ/J system composed of charmed quarks (Fig. 5b) and the ϒ system made up of b quarks, have been discovered. Both have level schemes characteristic of atomic spectra, which have been analyzed by using tools of nonrelativistic quantum mechanics developed for ordinary atoms. Neither system is so massive that the one-gluon-exchange description originally envisaged is entirely adequate, but the atomic analogy has proved extremely fruitful for studying the strong interaction. SEE CHARM; J PARTICLE; POSITRONIUM.

Future prospects. Quantum chromodynamics is an extremely promising candidate to be the correct theory of the strong interactions. In the regime in which asymptotic freedom renders some aspects of the strong interactions calculable, sharp predictions and incisive experiments are on the horizon. It is extremely important to seek verification that gluons interact among themselves. At the opposite extreme of very formidable strong interactions, the issue of permanent confinement of quarks and gluons remains to be resolved. A full understanding will necessarily bring with it solutions of two corollary problems: how quarks and gluons created in violent collisions materialize into the observed hadrons, and a complete description of the spectrum of hadrons.

To deal with the existence and properties of the hadrons themselves, it is necessary to devise a new computational approach which does not break down when the interaction becomes strong. The most promising approach has been the crystal lattice formulation of the theory, in which space-time is accorded a discrete, rather than continuous, structure. By considering the values of the color field only on individual lattice sites, it is possible to make use of many of the techniques developed in statistical physics for the study of spin systems such as magnetic substances.

FUNDAMENTAL INTERACTIONS AND FIELDS **331**

One of the most valuable methods has been the use of computer simulations in which different gluon configurations are explored by random sampling (Monte Carlo) techniques. This program makes extremely heavy demands on computer time, and has spurred the development and implementation of new computer architectures. Calculations of this sort have yielded suggestive evidence that quarks and gluons are indeed permanently confined in quantum chromodynamics. The eventual goal of this work is to compute the spectrum and properties of hadrons from first principles.

Attempts to understand confinement and the nature of the quantum chromodynamic vacuum have led to the prediction of new phenomena. It seems likely that when hadronic matter is compressed to very great densities and heated to extremely high temperatures hadrons will lose their individual identities. When the hadronic bubbles of the image discussed above overlap and merge, quarks and gluons may be free to migrate over great distances. A similar phenomenon occurs when atoms are squashed together in stars. The resulting new state of matter, called quark-gluon plasma, may exist in the cores of collapsing supernovae and neutron stars. The possibility of creating quantum chromodynamic plasma in the laboratory in collisions of energetic heavy ions is under active study.

Finally, because quantum chromodynamics and the unified theory of weak and electromagnetic interactions have the same gauge theory structure, it is interesting to contemplate the possibility that all three interactions have a common origin in a single gauge symmetry. The construction of such "grand unified theories" is a very active, though highly speculative, area of theoretical research. SEE ELEMENTARY PARTICLE; FUNDAMENTAL INTERACTIONS; GLUONS; GRAND UNIFICATION THEORIES; QUANTUM FIELD THEORY; QUARKS; STRONG NUCLEAR INTERACTIONS.

Bibliography. I. J. R. Aitchison and A. J. G. Hey, *Gauge Theories in Particle Physics*, 1982; N. Calder, *The Key to the Universe,* 1977; W. J. Marciano and H. Pagels, Quantum chromodynamics, *Phys. Rep.,* 36C:137–276, 1978; Y. Nambu, The confinement of quarks, *Sci. Amer.*, 235(5):48–60, November 1976; H. D. Politzer, Asymptotic freedom: An approach to strong interactions, *Phys. Rep.,* 14C:129–180, 1978; C. Quigg, *Gauge Theories of the Strong, Weak, and Electromagnetic Interactions*, 1983; C. Rebbi, Solitons, *Sci. Amer.,* 242(2):92–116, February 1979; F. J. Yndurain, *Quantum Chromodynamics: An Introduction to the Theory of Quarks and Gluons,* 1983.

GAUGE THEORY
C. N. YANG

In fundamental physics, theories conforming to the gauge principle of interaction between constituents of matter. It is the basis of all currently known interactions in physics.

Electromagnetism as a gauge theory. H. Weyl proposed in 1918 a theory of electromagnetism which he called gauge theory, because it has a gauge invariance. The idea was that electromagnetism causes a space-time–dependent scale change, so that if a particle is moved from the point x^μ to $x^\mu + dx^\mu$, its scale would change by $1 + S_\mu dx^\mu$, where S_μ is a space-time–dependent vector function. Weyl tried to identify S_μ with a numerical constant times the electromagnetic four-vector potential A_μ. A. Einstein raised the following objection to Weyl's idea: Consider two clocks brought around loops L_1 and L_2 as shown in the **illustration**. Einstein said that if Weyl's idea of space-time–dependent scale changes were right, these clocks, originally identical, would acquire different scales after they were brought around their respective loops. That implies also different time scales, so that the clocks would then go at different speeds. Einstein said, "The length of a common ruler (or the speed of a common clock) would depend on its history." That clearly would lead to absurdities, and the scale variation idea of Weyl could not be valid.

Few physicists took up Weyl's idea until 1927, two years after the development of quantum mechanics, when V. Fock observed that in quantum mechanics the expression for a charged particle in classical electrodynamics, $p_\mu - eA_\mu$, should be replaced as in Eq. (1). F. London then

$$p_\mu - eA_\mu \to -i\hbar\partial_\mu - eA_\mu = -i\hbar[\partial_\mu - ie\hbar^{-1}A_\mu] \tag{1}$$

pointed out that the expression in the square brackets would be the same as Weyl's $[\partial_\mu + S_\mu]$ if

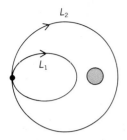

Configuration devised by Einstein to test Weyl's theory of electromagnetism. Clocks are brought around loops L_1 and L_2. This configuration also represents the Aharonov-Bohm experiment, in which electrons are brought around the loops and the shaded circle is a solenoid.

one makes the identification in Eq. (2). This is similar to Weyl's identification, but with the inser-

$$S_\mu = -ie\hbar^{-1}A_\mu \qquad (2)$$

tion of the important factor $i = \sqrt{-1}$. That, of course, implies that the scale change $1 + S_\mu dx^\mu$ becomes Eq. (3), which is a phase change, not a scale change. Thus was born the concept of

$$1 - ie\hbar^{-1}A_\mu dx^\mu = \exp[-ie\hbar^{-1}A_\mu dx^\mu] \qquad (3)$$

electromagnetism as a gauge theory in quantum mechanics. With the scale change replaced by a phase change, the Einstein objection is rendered inoperative, since the two clocks would come back after their loop motions with different electromagnetic phases but without changes of scale. Thus, they would still keep time at the same rate. One could ask whether the difference of their electromagnetic phases is observable. The answer is no, with present technology, but if the objects are electrons and not clocks, the difference is observable by interference experiments. This is, in fact, the important Aharonov-Bohm experiment, first proposed and studied in 1959, in which the phase difference is controlled by the strength of the magnetic flux in the solenoid indicated by the shaded circle in the illustration.

Nonabelian gauge theory. In 1954, in order to find new principles of interaction between fundamental particles, C. N. Yang and R. L. Mills generalized the electromagnetic gauge principle. The result is nonabelian gauge theory. In this theory, the phase [Eq. (3)] is replaced by a generalized phase, which is a concept developed since 1880 by S. Lie, E. Cartan, and others. Lie had explored the theory of Lie groups, the elements of which can be taken as generalized phases. Lie groups can be abelian or nonabelian. With a nonabelian Lie group, the resultant gauge theory is called a nonabelian gauge theory, and has a particularly intricate and beautiful structure. To appreciate the origin of this structure in general terms, it may be observed that in any gauge theory a conserved quantity serves as the source of the gauge field, in the same way that an electric charge (which is conserved) serves as the source of an electromagnetic field. For abelian gauge theories, such as electromagnetism, the field itself is not charged. But for a nonabelian theory, the field is itself charged with the same conserved quantity that generates the field. In other words, a nonabelian gauge field generates itself. It is this last feature that gives rise to the particularly intricate structure of the theory. In 1971, G. t'Hooft proved that nonabelian theories are renormalizable, a very important development in the possible structure of interactions. SEE RENORMALIZATION.

Symmetry dictates interactions. Theoretical and experimental developments since 1970 have established that the Weinberg-Salam model describes the weak interactions very well. This model is based on a nonabelian gauge theory with the added concept of symmetry breaking. Also, the generally accepted theory of strong interactions today is quantum chromodynamics, which is a nonabelian gauge theory. Gravity is also a nonabelian gauge theory, but in a sense which has not yet been totally clarified. These developments lead to the now universally accepted notion that all fundamental interactions of nature are gauge theories. Since gauge theories are based on the principle of local phase symmetry, the current notion about the origin of interactions is summarized by the statement: symmetry dictates interactions. This idea, that symmetry could be used as a guiding principle for formulating interactions in fundamental physics, was in fact first used and emphasized by Einstein and led to his theory of general relativity of 1915. SEE ELEMENTARY PARTICLE; FUNDAMENTAL INTERACTIONS; QUANTUM CHROMODYNAMICS; SYMMETRY BREAKING; SYMMETRY LAWS; WEINBERG-SALAM MODEL.

Bibliography. I. J. Aitchison, *An Informal Introduction to Gauge Field Theories*, 1982; E. Leader and E. Predazzi, *An Introduction to Gauge Theories and the "New Physics"*, 1982; G. t'Hooft, Renormalizable Lagrangians for massless Yang-Mills fields, *Nucl. Phys.*; B35:167–188, 1971; G. t'Hooft, Renormalization of massless Yang-Mills fields, *Nucl. Phys.*, B33:173–199, 1971; H. Weyl, *Sitzungsber. d.k. Preuss. Akad. d. Wiss.* p. 465, 1918; C. N. Yang, in *Proceedings of the 1983 Tokyo Conference on the Foundation of Quantum Mechanics*, 1984; C. N. Yang and R. L. Mills, Conservation of isotopic spin and isotopic gauge invariance, *Phys. Rev.*, 95:631, 1954, and 96:191–195, 1954.

INSTANTON
ROMAN JACKIW

A solution to the (imaginary-time) nonlinear field equation which arises in Yang-Mills field theory—a modern nonlinear generalization of Maxwell electromagnetic theory that is believed to give a fundamental description of elementary particles and forces. The instanton solution carries information about quantum tunneling. (Initially, the instanton was called a pseudoparticle.) SEE GAUGE THEORY.

A dynamical model when analyzed within quantum mechanics gives rise to processes different from those seen in an analysis based on classical mechanics. Of course, since nature is quantum mechanical, only the former processes are truly physical, but the latter can be used to give an approximate, semiclassical description. Of specific interest here is the tunneling phenomenon whereby a quantum-mechanical system evolves along paths that are classically forbidden, in that the classical equation of motion does not possess solutions which follow the forbidden paths. Nevertheless, there does exist in the classical theory a spur of quantum-mechanically allowed transitions: if, in the classical equation of motion, time t is replaced by imaginary time $(t \to \sqrt{-1} t)$, solutions which follow tunneling paths can be found. In other words, processes classically forbidden in real time become allowed in imaginary time. Moreover, the quantum-mechanical tunneling amplitude Γ is approximately given by the equation below, where I is the (imagi-

$$\Gamma \approx \exp\left(-\frac{1}{\hbar} I\right)$$

nary-time) action for the (imaginary-time) tunneling solution, and $\hbar$ is Planck's constant. Obviously only solutions of noninfinite action are relevant, since Γ vanishes if I diverges. Also, I must be positive, so that Γ disappears in the classical limit, $\hbar \to 0$.

While the above ideas, as they relate to point particles, were widely appreciated from the early days of quantum mechanics under the names semiclassical and WKB theory, their relevance to fields, such as the Maxwell electromagnetic field and its generalization the Yang-Mills field, was not realized until the mid-1970s, when an imaginary-time, finite-action solution to the Yang-Mills nonlinear wave equation was found. First called a pseudoparticle, then renamed an instanton, this solution was quickly interpreted as evidence for quantum tunneling in the Yang-Mills theory. This tunneling in turn served to explain the mass difference between the pi and eta mesons, and predicted that nucleons (protons and neutrons) are not stable, although their decay rate is exponentially small, hence negligible. [The instanton (tunneling) mediated proton decay is independent from that predicted by theories that try to unify all fundamental in-teractions.] SEE FUNDAMENTAL INTERACTIONS; GRAND UNIFICATION THEORIES; QUANTUM FIELD THEORY.

The instanton solution is also of interest to mathematicians, who independently had come to a study of Yang-Mills fields from a purely mathematical context called fiber-bundle theory, which combines differential geometry and topology. The original solution was generalized, and it was found that the (imaginary-time) Yang-Mills action of an arbitrary solution is an integral multiple of the action for the original solution. The integer is called instanton number by physicists and Chern-Pontryagin number by mathematicians. The reason for this quantization with classical field theory derives from the topological properties of Yang-Mills fields. The most general N-instanton solution has been found.

In addition to calling attention to unexpected physical processes, namely tunneling, the instanton solution has put into evidence the rich topological structure that is present in Yang-Mills theory. Contemporary research is concerned with understanding this structure, an activity in which both physicists and mathematicians are engaged in a rare collaborative effort. SEE ELEMENTARY PARTICLE.

Bibliography. M. Atiyah et al., Construction of instantons, *Phys. Lett.*, 65A:185–187, 1978; A. Belavin et al., Pseudoparticle solutions of the Yang-Mills equations, *Phys. Lett.*, 59B:85–87, 1975; G. 't Hooft, Symmetry breaking through Bell-Jackiw anomalies, *Phys. Rev. Lett.*, 37:8–11, 1976; D. S. Freed and K. K. Uhlenbeck, *Instantons and Four-Manifolds*, 1984; R. Jackiw, Introduction to the Yang-Mills quantum theory, *Rev. Mod. Phys.*, 52:661–673, 1980; R. Jackiw, Quantum meaning of classical field theory, *Rev. Mod. Phys.*, 49:681–706, 1977.

WEINBERG-SALAM MODEL
CHARLES BALTAY

A theory describing the electromagnetic and the weak forces between the elementary particles known as quarks and leptons that are presently believed to be the fundamental constituents of matter. Before this model there were believed to be four basic forces or interactions between the elementary particles: the strong nuclear forces, the electromagnetic forces, the weak interactions, and gravity. The model has succeeded in unifying the electromagnetic and the weak forces into so-called electroweak interactions, thus reducing the basic number of forces required to describe the interactions of the elementary particles to three. This consolidation may be similar in significance to the work of J. C. Maxwell and others in the nineteenth century that united the electric and the magnetic forces into a single electromagnetic theory.

Neutral currents. One of the most striking predictions of the Weinberg-Salam model was that of neutral current weak interactions. Until the development of the model, these neutral current interactions were generally believed to be absent, due to the very low experimental limits on some strangeness-changing neutral current processes such as those given by Eqs. (1) and (2). However,

$$K^{\pm} \rightarrow \pi^{\pm} + e^{+} + e^{-} \quad (1) \qquad K^{0} \rightarrow \mu^{+} + \mu^{-} \quad (2)$$

the existence of neutral currents was experimentally discovered in strangeness-conserving neutral current neutrino interactions, such as neutrino-electron elastic scattering, Eq. (3), and single-pion production by neutrinos, Eq. (4). The apparent dilemma, that is, the existence of strangeness-

$$\nu_{\mu} + e^{-} \rightarrow \nu_{\mu} + e^{-} \quad (3) \qquad \nu_{\mu} + p \rightarrow \nu_{\mu} + p + \pi^{0} \quad (4)$$

conserving neutral currents, while strangeness-changing neutral currents were absent, was reconciled by a new property of elementary particles called charm (similar to but distinct from the property called strangeness) postulated by S. Glashow, J. Iliopoulos, and L. Maiani. In this scheme, often called the GIM mechanism, the strangeness-changing neutral currents were exactly canceled out by the charm-changing neutral currents, while the strangeness- (and charm-) conserving neutral currents remain in existence. This postulate also predicted the existence of a new family of elementary particles, called the charmed particles, which were actually discovered in 1974. SEE CHARM; NEUTRAL CURRENT; NEUTRINO.

Intermediate bosons. The Weinberg-Salam model introduced a triplet of intermediate bosons, the W^{+}, W^{0}, and W^{-}, and a singlet, the B^{0}. The neutral bosons form a quantum-mechanical mixture given by Eqs. (5), where θ is a mixing angle called the Weinberg angle. The γ is the

$$\gamma = \cos \theta \, B^{0} - \sin \theta \, W^{0} \qquad Z^{0} = \sin \theta \, B^{0} + \cos \theta \, W^{0} \quad (5)$$

well-known photon that mediates the electromagnetic interactions, and the Z^{0} mediates the neutral current weak interactions. The $W^\pm$ mediate the charged current weak interactions. Thus the electromagnetic and the weak interactions are described by a single theory. SEE INTERMEDIATE VECTOR BOSON.

Predictions. The theory had great predictive power in that it predicted the cross sections (reaction rates) and other properties of a large variety of neutral current processes in terms of a single parameter, the mixing angle θ. An extensive experimental program was undertaken to study neutral current processes such as those given by Eqs. (6)–(9), the scattering of polarized

$$\nu + e \rightarrow \nu + e \quad (6)$$

$$\nu + p \rightarrow \nu + p \quad (7)$$

$$\nu + p \rightarrow \nu + p + \pi^0 \quad (8)$$

$$\nu + p \rightarrow \nu + \text{anything} \quad (9)$$

electrons on deuterium, parity-violating effects in atomic-level transitions, and so forth. The result of this experimental program was that all of the processes studied were consistent with the predictions of the model, and the value of the Weinberg angle was found to be given by Eq. (10).

$$\sin^2 \theta = 0.23 \pm 0.02 \quad (10)$$

Given this experimentally determined value of the mixing angle, the model was able to predict the masses of the $W^\pm$ and the Z^0 intermediate bosons to be around 80 and 90 GeV, respectively. These particles were indeed found in 1983 at the antiproton-proton collider at the CERN Laboratory, with masses consistent with the above prediction. These discoveries were considered a major triumph for the Weinberg-Salam model.

Standard model. This model, together with the Glashow-Iliopoulos-Maiani charm scheme, forms the basis of the current understanding of particle physics, called the standard model. The fundamental constituents in this model are six quarks (d, u, s, c, b, t) and six leptons (e, ν_e, μ, ν_μ, τ, ν_τ). The quarks interact via both the strong and the electroweak interactions, while the leptons have only electroweak interactions. The strong interactions are mediated by a neutral boson called the gluon (g); the theory that describes the strong interactions is called quantum chromodynamics (QCD). The electroweak interactions are mediated by the γ, the Z^0, and the $W^\pm$ bosons and are described by the Weinberg-Salam model. SEE ELEMENTARY PARTICLE; FUNDAMENTAL INTERACTIONS; GLUONS; LEPTON; QUANTUM CHROMODYNAMICS; QUARKS; WEAK NUCLEAR INTERACTIONS.

Bibliography. G. Arnison et al., Experimental observation of isolated large transverse energy electrons with associated missing energy at $\sqrt{s}$ = 540 GeV, *Phys. Lett.*, 122B:103–116, 1983; P. Becher, M. Bohm, and H. Joos, *Gauge Theories of Strong and Electroweak Interactions*, 1984; E. D. Commins and P. H. Bucksbaum, *Weak Interactions of Leptons and Quarks*, 1983; H. Georgi, *Weak Interactions and Modern Particle Theory*, 1984; S. Glashow, J. Iliopoulos, and L. Maiani, Weak interactions with lepton-hadron symmetry, *Phys. Rev.*, D2:1285–1292, 1970; C. H. Lai (ed.), *Gauge Theory of Weak and Electromagnetic Interactions*, 1981; S. Weinberg, A model of leptons, *Phys. Rev. Lett.*, 19:1264–1266, 1967.

NEUTRAL CURRENTS
CHARLES BALTAY

Exchange currents which carry no electric charge and mediate certain types of electroweak interactions. The discovery of the neutral-current weak interactions and the agreement of their experimentally measured properties with the theoretical predictions were of great significance in establishing the validity of the Weinberg-Salam model of the electroweak forces.

The neutral-current weak interactions are one subclass of the forces between the fundamental constituents of matter, the quarks and leptons, called the elementary particles. The three basic interactions between these particles are the strong nuclear forces, the electroweak forces, and gravity. The electroweak forces come in three subclasses: the electromagnetic interactions, the charged-current weak interactions, and the neutral-current weak interactions.

Electroweak interactions. The electroweak interactions are theoretically understood to be a current-current type of interaction, where the interaction is mediated by an exchange current or particle. The electromagnetic interaction is mediated by an exchanged photon γ (**illus.** a).

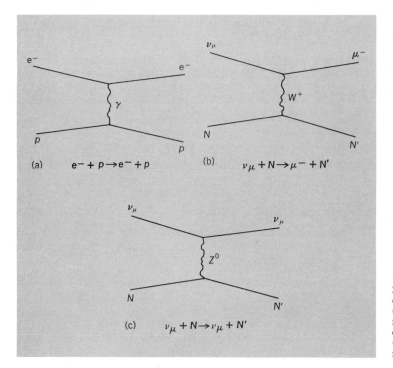

Diagrams for various electroweak interactions. (a) Electromagnetic interaction. (b) Charged-current weak interaction. (c) Neutral-current weak interaction.

Since the photon carries no electric charge, there is no change in charge between the incoming and the outgoing electron (the upper vertex in illus. a). The charged-current weak interaction is mediated by the exchange of a charged intermediate boson, the $W^\pm$, and thus, for example, an incoming neutral lepton such as the ν_μ in illus. b is changed into a charged lepton, the μ^-. In the neutral-current weak interactions, the exchanged intermediate boson, the Z^0, carries no electric charge (hence the name neutral-current interaction), and thus an incident neutral lepton, the ν_μ in illus. c, remains an outgoing neutral ν_μ. SEE ELECTRON; INTERMEDIATE VECTOR BOSON; LEPTON; NEUTRINO.

Discovery and investigation. The existence of the neutral-current weak interactions was predicted by the Weinberg-Salam model, which unified the electromagnetic and the weak nuclear forces into one basic interaction, now called the electroweak interactions. The neutral-current interactions were experimentally discovered in 1973, and have since been extensively studied, in neutrino scattering processes such as neutrino-electron elastic scattering, $\nu + e^- \to \nu + e^-$; neutrino-proton scattering, $\nu + p \to \nu + p$; single-pion production, $\nu + p \to \nu + p + \pi^0$; and inclusive neutrino scattering, $\nu + p \to \nu +$ any number of hadrons. Very important information about the properties of the neutral currents have been obtained by studying the interference effects between the electromagnetic and the neutral-current weak interactions in the scattering of polarized electrons on deuterium. Parity violating effects in atomic physics processes due to the neutral weak currents have been observed, and predicted parity-violating nuclear effects have been searched for. SEE PARITY.

Properties. As a result of the intensive experimental studies of neutral currents in the reactions mentioned above, the properties of the neutral currents have been fairly well determined. They are a mixture of vector (V) and axial-vector (A) currents (that is, the neutral currents transform like vectors and axial vectors under spatial rotations, and not like scalars, pseudoscalars, or tensors), and they are a mixture of isoscalar ($I = 0$) and isovector ($I = 1$) currents (that is, they transform like isoscalars and isovectors under isotopic spin rotations). The coupling constants that determine the relative strength of the vector, axial-vector, and isoscalar and isovector components have been measured, and are in very good agreement with the values predicted by the Weinberg-

Salam model. *See* ELEMENTARY PARTICLE; FUNDAMENTAL INTERACTIONS; ISOBARIC SPIN; QUARKS; SYMMETRY LAWS; WEAK NUCLEAR INTERACTIONS; WEINBERG-SALAM MODEL.

RENORMALIZATION
ITZHAK BARS

A program in quantum field theory consisting of a set of rules for calculating S-matrix amplitudes which are free of ultraviolet (or short-distance) divergences, order by order in perturbative calculations in an expansion with respect to coupling constants. *See* SCATTERING MATRIX.

Divergences in quantum field theory. To describe the nature of the problem, it is useful to consider the simple example of a ϕ^4 theory defined by the lagrangian density $L(x)$ in Eq. (1) in four-dimensional Minkowski space-time. Here, $\phi(x)$ is a quantum field operator depending

$$L(x) = \frac{1}{2}\partial_\mu\phi(x)\partial^\mu\phi(x) - \frac{1}{2}m^2\phi(x)^2 - \frac{1}{4!}\lambda\phi(x)^4 \qquad (1)$$

on the four-vector x, ∂_μ and ∂^μ represent differentiation with respect to space-time coordinates, and m and λ are parameters.

If one attempts to calculate any physical process in a perturbative expansion with respect to the coupling constant λ, in terms of Feynman diagrams, one has to confront divergent integrals. An example is the one-loop contribution to two-body scattering which is of order λ^2 (the second diagram in the **illustration**), and is proportional to the logarithmically divergent momentum integral given by expression (2), where p_1 and p_2 are the initial momenta of the two particles, and k is the loop momentum.

$$\lambda^2 \int d^4k [(p_1 + k)^2 - m^2 + i\epsilon]^{-1} [(p_2 - k)^2 - m^2 + i\epsilon]^{-1} \qquad (2)$$

These ultraviolet (large-momentum) divergences have their origin in short-distance singularities occurring in the product of the quantum field operators (or their matrix elements) such as $\phi(x_1)\phi(x_2)$ as x_1 approaches x_2 [see expression (5) below]. Therefore, the quantum theory is not well defined since the lagrangian (or hamiltonian) contains these fields multiplied at the same point $x_1 = x_2 = x$. Renormalization is the procedure for constructing a well-defined finite quantum field theory, and deals with the proper definition of such singular terms in the lagrangian.

Regularization procedure. First the nature and kind of singularities must be identified, and then they should be removed to define the physical theory. The theory is first rendered finite by introducing a regularization parameter so that as it approaches a limiting value the divergences appear as definite singularities in this parameter. This process is called regularization. There are many methods of regularization. For example, in the cutoff method the integral in Eq. (2) will be rendered finite if it is cut off at $k^2 = \Lambda^2$, where Λ is finite. The result of integration, which is a function of Λ, will be seen to diverge as log Λ when $\Lambda \to \infty$. Another, more modern method is dimensional regularization: The integral is first performed in n dimensions rather than four. It will converge for $n < 4$ and will have a definite dependence on n. An example of this dependence is given by Eq. (3), where $\Gamma(\alpha)$ is the Euler gamma function. All integrals, including the one in

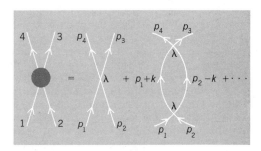

Two-body scattering in lowest-order perturbative expansion.

$$\int d^n k(-k^2 + c)^{-\alpha} = i\pi^{n/2} c^{(n/2)-\alpha} \Gamma[\alpha - (n/2)]/\Gamma(\alpha) \tag{3}$$

expression (2), can be put in the form of Eq. (3) by using Feynman parameters to combine denominators. The answer is then analytically continued to complex values of n and the four-dimensional theory is defined as the limit of n approaching 4. In this way, as long as $n \neq 4$ the gamma functions in Eq. (3) are finite, and only finite well-defined quantities are manipulated. In this method the singularities reappear as poles of the form $(n - 4)^{-k}$ as n approaches 4. The behavior of the integral of expression (2) is given by expression (4). The remainder of this article is restricted to dimensional regularization.

$$\frac{2i\pi^2}{n - 4} + (\text{finite}) + 0(n - 4) \tag{4}$$

Renormalization of parameters. In a regularized quantum field theory, one can, in principle, calculate finite expressions for all the Feynman diagrams that contribute to any physical process, to any desired order in perturbation theory. The answer will be a well-defined function of the momenta of the external legs, the parameters of the theory, such as m and λ in Eq. (1), and the number of dimensions n, which is taken as a complex number. As n approaches 4, the singularities can be studied and the method for removing them can be given. The degree of divergence of a graph is determined by Weinberg's theorem. In a field theory such as ϕ^4 of Eq. (1), one finds only two types of Feynman graphs that diverge as $n \to 4$. Those are the graphs with two external legs (propagator) and four external legs (scattering amplitude) and any other graph that contains these two types of graphs as subgraphs. Therefore, to further study the singularities it is sufficient to concentrate only on the two- and four-point functions of expression (5), whose Fourier

$$\begin{aligned} &<0|T(\phi(x_1)\phi(x_2))|0> \\ &<0|T(\phi(x_1)\phi(x_2)\phi(x_3)\phi(x_4))|0> \end{aligned} \tag{5}$$

transforms give the scattering amplitude. The singularities occur at short distances in the product of the field operators as points approach each other. This suggests that they can be removed from all physical processes by properly defining the local product of operators that appear in the lagrangian of Eq. (1). It turns out that indeed it is possible to absorb and remove all the infinities by a simple renormalization of the finite number of constants that define the theory. A theory is called nonrenormalizable when this procedure fails, and it can be rendered finite only by introducing an infinite number of renormalization parameters.

Nonphysical nature of parameters. The possibility of renormalization by a redefinition of parameters such as masses and coupling constants that appear in the lagrangian hinges on the fact that these are not the physical quantities that would be observed as the prediction of the theory. For example, the physical finite coupling constant is defined as the strength of the interaction that an experimentalist will observe in the process shown in the illustration. The strength of the interaction is defined by the measured value of the physical probability amplitude in the illustration, $A(p_1,p_2,p_4,\lambda,m)$, at an agreed value of the external momenta. This clearly is a complicated function of the parameter λ when the perturbation series is summed. The strength of the interaction is equal to λ only in lowest order and in general differs from it in the full theory. Similar remarks apply to the parameter m. The true mass of the particle in the theory differs from m. It is in general a complicated function of m and λ and is defined as the location of the pole in the full propagator. Similarly, the field $\phi(x)$ does not create the correctly normalized interacting particle when applied on the vacuum. It differs from the correctly normalized field ϕ_R by the multiplicative wave function renormalization constant $\sqrt{Z}$. Renormalization theory is the process of rewriting systematically the field theory in terms of the physical, finite, renormalized coupling constant λ_R, mass m_R, and field ϕ_R. If the theory is renormalizable and rewritten in terms of renormalized quantities, then all infinities cancel and all physical quantities, such as S-matrix elements, are finite and meaningful, as the regularization parameter is removed.

Power series in the renormalized parameters. An outline of this procedure can be given as follows. Since it has been agreed that λ, m and the renormalization of the field are not the observed physical quantities, they need not be finite. They are reexpressed as power series in terms of finite renormalized parameters λ_R and m_R and the regularization parameter. In the dimensional regularization scheme they take the form of Eqs. (6), where the renormalized field ϕ_R is

$$\lambda = \mu^{4-n}\left(\lambda_R + \frac{a_1(\lambda_R)}{n-4} + \frac{a_2(\lambda_R)}{(n-4)^2} + \cdots\right)$$

$$m^2 = m_R^2\left(1 + \frac{b_1(\lambda_R)}{n-4} + \frac{b_2(\lambda_R)}{(n-4)^2} + \cdots\right) \quad (6)$$

$$Z = 1 + \frac{c_1(\lambda_R)}{n-4} + \frac{c_2(\lambda_R)}{(n-4)^2} + \cdots$$

defined as $\phi = \sqrt{Z}\phi_R$. Here $a_i(\lambda_R)$, $b_i(\lambda_R)$, $c_i(\lambda_R)$ are power series in the finite parameter λ_R; that is, $a_i(\lambda_R) = \Sigma a_{ik}\lambda_R^k$, and so forth, where the coefficients a_{ik}, b_{ik}, c_{ik} are to be determined so as to render the theory finite as described below. Here μ is a mass parameter called the renormalization point, whose role is better understood by studying the renormalization group (discussed below). All Green's functions or probability amplitudes such as the one in the illustration will be renormalized by setting the momenta at $p_i^2 = \mu^2$. Furthermore, in Eq. (6), λ_R and m_R implicitly depend on μ, while λ and m are independent of it.

The perturbation series is now arranged as an expansion in the finite parameter λ_R and not in λ. This can be done by rewriting the lagrangian in terms of renormalized field ϕ_R and the parameters λ_R and m_R by substituting directly from Eqs. (6) into Eq. (1). Thus, the lagrangian takes the form of Eq. (7), where $\Delta L_{\text{counterterm}}$, the "counterterm" piece, has the same functional

$$L = \frac{1}{2}\partial_\mu\phi_R\partial^\mu\phi_R - \frac{1}{2}m_R^2\phi_R^2 - \frac{1}{4!}\mu^{4-n}\lambda_R\phi_R^4 + \Delta L_{\text{counterterm}} \quad (7)$$

dependence on ϕ_R as the explicitly written terms in Eq. (7) but with coefficients that correspond to the pole terms in Eq. (6). This piece would diverge as $n \to 4$ and is a power series in λ_R involving the coefficients a_{ik}, b_{ik}, c_{ik} mentioned above.

Minimal subtraction. These coefficients are determined in order to cancel all infinities by the following procedure: The amplitude $A(p_i,\lambda_R,m_R,\mu,n)$ in the illustration and the propagator $\Delta_F(p_i,\lambda_R,m_R,\mu,n)$ are calculated to any desired order in λ_R in terms of Feynman graphs [these are related to the Green's functions in expression (5)]. To these there are contributions from the explicit part of the lagrangian in Eq. (7) and also from the counterterms which involve the coefficients a_{ik}, b_{ik}, and c_{ik} up to the desired order in λ_R. Then, order by order, one requires that the poles $(n-4)^{-i}$ must completely cancel in the three measurable quantities given in Eq. (8). This

$$A(p_i,\lambda_R,m_R,\mu,n)|_{p^2=\mu^2} = \text{finite as } n \to 4$$

$$\Delta_F^{-1}(p,\lambda_R,m_R,\mu,n)|_{p^2=\mu^2} = \text{finite as } n \to 4 \quad (8)$$

$$\frac{\partial^2}{\partial p^2}\Delta_F^{-1}(p,\lambda_R,m_R,\mu,n)|_{p^2=\mu^2} = \text{finite as } n \to 4$$

is called the minimal subtraction scheme, and it completely fixes the coefficients a_{ik}, b_{ik}, and c_{ik}. There are other subtraction schemes which require the above quantities to be equal to a fixed observed value rather than being simply finite as $n \to 4$. The observable quantities such as the S-matrix elements are independent of the particular subtraction scheme adopted.

Finite S-matrix. As mentioned above, the only "primitive" infinities in the ϕ^4 theory appear in the Green's functions of expression (5) containing two or four external legs and in those graphs containing them as subgraphs. With the above renormalization procedure (choosing a_{ik}, b_{ik}, c_{ik}) all such infinities in these two- and four-point functions have been arranged to cancel. It remains to be shown that as $n \to 4$ (removal of cutoff) the S-matrix is finite. This is done by mathematical induction: Assume that it is true to order m in perturbation theory, then prove it to order $m+1$. This is carried out successfully if the theory is renormalizable. Therefore, all infinities can be removed by a proper definition of the products of the field operators appearing in the lagrangian. This amounts to properly identifying the observable finite quantities which led to a renormalization of the field, the coupling constant and the mass as in Eqs. (6). Any other physical quantity can now be calculated and will be a function of only the renormalized finite parameters λ_R and m_R, in addition to momenta.

Examples of renormalizable fields. So far the only field theories known to be renormalizable in four dimensions are those which include spin-0, spin-½ and spin-1 fields such that no term in the lagrangian exceeds operator dimension 4. The operator dimension of any term is calculated by assigning dimension 1 to bosons and derivatives ∂_μ, and dimension ³⁄₂ to fermions. Spin-1 fields are allowed only if they correspond to the massless gauge potentials of a locally gauge-invariant Yang-Mills-type theory associated with any compact Lie group. The gauge invariance can remain exact or can be allowed to break via spontaneous breakdown without spoiling the renormalizability of the theory. In the latter case the spin-1 field develops a mass. The successful quantum chromodynamics theory describing the strong forces and the SU(2) × U(1) Weinberg-Salam-Glashow gauge model of unified electroweak particle interactions are such renormalizable gauge models containing spin-0 −½, and −1 fields. The renormalization procedure in a gauge theory is much more complicated than in the simple ϕ^4 theory because of gauge fixing and lack of either manifest unitarity or Lorentz invariance, but gauge theories have been shown to be renormalizable. SEE FUNDAMENTAL INTERACTIONS; GAUGE THEORY; QUANTUM CHROMODYNAMICS; WEAK NUCLEAR INTERACTIONS; WEINBERG-SALAM MODEL.

Renormalization group. An important topic in renormalization theory is the renormalization group. This is the study of the dependence of the theory on the renormalization point μ that appeared in the subtraction procedure of Eqs. (6) and (8). Observable quantities such as the S-matrix elements, the measured strength of the interaction in the illustration, and so forth, do not depend on μ, but the finite expansion parameter λ_R or the mass parameter m_R do. All computations depend on the lagrangian in Eq. (1) which has no information about μ. Changing the value of the renormalization point in Eq. (6) induces new values of λ_R and m_R in such a way as to keep the measurable S-matrix elements unchanged. Since changing the mass parameter μ can be viewed as a change of scales, the renormalization group is intimately connected to scale transformations in the theory. It makes it possible to study the high- or low-energy behavior of the field theory by applying scale transformations to the momenta of the particles involved in a particular reaction. It is then found that the value of the effective expansion parameter $\lambda_R(\mu)$ of the theory depends on the energy scales of the reaction under study. It becomes convenient to choose the value of μ such that $\lambda_R(\mu)$ is small enough for a valid perturbative expansion. For certain gauge theories the effective coupling constant, which is a measure of the interaction, decreases as the energy scale increases. This behavior is called asymptotic freedom, and it explains why quarks act as free particles (small effective coupling constant) when they collide at very high energies and come to within very short distances of each other. SEE QUANTUM FIELD THEORY; QUARKS; SYMMETRY LAWS.

Bibliography. J. C. Collins, *Renormalization: An Introduction to Renormalization, the Renormalization Group, and the Operator-Product Expansion*, 1984; C. Itzykson and J. B. Zuber, *Quantum Field Theory*, 1980; W. Marciano and H. Pagels, *Phys. Rep.*, C36:137, 1978.

GRAND UNIFICATION THEORIES
CHUNG W. KIM

Attempts to unify three fundamental interactions—strong, electromagnetic, and weak—with a postulate that the three forces, with the exception of gravity, can be unified into one at some very high energy. The basic idea is motivated by the incompleteness of the electroweak theory of S. Weinberg, A. Salam, and S. Glashow, which has been extremely successful in the energy region presently accessible with the use of accelerators, and by the observation that the coupling constant for strong nuclear forces becomes smaller as energy increases whereas the fine-structure constant ($\alpha = 1/137$) for electromagnetic interactions is expected to increase with energy. The Weinberg-Salam-Glashow theory, which unifies electromagnetic and weak interactions, is incomplete in that strong interactions are not included; two coupling constants in the theory are unrelated; and in spite of many properties shared by them, quarks and leptons are unrelated, and their mass spectra and other properties remain unexplained. SEE STRONG NUCLEAR INTERACTIONS; WEAK NUCLEAR INTERACTIONS; WEINBERG-SALAM MODEL.

Models and successes. The simplest grand unification theory (GUT), proposed by H. Georgi and Glashow, is based on the assumption that the new symmetry that emerges when the three forces are unified is given by a special unitary group SU(5) of dimension 24. This symmetry is not observable in the low-energy region since it is badly broken. In this model, as in most GUTs, the coupling constants for the three interactions do actually merge into one at an energy of about 10^{14} GeV. Quarks (constitutents of the proton, neutron, pion, and so forth) and leptons (the electron, its neutrino, the muon, and so forth) belong to the same multiplets, implying that distinctons between them disappear at the energy of 10^{14} GeV or above. In addition to the known 12 quanta of strong, electromagnetic, and weak interactions (namely 8 gluons, the photon, and the W^+, W^-, and Z^0 particles), there appear, in this model, 12 new quanta with the mass of 10^{14} GeV. These generate new but extremely weak interactions that violate baryon- and lepton-number conservation. The most spectacular prediction of GUTs is the instability of the proton, which is a consequence of baryon-number (and lepton-number) violation. In the simplest SU(5) model, the decay $P \rightarrow e^+ + \pi^0$ is a dominant proton-decay mode, with a predicted proton lifetime of $10^{29 \pm 2}$ years. The baryon-number violation can also induce neutron-antineutron oscillations. SEE GLUONS; INTERMEDIATE VECTOR BOSON; LEPTON; NEUTRON-ANTINEUTRON OSCILLATIONS; PROTON; QUARKS; SYMMETRY BREAKING; SYMMETRY LAWS.

GUTs, in general, explain why the charge of the electron is precisely that of the proton with the opposite sign. In some GUTs with left-right symmetry, the lepton-number violation provides a novel way to generate the mass of neutrinos. Thus massive neutrinos are a distinct possibility in GUTs, and the smallness of their mass can also be understood. This leads to a possibility of neutrino oscillations which allow transformation of one neutrino species into another. Neutrinos with the appropriate mass may explain the missing mass in galaxies and clusters and perhaps the formation of galaxies. The prediction of the SU(5) theory of the ratio of the two unrelated constants in the Weinberg-Salam-Glashow model agrees impressively with observation, and the ratios of certain lepton and quark masses can also be calculated with reasonable success. A characteristic of the SU(5) model is the existence of the so-called desert between 100 GeV, which is the mass scale of the Weinberg-Salam-Glashow model, and 10^{14} GeV. This desert may bloom in more general GUTs. SEE NEUTRINO.

When applied to cosmology, GUTs have many implications. According to the scenario based on the GUTs, the universe underwent a phase transition when its temperature cooled to 10^{27} K which corresponds to 10^{14} GeV in energy and to the first 10^{-35} s after the big bang. This was when, for example, the SU(5) symmetry would have been broken. The phase transition caused an exponential expansion (10^{30}-fold in 10^{-32} s) of the universe, which explains why the observed 3 K microwave background radiation is uniform (the horizon problem), and why the universe behaves as if space is practically flat (the flatness problem). Baryon-number–violating interactions that were efficient in the early universe can qualitatively account for the observed ratio of the baryon-number density n_B and the photon-number density n_γ in the present universe, $n_B/n_\nu \simeq 10^{-10}$, which has been a long-standing puzzle in the standard hot big bang model of the universe. Another prediction common to many popular GUTs is the production of magnetic monopoles with a mass of 10^{16} GeV or less in the early universe.

Problems. In spite of its theoretical triumph and spectacular predictions, the simple SU(5) model is practically untested by experiment and appears to be incomplete or even incorrect. No experimental evidence of proton decay has been established. In fact, experiments have set a lower limit for the proton lifetime of 6×10^{31} years, which is somewhat larger than the SU(5) prediction. Although various modifications and generalizations of the simple SU(5) model can solve this problem, they do not, in general, provide significant improvements over the simple version. The problems which GUTs leave unsolved are numerous and include the following: (1) gravity is not included in the unification; (2) the number of unknown paremeters turns out to be unsatisfactorily large because of the complexity of symmetry-breaking mechanisms; (3) the mass spectra of the leptons and quarks remain unexplained; and (4) it is difficult to understand why there exist at least two vastly different but stabilized energy scales, one at 100 GeV and the other at 10^{14} GeV.

Prospects. Perhaps a true unification theory including gravity will emerge when the nature of symmetry breakings, the possible substructure of the leptons and quarks, and implications of supersymmetry, which is a new kind of symmetry relating matter (quarks and leptons) and quanta (such as the photon and the W^+, W^-, and Z^0 particles), are explored and properly understood. Unification theories with supersymmetry, known as supergravity, have been favored as candidates for such a theory. SEE ELEMENTARY PARTICLE; FUNDAMENTAL INTERACTIONS; SUPERGRAVITY; SUPERSYMMETRY.

Bibliography. A. Buras et al., Aspects of the grand unifications of strong, weak and electromagnetic interactions, *Nucl. Phys.*, B135:66–92, 1978; H. Georgi and S. Glashow, Unity of all elementary-particle forces, *Phys. Rev. Lett.*, 32:438–441, 1974; S. Glashow, Particle symmetries of weak interactions, *Nucl. Phys.*, 22:579–588, 1961; A. Guth, Inflationary universe: A possible solution to the horizon and flatness problems, *Phys. Rev.*, D23:347–356, 1981; C. Kounnas et al. (ed.) *Grand Unification With and Without Supersymmetry and Cosmological Implications*, 1984; G. Ross, *Grand Unified Theories*, 1985; A. Salam, Weak and electromagnetic interactions, *Nobel Symposium, Gothenburg*, 1968; S. Weinberg, A model of leptons, *Phys. Rev. Lett.*, 19:1264–1266, 1967.

NEUTRON-ANTINEUTRON OSCILLATIONS
GIUSEPPE FIDECARO

Periodic transitions between the state of a neutron and the state of an antineutron.

In 1970 the possible existence of neutron-antineutron oscillations and of other processes violating the conservation of baryon number B, which until then appeared to be a well-established law of nature, was first explicitly suggested by V. A. Kuz'min. In 1979 S. L. Glashow again raised this problem in the framework of theories unifying strong, weak, and electromagnetic interactions. These theories allow for processes violating baryon number conservation. Glashow pointed out that neutron-antineutron ($n\bar{n}$) transitions (with change in baryon number $\Delta B = 2$) would be mediated by a particle having a mass of the order of 10^6 GeV, intermediate between the mass of the $W^\pm$, Z^0 bosons (10^2 GeV) and the mass of the particle unifying the three interactions (10^{15} GeV). The search for $n\bar{n}$ oscillations would then also provide information on a range of energies not accessible to particle accelerators for some years to come. SEE BARYON; FUNDAMENTAL INTERACTIONS; GRAND UNIFICATION THEORIES; INTERMEDIATE VECTOR BOSON; SYMMETRY LAWS.

Phenomenology. The neutron and antineutron wave functions are written as linear combinations of the wave function of the mass eigenstates n_1 and n_2, as in Eq. (1), where θ is the mixing angle, and the finite lifetime of the neutron has been neglected.

$$\psi(n) = \psi(n_1) \cos\theta + \psi(n_2) \sin\theta \qquad (1)$$
$$\psi(\bar{n}) = \psi(n_1) \sin\theta - \psi(n_2) \cos\theta$$

In the case of free neutrons, when external fields are so weak that the energy difference $2\cdot\delta E$ between the two states n and $\bar{n}$ is negligible in comparison with the interaction energy $2 \cdot \delta m$, that is, with the mass difference $n_1 - n_2$ that induces $n\bar{n}$ transitions, $\sin\theta = \cos\theta = 1/\sqrt{2}$. This is very much the same situation as with the $K^0 - \bar{K}^0$ pair. The probability that a neutron at $t = 0$ is observed as an antineutron at the time t is given by Eq. (2). Thus, particles oscillate between the two states n and $\bar{n}$ with an oscillation time $\tau_{n\bar{n}} = \hbar/\delta m$, where $\hbar$ is Planck's constant divided by 2π. For very short times ($t \ll \tau_{n\bar{n}}$) Eq. (2) becomes Eq. (3), and the probability grows quadratically with time. SEE MESON.

$$P(t) = \sin^2\left(\delta m \cdot \frac{t}{\hbar}\right) \qquad (2) \qquad P(t) = \left(\delta m \cdot \frac{t}{\hbar}\right)^2 \qquad (3)$$

In the case of neutrons bound in nuclei, or traveling in a magnetic field, the maximum probability and the oscillation time are both very heavily reduced. However, if $t \ll \hbar/\delta E$, Eq. (3) is still valid and neutrons behave as if they were free. This case is very important since it makes possible oscillation experiments with stronger magnetic fields and a rougher vacuum than would otherwise be acceptable.

Experimental results. Neutrons oscillating inside a nucleus would occasionally annihilate with emissions of π mesons, and this would be another process that would make matter unstable in addition to proton decay ($\Delta B = 1$). From the corresponding proton lifetime τ_m, it is possible to obtain the oscillation time $\tau_{n\bar{n}}$ for free neutrons through Eq. (4), where M can be made

$$\tau_{n\bar{n}} = \left(\tau_m \cdot \frac{\hbar}{M}\right)^{1/2} \qquad (4)$$

equal to one nucleon mass. The current value, $\tau_m \gtrsim 10^{31}$ years, yields $\tau_{n\bar{n}} \gtrsim 10^7$ s corresponding to $\delta m \lesssim 10^{-22}$ eV. Although more accurate formulas have been worked out, this type of extrapolation remains uncertain, and indirect estimates of $\tau_{n\bar{n}}$ from the lifetime of matter cannot replace direct measurements using free neutrons. These measurements consist in allowing an intense beam of slow neutrons to travel in a vacuum for as long a time as possible, in a region where the magnetic field of the Earth has been reduced, and looking for antineutron annihilations on a target. SEE PROTON.

A direct measurement was carried out at the high-flux nuclear reactor of the Laue-Langevin Institute, Grenoble (France), with a neutron flight time of 0.03 s, in a magnetic field smaller than 10^{-7} tesla (1 milligauss), using a beam of 10^9 cold neutrons per second. A lower limit $\tau_{n\bar{n}} \gtrsim 10^6$ s was obtained. Limits in the region 10^8–10^9 s could be reached with a more sensitive experiment at the same reactor. Cosmic rays are the main limiting source of background. SEE ANTINEUTRON; NEUTRON.

Bibliography. M. Gell-Mann and A. Pais, Behaviour of neutral particles under charge conjugation, *Phys. Rev.*, 97:1387–1389, 1955; V. A. Kuz'min, CP non-invariance and baryon asymmetry of the Universe, *JETP Lett.*, 12:228–230, 1970; M. Lévy et al. (eds), *Quarks and Leptons: Proceedings of the 1979 Cargèse Summer Institute*, 1980; R. N. Mohapatra and R. E. Marshak, Local B-L symmetry of electroweak interactions, Majorana neutrinos, and neutron oscillations, *Phys. Rev. Lett.*, 44:1316–1319, 1980; H. A. Weldon et al. (eds.), *4th Grand Unification*, 1983.

SUPERGRAVITY
FREYDOON MANSOURI

A theory that attempts to unify gravitation with the other fundamental interactions. The notion of a unification theory of all interactions (forces) responsible for natural phenomena has fascinated physicists for over a century. The first, and only, completely successful unification theory was constructed by James Clerk Maxwell, in which the up-to-then unrelated electric and magnetic phenomena were unified in his electrodynamics. Early in the twentieth century, the problem of embedding electrodynamics and gravitation into a unified theory was pursued by A. Einstein, H. Weyl, Th. Kaluza, O. Klein, and others. Later, with the discovery of weak and strong nuclear forces, it became clear that the unification problem was more complex than had been anticipated. Moreover, experimental discoveries in nuclear and particle physics and the emergence of internal symmetry groups provided the hint that a more attainable goal on the way to full unification was to first unify electromagnetic, weak, and possibly strong forces. If successful, the next step would be unification with gravity into a superunification theory. Supergravity models are special superunification theories. To put the aims and the achievements of these models in a proper perspective, it will be helpful to briefly recount the earlier stages of unification. SEE FUNDAMENTAL INTERACTIONS; STRONG NUCLEAR INTERACTIONS; SYMMETRY LAWS; WEAK NUCLEAR INTERACTIONS.

Electroweak theory. The second stage of unification concerns the unification of electromagnetic and weak interactions, using Maxwell's theory as a guide. This was accomplished by S. L. Glashow, A. Salam, J. C. Ward, S. Weinberg, and others, making use of the non-Abelian gauge theories invented by C. N. Yang and R. L. Mills, and of spontaneous symmetry breaking pioneered by Y. Nambu and G. Goldstone. The symmetry of Maxwell's theory is very similar to spatial rotations about an axis, rotating the vector potentials while leaving the electric and magnetic fields unchanged. It is a local invariance because the rotations about a fixed axis can be made by different amounts at different points in space-time. Thus, Maxwell's theory is invariant under a one-parameter group of transformations U(1). In Yang-Mills theory this local invariance was generalized to theories with larger symmetry groups such as the three-dimensional rotation group SO(3) $\simeq$ SU(2) which has three parameters. The number of parameters of the local symmetry (gauge) group is also equal to the number of 4-vector potentials in the gauge theory based on that group. A detailed analysis of weak and electromagnetic forces shows that their description requires four 4-vector potentials (gauge fields), so that the gauge group must be a four-parameter group. In fact, it is the product SU(2) · U(1). Of the four gauge fields, one transmits electromagnetic force which is long-range indicating that the corresponding quanta (photons) are massless. The other three transmit short-range weak forces of the type responsible for the neutron beta decay, indicating that the quanta of short-range forces are massive. Instead of the ad hoc introduction of masses, which destroys the gauge invariance, a more satisfactory method of making weak forces short-range is by the spontaneous breaking of the SU(2) · U(1) symmetry down to $U_Y(1)$ of Maxwell's theory, thus ensuring that the photon remains massless. See GAUGE THEORY; SYMMETRY BREAKING; WEINBERG-SALAM MODEL.

Grand unification theories. In the third stage of unification, electroweak and strong forces are regarded as different components of a more general force which mediates the interactions of particles in a grand unification model. Strong forces are responsible for the interactions of hadrons and for keeping quarks confined inside hadrons. They are described by eight massless 4-vector potentials (gluons), the corresponding eight-parameter group being SU(3). This local symmetry is called color, and the corresponding theory quantum chromodynamics (QCD). Thus the gauge group of a grand unification theory must include SU(3) · SU(2) · U(1) as a subsymmetry. The most dramatic prediction of these theories is the decay of proton, first pointed out by J. C. Pati and A. P. Salam. See GLUONS; GRAND UNIFICATION THEORIES; PROTON; QUANTUM CHROMODYNAMICS; QUARKS.

Superunification theories and supersymmetry. The last, and the most ambitious, stage of unification deals with the possibility of combining grand unification and gravitation theories into a superunification theory, also known as supergravity. To achieve this, use is made of the dual role played by local symmetry groups. On the one hand, they describe the behavior of forces. On the other hand, they classify the elementary particles (fields) of the theory into multiplets: spin-zero fields in one multiplet, spin-½ fields in another multiplet, and so forth, but never fermions and bosons in one irreducible multiplet. This last restriction used to be a major obstacle on the way to superunification. This is because, of all the elementary particles, only the quanta of gravitational field (gravitons) have spin 2, so that a multiplet of elementary particles including the graviton must of necessity involve particles of different spin. But then by an internal symmetry transformation, which is by definition distinct from space-time (Lorentz) transformations, one can "rotate" particles of different spin into one another, thus altering their space-time transformation properties. This apparent paradox can be circumvented if both the internal symmetry and Lorentz transformations are part of a larger group which also includes the spin-changing transformations. This is how supersymmetry makes its appearance in supergravity theories. See GRAVITON; SUPERSYMMETRY.

Local supersymmetry transformations in field theory were first discussed by Y. A. Golfand and E. P. Likhtman. Under supersymmetry transformations, fermions and bosons mix. To visualize these transformations, it is useful to compare and contrast them to the transformations in the familiar three-dimensional euclidean space. In such a space, the coordinates (x_1, x_2, x_3) of a point are given by real numbers, so that they commute: $x_1 x_2 = x_2 x_1$, and so forth. Unlike real numbers, there is another class of numbers, known as Grassmann numbers, the elements of which do not commute: $\theta_1 \theta_2 = -\theta_2 \theta_1$. In particular, $\theta_1^2 = \theta_2^2 = 0$. Since their square yields a real number, zero, they may be thought of as square roots of real numbers in the same sense that the Dirac equation is the square root of the Klein-Gordon equation. The numbers θ_1 and θ_2 may be thought of as components of a two-dimensional (anticommuting) spinor $\theta = (\theta_1, \theta_2)$ in a spinor space. Now an even more general space can be envisioned, a superspace, in which some of the coordinates are x-like (commuting) and some are θ-like (anticommuting). Translations in superspace are examples of supersymmetry transformations. When $\vec{x}$ is changed by an amount $\vec{a}$, θ does not change. But when θ changes by an amount $\epsilon = (\epsilon_1, \epsilon_2)$, $\vec{x}$ also changes according to the rule given below,

$$\theta \to \theta + \epsilon$$
$$\vec{x} \to \vec{x} + \epsilon^+ \vec{\sigma} \theta$$

where ϵ^+ is the adjoint of ϵ, and $\vec{\sigma}$ consists of 2 × 2 Pauli matrices $(\sigma_1, \sigma_2, \sigma_3)$. Thus, by putting together a 3-vector $\vec{x}$ and a two-component spinor θ, it is possible to construct a supersymmetric multiplet $(\vec{x}, \theta)$ on which supersymmetry transformations can be defined. These transformations can also be defined by putting together multiplets of fields of different spin to form a supermultiplet. It is precisely this feature which is indispensable in the construction of supergravity theories.

Models. From the beginning, the development of superunification theories proceeded along two distinct but related directions: explicit supergravity models in four-dimensional space-time; general theories in superspace aimed at first constructing all the possibilities allowed by the invariances and then specializing to those with desirable particle spectra. The two approaches are complementary to each other since each has advantages which the other lacks.

In the first explicit supergravity model, the supermultiplet consisted of a spin-2 particle, the graviton, and a spin-$3/2$ particle, called gravitino. In a supersymmetric (super-Lie) group, internal and space-time symmetry transformations are bridged by θ-like transformations discussed above. As a result, the form of the internal symmetry group is determined by the particle content of a supermultiplet, which in turn is fixed by the requirement that it contain only one spin-2 particle, the graviton, and no particles with spin higher than 2. Then, depending on whether the gravitational sector is Einstein's theory or Weyl's theory, the internal symmetry groups turn out to be, SO(N) or SU(N), where $N = 1, \ldots, 8$. The SU(N) series are known as conformal supergravity theories. The spectrum of particles for each of these possibilities is given in the **table**. The largest internal symmetry group in the SO(N) series is thus SO(8), which does not contain SU(3) · SU(2) · U(1) of the grand unification theories as a subgroup. This means that it cannot accommodate the known quarks, leptons, gluons, and weak gauge bosons in a single supermultiplet. Opinions vary as to whether SO(8) should be regarded as a global symmetry or a local one. If it is only a global symmetry, a model has been constructed which has the spectrum of SO(8) and additional symmetries including a local SU(8) symmetry.

One-graviton supermultiplets of particles with spin equal to or less than 2

SO(N) or SU(N) internal symmetry	Number of particles of a given spin in a supermultiplet				
	Spin 0	Spin ½	Spin 1	Spin 3/2	Spin 2
$N = 0$	—	—	—	—	1
$N = 1$	—	—	—	1	1
$N = 2$	—	—	1	2	1
$N = 3$	—	1	3	3	1
$N = 4$	2	4	6	4	1
$N = 5$	10	11	10	5	1
$N = 6$	30	26	16	6	1
$N = 7$	70	56	28	7	1
$N = 8$	70	56	28	8	1

Superunification in superspace. One problem with supergravity models in four dimensional spacetime is that the transformation laws of the fields in a supermultiplet change from one model to the next, so that, in contrast to the Yang-Mills theory, the knowledge of known models is not very helpful in constructing others. This is because, as discussed above, the space-time coordinates $x^\mu = (t, x)$ by themselves do not form a supersymmetric multiplet, and in the absence of their partners $\theta = (\theta_1, \theta_2)$, the supersymmetry transformations enter the picture in an indirect manner. A superspace, on the other hand, has both x-like and θ-like coordinates, so that it is naturally suited for implementing local supersymmetry transformations of general supergravity theories. General theories in superspace and their physical interpretations have been studied from various points of view, and there is no consensus as to which set of criteria are the most reasonable ones. In one of the superspace approaches, it can be shown that a supergravity theory, like pure gravity, is a nonlinear realization of a gauge symmetry: The fields transform linearly in the internal symmetry (Yang-Mills) sector but nonlinearly in the space-time (gravity) sector. Then, SO(N) and SU(N) groups, respectively, appear as local symmetries, and it is possible to derive model-independent transformation laws for every field by suitable generalizing Yang-Mills transformations. Using these it is also possible to write down all the invariants that are needed for the construction of lagrangians.

In the same way that in a relativistic quantum they is often confronted with unphysical degrees of freedom such as timelike oscillations, which must be eliminated by imposing gauge conditions, superunified theories in superspace contain unphysical degrees of freedom which must be eliminated by imposing a suitable set of gauge conditions. This problem has been investigated, and no systematic method of dealing with it has been found.

Outlook. One of the initial motivations for constructing supergravity models was the hope of obtaining finite quantum theories of gravity coupled to matter. Arbitrary couplings of gravity to matter lead to quantum theories in which the amplitudes for various processes are infinite, and these infinities (divergences) cannot be isolated and eliminated from the theory in a process known as renormalization. In supergravity models constructed so far, the degree of renormalization is improved in the sense that these gravity-coupled-to-matter theories are as renormalizable as pure gravity. But the overall picture is not as hopeful as it was once thought. The reason for this can be seen from the point of view of the nonlinear realizations. Local supersymmetry transformations are nonlinear. Just as general coordinate invariance does not improve the renormalizability of pure gravity by completely eliminating the available invariants, local supersymmetry transformations do not appear to limit the construction of invariants to a finite number, an essential criterion for the elimination of divergences. This does not mean, however, that superunification theories are useless and that they must be abandoned. Even if these obstacles cannot be overcome within the framework of the currently popular models, the rich structure of these theories is likely to provide a point of departure for the construction of other physically acceptable theories.
SEE ELEMENTARY PARTICLE; QUANTUM FIELD THEORY; RENORMALIZATION.

Bibliography. L. Castellani et al., *Supergravity and the Geometry of Higher Dimensions*, 1986; D. Z. Freedman and P. van Nieuwenhuizen, Supergravity and the unification of the laws of physics, *Sci. Amer.*, 238(2):126–143, February 1978; G. Furiani et al. (eds.), *Superstrings, Supergravity, and Unified Theories*, 1986; H. Georgi and S. L. Glashow, Unified theory of elementary particle forces, *Phys. Today*, 33(9):30–39, September 1980; A. Salam, Gauge unification of fundamental forces, *Rev. Mod. Phys.*, 52(3):525–538, July 1980; A. Salam and E. Szegin (eds.), *Supergravity Theories, Anomalies and Compactification: Commentaries and Reprints*, 1986.

10
SYMMETRY LAWS AND CONSERVED QUANTITIES

Symmetry laws	348
Selection rules	354
Parity	358
Time reversal invariance	361
Spin	363
Helicity	363
CPT theorem	364
Isobaric spin	365
Unitary symmetry	367
Hypercharge	370
Flavor	370
Charm	371
Color	372
Symmetry breaking	372
Supersymmetry	373

SYMMETRY LAWS
CHARLES J. GOEBEL

The physical laws which are the expressions of symmetries. A conservation law results from each symmetry; that is, from each symmetry the existence of a quantity which is conserved (a constant of the motion) can be deduced. Selection rules result from conservation laws.

Space-time symmetries. A symmetry (or invariance) of the world exists whenever the description of the laws of physics is unaffected by a change in the frame of reference. For instance, the position of the origin of a space coordinate system is quite arbitrary; changing it makes no difference in the description of the motion of bodies because the forces between bodies depend only on their relative positions and not on any absolute position. Equivalently, a system of bodies behaves the same if translated to another place. This symmetry of space to translation implies the conservation of momentum.

Other symmetries of space-time are the irrelevance of (1) the origin of the time coordinate, (2) the orientation of a coordinate system in space, and (3) the velocity of a coordinate system (Lorentz invariance). Each of these implies a conservation law, as shown in the **table**. All these symmetries are termed continuous because the changes can be arbitrarily small; that is, a finite change can be made bit by bit. The resulting constants of the motion are classical quantities and are additive.

Discrete symmetries (reflections) also exist, for which the irrelevant change is not arbitrarily small. They imply constants of the motion (parities) in quantum mechanics. These parities are multiplicative. For instance, the direction of increasing time is irrelevant; the world is invariant to time reversal (microscopic reversibility). Although, macroscopically, future and past seem distinct, this is merely a result of the disposition of matter (a state of anomalously small entropy at some time in the past) in the same way that a point of space seems distinct by having a particular piece of matter there. Space is also symmetrical to reflection of space or to inversion, the reflection of all three directions of space; it is irrelevant which is the positive direction of a space axis or of all three axes. This amounts to the irrelevance of whether a right-handed or a left-handed coordinate system is used. The resulting conserved quantity (eigenvalue of space inversion) is (space) parity. Actually, the preceding statements about space inversion symmetry must be qualified. Although inversion symmetry is observed by the strong interactions (such as nuclear forces) and electromagnetic interactions, it is not observed by the weak interactions, such as decays of quasi-stable elementary particles, including beta decay. The description of a beta-decay event depends on the handedness of the coordinate system. SEE FUNDAMENTAL INTERACTIONS; PARITY; STRONG NUCLEAR INTERACTIONS; WEAK NUCLEAR INTERACTIONS.

Symmetry in quantum mechanics. In quantum mechanics a change in the frame of reference is described by the multiplication of the state vectors by a unitary operator $\psi \to A\psi$. This has no effect on an observable O if its operator commutes with A, $OA = AO$. In particular, energy eigenvalues are unchanged if the hamiltonian H commutes with A. The totality of operators which commute with H form a group, the symmetry group of H. Starting with an energy eigenstate ψ, the states $A\psi$, $B\psi$, ..., where $A, B, \ldots$ are the members of the symmetry group of H, all have the same energy eigenvalue (are degenerate). A linearly independent set of states which spans all the degenerate states formed from ψ is called a multiplet; its number is called the multiplicity of the multiplet. If ψ_n are the members of a multiplet, the matrices whose elements are the matrix elements of the symmetry operators $A_{mn} = (\psi_m, A\psi_n)$ form a matrix representation of the symmetry group. The importance of this is that all the possible matrix representations of a group can be determined abstractly from the multiplication table of the group, and therefore independently of the physical significance of the transformations. (Further, the matrix elements of other operators are determined by their commutation relations with the symmetry operators; this is the Wigner-Eckart theorem.) For example, if the symmetry group of the hamiltonian is the rotation group (rotations of the spatial coordinate frame about a point), the possible multiplets have multiplicities 1, 2, 3, ..., corresponding to angular momentum quantum number $j = 0, \frac{1}{2}, 1, \ldots$. The multiplets of the internal symmetry group SU_2 are the same; for SU_3 they are 1, 3, 6, 8, ...

Internal symmetries. Further symmetries, which are not space-time symmetries, are called internal symmetries. For example, the zero of both scalar and vector electromagnetic poten-

Invariances (symmetries) and conservation laws

Invariance to	Conserved quantity	Range of validity
Homogeneity of space-time		
Translation of space	Momentum, **p**	
Translation of time	Energy, E	
Isotropy of space-time	Angular momentum, **J**	
Rotation of space	Velocity of the	
Lorentz transformation	center of energy (center of mass)	Exact
Interchange of identical particles	Symmetry of the wave function (statistics)	
Inversion of time, space, and charge	CPT	
Gauge invariance		
Charge U_1	Net charge	
Color SU_3	Color[a]	
?	Net number of baryons, e-leptons, μ-leptons [b]	
Reversal of time	Time parity, T	
Reflection of space and charge	Product of parity and charge parity, CP [c]	
Reflection (or inversion) of space	Parity, P	
Reflection of charge (charge conjugation)	Charge parity, C[e]	Approximate
?	Net number of up quarks, down quarks, strange quarks, charm quarks [d]	
⋮	⋮	
Flavor SU_2	Isotopic spin I, isotopic parity G	
Flavor SU_3	"Unitary spin" [f]	
⋮	⋮ [g]	

[a]A hypothesized hidden symmetry.
[b]Conserved at the present level of detection, but possibly not exactly conserved.
[c]Violation seen only in K_L decay.
[d]Conservation of these three is equivalent to the conservation of baryon number, I_3 (the charge axis component of isotopic spin), and hypercharge.
[e]Violated by the weak interactions.
[f]Violated by electromagnetic interactions and by the up-down quark mass difference, $m_u - m_d \approx -5$ MeV.
[g]Violated by the strange-nonstrange quark mass difference, $m_s - m_u \approx m_s - m_d \approx 150$ MeV.

tials is irrelevant; the addition of a constant to an electromagnetic potentials is irrelevant; the addition of a constant to an electromagnetic potential is of no consequence (so-called gauge invariance of the first kind). In quantum mechanics this symmetry implies the conservation of charge. There is also a discrete symmetry: It is (nearly) irrelevant which sign of charge is called positive and which is called negative. The qualification "nearly" is necessary here, just as in space inversion, because the weak interactions do not observe the symmetry. Thus the world is (nearly) invariant to charge reversal, the interchange of positive and negative charge. At first sight this symmetry appears not to exist because one can distinguish positive charge as that which is carried by the heavy constituent of matter, the proton, whereas negative charge is that carried by the light constituent, the electron. However, antiprotons (negatively charged protons) and antielectrons (positrons) exist, and a world with (nearly) the same properties as Earth's would result if all electrons were replaced with positrons and all protons by antiprotons and, in general, if all particles were replaced by their antiparticles (the operation of charge conjugation). The resulting (nearly) conserved quantity is termed charge conjugation parity or charge parity, C. More precisely, charge parity is the eigenvalue of the operator of charge reversal. A system can be in an eigenstate of charge reversal only if it goes into itself under the operation, in other words, if it is self-charge conjugate. Such a system must be completely neutral, having no electric or magnetic moments; in fact, it must have no internal quantum numbers of any kind that change sign under charge conjugation. The neutron and the K^0 meson are examples of neutral but non-self-charge conjugate systems ($\bar{n} \neq n$; $\overline{K^0} \neq K^0$). The neutral π meson, the photon, and the graviton are self-charge-conjugate; their charge parities are the charge parities of their sources $+1$, -1, and $+1$, repsectively. A self-charge conjugate system of some interest is positronium, a bound state of an electron and a positron; it has $C = (-)^{s+l}$ in a state of orbital angular momentum l and spin s ($s = 0$ or 1). SEE ANTIMATTER; ANTIPROTON; ELEMENTARY PARTICLE; POSITRONIUM.

Although the weak interactions are not invariant under charge reversal or space inversion, they are invariant (with one observed exception, K^0 decay) to the combination of these reflections; they are also invariant under time reversal. Equivalently stated, all interactions appear to very nearly conserve CP and also T (C = charge conjugation, P = space inversion, T = time reversal). For instance, consider the decay of a muon into an electron and a pair of neutrinos. The probability that the electron's momentum $\vec{p}_e$ makes an angle θ with the direction of the muon's spin $\vec{s}$ is observed to be of the form given by Eq. (1), where the upper sign holds for a positive muon

$$P_{\pm} = a \pm b \cos \theta \tag{1}$$

and the lower for a negative muon. In an inverted coordinate system, $\vec{p}_e$ would be reversed in sign, but $\vec{s}$ would not (spin being an axial vector) and so $\cos \theta$ would reverse sign, resulting in Eq. (2). Hence the description of the decay is not invariant to space inversion. $P_{\pm}$ would also

$$P_{\pm} = a \mp b \cos \theta \tag{2}$$

change from Eq. (1) to Eq. (2) under charge reversal, that is, interchanging what is called positive and negative charge. Thus under both reflections together, $P_{\pm}$ is unchanged and so decay "conserves CP."

A sensitive test of T invariance is the size of a possible electric dipole moment of the neutron. The only direction provided by the neutron to direct such a moment is its spin; under time reversal this would reverse, and so the interaction energy of the moment with an electric field would also reverse, in contrast to all other energies. The electric dipole moment is less than 10^{-24} cm times e, the charge of the electron.

A very sensitive test of CP conservation is provided by K^0 decay. Neutral K mesons are produced by strong interactions, and so are produced as either K^0 or $\overline{K^0}$ mesons. But they decay differently, through weak interactions, and so it is linear combinations of K^0 and $\overline{K^0}$ which decay with definite lifetimes, namely K_L and K_S (L = long life, S = short life). [As an analogy, a light source might produce only right-hand circularly polarized photons (just as in the process $\pi^- p \rightarrow \Lambda K^0$, only K^0, not $\overline{K^0}$, mesons are produced); but if the light passes through Polaroid, it decomposes into two components with different decay (absorption) lengths, namely, photons which are linearly polarized, perpendicular or transverse to the optic axis.] If all interactions, including the weak ones, conserved CP, then K_L and K_S would be eigenstates of CP, and only one of them would decay into two pions, which is a CP eigenstate. In fact, K_S decays almost exclusively into two pions, but K_L is found to also have a nonvanishing, though small, probability of this decay. Thus CP is not exactly conserved in K^0 decay. The smallness of the violation contrasts with the maximal violation in weak interactions of C and P separately. No other system is as sensitive to CP violation as the neutral K mesons (they are a degenerate doublet, which is easily mixed by a small perturbation); no violation of either CP or T has been observed elsewhere. This makes it difficult to test the various detailed mechanisms which have been suggested for the violation. SEE MESON.

CPT theorem. The reflection symmetries are correlated by the CPT theorem of G. Lüders. This theorem states that a Lorentz invariant field theory is necessarily invariant to the product of the three reflections: charge conjugation C, space inversion P, and time reversal T. For example, this would mean that K^0 decay, which does not obey CP invariance, also should not obey T, time reversal invariance. In principle, the theorem can be tested experimentally, particularly in K-meson decay, but so far no stringent results have been obtained. Elementary particle theory would be enormously challenged if CPT invariance were found not to hold.

Invariances of strong interactions. The nuclear force between two protons is found to be identical to the force between two neutrons. This is called the charge symmetry of the nuclear force. More generally, the motion of a system composed of nucleous and pions (the lowest mass quanta of the nuclear force field) is invariant to the operation is $p \leftrightarrow n$, $\pi^+ \leftrightarrow \pi^-$, $\pi^0 \leftrightarrow \pi^0$. The combination of charge conjugation and the charge symmetry operation is called isotopic inversion, G, and carries each π-meson into itself. The π-meson thus has a G parity.

Further, the nuclear force between a neutron and a proton is identical to the force between two protons or two neutrons in the same orbital and spin state (charge independence). SEE NUCLEAR STRUCTURE; SCATTERING EXPERIMENTS (NUCLEI).

Isotopic spin. The foregoing symmetry can be expressed as the isotropy of a three-dimensional "isotopic space," which implies the conservation of "angular momentum," or isotopic spin **I**, in this space. The component of a particle's isotopic spin along the "third" axis I_3 is related linearly to the charge of the particle. The consequences of charge independence are formally very similar to the consequences of the conservation of angular momentum, except that nothing here corresponds to orbital angular momentum.

In terms of isotopic spin, the charge symmetry operation is a special rotation in isotopic spin space, namely, one which reverses the direction of the third axis. It follows from angular momentum calculus that a system having $I_3 = 0$ has a charge symmetry parity which is $(-)^i$, where i is the magnitude of $\bar{\mathbf{I}}$ (total isotopic spin) of the system. Thus the charge symmetry parity of the π^0 meson is -1. The π meson, therefore, has a G parity of -1; nucleonium, a system of nucleon and antinucleon, has $G=(-)^{l+s+i}$ in a state of orbital angular momentum l, spin s (0 or 1), and isotopic spin i (0 or 1).

All the strong interactions are isotropic in isotopic space and conserve isotopic spin; all the strongly interacting elementary particles carry isotopic spin. The anomalous stability of the heavier (the so-called strange) particles is explained by the fact that charge is conserved, and in both the strong and electromagnetic interactions I_3 is conserved (Gell-Mann–Nishijima scheme). SEE ISOBARIC SPIN.

Unitary symmetry. Charge independence, described above as the isotropy of a three-dimensional space, can also be described as symmetry with respect to arbitrary unimodular unitary transformations of the proton and neutron, that is, invariance of the strong interactions to transformations of the form $p \rightarrow \Sigma \cos \theta e^{i\Psi} p + i \sin \theta e^{i\Phi} n$, $n \rightarrow \cos \theta e^{-i\Psi} n + i \sin \theta e^{-i\Phi} p$, where p and n stand for the state vectors of a proton and a neutron, respectively. This transformation is called mixing p and n. This group of transformations is called SU_2. The analogous group of transformations on n particles is SU_n. SEE UNITARY SYMMETRY.

It appears that hadrons are well described as compounds of so-called quarks, of which there are n kinds (flavors): $u, d, s, c, b, \ldots$. The net number of quarks of each flavor is changed only in weak interactions. For example, in the strong reaction in Eq. (3) the net number of u, d, and s

$$\overline{K^0} p \rightarrow \Sigma^+ \pi^0 \tag{3}$$

quarks is 2, 0, and 1 respectively in both the initial and the final state. Further, the quarks of different flavors intereact in the same way with the fundamental strong interaction between them (the "glue" which binds them together to make hadrons). This means that the strong interactions would have a flavor SU_n internal symmetry, were it not for the fact that the quarks of different flavors have different masses; their masses seem to form roughly a geometrical progression. Aside from the effect of this on the internal symmetry of the strong interactions, this has the practical effect that the number of flavors n which are known increases as the maximum energy of particle accelerators increases; at present (1985) n is 5, and a sixth flavor of quark (t) may have been observed.

The masses of the lightest two quarks, u and d, are small compared to the zero-point energy associated with the binding of the quarks into hadrons, and so the properties of hadrons are rather insensitive to the mixing of u and d quarks; consequently the corresponding SU_2 symmetry is rather good. This is the isotopic spin symmetry described above; there it was described as an insensitivity to the mixing of p and n nucleons, which is equivalent to the mixing of u and d quarks. The next higher mass quark s has a mass which is not much smaller than the zero-point energy, and therefore the mixing of s with u or d quarks affects the properties of a hadron much more than does the mixing of u and d quarks. Hence the corresponding SU_3 symmetry is not as good a symmetry as the SU_2 (isotopic spin) symmetry. The masses of the remaining quarks are so large that it is useful, not to describe the relationship of hadrons which contain various numbers of these heavy quarks in terms of SU_4 and SU_5 symmetries, but rather to calculate the properties of these hadrons by using directly the symmetry principle that the glue couples equally to all flavors of quarks.

In addition to this flavor SU_n symmetry of strong interactions, it is believed that there is also a color SU_3 symmetry which is exact but hidden. *See* COLOR; FLAVOR; QUARKS.

Chiral symmetry. To describe this symmetry, one must first define helicity, η; η is the projection of the spin of a particle on the direction of its motion, that is, $\eta = \vec{S} \cdot \hat{p}$, where $\vec{S}$ is the particle's spin (in units of $\hbar$, Planck's constant divided by 2π) and $\hat{p} = \vec{p}/|p|$ is its unit momentum vector. In quantum mechanics, η has the quantized values $s, s-1, \ldots, -s$, where s is the magnitude of the spin of the particle. Thus a spin-½ particle, such as e, p, or n, can have just two eigenvalues of η, namely ½ and $-½$; these polarization states are called right-hand and left-hand, respectively. Usually, the value of η is not an intrinsic property of a particle because it is not a Lorentz invariant; that is, a particle which is, say, right-handed as observed in one space-time frame will not be as observed in another. But, exceptionally, the helicity of a massless particle is an intrinsic property, independent of the choice of space-time frame (reflections of space excluded), just as much as properties such as charge or i-spin. Hence, a system of massless particles could obey laws such as "the number of left-hand and right-hand particles are each conserved" or "the i-spins carried by the left-hand particles and by the right-hand particles are each conserved." (Relativistic invariance requires that the category "left-hand particles" must also include their charge conjugates, namely, right-hand antiparticles.) Such conservation laws follow from symmetries of the form of a product of an internal symmetry group for the left-hand particles (that is, their field) and a similar one for the right-hand particles. Such a symmetry group is called chiral (meaning "handed"). It is well known that the electromagnetic and weak interactions are chiral-symmetric. But it is obvious that the world as a whole is not chiral-symmetric, because not all (in fact very few) particles are massless. Many scientists believe that the world is fundamentally chiral-symmetric, but that this symmetry is "spontaneously broken." *See* HELICITY.

Gauge invariance. A gauge invariance is an internal symmetry which is local; that is, the symmetry operation is a rotation of the internal coordinates independently at each point of space and time. This symmetry is possible only if there exist "gauge fields," massless vector fields (that is, boson fields whose quanta are massless particles with spin-parity $J^P = 1^-$), one for each independent internal symmetry "rotation." The internal symmetry whose conserved quantum number is charge has long been known to be guage-invariant; its vector field is the electromagnetic field, whose quanta are photons. The theory of a gauge-invariant SU_2 internal symmetry was first considered by C. N. Yang and R. Mills; the Yang-Mills vector field is an isovector ($i = 1$) field.

There is great interest in gauge-invariant field theories, both for strong and weak interactions: (1) They are the only renormalizable field theories (with one uninteresting exception) which contain vector fields; they can therefore describe weak interactions, which are believed to be mediated by the exchange of intermediate vector bosons. (2) They are the only theories with asymptotic freedom, a term which means roughly that interactions become weaker for larger momentum transfers. As far as known, the "scaling" observed in deep inelastic electron scattering requires the strong interactions to be asymptotically free.

Spontaneous symmetry breaking. At first sight, however, the masslessness of the gauge fields would appear to rule out gauge-invariant theories of the weak and strong interactions, because the only observed massless $J^P = 1^-$ boson is the photon. But a phenomenon called spontaneous symmetry breaking can occur, in which the states of the system do not have all of the symmetry of the equations of motion. A particular consequence can be that some or all of the gauge field quanta are massive (this is known as the Higgs mechanism). Remarkably, the good properties of renormalizability and asymptotic freedom remain true. Further, the algebraic (commutation) relations between the operators which generate the group of internal symmetries remain true, despite the lack of symmetry in the states. Such relations are called a current algebra. An example is the chiral i-spin current algebra; although the chiral symmetry is broken and particle states are not chiral eigenstates, the current algebra appears to be valid and a matrix element yields correctly the ratio of the axial vector (Teller) to vector (Fermi) beta-decay constants of the nucleon, the Adler-Weisberger relation. SEE QUANTUM FIELD THEORY; RENORMALIZATION; SYMMETRY BREAKING.

Baryonic and leptonic charges. An important feature of a gauge field with unbroken gauge invariance is that the internal symmetry, and its corresponding conservation laws, is forced to hold. Electromagnetism provides an example: Charge conservation is a consequence of Maxwell's equations. According to experiment, not only is electric charge conserved, but also three other "charges," namely, baryon number, electron number, and muon number. (These are net numbers; for example, electron number, or better, e-lepton number, means the number of electrons and electron neutrinos minus the number of positrons and antielectron neutrinos.) However, these three charges are not the source of any known massless (that is, unbroken) gauge field, and thus there is no known reason for these numbers to be exactly conserved. It is speculated that they are in fact not conserved, and that processes as in Eqs. (4) can occur, though at a rate too

$$\mu^- \to e^- \gamma \qquad p \to e^+ \pi^0 \qquad (4)$$

slow to have been yet detected. These speculations are based on so-called grand unified theories in which all particles (quarks and leptons) are treated as basically equivalent. The following is a simple qualitiative reason for the existence of such processes: It is known to a very high accuracy that the charge of the proton and the positron are equal, or equivalently that the proton and electron have equal but opposite charges, or that the hydrogen atom is neutral. Why should this be? If baryons and electrons are separately conserved, there is no reason for there to be any relation between the charge of the proton and the electron (although the existence of beta decays like $n \to pe\bar{\nu}_e$ requires that the charge differences $Q_p - Q_n$ and $Q_{e^+} - Q_{\bar{\nu}_e}$ must be equal). It is only the existence of a reaction like $p \to e^+ \pi^0$ (where π^0 is certainly neutral, because it is self-charge-conjugate) which forces Q_p to equal Q_{e^+}. SEE BARYON; GRAND UNIFICATION THEORIES; LEPTON.

Quantum chromodynamics. In the quark theory, which seems highly successful in describing hadrons and their interactions, the glue which binds quarks together to form hadrons is a gauge field of Yang-Mills type in which the internal symmetry is SU_3, and each flavor quark, u, d, ..., is a triplet, the defining representation of the SU_3 ("each flavor of quark comes in three colors"). This symmetry is called color SU_3 to distinguish it from the flavor SU_n groups. The glue field couples equally to quarks of all flavors, and this is responsible for the flavor SU_n symmetry. (There is a close parallel to electromagnetism: The electromagnetic field is a gauge field which binds electrons to nuclei to form atoms; the electromagnetic field is coupled to electric charge, which is consequently forced to be conserved, and is indifferent to any other properties of the particle which carries the charge.) This gauge field theory of the glue is called quantum chromodynamics (QCD).

It is believed that color SU_3 is an exact, unbroken symmetry, despite the fact that hadrons do not occur in color multiplets, nor is there a massless quantum (gluon) of the glue field observed or a long-range (inverse-square) force, analogous to the photon and the Coulomb force. But the analogy to electromagnetism is imperfect, because the glue field is coupled to itself: The glue field is coupled to color, but it itself carries color since it is an octet representation of color SU_3. A consequence is asymptotic freedom, one aspect of which is that the force between two color charges brought close together rises more slowly than inverse square, in contrast to the electric force between two charges which becomes stronger than inverse square, due to vacuum polarization. Conversely, as two color charges are moved farther apart, the force between them falls more slowly than inverse square. At a large enough separation, the well-understood method of calculation, perturbation theory, fails; it is conjectured that the force falls asymptotically to a nonvanishing constant value (roughly 10^5 newtons = 10 tons weight). This is known as the confinement conjecture, since the consequence is that color is confined: Two bodies can separate freely to large distances only if they do not carry color charge, that is, are singlets of color SU_3; hence all hadrons must be colorless, that is, color singlets. Since the gluon is colored, it cannot appear as a free particle. Thus the consequence of confinement—that is, the color force is super-long-range—is that color SU_3, although exact, is a hidden symmetry, not directly observed in hadrons the way that the flavor SU_n symmetries are.

But there are a number of ways that the hidden color does reveal itself. The most dramatic is in very high-energy collisions involving large momentum transfers, that is, short-distance collisions; these create high-energy quarks and gluons as though these were physical particles (since at short distances confinement is irrelevant). As these colored particles separate from one another, confinement becomes important and each quark or gluon becomes converted into a narrow spray of hadrons, a jet. From the properties of the jets, properties of their progenitors can be deduced. SEE GAUGE THEORY; GLUONS; QUANTUM CHROMODYNAMICS.

Selection rules. As stated earlier, selection rules are an important result of conservation laws; They express whether or not particular reactions can satisfy the conservation laws. A few examples follow.

The conservation of angular momentum, parity, and statistics implies, for example, that unless a level of Be^8 has even angular momentum and positive parity, it cannot decay into two alpha particles since these two identical spinless bosons can only be in such states.

The conservation of angular momentum and parity implies the selection rules for the emission of radiation. For instance, the selection rules for the emission of electric dipole radiation [$\Delta J = 0, \pm 1$ (but not $J = 0$ to another state with $J = 0$) and parity change] are a consequence of the vector addition rules of angular momentum plus the fact that the electric dipole field is 1^-; that is, its angular momentum is one unit of $\hbar$ and its parity is -1.

The conservation of charge parity C implies that a given state of positronium cannot decay into both an even number and an odd number of photons. Similarly, the conservation of isotopic spin parity G implies that a state of nucleonium cannot decay into both an even number and an odd number of π^- mesons. SEE NUCLEAR REACTION; SELECTION RULES.

Bibliography. I. Bars et al. (eds.), *Symmetries in Particle Physics*, 1984; J. W. Cronin, CP symmetry violation, *Rev. Mod. Phys.*, 53(3):378–383, 1981; J. J. Sakurai, *Invariance Principles and Elementary Particles*, 1964; J. P. Elliot and P. G. Dawber, *Symmetry in Physics*, 2 vols., 1979, paper 1984; G. 't Hooft, Gauge theories of the forces between elementary particles, *Sci. Amer.*, 242(6):104–138, 1980; C. N. Yang, Law of parity conservation and other symmetry laws, *Science*, 127(3298):565–569, 1958.

SELECTION RULES
J. B. FRENCH

General rules concerning the transitions which may occur between the states of a quantum-mechanical physical system. They derive in almost all cases from the symmetry properties of the states and of the interaction which gives rise to the transitions. The system may have a classical (nonquantum) counterpart, and in this case the selection rules may often be related to the classical conserved quantities. A first use of selection rules is in determining the symmetry classes of the states, but in a great variety of ways they may yield other information about the system and the conservation laws. SEE SYMMETRY LAWS.

Angular momentum and parity rules. For an isolated system the total angular momentum is a conserved quantity; this fact derives from a fundamental fact of nature, namely, that space is isotropic. Each state is then classifiable by angular momentum J and its z component M ($= -J, -J + 1, \ldots, +J$). Angular momenta combine in a vectorial fashion. Thus, if the system makes a particle-emitting transition $J_1, M_1 \rightarrow J_2, M_2$, the emitted particles must carry away angular momentum (j, μ), where $\mathbf{j} = \mathbf{j}_1 - \mathbf{j}_2$. This implies that $\mu = M_1 - M_2$ and that j takes on values $J_1 - J_2, J_1 - J_2 + 1, \ldots, J_1 + J_2$. Thus in transitions ($J = 4 \leftrightarrow J = 2$) the possible j values comprise only 2, 3, 4, 5, 6, and, if it is also specified that $M_1 - M_2 = \pm 4$, only 4, 5, 6. Observe that J_z is additive.

Another fundamental symmetry, the parity, which determines the behavior of a system (or of its description) under inversion of the coordinate axes, is conserved by the strong and electromagnetic interactions, and gives a classification of systems as even ($\pi = +1$) or odd ($\pi = -1$). Under combination the parity combines multiplicatively. Thus, if the transition above is $4^\pm \rightarrow 2^\mp$, it follows that $j^\pi = 2^-, \ldots, 6^-$, while $4^\pm \to 2^\pm$ would give $j^\pi = 2^+, \ldots, 6^+$. The angular momentum **j** may be a combination of intrinsic spin **s** and orbital angular momentum **l**. Scalar, pseudoscalar, vector, and pseudovector particles are respectively characterized by $s^{\pi_s} = 0^+, 0^-, 1^-, 1^+$, where π_s is the "intrinsic" parity, while l always carries $\pi_l = (-1)^l$. SEE PARITY; SPIN.

Electromagnetic transitions. The photon is a transverse vector particle; this implies that its electric field vector (classically) or its intrinsic spin vector (quantum-mechanically) lies in the plane normal to the propagation direction. Thus, while an ordinary vector particle with total angular momentum j may, for $l > 0$, have $l = j, j - 1, j + 1$, corresponding to the three internal degrees of freedom, a transverse particle may have $l = j$ or else a particular linear combination of $l = j \pm 1$ (two internal degrees). An $l = j$ photon ("magnetic" type, or Mj) then carries parity $(-1)^{j+1}$, while the $l = j \pm 1$ combination (electric type, or Ej) carries $\pi = (-1)^j$; the magnetic-electric nomenclature is appropriate because, in the long-wavelength (LWL) limit (that is, when the photon wavelength is much greater than the radius of the source) the radiation is generated by a magnetic-type interaction (for example, $\mathbf{L} \cdot \mathbf{H}$ or $\mathbf{S} \cdot \mathbf{H}$ for M1, with **H** the magnetic field) or electric-type interaction ($\mathbf{P} \cdot \mathbf{E}$ for E1, with **P** the dipole operator and **E** the electric field). For $j = 0$, the only combination of s and l is $(s, l, j) = (1, 1, 0)$, which describes an oscillating E0 field, which however vanishes outside the source and hence does not radiate. Thus, for transitions of the form $J_1^\pm \leftrightarrow J_2^\pm$, photons of type M1, E2, M3, ... may be emitted, and, for $J_1^\pm \leftrightarrow J_2^\mp$, E1, M2, E3, ..., the j values being restricted as above. If M_1, M_2 are also specified, as in Zeeman transitions, there are further restrictions. For dipole transitions ($j = 1$), only $\mu = 0$ (π lines) and $\mu = \pm 1$ (σ lines) occur, the former not being radiated parallel to the magnetic field because of the photon transversality. If $M_1 = M_2 = 0$, the transition is forbidden unless $J_1 + J_2 + j$ is even; this inversion rule, not restricted to electromagnetic transitions, arises from the behavior of states and operators under a rotation through π radians about an axis in the x-y plane. It is important in discussing molecular symmetries and isospin selection rules. SEE MULTIPOLE RADIATION.

Approximate rules. The "exact" selection rules are to be supplemented by approximate rules which depend on the internal structure of the system. To the extent that L, S, the total orbital and spin angular momenta, are good quantum numbers, as happens in many atoms and a few nuclei, the angular momentum radiated in a long-wavelength E1 or E2 transition (whose transition operators are spin-independent) must come from the orbital structure; thus $(\Delta\mathbf{L}, \Delta\mathbf{S}) = (1, 0), (2, 0)$, respectively. For the M1 operator, which has the form $\alpha\mathbf{L} + \beta\mathbf{S}$, $(\Delta\mathbf{L}, \Delta\mathbf{S})$ can have the values $(1, 0)$ or $(0, 1)$ but not $(0,0)$ or $(1, 1)$; however, beyond that, since **L** and **S** can generate no orbital, spin, or radial excitations, M1 transitions in LS-coupling occur only between two levels of the same term (although this is not true for Mj with $j > 1$).

Many processes, including photon emission and beta decay, occur via one-body operators [for example, by

$$\mathbf{P} = \sum_i e_i \mathbf{r}(i)$$

for E1, where the sum is over particles, with the ith particle having charge e_i and position operator $\mathbf{r}(i)$]. These generate only single-particle transitions, in which at most one particle changes orbit. (The number of such particles is $\delta n = 1$ for odd parity, as in

$$p^2 \xrightarrow{E1} sp, pd$$

$\delta n = 0, 1$ for even, as in

$$d^2 \xrightarrow{E2} d^2, sd, dg$$

$\delta n = 0$ for M1 in LS-coupling, but not in jj-coupling, since, for example, $f_{5/2} \xrightarrow{M1} f_{7/2}$ is allowed). Because of this "$\delta n < 1$" rule, the ($J\pi$) rules apply to individual particles (for example, $\Delta l^\pi = 1^-$ for E1) and thus greatly facilitate study of the microscopic structures of the states involved. SEE NUCLEAR STRUCTURE.

Forbidden transitions. Forbidden transitions are those which are hindered by a significant selection rule or combination of them (the term is often applied in a more restricted sense in some domains, for example, in atomic physics, to all photon transitions except E1). Several examples will be used to illustrate the remarkable variety of information which they yield.

Parity violation. The alpha-decay process $A(J^\pi) \to B(0^+) + \alpha(0^+)$ is allowed only if $\pi = (-1)^J$ since the final angular momentum J is orbital only. The forbidden process, with A a $2-$

state in ^{16}O and B the ^{12}C ground state, has been observed, with a lifetime on the order of 10^{13} larger than that of a nearby 2+ state in ^{16}O. Since neither of the decay products has a nearby 0$^-$ companion, the decay can be ascribed to a small 2$^+$ admixture in the 2$-$ state, the admixing "intensity" (square of the 2$^+$ amplitude) being then on the order of 10^{-13}. This gives a measure of a fundamental symmetry breaking, the violation of parity by the weak interaction. This example, like several of those following, involves symmetry breaking in a two-level substructure, the simple quantum mechanics of which is explained in the **illustration**.

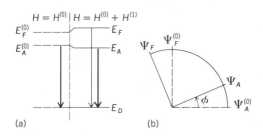

Symmetry admixing in a two-level substructure. (a) Energy-level diagram. (b) Rotation of states.

Two close-lying states Ψ_A and Ψ_F are eigenfunctions of a hamiltonian $H = H^{(0)} + H^{(1)}$ with energies E_A and E_F (illus. a), where $H^{(0)}$ preserves a symmetry and the small $H^{(1)}$ breaks it. Transitions occur to a distant state Ψ_D (with energy E_D), in which the symmetry admixing is negligible. $\Psi_A^{(0)}$ and $\Psi_F^{(0)}$, eigenfunctions of $H^{(0)}$, have different symmetries. The selection rules are such that the transition $\Psi_A^{(0)} \rightarrow \Psi_D$ is allowed, while $\Psi_F^{(0)} \rightarrow \Psi_D$ is forbidden. The effect of $H^{(1)}$, which has no diagonal matrix elements in the $\Psi^{(0)}$ states, is to "rotate" the states through a small angle ϕ (illus. b), thereby admixing the symmetries, generating a weak transition for Ψ_F, and slightly shifting the energies. The ratio of the transition strengths S determines ϕ, which in turn fixes the admixing intensity and the level shifts. Diagonalizing the 2 × 2 matrix gives: $S_F/S_A = \tan^2 \phi$; $(\Psi_F^{(0)} H^{(1)} \Psi_A^{(0)}) = (E_F - E_A) \sin \phi \cos \phi$; $(E_F - E_F^{(0)}) = -(E_A - E_A^{0}) = (E_F - E_A) \sin^2 \phi$. The perturbation solution (small ϕ) is often adequate.

Multiple-photon transitions. Because the M1 operator has no radial dependence in the long-wavelength limit, the single-photon transition between the $2S_{1/2}$ and $1S_{1/2}$ states in hydrogen is strongly inhibited; the lifetime should be on the order of 10^5 s. The very low-energy E1 transition to $2P_{1/2}$ (which lies below $2S_{1/2}$ by the Lamb shift energy), followed by E1 to ground, would take on the order of 10^8 s. The dominant process instead, with a lifetime on the order of 10^{-1} s, is the simultaneous emission of two photons, which share the energy to produce a continuous photon spectrum. In nuclei the two-photon $0^+ \rightarrow 0^+$ transition has also been observed, as well as E0 deexcitation by ejection of a penetrating atomic electron ("complete internal conversion") and by the closely related emission of an e^+-e^- pair. All of these processes are forbidden for $0^+ \leftrightarrow 0^-$, which has not been observed.

LS-forbidden transitions. Early examples discovered in optical astronomy involve $6s6p\ ^3P_0 \leftarrow 6s^2\ ^1S_0 \rightarrow 6s6p\ ^3P_2$ in HgI. The first is absolutely forbidden, while the second would require that highly improbable M2, with wave-function corrections (since $s \rightarrow p$ would require E1). But the decoupling effect of the hyperfine interaction between valence electrons and the nuclear magnetic moment generates a small $^3P_1$ amplitude in the $^3P_0$ state (the total angular momentum being $F = \frac{1}{2}$ in each case) and thereby, as in the illustration, an allowed but greatly retarded $^1S_0 \rightarrow ^3P_1$ transition. A similar process occurs for the second case. Astronomical observation of such forbidden lines indicates that particle densities in the source are low; otherwise deexcitation would occur via collisions.

Spherical shell-model orbits in nuclei. The process of adding a neutron in orbit (l, j) to a nucleus, $A + n(l, j) \rightarrow B$, is realizable in ($d, p$) "direct" reactions, whose angular distribution determines l and whose magnitude gives the intrinsic neutron-capture probability. If one of the nuclear states involved is 0^+, the j value must equal the angular momentum of the other. For example, if A is describable as $(j^{2n})_{J=0^+}$, there should be large cross sections to $(j^{2n+1})_{J=j}$ and

$((j^{2n})_0 j')_{J=j'}$, in which the particles in orbit j are themselves coupled to zero. Forbidden transitions to $(j^{2n+1})_{J=j'\neq j}$ are, however, often observed. They arise via admixtures (on the order of a few percent) of $((j^{2n})_0 j')$ in the final state which, by the analysis associated with the illustration, give measures of the orbital "purity." SEE NUCLEAR REACTION.

K selection rules in deformed nuclei. There is a J_z splitting of single-particle orbits in many heavy nuclei; this is produced by the large permanent quadrupole deformation of the nucleus which gives rise to a noncentral single-particle potential. K labels the angular-momentum component along the symmetry axis, whose rotation then generates from each intrinsic state a band of levels with $J \geq K$, all with the same parity. If two levels belong to bands with very different K values (specifically if $\Delta = |K_1 - K_2| - j$ is large, where the multipolarity j follows from the J^π values), electromagnetic transitions between them will be very highly forbidden. This is so formally because the j-multipole transition operator can only generate $|\Delta K| \leq j$, and more physically because the very different rotations involved in the two states generate large differences between all the corresponding orbits which play a role in the rotation. Empirically one encounters Δ values as high as 8 and finds retardations on the order of $10^{2\Delta}$, the process itself occurring via Coriolis-type admixtures. Similar multiparticle forbiddenness would obtain for transitions between states built on the two potential minima observed in fission studies for many nuclei. SEE NUCLEAR FISSION.

Molecular selection rules. The example of K selection rules in deformed nuclei involves both single-particle and rotational motion in these nuclei. Vibrational motion, especially quadrupole, can be found also. The classification of the motions derives from molecular physics and is most developed in that domain. To the extent that the molecular motions can be well separated, the angular momentum transferred in a transition must come from one or another of them. Along with the selection rules implied by this (the rule for harmonic vibrations is $n_v \to n_v \pm 1$), the various coupling possibilities and any special symmetries displayed by the intrinsic structure must be considered. Let N, Λ, Σ be the projection onto the symmetry axis (the line joining the two nuclei for the diatomic molecules to which the discussion is now restricted) of the collective rotational, the electronic orbital, and the electronic spin angular momenta. Λ and Σ may combine to Ω, the total intrinsic angular momentum along the symmetry axis; then the rotation gives $J = \Omega, \Omega + 1, \ldots$ (Hund's case A). Alternatively (case B), $\Lambda + N = K$, which combines with S to give $\mathbf{J} = \mathbf{S} + \mathbf{K}$, so that $|K - S| \leq J \leq K + S$. In both cases the supplementary rules for $E1$ (with natural extensions to other multipoles) that $\Delta\Lambda = 0, \pm 1$ and $\Delta S = 0$ are valid, while for case A, $\Delta\Sigma = 0$, $\Delta\Omega = 0 \pm 1$, and for case B, $\Delta\mathbf{K} = 1$; these being valid if the spin-orbit interactions are weak.

The special symmetries of the intrinsic structure refer to its behavior under a reflection through the symmetry center and through a plane containing the symmetry axis. The first operation, $\mathscr{P}$, defines the intrinsic parity, labeled as g (for $\pi = +1$) and u ($\pi = -1$). The second, $\mathscr{S}$, defines a symmetry $\pm$; in nuclear physics one commonly uses $\mathscr{R} = \mathscr{PS}$, with eigenvalue $r = \pm 1$, which generates a rotation through π radians about an axis normal to the symmetry axis. For homonuclear molecules the interchange of the two nuclei defines states as symmetric (s) or antisymmetric (a). Then for $E1$ the rules $g \leftrightarrow u$, $+ \leftrightarrow -$ are valid, and with homonuclear molecules $s \to s$, $a \to a$.

Further symmetries. The isospin symmetry of the elementary particles is almost conserved, being broken by electromagnetic and weak interactions. It is described by the group SU(2), of unimodular unitary transformations in two dimensions. Since the SU(2) algebra is identical with that of the angular momentum SO(3), isospin behaves like angular momentum with its three generators $\mathbf{T}$ replacing $\mathbf{J}$. For a nucleon $t = \frac{1}{2}$, the values $t_z = \frac{1}{2}, -\frac{1}{2}$ may be assigned for protons (p) and neutrons (n), respectively. Then, for nuclear states, $T_z = (Z - N)/2$ (where Z is the atomic number and N is the neutron number) and $T (\geq|T_z|)$ is an almost good quantum number. The argument associated with the illustration can often be used to measure the isospin admixing; for example, with the pair of decays ^{15}N ($T = \frac{1}{2}, \frac{3}{2}$) $\to$ ^{14}N ($T = 0 + n$), the second of which is forbidden by the selection rule $\Delta T = \frac{1}{2}$, one finds about 4% admixing. Since the nucleon charge is $|e|\{\frac{1}{2} + t_z\}$, the electromagnetic transition operators split into an isoscalar part (for which $\Delta \mathbf{T} = 0$) and an isovector ($\Delta \mathbf{T} = 1$). Since the isoscalar part of the (long-wavelength) $E1$ operator

$$|e|\left\{\frac{1}{2}\sum_i \mathbf{r}(i) + \sum_i t_z(i)\mathbf{r}(i)\right\}$$

is ineffective, being proportional to the center-of-mass vector, and the inversion rule described earlier forbids isovector transitions in self-conjugate nuclei ($T_z = 0$, $N = Z$), it follows that $E1$ transitions between states of the same T are forbidden in such nuclei. SEE ISOBARIC SPIN.

The isospin group is a subgroup of SU(3) which defines a more complex fundamental symmetry of the elementary particles. Two of its eight generators commute, giving two additive quantum numbers, T_z and strangeness S' (or, equivalently, charge and hypercharge). The strangeness is conserved ($\Delta S' = 0$) for strong and electromagnetic, but not for weak, interactions. The selection rules and combination laws for SU(3) and its many extensions, and the quark-structure ideas underlying them, correlate an enormous amount of information and make many predictions about the elementary particles. SEE BARYON; ELEMENTARY PARTICLE; MESON; QUARKS; UNITARY SYMMETRY.

The fundamental beta-decay transition is $n \rightarrow p + e^- + \bar{\nu}$. For the Gamow-Teller decay mode (electron and antineutrino in a triplet state) and the Fermi mode (singlet), the transition operators are respectively components of

$$\sum_i \mathbf{s}(i)\,\mathbf{t}(i)$$

and of **T**. When these decays are realized in a nucleus, in which also $p \rightarrow n + e^+ + \nu$ may occur, the rules $\Delta \pi = 0$, $\Delta \mathbf{T} = 1$, $\Delta T_z = \pm 1$ clearly hold, with $\Delta \mathbf{J} = 1, 0$ for Gamow-Teller and Fermi, respectively. But for Fermi decay there is also the very strong rule (which includes $\Delta T = 0$) that the decay takes place only to a single state, the isobaric analog, which differs from the parent only in T_z, the transition rate being then easily calculable. This special feature, analogous to that for M1 transitions in LS coupling, arises from the fact that the transition operator is a generator of an (almost) good symmetry. Besides superallowed Fermi transitions between $J = 0$ analog states (forbidden for Gamow-Teller), there are also forbidden Fermi decays between non-analog states which, as in the illustration, arise from isospin admixing and have been used to give measures for it. Special rules, by no means as well satisfied, arise also for Gamow-Teller decays from the fact that the 15 components of the basic transition operators are the generators of an SU(4) group whose symmetry is fairly good in light nuclei. (For these same nuclei a different SU(3) realization than above gives a model for rotational bands.) Besides the allowed transitions, there is also a hierarchy of forbidden transitions in nuclei which, through recoil (retardation) effects, involve the orbital angular momentum and have therefore less restrictive selection rules. SEE NEUTRINO; RADIOACTIVITY.

A great variety of other groups have been introduced to define relevant symmetries for atoms, molecules, nuclei, and elementary particles. They all have their own selection rules, representing one aspect of the symmetries of nature.

Bibliography. A. Bohr and B. R. Mottelson, *Nuclear Structure*, vol. 2, 1975; E. U. Condon and H. Odabasi, *Atomic Structure*, 1980; R. D. Cowan, *The Theory of Atomic Structure and Spectra*, 1981; R. H. Garstang, Forbidden transitions, in D. R. Bates (ed.), *Atomic and Molecular Processes*, pp. 1–46, 1962.

PARITY
Charles J. Goebel

A physical property of a wave function which specifies the wave function's behavior under simultaneous reflection of all spatial coordinates of the wave function through the origin, that is, when x is replaced by $-x$, y by $-y$, and z by $-z$. If the wave function ψ satisfies Eq. (1), it is

$$\psi(x,y,z) = \psi(-x,-y,-z) \qquad (1)$$

said to have even parity. If, on the other hand, Eq. (2) holds, the wave function is said to have

$$\psi(x,y,z) = -\psi(-x,-y,-z) \qquad (2)$$

odd parity. These two expressions can be combined in Eq. (3), where $P = \pm 1$ is a quantum

$$\psi(x,y,z) = P\psi(-x,-y,-z) \qquad (3)$$

number having only the two values $+1$ (designated as even parity) and -1 (odd parity). The physical property defined by P is quantized and is called parity. More precisely, parity is defined as the eigenvalue of the operation of space inversion. Parity is a concept that has meaning only for waves and therefore has no meaning in classical particles physics.

Parity does have a meaning for the Schrödinger wave function of a particle in quantum mechanics. It likewise has meaning for the wave function of any system. Corresponding to the fact that the wave function of a complex system is the product of the wave function of the coordinates of the subsystems into which the system may be subdivided times the internal wave functions of those subsystems, the parity of the system is the product of the parity of the wave function of the coordinates of the subsystems multiplied by the intrinsic parities of these subsystems.

Conservation. The conservation of parity is a consequence of the inversion symmetry of space. To show this formally, let $\mathcal{P}$ be the parity operator which inverts space; that is, $\mathcal{P}$ acting on a wave function yields the wave function at the inverse point of space, $\mathcal{P}\psi(\mathbf{r}) = \psi(-\mathbf{r})$. Similarly, for an operator A, $\mathcal{P} A(\mathbf{r})\mathcal{P}^{-1} = A(-\mathbf{r})$. The statement that the world is symmetrical to inversion means that the Hamiltonian H after inversion is the same as before, that is, $\mathcal{P}H\mathcal{P}^{-1} = H$; and thus $\mathcal{P}H - H\mathcal{P} = [\mathcal{P},H] = 0$. Since $\mathcal{P}$ commutes with H, it is a constant of the motion. Further, H and $\mathcal{P}$ can be simultaneously diagonal, that is, the eigenfunctions of H can be simultaneously eigenfunctions of $\mathcal{P}$. In fact, if for an eigenvalue E of H there is only a single eigenfunction (nondegenerate level), this eigenfunction must be an eigenfunction of $\mathcal{P}$. As for the eigenvalues of $\mathcal{P}$, note that $\mathcal{P}^2 = 1$, from which it follows that the possible eigenvalues of $\mathcal{P}$ are $+1$ or -1. That is, an eigenfunction of $\mathcal{P}$ satisfies $\mathcal{P} \psi_{\pm}(\mathbf{r}) = \psi_{\pm}(-\mathbf{r}) = \pm\psi_{\pm}(\mathbf{r})$, where the upper (lower) sign indicates an eigenfunction of positive (negative) parity, also known as even (odd) parity, as stated above.

Thus, parity would be conserved if the statement of physical laws were independent of the handedness of the coordinate system used. Of course, the fact that most people are right-handed is not a physical law but an accident of evolution; there is nothing in the laws of physics which favors a right-handed to a left-handed human. The same holds for optically active organic compounds, such as the amino acids. However, the statement that the neutrino is left-handed *is* a physical law. SEE NEUTRINO.

All the strong interactions between elementary particles (for example, nuclear forces) and the electromagnetic interactions are symmetrical to inversion, so that parity is conserved by these interactions. As far as is known, only the β-interactions (which involve neutrinos) and the other weak interactions contribute little to all processes except the decays of elementary particles (including beta decay of nuclei), so that in all other processes parity is very nearly conserved.

Atomic and nuclear energy states are characterized by a definite parity (which may be different for different energy states of the same nucleus), and the conservation of parity has an important bearing on atomic and nuclear reactions. Operators representing dynamical variables may also be classified in terms of the parity concept, depending upon how they are affected by an inversion of their spatial coordinates.

Orbital parity. Since parity is conserved in strong and electromagnetic interactions, it is termed a good quantum number, and an energy eigenstate (unless it is degenerate) must be an eigenstate of parity. The parity of a one-particle state of orbital angular momentum l is given by $P = (-)^l$, that is, even $(+1)$ for s, d, . . . , states, and odd (-1) for p, f, . . . , states. Thus the deuteron, whose state is a linear combination of $^3S_1$ and $^3D_1$, has even parity; there cannot be any admixture of $^3P_1$. The orbital parity of an n-particle system is the product of the parities of the $n-1$ relative orbital angular momentum states: $P_{\text{orb}} = (-)^{l+\cdots+l_{n-1}}$. Thus the parity of an atom is the product of the parities of the one-electron orbital wave function; all configurations which mix must have the same parity. The Laporte rule of atomic spectroscopy, which states that an electric dipole transition can occur only between states of opposite parity, depends on the fact that the electric dipole radiation field has odd parity.

Intrinsic parity. The intrinsic parities of the particles composing a system must be multiplied by the orbital parity to yield the total parity. But the intrinsic parity of a conserved particle is irrelevant and can be omitted. For if the particle is conserved in a reaction, so is the contribution of its intrinsic parity to the total parity, so that its intrinsic parity is irrelevant to the balance of parity in the reaction. In fact, if a particle is conserved in all reactions, its intrinsic parity can never be determined. The photon is an unconserved particle; its intrinsic parity is odd. The parity of the π^0 meson (a pseudoscalar) is odd, so that to conserve parity it must be emitted by a nucleon into a P state. By charge independence, the charged π meson must also be emitted in a P state; it is natural to call the parity of the charged π meson odd also, which amounts to defining the parity of the neutron and proton to be the same. An electron by itself is conserved, but an electron plus a positron can annihilate. Thus the product of the parities of an electron and a positron must be well defined. According to the Dirac equation of relativistic quantum theory, the product of their parities is -1. The same result holds for any fermion particle-antiparticle pair. Thus the parity of positronium is -1 times its orbital parity, that is, $-(-)^l$. SEE ELEMENTARY PARTICLE.

Spin and momentum correlations. The symmetry of the strong and electromagnetic interactions with respect to inversion implies statements about possible correlations of momenta and spins of the particles emitted as a result of such reactions. The principle is that the probability of a configuration of momenta and spins must be a scalar, in order that it not change under inversion of the coordinate system. Thus in a reaction yielding three particles with momenta, $\mathbf{p}_1$, $\mathbf{p}_2$, $\mathbf{p}_3$, the angular distribution might be of the form $a + b\mathbf{p}_1 \cdot \mathbf{p}_2$ but not $a + b\mathbf{p}_1 \cdot \mathbf{p}_2 \times \mathbf{p}_3$, for under inversion the last term changes sign as in Eq. (4). This triple product is a pseudoscalar, and

$$\mathbf{p}_1 \cdot \mathbf{p}_2 \times \mathbf{p}_3 \to (-\mathbf{p}_1) \cdot (-\mathbf{p}_2) \times (-\mathbf{p}_3) = -\mathbf{p}_1 \cdot \mathbf{p}_2 \times \mathbf{p}_3 \qquad (4)$$

the description of the angular distribution would not be independent of the handedness of the coordinate system, because the coefficient b would appear to change sign. Orbital angular momentum $\mathbf{L} = \mathbf{r} \times \mathbf{p}$ is a pseudovector, since under inversion $\mathbf{L} \to +\mathbf{L}$; the same must hold for spin angular momentum $\mathbf{S}$. Thus $\mathbf{S} \cdot \mathbf{p}$ is a pseudoscalar, and so such a term cannot occur in the angular distribution of a parity conserving process. This term, $\mathbf{S} \cdot \mathbf{p}$, in an angular distribution, would correlate a particle's spin with its momentum; that is, it would imply a polarization in the momentum direction, or longitudinal polarization, which is accordingly absent in strong and electromagnetic reactions. Transverse polarizations, indicated by terms such as $\mathbf{S}_1 \cdot \mathbf{p}_1 \times \mathbf{p}_2$, are of course always possible.

Nonconservation. One of the selection rules which follows from parity conservation is the following: The same spin zero boson cannot decay both into two π mesons and three π mesons, because these final states have opposite parities, even and odd respectively. But the positive K meson is observed to do just this: It has both the $K_{\pi 2}$ and $K_{\pi 3}$ decay modes, and its spin is zero as deduced from the distribution of momenta in the $K_{\pi 3}$ mode. Thus the conclusion is that parity is not conserved in this decay. In 1956, T. D. Lee and C. N. Yang made the bold hypothesis that parity also is not conserved in beta decay. They reasoned that the magnitude of the beta decay coupling is about the same as the coupling which leads to decay of the K meson, so these decay processes may be manifestations of a single kind of coupling. Also, there is a very natural way to introduce parity nonconservation in beta decay, namely by assuming a restriction on the possible states of the neutrino (two-component theory). They pointed out that no beta-decay experiment had ever looked for the spin-momentum correlations, which would indicate parity nonconservation; they urged that these correlations be sought.

In the first experiment to show parity nonconservation in beta decay, the spins $\mathbf{S}_{Co}$ of the beta-active nuclei cobalt-60 were polarized with a magnetic field $\mathbf{H}$ at low temperature; the decay electrons were observed to be emitted preferentially in directions opposite to the direction of the ^{60}Co spin (see **illus.**). Thus a $\mathbf{S}_{Co} \cdot \mathbf{P}_e$ correlation was found or, in terms of macroscopic quantities, an $\mathbf{H} \cdot \mathbf{p}_e$ correlation. The magnitude of this correlation shows that the parity-nonconserving and parity-conserving parts of the beta interaction are of equal size, substantiating the two-component neutrino theory.

It is believed that parity conservation fails in all the weak decays, which includes all decays of the quasi-stable elementary particles (except the electromagnetic decays $\pi^0 \to 2\gamma$ and $\Sigma^0 \to \Lambda + \gamma$).

It was at first somewhat disconcerting to find parity not conserved, for that seemed to imply a handedness of space which would then not be the empty thing which (since the demise of the ether hypothesis) most physicists think it to be. That is, an ether would be needed to provide a standard of handedness at each point of space, to tell ^{60}Co which direction to decay into. But this is not really the situation; the saving thing is that anti-^{60}Co decays in the opposite direction. Thus, after all, there is nothing intrinsically left-handed about the world, just as there is nothing

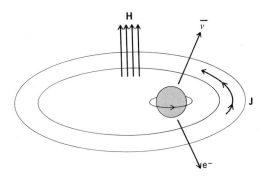

Beta decay from polarized cobalt-60 nuclei. When the spin axes of the cobalt nuclei are not polarized, the preferential emission of the electrons and antineutrinos in the directions shown is not detectable.

intrinsically positively charged about nuclei. What really exists here is a correlation between handedness and sign of charge. *See* SELECTION RULES.

Bibliography. M. A. Bouchiat and L. Potier, An atomic preference between left and right, *Sci. Amer.*, 250:100–111, June 1984; S. L. Gilbert and C. E. Wieman, Atomic-beam measurements of parity nonconservation in atomic cesium, *Phys. Rev. A*, 34:792–803, 1986; M. Gardner, *The Ambidextrous Universe*, 2d ed., 1979; T. D. Lee, Weak interactions and nonconservation of parity, *Science*, 127(3298):569–573, 1958; P. Morrison, The overthrow of parity, *Sci. Amer.*, 196(4):45–53, 1957; C. N. Yang, Law of parity conservation and other symmetry laws, *Science*, 127(3298):565–569, 1958.

TIME REVERSAL INVARIANCE
CHARLES J. GOEBEL

A symmetry of the fundamental (microscopic) equations of motion of a system; if it holds, the time reversal of any motion of the system is also a motion of the system. With one exception (K_L meson decay), all observations are consistent with time reversal invariance (T invariance).

Time reversal invariance is not evident from casual observation of everyday phenomena. If one takes a movie of a phenomenon, the corresponding time-reversed motion can be exhibited by running the movie backward. The result is usually strange. For instance, water in the ground is not ordinarily observed to collect itself into drops and shoot up into the air. However, if the system is sufficiently well observed, the direction of time is not obvious. For instance, a movie which showed the motion of the planets would look just as right run backward or forward. The apparent irreversibility of everyday phenomena results from the combination of imprecise observation and starting from an improbable situation (a state of low entropy, to use the terminology of statistical mechanics).

The known fundamental equations of motion are all time reversal–invariant. For instance, suppose that $\vec{r}_j(t)$ and $\vec{p}_j(t)$, the coordinates and momenta of charged particles, and $\vec{E}(\vec{r},t)$ and $\vec{B}(\vec{r},t)$, the electric and magnetic fields, satisfy the equations of motion of the particles and the fields (Maxwell's equations); that is, these functions of time describe a motion of the system. Then the functions $\vec{r}_{j\mathrm{REV}}(t) = \vec{r}_j(-t)$, $\vec{p}_{j\mathrm{REV}}(t) = -\vec{p}_j(-t)$, $\vec{E}_{\mathrm{REV}}(r,t) = \vec{E}(r,-t)$, and $\vec{B}_{\mathrm{REV}}(\vec{r},t) = -\vec{B}(\vec{r},-t)$, which describe the time reversal of the original motion, also satisfy the equations of motion. Thus the system is time-reversible.

Quantum mechanics. In quantum mechanics, if the hamiltonian operator H is independent of time (energy-conserving) and real, and if the wave function $\psi(t)$ is a solution of the Schrödinger equation below, where $\hbar$ is Planck's constant divided by 2π and ∂_t is partial differentiation

$$i\hbar \partial_t \psi = H\psi$$

with respect to time, then the function $\psi_{\mathrm{REV}}(t) = \psi^*(-t)$, where the asterisk indicates complex conjugation, is also a solution of the Schrödinger equation. The function ψ_{REV} is the wave function of the time reversal of the motion described by ψ; hence the Schrödinger equation is time reversal–invariant. In the time-reversed motion, other kinds of amplitudes, for example transition amplitudes, are also complex-conjugated.

Tests. If time reversal invariance holds, no particle (a system with a definite mass and spin) can have an electric dipole moment D, that is, an interaction energy of the form $-\vec{D}\cdot\vec{E}$, where $\vec{E}$ is the electric field. This is because the only vector quantity which can be defined by such a system (that is, the only vector quantity with nonvanishing expectation value) is the spin $\vec{S}$; thus an electric dipole moment $\vec{D}$ would have to be a multiple of $\vec{S}$, resulting in an interaction energy of the form $c\vec{S}\cdot\vec{E}$, where c is a constant. This energy would change sign under time reversal, because $\vec{S}$ does but $\vec{E}$ does not. (Since spin is an axial vector, an electric dipole moment of a particle would also violate space inversion symmetry or parity.) A polar body, for example, a water (H_2O) molecule, has an electric dipole moment, but its energy and spin eigenstates (which are particles) do not. No particle has been observed to have an electric dipole moment; for instance, the present experimental upper limit on the electric moment of the neutron is approximately 10^{-24} cm times e, where e is the charge of the proton. SEE NEUTRON; PARITY; SPIN.

Another test of time reversal invariance is to compare the cross sections for reactions which are inverse to one another, for example, the reactions $^{16}O + d \longleftrightarrow\,^{14}N + \alpha$. The present experimental upper limit on the relative size of the time reversal invariance–violating amplitude of such reactions is approximately 3×10^{-3}. SEE NUCLEAR REACTION.

Another class of tests involves looking at the relative phases of amplitudes. For instance, if a nuclear transition can emit both electric quadrupole and magnetic dipole electromagnetic radiation (gamma rays), certain interference terms in the angular distribution of the radiation cannot occur, because of the relative phase of the amplitudes for these emissions which is imposed by time reversibility. Experiments looking for such effects put an upper limit of approximately 10^{-3} on the relative size of a time reversal invariance–violating amplitude. SEE MULTIPOLE RADIATION.

Evidence for violation. If time reversibility holds, then by the CPT theorem CP invariance must hold, that is, invariance of the fundamental equations under the combined operations of charge conjugation C and space inversion P. Consequently if time reversal invariance holds, a self-conjugate particle must be a CP eigenstate. The K_L meson is self-conjugate, but it or its decay state is not a CP eigenstate. It is nearly a CP = -1 eigenstate, but has a small probability of decaying into a pion pair, $\pi^+\pi^-$ or $\pi^0\pi^0$, which states have CP = $+1$. Further, the probabilities of its decay into the charge-conjugate channels $\pi^+e\bar{\nu}_e$ and $\pi^-\bar{e}\nu_e$ (and similarly for channels with μ in place of e) are not the same. SEE MESON.

If this observed CP violation came from the interactions responsible for the K_L decay, they should produce other observable CP-violating effects as well, namely an electric dipole moment of the neutron and a change in the relative phase of the amplitudes for K_L decay into pions; neither is seen. Apparently the interactions which violate time reversibility are very weak, and for this reason they are called superweak. They produce an observable effect in K_L decay only because of the very special circumstance that the masses of the two K_0 mesons are very nearly the same (in the absence of the weak interaction responsible for their decays, they would have the same mass), with the consequence that a very weak interaction can have appreciable effect by mixing the two. The superweak interaction that violates time reversal invariance mixes a little CP = $+1$ amplitude into the K_L (and correspondingly a little CP = -1 into the K_S, although this is harder to observe). It is an open question just what these superweak interactions are, and whether the violation is due to a spontaneous symmetry breaking. SEE FUNDAMENTAL INTERACTIONS; SYMMETRY BREAKING; WEAK NUCLEAR INTERACTIONS.

In addition to the direct but very subtle evidence for violation of time reversal invariance seen in K_L decay, there is an indirect but very prominent evidence, namely the baryon asymmetry of the universe, that is, the difference in the numbers of baryons and antibaryons. (There is likewise a lepton asymmetry.) In concrete terms, nearly all the positively charged "elementary" particles in the universe are protons (not positrons) and the negative particles are electrons (not antiprotons). Put quantitatively, early in the history of the universe, when the value of kT (the product of Boltzmann's constant and the thermodynamic temperature) was much larger than 1 GeV, the ratio of the difference of the numbers of baryons and antibaryons to the sum was of the order of 10^{-9}. (This is roughly the same as the present ratio of the numbers of protons to cosmic

3 K blackbody photons.) If this ratio is not accidental or arbitrary but is a result of the evolution of the universe from an earlier state, it implies the operation of processes which violated not only baryon number conservation but also CP symmetry, and time reversal symmetry. See ELEMENTARY PARTICLE; SYMMETRY LAWS.

Bibliography. J. P. Elliot and P. G. Dawber, *Symmetry in Physics*, 2 vols., 1979, paper 1984; W. M. Gibson and B. R. Pollard, *Symmetry Principles in Elementary Particle Physics*, 1976, paper 1980; K. Gottfried, *Quantum Mechanics*, vol. 1: *Fundamentals*, 1966; E. M. Henley, Parity and time-reversal invariance in nuclear physics, *Annu. Rev. Nucl. Sci.*, 19:367–432, 1969; E. Merzbacher, *Quantum Mechanics*, 2d ed., 1970; D. Park, *Introduction to the Quantum Theory*, 2d ed., 1974; L. I. Schiff, *Quantum Mechanics*, 3d ed., 1968; S. Weinberg, The decay of the proton, *Sci. Amer.*, 244(5):64–75, 1981.

SPIN
Charles J. Goebel

The intrinsic angular momentum of a particle. It is that part of the angular momentum of a particle which exists even when the particle is at rest, as distinguished from the orbital angular momentum. The total angular momentum of a particle is the sum of its spin and its orbital angular momentum resulting from its translational motion. The general properties of angular momentum in quantum mechanics imply that spin is quantized in half integral multiples of $\hbar$ ($\hbar = h/2\pi$, where h is Planck's constant); orbital angular momentum is restricted to half *even* integral multiples of $\hbar$. A particle is said to have spin 3/2, meaning that its spin angular momentum is 3/2$\hbar$.

A nucleus, atom, or molecule in a particular energy level, or a particular elementary particle, has a definite spin; for instance, a deuteron has spin 1, a $_3Li^5$ nucleus in its ground state has spin 3/2, and an electron has spin 1/2. The spin is an intrinsic or internal characteristic of a particle, along with its mass, charge, and isotopic spin. See ISOBARIC SPIN; SYMMETRY LAWS.

A particle of spin s has $2s + 1$ spin states, since according to quantum mechanics the projection of an angular momentum of magnitude j along an axis can have the $2j + 1$ integrally spaced values $j, j - 1 \ldots -j + 1, -j$. These spin states represent an internal degree of freedom of a particle, in addition to its external freedom of motion in three-dimensional space.

In field theory, in which particles are regarded as quanta of a field, the spin of the particle is determined by the tensor character of the field. For instance, the quanta of a scalar field have spin 0 and the quanta of a vector field have spin 1. A celebrated theorem of quantum field theory, proved first by W. Pauli, states a connection between spin and statistics: A particle with half even integral spin obeys Bose-Einstein statistics and is called a boson; a particle with half odd integral spin obeys Fermi-Dirac statistics and is called a fermion. See QUANTUM FIELD THEORY.

HELICITY
D. Allan Bromley

A fundamental quantized variable used in quantum mechanics to specify the relative orientations of spin and linear momentum of massless particles. It is a requirement of fundamental Dirac quantum mechanics that such particles have their spins aligned either parallel or antiparallel to their linear momentum. Particles having parallel alignment are arbitrarily assigned helicity $+1$; those have antiparallel alignment, -1. See SPIN.

In a classic experiment on K electron capture by ^{152}Eu, M. Goldhaber, L. Grodzins, and A. Sunyar first showed that the neutrino emitted in the weak nuclear interaction had negative helicity—that is, its spin was aligned antiparallel to its momentum. An equivalent description of this situation is that these neutrinos are left-handed. Symmetry requires that antineutrinos be right-handed and have positive helicity. See ELECTRON CAPTURE; NEUTRINO; SYMMETRY LAWS.

This discovery also requires that the Gamow-Teller beta decay proceed predominantly by means of an axial vector interaction. See SELECTION RULES; WEAK NUCLEAR INTERACTIONS.

The questions of whether the neutrino is truly massless and of helicity and lepton number conservation are intimately related. One of the main points at issue is whether the lepton number of the neutrino is uniquely related to its helicity; usually it is taken as the negative of the helicity. Other theories have been suggested wherein the two are completely dissociated and wherein the usual additive lepton number conservation law is replaced by a multiplicative one. SEE ELEMENTARY PARTICLE.

Bibliography. E. Commins and P. H. Bucksbaum, *Weak Interactions of Leptons and Quarks*, 1983; M. Goldhaber, L. Grodzins, and A. Sunyar, *Phys. Rev.*, 109:1015–1017, 1958.

CPT THEOREM
VAL L. FITCH

A fundamental ingredient in quantum field theories, which dictates that all interactions in nature, all the force laws, are unchanged (invariant) on being subjected to the combined operations of particle-antiparticle interchange (so-called charge conjugation, C), reflection of the coordinate system through the origin (parity, P), and reversal of time, T. In other words, the CPT operator commutes with the hamiltonian. The operations may be performed in any order; TCP, TPC, and so forth, are entirely equivalent. If an interaction is not invariant under any one of the operations, its effect must be compensated by the other two, either singly or combined, in order to satisfy the requirements of the theorem. SEE QUANTUM FIELD THEORY.

The CPT theorem appears implicitly in work by J. Schwinger in 1951 to prove the connection between spin and statistics. Subsequently, G. Lüders and W. Pauli derived more explicit proofs, and it is sometimes known as the Lüders-Pauli theorem. The proof is based on little more than the validity of special relativity and local interactions of the fields. The theorem is intrinsic in the structure of all the successful field theories. SEE SPIN.

Significance. CPT assumed paramount importance in 1957, with the discovery that the weak interactions were not invariant under the parity operation. Almost immediately afterward, it was found that the failure of P was attended by a compensating failure of C invariance. Initially, it appeared that CP invariance was preserved and, with the application of the CPT theorem, invariance under time reversal. Then, in 1964 an unmistakable violation of CP was discovered in the system of neutral K mesons. SEE PARITY.

One question immediately posed by the failure of parity and charge conjugation invariance is why, as one example, the π^+ and π^- mesons, which decay through the weak interactions, have the same lifetime and the same mass. It turns out that the equality of particle-antiparticle masses and lifetimes is a consequence of CPT invariance and not C invariance alone. A casual proof can be obtained by applying the CPT operation to a free particle. This results in an antiparticle with precisely the same momentum and total energy and therefore, from special relativity, the same rest mass as the original particle. This obvious result can be generalized heuristically to include equality of lifetimes by including an imaginary component in the mass.

Experimental tests. The validity of symmetry principles such as the CPT theorem should not be assumed beyond the limits of the experimental tests. In that the CPT theorem predicts the equality of particle-antiparticle properties such as mass, lifetime, and the magnitude of the magnetic moment, measurements which compare these quantities constitute tests of the theorem. For example, it is known from studies of the neutral K-meson system that the mass of K^0 is the same as the $\overline{K^0}$ to within 1 part in 10^{18}. The magnitude of the magnetic moment of the electron is known to be the same as the positron to 1 part in 10^{10}. The lifetime of the π^+ and π^- mesons is the same to better than 0.1%, and so with the μ^+ and μ^- mesons.

The discovery of CP violations in 1964 in the system of neutral K mesons reopened the question of CPT invariance. That system has been analyzed in great detail, and it is known now that at least 90% of the observed CP violation is compensated by a violation of time-reversal invariance. That is, not more than 10% of the observed CP violation can be credited to a violation of CPT. These are all experimental limits and do not in any way suggest a violation of CPT.

The remarkable sensitivity of the K mesons to possible departures from CPT invariance, as demonstrated by the precision with which the masses of the K^0 and $\overline{K^0}$ are known to be the same, suggests that this system is perhaps the best one for further experimental tests of the validity of the CPT theorem. SEE ELEMENTARY PARTICLE; MESON; SYMMETRY LAWS.

ISOBARIC SPIN
Robert K. Adair

A quantum-mechanical variable or quantum number applied to hadrons, the strongly interacting fundamental particles, and compounds of hadrons (such as nuclear states) to facilitate consideration of the consequences of the charge independence of the strong (nuclear) forces. This variable is also labeled isotopic spin or isospin, and most commonly I spin.

The many strongly interacting particles (hadrons) and the compounds of these particles, such as nuclei, are observed to form sets or multiplets such that the members of the multiplet differ in their electromagnetic charge and electromagnetic properties but are otherwise almost identical. For example, the neutron and proton, with electric charges of zero and plus one fundamental unit (of the magnitude of the electronic charge), form a set of two such states. The pions, one with a unit of positive charge, one with zero charge, and one with a unit of negative charge, form a set of three. It appears that if the effects of electromagnetic forces and the closely related weak nuclear forces (responsible for beta decay) are neglected, leaving only the strong forces effective, the different members of such a multiplet are equivalent and cannot be distinguished in their strong interactions. The strong interactions are thus independent of the different electric charges held by different members of the set; they are charge-independent. *See* ELEMENTARY PARTICLE; FUNDAMENTAL INTERACTIONS; HADRON; STRONG NUCLEAR INTERACTIONS.

The isobaric spin I of such a set or multiplet of equivalent states is defined such that Eq. (1) is satisfied, where N is the number of states in the set. Another quantum number I_3, called the

$$N = 2I + 1 \tag{1}$$

third component of isobaric spin, is used to differentiate the numbers of a multiplet where the values of I_3 vary from $+I$ to $-I$ in units of one. The charge Q of a state and the value of I_3 for this state are connected by the Gell-Mann–Okubo relation, Eq. (2), where Y, the charge offset, is called

$$Q = I_3 + Y/2 \tag{2}$$

hypercharge. For nuclear states, Y is simply the number of nucleons. Electric charge is conserved in all interactions; Y is observed to be conserved by the strong forces so that I_3 is conserved in the course of interactions mediated by the strong forces. *See* HYPERCHARGE.

Similarity to spin. This description of a multiplet of states with isobaric spin I is similar to the quantum-mechanical description of a particle with total angular momentum or spin of j (in units of $\hbar$, Planck's constant divided by 2π). Such a particle can be considered as a set of states which differ in their orientation or component of spin j_z in a z direction of quantization. There are $2j + 1$ such states, where j_z varies from $-j$ to $+j$. To the extent that the local universe is isotropic (or that there are no external forces on the states which depend upon direction), the components of angular momentum or spin in any direction are conserved, and states with different values of j_z are dynamically equivalent.

There is then a logical or mathematical equivalence between the descriptions of (1) a multiplet of states of definite isobaric spin I and different values of charge and I_3 with respect to charge-independent forces and (2) a multiplet of states of a particle with a definite spin j and different values of j_z with respect to direction-independent forces. In each case, the members of multiplet with different values of the conserved quantity I_3 on the one hand and j_z on the other are dynamically equivalent; that is, they are indistinguishable by any application of the forces in question. *See* SPIN.

Importance in reactions and decays. The charge independence of the strong interactions has important consequences defining the intensity ratios of different charge states produced in those particle reactions and decays which are mediated by the strong interactions. As a simple illustration, consider the virtual transitions of a nucleon to a nucleon plus a pion, transitions which are of dominant importance in any consideration of nuclear forces. A proton may undergo a transition to a proton and neutral pion or to a neutron and positive pion and conserve charge; a neutron can go to a neutron and neutral pion or to a proton and negative pion. The neutron and proton form a nucleon isobaric spin doublet with $I = \frac{1}{2}$; the pions constitute an isobaric spin triplet with $I = 1$. If one starts with an initial set of one proton and one neutron, with no bias in charge state or I_3, and the strong forces responsible for the virtual transitions do not discriminate between states with different charge or different values of I_3, it follows by inspection that the ratio of transition intensities will be defined as shown in the **table**.

If the forces cannot distinguish between charge states and there is no initial charge asymmetry, there can be no charge asymmetry in the final sets of states. If initially there are one neutron and one proton, equal numbers of each charge member of the nucleon isobaric spin doublet, in the final system there must be equal probabilities of finding each charge member of the pion triplet and equal probabilities of finding a neutron or proton. This condition, that the strong interactions cannot differentiate among the members of an isobaric spin multiplet, defines the relative intensities given in the table. These arguments hold equally well for real decays.

Relative intensities of virtual transitions determined by isobaric spin symmetry

Transition	Relative intensity
$p \to n + \pi^+$	2/3
$p \to p + \pi^0$	1/3
$n \to n + \pi^0$	1/3
$n \to p + \pi^-$	2/3

The above demonstration considers the decay of an initial (nucleon) doublet with $I = \frac{1}{2}$, to a final state of a (nucleon) doublet and a (pion) triplet with $I = 1$. Using the same kind of argument, it is easy to see that the conditions of equal intensity of each member of a multiplet cannot be fulfilled in a transition from an initial doublet to a final state of a doublet and quartet. Therefore, none of the individual transitions is allowed by charge independence, though charge or I_3 is conserved in the decays. In general, decays are allowed for a transition $A \to B + C$ only if inequality (3) is satisfied. This is analogous to the vector addition rule for spin or angular mo-

$$|I(B) + I(C)| \geq |I(A)| \geq |I(B) - I(C)| \qquad (3)$$

mentum; the strong interactions conserve isobaric spin in the same manner as angular momentum is conserved. This is an example of the general rule that the whole content of the description of spin or angular momentum can be taken over for isobaric spin. SEE NUCLEAR REACTION; SELECTION RULES.

Classification of states. Isobaric spin considerations provide insight into the total energies or masses of nuclear and particle states. The fundamental constituents of nuclei are the nucleons, the neutron and proton, spin ½ fermions which must obey the Pauli exclusion principle to the effect that the sign of a wave function which describes a set of identical fermions must change sign upon exchange of any two fermions. Similarly, hadrons are now described as compounds of quarks, which are also spin ½ fermions. The two fermions which belong to an isobaric spin doublet can be considered as different charge states of a basic fermion, even as states with spin in the plus and minus direction of quantization are considered as different spin states of the fermion. The extended Pauli exclusion principle then requires that the wave function amplitude change sign upon exchange of spin, charge, and spatial coordinates for two fermions.

A space state $u(r)$ of two fermions, where r is the vector distance between the two particles, will be even upon exchange of the two particles if $u(r)$ has an even number of nodes, and will be odd under exchange if there is an odd number of nodes. With more nodes, the space wavelength is smaller, and the momentum and energy of the particles are larger. The lowest energy state must then have no spatial nodes and must be even under spatial interchange. From the Pauli principle, the whole wave function must be odd, and then the exchange under spin and isobaric spin coordinates must be odd. Using this kind of argument, Eugene Wigner was able to classify the low-mass (low-energy) states of light nuclei in terms of their isobaric spin symmetries.

An application of the same principle was an important element in the discovery of a new quantum number (labeled color) for quarks, the elementary constituents of the hadrons. The lightest, least energetic baryon states, such as the neutron and proton, were shown to be even under the exchange of quark spin and quark isobaric spin, a result which would seem to require that the space wave function was odd, violating the general energy argument. It is now known that the states are odd under the additional color exchange, allowing the space function to be even, as expected, for the lightest states. SEE COLOR; QUARKS; SYMMETRY LAWS.

Bibliography. S. deBenedetti, *Nuclear Interactions*, 1964; H. A. Enge, *Introduction to Nuclear Physics*, 1966; M. Gell-Mann and Y. Neeman, *The Eightfold Way*, 1964; D. Lurie, *Particles and Fields*, 1968; L. H. Ryder, *Elementary Particles and Symmetries*, 1986.

UNITARY SYMMETRY
CHARLES J. GOEBEL

One of the approximate internal symmetry laws obeyed by the strong interactions of elementary particles. A system of particles has an SU_n internal symmetry if all of the particles can be described as compounds of a fundamental multiplet of n particles, and if all physical properties of the system are unchanged by an arbitrary unitary transformation of the fundamental multiplet. SEE SYMMETRY LAWS.

An analog is the approximate spin independence of electrons under electrostatic forces (as in an atom): There is a fundamental doublet, namely the spin-up electron and the spin-down electron. Denoting these two states by $|u\rangle$ and $|d\rangle$, all physical properties (energy eigenvalues, charge density, and so on) are unchanged by the replacements shown in Eqs. (1), where α and β

$$|u\rangle \to \alpha|u\rangle + \beta|d\rangle \qquad |d\rangle \to -\beta^*|u\rangle + \alpha^*|d\rangle \qquad |\alpha|^2 + |\beta|^2 = 1 \qquad (1)$$

are complex numbers. This transformation corresponds to a rotation of space (α and β can be expressed in terms of the three numbers which describe the rotation). It is easily seen that states of several electrons decompose under the rotation, Eqs. (1); for example, the two-electron state $(|u,d\rangle - |du\rangle)/\sqrt{2}$ is unchanged by the rotation, and the three remaining two-electron states, $|u,u\rangle$, $|d,d\rangle$, and $(|u,d\rangle + |d,u\rangle)/\sqrt{2}$, transform to linear combinations of themselves. This is the decomposition into singlet and triplet spin states, that is, into total spin $S = 0$ and 1 respectively; the nonmixing between them is equivalent to the invariance of S to rotation. The group of all the transformations of two states which preserves their scalar products [$\langle u|d\rangle = 0$, $\langle u|u\rangle = \langle d|d\rangle = 1$] is known as the two-dimensional unitary group, U_2; the transformations of Eqs. (1) form a subgroup known as SU_2 which merely lacks the uninteresting transformations of the form $|u\rangle \to e^{i\varphi}|u\rangle$ and $|d\rangle \to e^{i\varphi}|d\rangle$, that is, an equal change of phase of the two states.

Charge independence. The strong interactions are approximately invariant to such a group; the fundamental doublet can be taken to be the nucleon, with the up and down states proton and neutron. This SU_2 symmetry is known as charge independence, or, loosely, as i-spin conservation, the analog to the electron spin being known as i-spin **I**. The nucleon has i-spin ½; the pion has i-spin 1. Although the transformation of the pion states under i-spin rotations follows correctly from regarding the pion as the triplet state of a pair of nucleons, a further symmetry, nucleon conservation, requires the pion to be regarded instead as a compound of a nucleon and antinucleon. However, the statement "$|\pi\rangle$ can be regarded as $|N\bar{N}\rangle$" is only a statement about the transformation of pion states under i-spin rotations; it is not a statement about the physical structure of the pion. It is like saying that a triplet state of a calcium atom transforms under rotations like the triplet state of two electrons; this does not mean that the atom is a compound of two electrons only. SEE ISOBARIC SPIN.

SU_3 symmetry. The existence of strange particles shows that the symmetry group of the strong interactions is yet larger; an additional quantum number strangeness is conserved equally well as I_3. This can be achieved by adding a third fundamental particle, an i-spin singlet $|\lambda\rangle$ to carry strangeness; the additional symmetry is invariance to a relative phase change of $|N\rangle$ and $|\lambda\rangle$, that is, $|N\rangle \to e^{i\psi}|N\rangle$ and $|\lambda\rangle \to e^{-i\psi}|\lambda\rangle$, a U_1 group. When a sufficient number of strange particles had been observed, it was seen that they, together with the old nonstrange particles, were grouped into multiplets whose members had the same space-time quantum numbers (except for mass; the masses of the members are only similar, not equal). This suggested the existence of a yet larger symmetry; it has turned out that this symmetry is the group of unitary transformations of a triplet of fundamental particles, SU_3. (The i-spin plus strangeness symmetry of the triplet described above, that is, $SU_2 \times U_1$, is a subgroup of SU_3.) This symmetry is often loosely called unitary symmetry. SEE STRANGE PARTICLES.

A striking difference in the manifestations of SU_2 and SU_3 is that whereas all possible multiplets of the former appear in nature, only those multiplets of the latter appear which can be regarded as compounds of the fundamental triplet in which the net number of component fundamental particles (number of particles minus number of antiparticles) is an integral multiple of 3. (An analog in SU_2 would be compounds of an even number of the fundamental doublet particles; these states would have integral, never half-odd-integral, i-spin.) In particular, no particle which could be regarded as the fundamental triplet is found. Despite this nonappearance, it turns out that a great deal about the strongly interacting particles (hadrons) is at least qualitatively explained if they are regarded as physical compounds of a fundamental triplet of particles, to which the name quark has been given. The color theory (quantum chromodynamics) of strong interactions explains why single quarks are never observed. SEE HADRON; QUANTUM CHROMODYNAMICS; QUARKS.

The **illustration** shows the values of the additive quantum numbers I_3 and Y of the states of some SU_3 multiplets. Y (hypercharge) is essentially the same as strangeness. The states of the triplet labeled d, u, s correspond to the triplet n, p, λ discussed above [d = down, u = up (referring to I_3), and s = strange]. By assigning appropriate values of charge Q to the quarks, one gets the correct charges of hadrons, which obey $Q = I_3 + \frac{1}{2}Y$.

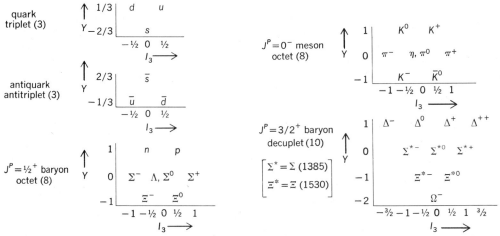

Weight diagrams of some SU_3 multiplets showing the values of I_3 and Y (hypercharge) of the members.

Unitary symmetry is only approximate. Its "breaking" appears to obey octet dominance. This means, for example, that the part of a mass operator which is not invariant under SU_3 (so that its matrix elements are the mass deviations of the members of an SU_3 multiplet) transforms as an octet. The consequence is the mass formula of M. Gell-Mann and S. Okubo, given by Eq. (2), where a, b, and c are constants, different for each multiplet, generally. (An analogy to this

$$m(I, Y) = a + bY + c[I(I + 1) - 1/4Y^2] \quad (2)$$

would be that the energy operator of atoms in a weak magnetic field transforms as a vector, and therefore the energy of an atom has the form $a + bJ_3$, where J_3 is the projection of the spin of the atom along the direction of the magnetic field.) This implies that the masses of the four i-spin multiplets of an octet have one relation between them, given by Eq. (3). This holds well for the

$$m(1/2, +1) + m(1/2, -1) = 3/2 m(0, 0) + 1/2 m(1, 0) \qquad (3)$$

$1/2^+$ baryon multiplet (i-spin multiplets N, Ξ, Λ, Σ) and also for the 0^- meson octet (K, $\overline{K}$, η, π) if m in the formula is replaced by squares of masses (field theory gives some reason for this, for mesons). For a decuplet, such as the $3/2^+$ baryons [Δ, $\Sigma(1385)$, $\Xi(1530)$, Ω], the GM-O formula, Eq. (2), predicts equal spacing of the masses; this served as a prediction of the Ω, at its correct mass. The GM-O formula does not hold for the eight 1^- mesons [$K(890)$, $\overline{K}(890)$, ω, ρ]. This is explained by the existence of a ninth 1^- meson of similar mass, the ϕ; the SU_3 breaking interaction has strongly mixed the original I, $Y = 0$, 0 "ϕ_0" member of the octet with the original singlet "ω_0," and pushed their masses apart. This is also true of the nine 2^+ mesons. In both cases, the lower I, $Y = 0$, 0 meson is nearly degenerate in mass with the I, $Y = 1$, 0 meson (that is, ω with ρ; f' with A_2), and the upper one is strongly coupled to $K\overline{K}$ (the major decay mode of both ϕ and f') but nearly decoupled from pions. The quark model explains these observations.

SU_3 symmetry, of course, implies relations between strong coupling constants, for example, $\pm\sqrt{3}\, g_{KN\Lambda} \pm g_{KN\Sigma} = 2g_{\pi NN}$, and between cross sections; these, when testable experimentally, seem to agree. With assumptions about the SU_3 transformation behavior, namely, as octets, of electromagnetic and weak current operators, there are many relations between their matrix elements, which agree well. For instance, the magnetic moments of the $1/2^+$ baryon octet are predicted to have the relations given by Eqs. (4). The relations among the weak couplings are de-

$$\begin{aligned} \mu(\Sigma^+) &= \mu(p) \\ 2\mu(\Lambda) = \mu(\Xi^0) &= -2\mu(\Sigma^0) = \mu(n) \\ \mu(\Sigma^-) = \mu(\Xi^-) &= -[\mu(p) + \mu(n)] \end{aligned} \qquad (4)$$

scribed by the Cabibbo theory. *See Baryon; Meson; Weak nuclear interactions.*

SU_N symmetry. According to the argument given above, hadrons have the approximate symmetry SU_N, where N is the number of kinds of quarks, or flavors. Five flavors of quark are known; in addition to the quarks with the flavors up, down, and strange described above, two more quarks, charm and bottom, have been found. There is reason to think that the number of flavors is even, and that a top partner to the bottom quark will be found; in fact, there is some evidence for its existence. But it is totally unknown how many yet heavier quarks there are. Parallel to the quarks, there are six kinds of leptons currently known. The resulting SU_6 symmetry plays a role in their weak interactions. *See Charm; Flavor; J particle; Lepton; Upsilon particles.*

Color SU_3. In the quark model of hadrons, baryons are correctly described only if quarks carry, in addition to spin and flavor, another quantum number, color, which can take on three values. The resulting symmetry, color SU_3, is thought to be exact; that is, the differently colored quarks of a given flavor are thought to be absolutely equivalent. However, this exact symmetry is rather well hidden, because apparently all free particles belong to the singlet representation of the symmetry. *See Color.*

Nonrelativistic symmetries. There are unitary symmetries which are not purely internal, in the strict sense, namely those in which it is supposed that the interaction between the fundamental particles is spin-independent (as for nonrelativistic electrons interacting with electrostatic forces). This was suggested by E. Wigner as an approximate symmetry in light nuclei; there the fundamental particle is a quartet, with states $|p\uparrow\rangle$, $|p\downarrow\rangle$, $|n\uparrow\rangle$, $|n\downarrow\rangle$ (where $p\uparrow\rangle$ means a proton with spin up, and so on), and the symmetry group is therefore SU_4. Similarly, if the interactions of the SU_3 quarks are supposed to be spin-independent, the symmetry group is SU_6. It is known that such unitary symmetries can never be exact in a relativistic theory, but that fact does not prevent them from being approximately true, at least for low-lying states. *See Elementary particle.*

Bibliography. J. P. Elliot and P. G. Dawber, *Symmetry in Physics*, vol. 1: *Principles and Simple Applications*, 1979, paper 1984; M. Gell-Mann and Y. Ne'eman (eds.), *The Eightfold Way*, 1964; F. Halzen and A. D. Martin, *Quarks and Leptons: An Introductory Course in Modern Particle Physics*, 1984; D. B. Lichtenberg, *Unitary Symmetry and Elementary Particles*, 2d ed., 1978; L. H. Ryder, *Elementary Particles and Symmetries*, 1986; N. P. Samios, M. Goldberg, and B. T. Meadows, Hadrons and SU(3): A critical review, *Rev. Mod. Phys.*, 46:49–81, 1974.

HYPERCHARGE
D. ALLAN BROMLEY

A quantized attribute, analogous to electric charge, introduced in the classification of a subset of elementary particles—the so-called baryons—including the proton and neutron as its lightest members. As far as is known, electric charge is absolutely conserved in all physical processes. Hypercharge was introduced to formalize the observation that certain decay modes of baryons expected to proceed by means of the strong nuclear force simply were not observed.

Unlike electric charge, however, the postulated hypercharge was found not to be conserved absolutely; the weak nuclear interactions do not conserve hypercharge—and indeed can change hypercharge by ± 1 or 0 units. These observations establish certain relationships but do not define the hypercharge. The simplicity of the relationships suggested a hypercharge (symbol Y) as follows: for the proton and neutron, $Y = +1$; for the pion, $Y = 0$.

For example, consider the heavier Λ^0 baryon, which in principle is energetically unstable against decay into a proton and a pion in a characteristic time of about 10^{-23} s. But it is observed that the Λ^0 lives about 10^{13} times longer than this characteristic time, suggesting that, for this baryon, $Y = -1$, and thus it can decay into a proton and a pion only by means of the weak nuclear interaction. Systematic observations of this kind have permitted assignment of hypercharge values to all the baryons and antibaryons.

When the known baryons are classified according to their electric charge and their hypercharge, they naturally group into octets in the scheme first proposed by M. Gell-Mann and K. Nishijima. The quarks, hypothesized as the fundamental building blocks of matter, must have fractional hypercharge as well as electrical charge; the simplest quark model suggests values of $\frac{1}{3}$ and $\frac{2}{3}$, respectively. SEE BARYON; ELEMENTARY PARTICLE; QUARKS; SYMMETRY LAWS; UNITARY SYMMETRY.

FLAVOR
VAL L. FITCH

A generic term referring to quarks, the most elementary constitutents of matter. The evidence is conclusive that all matter is constituted of quarks of at least five different types of flavors, with a sixth predicted. They are labeled u, d, s, c, b, and t. The last, the t quark, is predicted from symmetry considerations. Corresponding mnemonics are up, down, strange, charmed, bottom, and top. The u, c, and t flavors each carry a positive electric charge equal in magnitude to two-thirds that of the electron, and the d, s, and b flavors have a negative charge one-third that of the electron.

The proton is composed of two u- and one d-flavored quarks; the neutron of two d and one u. Strange particles contain an s; charmed particles contain a c-flavored quark. The strange lambda particle is similar to a neutron except that one of the d quarks has been replaced by an s quark. SEE BARYON; CHARM; NEUTRON; PROTON; STRANGE PARTICLES.

Mesons are composed of quark-antiquark ($q,\bar{q}$) combinations. For example, the ρ meson is made of $u\bar{u}$ or $d\bar{d}$, the φ is $s\bar{s}$, the ψ/J is $c\bar{c}$, and the Υ is $b\bar{b}$. The particle corresponding to $t\bar{t}$ has yet to be discovered. It is presumed that this is because its rest mass is so great as to preclude its production at the highest-energy accelerators currently available. SEE J PARTICLE; MESON; UPSILON PARTICLE.

Interquark forces are produced by the exchange of massless, spin-one bosons called gluons. Evidence exists that these forces are flavor-independent. SEE GLUONS.

A subdivision of flavors into colors is required from statistical considerations. Each quark flavor occurs in three colors. The color of the quarks in particles is such that the particles are colorless. SEE COLOR.

Whether quarks can exist in isolation is an open question. Various reports of their detection remain to be confirmed. A major theoretical problem lies in understanding why quarks appear to be always confined, and always appear in combination with other quarks to compose particles which have integral multiples 0, ± 1, ± 2, and so forth, of the electron charge. SEE ELEMENTARY PARTICLE; QUARKS.

CHARM
Nicholas P. Samios

A term used in elementary particle physics to describe a class of elementary particles.

Theory. Ordinary atoms of matter consist of a nucleus composed of neutrons and protons and surrounded by electrons. Over the years, however, a host of other particles with unexpected properties have been found, associated with both electrons (leptons) and protons (hadrons).

Leptons. The electron has as companions the mu meson (μ) and the tau meson (τ), approximately 200 times and 3700 times as heavy as the electron, respectively. These particles are similar to the electron in all respects except mass. In addition, there exist at least two distinctive neutrinos, one associated with the electron, v_e, and another with the mu meson, v_μ. In all, there are five or six fundamental, distinct, structureless leptons. See Lepton.

Hadrons. A similar but more complex situation exists with respect to the hadrons. These particles number in the hundreds, and unlike the leptons they cannot be thought of as fundamental. In fact, they can all be explained as composites of more fundamental constitutents, called quarks. It is the quarks which now seem as fundamental as the leptons, and the number of quark types has also increased as new and unexpected particles have been experimentally uncovered. The originally simple situation of having an up quark (u; charge $+2/3$), and a down quark (d; charge $-1/3$) has evolved as several more varieties or flavors have had to be added. These are the strange quark (s; charge $-1/3$), with the additional property or quantum number of strangeness ($S = -1$), to account for the unexpected characteristics of a family of strange particles; the charm quark (c; charge $+2/3$), possessing charm ($C = +1$) and no strangeness, to explain the discovery of the J/ψ particles, massive states three times heavier than the proton; and a fifth quark (b; charge $-1/3$) to explain the existence of the even more massive upsilon (Υ) particles. See Hadron; J particle; Quarks.

The quarks and leptons discovered so far appear to form a symmetric array (see **table**). Both the leptons and quarks come in pairs, although the anticipated partner (t) of the b quark has not yet been found.

Fundamental constituents of matter		
Family of particles	Charge	Particles
Leptons	0	$v_e, v_\mu,$
	-1	e, μ, τ
Quarks	$+2/3$	$u, c, (t)^*$
	$-1/3$	d, s, b

*Existence uncertain.

Observations. The members of the family of particles associated with charm fall into two classes: those with hidden charm, where the states are a combination of charm and anticharm quarks ($c\bar{c}$), charmonium; and those where the charm property is clearly evident, such as the D^+ ($c\bar{d}$) meson and Λ_c^+ (cud) baryon.

Charmonium. In the charmonium family, six to seven states with various masses and decay modes have been identified. Although a detailed understanding of all these experimentally measured properties has not yet been achieved, everything seems to be in qualitative agreement with theoretical expectations.

Bare charm states. There are several identified bare charm states, including the D ($c\bar{d}$) mesons in both the $J^P = 0^-$ and $J^P = 1^+$ categories (where J is spin, and P is parity), and the Σ_c^{2+} (cuu) and Λ_c^+ (cud) charmed baryons. Information about the lifetimes of these states has been derived from experiments utilizing the high resolution of emulsions to measure the finite distance traveled by charmed particles. The lifetimes of the Λ_c^+ and D^0 have been determined to be on the order of 10^{-13} s with that of the D^+ about a factor of 7 longer. These values are in good agreement with theoretical expectations.

The Λ_c^+ charmed baryon has been observed to be produced in a variety of interactions, including neutrino-proton, proton-proton, electron-positron, and neutrino-neon reactions. A large number of decay modes have been observed, including Kp, $Kp\pi$, $\Lambda\pi$, $\Lambda\pi\pi\pi$, and $Kp\pi\pi$.

Prospects. Although reasonable progress has been made in the study of charmed states, only a handful of states has been observed. Just as the basic SU(3) symmetry arose from a study of the numerous hadron strange and nonstrange resonances, the complete understanding of charm awaits the uncovering of additional states. SEE ELEMENTARY PARTICLE.

Bibliography. W. Chinowsky, Psionic matter, *Ann. Rev. Nucl. Sci.*, 27:393–464, 1977; S. L. Glashow, Quarks with color and flavor, *Sci. Amer.*, 233(4):38–50, October 1975; F. Halzen and A. D. Martin, *Quarks and Leptons: An Introductory Course in Modern Particle Physics*, 1984; R. F. Schwitters, Fundamental particles with charm, *Sci. Amer.*, 237(4):56–70, 1977.

COLOR
THOMAS APPELQUIST

A term used to describe a hypothetical quantum number carried by the quarks which are thought to make up the strongly interacting elementary particles. It has nothing to do with the ordinary, visual use of the word color.

The quarks which are thought to make up the strongly interacting particles have a spin angular momentum of one-half unit of $\hbar$ (Planck's constant). According to a fundamental theorem of relativity combined with quantum mechanics, they must therefore obey Fermi-Dirac statistics and be subject to the Pauli exclusion principle. No two quarks within a particular system can have exactly the same quantum numbers.

However, in making up a baryon, it often seemed necessary to violate this principle. The Ω-particle, for example, is made of three strange quarks, and all three had to be in exactly the same state. O. W. Greenberg was responsible for the essential idea for the solution to this paradox. In 1964 he suggested that each quark type (u, d, and s) comes in three varieties identical in all measurable qualities but different in an additional property, which has come to be known as color. The exclusion principle could then be satisfied and quarks could remain fermions, because the quarks in the baryon would not all have the same quantum numbers. They would differ in color even if they were the same in all other respects.

The color hypotheses triples the number of quarks but does not increase the number of baryons and mesons. The rules for assembling them ensures this. Tripling the number of quarks does, however, have at least two experimental consequences. It triples the rate at which the neutral π meson decays into two photons and brings the predicted rate into agreement with the observed rate.

The total production cross section for baryons and mesons in electron-positron annihilation is also tripled. The experimental result at energies between 2 GeV and 3 GeV is in reasonable agreement with the color hypothesis and completely incompatible with the simple quark model without color. SEE BARYON; ELEMENTARY PARTICLE; GLUONS; MESON; QUARKS.

Bibliography. H. Georgi, A unified theory of elementary particles and forces, *Sci. Amer.*, 244(4):48–63, 1981; S. L. Glashow, Quarks with color and flavor, *Sci. Amer.*, 233(4):38–50, 1975; O. W. Greenberg, Spin and unitary spin independence in a paraquark model of baryons and mesons, *Phys. Rev. Lett.*, 13, 598–602, 1964; F. Halzen and A. D. Martin, *Quarks and Leptons: An Introductory Course in Modern Particle Physics*, 1984.

SYMMETRY BREAKING
PIERRE RAMOND

A generic term describing the deviation from exact symmetry exhibited by many physical systems. It can occur either explicitly (explicit symmetry breaking) or spontaneously (spontaneous symmetry breaking), with distinct observable consequences characterizing the two cases.

Explicit symmetry breaking. In the explicit case, the system is "not quite" the same for two configurations related by exact symmetry. A simple example is a bicycle wheel with the valve stem sticking out: it is almost symmetric with respect to rotations about the bicycle axis, but the symmetry is explicitly broken by the valve stem. In quantum mechanics the system is characterized by its energy (hamiltonian), and in the case of explicit symmetry breaking the hamiltonian is almost symmetric. For example, the hamiltonian describing electrons inside a spherical cavity is symmetric with respect to rotations. If a weak magnetic field is applied, the electron spin (magnetic moment) will interact with the magnetic field and the energy will be different for electrons with spin up or down, thus breaking the rotational symmetry. The consequence will be a symmetrical splitting of the allowed energy levels about each energy level with no magnetic field. Usually explicit symmetry breaking is useful in explaining numerical relations between the energy levels (or masses of nuclear particles) of certain physical systems.

Spontaneous symmetry breaking. On the other hand, in the case of spontaneous symmetry breaking, the hamiltonian always displays the exact symmetry, but the state of lowest energy of the system does not share this symmetry. This case can be completely understood only within the framework of quantum mechanics, although a simple classical analog exists: a spinning roulette wheel settling into a state with the ball in one slot, thus breaking the rotational symmetry. The physical consequences of spontaneous symmetry breaking are very different from the explicit case in that they predict domain structure when the broken symmetry is discrete, and new gapless (massless) modes of excitations, called Nambu-Goldstone modes, when the broken symmetry is continuous.

Spontaneous symmetry breaking also serves to describe the different phases of physical systems such as the superconducting phase of conductors, the various phases of ^{3}He, and so forth. The superconducting phase is achieved when it becomes energetically favorable for two conduction electrons to bind due to the effect of lattice interactions. These Cooper pairs of electrons can easily be unbound by small thermal effects. In a superconductor the state of lowest energy is filled with these Cooper pairs, and the conduction electrons are said to have condensed into Cooper pairs. The vacuum state contains an indefinite number of Cooper pairs, with the result that charge conservation is spontaneously broken by the superconducting ground state. This has many physical consequences such as the Meissner effect. In ^{3}He, a similar pairing occurs between ^{3}He atoms, and the possible ways in which this pairing occurs describes the observed phases of ^{3}He.

In particle physics, spontaneous symmetry breaking occurs in many contexts. The condensation of quark pairs into the vacuum (due to quantum chromodynamics) breaks chiral symmetry with the result of creating Nambu-Goldstone particles called pions (which would be massless were it not for small explicit breaking of chiral symmetry). The most spectacular effects of spontaneous symmetry breaking occur in the electroweak theory. The electroweak gauge symmetry is broken spontaneously by the vacuum state chosen by nature. As a result, some gauge particles acquire a mass by "eating" the Nambu-Goldstone boson produced by the spontaneous symmetry breaking through a process known as the Higgs mechanism. These massive gauge particles, called W^+, W^-, and Z, were first observed in 1982 and 1983. SEE GAUGE THEORY; INTERMEDIATE VECTOR BOSON; QUANTUM CHROMODYNAMICS; QUARKS; SYMMETRY LAWS; WEAK NUCLEAR INTERACTIONS; WEINBERG-SALAM MODEL.

SUPERSYMMETRY
FRANCESCO IACHELLO

A symmetry that relates bosons to fermions, and vice versa. Since the 1960s there has been an unprecedented expansion in the use of symmetry considerations in the study of problems in physics. Supersymmetry is the latest and most elaborate application. Although supersymmetry was originally introduced in the early 1970s for studying problems in elementary particle physics, its use and applications have spread to other fields of physics. Experimental examples of symmetry have been discovered in the spectra of atomic nuclei.

Definition. Physicists believe that all constituents of matter have a property called angular momentum. According to quantum theory, angular momentum can be either an integer or half-integer multiple of Planck's constant, $\hbar$. Particles for which the angular momentum is an integer multiple of $\hbar$ are called bosons; those for which the angular momentum is a half-integer multiple of $\hbar$ are called fermions. Up to the early 1970s, when applying symmetry considerations to the study of phenomena in physics, physicists always considered symmetry transformations relating either bosons to bosons or fermions to fermions. In the early 1970s it was suggested that there may exist symmetries relating bosons to fermions, and vice versa. Because of their remarkable properties, and in order to distinguish them from the usual symmetries, these symmetries were called supersymmetries.

Mathematical formalism. The study of symmetries requires the use of a mathematical formalism, known as theory of group transformations. Most applications in physics rely on the theory of Lie groups, which constitute a special type of group of transformations first studied by S. Lie and others at the end of the nineteenth century. The mathematical properties of these groups can be concisely stated by writing down the commutation relations satisfied by the generators G_α of the corresponding infinitesimal transformations, Eq. (1), where the commutator of

$$[G_\alpha, G_\beta] = \sum_\alpha c_{\alpha\beta}^\gamma G_\gamma \qquad (1)$$

two operators A and B is defined by $[A,B] = AB - BA$, and the coefficients $c_{\alpha\beta}^\gamma$ specify the group. The study of supersymmetries requires the introduction of a new mathematical formalism, the theory of graded groups of transformations or supergroups. While the mathematical theory of Lie groups has been known for more than 80 years, that of graded Lie groups is still being developed. The main difference between graded and nongraded Lie groups is that in the graded case there are two types of generators, the bosonic, G_α, and the fermionic, F_i. These generators satisfy commutation relations more elaborate than those of nongraded groups, Eqs. (2). In these expressions,

$$[G_\alpha, G_\beta] = \sum_\gamma c_{\alpha\beta}^\gamma G_\gamma \qquad [G_\alpha, F_i] = \sum_j f_{\alpha i}^j F_j \qquad \{F_i, F_j\} = \sum_\alpha g_{ij}^\alpha G_\alpha \qquad (2)$$

in addition to commutators, there now appear anticommutators. The anticommutator of two operators A and B is defined as $\{A,B\} = AB + BA$. The coefficients $c_{\alpha\beta}^\gamma$, $f_{\alpha i}^j$, g_{ij}^α again specify the group.

Supersymmetry in particle physics. Supersymmetry was originally formulated for applications to this field of physics. Early ideas about supersymmetry, put forward in 1971, went to a large extent unnoticed. The subject was rediscovered in 1973 and placed into a systematic perspective in 1973 and 1974, in an approach that led to the construction of theories of elementary particles with well-defined relations among masses and coupling constants of the fundamental particles in the theory. Subsequently, supersymmetries were used in attempts to construct a unified theory of all known interactions in physics: the gravitational, the weak, the electromagnetic, and the strong interactions. Particularly important here is the attempt to include the gravitational interaction, since there is at present no workable theory of gravitation consistent with the principles of quantum mechanics. Theories of gravitation that make use of supersymmetry have been called supergravity theories. The most elementary example of a supergravity theory, constructed in 1976, assumes the existence of a field particle with angular momentum 2 (graviton) and a second one with angular momentum $\frac{3}{2}$ (gravitino). Many other theories have been constructed since then. Despite all these attempts, no experimental verification of supersymmetric theories in elementary particles physics has yet been found. SEE ELEMENTARY PARTICLE; FUNDAMENTAL INTERACTIONS; SUPERGRAVITY.

Supersymmetry in nuclear physics. Supersymmetry has also been applied to the study of problems in nuclear physics. The early ideas on this subject were presented in 1980 and 1981. Atomic nuclei are composed of neutrons and protons. In medium-mass and heavy nuclei, two protons or two neutrons appear to bind into pairs. The pairs have angular momenta that are integer multiples of $\hbar$ and thus are bosons. In nuclei with an even number of protons and neutrons (even-even nuclei), all particles are paired. However, in nuclei with an even number of protons and an odd number of neutrons, or vice versa (even-odd nuclei), complete pairing is impossible and in addition to pairs (bosons) there are unpaired neutrons or protons. These unpaired particles are fermions. If the theory describing these nuclei is required to have supersymmetry, a set of relations is obtained linking together properties of even-even and even-odd nuclei. The most important relation is that linking the excitation spectra of atomic nuclei, that is, the set of their allowed quantum-mechanical energy levels. The first experimentally found example of supersymmetry in nuclei is shown in the **illustration**. Since the discovery of this example, several others

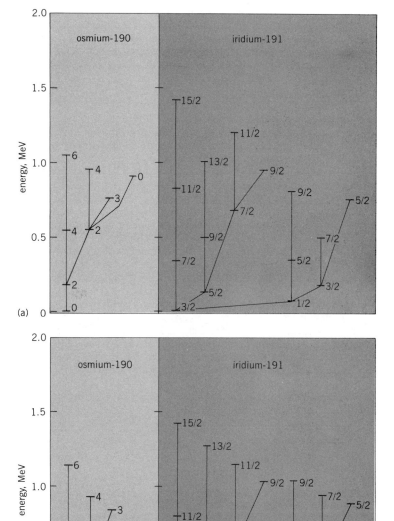

Examples of a dynamic supersymmetry in nuclear physics: energy spectra of osmium-190, composed of 76 protons and 114 neutrons, and iridium-191, composed of 77 protons and 114 neutrons. (*a*) Observed spectra; (*b*) theoretically predicted spectra. Each allowed energy level is labeled by the value of the total angular momentum *J*. The lines between levels indicate decay from the upper to the lower level by emission of intense electromagnetic radiation. The extent to which the observed and predicted spectra agree is evidence of supersymmetry in these nuclei; the extent to which they disagree is indicative of partial breaking of the supersymmetry.

have been found, in the same and in other regions of the periodic table. A general feature of the experimental examples found so far is that supersymmetry in nuclei appears to be somewhat broken (see illus.). This is not unexpected, since supersymmetry is a rather complex type of symmetry requiring very stringent conditions on the interactions between the constituent particles. Despite the partial breaking, the fact that the experimental situation can be described, at least approximately, by invoking purely symmetry concepts is a remarkable result. This result has been exploited systematically in the study of complex atomic nuclei. SEE NUCLEAR STRUCTURE.

Outlook. Experimental examples of supersymmetry have been found in nuclear physics. The implications of this discovery in the search for supersymmetry in other fields of physics, especially in elementary particle physics, are not yet clear. Supersymmetries observed in nuclei are of a different type than those mostly sought in elementary particle physics, since the bosons employed in the description of atomic nuclei are composite particles (pairs). Most supersymmetric theories in elementary particle physics employ bosons which are not composite. Supersymmetry has also been applied to the problem of electrons moving in a type II superconductor. In this application, the bosons are also composite particles (pairs of electrons). The fact that no experimental verification of supersymmetric theories with noncomposite bosons at present exists indicates either that the experimental search has not proceeded far enough or that the only realizable types of supersymmetry are those with composite bosons.

In conclusion, the discovery of supersymmetry in nuclear physics has led to major advances in this field and has shown that this new and elaborate type of symmetry exists in nature. However, the scope and limitation of the concept of supersymmetry remain to be determined. SEE SYMMETRY LAWS.

Bibliography. D. Z. Freedman and P. van Nieuwenhuizen, Supergravity and the unification of the laws of physics, *Sci. Amer.*, 238(2):126–143, February 1978; F. Iachello, Supersymmetry in nuclei, *Amer. Sci.*, 70(3):294–299, May 1982; M. Jacob (ed.), *Supersymmetry and Supergravity: Collected Articles From Physics Reports*, 1986; B. G. Levi, Nuclei may exhibit supersymmetry, *Phys. Today*, 33(9):21–22, September 1980; G. B. Lubkin, Interacting-boson model emphasizes symmetry group, *Phys. Today*, 31(7):17–20, July 1978; P. West, *Introduction to Supersymmetry and Supergravity*, 1986.

11
KINEMATICS AND SCATTERING THEORY

Dalitz plot	**378**
Goldhaber triangle	**381**
Scattering matrix	**384**
Dispersion relations	**386**
Causality	**391**
Regge pole	**392**

DALITZ PLOT
RICHARD H. DALITZ

Pictorial representation in high-energy nuclear physics for data on the distribution of certain three-particle configurations. Many elementary-particle decay processes and high-energy nuclear reactions lead to final states consisting of three particles (which may be denoted by a, b, c, with mass values m_a, m_b, m_c). Well-known examples are provided by the K-meson decay processes, Eqs. (1) and (2), and by the K- and $\bar{K}$-meson reactions with hydrogen, given in Eqs. (3) and (4). For definite

$$K^+ \to \pi^+ + \pi^+ + \pi^- \qquad (1) \qquad\qquad K^+ \to \pi^0 + \mu^+ + \nu \qquad (2)$$

$$K^+ + p \to K^0 + \pi^+ + p \qquad (3) \qquad\qquad K^- + p \to \Lambda + \pi^+ + \pi^- \qquad (4)$$

total energy E (measured in the barycentric frame), these final states have a continuous distribution of configurations, each specified by the way this energy E is shared among the three particles. (The barycentric frame is the reference frame in which the observer finds zero for the vector sum of the momenta of all the particles of the system considered.) SEE ELEMENTARY PARTICLE.

Equal mass representation. If the three particles have kinetic energies T_a, T_b, and T_c (in the barycentric frame), Eq. (5) is obtained. As shown in **Fig. 1**, this energy sharing may be

$$T_a + T_b + T_c = E - m_a c^2 - m_b c^2 m_c c^2 = Q \qquad (5)$$

represented uniquely by a point F within an equilateral triangle LMN of side $2Q/\sqrt{3}$, such that the perpendiculars FA, FB, and FC to its sides are equal in magnitude to the kinetic energies T_a, T_b, and T_c. This exploits the property of the equilateral triangle that $(FA + FB + FC)$ has the same value (Q, the height of the triangle) for all points F within it. The most important property of this representation is that the area occupied within this triangle by any set of configurations is directly proportional to their volume in phase space. In other words, a plot of empirical data on this diagram gives at once a picture of the dependence of the square of the matrix element for this process on the a, b, c energies.

Not all points F within the triangle LMN correspond to configurations realizable physically, since the a, b, c energies must be consistent with zero total momentum for the three-particle system. With nonrelativistic kinematics (that is, $T_a = p_a^2/2m_a$, etc.) and with equal masses m for a, b, c, the only allowed configurations are those corresponding to points F lying within the the circle inscribed within the triangle, shown as (i) in Fig. 1. With unequal masses the allowed domain becomes an inscribed ellipse, touching the side MN such that $MP:PN$ equals $m_b:m_c$ (and cyclical for NL and LM). More generally, with relativistic kinematics ($T_a = \sqrt{(m_a^2 c^4 + p_a^2 c^2)} - m_a c^2$,

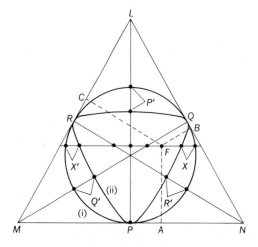

Fig. 1. A three-particle system (abc) in its barycentric frame is specified by a point F so that perpendiculars FA, FB, and FC to the sides of an equilateral triangle LMN (of height Q) are equal in magnitude to the kinetic energies T_a, T_b, T_c, where Q denotes their sum. See Eq. (6). Curve (i) encloses all points F which correspond to physically allowed configurations, for equal masses and nonrelativistic kinematics; curve (ii) corresponds to curve (i) when relativistic kinematics are used, appropriate to the decay process $\omega(785 \text{ MeV}) \to \pi^+ \pi^- \pi^0$.

etc.), the limiting boundary is distorted from a simple ellipse or circle. This is illustrated in Fig. 1 by the boundary curve (ii), drawn for the $\omega \to 3\pi$ decay process, where the final masses are equal. This curve was also calculated by E. Fabri for Eq. (1), and this plot is sometimes referred to as the Dalitz-Fabri plot. In the high-energy limit $E \to \infty$, where the final particle masses may be neglected, the boundary curve approaches a triangle inscribed in LMN.

The following points (and the regions near them) are of particular interest:

1. All points on the boundary curve. These correspond to collinear configurations, where a, b, c have parallel momenta.

2. The three points of contact with the triangle LMN. For example, point P corresponds to the situation where particle c is at rest (and therefore carries zero orbital angular momentum).

3. The three points which are each farthest from the corresponding side of the triangle LMN. For example, point P' on Fig. 1 corresponds to the situation where b and c have the same velocity (hence zero relative momentum, and zero orbital angular momentum in the bc rest frame).

If the process occurs strongly through an inter- mediate resonance state, say $a+(bc)^*$ where $(bc)^* \to b+c$, there will be observed a "resonance band" of events for which T_a has the value appropriate to this intermediate two-body system. Such a resonance band runs parallel to the appropriate side of the triangle [the side MN for the case $(bc)^*$, and cyclically] and has a breadth related with the lifetime width for the resonance state.

Antiproton annihilation. The Dalitz plot shown for equal masses in Fig. 1 has been especially useful for three-pion systems, since it treats the three particles on precisely the same footing. Points placed symmetrically with respect to the symmetry axis PL represent configura-

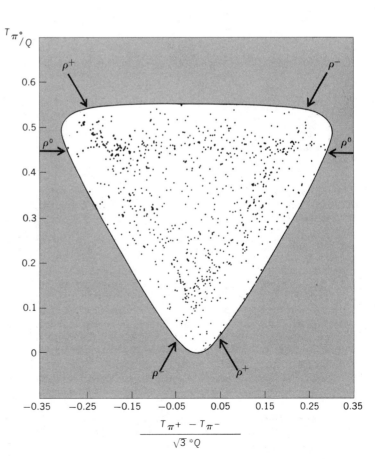

Fig. 2. Dalitz plot for 823 examples of the antiproton annihilation process, $\bar{p} + p \to \pi^+ + \pi^+ + \pi^0$. Arrows show positions expected for $\rho(765)$-meson resonance bands, which appear clearly for the $(\pi^+\pi^-)$, $(\pi^-\pi^0)$, and $(\pi^0\pi^+)$ systems. Distribution is symmetrical between the six sectors obtained by drawing the three axes of symmetry. The authors interpret this distribution as being due to (1) the reaction $\bar{p}p \to \rho\pi$ occurring in the $I = 0$ S_1^3 initial state, (2) the reaction $\bar{p}p \to 3\pi$ (s-wave pions) occurring in the $I = 1$ S_0^1 initial state, with roughly equal intensities. (After C. Baltay et al., Annihilation of Antiprotons in Hydrogen at Rest, Phys. Rev., 140:B1039, 1965)

tions related by the interchange of π_b and π_c. The symmetry axes PL, QM, and RN divide the allowed regions into six sectors; the configurations in each sector can be obtained from those corresponding to one chosen sector (for example, the sector such that $T_a \geq T_b \geq T_c$) by the six operations of the permutation group on three objects. These operations are of particular interest for three-pion systems, since pions obey Bose statistics; the intensities in the six sectors are related with the permutation symmetry of the orbital motion in the three-pion final state. The Dalitz plot shown in **Fig. 2**, for the antiproton capture reaction $\bar{p}p \to \pi^+\pi^-\pi^0$, illustrates these points. Three ρ-meson bands [ρ(765 MeV)$\to\pi\pi$] are seen, corresponding to intermediate states $\pi^+\rho^-$, $\pi^0\rho^0$, and $\pi^-\rho^0$; the six sectors have equal intensity.

Relativistic three-particle system. Figure 3 depicts a less symmetric specification for a three-particle system. The momentum of c in the (bc) barycentric frame is denoted by **q**, the momentum of a in the (abc) barycentric frame by **p**, and the angle between **p** and **q** by θ. For

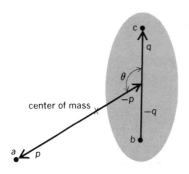

Fig. 3. Coordinate system for relativistic three-particle system. For given total energy E, the two momenta, q and p, are related in magnitude by the following equations: $E = \sqrt{(m_a^2 + p^2)} + \sqrt{(M_{bc}^2 + p^2)}$ and $M_{bc} = \sqrt{(m_b^2 + q^2)} + \sqrt{(m_c^2 + q^2)}$.

fixed energy T_a, the points F on Fig. 1 lie on a line parallel to MN; as $\cos\theta$ varies from $+1$ to -1, the point F representing the configuration moves uniformly from the left boundary X' to the right boundary X. If cartesian coordinates are used for F, with origin P and y axis along NM (as has frequently been found useful in the literature), then Eqs. (6) hold. Note that, for fixed x, y is linearly related with $\cos\theta$.

$$x = T_a/Q \qquad y = (T_b - T_c)/Q\sqrt{3} \qquad (6)$$

Unsymmetrical plot. The Dalitz plot most commonly used is a distorted plot in which each configuration is specified by a point with coordinates (M_{ab}^2, M_{bc}^2) with respect to right-angled axes. This depends on the relationship given in Eq. (7) and its cyclic permutations, for the total

$$M_{bc}^2 = (E - m_a c^2)^2 - 2ET_a \qquad (7)$$

barycentric energy M_{bc} of the two-particle system bc. This plot may be obtained from Fig. 1 by shearing it to the left until $\overline{LM}$ is perpendicular to MN, and then contracting it by the factor $\sqrt{3/2}$ parallel to MN [which leads to a cartesian coordinate system (T_c, T_a)], and finally reversing the direction of the axes [required by the minus sign in Eq. (7)] and moving the origin to the point $M_{ab}^2 = M_{bc}^2 = 0$. The plots shown in **Fig. 4** correspond in this way to the relativistic curve (ii) in Fig. 1 for two values of the total energy E. This distorted plot retains the property that phase-space volume is directly proportional to the area on the plot. As shown in Fig. 4, the $(ab)^*$ and $(bc)^*$ resonance bands have a fixed location on this plot; data from experiments at different energies E can then be combined on the same plot to give a stronger test concerning the existence of some intermediate resonance state. On the other hand, the $(ca)^*$ resonance bands run across the plot at 135° and move as E varies, so that a different choice of axes [say (M_{ca}^2, m_{ab}^2)] is more suitable for their presentation.

Conclusion. It must be emphasized that the Dalitz plot is concerned only with the internal variables for the system (abc). In general, especially for reaction processes such as Eqs. (3) and (4), there are other variables such as the Euler angles which describe the orientation of the plane

KINEMATICS AND SCATTERING THEORY **381**

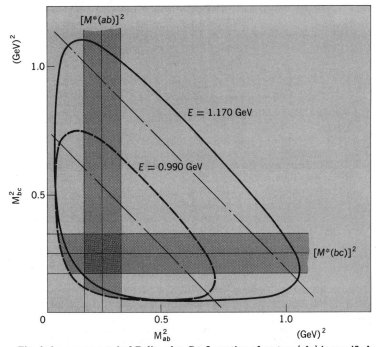

Fig. 4. An unsymmetrical Dalitz plot. Configuration of system (*abc*) is specified by a point (M_{ab}^2, M_{bc}^2) in a rectangular coordinate system, where M_{ij} denotes the barycentric energy of the two-particle system (*ij*). Kinematic boundaries have been drawn for equal masses $m_a = m_b = m_c = 0.14$ GeV and for two values of total energy E, appropriate to a three-pion system ($\pi^+\pi^-\pi^+$). Resonance bands are drawn for states (*ab*) and (*bc*) corresponding to a (fictitious) π-π resonance mass 0.5 GeV and full width 0.2 GeV. The dot-dash lines show the locations a (*ca*) resonance band would have for this mass of 0.5 GeV, for the two values of the total energy E.

of (*abc*) relative to the initial spin direction or the incident momentum, which may carry additional physical information about the mechanism for formation of the system (*abc*). The Dalitz plots usually presented average over all these other variables (because of the limited statistics available); this sometimes leads to a clearer picture in that there are then no interference terms between states with different values for the total spin-parity. However, it is quite possible to consider Dalitz plots for fixed values of these external variables, or for definite domains for them. SEE GOLD-HABER TRIANGLE.

Bibliography. R.H. Dalitz, *K*-mesons and hyperons: Their strong and weak interactions, *Rep. Progr. Phys.*, 20:163, 1957; G. Kallen, *Elementary Particle Physics*, 1964; G. Kopilov, *Elementary Kinematics of Elementary Particles*, 1983; C. Zemach, Three-pion decays of unstable particles, *Phys. Rev.*, 133:B1201, 1964.

GOLDHABER TRIANGLE
GERSON GOLDHABER

The phase space triangle, or Goldhaber triangle, corresponds to the kinematically allowed boundary for a high-energy reaction leading to four or more particles. In a high-energy reaction between two particles *a* and *b* yielding four particles 1, 2, 3, and 4 in the final state (*a* + *b* → 1 + 2 +

3 + 4), it is convenient to consider the reaction in terms of the production of two intermediate-state quasi-particle composites x and y, which then decay into two particles each, as in expression (1).

$$a + b \to x + y$$
$$ \hookrightarrow 1 + 2 \quad \hookrightarrow 3 + 4 \tag{1}$$

Most high-energy interactions indeed proceed through such intermediate steps, in which, for specific values of the invariant masses $m_x = M_x^*$ and $m_y = M_y^*$, the quasi-particle composites may form resonances. However, the description in terms of the composites x and y, with the invariant masses m_x and m_y as variables, is valid irrespective of whether or not these composites form resonances.

Kinematical limits. The kinematical limits in this representation are particularly simple, namely, they form a right-angle isosceles triangle. A Goldhaber triangle plot corresponds to plotting a point (m_x, m_y) for each event occurring in the above high-energy reaction. Because of the kinematical constraints, these points must all lie inside the triangle.

If one considers the general reaction given in Eq. (1), then the length of each of the two equal sides of the triangle is Q, defined in Eq. (2). Here W is the total energy in the center of mass

$$Q = W - \sum_{i=1}^{4} m_i \tag{2}$$

of particles a and b, and m_i, for $i = 1$ to 4, is the mass of the particles 1 to 4. Hence Q corresponds to the total kinetic energy available in the reaction. All quantities are in energy units of millions or billions of electronvolts.

The values of m_x and m_y run over the intervals $m_1 + m_2 \leq m_x \leq m_1 + m_2 + Q$ and $m_3 + m_4 \leq m_y \leq m_3 + m_4 + Q$, respectively. The effect of changing the incident momentum, and thus Q, is then simply to move the hypotenuse of the triangle, leaving the two sides, as well as the location of any resonances which may occur for certain mass values of the composites x and y, fixed.

In the triangle corresponding to the general reaction (**Fig. 1**), the vertical and horizontal bands indicate resonances at masses M_x^* and M_y^* with full width at half-maximum height or Γ_x and Γ_y, respectively. The bands shown of width 2Γ represent the regions usually chosen if the events corresponding to a given resonance are selected.

Phase space distribution. The phase space is given by $\Phi \propto 1/W \int k_x k_y p_{xy} dm_x dm_y$ where the integral extends over the triangle. Here k_x and k_y are the momenta in the center of mass of the composites x and y, respectively, and p_{xy} is the outgoing momentum of each of the composites in the overall center of mass of particles a and b. It is noteworthy that, along each of the three sides of the triangle, one of the factors in the integrand vanishes. This corresponds to the fact that, along the m_x axis, $m_y = m_3 + m_4$; thus there is no internal kinetic energy in the y

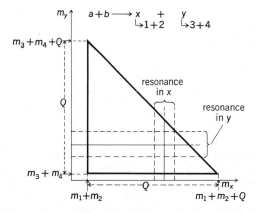

Fig. 1. Definition of the kinematical boundary of the Goldhaber triangle for four particles.

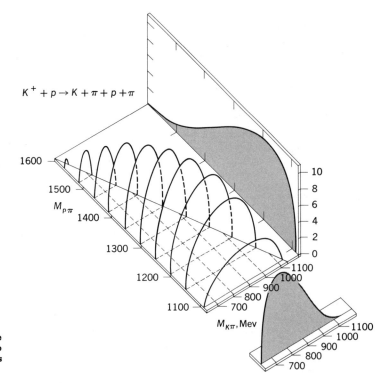

Fig. 2. Phase space distribution over the Goldhaber triangle for the example $K^+p \to (K^+\pi^-) + (p\pi^+)$ at 2 GeV/c incident laboratory momentum. The shaded areas correspond to the phase space projections onto the two mass axes.

composite and hence $k_y = 0$. Similarly $k_x = 0$ along the m_y axis. The hypotenuse corresponds to the situation for which the entire available energy is in the mass of the two composites; thus $W = m_x + m_y$. Hence here one gets $p_{xy} = 0$. The phase space distribution over the triangle is illustrated in **Fig. 2** for the example $K^+p \to K^+\pi^- p\pi^+$ for an incident laboratory momentum of 2 GeV/c. The shaded areas correspond to the projections on the two mass axes.

Comparison with Dalitz plot. Superficially there is a great similarity between the Dalitz plot and the Goldhaber triangle plot. There are, however, several important differences: (1) The Dalitz plot applies to three particles in the final state and corresponds to plotting m_x^2 versus m_y^2, however, in this case the two composites x and y have one particle in common. (2) The Dalitz plot has the advantage that the phase space distribution is uniform but the kinematical boundary is a more complicated function of the variables. In the Goldhaber triangle plot the phase space distribution is more complicated, as shown in Fig. 2, but the kinematical boundary is very simple—a triangle. (3) If two resonances overlap on the plot the interpretation is completely different. In the Dalitz plot the overlap corresponds to interference between the two resonances; in the Goldhaber triangle plot the overlap corresponds to double resonance formation, that is, both the x and y composites form essentially independent resonances at the same time. *See* Dalitz plot.

Example of triangle plot. To illustrate the appearance of the Goldhaber triangle plot, **Fig. 3** shows an example from the reaction $K^+p \to (K^+\pi^-) + (p\pi^+)$ at a laboratory momentum of 9 GeV/c. The two sets of particles in parentheses correspond to the choice of x and y composites, respectively. The clear-cut vertical band corresponds to the $K^{*0}(890)$ meson resonance with $K^+\pi^-$ decay. The horizontal band corresponds to the $\Delta^{++}(1238)$ baryon resonance with $p\pi^+$ decay. The overlap region of these two corresponds to $K^{*}(890) + \Delta^{++}(1238)$ double resonance formation. A second faint vertical band corresponds to the $K^{*0}(1420)$ meson resonance again with $K^+\pi^-$ decay. This resonance is most apparent in the $K^{*0}(1420) + \Delta^{++}(1238)$ double-resonance region. The projections on the two mass axes are also shown.

Extension to five particles. The concepts discussed here can be extended to five particles in the final state; here the composite x corresponds to three particles: 1, 2, and 3; and the

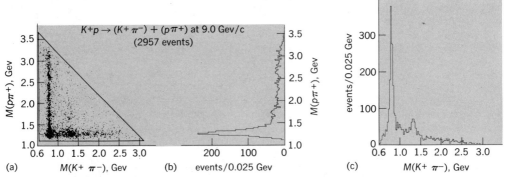

Fig. 3. (a) Example of a triangle plot. (b) Projection for $M(p\pi^+)$. (c) Projection for $M(K^+\pi^-)$. (*After A. Firestone, G. Goldhaber, and B. C. Shen, Lawrence Radiation Laboratory, Berkeley, California*)

composite y to two particles: 4 and 5. The kinematic limits will still be determined, as before, by an isosceles triangle where the generalization is straightforward. For example, in reaction (3) a double resonance formation occurs for ω^0 and $\Delta^{2+}(1238)$ production according to reaction (4).

$$\pi^+ p \to (\pi^+ \pi^- \pi^0) + (p\pi^+) \tag{3}$$

$$\pi^+ p \to \underset{\hookrightarrow \pi^+ \pi^- \pi^0}{\omega^0(890)} + \underset{\hookrightarrow p\pi^+}{\Delta^{++}(1238)} \tag{4}$$

SCATTERING MATRIX
Loyal Durand

A matrix which expresses the initial state in a scattering experiment in terms of the possible final states, and hence enters the calculation of the probabilities that certain reactions will occur in a collision of two or more particles. The scattering matrix was introduced by J. A. Wheeler in 1937 in the discussion of the theory of nuclear reactions. Previous work on scattering theory (applied to atomic collisions) had been based for the most part on the use of the Schrödinger equation for the direct calculation of the scattering amplitude. However, the multiplicity of different reactions in a typical nuclear collision and the uncertainties in the form of the nuclear forces made a more general approach to reaction theory desirable.

The problem is compounded in the relativistic domain, in which particles may be created or destroyed in a collision and the forces between elementary particles are known only approximately. There is, furthermore, no useful relativistic analog of the Schrödinger equation. It was therefore suggested by W. Heisenberg in 1943 that the S matrix should play a fundamental rather than a subsidiary role in relativistic quantum mechanics, and considerable progress has been made toward this goal. SEE ELEMENTARY PARTICLE; NUCLEAR REACTION; SCATTERING EXPERIMENTS (NUCLEI).

Definition and properties. The initial state of a system of particles specified by the set of quantum numbers γ may be described by an ingoing wave function $|\gamma, \text{in}\rangle$. Ingoing functions satisfy boundary conditions such that it is possible to construct from them wave functions in which the individual particles are localized and converge toward the region of interaction prior to their collision. The final states of the system are described by a similar set of outgoing wave functions $|\beta, \text{out}\rangle$, from which wave functions can be constructed which describe particles that diverge from the interaction region at times long after the collision. It is convenient to normalize the in- and out-states to unit ingoing or outgoing particle flux in the center-of-mass coordinate system. An in-state $|\lambda, \text{in}\rangle$ may be reexpressed in terms of the out-states as in Eq. (1). The matrix

$$|\gamma, \text{in}\rangle = \Sigma_\beta |\beta, \text{out}\rangle S_{\beta\gamma} \qquad (1)$$

of probability amplitudes $S_{\beta\gamma}$, which relates all possible initial and final states, is the scattering or S matrix.

Conservation of probability in the possible reactions requires that S be a unitary matrix, that is, that $\Sigma_\gamma S^*_{\gamma\alpha} S_{\gamma\beta} = \Sigma_\gamma S_{\alpha\gamma} S^*_{\beta\gamma} = \delta_{\alpha\beta}$, where $\delta_{\alpha\beta}$ has the value 1 if $\alpha = \beta$ and 0 if $\alpha \neq \beta$. Additional restrictions on S may be deduced if it is assumed that the interactions are invariant under Lorentz transformations, discrete operations of reflection of space or time axes, particle-antiparticle interchange (charge conjugation), or such internal symmetries as isobaric-spin and unitary symmetry. SEE ISOBARIC SPIN; PARITY; SYMMETRY LAWS; TIME REVERSAL INVARIANCE; UNITARY SYMMETRY.

The expansion of the ingoing states in terms of the outgoing states provides a clear physical picture of the results to be expected from a collision of the particles in the initial state. The scattered wave is obtained by subtracting from the final wave those components which did not interact. The resulting probability amplitude for finding the state $|\beta, \text{out}\rangle$ in the scattered wave is given by $S_{\beta\gamma} - \delta_{\beta\gamma}$. The scattering amplitude $f_{\beta\gamma}$ is obtained by multiplying the probability amplitude by the flux in the actual incident state. For a two-particle plane-wave state, this factor is $\hbar/2p$, where p is the momentum of either incident particle in the center-of-mass system and $\hbar$ is Planck's constant divided by 2π. Thus Eq. (2) holds. The scattering cross section for a transition

$$f_{\beta\gamma} = \frac{\hbar}{2ip}(S_{\beta\gamma} - \delta_{\beta\gamma}) \qquad (2)$$

to a two-particle final state is then given by $4\pi |f_{\beta\gamma}|^2$.

The mathematical properties of the S matrix have been thoroughly explored in the nonrelativistic Schrödinger theory, and the analytic properties of scattering amplitudes as functions of energy and scattering angle can be given precisely for physically interesting interactions. Less is known about relativistic scattering theory. The S-matrix elements of a quantum field theory can be related by reduction formulas to the basic Green's functions of the theory (vacuum expectation values of time-order products of quantized fields). A number of exact properties of relativistic scattering amplitudes have been derived, for example, bounds on the growth of the amplitudes with increasing energy, limits on their variation with angle, and the relations between the real and imaginary parts of amplitudes and between amplitudes for different processes described by dispersion relations. The dispersion relations, which are a consequence of the unitarity and analyticity of the S matrix, have been particularly useful in problems which involve strong nuclear forces. SEE DISPERSION RELATIONS; QUANTUM FIELD THEORY.

Calculation. It is usually not possible to calculate S exactly for a complex process, for example, atomic or nuclear scattering or high-energy scattering with particle production, and it is necessary to resort to approximate methods. An exception is the case of a finite set of coupled two-body channels in the Schrödinger theory. The S matrix for a given energy and total angular momentum (or given partial wave) can be calculated in this case by solving the coupled Schrödinger equations numerically. The complete scattering amplitude is then constructed as a sum of partial-wave scattering amplitudes. This approach was extended by L. D. Faddeev, who developed an exact method applicable to multiparticle Schrödinger problems in which, for example, a three-body scattering amplitude is related through an integral equation to the two-body scattering amplitudes for the pairs of particles and the three-body interaction. This method has been used successfully in some problems in nuclear physics, but rapidly becomes intractable as the number of particles is increased.

A variety of approximation methods have been developed which are appropriate to different physical situations: perturbative methods for weak potentials or high energies, eikonal methods useful for small-angle scattering, distorted wave approximations in which part of the interaction is treated exactly and part perturbatively, optical approximations in which the effects of many scatterers are described by an averaged interaction, variational methods, and a number of methods valid for low energies.

The most commonly used methods are perturbative. A Lorentz covariant perturbation theory for the calculation of S in ordinary quantum field theories was developed in 1948–1949 by J. Schwinger, R. P. Feynman, and F. J. Dyson. This approach was extended around 1970 to the nonabelian gauge theories thought to describe the strong nuclear forces and the combined electromagnetic and weak forces. The S-matrix element $S_{\beta\gamma}$ may be written as the matrix element of a time-ordered exponential operator between states $|\gamma\rangle$ and $|\beta\rangle$ of the noninteracting system as in Eq. (3). Here $H'(x)$ is the interaction hamiltonian, and the $+$ on the bracket indicates that the

$$S_{\beta\gamma} = \left\langle \beta \left| \exp\left[-\left(\frac{i}{h}\right) \int d^4x H'(x) \right] \right| \gamma \right\rangle^+ \quad (3)$$

operators $H'(x)$, $H'(x')$, ... in the Taylor series expansion of the exponential are to be arranged from left to right in order of decreasing time variables. The terms in this Taylor series can be represented by means of the diagrammatic technique introduced by Feynman (see **illus**.). Each vertex and leg in the diagrams is associated with the component of the matrix element corresponding to the process depicted. The set of Feynman diagrams thereby provides a convenient algorithm for the construction of S to any order in H'. SEE FEYNMAN DIAGRAM.

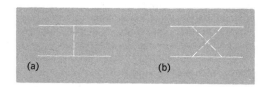

Typical Feynman diagrams for two-particle scattering: (*a*) second-order diagram; (*b*) fourth-order diagram. Solid lines represent scattered particles; broken lines, particles which transmit force between them.

The covariant perturbation techniques have been remarkably successful in quantum electrodynamics; the fine structure constant is small, and provides a natural expansion parameter for the perturbation series. In contrast, it was long thought that the strength of the specifically nuclear forces between strongly interacting particles (hadrons) prevented a meaningful perturbation expansion of S for hadronic reactions. This view changed with the development of the modern theory of strong interactions, in which hadrons are described as composites of elementary quarks which interact through the exchange of gluons as described by a nonabelian gauge theory. It was realized in 1973 that the strength of the elementary interaction becomes weak at short space-time distances, hence that one could use perturbation theory to describe the hard scattering processes which involve large momentum transfers between the constituents which interact in a hadronic collision. The soft parts of a hadronic scattering amplitude, including the wave functions of the constituents in a hadron, were described phenomenologically. This approach has been quite successful in explaining the main features of hadronic processes at high energies. While it is still not possible to calculate the soft parts of a hadronic scattering amplitude reliably, some progress has been made, and the quark-gluon description is now applied to problems in nuclear scattering. SEE GAUGE THEORY; GLUONS; QUANTUM CHROMODYNAMICS; QUANTUM ELECTRODYNAMICS; QUARKS; WEAK NUCLEAR INTERACTIONS; WEINBERG-SALAM MODEL.

Bibliography. W. O. Amrein et al., *Scattering Theory in Quantum Mechanics*, 1977; J. E. Farina, *Quantum Theory of Scattering Processes*, pt. 1, 1976, pt. 2, 1973; C. Itzykson and J. B. Zuber, *Quantum Field Theory*, 1980; R. G. Newton, *Scattering Theory of Waves and Particles*, 2d ed., 1982; M. Reed and B. Simon, *Methods of Modern Mathematical Physics*, vol. 3: *Scattering Theory*, 1979; J. R. Taylor, *Scattering Theory*, 1972, reprint 1983.

DISPERSION RELATIONS
CHARLES J. GOEBEL

Relations between the real and imaginary parts of a response function. The term dispersion refers to the fact that the index of refraction of a medium is a function of frequency. In 1926 H. A. Kramers and R. Kronig showed that the imaginary part of an index of refraction (that is, the absorptivity) determines the real part (that is, the refractivity); this is called the Kramers-Kronig relation. The term dispersion relation is now used for the analogous relation between the real and imaginary parts of any response function, such as Eq. (14) below.

Response function. Consider a system in which a cause $C(t)$ (for example, a force) and its effect $E(t)$ (for example, a displacement) are related by Eq. (1), where $G(t)$ is called the response

$$E(t) = \int_{-\infty}^{t} dt' \, G(t - t') C(t') \tag{1}$$

function. Because this relation is linear in both $C(t)$ and $E(t)$, the response is said to be linear: The superposition of two causes results in the sum of their effects. Because the response function G is a function only of $t - t'$, the response is said to be time-independent: The effect of time t of a cause at time t' depends only on the time difference $t - t'$. Because the upper limit of integration is t [or equivalently, $g(t)$ can be said to vanish for negative argument, Eq. (2)], the response is said

$$G(t) = 0 \quad \text{if } t < 0 \tag{2}$$

to be causal: the cause has no effect at earlier times. *See Causality.*

Many examples can be given. A few pairs of cause and effect are: force and spatial displacement, electric field and polarization (G is electrical susceptibility), and incident wave and scattered wave (G is scattering amplitude).

Pure-tone response. The Fourier transform $g(\omega)$ of the response function $G(t)$ is defined in Eq. (3) and the inverse relation in Eq. (4). The function $g(\omega)$ is complex-valued; however, if $G(t)$

$$g(\omega) = \int_0^{\infty} dt \, e^{i\omega t} G(t) \tag{3} \qquad G(t) = \frac{1}{2\pi} \int_{-\infty}^{\infty} d\omega \, e^{-i\omega t} g(\omega) \tag{4}$$

is real [which it must be if $C(t)$ and $E(t)$ of Eq. (1) are real], the crossing relation, Eq. (5), is valid,

$$g^*(\omega) = \int_0^{\infty} dt \, e^{-i\omega * t} G(t) = g(-\omega^*) \tag{5}$$

where an asterisk indicates complex conjugation. The function $g(\omega)$ is called the frequency-dependent response; it describes the response to a simple harmonic (pure tone) cause: if the cause is given by Eq. (6), then Eq. (1) implies that its effect is given by Eq. (7), where c.c. means complex conjugate and $|g|$ and δ are given by Eq. (8). One sees that the magnitude $|g|$ and phase δ of $g(\omega)$

$$C(t) = \cos(\omega t - \eta), \; \omega \text{ real} \tag{6}$$

$$E(t) = \int_{-\infty}^{\infty} dt' \, G(t - t') \cos(\omega t' - \eta)$$

$$= \tfrac{1}{2} \int_{-\infty}^{\infty} dt' \, G(t - t') e^{-i(\omega t' - \eta)} + \text{c.c.}$$

$$= \tfrac{1}{2} e^{-i(\omega t - \eta)} \int_{-\infty}^{\infty} dt' \, G(t - t') e^{i\omega(t - t')} + \text{c.c.}$$

$$= \tfrac{1}{2} e^{-i(\omega t - \eta)} g(\omega) + \text{c.c.}$$

$$= \tfrac{1}{2} |g(\omega)| e^{-i(\omega t - \eta - \delta(\omega))} + \text{c.c.}$$

$$= |g(\omega)| \cos(\omega t - \eta - \sigma(\omega)) \tag{7}$$

$$g(\omega) = |g(\omega)| e^{i\delta(\omega)} \tag{8}$$

are the amplitude and phase shift, respectively, of the simple harmonic effect relative to the cause.

Causality. Causality, Eq. (2), has an important consequence for $g(\omega)$, namely the dispersion relations given in Eq. (14) below. Their derivation will be outlined, using Cauchy's formula and other properties of analytic functions of a complex variable. Let ω take on complex values, $\omega = \text{Re } \omega + i \, \text{Im } \omega$. Then $g(\omega)$ as given by Eq. (3) is an analytic function of ω as long as the integral, and its derivative with respect to ω, exist (that is, are finite). For any stable system, $G(t)$ is bounded in magnitude for all values of t. It follows that the integral in Eq. (3) exists for all ω such that $\text{Im } \omega > 0$ ("in the upper half ω-plane") because then the magnitude of the factor $e^{i\omega t}$ of the integrand falls exponentially with increasing t according to Eq. (9). Since, on the other hand,

$$|e^{i\omega t}| = e^{-(\text{Im } \omega)t} \tag{9}$$

for decreasing t this factor rises exponentially, the existence of the integral depended on causality, Eq. (2). So, causality implies that $g(\omega)$ is an analytic function in the upper half ω-plane.

Consequently one may write Cauchy's formula for $g(\omega)$, Eq. (10), where the integration

$$g(\omega) = \frac{1}{2\pi i} \int_C d\omega' \, g(\omega')/(\omega' - \omega) \tag{10}$$

contour C is any simple loop, traversed counterclockwise, which surrounds the point ω and which lies in the upper half ω'-plane (the region where $g(\omega)$ is an analytic function). Letting the point ω approach the contour C, the Cauchy formula comes to the form Eq. (11), where P means principal

$$g(\omega) = \frac{1}{2\pi i} \left[P \int_C \frac{d\omega'}{\omega' - \omega} g(\omega') + i\pi g(\omega) \right] \tag{11}$$

value. [Roughly speaking, it means: omit from the range of integration the interval of C which lies within a distance ϵ on either side of the singular point $\omega' = \omega$, and then take the limit $\epsilon \to 0$.] Solving for $g(\omega)$ gives Eq. (12), and then taking real and imaginary parts gives Eqs. (13). HT means

$$g(\omega) = \frac{-i}{\pi} P \int_C \frac{d\omega'}{\omega' - \omega} g(\omega) \equiv -i\text{HT}[g] \quad (12) \qquad \text{Re } g = \text{HT}[\text{Im } g] \quad \text{Im } g = -\text{HT}[\text{Re } g] \quad (13)$$

Hilbert transform (using the contour C), and Re g and Im g are the real and imaginary parts of g; these relations hold for any function $g(\omega)$ which is analytic inside the contour C. Let now the contour C be expanded until it encloses the entire upper half ω'-plane. If $g(\omega)$ vanishes at large ω the only part of the contour which contributes to the HT integral runs along the real ω'-axis. Finally, using crossing, Eq. (5), the integral can be expressed as an integral over just positive ω', Eqs. (14). These are the dispersion relations for the frequency-dependent response function $g(\omega)$.

$$\text{Re } g(\omega) = \frac{2}{\pi} P \int_0^\infty d\omega' \, \omega' \, \text{Im } g(\omega')/(\omega'^2 - \omega^2)$$
$$\text{Im } g(\omega) = \frac{-2\omega}{\pi} P \int_0^\infty d\omega' \, \text{Re } g(\omega')/(\omega'^2 - \omega^2) \tag{14}$$

They say that Re g and Im g are not independent functions; given one of them, the other is determined.

Resonance. Although a causal $g(\omega)$ is analytic in the upper ω-plane, it can, and usually does, have singularities in the lower ω-plane. For instance, $g(\omega)$ may have a pole at $\omega = \omega_0$ (and to satisfy Eq. (5) likewise at $\omega = -\omega_0^*$); causality requires that Im $\omega_0 < 0$. In Eq. (15) such a pole

$$g(\omega) = \frac{A}{\omega - \omega_0} + \frac{A}{-\omega - \omega_0^*} + \cdots = \frac{\text{Re } A + i \, \text{Im } A}{\omega - \omega_R + i^1/2\gamma} + \frac{\text{Re } A - i \, \text{Im } A}{-\omega - \omega_R - i^1/2\gamma} + \cdots \tag{15}$$

term of $g(\omega)$ is written out, with the notation Re $\omega_0 = \omega_R$, Im $\omega_0 = -\frac{1}{2}\gamma$. If γ is small, then $|g(\omega)|^2$ as a function of real ω peaks sharply at $\omega = \omega_R$ with a width (defined as the interval of ω between the half height points, where $|g(\omega)|^2$ is half as big as the peak) equal to γ. This is called a resonance of the response function $g(\omega)$; ω_R is its position and γ is its width.

The consequence of a resonance for the effect $E(t)$ is found by using Eq. (15) in Eqs. (1) and (4). If the cause $C(t)$ turns off rapidly enough, then at large t the exponential factor in the integrand of Eqs. (16) decreases into the lower ω-plane, and so the ω-integration can be done by

$$E(t) = \int dt' \, C(t') \frac{1}{2\pi} \int d\omega \, e^{-i\omega(t-t')} \frac{A}{\omega - \omega_0} + \text{c.c.} + \cdots$$
$$\xrightarrow{t \to \infty} -iA \int dt' \, C(t') e^{-i\omega_0(t-t')} + \text{c.c.} + \cdots$$
$$= [-iAe^{-i\omega_R t} \int dt' \, C(t') e^{i\omega_0 t'} + \text{c.c.}] e^{-1/2\gamma t} + \cdots$$
$$= C_1 \cos(\omega_R t + C_2) e^{-1/2\gamma t} + \cdots \tag{16}$$

(where C_1, C_2 are constants)

residues. The result is that the effect $E(t)$ has an exponentially decaying term whose time constant (for E^2) is γ^{-1}; that is, the width of the resonance peak of $|g(\omega)|^2$ and the decay constant of $E(t)^2$ are reciprocals of one another. In more physical terms, the decaying oscillation of $E(t)$ reflects a (decaying) normal mode of oscillation, a "ringing," of the system.

If Im ω_0 were positive, contrary to causality, the calculation of Eqs. (16) would show the effect $E(t)$ to have an exponentially growing term at an early time, preceding the cause. As for $g(\omega)$, the qualitative effect of causality is that at the resonance, where the magnitude $|g(\omega)|$ has a narrow peak, the phase of $g(\omega)$ [$\delta(\omega)$ in Eq. (7)] rises rapidly; if Im ω_0 were positive, contrary to causality, the phase would instead fall rapidly.

Dielectric constant and refractive index. The response of a macroscopically homogeneous dielectric (polarizable) medium to an electromagnetic plane wave of frequency ω is described by the dielectric constants $\epsilon_\parallel(\omega)$ and $\epsilon_\perp(\omega)$. The subscripts distinguish whether the electric field is parallel or perpendicular to the direction of propagation of the plane wave. The discussion will be limited to $\epsilon_\perp(\omega)$, which will be referred to as simply $\epsilon(\omega)$. The refractive index $n(\omega)$ is simply related: $n^2 = \epsilon$. But $n(\omega)$ itself will not in general satisfy dispersion relations because if $\epsilon(\omega)$ has a zero in the upper ω-plane, $\sqrt{\epsilon} = n$ will have a square root singularity.

At $\omega = \infty$, $\epsilon(\omega)$ has the value 1 (the vacuum value). At $\omega = 0$, $\epsilon(\omega)$ is singular if the medium is conductive (that is, contains charges which are free to move), according to Eq. (17), where σ_0

$$\epsilon(\omega) \xrightarrow{\omega \to 0} \frac{4\pi i \sigma_0}{\omega} - \frac{\omega_0^2}{\omega^2} \qquad \omega_0^2 = n_v 4\pi e^2/m \qquad (17)$$

is the static conductivity and ω_0 is the plasma frequency; n_v is the number density of free charges of charge e and mass m. Thus the function given in Eq. (18) has the properties assumed for $g(\omega)$

$$g(\omega) = \epsilon(\omega) - 1 - \frac{4\pi i \sigma_0}{\omega} + \frac{\omega_0^2}{\omega^2} \qquad (18)$$

above and can be substituted for it in Eq. (14), resulting in the dispersion relations, Eqs. (19). In

$$\text{Re } \epsilon(\omega) = 1 - \frac{\omega_0^2}{\omega^2} + \frac{2}{\pi} \int_0^\infty d\omega' \, \frac{\omega' \, \text{Im } \epsilon(\omega') - \omega \, \text{Im } \epsilon(\omega)}{\omega'^2 - \omega^2} \qquad (19a)$$

$$\text{Im } \epsilon(\omega) = \frac{4\pi \sigma_0}{\omega} - \frac{2}{\pi} \omega \int_0^\infty d\omega' \, \frac{\text{Re } \epsilon(\omega') + \omega_0^2/\omega'^2 - \text{Re } \epsilon(\omega) - \omega_0^2/\omega^2}{\omega'^2 - \omega^2} \qquad (19b)$$

Eqs. (19), terms have been added to the integrands in order to make them nonsingular; these terms do not alter the integrals because $P \int d\omega'/(\omega'^2 - \omega^2) = 0$.

By setting ω equal to special values, interesting results are obtained; for simplicity these will be given for the case of a nonconductor, where $\sigma_0 = \omega_0 = 0$. Setting $\omega = 0$ in Eq. (19a) results in an expression for the static dielectric constant, Eq. (20). Taking ω large in Eqs. (19)

$$\epsilon(0) = 1 + \frac{2}{\pi} \int_0^\infty d\omega \, \text{Im } \epsilon(\omega)/\omega \qquad (20)$$

gives Eqs. (21), which are sum rules. The first of these relates an integral over all frequencies of

$$\text{Re } \epsilon(\omega) \longrightarrow 1 - \omega_\infty^2/\omega^2 \qquad \text{where} \qquad \omega_\infty^2 = \frac{2}{\pi} \int_0^\infty d\omega \, \omega \, \text{Im } \epsilon(\omega) \qquad (21a)$$

$$\text{Im } \epsilon(\omega) \to 4\pi \sigma_\infty/\omega \qquad \text{where} \qquad 4\pi \sigma_\infty = \frac{2}{\pi} \int_0^\infty d\omega \, [\text{Re } \epsilon(\omega) - 1] \qquad (21b)$$

the absorptivity, Im ϵ, to the quantity ω_∞^2, for which a direct calculation gives the value $\omega_\infty^2 = \Sigma \, n_v 4\pi e^2/m$; where the sum is over all kinds of charges in the medium, in contrast to the formula for the plasma frequency ω_0, Eq. (17); at infinite frequency all particles of the medium behave as though free.

Forward-scattering amplitude. It can be shown by arguments like those used above that the forward-scattering amplitude $f(\omega)$ of light incident on an arbitrary body satisfies the dispersion relation given in Eq. (22) [the less interesting relation giving Im f in terms of Re f has been

$$\text{Re } f(\omega) = f(0) + \frac{\omega^2}{2\pi^2 c} \int_0^\infty d\omega' \, \frac{\sigma_T(\omega') - \sigma_T(\omega)}{\omega'^2 - \omega^2} \quad \text{where} \quad f(0) = -Q^2/Mc^2 \quad (22)$$

omitted], where Q is the charge, M is the mass of the body, and c is the speed of light. In Eq. (22) Im $f(\omega)$ has been expressed in terms of the total cross section σ_T by the optical theorem, Eq. (23); this formula is true for the scattering of waves of any sort.

$$\text{Im } f(\omega) = (k/4\pi)\sigma_T(\omega) \quad \text{where} \quad k = \text{wave number} \quad (23)$$

Equations (22) and (14) can be related by using the "optical potential" formula, Eq. (24),

$$k^2(\text{in medium})/k^2(\text{in vacuum}) = n^2 - 1 + 4\pi n_v f(\omega)/k^2 \quad (24)$$

which relates the forward scattering amplitude $f(\omega)$ of a wave (of any sort) of frequency ω and wave number k incident on a body to the index of refraction n of that wave in a medium consisting of a random arrangement of such bodies with mean number density n_v. However, Eq. (24) is valid only for a dilute medium, whereas Eqs. (14) and (22) are both exact if causality is true.

An interesting consequence of Eq. (22) is that a system consisting of the electromagnetic field and point charges, obeying classical mechanics, cannot be causal. For such a system it can be shown that $f(\omega)$ vanishes for $\omega = \infty$; using this in Eq. (22) gives Eq. (25), which is untrue since

$$0 = f(0) - (2\pi^2 c)^{-1} \int_0^\infty d\omega \sigma_T \quad (25)$$

both terms on the right-hand side are negative. In the quantum theory, by contrast, $\sigma_T > 0(1/\omega)$ and so no formula like Eq. (25) follows from causality.

Elementary particle theory. In the quantum theory the scattering amplitudes of matter waves satisfy dispersion relations similar to Eq. (22) if causality holds. This gives the possibility of experimentally testing causality up to very high frequencies (corresponding to very short time intervals) by using, for instance, protons of energy up to 500 GeV, incident on protons in a target, at Fermilab. Since the right-hand side of Eq. (22) requires knowledge of σ_T at all ω, one can strictly say only that the observed $f(\omega)$ is consistent with causality and a reasonable extrapolation of σ_T beyond observed energies.

Not only is there no experimental evidence against causality, but causality has always been a property of relativistic quantum field theory (elementary particle theory). The following is essentially equivalent to causality: Considered as functions of the Lorentz invariant scalars formed from the 4-momenta of the interacting particles (for instance, the total energy and the momentum transfer in the center-of-mass frame, in the case of a two-body scattering amplitude), the S-matrix elements (scattering amplitudes) of quantum field theory are analytic functions. More precisely, they are analytic except for singularities on Landau surfaces in the space of the (complex) arguments, the Lorentz scalars. These singularities are branch points, and if the amplitude is analytically continued around one, the amplitude becomes changed by a certain amount.

On the one hand, knowing this change allows the amplitude itself to be determined, by use of Cauchy's formula [a dispersion relation like Eq. (22) is a simple example of this]. On the other hand, the change is given in terms of other S-matrix elements. [The optical theorem, Eq. (23), is a simple example of this; the point is that σ_T is a sum (and integral) over products of two amplitudes.] These S-matrix equations, resulting from the combination of the analyticity and the unitarity of the S-matrix, are the subject of S-matrix theory. SEE RELATIVITY.

These equations can to some extent replace the equations of motion of quantum field theory. They have the advantage over the only known systematic solution of the equations of motion, namely perturbation theory (Feynman diagrams), that they involve only S-matrix elements, that is, the amplitudes of real, observable processes. However, there is no known way to find the solution of the S-matrix equations which corresponds to a given set of field theory equations of motion.

It was the original hope of many workers in S-matrix theory that the equations of motion were in fact irrelevant and that the S-matrix was determined by a few properties of low-mass particles (in fact perhaps by none, in the extreme view termed bootstrapping). However, the successes of quantum chromodynamics (QCD) in describing and predicting many phenomena in high-energy physics seems to show that a specific field theory, whose elementary quanta (quarks and gluons) are not observable as particles, underlies the S-matrix of "elementary" particle interactions. The S-matrix equations are true in any case, and will remain an important tool in the description and correlation of experimental data. SEE ELEMENTARY PARTICLE; QUANTUM CHROMODYNAMICS; QUANTUM FIELD THEORY; SCATTERING EXPERIMENTS (NUCLEI); SCATTERING MATRIX.

Bibliography. G. F. Chew, *The Analytic S Matrix*, 1966; R. V. Churchill and J. W. Brown, *Complex Variables and Applications*, 4th ed., 1984; R. J. Eden, *High Energy Collisions of Elementary Particles*, 1967; R. J. Eden et al., *The Analytic S Matrix*, 1966; R. Good and T. Nelson, *Classsical Theory of Electric and Magnetic Fields*, 1971; A. Martin and T. Spearman, *Elementary Particle Theory*, 1970; R. G. Newton, *Scattering Theory of Waves and Particles*, 2d ed., 1982; P. Roman, *Advanced Quantum Theory*, 1965.

CAUSALITY
GORDON L. SHAW

In classical mechanics causality has been taken to mean that all the dynamical variables of a system can be precisely measured and their evolution in time is strictly determined by the forces. Thus, specifying the initial variables of position x, y, z and momentum p_x, p_y, p_z at time $t = t_0$ completely determines the future behavior of the system for $t > t_0$. Therefore, classically, one can repeatedly prepare a system in the same specific initial state with arbitrary precision and then measure the same final state after a given lapse in time.

In quantum mechanics there is also a rigorous law of causality. However, there are important differences. All the variables cannot be completely specified; there are inherent uncertainties δ such that, for example, $(\delta x)(\delta p_x) \gtrsim$ Planck's constant. This Heisenberg uncertainty relation is completely built into the structure of quantum mechanics. Hence, one prepares an initial state and specifies a wave function $\psi(x,y,z,t_0)$ whose (absolute) square is interpreted as a probability distribution of finding the system at time t_0 at position x,y,z. The law of causality in quantum mechanics states that for an isolated system the initial state $\psi(x,y,z,t_0)$ completely determines, by the dynamics of the Schrödinger equation, the future state $\psi(x,y,z,t)$, which again leads to a probability distribution. To understand this physically, consider the scattering of protons off a target nucleus. Quantum mechanics can predict only an average distribution for the scattering process; thus it is the average of a large number of measured events that one accurately compares with theory. One also notes that in accord with the uncertainty principle there will always be some spread in momentum for the initial proton states. Furthermore, the detection of the protons disturbs the system (again because of the uncertainty principle), destroying the exact relation between ψ before and after the measurement.

Causality has also been used in reference to the principle that an event cannot precede its cause. Consider, for example, a packet of light (or electromagnetic radiation) having a sharp wave front incident on an obstacle. After the incident wave reaches the obstacle or scattering center (located at the origin) at time $t = 0$, scattered light moves out in all directions. Since light travels with the finite velocity c, causality requires that no scattered light be observed at any point P at a distance r from the scattering center until there is sufficient time for the incident light to reach P, that is, until $t > r/c$. This requirement places important restrictions on the behavior of the scattering process as a function of frequency ω. One writes the incident wave as expression (1).

$$A_I(z,t) \propto \int_{-\infty}^{\infty} d\omega\, a\,(\omega) e^{j\omega(z/c - t)} \tag{1}$$

The statement that the incident wave does not reach the scattering center at the origin before $t = 0$ means that $A_I(0,t) = 0$ for $t < 0$.

Using Cauchy's residue theorem to evaluate (1) for $A_I(0,t)$ by contour integration, completing the contour for $t < 0$ by an infinite semicircle in the upper-half complex ω plane where

$Im\omega > 0$, it follows that $a(\omega)$ has no poles for $Im\omega > 0$. The scattered wave in the forward direction is given by expression (2), where $f(\omega)$ is the forward scattering amplitude. Causality requires that

$$A_s(z,t) \propto \int_{-\infty}^{\infty} d\omega f(\omega) a(\omega) e^{i\omega(z/c - t)} \tag{2}$$

$A_S(z,t) = 0$ for $(z/c - t) < 0$. It again follows (by a similar argument) that $f(\omega)$ has no poles for $Im\omega > 0$. This analyticity for $f(\omega)$ can be expressed in integral form, again using Cauchy's theorem, as in Eq. (3), where P signifies that it is a principal value integral. Equation (3), which follows

$$Ref(\omega) = \frac{1}{\pi} P \int_{-\infty}^{\infty} d\omega' \frac{Imf(\omega')}{\omega' - \omega} \tag{3}$$

from causality [plus the assumption that $f(\omega) \to 0$ as $|\omega| \to \infty$], forms the basis of a large number of applications ranging from classical physics (for example, electric circuit theory) to relativistic quantum physics. (The causality principle in relativistic quantum field theory is expressed by the condition that the fields at different space-time points do not interfere with each other if the separation of the points is spacelike.) SEE RELATIVITY.

In particular, consider the scattering of light in an optical medium. Let $n(\omega)$ be the usual (real) geometrical index of refraction and $\gamma(\omega)$ the absorption coefficient of the medium. Here, one has for Eq. (3) the relationship shown as Eq. (4) [Kramers-Kronig relation], relating the absorption

$$n(\omega) = 1 + \frac{c}{\pi} P \int_0^{\infty} \left(\frac{\omega'}{\omega}\right)^2 \frac{d\omega'}{\omega'^2 - \omega^2} \gamma(\omega') \tag{4}$$

process γ to the dispersive process n. Thus Eq. (4) is referred to as a dispersion relation, and the general equation, Eq. (3), is commonly called a dispersion relation even in its applications in high-energy physics. The forward dispersion relations for pion-nucleon scattering have been verified experimentally up to very high energies. SEE DISPERSION RELATIONS.

Bibliography. J. Bjorken and S. Drell, *Relativistic Quantum Fields*, 1965; P. Fong, *Elementary Quantum Mechanics*, 1962; S. Gasiorowicz, *Elementary Particle Physics*, 1966; R. G. Newton, *Scattering Theory of Waves and Particles*, 2d ed., 1982; R. Swinburne, *Space, Time and Causality*, 1983.

REGGE POLE
LUIGI SERTORIO

Pole singularity of the scattering amplitude in the complex angular momentum plane. The concept of Regge pole arises in the theory of nonrelativistic two-body quantum dynamics. To understand the rather complex ideas which follow, it is necessary to begin by reviewing a few elements of potential scattering.

Theory. A system composed of two elementary spinless particles is described by the wave function $\psi(\vec{x})$ ($\vec{x}$ is the relative distance between the two particles), which is the solution, with appropriate asymptotic conditions, of the Schrödinger equation, Eq. (1), where $\hbar$ is the Planck

$$-\frac{\hbar^2}{2M} \Delta \psi(\vec{x}) + U(\vec{x}) \psi(\vec{x}) = k^2 \psi(\vec{x}) \tag{1}$$

constant, M the reduced mass, and U the potential.

For the sake of simplicity assume $U(\vec{x}) = U(x)$.

The dynamics of the system is fully specified by the interaction potential U.

The solution ψ, which describes the collision of a plane wave with the scattering center, satisfies the asymptotic conditions given by Eq. (2).

$$\psi(\vec{x}) \xrightarrow[x\text{large}]{} e^{i\vec{k}\cdot\vec{x}} + \frac{e^{ikx}}{x} F(k, \cos\theta) \quad \text{where} \quad \cos\theta = \frac{\vec{k}\cdot\vec{x}}{kx} \tag{2}$$

This formula, in conjunction with Eq. (1), defines the scattering amplitude F. The knowledge of F is equivalent to the knowledge of the Heisenberg S matrix. In turn the S matrix gives a complete description of any scattering process associated with Eq. (1). In order to compute F one needs to find an explicit solution of the Schrödinger equation. F is directly connected with the quantum flux of scattered particles according to well-known formula (3). This flux (differential cross

$$\frac{d\sigma}{d\Omega} = |F|^2 \tag{3}$$

section) is directly measurable in a scattering experiment. SEE SCATTERING MATRIX.

The modern quantum theory of scattering has recognized the importance of analyzing the mathematical properties of the scattering amplitude for wide classes of potentials. The general goal is the study of the properties (analytic properties) of the amplitude $F(k, \cos\theta)$ defined in the complex plane of the variables k and $\cos\theta$. The extension to the complex domain provides the physicist with a comprehensive algorithm; general properties of class of $U(x)$ reflect themselves in general analytic properties of the function F. In particular, singularities in the k plane represent definite features of the physical system and they are directly related to the properties of $U(x)$. According to W. Heisenberg, a pole of F in the k plane can be associated with either a resonant state or a bound state of the system. Other analytic properties also admit related interpretation, but for the sake of simplicity they are not discussed here.

The theory is considerably simplified if the partial wave amplitude defined by Eq. (4) is

$$a_l(k) = 1/2 \int_{-1}^{1} F(k, \cos\theta) \, P_l(\cos\theta) \, d\cos\theta \tag{4}$$

used. Here the $P_l(\cos\theta)$ are the standard spherical harmonics.

The $a_l(k)$ depend on the angular momentum l instead of the angle θ. In this way the symmetry properties of the scattering amplitude under rotations are explicitly exhibited. SEE SYMMETRY LAWS.

The study of $a_l(k)$ is related to the solution of an ordinary differential equation, the radial Schrödinger equation, instead of the three-dimensional one.

The connections between the radial Schrödinger equation and the partial wave amplitude are expressed by Eqs. (5)–(7), where $\delta_l(k)$ is the well-known phase shift. Given $a_l(k)$, the full

$$\phi''(x) + k^2 \phi(x) - \frac{l(l+1)}{x^2} \phi(x) - U(x) \phi(x) = 0 \tag{5}$$

$$\lim_{x \to 0} \phi(x) = 0$$

$$\phi(x) \underset{x \to \infty}{=} \sin\left[kx - \frac{\pi l}{2} + \delta_l(k)\right] \tag{6} \qquad a_l(k) = \frac{1}{k} e^{i\delta_l(k)} \sin \delta_l(k) \tag{7}$$

$F(k, \cos\theta)$ can be reconstructed through the partial wave expansion in Eq. (8). It is obvious that this

$$F(k, \cos\theta) = \sum_{0}^{\infty} (2l+1) \, a_l(k) \, P_l(\cos\theta) \tag{8}$$

expansion is particularly useful when only a few partial waves are stimulated. This situation certainly arises when the energy $E = k^2$ is close to a pole of $a_l(k)$. If this is the case, the system is resonating in a state of angular momentum l, and the pole in the lth component in the expansion is the predominant contribution to the amplitude $F(k, \cos\theta)$.

T. Regge extends the study of the partial wave amplitude a_l by generalizing it to be a function of l as a complex variable. This extension is made by considering Eqs. (5)–(7) for complex values of l.

The study of the analytic properties of $a(l,k)$ for complex l is particularly interesting for the restricted class of Yukawian potentials given by Eq. (9). Indeed, these potentials yield amplitudes

$$U(x) = \int_{m}^{\infty} \frac{e^{-\mu x}}{x} \sigma(\mu) \, d\mu \tag{9}$$

$a(l,k)$ with remarkably simple properties: In the right-hand plane Re $l \geq -1/2$, $a(l,k)$ is meromorphic; that is, it has only simple poles. These are called Regge poles. The position and residues of Regge poles are functions of the variable k^2. In the formula $a(l,k) \simeq \beta(k^2)/l - \alpha$ the function $\alpha(k^2)$ is called the Regge trajectory and $\beta(k^2)$ is called the Regge residue.

In the region Re $l < 1/2$ the analytic properties of $a(l,k)$ are complicated and depend on the interaction U in a somewhat unstable way.

Note that the values of $a(l,k)$ for noninteger l do not enter directly into the Rayleigh-Faxen formula, Eq. (8).

By the Watson-Sommerfeld transformation a convenient substitute can be obtained for Eq. (8) which at the same time provides a physical interpretation of the theory.

Suppose that the amplitude $a(l,k)$ in Re $l > -1/2$ has only a simple pole for $l = \alpha(k^2)$, the Watson transformation implies that Eq. (10) is valid. The term $\beta P_\alpha/\sin \pi\alpha$ is the contribution to F

$$F(k, \cos \theta) = -i \int_{-1/2 - i\infty}^{-1/2 + i\infty} (2l + 1) \frac{a(l,k)}{\sin \pi l} P_l(-\cos \theta) + \frac{\beta(k^2)}{\sin \pi\alpha(k^2)} P_{\alpha(k^2)}(-\cos \theta) \quad (10)$$

of the pole of $a(l,k)$ at $l = \alpha$. The integral between $-1/2 - i\infty$ and $-1/2 + i\infty$ is called the background integral. It takes into account the contribution to $F(k, \cos \theta)$ of all the singularities of $a(l,k)$ on the left plane Re $l < -1/2$.

Consider the behavior of $F(k, \cos \theta)$ for $E = k^2$ real and increasing. The simple pole of $a(l,k)$ moves along the trajectory $\alpha(E)$. Suppose, moreover, that for E close to a value E_0 Eq. (11) holds,

$$\alpha = n + \epsilon(E) + i\gamma(E) \quad (11)$$

where n is an integer and ϵ, γ are small real numbers. If ϵ, γ were zero at $E = E_0$, F would have a pole for $E = E_0$ of residue $\beta(E_0) P_n(\cos \theta)$. This would correspond to a bound state of physical angular momentum n. However, it should be noted that this phenomenon occurs only when $E < 0$.

In general, for $E > 0$, γ never vanishes, and whenever the ϵ, γ are small the scattering can be described rather as occurring via an intermediate resonance. Near the point $E = E_0$, where $\epsilon, \gamma \ll 1$, Eq. (12) holds. Supposing $\epsilon(E_0) = 0$ one may keep the linear terms in $E - E_0$ only, in the

$$a(l,E) \simeq \frac{\beta(E_0)}{1 - \alpha(E)} \simeq \frac{\beta(E_0)}{1 - n - \epsilon(E) - i\gamma(E)} \quad (12)$$

denominator. From the general theory, γ and β can be taken as constant. It follows approximately that Eqs. (13) and (14) are valid. If $l = n$ it becomes obvious that $a(l,E)$ has a complex pole in E,

$$1 - n - \epsilon - i\gamma = 1 - n - (E - E_0)\frac{d\epsilon}{dE} - i\gamma \quad (13)$$

$$a(l,E) \simeq \frac{\beta}{1 - n - (E - E_0)\frac{d\epsilon}{dE} - i\gamma} \quad (14)$$

as shown in Eq. (15), Γ being defined by Eq. (16). This corresponds to the classical picture of a

$$E = E_0 - \frac{i\Gamma}{2} \quad (15) \qquad \Gamma = \frac{2\gamma}{d\epsilon/dE} \quad (16)$$

resonance of width Γ and life $\hbar/\Gamma$ at the energy E_0. If instead E is real, $a(l,E)$ is singular in l at a value given by Eq. (17), and for $E = E_0$ the angular momentum has integer real part plus an imaginary part γ.

$$\begin{aligned} l &= n + (E - E_0)\frac{d\epsilon}{dE} + i\gamma \\ &= n + \epsilon(E) + i\gamma \end{aligned} \quad (17)$$

The angular life of the system can be considered to be $1/\gamma$. As E varies, $\alpha(E)$ may pass close to different integers, and therefore it may originate several different resonances. In this way resonances and bound states are grouped into families given by the same trajectory $\alpha(E)$.

As clearly seen by the Watson transformation in Eq. (10), knowledge of the family of resonances is not enough to reconstruct the full $F(k, \cos \theta)$ because of the contribution of the background integral. Nevertheless, the properties of the background integral are sufficiently known to allow one to extract from the inversion formula (10) the important asymptotic result in Eq. (18),

$$F(k, \cos \theta) \underset{\cos \theta \to \infty}{\simeq} (\cos \theta)^{\alpha(E)} + O(\cos \theta^{-1/2}) \qquad (18)$$

which gives a precise prediction for the asymptotic behavior of $F(k, \cos \theta)$ in the large $\cos \theta$ limit (unphysical region) and which is important in the theory of potential scattering as a step in the proof of the Mandelstam representation.

Hypothesis. The Regge pole hypothesis can be stated as follows: The properties of the scattering amplitude which have been proved by Regge in potential theory represent a deep property of nature and are indeed true also for physical relativistic amplitudes. The consequences of the hypothesis are far-reaching.

The relativistic theory of strong interacting particles shares with potential theory the general concept of S matrix but does not have the counterpart of the Schrödinger equation.

It is generally agreed that elementary particles, bound states, and resonances are described by poles of the S-matrix elements as functions of the energy. Researchers have obtained experimental evidence for the existence of hundreds of new particles. As it is extremely difficult to establish a clear-cut distinction between elementary and compound particles, the very idea of elementary particles has become the target of mounting criticism. Heisenberg, and later G. F. Chew, proposed theories in which the idea of elementarity is altogether abolished. This is not the place to discuss the merit of such theories. The conjecture that the mechanism of Regge poles holds for relativistic amplitudes has provided physicists with a workable criterion by which it will be possible to reach a decision on whether elementarity of a given particle exists. Moreover, the Regge pole hypothesis leads to extremely interesting predictions on the high-energy behavior of scattering processes (see **illus.**). Consider the scattering process $A + B \to C + D$ symbolized by the diagram. Here p_a, p_b, $-p_c$, $-p_d$ are the energy-momentum four vectors associated with A, B, C, D respectively. For simplicity assume $A, \ldots, D$ to have equal mass M and to be spinless. The invariants are defined by Eqs. (19), where $p^2 = -(\vec{p})^2 + E_p^2$.

$$(p_a + p_b)^2 = s \qquad\qquad (p_a + p_c)^2 = t \qquad (19)$$

Also, $s = (p_c + p_d)^2$ because of the conservation of energy-momentum as in Eq. (20). The

$$p_a + p_b + p_c + p_d = 0 \qquad (20)$$

scattering amplitude will be a function of s, t, only, as in Eqs. (21).

$$A = A(s, t) \qquad s = 4M^2 + k_s^2 \qquad t = -2k_s^2(1 - \cos \theta_s) \qquad (21)$$

Here k_s, $\sqrt{s}$ are respectively the relative momentum and total energy in the center of mass system of A and B, and θ_s is the scattering angle. For a physical scattering process one has $s > 4M^2$, $t < 0$. Consider now the scattering process in relation (22), where $\bar{C}, \bar{B}$ are the antiparticles

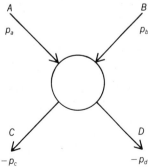

Scattering process.

$$A + \bar{C} \to \bar{B} + D \tag{22}$$

of C, B. Let p'_a, $-p'_b$, p'_c, $-p'_d$ be the energy momentum vectors of the particles $\bar{ABCD}$ respectively, and let t' and s' be defined by Eqs. (23). The corresponding amplitude is written as $A'(s', t')$.

$$t' = (p'_a + p'_c)^2 \qquad\qquad s' = (p'_a + p'_b)^2 \tag{23}$$

Notice that here t' plays the role of (square) energy in the CM system. It was long conjectured and subsequently proved by V. Glaser and coworkers that $A(s, t)$ and $A'(s', t')$ are actually the same analytic function in different domains, for the $A + B \to C + D$ process $s > 4M^2$, $t < 0$ and for $A + \bar{C} \to \bar{B} + D$, $t' \geq 4M^2$, $s' < 0$. Accordingly the distinction between A', s', t' and A, s, t is dropped. Suppose now that in the channel $A + \bar{C} \to \bar{B} + D$, a scattering angle θ_t is defined by Eq. (24). If the Regge pole hypothesis holds, then for $\cos \theta_t \to \infty$ one should have $s \to \infty$ and

$$s = -2k_t^2(1 - \cos\theta_t) \tag{24}$$

$A(s,t) \cong \beta(t)s^{\alpha(t)}$, where $\alpha(t)$ is a pole in the partial scattering amplitude $a_l(t)$ as a function of t, l defined by Eq. (25). Although the region $s > 0$ is unphysical for the process $A + \bar{C} \to \bar{B} + D$, it

$$A(s,t) = \int_{-\infty}^{\infty} \frac{dl(2l+1)}{\sin \pi l} a_l(t) P_l(-\cos\theta_t) \tag{25}$$

is not so for $A + B \to C + D$, and this yields a definite prediction for the asymptotic behavior of $A(s,t)$ for large s and constant t of the kind given by Eq. (26).

$$A(s,t) \simeq \beta(t) s^{\alpha(t)} \tag{26}$$

The Regge pole hypothesis essentially assumes that this behavior is valid for $t < 0$ and not only for $t > 4M^2$, as implied analogously by the $A + \bar{C} \to \bar{B} + D$ reaction.

In this sense the conjecture is a definite statement on the behavior of the scattering amplitude for constant transmitted momentum t and high energies. The exponent $\alpha(t)$ is tied to an angular momentum of an exchanged Regge pole of $(mass)^2 = t (t < 0)$. Since $\alpha(t)$ is not integer in this region, the pole does not correspond to any physical particle. The trajectory $\alpha(t)$ may, however, assume integer values for $t > 0$, and here it may correspond to physical particles of angular momentum $\alpha(t)$ and $(mass)^2 = t$. In this way the outcome of high-energy scattering experiments may be related to $\alpha(t)$ and therefore to a family of particles. As the existence of a trajectory is inferred from the particular potential scattering mechanism, it is clear that a Regge pole ought always to be interpreted as a compound state.

If now a particle lies on a trajectory, there is a strong indication that the particle is not elementary. It has been conjectured by Chew and others that all particles lie on Regge trajectories (or that at least the strongly interacting particles do so) and that the concept of elementarity should be abandoned. Certainly, if the Regge pole hypothesis is correct, physicists will have at least one objective criterion for elementarity. Elementary particles will correspond in general to fixed poles, that is, poles of $a_l(t)$ whose position does not depend on t. SEE ELEMENTARY PARTICLE.

Considerable theoretical and experimental research has been carried out on basis of this conjecture, and although the basic elements for an axiomatic foundation of such a dynamics have not been formulated, from a pragmatic point of view much has been achieved. Much mathematical work has been done to clarify the concept of Regge pole contribution to high-energy scattering for any reaction involving spinning particles, and an equally great effort has been made in the statistical analysis of the experimental data in this scheme by using the most powerful computers available.

It is hoped that this great amount of work will be the basis for a future theory. SEE DISPERSION RELATIONS; SCATTERING EXPERIMENTS (NUCLEI).

Bibliography. P. D. Collins, *An Introduction to Regge Theory and High Energy Physics*, 1977; R. J. Eden et al., *The Analytic S-Matrix*, 1966; R. Omnes and M. Froissart, *Mandlestam Theory and Regge Poles: An Introduction for Experimentalists*, 1963.

PARTICLE ACCELERATORS

Particle accelerator	398
Resonance transformer	441
Tandetron	441
Dynamitron accelerator	443
Cockcroft-Walton accelerator	445
Van de Graaff generator	447
Pelletron accelerator	449
Betatron	451
Alternating gradient focusing	452

PARTICLE ACCELERATOR

Harvey E. Wegner, David J. Clark, Helmut Wiedemann, Hermann A. Grunder, Kenneth Batchelor, Harold E. Jackson, Jr., Karl Strauch, Mark Q. Barton, and Uwe Schumacher

H. E. Wegner wrote the first sections and Electrostatic Accelerators; D. J. Clark, Cyclotrons; H. Wiedemann, Electron Synchrotron; H. A. Grunder, Proton Synchrotron and Heavy-Ion Synchrotron; K. Batchelor, Linear Accelerator; H. E. Jackson, Jr., Microtron; K. Strauch, Principles of Colliding-Beam Systems and Positron-Electron Storage Rings; M. Q. Barton, Proton and $\bar{p}p$ Storage Rings; and U. Schumacher, Electron Ring Accelerator.

An electrical device which accelerates charged atomic or subatomic particles to high energies. The particles may be charged either positively or negatively. If subatomic, the particles are usually electrons or protons and, if atomic, they are charged ions of various elements and their isotopes throughout the entire periodic table of the elements. Before the advent of accelerators the only source of energetic particles for research was the naturally occurring radioactive atoms that emit subatomic particles such as electrons and alpha particles at various energies ranging from kilovolts to over 8 MeV.

The energy of an accelerated atomic or subatomic particle is usually expressed in units of electronvolts (eV). An electronvolt is the amount of energy that a particle with unit charge, such as an electron or proton, receives when accelerated through a potential difference of 1 V. Therefore, when a proton is accelerated with an electrostatic accelerator operating at 1,000,000 V, its energy will be 1,000,000 eV; equivalently, if the particle has q units of charge, its resultant energy in this example will be q MeV. Commonly used multiple units are kiloelectronvolts (keV), megaelectronvolts (MeV), gigaelectronvolts (GeV), or teraelectronvolts (TeV).

Very high-energy atomic and subatomic particles are accelerated in outer space, and constantly bombard the Earth in the form of cosmic rays ranging in energy from hundreds of keV to hundreds of thousands of MeV. These natural sources of particles are uncontrollable and must be used for research in whatever energy, intensity, and kind of natural radiation is available. Particle accelerator devices, on the other hand, allow these same particles to be accelerated to precise energies with complete control of the intensity and energy over wide ranges. The particles may also be directed to collide with specific target materials in whatever way desired so as to greatly expand knowledge of the fundamental interactions of charged particles with each other and with other materials. Moreover, with accelerators it is possible to produce a wide variety of secondary beams of exotic particles with such short half-lives that they would otherwise be unavailable for controlled experimentation. See ELEMENTARY PARTICLE.

Accelerators that produce various subatomic particles at high intensity have many practical applications in industry and medicine as well as in basic research. Electrostatic generators, pulse transformer sets, cyclotrons, and electron linear accelerators are used to produce high levels of various kinds of radiation that in turn can be used to polymerize plastics, provide bacterial sterilization without heating, and manufacture radioisotopes which are utilized in industry and medicine for direct treatment of some illnesses as well as research. They can also be used to provide high-intensity beams of protons, neutrons, heavy ions, pi mesons, or x-rays that are used for cancer therapy and research. The accelerator-produced x-rays are also used for radiographic determination of flaws and structural problems in heavy industrial steel castings and other types of structures.

Particle accelerators fall into two general classes—electrostatic accelerators that provide a steady dc potential, and varieties of accelerators that employ various combinations of time-varying electric and magnetic fields.

Electrostatic accelerators. Electrostatic accelerators in the simplest form either accelerate the charged particle from the source of high voltage to ground potential or from ground potential to the source of high voltage. The high-voltage dc potential may be either positive or negative, and consequently positive or negative particles will be accelerated by being either attracted to or repelled from the high voltage.

All particle accelerations are carried out inside an evacuated tube so that the accelerated particles do not collide with air molecules or atoms and may follow trajectories characterized specifically by the electric fields utilized for the acceleration. Usually the evacuated tube is provided with a series of electrodes arranged to have gradually increasing potentials from ground to the maximum high-voltage potential. In this way, the high voltage is distributed uniformly along the acceleration tube, and the acceleration process is thereby simplified and better control permitted. The maximum energy available from this kind of accelerator is limited by the ability of the voltage generator to provide some maximum high voltage. This limitation in energy was a severe problem in the early days of nuclear and atomic research, when electrostatic accelerators were used, and led to the development of the second kind of accelerator, which uses time-varying electric or magnetic fields, or both, and is not restricted by any particular limit of potential that can be maintained. SEE COCKCROFT-WALTON ACCELERATOR; DYNAMITRON ACCELERATOR; PELLETRON ACCELERATOR; RESONANCE TRANSFORMER; TANDETRON; VAN DE GRAAFF GENERATOR.

Time-varying field accelerators. In contrast to the high-voltage-type accelerator which accelerates particles in a continuous stream through a continuously maintained potential, the time-varying accelerators must necessarily accelerate particles in small discrete groups or bunches. Since the voltage on any given electrode is varying in time, at certain times the voltage will be suitable for acceleration, while at other times it would actually decelerate the particles. For this reason the electrodes must be arranged so that the particle bunches appear in their vicinity only when the voltage is correct for acceleration.

Linear accelerators. An accelerator that varies only in electric field and does not use any magnetic guide or turning field is customarily referred to as a linear accelerator or linac. In the simplest version of this kind of accelerator, the electrodes that are used to attract and accelerate the particles are connected to a radio-frequency (rf) power supply or oscillator so that alternate electrodes are of opposite polarity. In this way each successive gap between adjacent electrodes is alternately accelerating and decelerating. If these acceleration gaps are appropriately spaced to accommodate the increasing velocity of the accelerated particle, the frequency can be adjusted so that the particle bunches are always experiencing an accelerating electric field as they cross each successive gap. In this way modest voltages can be used to accelerate bunches of particles indefinitely, limited only by the physical length of the accelerator construction.

All research linacs usually are operated in a pulsed mode because of the extremely high rf power necessary for their operation. The pulsed operation can then be adjusted so that the duty cycle or amount of time actually on at full power averages to a value that is reasonable in cost and practical for cooling. This necessarily limited duty cycle in turn limits the kinds of research that are possible with linacs; however, they are extremely useful (and universally used) as pulsed high-current injectors for all electron and proton synchrotron ring accelerators. Superconducting linear accelerators have been constructed that are used to accelerate electrons and also to boost the energy of heavy ions injected from electrostatic machines. These linacs can easily operate in the continuous-wave (cw) rather than pulsed mode, because the rf power losses are only a few watts.

Circular accelerators. As accelerators are carried to a higher and higher energy, a linac eventually reaches some practical construction limit because of length. This problem of extreme length can be circumvented conveniently by accelerating the particles in a circular path maintained by either static or time-varying magnetic fields. Accelerators utilizing steady magnetic fields as guide paths are usually referred to as cyclotrons or synchrocyclotrons, and are arranged to provide a steady magnetic field over relatively large areas that allow the particles to travel in a circular orbit of gradually increasing diameter as they increase in energy. After many accelerations through various electrode configurations, the particles eventually achieve an orbit as large as the maximum diameter of the magnetic field and are extracted for use in research and other kinds of applications.

A special kind of circular accelerator for electrons using a static magnetic field is called a microtron. A constant frequency in a small acceleration gap or cavity near the edge of a circular-shaped uniform magnetic field is arranged so that it accelerates the electrons tangentially. Electrons are provided at an energy of one or a few MeV so that they are moving close to the velocity of light. This means that when the electrons are accelerated their velocity can only increase slightly because the velocity of light cannot be exceeded according to the laws of special relativity. The acceleration and increase in energy is consequently accomplished by increasing their mass instead of their velocity. With constant velocity, the acceleration frequency can also be constant and arranged in such a way that the circular orbits increase in circumference by one or more beam bunches or wavelengths on each revolution of larger diameter. Much larger microtrons in a racetracklike configuration of two opposed half-moon-shaped magnets have been built with an electron linac between them and aligned with the outermost racetrack orbit.

A circular accelerator utilizing a time-varying magnetic field that in turn produces an electric field for an acceleration through the induction principle is called a betatron and has been used only for the acceleration of electrons. Betatrons are limited to the acceleration of electrons to energies not much in excess of 300 MeV and have largely been replaced by synchrotron accelerators, which use a time-varying electric field for acceleration rather than a field produced by induction. A few small betatrons designed for 20 to 30 MeV are still used for producing extremely hard x-rays, which are employed in radiographic testing of thick steel castings in industry and other similar applications. S*EE* B*ETATRON*.

Early electron synchrotrons used betatron acceleration in the initial phase of acceleration of each bunch of particles and subsequently continued the acceleration process through energizing of one or more rf cavities through which the particles passed in each orbit. Although electron synchrotrons were formerly used in high-energy physics research and as synchrotron radiation sources, they are now used exclusively as injectors for storage rings.

Practical limitations of magnet construction and cost have kept the size of circular proton accelerators with static magnetic fields to the vicinity of 100 to 1000 MeV. Most circular accelerators can easily operate over a range of 10 to 1 and some as high as 30–50 to 1. However, increasing the size and coupling of smaller to larger machines is necessary for larger ranges in energy. For even higher energies, up to 400 GeV per nucleon in the maximum-size proton accelerator in operation, it is necessary to vary the magnetic field as well as the electric field in time. In this way the magnetic field can be of a minimal practical size, which is still quite extensive for a 400-GeV accelerator (6500 ft or 2000 m in diameter). This circular magnetic containment region, or "race track," is injected with relatively low-energy particles that can coast around the magnetic ring when it is at minimum field strength. The magnetic field is then gradually increased to stay in step with the higher magnetic rigidity of the particles as they are gradually accelerated with a time-varying electric field. Again, when the particles achieve an energy corresponding to the maximum magnetic field possible in the circular guide field, they are extracted for utilization in research programs, and the magnetic field is cycled back down to its low or near-zero value and the acceleration process is repeated.

Focusing. One of the chief problems in any accelerator system is that of maintaining spatial control over the beam; this requires some form of focusing. In addition, the presence of focusing elements makes it possible to trade off beam cross-sectional area and angular divergence as the specific utilization may require.

A wide variety of focusing principles and devices are used in different accelerator types. The natural bowing-out of magnetic field lines in a cyclotron magnet results in a net focusing action as will be discussed below; so also does the electrostatic field line configuration between drift tubes of many simple linear accelerators. But these systems are classed generally as weak-focusing approaches. The particle beams under their control can make relatively large excursions from the desired equilibrium orbits with the consequence that large vacuum envelopes are essential if the beams are not to be lost through collision with the envelope walls.

Fortunately, an alternate strong-focusing approach has made possible very substantial improvements in this area. This approach was described in 1952 by E. Courant, M. S. Livingston, and H. S. Snyder, who reported the invention of electrostatic and magnetic lens systems arranged alternately positive and negative (in the sense of convex and concave optical elements) so that, as in the optical analog, they have a net- and strong-focusing action on charged particle beams. This alternation of positive and negative elements has been implemented in many ways, with many devices and for a variety of geometries. Perhaps its most frequent appearance is in quadrupole (four-pole) or hexapole (six-pole) magnets used as variable-focal-length, variable-astigmatism elements in beam transport both within large accelerators themselves and in extensive systems external to the accelerator.

Superconducting magnets. The study of the fundamental structure of nature and all associated basic research require an ever increasing energy in order to allow finer and finer measurements on the basic structure of matter. Since the voltage-varying and magnetic-field-varying accelerators also have limits to their maximum size in terms of cost and practical construction problems, the only way to increase particle energies even further is to provide higher-varying magnetic fields through superconducting magnet technology, which can extend electromagnetic capability by a factor of 4 to 5.

A large superconducting cyclotron is in operation at Michigan State University, and several others are under construction. A superconducting synchrotron called the tevatron, which accelerates protons to 1000 GeV or 1 TeV, has also been constructed at Fermilab, Batavia, Illinois. An extremely large superconducting supercollider (SSC) is undergoing intensive design studies in the United States, and a similar but smaller machine is being considered in Europe.

Storage rings. Beyond this limit the only other possibility is to accelerate particles in opposite directions and arrange for them to collide at certain selected intersection regions around the accelerator. The main technical problem is to provide adequate numbers of particles in the two colliding beams so that the probability of a collision is moderately high. This intensity is usually expressed as luminosity and is in units of number of particles per square centimeter per second. Such storage ring facilities are in operation for both electrons and protons. Besides storing the particles in circular orbits, the rings can operate initially as synchrotrons and accelerate lower-energy injected particles to much higher energies and then store them for interaction studies at the beam interaction points.

Electron-positron storage rings are also used for the production of synchrotron radiation, ranging from ultraviolet to x-rays, depending on the size of the machine. This unique source of radiation is used in the study of biology, solid-state physics, and atomic and molecular physics.

Collective accelerators. Development of a completely different kind of accelerator, the collective-effect accelerator, has been undertaken in several countries. The idea is based on the fact that electrons, because of their extremely low mass, can be accelerated close to the velocity of light with very modest voltages compared with those required for a particle approximately 2000 times heavier, such as the proton, or for much heavier particles, such as atomic ions. An acceleration voltage of only 0.5 megavolts (MV) will accelerate an electron to seven-tenths the velocity of light, while an effective voltage of over 900 MV is required to achieve the same velocity for protons.

The idea of a collective accelerator is to accelerate a cloud or bunch of electrons containing within the electrostatic well established by its cloud structure one or a few protons or heavier atoms in such a way that the electrical forces providing the containment are sufficiently strong to literally drag the heavy particle along. An electron bunch containing a number of protons or other atoms, when accelerated to 7/10 the velocity of light by 0.5 MV, would carry along the heavier particles at the same velocity equivalent to many hundreds of MeV. The idea is very appealing because such an accelerator would be much simpler than conventional acceleration techniques; however, the system has not yet been proved practical although some devices have accelerated various heavy ions to a few tens of MeV.

Laser-driven accelerators. A much newer accelerator concept based on laser-produced electric fields is under intense study and has largely eclipsed the older collective-effect accelerator work. Basically, an intense laser beam is used to produce an intense alternating electric field in close proximity to an optical grating, or else a beat frequency produced by two lasers produces similar electric fields in a plasma. These electric fields can then in principle be used to accelerate electrons or heavier particles to high energies. While conventional acceleration systems use rf-driven (copper) cavities with dimensions of several inches (many centimeters) to produce the particle-accelerating electric fields, the laser-produced alternating electric fields are produced on a scale of 10^{-10} m. This means that accelerators utilizing this principle, if successfully developed, could accelerate protons to energies of 10–100 GeV in a distance of a few feet (1 m) or so.

Performance characteristics. This introduction has shown that there are many kinds of accelerators, varying in size and performance characteristics and capable of accelerating many kinds of particles at varying intensities and energies, in the form of bunched or pulsed beams of particles or as a steady dc current. **Table 1** lists a few of the more common types of accelerators and their basic performance characteristics. This table is not intended to be complete. The machine characteristics listed are for the largest machine in each class, and many accelerators of a given type can have widely different performance characteristics depending on their use. Accelerators used principally for industrial applications are not listed.

Table 1. Operating characteristics of particle accelerators

Accelerator type	Particle accelerated	Energy range*	Beam current (average; peak), or intensity, luminosity	Duty cycle	Energy spectrum†	Beam geometry	Development status (1985)
ELECTROSTATIC ACCELERATORS							
Tandetron	p,d,α, heavy ions	To 3 MV	1 μA; 10 μA	Continuous	~0.01%	Small focal spot	1–3 MV operating
Cockcroft-Walton	p, d, α, e, heavy ions	To 4 MV	1 mA; 10 mA	Continuous	~0.01%	Small focal spot	4-MeV operating
Dynamitron	p, d, α, e, heavy ions	To 4 MV	1 μA; 50 mA	Continuous	~0.01%	Small focal spot	4-MV operating
Tandem Van de Graaff	p, d, α, e, heavy ions	To 25 MV	1 μA; 50 μA	Continuous	~0.01%	Small focal spot	18-MV operating; 25-MV under construction
Tandem pelletron	p, d, α, e, heavy ions	To 25 MV	1 μA; 50 μA	Continuous	~0.01%	Small focal spot	22.5-MV operating; 25-MV under construction
Vivitron	p, d, α, e, heavy ions	To 35 MV	1 μA; 50 μA	Continuous	~0.01%	Small focal spot	35-MV under construction
TIME-VARYING FIELD ACCELERATORS							
Circular magnetic types (radio-frequency resonance accelerators)							
Microtron	e⁻	To 200 MeV	10 μA; 100 mA	0.1%	0.1%	Small focal spot	30–200 MeV operating
Sector or isochronous cyclotron	p, d, α, heavy ions	To 590 MeV (p)	20 μA; 2 mA	Continuous	~0.01%	Internal target or external beam with small focal spot	590-MeV (p), 30–50 MeV/amu (heavy ions), operating; 200 MeV/amu under construction
Synchrocyclotron	p, d, α, heavy ions	To 1 GeV (p) 50–100 MeV/amu	1 μA; depends on duty cycle	< 10^{-2}	0.1%	External beam of fair collimation at lower intensity	1-GeV (p), mostly closed down
Synchrotron (weak focusing)	p, e, heavy ions	1–6 GeV (p) 2 GeV/amu	0.1 μA; depends on duty cycle	30%	0.1%	Internal targets; external beam of fair collimation at lower intensity	2-GeV/amu heavy ion operating
Alternating-gradient synchrotron	p, e⁺, e⁻, heavy ions: mass 12–32 mass 12–16	10–800 GeV (p) 22 GeV (e⁺, e⁻) 15 GeV/amu 200 GeV/amu	1.0 μA (p) 10^8–10^{10} ions/s 10^8–10^{10} ions/s	30% 30% 30%	0.1% 0.1% 0.1%	(p) internal targets; external beam of fair collimation; external secondary-particle beams	800 GeV (p) operating; 1000 GeV (p), 45 GeV (e⁺, e⁻), heavy-ion machines under construction
Linear accelerators							
Heavy-ion linear accelerator	p, d, α, heavy ions	To 30 MeV/amu	10 μA; 130 mA	~10%	0.5%	Well-collimated and well-focused external beam	30-MeV/amu (heavy ion) operating
Linear accelerator	p	50–800 MeV	1 mA; 100 mA	~10%	0.1%	Well-collimated and well-focused external beam	800-MeV (p) operating
Electron linear accelerator	e⁺, e⁻	6 MeV to 50 GeV 100 GeV (CM) in e⁺, e⁻ collider mode	60 μA; 400 μA 10^{29}/cm²s	~6%	~0.2% 0.06%	Well-collimated and well-focused external beam	50-GeV under construction (SLED mode)
Colliding-beam storage rings							
Electron storage ring	e⁺, e⁻	0.3–45 GeV (CM)	~10^{31}/cm² s	Continuous	0.1%	Small-diameter internal beam	45 GeV operating
Proton storage ring	p, d, α	10–31 GeV	~10^{32}/cm² s	Continuous	0.1%	Small-diameter internal beam	31-GeV (CM) closed down
Proton-antiproton storage ring collider	(pp̄)	540 GeV (CM)	~10^{29}/cm² s	Continuous	0.1%	Small-diameter internal beam	2-TeV (CM) under construction

*Voltage range is given for electrostatic accelerators. amu = atomic mass unit. CM = center of mass.
†Spread in energy of beam expressed as a percentage of total energy of beam; that is, 1% for the cyclotron means for a 1-MeV beam a spread in energy of 0.01 MeV.

ELECTROSTATIC ACCELERATORS

Electrostatic accelerators are used to accelerate atomic or subatomic particles to high energies by means of a high voltage potential that is maintained through some electrical or mechanical means of transporting charge from ground to the high voltage potential. Modern machines of larger size all use mechanical charge transport, but electrical transport systems are widely used in small Cockcroft-Walton or dynamitron-type electrostatic accelerator units. The most commonly used electrostatic generator was invented in the 1930s by R. J. Van de Graaff and utilizes a high-speed rubberized-fabric belt to transport the charge. Many machines utilizing this principle have been homemade or built by various research groups throughout the world.

Most electrostatic accelerators are housed inside a large high-pressure vessel that is filled to a pressure as high as 15 atm (220 psi or 1.5×10^6 pascals) with very dry insulating gas. The insulating gas may be pure sulfur hexafluoride (SF_6) or a mixture with carbon dioxide (CO_2) and nitrogen (N_2) at a dew point of $-60°F$ ($-51°C$) or less. This high-pressure insulating gas allows the machine to be housed in a much smaller space than would be possible in air at atmospheric pressure (for example, 6-ft or 1.8-m clearance from the high-voltage terminal to ground has been necessary to insulate successfully up to 17 MV).

The pelletron electrostatic accelerator utilizes a chain-charging system instead of the rubberized-fabric belt. Instead of the charge being sprayed onto the charging belt system from sharp corona points, each metal link is charged and discharged by induction methods so that no electrical sparking or erosion is involved.

Tandem design. These large electrostatic accelerators are constructed in the form of tandem electrostatic accelerators that utilize the high voltage potential for acceleration more than once; the first of these was installed in Chalk River Nuclear Laboratories of Atomic Energy of Canada in the late 1950s, although the concept in its present form was presented by Willard Bennett before World War II and independently by Luis Alvarez after World War II.

The tandem electrostatic accelerator gets its name from the fact that it is effectively two electrostatic accelerators back to back, arranged so that the high voltage potential is first used to accelerate negative ions from ground potential up to the high positive potential. The charge of the particle is then switched from negative to positive, and the positive ions are additionally accelerated from the high voltage potential back to ground, continuing in the same direction as the original acceleration. The charge-changing inside the high-voltage terminal allows the high voltage to be used to accelerate the particles twice and consequently produce much higher energies than would be possible in the conventional single-stage electrostatic accelerator. A third stage of acceleration can be accomplished by first accelerating negative particles from high voltage to ground and then injecting them into a conventional tandem instead of utilizing negative particles injected directly from an ion source at low energy as is more customary.

The charge-changing in the high-voltage terminal is accomplished by either a gas or foil stripper. The negative ions which are accelerated from ground potential to the high-voltage terminal have enough energy to penetrate either thin foils or a high-pressure gas region within an open-ended tube. In so doing, the collisions with other atoms violently strip away many electrons from the negative ion, leaving the ion with a net positive charge. In cases of the lighter heavy ions such as carbon or oxygen, it is possible to strip away all the electrons from the atom, leaving the bare nucleus of the atom, and thereby accomplish the maximum possible acceleration, represented by the high charge state of the accelerated ion. Each charge provides a particle with an energy in MeV equivalent to the voltage of the high-voltage terminal in MV; therefore an oxygen ion with charge 8+, $^{16}O^{8+}$, receives an energy of 80 MeV from its acceleration through a potential of 10 MV. Secondary strippers can be placed farther down the acceleration tube from the high-voltage terminal to produce even higher charge states for heavier ions and consequently higher energies. Zirconium atoms, for example, have been accelerated to energies in excess of 350 MeV by such multiple stripping techniques.

Folded tandem design. The folded tandem design has been developed for large electrostatic accelerators. They are constructed like a large single-stage electrostatic accelerator; however, the column which supports the high-voltage terminal contains two acceleration tubes—one for the acceleration of negative ions from ground potential to the high-voltage terminal, and the other for acceleration of positive ions from the high-voltage terminal down to ground potential. In **Fig. 1** negative ions produced in the ion source on the left are directed vertically by the magnet and then accelerated up to the high-voltage terminal. The large magnet inside the terminal bends the ions 180° to a downward direction after they have been stripped to high positive-charge states. They are then accelerated in the second acceleration tube on the right, back down to ground potential to achieve their final energy, and then directed horizontally by a final magnet to experimental stations. The advantage of the folded design is that both the accelerator and building enclosure are much shorter and the complex negative-ion source is at ground level instead of several stories above the top of a very tall accelerator. However, the diameter does have to be increased somewhat, at least in the vicinity of the high-voltage terminal. The added problem of

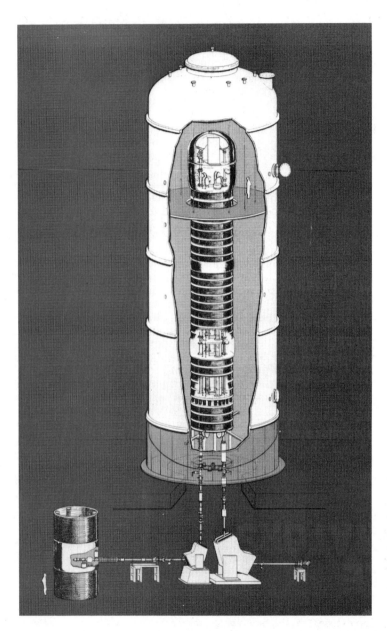

Fig. 1. The 25-MV pelletron (up-down) tandem accelerator designed for the Oak Ridge National Laboratory. The people on the left of the ion source and on the elevator platform near the high-voltage terminal indicate the large size of this machine. (*National Electrostatics Corp.*)

providing a reliably operating large bending magnet inside the high-voltage terminal is a complexity not found in the conventional tandem, in which the particles pass from ground-to-ground potential in a straight line.

Large machines. The first of the large electrostatic accelerator facilities, completed in 1970, is the three-stage tandem facility at Brookhaven National Laboratory in Upton, New York. It comprises two of the largest tandem Van de Graaff machines available at the time connected together to operate in a three-stage mode. The three-stage operation provides heavy ions at energies comparable to a larger two-stage tandem operating at 18–19 MV.

These two machines also have a unique operating capability in a four-stage mode where the last stage is used to decelerate the ions. The first machine operates as a conventional tandem, injecting partially stripped positive heavy ions into the second machine. The second machine is operated with a negative terminal so that positive ions are accelerated to much higher energies at the terminal, where they are then completely stripped to "bare" nuclei. These highly charged bare nuclei then decelerate back to ground potential, coming out of the second machine with very low energies that can be adjusted by the specific positive and negative potentials of the two tandem accelerators. In this way totally stripped ions up to chlorine (17+) have been produced at energies as low as 2.5 MeV. These unusual ions are used in a variety of atomic physics measurements.

Other large machines in operation are the 20-MV folded tandem at the Japan Atomic Energy Research Institute in Tokai that operates up to 18 MV; the 30-MV vertical tandem at the Daresbury Nuclear Physics Laboratory in Daresbury, England, that operates up to 20 MV; and the 25-MV folded tandem at the Holifield National Laboratory in Oak Ridge, Tennessee, that operates up to 22.5 MV (Fig. 1). None of these machines has been able to operate at its maximum design voltage; however, the machines are all used continuously in research programs, and their maximum voltage capabilities have also been upgraded and improved. The Oak Ridge machine is also the first large tandem used to inject a cyclotron for a boost in heavy-ion energy of 3 to 4, and operates more than half the time as an injector rather than directly as a research machine.

The 35-MV Vivitron tandem, under construction at the Center of Nuclear Research in Strasbourg, France, is a completely new kind of electrostatic structure, invented by Letournel, that greatly reduces the size and cost as compared to conventional machines. The machine is operated with a charging belt and is long enough (164 ft or 50 m) to accommodate a reasonable voltage gradient on the acceleration tube; however, it is much smaller in diameter (25 ft or 7.6 m) than would normally be required (40–50 ft or 12–15 m). It contains a series of seven concentric, cage-like, conducting shells, each of which is connected to uniform symmetric positions along the high-voltage columns from the central high-voltage terminal to ground. The concentric shells, called porticoes, are in turn supported by specially designed insulators between the shells from the terminal and column to the outer wall of the enclosing pressure vessel.

Control and adjustment. These large electrostatic accelerators can be controlled very precisely in terms of the absolute voltage of the high-voltage terminal, so that particles can be accelerated with an accuracy of better than 1 part in 10^4 on a routine basis and better than 1 part in 10^5 when additional control measures are invoked.

Since singly charged negative ions can be made of most elements or some compound of the elements, it is a straightforward procedure to accelerate almost any atom throughout the periodic table from hydrogen to uranium. The large electrostatic accelerator facility at Brookhaven National Laboratory has accelerated as many as 40 different elements and 50 different isotopes throughout the periodic table. Since the electrostatic accelerator identically accelerates different elements or mass particles having the same charge, only minimal adjustments are required in order to change from the acceleration of one mass particle to another. In contrast, the accelerators that utilize varying magnetic and electric fields require considerably more adjustment because of the different velocities encountered as the heavy ions are changed. Although the adjustments are more complex, the field-varying accelerators are not limited in the maximum energy obtainable by the high voltage potential, unlike the electrostatic accelerators.

Dual accelerator systems. Plans have been undertaken at several electrostatic accelerator facilities for various kinds of field-varying accelerators as energy booster accelerators, so that their heavy-ion energy range can be greatly extended. In this dual accelerator arrangement, the tandem accelerator is arranged to inject a pulsed beam matched to the timing requirements of the field-varying accelerator. At present, both a small, single-stage and tandem electrostatic accelerators are used to inject an open sector cyclotron at the Hahn-Meitner Institut in Berlin. Tandem accelerators are also used to inject superconducting linacs at the Argonne National Laboratory in Illinois and at the State University of New York at Stony Brook, and a room temperature linac at the Max Planck Institut für Physik in Heidelberg, Germany. Other tandem facilities also have superconducting linacs and cyclotrons under construction for injection and subsequent energy boosting of their heavy ions. Also under construction is a 2000-ft (610-m) beam transport line that will allow the three-stage tandem accelerators at Brookhaven National Laboratory to inject the alternating gradient synchrotron (AGS). The AGS normally provides 30 GeV protons; however, when injected by the tandems, it will provide heavy ions ranging from carbon (mass 12) to sulfur (mass 32) at energies of 15 GeV per mass unit (that is, 180 GeV for carbon and 480 GeV for sulfur).

Applications. The larger electrostatic accelerators are used primarily for basic research in the study of nuclear structure and heavy-ion reactions, including fusion and fission processes. However, the heavy-ion research interests have moved rapidly to much higher energies that are beyond the capability of electrostatic machines as indicated by the fact that most large machines are now injecting, or are planned for future injection, of other linacs or circular machines to provide an energy boost to the region of research interest. Although the role of large electrostatic accelerators in the forefront of heavy-ion research accelerators has waned, they are still the most versatile accelerators in operation in their energy region for any kind of research from solid-state to atomic and nuclear physics. They also perform outstandingly well as heavy-ion injectors for other kinds of accelerators.

The smaller machines are used throughout industry for many applications, including neutron and x-ray production for diagnostic measurements, polymerization and other treatments of plastics and various materials, ion implantation, thin-film analysis, and many other related applications.

CIRCULAR ACCELERATORS

Circular accelerators utilize a magnetic field to bend charged-particle orbits and confine the extent of particle motion.

Cyclotrons. The cyclotron is a circular accelerator in which the particle orbits start at the center and spiral outward in a guide magnetic field which is constant in time. The cyclotron concept was put forward by E. O. Lawrence in 1930; the exploitation of this concept is still the basis for major developments in accelerator science. The survival of the cyclotron idea has resulted from successive blending of the basic concept with a series of later ideas (phase stability, strong focusing, and superconductivity). In this process the classic Lawrence cyclotron of the 1930s evolved into the synchrocyclotron of the 1950s, then into the isochronous cyclotron of the 1960s, and finally into the superconducting cyclotron of the late 1970s.

Basic principles. The basic cyclotron concept is given by Eq. (1), the Lawrence equation

$$\omega_c = \frac{qB}{m} \qquad (1)$$

for the angular frequency of rotation of a particle in a plane perpendicular to a magnetic field, where q and m are the charge and mass of the particle and B is the magnetic induction of the perpendicular magnetic field. (SI or mksa units are used throughout this section of the article.) This equation can be derived by noting that (1) $\omega = v/r$, where v is the linear velocity and r is the orbital radius; (2) the magnetic force is perpendicular to the velocity and is given by qvB; (3) this perpendicular force produces a centripetal acceleration and, from Newton's second law, gives $qvB = mv^2/r$. Solving the last equation for v/r leads directly to Eq. (1). The striking statement of Eq. (1) is that if field, charge, and mass are all constant, then angular velocity is likewise constant and, in particular, does not depend on the linear velocity. Faster-moving particles travel on circles of larger radius—the increase in the length of a revolution due to the larger radius just matches the increase in speed, and fast or slow particles thus require the same amount of time to make a 360° revolution. Thus, an accelerating electrode operating at constant frequency can be used. This is the key idea of the Lawrence cyclotron.

The essential features of such a cyclotron are diagrammed in **Fig. 2**. A pair of D-shaped accelerating electrodes, called dees or D's, is electrically attached to an rf voltage source whose frequency is matched to the rotation frequency of the particle in a surrounding magnetic field. Ions are formed in the ion source at the center and are drawn out by the high voltage on the facing dee. Particles which cross the gap between the two dees at a time when the rf voltage produces an accelerating electric field are accepted for acceleration; the remainder are lost. Beyond the accelerating gap the particle passes inside the dee, which is an electrically shielded region; at the same time it is pulled into a circular path by the perpendicular magnetic field, and after 180° it arrives again at the gap between the dees. Since the frequency of the rf source which drives the dees was selected to match the rotational frequency of the particle, the voltage between

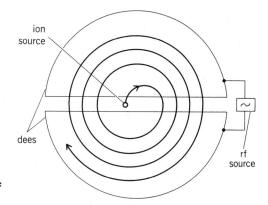

Fig. 2. Principle of the cyclotron. An ion in the accelerator follows the arrowed path. The magnetic field is perpendicular to the page.

the dees will have reversed while the particle was making its 180° turn, and therefore the particle is again accelerated as it crosses from dee to dee. This process can clearly be repeated as often as desired. The particle will thus be repetitively accelerated and, as it speeds up, will gradually move on circles of larger and larger radius. Finally the particle will come to the limit of the magnetic field. At this point an additional electrode—referred to as a deflector—can be inserted to direct all particles out along a single path. **Figure 3** shows the dees in the magnetic field provided by an electromagnet.

Limitations. This theoretical picture of exactly synchronized orbits and voltage is an oversimplification, and more detailed investigation by H. A. Bethe and M. E. Rose in 1937 revealed

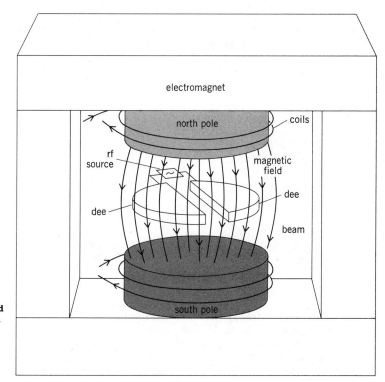

Fig. 3. Cyclotron, showing a particle beam accelerated from the center by dees. A vacuum chamber (not shown) encloses the dee system and beam. (*Lawrence Berkeley Laboratory*)

two fundamental problems with the cyclotron concept. First of all, as any particle speeds up, its mass increases according to Albert Einstein's equation, Eq. (2), where m_0 is the particle's rest

$$m = \frac{m_0}{\sqrt{1 - v^2/c^2}} \qquad (2)$$

mass and c is the speed of light. When this accurate expression is inserted into Eq. (1), the rotation frequency then depends on velocity through the v^2/c^2 term, although for low velocities the dependence is weak because the factor v/c is small. Nevertheless, in a constant magnetic field, the gradual increase in mass causes the rotation frequency to steadily decrease as the particle speeds up, and the particle then gradually lags behind a constant rf frequency. Ultimately the particle will reach an accelerating gap so late that the voltage will have already shifted to the reverse direction, and thereafter the particle will steadily slow down at successive gaps. The mass increase thus sets a limit on the acceleration process.

A second limitation comes from the magnetic field factor in Eq. (1). Thus far, motion in a plane has been assumed; however, to prevent particles from drifting away from the plane due to their initial velocities there must be a restoring force which pushes particles back toward the plane. This is called axial focusing. A way to obtain the needed restoring force is to make the magnetic field lines bend (**Fig. 4**). The magnetic force (always perpendicular to the field lines) slants downward for an orbit above the midplane and upward for an orbit below the midplane, so that a particle displaced from the desired plane always feels a component of the magnetic force pushing it back toward the plane. From Fig. 4 it can be seen that such a restoring force requires the magnetic field lines to bend toward the axis of rotation of the particle. However, allowed magnetic field shapes are limited by Maxwell's equations; applying these equations, one finds that when lines of force bend toward the axis as in Fig. 4, the field strength perpendicular to the plane of motion must decrease as the radius becomes larger, producing a slowing-down according to Eq. (1). This effect adds to that of the mass increase. In their 1937 paper, Bethe and Rose calculated both of these effects for the Lawrence cyclotron and concluded that such a cyclotron could not achieve an energy greater than 10 MeV per nucleon, corresponding to a mass increase of about 1%. A Lawrence cyclotron, constructed at Oak Ridge, in fact achieved an energy of 23 MeV per nucleon (23-MeV protons) by raising the voltage on each accelerating electrode to 200 kV; this energy was much higher than Bethe and Rose thought possible, but actually in precise accord with the Bethe-Rose prediction when corrected for the higher dee voltage.

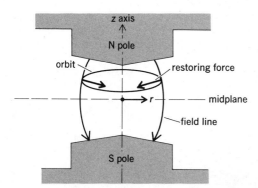

Fig. 4. Focusing forces in an axially symmetric magnetic field. If the field lines curve as shown, toward the axis, the magnetic force has a component parallel to the midplane (which holds the particle in its circular orbit) and perpendicular to the midplane (which pulls the particle back toward that plane).

Synchrocyclotron. A way to avoid the Bethe-Rose energy limit was proposed in 1946 in independent papers by E. M. McMillan in the United States and V. Veksler in the Soviet Union: namely, to slow down the frequency of the rf power source at the same rate as the slowing down of the rotational frequency of the particle. The usefulness of this process depends on the question of stability. Particles which are not exactly at the design values of energy and phase will either oscillate about the design values (stability) and accelerate, or diverge from the design values (instability) and not accelerate. **Figure 5** sketches the ideas involved in establishing that the accelerating process is stable. The frequency of the driving oscillator is adjusted to just match the rotation frequency of a reference particle, shown in Fig. 5. Its phase is called the synchronous phase ϕ_s.

Particles which arrive at the accelerating gap before the synchronous particle receive a larger accelerating voltage and energy gain than the reference particle; this causes their mass to increase faster than that of the reference particle, which causes them to rotate more slowly according to Eq. (1). The larger energy gain causes the early particles to reach a larger radius and thus a lower magnetic field than that of the reference particle. This also causes them to rotate more slowly according to Eq. (1). They thus wait for the reference particle to catch up and are gradually restored to the synchronous phase. Similarly, particles which reach the accelerating gap late receive a smaller energy increase than the reference particle. The smaller mass increase and higher magnetic field cause them to rotate faster and enable them to catch up with the reference particle. This restoring action relative to the phase of the synchronous particle is referred to as phase focusing, and the idea is referred to as the principle of phase stability. This is the basic concept in the frequency-modulated cyclotron or synchrocyclotron. **Figure 6** shows the cyclic nature of the acceleration process in a synchrocyclotron. This concept provided the leading advances in energy in the post-World War II years. The size of such cyclotrons is limited only by cost and construction difficulty. The largest which have been built achieve an energy of about 1000 MeV for protons.

In changing from the Lawrence cyclotron to the synchrocyclotron, however, a very valuable beam property is given up, namely, beam intensity. In the Lawrence cyclotron, particles can leave the ion source on every rf cycle and start their acceleration journey, because every rf cycle is exactly identical. The synchrocyclotron is, in contrast, a batch system (Fig. 6). A group of particles leave the source, and the frequency of the accelerating system is then steadily lowered to stay matched with those particles while they are being accelerated. During this time other particles "wanting" to leave the ion source "see" a frequency which no longer matches their rotation frequency and must therefore "wait" until the reference group has been accelerated all the way to full energy and the rf returned to the frequency corresponding to the rotation of particles just leaving the source. This results in a large intensity loss. The beam coming from a synchrocyclotron typically has about 1/100 as many particles per second as the beam from a Lawrence cyclotron under comparable conditions. This low beam intensity became the limiting factor in a major class

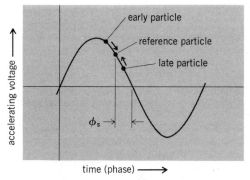

Fig. 5. Principle of phase stability in the synchrocyclotron. The reference particle has resonance energy and synchronous phase. The early particle accelerates faster to larger radius, where it has a lower rotation frequency which moves it later in time toward the central particle. The late particle is correspondingly moved earlier in time toward the central particle.

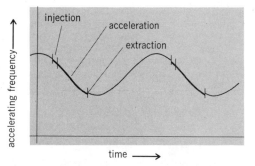

Fig. 6. Accelerating frequency as a function of time in the synchrocyclotron. The frequency decreases during acceleration. It matches the particle's frequency, which decreases because of mass increase and magnetic field reduction with increasing radius.

of precise physics experiments which require stringent filtering of beams or which use primary proton beams for production of secondary beams such as pions or muons.

Isochronous cyclotron. Responding to this experimental need, the isochronous or sector cyclotron was introduced in the late 1950s. Although the basic ideas for the isochronous cyclotron had actually been put forward in 1938 by L. H. Thomas (immediately after the Bethe-Rose paper), its practical application was delayed because of the complicated equations involved. Computers were not then available to solve the equations numerically and, as a result, Thomas's idea was generally overlooked for nearly two decades. Thomas's idea can be understood by starting with Eq. (1). As the particle speeds up, its mass increases, as discussed earlier, and this contributes a slowing-down effect which cannot be avoided; however, if magnetic field B is increased in a compensating way, then the frequency is again constant. Since faster (heavier) particles move in circles of larger radius, a magnetic field is required whose strength increases with the radius. Such a field can readily be built, but the discussion of Fig. 4 applies: if the strength of the magnetic field increases with radius, then, according to Maxwell's equations, the field lines must curve away from the axis and therefore produce force components pushing displaced particles further away from the central plane. In the absence of other restoring forces, the number of particles surviving the acceleration process would then be extremely small.

The discussion in Fig. 4, however, is specifically applicable only to fields which are axially symmetrical. If azimuthal variations are included in the magnetic field, additional terms in Maxwell's equations come into play, and upon analysis one finds additional focusing forces. The origin of these forces is shown in **Fig. 7**, which shows a particularly simple version of the azimuthally varying field, or isochronous, cyclotron. In this example the magnetic field comes from three wedge-shaped magnets which meet at a point in the center. In the region between the magnets, the orbits are an approximate straight line (since very little magnetic field is present in this region). They bend sharply through 120° in the region within the magnets and continue to repeat this pattern. The orbit crosses the edge of each sector at a nonperpendicular angle denoted by ϕ in Fig. 7a. As a result, the orbit velocity at the field edge has a component $v_r = |v| \sin \phi$ in the radial direction. As indicated in Fig. 7b, the magnetic field at this point has an azimuthal component B_θ due to the inevitable bowing of the field lines at the edge of any magnet. This component is pointed away from the magnet above the median plane and toward the magnet below the median plane. Taking the vector product of the r component of velocity times the θ component of the field gives an axial force F_z toward the median plane which pushes particles which are out of the median plane back into the plane. It is easy to verify that the z force is likewise focusing at the point where the orbit leaves the wedge since both the r component of velocity and the θ component of the field have reversed direction. This "edge" focusing is then an additional force that is introduced by the sector structure; it can be used to override defocusing which comes from having the average field increase with radius in order to maintain constant angular frequency in Eq. (1).

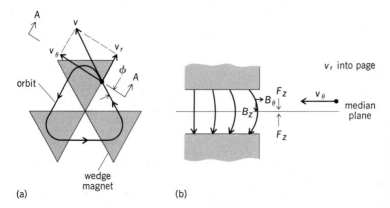

Fig. 7. Additional axial focusing in an azimuthally varying field. (*a*) View along axis of cyclotron. (*b*) Section *A-A* through median plane of *a*.

PARTICLE ACCELERATORS 411

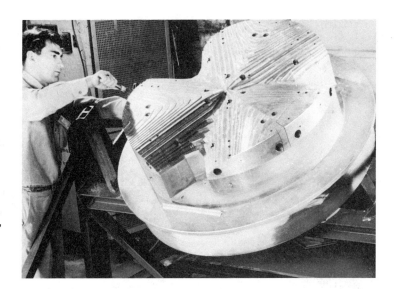

Fig. 8. Magnet pole for the first isochronous cyclotron, which operated at the Lawrence Berkeley Laboratory in 1950. (*Lawrence Berkeley Laboratory*)

Figures 8, **9**, and **10** give an idea of the broad spectrum of sizes and types of isochronous cyclotrons. Figure 8 shows the pole tip of the first isochronous cyclotron, which was put into operation in 1950 in a classified program at the Lawrence Berkeley Laboratory, but not generally known until declassified and published in 1956. The elaborate shaping of the pole gave an approximately sinusoidal variation of field strength with angle, the specific shape studied by Thomas. (Later, computer studies indicated that almost any azimuthal form for the field variation would produce focusing, and elaborate pole tip contours were thereafter unnecessary.) Figure 9 shows the basic magnet core of a typical isochronous cyclotron for nuclear physics research, the

Fig. 9. Workers assembling the yoke for a typical isochronous cyclotron, the 50-MeV variable-energy cyclotron at Michigan State University. (*Michigan State University*)

Fig. 10. Partially assembled magnet for the TRIUMF cyclotron, showing the spiral iron sectors of the lower pole. (*TRIUMF, University of British Columbia*)

50-MeV machine at Michigan State University. This cyclotron has achieved an overall energy resolution of 1 part in 20,000. Figure 10 shows the magnet for the largest isochronous cyclotron (56 ft or 17 m in diameter), that of TRIUMF Laboratory, Vancouver, British Columbia, Canada. The sectors in this design have a spiral shape to increase the edge axial focusing. Many isochronous cyclotrons use the spiral concept. This cyclotron accelerates negative hydrogen ions to an energy of 500 MeV. The cyclotrons at TRIUMF and at Zurich, Switzerland, have produced over 100 microamperes of proton beams, and are known as meson factories, since large quantities of mesons are produced when these beams hit a target. A high-intensity injector cyclotron, under construction at Zurich, will increase the proton current from 100 μA to 1 mA at 590 MeV. *See Meson*.

Several laboratories have multistage cyclotron systems. At the GANIL national laboratory in Caen, France, a small cyclotron injects a large 400-MeV cyclotron, which then injects another 400-MeV cyclotron. This system is designed for heavy-ion beams, using stripping to higher charge state between the large cyclotrons to produce higher energies. Other laboratories inject a cyclotron with an electrostatic accelerator by using stripping between accelerators.

Superconducting cyclotron. The superconducting cyclotron is another major evolution of the cyclotron concept. In contrast to earlier cyclotron concepts, the superconducting cyclotron is mainly a technical change rather than a conceptual one. The only superconducting element in the superconducting cyclotron is in fact the main coil, which is typically housed in an annular cryostat (**Fig. 11**). Conventional room-temperature components, including pole tips, accelerating system, vacuum system, and ion source, are inserted in the warm bore of the cryostat from top

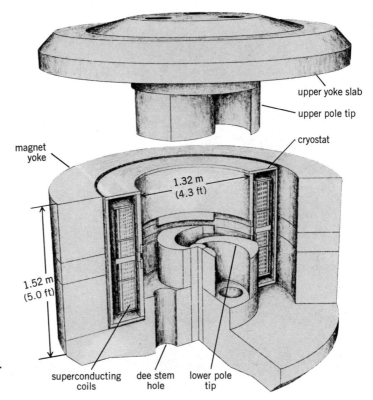

Fig. 11. Conceptual drawing of $K = 500$ superconducting cyclotron at Michigan State University. The upper pole cap is shown in the raised position used for maintenance. (After H. G. Blosser, The Michigan State University superconducting cyclotron program, IEEE Trans. Nucl. Sci., NS-26(2), pt. 1:2040–2047, 1979)

and bottom. The superconducting coil allows the strength of the magnetic field to be greatly increased up to the level of approximately 50,000 gauss (5 teslas), or three times higher than was previously typical of cyclotrons. This increase in field reduces the linear size of the cyclotron to approximately one-third, areas to approximately one-ninth, and so on, compared to a normal cyclotron of the same energy. The result is a large reduction in the cost of many cyclotron components. The first of the superconducting cyclotrons to come into operation is the 500-MeV machine at Michigan State University (Fig. 11 and **Fig. 12**). A larger superconducting cyclotron is under construction that will be injected by the first, and will boost heavy ions to energies as high as 200 MeV per nucleon. Several other such cyclotrons are being constructed at other laboratories. The superconducting cyclotron has a variety of potential applications, particularly for accelerating heavy nuclei to energies needed for nuclear reactions. SEE NUCLEAR REACTION.

Ion sources. An important property of an accelerated ion is its charge state, Q. The maximum energy E of an ion in a cyclotron is given by Eq. (3), where K is the energy constant (usually

$$E = \frac{Kq^2}{m} \quad (3)$$

in MeV) and q and m are in units of proton charge and mass. This equation comes from the relations $E = \frac{1}{2} mv^2$, $v = \omega r$, and Eq. (1), setting B equal to the maximum field available, and r equal to the maximum radius. For protons, $q = m = 1$ and thus $E = K$. For heavier ions there is a strong incentive to produce high q ions with the ion source, since $E \sim q^2$.

The standard ion source is based on the principle of the Penning ion gauge (PIG), in which electrons oscillating in a magnetic field ionize the source gas. The highest-charge-state ions produced by a PIG source in useful intensities are, for example, $q = 5$ for nitrogen (N^{5+}). New concepts in ion sources have been developed which will produce N^{7+} (fully stripped) beams,

Fig. 12. Magnet of $K = 500$ superconducting cyclotron at Michigan State University with upper pole cap raised. (*From H. G. Blosser, The Michigan State University superconducting cyclotron program, IEEE Trans. Nucl. Sci., NS-26(2), pt. 1:2040–2047, 1979*)

doubling the nitrogen energy according to Eq. (3). These advanced sources are called the electron cyclotron resonance (ECR) source and the electron-beam ion source (EBIS). The ECR source uses microwave-heated electrons in a magnetically confined plasma to produce high-charge-state ions. The EBIS uses an electron beam to ionize the gas to very high charge states (Xe^{52+}). These advanced sources offer important opportunities to increase the heavy-ion energies of existing cyclotrons, and are also of interest for use at other types of accelerators such as synchrotrons. SEE ION SOURCES.

Applications. There are approximately 100 cyclotrons in operation. They are used principally for basic research in nuclear physics and chemistry, aimed at understanding the structure of nuclei, nuclear forces, and nuclear reactions. Practical applications are also pursued in many laboratories and hospitals including isotope production for medical use, secondary neutron beams for cancer therapy, and analysis of environmental material. More than 30 cyclotrons are used primarily for medical applications. SEE NUCLEAR STRUCTURE.

Electron synchrotron. An electron synchrotron is a circular accelerator optimized to accelerate electrons or positrons to high energies. The acceleration is achieved as the particles pass through electric fields created in resonant rf cavities. A string of bending magnets along a circular path forces the particles to travel through the accelerating cavities at each revolution. In this way each particle is accelerated many times during the accelerating cycle by the same cavity, and even a modest accelerating field will eventually lead to a high particle energy.

To make this acceleration mechanism function properly, certain conditions have to be met. The oscillating field in the cavity has to be in the accelerating phase every time a bunch of particles arrives. Thus the time it takes the particles to orbit once around the ring must be an integral multiple of the rf period (the synchronicity condition). In an electron synchrotron the particles are injected at some tens of MeV and therefore always have a velocity close to the speed of light. As a consequence the revolution frequency of the particles is constant, and it is possible to use a fixed-frequency accelerating cavity. This is the primary difference between electron synchrotrons and proton synchrotrons or cyclotrons, where the frequency of the accelerating radio-frequency fields must be adjusted during acceleration by factors of 10 or more. A linear accelerator or a microtron is usually used as a preaccelerator to inject the particles into the synchrotron at an energy of tens or hundreds of MeV.

In the past, electron synchrotrons have been used extensively for research in high-energy physics and as sources of synchrotron radiation. However, full-time synchrotrons are now constructed and used exclusively as injectors into storage rings. (Storage rings may also operate as synchrotrons for a few minutes while they are being filled.) Electron synchrotrons have been built in sizes from 6.5 ft (2 m) in diameter for an energy of 100 MeV, up to 650 ft (200 m) in diameter at Cornell University (**Table 2**) for a maximum particle energy of 12 GeV.

Table 2. Some parameters of electron synchrotrons

Parameters	Cornell University	DESY II*	National Synchrotron Light Source (NSLS)
Energy, GeV	12	9.2	0.75
Radius, ft (m)	328 (100)	153 (46.6)	14.8 (4.51)
Cycling rate, Hz	60	12.5	1
Maximum magnet field, teslas	0.4	1.13	1.31
Injection energy, MeV	150	55 or 200	70
Particles per pulse	3×10^{10}	4×10^{10}	10^{10}

*Deutsches Elektronen Synchrotron, located in Hamburg, Germany.

Acceleration and beam control. The operation of the electron synchrotron is determined by its accelerating cycle, which may range from a few milliseconds up to seconds. At the beginning of a cycle the preaccelerator is triggered to produce a particle beam that is injected into the synchrotron. The beam from the injector will have certain characteristics as required for optimum acceptance by the synchrotron. The synchronicity condition is fullfilled only for particles which arrive at the right time for acceleration in the cavity. The ratio of the radio frequency to the revolution frequency is called the harmonic number and is equal to the maximum number of places (called buckets) around the ring where particles can be placed for acceleration. While these buckets rotate around the ring at nearly the speed of light, the injector has to deliver a beam structured and timed so that the particles arrive at the injection point at the same time as the buckets arive. Either one or all of the buckets can be filled with particles.

While the synchronicity condition is very important, it has to be fulfilled only approximately. The principle of phase stability ties together the characteristics of the magnetic guide field and the synchronicity condition in such a way that the particles stay trapped close to and oscillating about the ideal equilibrium position. This is important during the accelerating process. As the magnetic guide field is increased, the principle of phase stability adjusts the particle energy to be proportional to the rising magnetic field. After executing many turns during which the particle energy experiences increments of up to a few MeV per turn, the particles reach the maximum design energy.

At this point the particle beam is extracted and guided to an experimental station or to a storage ring. The extraction is commonly achieved by the use of a pulsed high-field kicker magnet that deflects the beam out of the synchrotron. After extraction the magnetic guide field is reduced again to the injection condition, and a new accelerating cycle can begin.

The maximum energy that can be achieved in a synchrotron is limited by either the maximum magnetic field necessary to keep the beam on its orbit or by the effect of synchrotron radiation. As the particles are made to follow a closed orbit, they lose energy by way of this radiation. The energy loss per turn increases as the fourth power of the energy, and the maximum energy is reached when this energy loss is as large as the maximum energy gain per turn in the accelerating cavities.

While a stronger magnet and rf system could increase the maximum energy somewhat, there is another fundamental limit on this energy. During the first phase of an accelerating cycle the beam cross section shrinks from its value at injection (typically a few square millimeters), inversely proportional to its energy. At some point, however, the statistical emission of synchrotron radiation photons causes a rapid increase in the beam cross section as the energy is raised. The limit is reached when the beam fills the available aperture in the vacuum pipe.

The maximum intensity in a synchrotron is determined by the performance of the preinjector as well as by beam instabilities. Individual particles can be lost due to collisions with other particles or due to the electromagnetic interaction of all particles in one bunch with the surrounding vacuum chamber or with other bunches. In all cases the limitation on the beam current is more severe at low energies than at high energies, which is why synchrotrons are used as a step-up injector between the preaccelerator and a storage ring.

Magnet system. The magnetic guide field in a synchrotron consists of dipole fields to keep the particles on a circular path and a focusing field generated by quadrupoles to keep the beam from diverging. In the older synchrotrons both bending and focusing were performed by one type of magnet, and the resulting magnet lattice was called a combined-function lattice. As magnet technology progressed and more flexibility was desired, electron synchrotrons began to employ the so-called separated-function lattice, where bending and focusing is done by different magnets with separate power supplies. To enhance the beam current capability, the newer synchrotrons also use sextupole magnets to correct for effects caused by the energy spread in the beam. In analogy to light optics, this procedure is called the chromatic correction of particle beam focusing.

Vacuum system. To avoid loss of particles during acceleration, the particles circulate in a vacuum pipe embedded in the lattice magnets. In older synchrotrons this vacuum pipe had to be made from an insulating material like glass, epoxyrecin, and more recently ceramic. This was required to allow the rapid cycling of synchrotrons at 50 or 60 Hz without creating problems due to eddy currents caused by the magnetic fields in the vacuum chamber.

Fig. 13. The 750-MeV electron synchrotron at the National Synchrotron Light Source (NSLS) at Brookhaven.

In the newer synchrotrons used exclusively as injectors for storage rings, the cycling rate is about 10 Hz or less. This makes it possible to use thin stainless steel vacuum pipes, thus reducing cost and improving reliability.

Examples. In Table 2 some parameters of the largest full-time electron synchrotron located at Cornell University are compared with two newer synchrotrons. The DESY II is a synchrotron that was constructed in 1984 and 1985 to serve as an improved injector for the three storage rings DORIS (5.5 GeV), PETRA (23 GeV), and HERA (30 GeV). The 750-MeV synchrotron in Table 2 is used as an injector into a 750-MeV and a 2.5-GeV storage ring at the National Synchrotron Light Source (NSLS) at Brookhaven. **Figure 13** shows the bending magnets, the quadrupoles in the middle of the straight sections, and the smaller sextupoles of this ring. Also shown is the rather flat vacuum chamber running through the magnets. In the foreground is the beam line leading from the synchrotron to the storage rings. The injection from a 70-MeV linear accelerator at the left corner is not visible.

Proton synchrotron. High-energy protons are a favorite probe for investigating the basic structure of matter. In collisions with other protons (in proton-proton colliders, for example) or with the subatomic constituents of fixed targets, they are used to study nature's elemental forces. In some instances, protons are merely intermediaries, which are used to produce secondary beams of mesons, neutrinos, and other particles. Like the protons that produced them, these particles then become the tools of the experimental physicist.

In the pursuit of higher proton energies, the proton synchrotron was developed as successor to the cyclotron and synchrocyclotron. Like the electron synchrotron, its essential feature (**Fig. 14**) is a roughly circular ring of variable-field magnets, arranged to confine the accelerated particles to an evacuated tube, or "racetrack," often having a cross-sectional diameter of only a few inches. In addition to dipole bending magnets, two types of quadrupoles are distributed around the ring, one focusing the beam in the horizontal plane, the other in the vertical plane. Injection and extraction magnets are energized only at the beginning and end, respectively, of an acceleration cycle. Protons are injected into the synchrotron ring at relatively low energy, requiring that the dipole bending magnets be adjusted initially to produce a correspondingly weak field. After injection, the strength of the magnetic field is steadily increased, but at the same time the particles gain energy with each pass through the rf cavity. A balance is maintained between field

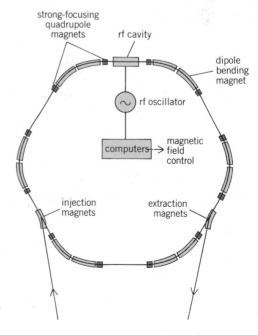

Fig. 14. Schematic diagram of an idealized proton synchrotron.

strength and particle energy such that the proton bunch is constrained to a nearly constant orbit. As the energy increases, the particle velocity (hence the orbital frequency) also increases, so the frequency of the rf cavity must rise to keep in step. (At relativistic speeds, energy increases manifest themselves mainly as increases in mass. In large synchrotrons, therefore, velocity changes are small, since protons are injected near the speed of light.) Once the desired energy is reached, the protons either are allowed to collide with targets in the ring (usually jets of gas) or, more often, are extracted from the ring and sent to experimental areas as much as a mile (1.6 km) away. The world's largest fixed-target proton synchrotrons are listed in **Table 3**.

Table 3. Large proton synchrotrons currently in operation or under construction.

Synchrotron	Date of Commissioning	Approximate radius, ft (m)	Beam energy, GeV
Proton synchrotron (PS), CERN, Switzerland*	1959	330 (100)	28
Alternating-gradient synchrotron (AGS), Brookhaven, Upton, New York	1961	410 (125)	33
IHEP, Serpukhov, Soviet Union	1967	770 (235)	76
Main ring, Fermilab,* Batavia, Illinois	1972	3300 (1000)	400
Superproton synchrotron (SPS), CERN	1976	3600 (1100)	400
Tevatron, Fermilab	1983	3030 (920)	1000
UNK, Soviet Union	1990s?	10,040 (3060)	3000

*The CERN PS and the main ring at Fermilab are now used principally as injectors for the SPS and the tevatron, respectively.

Proton sources and injectors. All of the facilities listed in Table 3 operate at energies too high to be reached economically in a single ring. Protons at high energies can be confined to a narrower beam (and in larger numbers) than can those at lower energies; therefore, a single machine designed for a wide range of energies would require both a large cross section (for low-energy operation) and a large radius (for high-energy operation). As a consequence, all large proton synchrotron facilities comprise a cascade of accelerators, each of which is responsible for some fraction of the total increase in proton energy. The sequence of accelerators at Fermilab near Batavia, Illinois, is shown in **Fig. 15**.

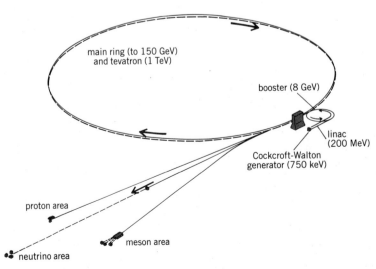

Fig. 15. Diagram of the Fermilab facility, showing sequence of accelerators used to produce 8-GeV protons for injection into the main ring. Extracted protons are routed to three experimental areas by a "proton switchyard." Part of the extracted beam is used to produce beams of secondary particles.

The first step in the Fermilab cascade is a Cockcroft-Walton generator, where H⁻ ions, produced by adding electrons to hydrogen atoms, are accelerated across a potential of 750 kV. A linac then increases the energy of the H⁻ ions to 200 MeV. Now sorted into discrete bunches, the ions are next stripped of their electrons, and the resulting protons are injected into the booster synchrotron, a ring with a radius of about 246 ft (75 m). There, the protons are accelerated to 8 GeV; they are now traveling at over 99% of the speed of light. From the booster, the protons go to the so-called main ring which, until 1983, was the final element in the cascade. Current would subsequently be increased in the main dipoles until beam energy reached 400 (rarely, 500) GeV. Now, however, at 150 GeV the protons in the main ring are injected into the superconducting tevatron, where they can be accelerated to 1000 GeV (1 TeV). A similar cascade (minus the superconducting ring) is in operation at CERN, where a Crockcroft-Walton generator, a linac, a booster synchrotron, and the 28-GeV proton synchrotron (PS) are used as the principal injector sequence for the 400-GeV super proton synchrotron (SPS).

Acceleration and beam dynamics. In the tevatron at Fermilab, the rf cavities typically add just over 1 MeV to the proton energy during each orbit. After close to a million revolutions, requiring about 17 s, the energy reaches 1 TeV. (Upgrades to the rf system will decrease this "ramp-up" time.) During this time, the dipole magnetic fields have been smoothly increased from 0.67 tesla at injection to 4.44 teslas. (A tesla is equal to 10^4 gauss.)

The simplicity of this picture ignores many complications, some of which are central to an understanding of particle motion in a synchrotron. A more detailed study of this motion begins with the definition of a reference particle (as in the discussion of synchrocyclotrons above). This particle, nominally at the center of the beam tube and of just the right energy, follows an "ideal" closed orbit around the circumference of the synchrotron. Practically all particles possess some transverse momentum, however, and thus oscillate around this ideal orbit with a frequency determined by the focusing magnetic fields. This motion (**Fig. 16**) is known as betatron motion, and the frequency is the betatron frequency (typically about 10^6 Hz). To avoid destabilizing resonances, the machine is usually tuned so that a given particle does not often repeat precisely the same orbit around the circumference. These resonances, which might be likened to the resonant vibrations of a poorly balanced automobile wheel, can cause uncontrolled growth of the beam. In early, weak-focusing proton synchrotrons, the focusing fields that controlled the betatron motion were weak quadrupole fields superimposed on the dipole magnetic fields. However, most modern machines rely on strong focusing, usually by means of separate quadrupole magnets that constrain the particles to a narrow path. A pair of such magnets can exert a net focusing force in both the vertical and horizontal directions. This is shown in **Fig. 17,** where the z axis is taken as passing through the centers of the quadrupoles. Each magnet focuses particles in either the xz or yz plane, but defocuses particles in the other. The resultant effect, however, is net focusing in both planes. After passing through two quadrupoles, charged particles that were initially moving parallel to the z axis converge. These strong-focusing machines are also known as alternating-gradient accelerators.

A proton in the synchrotron may also differ from the reference particle in the amount of kinetic energy it possesses. At relatively low energies, this difference is reflected mainly in a difference in particle velocity; thus, the more energetic particle moves ahead. At high energies,

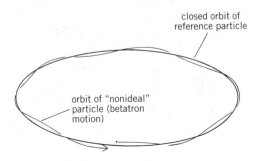

Fig. 16. Schematic illustration of betatron motion in a proton synchrotron. A proton with transverse momentum follows a path that oscillates around the ideal closed orbit.

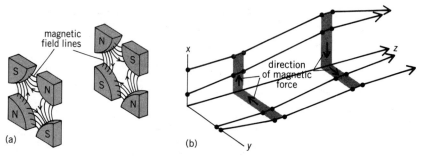

Fig. 17. Net focusing effect of paired quadrupole magnets. (a) Magnet geometry. (b) Particle trajectories.

energy manifests itself mainly as mass; therefore, the more energetic particle does not move appreciably faster, but rather in a longer, wider orbit than the reference particle. Thus, the energetic particle falls behind. In either case, the accelerating rf field can be tuned to give the more energetic particles a weaker-than-average push, thus allowing the less energetic ones to catch up. As the energy of the beam increases, then, the energy of each particle oscillates about a steadily increasing mean energy. The result is a longitudinal, oscillatory synchrotron motion (of about 10^3 Hz) that bunches the particles and ensures phase stability. At some intermediate energy, however, protons with different energies circulate with the same orbital frequency. At this transition energy, the rf field cannot sort out particles with different energies, and phase stability is momentarily lost.

Extraction. Once the protons in the synchrotron reach the energy required by the experimentalists, the particles are most often extracted and directed to experimental areas outside the ring. There, interactions with various fixed targets are studied, and beams of secondary particles, such as mesons and neutrinos, are created. By far the most common mode of extraction is a slow process (requiring one to several seconds) called resonant extraction. In this process, the dipole magnetic field is usually held constant and the accelerating rf field is turned off. The quadrupole magnets are then tuned until the betatron oscillations are very near a destabilizing resonance. At the same time, a field perturbation is imposed locally, often with a sextupole magnet, then varied slowly to sweep the beam through the resonance. During this sweep, particles are brought sequentially and continuously into resonance, resulting in a controlled growth of the beam. The actual extraction then involves the use of a knifelike septum to physically "scrape" the most errant protons from the periphery of the expanding beam, after which they can be guided in a continuous stream down a separate beam pipe to the experimental regions. **Figure 18** shows the dipole

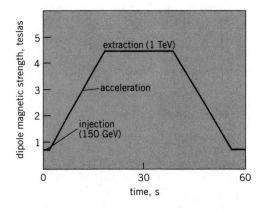

Fig. 18. Dipole magnet field strength during a typical maximum-energy acceleration cycle at the tevatron. The field increases from 0.67 T at injection to a 4.44 T "flat top" during extraction.

magnet field strength during a typical acceleration and extraction cycle at Fermilab's tevatron. The duration of the flat top of the field during extraction, as a fraction of the total cycle, determines the duty cycle of the machine.

Methods of fast extraction are also available, either for special experimental needs or for beam transfer between synchrotron rings. Such methods entail special "kicker" magnets that either remove individual pulses from the ring or distort the beam orbit so that the entire beam can be extracted during a single turn. Single-turn extraction is also used when an operational problem demands an abort. In such instances, the beam, which may have the kinetic energy of an artillery shell, is directed to a specially designed beam dump.

Tevatron and UNK. Perhaps the most obvious limitation on the energy that can be reached in a proton synchrotron is the field strength of the dipole magnets used to constrain the particles to a circular path. The stronger the magnets, the smaller the ring size and the less real estate required to achieve a given energy $= m$; or, the greater the energy that can be achieved in a tunnel of given circumference. Experimental accelerator magnets have produced fields in the neighborhood of 9 T, and magnets designed for mass production go up to about 6.5 T. All such magnets are wound with superconducting alloys (usually Nb-Ti or Nb_3Sn) and must be cooled to near 4.5 K ($-452°F$) by liquid helium.

The first superconducting proton synchrotron, Fermilab's tevatron, has been in operation since 1983. It comprises 774 superconducting dipole magnets, each 21 ft (6.4 m) long and capable of a maximum field strength of 4.44 T. It is laid out directly beneath the 400-GeV main ring (**Fig. 19**), which serves as its injector. When operated at maximum energy, the tevatron accelerates protons to 1 TeV. A second superconducting synchrotron, known as UNK and under construction in the Soviet Union, aims at even higher energies. The 76-GeV proton synchrotron at the Serpukhov Institute of High Energy Physics is to be used as the injector for a 400–700-GeV ring, which will, in turn, inject a 3-TeV ring of superconducting magnets, occupying the same 12-mi-circumference (19.2-km) tunnel.

Fig. 19. Tunnel of the main accelerator at Fermilab. The upper magnets are part of the 400-GeV main ring. The lower ones are superconducting magnets for the tevatron. (*Fermi National Accelerator Laboratory*)

All such superconducting accelerators face problems not encountered in conventional synchrotrons. The most obvious new requirements are reliable refrigerators for the string of magnets and appropriate couplings and insulators to accommodate the effects of extremely low temperatures. The length of a tevatron magnet, for example, decreases by about 0.8 in. (2 cm) when it is cooled to its operating temperature. The most severe problems, however, are associated with quenches, events in which a local hot spot (usually the result of a tiny movement in the superconducting wire) causes the magnet winding to lose its superconductivity and thus "go normal." The current flowing through the windings thus encounters local resistance, causing further heating and producing the danger of melted conductor. The magnet design must accommodate the heating caused by the field-induced current (passive protection), or active steps must be taken to heat the entire magnet, thus spreading out the region of high resistance (active protection). In addition, immediate steps must be taken to dump the proton beam harmlessly, lest it go out of control and damage the magnets.

Heavy-ion synchrotron. The development of heavy-ion accelerators is closely tied to that of proton accelerators. Indeed, most heavy-ion accelerators, including cyclotrons, linacs, and electrostatic machines, are used to accelerate both protons and heavier nuclei. Converted proton synchrotrons at Princeton and Berkeley were first used to accelerate heavy ions to relativistic energies in 1970. The Berkeley facility is the only heavy-ion synchrotron operating in the United States (**Fig. 20**). The synchrotron itself is linked to a high-intensity linac (the superHILAC), which

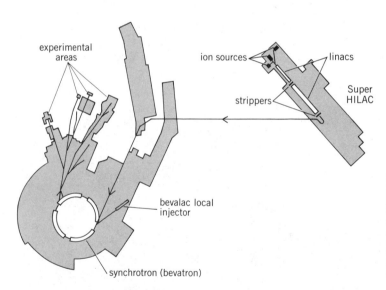

Fig. 20. Plan view of the bevalac complex at the Lawrence Berkeley Laboratory, showing the superHILAC with its two Alvarez linear accelerators and two strippers, the transfer line, and the bevatron, originally built as a proton synchrotron.

serves as one of its injectors. The combined facility (the bevalac) can produce beams of ions as heavy as uranium, stripped of some or all of its electrons. The principal heavy-ion synchrotrons are listed in **Table 4**.

Basic considerations. Because of their many similarities, it is useful to focus on the differences between synchrotrons designed or modified for the acceleration of heavy ions and the proton machines discussed above. A concept that underlies many of these differences is the charge-to-mass ratio q/A (the ratio of charge state q to mass number A) of the ion species being accelerated. For protons this ratio is fixed, but for heavy ions it depends on how many electrons have been "stripped" from the nucleus. For a given accelerating potential, the increase in energy per nucleon (E_n) is directly proportional to q/A. (When working with heavy ions, it is usually E_n, rather than the total energy E, that is important.) The behavior of the ions in a magnetic field also

Table 4. Major heavy-ion synchrotrons currently in operation or near completion.

Facility	Maximum beam energy,* GeV per nucleon	Maximum ion mass
Saturne II, Saclay, France	1.2	40 (Ar)
Bevalac, Berkeley, California	2.1	238 (U)
Synchrophasotron, Dubna, Soviet Union	4.6	20 (Ne)
AGS, Brookhaven, Upton, New York	15.6	32 (S) [197 (Au)]†
SPS, CERN, Switzerland	200	16 (O)

*The maximum energy is rarely achieved with the heaviest ions.
†Projected capability, pending new construction.

depends on q/A: Of two ions with the same E_n, the one with the lower charge-to-mass ratio will require a higher field strength for beam handling. Thus, where magnetic field strengths and accelerating potentials are limiting factors, high values of q/A are desirable. It is this observation, together with the fact that q/A changes if the ion gains or loses electrons during the acceleration process, that dictates many of the features of heavy-ion synchrotrons.

Ion sources and injectors. One of the most important differences between proton machines and heavy-ion accelerators is the source of ions. In the latter, simple sources of ionized hydrogen or H⁻ ions give way to complex devices, such as the Penning ion gage, that ideally produce pure beams of any atomic species, at high intensity and in the highest possible charge state (high q/A).

The first stages of acceleration are similar for heavy ions and protons. The beam passes first through an electrostatic accelerator (a Cockcroft-Walton generator or tandem Van de Graaff accelerator), then through successive linacs. What distinguishes this series of accelerators from its proton analog is the presence of one or more strippers. As ions pass through these strippers—typically comprising carbon foils or streams of liquid or vapor—they lose electrons and are thus elevated to higher charge states. Unfortunately, beam intensity is usually much attenuated at the same time, since various charge states are produced, of which only one is usually selected for further acceleration.

Acceleration in the synchrotron. The acceleration cycle in the heavy-ion synchrotron itself is exactly analogous to that in a proton synchrotron (Fig. 18); however, a few new considerations come into play. First, the injected ions are not yet near the speed of light, hence their velocity will increase substantially as their energy increases. The frequency range of the rf system must accommodate this increase. Second, the beam intensities are typically much lower than in proton machines; therefore, beam diagnostic instrumentation must be especially sensitive. Finally, partially stripped ions are susceptible to changes in charge state during collisions with residual gas molecules in the beam tube. When this happens during the acceleration cycle, the ion is lost; consequently, an extremely good vacuum must be maintained in heavy-ion synchrotrons. Pressures near 10^{-10} torr (10^{-7} pascal) are typically required to ensure survival of partially stripped ions.

Prospects. In December 1983, an advisory committee to the Department of Energy and the National Science Foundation assigned the highest priority to the construction of a relativistic heavy-ion collider. Such a machine might be capable of studying extreme states of nuclear matter, including a plasma of quarks and gluons. The envisioned collider would provide beams of very heavy ions of at least 30 GeV per nucleon.

LINEAR ACCELERATOR

A linear accelerator accelerates particles in a straight line by means of electric fields developed across a series of accelerating gaps in sequence. Each gap is transversed once and the electric fields must be produced along the entire orbit. Extremely high field strengths and hence high radio-frequency power levels are required for room-temperature structures and high beam currents. Superconducting structures requiring low power have been developed for certain applications.

Principles of operation. The synchronism between particles and accelerating field may be achieved by either of two methods: traveling-wave acceleration, whereby a wave with an accelerating field component whose phase velocity is equal to the particle velocity is utilized; or standing-wave acceleration, whereby a standing-wave pattern is produced by the superposition of forward and backward waves in the structure. In the latter case, either the forward-wave phase velocity is made equal to the particle velocity or, more frequently, a set of drift tubes is introduced into the structure to shield the particles from the fields when they are passing through regions where the fields would otherwise be decelerating.

The energy W to which a particle of charge q can be accelerated in a structure of length L fed by rf power sources of total power P and wavelength λ is given by Eq. (4), where K is a

$$W = qK(PL)^{1/2}\lambda^{-1/4} \qquad (4)$$

constant depending on the structure and mode of acceleration used. From this equation it can be seen that length and power are equally instrumental in attaining high energies and that short wavelengths are also advantageous in terms of power economy. Countering this are the following disadvantages: (1) higher power levels are generally available from power tubes operating at longer wavelengths and the accelerator guide is capable of high power handling; (2) the absolute tolerances of the mechanical structure are larger at larger wavelengths; (3) longer wavelengths allow for higher beam currents. In practice, electron accelerators use wavelengths of 3–20 cm, proton accelerators 30–200 cm, and heavy-ion accelerators 300–1000 cm. In general, the high power levels required for room temperature operation result in pulsed rather than continuously operating machine design, but there is no fundamental reason for pulsed operation and some continuously operating machines are under design.

Successful acceleration of particles requires that acceleration be phase-stable and that the beam of particles remain focused along the orbit. Consider a field region where the accelerating field increases in time as the particle crosses the region (**Fig. 21***a*). It is clear that there will be a net defocusing action since, in an increasing field, the defocusing transverse momentum imparted at B is greater than the focusing momentum imparted at A. However, the action is phase-stable,

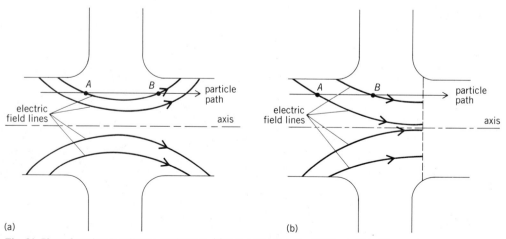

Fig. 21. Phase focusing in a linear accelerator. (*a*) Gap without grids. (*b*) Gap with grids.

since a particle arriving late will experience a larger acceleration than a faster-traveling particle which arrives early. If the field decreased in time during particle passage, there would be transverse focusing but longitudinal phase instability. Thus focusing and phase stability are incompatible, and this appeared to be a serious obstacle to the design of long linear accelerators. It has been circumvented in various ways as follows: (1) Charge can be included in the beam to terminate field lines inside the beam and cause a convergent action even at phase-stable transit phases. This charge has typically been induced by grids placed in the field to produce field lines similar to those shown in Fig. 21b. In this case there is focusing momentum at both A and B, and in an increasing field the acceleration is also phase-stable. Unfortunately, the grids also intercept some beam and therefore prohibit high-current operation. (2) As the particle velocity approaches c, the incompatibility becomes irrelevant because, for relativistic velocities, the action of the radial electric time-varying field is almost canceled by the accompanying time-varying magnetic field; also, because the velocity is almost invariable, the particles are in neutral equilibrium longitudinally. (3) External magnetic fields (solenoidal or strong-focusing) or electrostatic lenses have been used. (4) By alternating the phase difference, or by suitable field shaping, a region of limited phase stability is attainable.

Electron accelerators. Virtually all electron linear accelerators are of the traveling-wave type. The wave is produced in a loaded waveguide excited in a transverse magnetic (TM) mode which has a longitudinal electric field component. The loading is typically achieved by means of disks which are sized to produce the correct phase velocity to match the particles. A typical field configuration is shown in **Fig. 22**; the field pattern may be visualized as translating along the axis with a velocity equal to that of the electron beam. Typically solenoid focusing is used over the first few feet of the accelerator, after which the electrons approach the velocity of light whereby all radial forces may be neglected. The momentum component transverse to the axis remains constant, and since the longitudinal component increases continuously, the angle of divergence of the beam decreases continuously. For uniform energy gain per unit length, this corresponds to a beam angle varying as the inverse of distance along the machine and to a beam radius increasing logarithmically. A set of weak magnetic lenses can be used to produce a decrease in beam diameter.

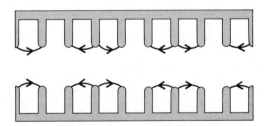

Fig. 22. Electric field configuration in a traveling-wave electron linear accelerator operating in the most commonly used $2\pi/3$ mode, where the waveguide is loaded by three disks per wavelength to obtain a phase velocity matching the particle velocity.

Accelerator guides must be manufactured to very close tolerances in order to control the phase velocity to the required accuracy, since there is no phase stability once the energy has grown to several times the rest energy of the electron. However, in practice the accelerator is divided into many separate accelerator guides, each with its own power source. These separate sections may be individually tuned and phased during operation, thus somewhat relieving the tolerance requirements.

Electrons are injected into the accelerator with an energy of typically 80–120 keV, corresponding to a velocity of approximately $c/2$, and the first accelerator section is often a special bunching device to concentrate the electrons near the crest of the traveling wave in the succeeding sections.

The total energy capability from a linear accelerator has no fundamental limit since electrons accelerated in a straight line do not lose any appreciable energy due to radiation. The performance of a given accelerator is closely tied to the available power source and the voltage breakdown capability of the accelerating guide. Most machines operate at microwave frequencies and utilize klystron amplifiers driven from a common master oscillator as power sources.

The acceleration of electrons at room temperature requires peak radio-frequency power levels of tens of megawatts per source. Costs of power and available power sources limit the beam duty cycle to a few tenths of a percent. At the Massachusetts Institute of Technology and Saclay, France, 400–500-MeV linacs of higher duty factor have been constructed with long pulses and fast repetition rates achieved at the expense of low acceleration rates. A superconducting electron linac which can operate continuously with dissipation of about 1 W/m has been constructed at Stanford University, utilizing standing-wave structures of niobium metal and operated at 1 to 4 K (−458 to −452°F). However, metallurgical and electron loading problems limited the accelerating gradients to about 3 MeV/m. At this level these accelerating guides are useful for use with recirculating beams as in the microtron at the University of Illinois.

The largest electron linac is at Stanford University and is 10,000 ft (3050 m) long (**Fig. 23**). Besides producing electrons directly for basic research, it injects two different synchrotron storage rings with electrons and positrons, the 4.2-GeV SPEAR storage ring and the 18-GeV PEP storage ring. The Stanford Linear Accelerator Center (SLAC) has increased the output energy of this 20-GeV machine to 32 GeV by means of an SLED scheme whereby the amplitude of the rf pulse is increased at the expense of the pulse length. New higher-power klystrons have been installed to further increase the energy to 50 GeV. A further extension of this facility involves the SLAC-Linac-Collider (SLC) Project, which will utilize the existing 50-GeV linac to accelerate both electrons and positrons in the same rf pulse. At the end of the linac both beams will be separated and sent through long arcs until they aim at each other. Then a final focus system will compress the transverse sizes of the beams to about 2 micrometers at the electron-positron interaction point, giving an energy of 100 GeV in the center-of-mass system.

Fig. 23. Stanford Linear Accelerator Center. (*a*) Aerial view showing the 2-mi (3.2-km) accelerator and associated buildings and storage rings. (*b*) The accelerator installed in its underground tunnel.

Proton accelerators. Proton linear accelerators are of the standing-wave type since protons do not reach relativistic velocities at typical injection voltages which are in the range of 500 to 2000 keV. The most usual structure is that due to Alvarez (**Fig. 24**), whereby the particles pass inside copper drift tubes where no electric field is present and are accelerated in the region between the drift tubes. The distance between the center of each drift tube, or cell length, is

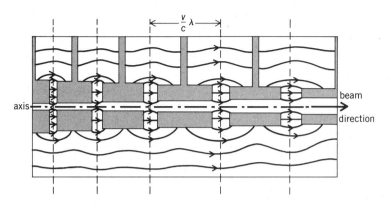

Fig. 24. Electric field configuration in a standing-wave proton accelerator employing a cavity utilizing an Alvarez structure operating in the TM_{101} mode. Drift tubes supported on stems shield the particles when the phase is decelerating. Particles cross gaps at approximately the same phase.

equal to $v\lambda/c$, where λ is the free-space exciting wavelength and v is the particle velocity at the cell. The particles cross each gap at the same phase, usually 20 to 30° before the time maximum of the electric accelerating field. Focusing is achieved by means of magnetic quadrupoles situated inside the drift tubes. A series of cells is a resonant cavity if each cell resonates at the same frequency. Resonators are sized to match available power sources at the typical operating frequency (around 200 MHz). Special resonant posts or stems are used to improve the energy transfer along the resonator and ease the mechanical and beam loading tolerances. Resonators are placed in tandem to achieve energies up to 200 MeV with the Alvarez structure at Brookhaven and Fermi National Accelerator Laboratories. Above 200 MeV, the efficiency of the Alvarez structure falls off, and special coupled cavity structures of the type used at the 800-MeV proton accelerator at the Los Alamos Meson Physics Facility (LAMPF) are necessary above this energy.

The 800-MeV machine at Los Alamos is the largest proton linac in operation and is 2870 ft (874 m) long. Besides producing intense meson beams for basic research and medical studies from proton collisions with nuclei, it injects a magnetic storage ring that in turn produces intense bursts of neutrons for basic and applied research.

In some accelerators a new structure called the radio-frequency quadrupole (RFQ) is used as an injector instead of the dc injectors previously favored. This device utilizes a special vane-type accelerating structure which also provides quadrupole focusing by electric fields near the axis (**Fig. 25**). Injection energies of 20–50 keV and output energies of 1–2 MeV are typical for such structures.

Fig. 25. Inside view of the radio-frequency quadrupole structure of the Brookhaven proton linear accelerator.

Heavy-ion accelerators. Heavy-ion linacs are similar in design to low energy proton accelerators. Because of the relatively low charge-to-mass ratios available from ion sources, with available electric fields, the ion velocities are lower than for protons, so practical cell lengths require operation at frequencies of typically 30–100 MHz. Typically a dc injector of 300 to 500 kV is followed by a series of accelerating cavities of the Wideröe or Alvarez type to increase the particle energy to about 5% of c, corresponding to energies of 1–1.4 MeV per atomic mass unit (amu). [Accelerators of the Wideröe type have been used for low values of v/c. The alternate drift tubes are connected to opposite sides of an alternating-current power supply (**Fig. 26**) so that the spacing between drift tube centers is $v/2c$ instead of $v\lambda/c$ as in the Alvarez structure.] The ions are then passed through a stripper (gas jet or carbon foil) where the ionic charge is increased by a factor of 2 to 4, and about half the ions are in a single-charge state. Subsequent Alvarez cavity sections are used, with a higher acceleration rate to increase the velocity to about 10% of c (6 MeV/amu). At certain laboratories, such as the unilac at Darmstadt, West Germany, further acceleration is achieved by the use of a number of independent, individually controlled, single cell, resonant cavities which may be used to increase the energy to as much as 13 MeV/amu.

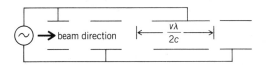

Fig. 26. Wideröe structure, used for linear accelerators with low values of v/c.

There has been considerable work on the development of a practical superconducting structure for heavy ions; spiral, split-ring or quarter-wave-type structures have been proposed and tested at various laboratories and work has started on the development of a radio-frequency quadrupole structure for heavy-ion acceleration. These structures are proposed as a method of increasing the capability of tandem accelerators by using such devices as postaccelerators. The Atlas project at the Argonne National Laboratory utilizes superconducting niobium split-ring resonators to achieve energies of up to 30 MeV/amu.

A similar machine at the State University of New York at Stony Brook (SUNYLAC), will accelerate ions up to 10 MeV/amu by means of superconducting lead-plated copper split-ring accelerating cavities following a tandem Van de Graaff injector and special prebunching system.

At Berkeley, California, the heavy-ion linear accelerator (HILAC) has been upgraded by adding a $\pi - 3\pi$ Wideröe-type injector (ABEL) for acceleration of particles with atomic number greater than 40 to the existing 750-kV Cockcroft-Walton injector used for particles with atomic mass less than 40. This so-called superHILAC accelerates ions to a maximum energy of 8.5 MeV/amu. As discussed above, a transfer line between the superHILAC and the bevatron synchrotron allows for acceleration of uranium ions to 1 GeV/amu. This facility known as the beVALAC may be further upgraded by following the superHILAC with two rings of 6-T superconducting magnets located one above the other in a tunnel that surrounds the existing bevatron site. This project, which has been called the teVALAC, would increase the beam current intensities available by a factor between 100 and 1000.

MICROTRON

The microtron is an electron accelerator in which the beam energy is raised in steps by recirculating the electrons through the same radio-frequency accelerating sections. A magnetic guide field provides stable orbits which the electrons follow as they are accelerated. The number of orbits required to reach a given energy is determined by the magnet geometry and is usually in the range of 10–100. Only moderate rf power is required because of beam recirculation.

Principle of operation. The basic principle of operation for all microtrons is embodied in the coherence condition which requires that the geometry of the electron orbits be adjusted to ensure that the transit time, from exit from the rf accelerating section to entrance on the next traversal of an rf section, be an integral multiple of the rf period. Under these conditions, the electron beam will be uniformly accelerated from injection to extraction energy. For particles traveling at close to the speed of light, successive orbits increase in length by an integral number ν of rf wavelengths λ. Possible microtron configurations are shown in **Fig. 27**, where the resulting restriction on the change in orbit radius, ΔR, is also shown for each geometry. In every case the fields in the bending magnets are assumed to be uniform. The condition on ΔR is met by adjusting the energy gain per turn, ΔW, and the magnetic field in the sector magnet B to give the prescribed ΔR, according to Eq. (5). The circular geometry of the "classical" microtron (Fig. 27a) was first

$$\Delta R = \frac{\Delta W}{qBc} \quad (5)$$

proposed by the Soviet physicist V. I. Veksler. The maximum energy attainable in this version is limited by the finite size of the rf cavity. In the "racetrack" geometry (Fig. 27b), developed to avoid this constraint, the magnetic field is split into two half-circular sectors separated by sufficient drift space to allow insertion of a linac section of substantial accelerating power. This geometry is used to accelerate electrons to energies of a few hundred MeV. To achieve higher energies, higher-order variants of the microtron geometry, the double-sided microtron (Fig. 27c) and the hexatron (Fig. 27d), have been proposed.

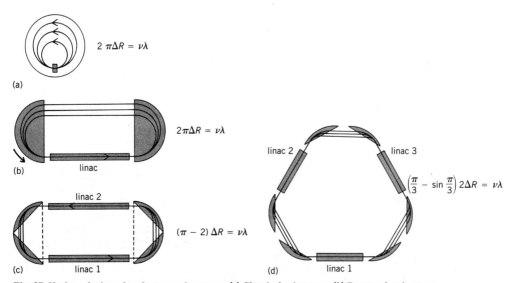

Fig. 27. Various designs for electron microtrons. (a) Classical microtron. (b) Racetrack microtron. (c) Double-sided microtron. (d) Hexatron.

Applications. The "classical" microtron finds application in medicine and industry, furnishing pulsed beams of electrons in the energy range of 10–50 MeV. Microtrons are also used in nuclear physics and other research fields, and as injectors for larger microtrons and larger machines such as electron synchrotrons.

Design. The MAMI system is illustrative of the most advanced designs. The cascaded system of racetrack microtrons shown in **Fig. 28** is designed to furnish 100 microamperes of electrons at 175 MeV. A small 20-turn 14-MeV microtron injects into the second stage, which accelerates the electrons to 175 MeV in 51 turns. Use of a high-efficiency wave guide for the rf accelerating sections in this design makes possible continuous acceleration of electrons at the 100 µA level. With the addition of a third racetrack microtron, MAMI will be capable of continuous operation at energies up to 800 MeV.

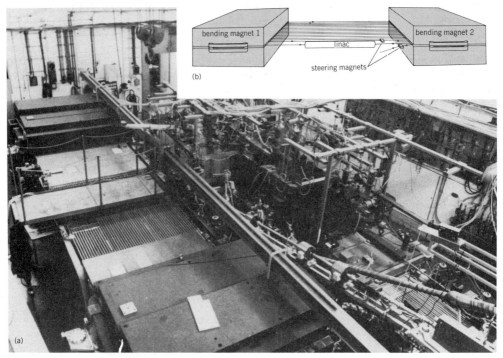

Fig. 28. MAMI cascaded racetrack microtron system at the University of Mainz. (a) System under construction. A small 14-MeV racetrack microtron at the right serves as injector to a second microtron which accelerates the electron beam to 175 MeV. (b) Basic microtron design.

Comparison with other accelerators. All high-energy electron accelerators increase the beam energy by repeated traversal of an rf accelerating cavity operating at high voltage. In cyclic machines such as electron synchrotrons, acceleration is accomplished by a very large number of passages through a relatively modest accelerating structure. Such accelerators are limited to maximum-beam currents in the 1–10-μA range. Acceleration in linear accelerators is accomplished by operating a long array of waveguide sections at very high voltage gradients. To avoid prohibitive losses due to power dissipation in the cavity walls, linacs are restricted to operation in a short pulse mode. The microtron can be viewed as a folded linac, with each traversal of the beam through the same linac section corresponding to passage of the beam through a successive section of a conventional linac. Consequently, the conversion efficiency of input to beam power can be made very much higher than for a linac. An advantage of the microtron is the ease with which the beam can be extracted on any turn by magnetic deflection of the electrons in the region of the accelerator where the orbits are dispersed. Widespread interest in microtrons for research in nuclear physics is due to excellent beam quality and stability and to the high power efficiency which characterizes their operation.

STORAGE RINGS

Storage rings consist of annular vacuum chambers embedded in a ring of magnets in which beams of high-energy charged particles can be stored and caused to collide in near head-on collisions.

Principles of colliding-beam systems. The motivation for using this technique is found in the following kinematic considerations. Unlike fixed-target systems (cyclotrons, synchrotrons, linear accelerators) in which a beam of accelerated particles traverses a fixed target, colliding-beam systems use the full energy of each particle to produce reactions. When two particles of mass m_1 and m_2 and energy E_1 and E_2 interact ($E = T + mc^2$, where T is the kinetic energy and

mc^2 the rest-mass energy), the center-of-mass energy E_{cm} measures the energy available for reactions. For a fixed-target accelerator which produces a beam of high-energy relativistic particles 1 ($E_1 > m_1c^2$) hitting stationary target particles 2, this energy is given approximately by Eq. (6).

$$E_{cm} = [2(m_2c^2)E_1]^{1/2} \qquad (6)$$

The useful energy is only a fraction of the available beam energy and increases slowly as its square root. By contrast, when two particles of the same mass and energy E_1 collide head-on in a storage ring, E_{cm} is given by Eq. (7). All of the available energy is useful. For instance, the CERN e^+e^-

$$E_{cm} = 2E_1 \qquad (7)$$

storage ring LEP (Large Electron Positron Storage Ring) is designed to operate with beams of up to 90 GeV yielding E_{cm} = 180 GeV. With a fixed e^- target, an e^+ beam of 32,000,000 GeV would be required to reach the same value of E_{cm}.

Low reaction rate. The high value of E_{cm} of colliding-beam systems is obtained at the expense of reaction rate, which is proportional to the density of target particles and to the beam intensity. It is helpful to think of one stored beam as the target for the other, and to consider the collision of a "beam bunch" with a "target bunch." In a typical e^+e^- storage ring the beam density is about 10^{14} particles/cm^3; a fixed target contains about 10^{24} electrons/cm^3. This huge difference is partially compensated by (1) the fact that at the interaction point each beam bunch collides with each target bunch f times per second (where f is the number of stored particle revolutions per second) and (2) the large intensity of the stored beams compared to accelerator beams which is the result of the accumulation of particles injected over a period of time. Because of the importance of these two factors for obtaining useful reaction rates, beam storage was used in all high-energy colliding beam systems placed in operation before 1987. However, there are limits to the density and intensity of beams that can be stored and made to collide, and thus colliding-beam reaction rates are always much lower than fixed-target reaction rates. Storage ring experiments are designed to take full advantage of the high center-of-mass energy within the limitations imposed by the relatively low reaction rates.

Luminosity. The performance of storage-ring colliding-beam systems is measured by the luminosity L defined by Eq. (8), where R = reaction rate or number of interactions per second;

$$L \equiv \frac{R}{\sigma} = \frac{N_1 N_2 f}{Ab} \qquad (8)$$

σ = interaction cross section; N_1, N_2 = number of particles in each circulating beam; A = beam cross-sectional area at the interaction point; f = number of stored particle revolutions per second; b = number of bunches in each beam. Thus, the reaction rate is given by $R = L\sigma$. Typical values are in the $L = 10^{29}-10^{32}$ cm^{-2} s^{-1} range.

To reach much higher energies than are now planned for LEP, the size of the storage ring would have to become unreasonably large. Instead, it appears possible to obtain useful luminosities in direct head-on collisions between two beams from two very high-energy linear accelerators by focusing each beam to a very small cross-sectional area A. The SLAC-Linac-Collider (SLC) project is a prototype scheduled to start operation in 1987 at E_{cm} = 100 GeV.

Positron-electron storage rings. A positron-electron (e^+e^-) storage ring consists of a ring of bending and focusing magnets enclosing a doughnut-shaped vacuum chamber in which counterrotating beams of e^+ and e^- are stored for periods of several hours. The two beams are made to collide with each other about 10^6 times per second in straight interaction sections of the vacuum chamber (**Fig. 29**). These are surrounded with detectors for the observation of collision products. An alternate design consists of two separate but intersecting rings, one to store e^-, the other to store e^- or e^+, with collisions taking place at the intersection points.

Collisions between antiparticle and particle such as e^+e^- are of particular interest since most of the resulting processes proceed through annihilation which forms a particularly simple final state with the quantum numbers of the photon. This is one of the main reasons why e^+e^- collisions have proved to be particularly fruitful in deepening understanding of the fundamental structure of matter.

Much has been learned about the charm-anticharm quark system (the J/ψ meson family) and other mesons containing charmed quarks with the e^+e^- storage rings SPEAR (Stanford Pos-

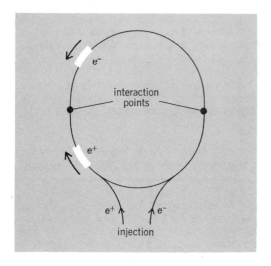

Fig. 29. Schematic diagram of e^+e^- storage ring.

itron-Electron Accelerating Ring) at Stanford Linear Accelerator Center (SLAC), in California, and DORIS at DESY (Deutsches Elektronen Synchrotron), Hamburg, Germany, where beams are stored up to 5 GeV. As a result, new, higher-energy storage rings have been constructed. PETRA (positron electron tandem ring accelerator) at DESY started operation in 1978 and has operated at up to $E_{cm} = 45$ GeV, the highest energy reached to date in e^+e^- collisions. With this ring, hadron jets produced by gluon decays were observed for the first time, as were interference effects between electromagnetic and weak interactions in $\mu^+\mu^-$ production. PEP (Positron-Electron Project) at SLAC started operation in 1980 and has run up to $E_{cm} = 30$ GeV. Important results on the properties of jets that are produced by quark and gluon decays have been obtained with these two storage rings. The TRISTAN storage ring at the KEK Laboratory in Japan is designed to reach $E_{cm} = 60$–70 GeV. S*ee* C*harm*; G*luons*; J *particle*; Q*uarks*; U*psilon particles*.

The LEP storage ring, under construction at CERN, will have a circumference of 17 mi (27 km) and will store e^+e^- beams reaching $E_{cm} = 180$ GeV. It is designed to produce Z^0 particles copiously and to explore the energy region below and above the Z^0 mass energy. S*ee* I*ntermediate vector boson*.

Beam instabilities. When particles from beam 1 (e^+ or e^-) pass through a more intense beam 2 (e^- or e^+) in the interaction region, they feel a strong force due to the electromagnetic field set up by beam 2. Thus, beam 2 acts as a very nonlinear lens on beam 1 and tends to cause a diffusive growth in the size of the latter. The force increases rapidly with the density of particles in beam 2 until it becomes so strong that the beam 1 area becomes suddenly very large and the luminosity decreases drastically; this is beam-beam instability. There also exist single-beam instabilities which are caused by electromagnetic forces between particles in the same beam or by fields set up by induced currents in the walls of the vacuum chamber. Maximum luminosity is achieved by operating with two beams of equal intensity just below the beam-beam instability limit, which in a well-designed storage ring is reached before single-beam instabilities become a problem.

Synchrotron radiation. The limitations discussed so far apply equally well in principle to e^+e^- as to p-p or $\bar{p}$-p colliding-beam systems. There are important differences, however. Because they have a small mass, e^+ and e^- in circular orbits radiate a substantial amount of energy in the form of synchrotron radiation, just as they do in electron synchrotrons. The power P_{rad} radiated by N_1 positrons and N_2 electrons of energy E revolving in circular orbits of radius r is given by Eq. (9), where e and mc^2 are the electron charge and rest-mass energy and $\epsilon_0 = 8.854 \times 10^{-12}$

$$P_{rad} = \left(\frac{1}{4\pi\epsilon_0}\right)\frac{2}{3}(N_1 + N_2)e^2c\left(\frac{E}{mc^2}\right)^4\frac{1}{r^2} \qquad (9)$$

farads/m. This power, which has to be supplied by the rf system to keep particles on the design orbit, is appreciable and costly; at the peak operating energy it amounts typically to several megawatts in multi-GeV storage rings. In the design of a storage ring for a given energy E and luminosity L, the most economic solution requires a careful balance between the cost of magnets and buildings (which increases with increasing r) and the cost of the rf system (which decreases with increasing r since less power is needed). At the highest operating energies the amount of available rf power limits the luminosity, since for fixed values of P_{rad} and r only a steep decrease in $(N_1 + N_2)$ can compensate for an increase of E. **Figure 30** shows the variation with E of the maximum luminosity L that is theoretically possible in an e^+ storage ring through adjustment of the focusing magnets. E_0 is the energy at which L is designed to have its maximum value L_0. L is limited by beam-beam instability for $E < E_0$ and by available rf power for $E > E_0$. At $E = E'$ the focusing magnets become saturated and can no longer be adjusted for maximum luminosity.

There are also very beneficial consequences from the emission of synchrotron radiation. This emission dampens the amplitude of e^- and e^+ oscillations around the equilibrium orbit, and produces a gradual transverse polarization of the stored beams. As in electron synchrotrons, the radiation itself constitutes a very intense source of x-rays which has proved very useful in various branches of biology, chemistry, and physics.

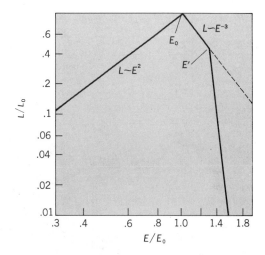

Fig. 30. Variation of luminosity with energy in an e^+e^- storage ring (log-log scale).

Ring. The ring consists of circular sectors of bending and focusing magnets and of straight sections for rf cavities and interaction regions. The bending magnets guide the beam particles around the ring, and the focusing magnets drive them toward the equilibrium orbit around which they execute radial and vertical oscillations. The use of separate magnets for guiding and for focusing permits tight control of beam size to maximize the luminosity as the energy of operation is changed.

Injector. To achieve high luminosity, it is desirable to inject and accumulate e^- and e^+ with energies close to the desired collision energy. Linear accelerators and synchrotrons are used as injectors. The e^- are accelerated directly; e^+ are collected from a target bombarded with high-energy e^- and then accelerated. Accumulation occurs through the addition of particles first to the circulating e^+ beam, then to the circulating e^- beam until desired intensities have been reached. Typical accumulation times are seconds for e^- and minutes for e^+.

Vacuum chamber and system. Very low pressures of 10^{-9} and 10^{-10} torr (10^{-7} and 10^{-8} Pa) are required to prevent the outscattering of beam particles by collisions with gas molecules, which reduces storage times below useful values. Special techniques are used to achieve such high vacuum in the presence of high-power synchrotron radiation hitting the chamber walls.

The rf power system. The power from the rf system replaces the energy emitted as synchrotron radiation and accelerates the stored beams to the desired energy. In a multi-GeV storage ring, several megawatts of power at 400–500 MHz are typically delivered to several rf cavities located in the ring.

Beam structure. The stored beams consist of bunches (typically 2 in. or 5 cm long), since rf power is being supplied continuously. For a given number of stored particles, maximum luminosity is achieved with the smallest number of bunches in each beam according to Eq. (8). This number equals half the number of interaction regions. (For example, in Fig. 29 there are two interaction regions and one bunch circulating in each direction.)

Interaction region. Magnet-free straight sections at the center of which e^+e^- collisions occur are provided for the installation of detectors. Special magnets are usually installed on each side of the interaction region to focus the beams to a very small area at the interaction point so as to maximize the luminosity; this is called low beta insertion.

Proton and $\bar{p}p$ storage rings. Proton storage rings consist of a pair of annular vacuum chambers in which high-energy protons can be stored and caused to collide nearly head-on. The motivation for using this technique is the same as for electron-positron storage rings—namely, the colliding-beam storage ring permits studying reactions at energies that cannot be achieved with economically feasible conventional accelerators.

Only one proton storage ring device has been constructed and operated—the intersecting storage rings (ISR) at CERN near Geneva, Switzerland. The ISR consists of two nearly circular rings arranged as in **Fig. 31**. Each ring consists of 132 magnets of a type similar to those used in

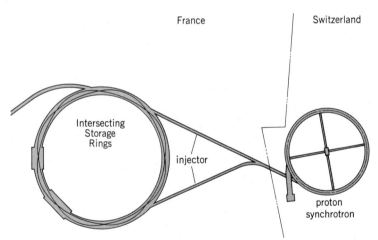

Fig. 31. Schematic layout of the intersecting storage rings at CERN and the injector.

a large proton synchrotron. The total circumference of each ring is nearly 0.6 mi (1 km). Protons of 26 GeV from the CERN proton synchrotron (PS) are injected into the rings. A stacking procedure permits accumulating protons from many PS pulses into the rings until currents of up to 50 A are stored. The entire stack of protons can then be slowly accelerated to an energy of 32 GeV. Collisions occur in all eight intersection regions (**Fig. 32**), permitting an extensive and varied research program. Experiments at this facility have made important contributions to high-energy physics, particularly in the study of total cross sections, the production of particles which have large transverse momenta, direct lepton production, and scaling laws in weak and electromagnetic interactions. The rings have also been used to study antiproton-proton, deuteron-deuteron, and alpha-alpha collisions. The ISR has now been decommissioned and its facilities and resources allocated to other CERN projects.

In 1978, an ambitious effort was started at Brookhaven National Laboratory to construct a proton storage ring of 400 GeV in each ring. This facility, named ISABELLE, was designed to

Fig. 32. Inside of the ISR tunnel, with most equipment installed during a late stage of construction. One of the intersection points is clearly visible. (*From K. Johnsen, Nucl. Instrum. Meth.*, 108(2):205–223, 1973; Photo CERN)

have six intersection regions in which collisions would occur with a luminosity at least an order of magnitude higher than that available at the ISR and at energies appropriate to study the wide range of physics questions expected in the energy range of the masses of the intermediate vector bosons, the Z^0 and the $W^{\pm}$. Superconducting magnets were fundamental to the design of this facility in order to reduce construction and operating costs and to reduce the circumference of the machine, a consideration for controlling beam instabilities which could limit the luminosity. Technical difficulties with the superconducting magnets forced some delay and cost escalation in the project, and in 1983 the project was canceled even though the technical problems had been solved and the project was half completed. This decision was made because the high-energy physics community believed that the remarkable success of the antiproton-proton ($\bar{p}p$) collider at CERN demonstrated that the physics questions to which ISABELLE was addressed could be explored with the $\bar{p}p$ technique and the funds and effort for ISABELLE construction could be better used as part of an effort to construct an even larger proton storage ring, the superconducting supercollider (SSC).

The antiproton-proton collider exploits the same feature that is used in the electron-positron storage ring; that is, protons and antiprotons having opposite electrical charge but identical mass can be stored circulating in opposite directions in the same ring. Thus, a large synchrotron such as the CERN 400-GeV superproton synchrotron (SPS) can be converted into a powerful collider facility. The chief problem, accumulating enough antiprotons to achieve an adequate luminosity, was neatly solved at CERN. The components of their collider systems are shown in **Fig. 33**. Antiprotons are created in collisions of protons from the PS with nuclei in a target. These antiprotons are focused and collected in an accumulator ring. When the antiprotons are created, they do not all emerge from the target with the same energy and direction. In the focusing fields of the accumulator ring, these variations in energy and angle correspond to large spreads in the velocities of the particles relative to the average velocity; that is, the antiprotons are "hot" and must be "cooled" in order to reduce the beam size enough for the beam to be used in the collider. In the accumulator ring, elaborate feedback loops are employed which damp the average motion arising from the statistical fluctuations in the beam population. Because of the relative motion of the particles, the feedback loops are continuously exposed to new samples of this population. This combination of damping of average motion and mixing to generate new averages results in a net cooling effect. The theory of this stochastic cooling process was developed at CERN, and the high bandwidth electronic detectors and beam deflectors for the feedback loops have been successfully deployed. In operation, a period of about 24 h is required for 10^{11} antiprotons to be accumulated and cooled. These are then injected into the PS, accelerated to about 10 GeV, and injected into the SPS. A bunch of protons is then accelerated in the PS and also injected into the SPS. The counterrotating protons and antiprotons are then accelerated to 270 GeV. At this final energy, the protons and antiprotons are made to collide in special straight sections of the ring where elaborate detectors are placed. A luminosity in excess of 10^{29} $(cm^2 \, s)^{-1}$ has been achieved. This overall configuration, while complicated, has been spectacularly successful. The discovery of the charged intermediate bosons $W^{\pm}$ was announced in 1982 and the Z^0 was reported in 1983.

At the Fermi National Accelerator Laboratory, an even more ambitious $\bar{p}p$ collider has been constructed. The design of this facility is based on the same fundamental principles as the CERN machine but uses the Energy Saver/Doubler Synchrotron to achieve energies of 1000 GeV in each beam. By exploiting improvements in the antiproton accumulation and cooling techniques, this facility is also designed to have a higher luminosity of 10^{30} $(cm^2 \, s)^{-1}$.

The superconducting supercollider (SSC) is visualized as a proton-proton storage ring with energies up to 20 TeV in each beam. The operation of the machine would be very similar to that of the ISR. Very high luminosity should also be possible. The circumference will be at least 40 mi (65 km) and might be more than twice that large. The problem which dominates the conceptual design of this device is how to construct the magnet system cheaply enough to make the construction of the machine financially feasible.

Superconducting magnets with very small beam aperture (perhaps as small as 1 in. or 2.5 cm in diameter) are being considered. The magnetic field strength chosen may impact the cost of the project. Low-field magnets are cheap to construct, but they force the circumference and circumference-related costs to be higher. Higher-field magnets operating at fields up to 8 T may

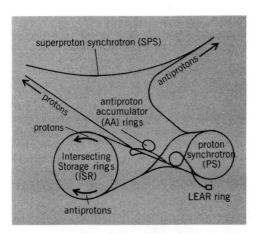

Fig. 33. Layout of the various accelerator rings of the CERN laboratory. Antiprotons are collected and cooled in the antiproton accumulator (AA) for subsequent injection into the superproton synchrotron (SPS). (*After D. B Cline, C. Rubbia, and S. van der Meer, The search for intermediate vector bosons, Sci. Amer., 246(3):48–59, March 1982*)

be possible to reduce circumference-related costs, but such magnets may require extensive research and development to realize. These high-field magnets might use niobium-tin instead of niobium-titanium as the superconductor. Niobium-tin has superior superconducting properties, but it is very brittle and requires expensive fabrication techniques.

Limitations. The foremost technical problem facing the builders of proton storage rings is achieving currents of sufficient intensity and density to provide adequate reaction rates for the experiments. There are fundamental considerations which limit the current that can be stacked. First, the current density of any accelerator is limited by fundamental thermodynamic arguments which apply to the source of ions in the injector. Given economic limits on size of magnets and vacuum chambers, the source brightness then becomes a limit on the current which can be stacked in the storage ring. The second limit is imposed by self-fields from the beam. The large charge density of the beam of a proton storage ring generates electrostatic fields of several kilovolts per meter and magnetic fields of several hundred amperes per meter (space charge fields). These electromagnetic fields are sufficient to cause deterioration of the beam quality or even sudden loss of the beam through collective oscillations of the beam particles forced by the self-fields. The third limit is set by a beam-induced pressure rise in the vacuum chamber. Ions, formed by collisions of beam particles with molecules of the residual gas, are driven into the vacuum chamber walls by the electrostatic potential of the beam. The ion impact causes desorption of molecules from the wall surface. At a sufficiently large beam current, the result is an avalanche increase in pressure. Loss of the beam or, at least, deterioration of its quality then results from beam-gas scattering. A fourth limit is imposed by the large stored energy of the beam. The stored energy in each of the beams in a proton storage ring can be many megajoules and can cause serious damage to components of the facility if the beam is dumped in an uncontrollable fashion. A fifth limit is determined by the electromagnetic interaction of one beam with the other. The proton storage ring is more sensitive to the beam-beam instability than is the electron storage ring because in the latter device the diffusive growth of the beam size from the beam-beam interaction is ameliorated by radiation damping.

Accelerator specialists have developed effective techniques for understanding these limitations and, in some cases, avoiding them. The ISR, for example, has realized a luminosity of 10^{32} $(cm^2\ s)^{-1}$, which exceeds the design luminosity by more than an order of magnitude. Special magnets can be deployed in the intersection regions for reducing the beam size so that the luminosity for a given available current can be enhanced. This technique (called the low beta insertion) is common to all newer colliders. Similarly, the beam can be tightly bunched azimuthally. The use of bunched beams has proved remarkably successful in the $\bar{p}p$ collider at CERN, and this feature will almost certainly be a part of future colliders.

Technology. The technology of constructing a proton storage ring is very similar to that for a proton synchrotron but with three important differences. The magnets must be of greater precision and stability and have an elaborate set of corrections to optimize stability of the proton beam against perturbations by space charge forces and imperfections in the magnets. Second, the vacuum requirements are very severe. A pressure of 10^{-10} torr (10^{-8} Pa) or less is required to prevent excessive beam loss and to minimize the background in the experiments caused by beam particles colliding with residual gas molecules. The vacuum chamber also must be designed and constructed to minimize the pressure bump phenomenon described earlier. Finally, in a storage ring there is a much more intimate connection between the machine and the experiments than is the case for conventional accelerators. The experiments must be designed about the specific properties of the beams in the interaction region, and the machine builders must build the storage rings as if they are part of the experiments. In particular, the machine builders must pay particular attention to the problem of maintaining good access to the intersection regions and keep these areas free of magnets, pumps, and other equipment incompatible with the experiments. The beam losses must be kept very small from all causes because the experiments, which are typically designed to study very low-event rates, are extremely sensitive to background events associated with the lost particles. Loss rates of less than a part per million per minute have been achieved at the ISR. At this rate nearly half of the loss is associated with real beam-beam collisions. This loss rate is so low that it is possible to have continuous experimental physics runs of several days without refilling the rings.

Superconducting magnets will almost certainly be used in future colliders. While this technology caused some difficulty for the ISABELLE project, the problems were ultimately resolved and completely satisfactory magnet models were produced. At the Fermi National Accelerator Laboratory, the successful operation of the superconducting energy saver/doubler ring has shown that the performance is satisfactory for collider-mode operation for the $\bar{p}p$ project.

ELECTRON RING ACCELERATOR

An electron ring accelerator (ERA) is a device for imparting to protons or other positive ions kinetic energies of several to several hundred MeV per nucleon by the electron drag force of stable ringlike configuration of electrons that is accelerated to the requisite velocity.

Collective acceleration. Conventional particle accelerators are subject to limitations in the strengths of the fields that can feasibly be employed to guide and accelerate the desired particles, and accordingly may be massive and expensive. This situation has motivated the development of techniques for accelerating a small group of particles by means of their interaction with another group of charges or with some form of plasma wave. This collective acceleration employs fields developed by charged particles within the accelerator itself. Such fields can be quite large and are not subject to the technological limitations that restrict externally applied fields.

Electron ring collective field. A direct and attractive method of collective acceleration is employed in the electron ring accelerator. The fields of significance in this machine are produced by a slender ring of electrons circulating in a magnetic field. The electron ring is formed by injecting an electron beam of several hundred amperes into a magnetic field that is rapidly deformed to trap one or more injected turns into closely spaced circular orbits and that then is further pulsed to shrink the ring dimensions to suitable size. The pulsed magnetic field serves both to increase the energy of the circulating electrons (through the transformer action of induced electric fields) and to reduce the major and minor dimensions of the ring (compression). In alternative experimental devices (at the University of Maryland) these rings were formed by injecting a hollow relativistic electron beam into a cusp-shaped magnetic field, wherein the forward motion of the electrons was converted into circulatory motion.

At the edge of the ring of major and minor radii R and a (R much greater than a), formed of N circulating electrons of charge e, electrostatic fields as great as those given by Eq. (10) arise.

$$E = \frac{Ne}{4\pi^2 \epsilon_0 R a} \tag{10}$$

Thus a field of 7.3×10^7 V/m is obtained at the edge of a ring of 10^{13} electrons, with $R = 0.025$ m (1 in.) and $R/a = 10$. This electrostatic field binds the ions to the ring, since they are formed in the ring volume by electron collisional ionization of neutral gas introduced into the vacuum chamber. The trapped ions can be accelerated with the electron ring, if this is accelerated in a direction along the magnetic field lines. The ions obtain substantial kinetic energies, which are in the initial stages and for small ion loading larger by a factor $M/\gamma m_0$ than the energy associated with the transverse velocity imparted to the individual electrons, where M/m_0 denotes the ratio of ion to electron rest mass and γ is the relativistic factor for the electrons. Hence the effectiveness of the externally applied acceleration field can be amplified by this factor—for example, by 60.4 for proton acceleration in a ring of 15-MeV electrons ($\gamma = 30.4$). The acceleration mechanism, moreover, is applicable to ions of any desired type, and indeed ions of differing charge-to-mass ratio can be simultaneously accelerated.

Figure 34 shows the cross section of an experimental electron ring compressor. The compression of the electron rings, which are formed by injecting electrons perpendicular to the plane of the figure into the median plane of the compressor, is obtained by the sequentially pulsed coil sets 1–3. The magnetic field thus forms a "magnetic bottle." The equilibrium and the single-particle stability of the ring are ensured by the focusing force that is achieved by the field index n, defined by Eq. (11), where r is the distance from the center of the ring and B_z is the magnetic

$$n = -\frac{r}{B_z}\frac{\partial B_z}{\partial r} \tag{11}$$

induction perpendicular to the ring. The focusing is necessary to compensate for the difference between electrostatic repulsion and magnetic attraction of the ring electrons. The field index n

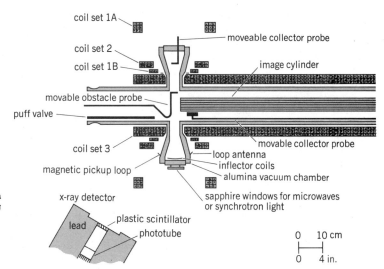

Fig. 34. Cross section of an experimental electron ring compressor employed at the Lawrence Berkeley Laboratory. (*Lawrence Berkeley Laboratory Rep. 2090*)

must lie between 0 and 1. The electron ring is released from the magnetic bottle by creating an unbalance between the currents in the left- and right-hand portions of the coils, and gets into the long solenoid to permit study of collective acceleration. The puff valve introduces a short gas burst for ion loading of the electron rings. The probes, loops, and x-ray detector provide information concerning the performance of the entire device.

Ring stability. Potentially destructive instabilities can arise in the motion of the individual electrons and in collective motion of the ring as a whole. The resonant excitation of single-particle transverse oscillations can be avoided during compression by selecting suitable values for the field index n, by controlling other characteristics of the field, and by passing rapidly during compression through the most dangerous values of the field index. When the ring is released into the acceleration column, the axial magnetic field no longer provides a focusing action, and the beam must then be preserved by the agency of image fields arising from nearby conducting (or dielectric) material.

Collective instabilities have proven to be more troublesome in achieving high-quality electron rings. An azimuthal instability, which can lead to bunching of the electron beam, is of particular importance. Effective means for the suppression of this instability include the provision of conducting surfaces close to the ring and the presence of an appreciable spread in the energies of the circulating electrons (Landau damping). An energy spread acts to increase the radial minor dimension of the ring (thus diminishing its maximum electric field), however, and may preclude the placement of conducting surfaces close to the ring. Additional instabilities may result from oscillations of the ring with respect to the trapped ions, but are subject to some control through adjustment of the ion abundance. A theoretical analysis of instabilities suggests that the dominant azimuthal instability should not preclude the attainment of peak fields of approximately $80B$ MV/m (effective acceleration fields of about $40B$ MV/m) in a ring situated in a guide field of B teslas. It is noteworthy that this limit increases in direct proportion to B.

Developmental work with electron rings has employed techniques that act to suppress instabilities in an electron ring compressor. The effect of single-particle resonances has been greatly reduced by the use of a fast compression cycle (compression time less than 10 microseconds, and all dangerous resonances traversed in less than 100 nanoseconds). Collective radial ion-electron oscillations also were suppressed by use of shaped conducting surfaces designed to modify the field index n and the derivative $\partial n/\partial r$ at the end of compression. An image cylinder (Fig. 34), a "squirrel-cage" structure formed of longitudinal conducting strips near the ring, has been used to provide image fields designed to preserve the integrity of the beam during the initial portion of the acceleration period. The longitudinal bunching instability has been greatly reduced by carefully designed conducting surfaces close to the ring.

Methods of acceleration. The acceleration of the electron ring and ions attached to it requires the release of the ring from the magnetic bottle in which it is held and the provision of fields that will accelerate the ring axially down an acceleration column. An axial magnetic field, together with suitable provisions for maintenance of ring stability, must be present throughout the length of this column in order to preserve the electron ring configuration. For acceleration to high energies the use of a sequence of rf or pulsed cavities appears necessary. Initial experiments have focused, however, on a simpler method of acceleration wherein the force that accelerates the ring arises from a small radial component B_r of the static magnetic field in the acceleration column (magnetic expansion acceleration). With this method the axial velocity imparted to the electrons arises specifically from the Lorentz force $ev_\phi B_r$ (where v_ϕ is the azimuthal component of the velocity) acting on them, and the axial motion of the electrons and ions arises entirely at the expense of the energy initially associated with the azimuthal motion of the electrons.

Results of experiments. Two groups have reported evidence for ion acceleration in short magnetic-expansion acceleration columns. A group at Dubna, in the Soviet Union, reported in 1971 the production of He^{2+} ions of 30 MeV energy by magnetic-expansion acceleration through a distance estimated as 16 in. (0.4 m). In 1978 this group succeeded in accelerating nitrogen ions with heavily loaded electron rings. An ion energy gain of about 4 MeV/nucleon-m over an acceleration length of about 20 in. (0.5 m) for ion numbers of as much as $(5\pm 2)\times 10^{11}$ ions per cycle was achieved. The acceleration parameters were measured by the method of activation analysis. **Figure 35** shows the dependence of the target activity ΔN of a deuterium target on the partial pressure of nitrogen, argon, and xenon mixtures, indicating that a certain pressure and hence ion loading range resulted in sufficient focusing and collective acceleration conditions. A 20 MeV/nucleon heavy-ion electron ring accelerator will be constructed as an injector for a high-energy heavy-ion accelerator.

The well-monitored experiments at the Max Planck Institute (West Germany) with rings of relatively low holding power produced both hydrogen and helium ions of approximately 400 keV kinetic energy by acceleration through a few centimeters. The acceleration of the electron rings with and without hydrogen or helium loading was measured with the aid of magnetic probes. **Figure 36** shows measurements of the acceleration of electron rings with and without hydrogen loading over the axial distance z from the compression plane. The increase of the acceleration with z can be used to determine the holding power of the electron rings from the maximum acceleration up to which the difference in the ring inertia persists. The holding power of this experiment (performed in 1975) corresponds to 3 to 4 MV/m, a value which was subsequently

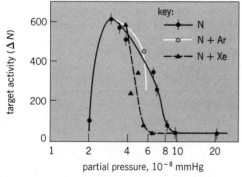

Fig. 35. Dependence of the target activity of deuterium on the partial pressures of nitrogen, argon, and xenon. 10^{-8} mmHg = 1.33×10^{-6} Pa. (*After V. P. Sarantsev and I. N. Ivanov, Development of collective acceleration methods, in H. J. Doucet and J. M. Buzzi, eds., Joint Institute for Nuclear Research, Proc. High-Power Beams 81, Ecole Polytechnique, Palaiseau, France, pp. 691–698, 1981*)

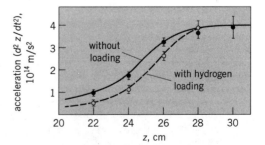

Fig. 36. Acceleration of the electron rings with and without hydrogen loading measured along the distance z from the compression plane. (*After U. Schumacher et al., Collective acceleration of protons and helium ions in the Garching ERA, IEEE Trans. Nucl. Sci., NS-22(3):989–991, 1975, and Phys. Lett. 51A:367–369, 1975*)

slightly improved. The collective acceleration of helium ions was confirmed by nuclear track registration in thin cellulose nitrate foils.

In addition to its potential for achieving collective-field acceleration of ions, a compressed electron ring offers a unique means for the production and spectroscopic study of highly stripped ions as well as for an intense synchrotron radiation source.

Bibliography. V. D. Barger and F. Halzen (eds.), *From Colliders to Supercolliders*, 1987; P. Blasi and R. Ricci (eds.), *Proceedings of the International Conference on Nuclear Physics*, Florence, 1983; D. B. Cline, C. Rubbia, and S. van der Meer, The search for intermediate vector bosons, *Sci. Amer.*, 246(3):48–59, March 1982; F. T. Cole (ed.), *Proceedings of the 12th International Conference on High Energy Accelerators*, Fermilab, August 11–16, 1985; H. Edwards, The tevatron energy doubler: A superconducting accelerator, *Ann. Rev. Nucl. Part. Sci.*, 35:605–660, 1985; S. Humphries, *Principles of Charged Particle Accleration*, 1986; J. D. Jackson, Physics at the superconducting supercollider, *Amer. Sci.*, 72:151–155, 1984; C. Joshi and T. Katsouleas (eds.), Laser Acceleration of Particles, AIP Conf. Proc. 130, 1985; I. M. Kapchinskiy, *Theory of Linear Resonance Accelerators*, 1985; S. P. Kapitza and V. N. Melekhin, *The Microtron*, 1978; R. D. Kohaupt and G. A. Voss, Progress and problems in performance of e^+e^- storage rings, *Annu. Rev. Nucl. Part. Sci.*, 33:67–104, 1983; E. M. McMillan, A history of the synchrotron, *Phys. Today*, 37(2):31–37, February 1984; M. Month et al. (eds.), *Physics of High-Energy Particle Accelerators*, American Institute of Physics Conf. Proc. 127, 1985; 1965 Particle Accelerator Conference, *IEEE Trans. Nucl. Sci.*, NS–12, no. 3, 1965, and similar conferences every 2 years following; C. L. Olson and V. Schumacher, *Collective Ion Acceleration*, Springer Tracts in Modern Physics, vol. 84, 1979; S. Penner, CW electron accelerators for nuclear physics, *IEEE Trans. Nucl. Sci.*, NS–28:2067–2073, 1981; *Proceedings of the 9th International Conference on Cyclotrons and Their Applications*, 1981; R. E. Rand, *Recirculating Electron Accelerators*, 1984; N. Rostoker and M. Reiser (eds.), *Collective Methods of Acceleration*, 1979; W. Scharf, *Particle Accelerators and Their Uses*, 2 vols., 1986; E. Ventura and P. Thieberger (eds.), 4th International Conference on Electrostatic Accelerator Technology, Buenos Aires, April 15–19, 1985, *Nucl. Instrum. Meth.*, 1986.

RESONANCE TRANSFORMER
Raymond G. Herb

An electrostatic particle accelerator in which the high-voltage terminal oscillates over the voltage range $\pm V$. These machines are used principally for acceleration of electrons. Electron current is allowed to pass only when the terminal voltage is near its peak value of $-V$. Thus the energy spread of the electrons can be held to a moderate value.

The resonant transformer consists of a low-voltage primary winding which surrounds the lower end of a high-voltage coil stack. The stack is made up of a number of thin flat windings called pancake coils, in which the multisection vacuum tube is coaxially mounted. The inductance and capacitance of the high-voltage secondary have values such that the resonant frequency of this circuit is equal to the frequency at which primary power is supplied. These machines utilize 180-Hz primary power. The resonant transformer and accelerating tube are housed in a steel tank, and high-voltage insulation is provided by the use of sulfur hexafluoride gas. *See* Particle accelerator.

Resonance transformers are manufactured in two different sizes: a 1-MeV, 5-mA unit, and a 2-MeV, 6-mA unit. They are used for industrial radiography, for x-ray therapy, for food and drug sterilization, and for the processing of plastics. *See* Cockcroft-Walton accelerator.

TANDETRON
Kenneth H. Purser

A constant-voltage (dc) particle acceleration system which produces a directed stream of high-energy atoms. The species of these atoms can be from most parts of the periodic table. These accelerators are used to analyze the composition of materials and their structure, and to change the physical and electrical properties of materials.

Applications. Most analyses are accomplished by impinging the fast atoms onto a sample and by detecting the energy of those atoms that are scattered in the direction opposite to that of the incident high-velocity ions. Another analytical procedure involves nuclear reactions that can be induced between atoms in the incoming stream and trace nuclei present in the sample. Both methods are used to measure the depth distribution of specific impurities with very high sensitivity. In addition, the physical and electrical properties of materials, including the semiconductors silicon and gallium arsenide, can be dramatically changed by injecting energetic atoms from the tandetron directly into the crystal lattice. Often this produces new materials that cannot be created by any other technique.

Acceleration principle. The acceleration principle used to produce these streams of high-energy atoms is based upon a tandem acceleration technique (see **illus.**). For example, to accelerate helium atoms to an energy of 3 MeV, a well-collimated stream of positively charged He^+ ions is extracted from an ion source and directed through a region containing lithium vapor at a pressure of about 7 pascals (50 millitorr). Here, the He^+ ions from the source interact with lithium atoms and pick up electrons to produce He^- ions. For each incident He^+ ion there is a probability of approximately 1% that two electrons will be transferred to the He^+ ion, converting it to the negative He^- species. These negative ions are analyzed by deflection in a magnetic field so that all unwanted ions are rejected and only the uncontaminated He^- particles are directed into the acceleration section. SEE ION SOURCES.

Within the acceleration region, the negative He^- ions are attracted toward the positive-polarity voltage terminal and are focused to a narrow waist so that they will pass through a small-diameter stripping tube. Here, oxygen gas is introduced to create a region of comparatively high pressure (0.7–1.3 Pa or 5–10 millitorr) where interactions between the 1-MeV He^- ions and gas molecules cause electrons to be stripped from the negative ions, changing back their overall

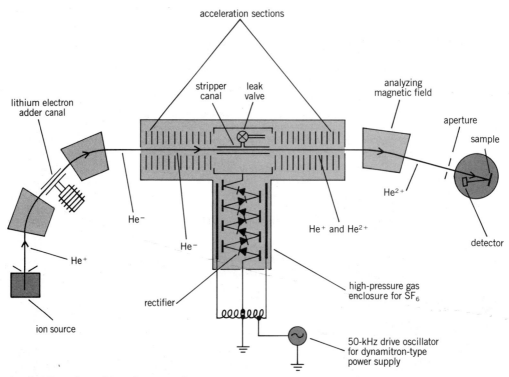

Acceleration scheme for tandetron accelerator.

charge from negative to positive. For helium, the maximum charge on each ion is +2 and, as these ions are repelled from the terminal in the second stage of the acceleration, they receive an additional 2 MeV so that at the exit from the accelerator they have a total kinetic energy of 3 MeV.

Reliability and safety. A unique feature of this acceleration scheme is that the ion source is located at ground potential outside of the high-pressure enclosure, rather than within a high-voltage terminal where electrical sparks can badly damage the equipment. Also, in this scheme, the ion source is always available for adjustment and maintenance without removing pressurized gas from the enclosing tank. For additional reliability, only simple nonmechanical components are located within the insulating sulfur hexafluoride (SF_6) gas pressurized to 600 kilopascals (90 lb/in.^{2} or 6 atm): an all-solid-state voltage generator, the positive- and negative-acceleration tubes, the stripping canal and its variable leak.

The accelerator system has a very low production rate for nuclear radiations and x-rays. Bremsstrahlung and x-rays, always produced during the stopping of fast electrons at surfaces within the accelerator, are kept to low levels by using high-vacuum techniques which minimize the production of the offending secondary electrons and by the use of transverse and magnetic suppression fields which force electrons away from the acceleration axis and cause them to impinge at low energy onto titanium surfaces. This design ensures that the x-ray flux is negligible in the operating area outside of the accelerator pressure vessel. This feature makes it possible to use the tandetron in an ordinary analytical laboratory without the need for concrete shielding blocks or walls.

Accelerating voltage. The 1-MV accelerating voltage is developed by solid-state rectification of a 50-kHz radio-frequency voltage. The solid-state de- vice substitutes for the mechanical charge transport system used in many electrostatic accelerators. The design of this power supply is based on the dynamitron principle, an arrangement which allows over 100 milliamperes of charging current to be reliably produced at voltages as high as several million volts. Such high current capacity is important when the tandetron is used for the production doping of semiconductors and for materials modification. Presently, 4-in.-diameter (100-mm) semiconductor wafers can be implanted with boron atoms to a concentration of 2×10^{13} atoms per square centimeter at the rate of about 100 per hour. SEE DYNAMITRON ACCELERATOR; PARTICLE ACCELERATOR.

DYNAMITRON ACCELERATOR
HARVEY E. WEGNER

A particle accelerator that utilizes a steady high-voltage direct-current (dc) potential to accelerate charged particles. The process is similar to that used in electrostatic accelerators; however, the high-voltage potential of the Dynamitron accelerator is produced by electrical means rather than by the mechanical charge transport system utilized in electrostatic accelerators. The high voltage is produced by a large number of dc voltage sources, all of which are connected in series. Each of these dc sources receives its power by means of individual parallel capacitive coupling to a radio-frequency (rf) power system operating at 100 kHz. This high voltage can then be used for the acceleration of particles in a manner similar to that used in any other kind of dc accelerator. Although other accelerators, for example, the Cockcroft-Walton machines, use various types of cascade voltage-multiplying circuits, the Dynamitron principle makes possible considerably higher operating particle currents. The advantages of the Dynamitron arise from its use of high frequency and low capacitance; other systems use low frequency and high capacitance, the high-energy storage of which sometimes leads to component failure. Invented by M. R. Cleland in 1956, Dynamitron accelerators are used in industrial, medical, and basic research applications. SEE COCKCROFT-WALTON ACCELERATOR; VAN DE GRAAF GENERATOR.

The voltage generator and acceleration column are housed within a high-pressure vessel filled to about 7 atm (700 kilopascals) pressure with sulfur hexafluoride (SF_6) gas, which provides the necessary insulation for the high-voltage dc potential and rf driving voltage.

Dynamitron accelerators can produce electron currents as high as 25–100 mA at their rated maximum voltage; electrostatic generators are capable of only 1 mA, or 2, at most. Although electrostatic accelerators can be very large, up to rated voltages of 25–30 MV; the practical limit of size for a Dynamitron-type accelerator is probably 8–10 MV; the largest machines constructed are rated at 5 MV.

The operating principle is best seen in the **illustration**, which shows the basic components. The two large, semicylindrical rf electrodes operate at approximately 150 kV at 100 kHz, with the resonant inductor and oscillator system as shown. Each split corona shield on the high-voltage column of the accelerator acts as an rf pickup electrode, which in turn drives an individual rectifier. As a pair, the split corona shields also help provide a smooth dc potential distribution, just as do the circular equipotential rings in an electrostatic accelerator.

As originally conceived, the Dynamitron accelerators utilized a high-voltage rectifier tube for the rf rectification states; however, modern units use a special solid-state rectifier which simplifies the design and increases the maximum current capability. Power is coupled into each rectifier through the shield-to-electrode capacitance, as indicated.

The Dynamitron-type accelerator is also produced in a two-stage tandem configuration, as are electrostatic accelerators. Two such machines, each rated at 4 MV (8-MeV proton beams), are in use in basic research in the United States and in Germany. As an electron accelerator, the Dynamitron is extremely powerful and is used extensively for the cross-linking of plastics, vulcanization of rubber, and sterilization of medical products, as well as other industrial applications.

The smaller, less powerful version of the Dynamitron tandem accelerator is called a Tandetron. These tandem accelerators are made in different sizes varying from 1 to 3 MV and have a

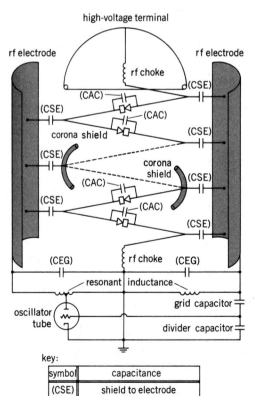

Radio-frequency coupling and direct-current rectification system utilized in the Dynamitron accelerator. (*After R. F. Shea, ed., 1st National Particle Accelerator Conference, IEEE Trans. Nucl. Sci., NS-12(3): 227-234, 1965*)

power supply capable of only 1 mA, in contrast to the 100 mA or more of a Dynamitron. The Tandetron is used as an analytical system by the semiconductor industry and in ^{14}C analysis as an ultrasensitive mass spectrometer. SEE CHARGED PARTICLE BEAMS; PARTICLE ACCELERATOR.

COCKCROFT-WALTON ACCELERATOR
J. C. OVERLEY

An electrostatic particle accelerator characterized by the method used to obtain the dc accelerating voltage. This type of machine was used in 1932 by J. D. Cockcroft and E. Walton in studying the first nuclear reactions induced by an accelerated ion beam. Since that time, a number of variations have been developed and applied to a wide range of problems in both science and industry.

In its most rudimentary form, a Cockcroft-Walton power supply contains a high-voltage transformer and a voltage multiplying circuit (**Fig. 1**). The multiplier circuit consists of two stacks of series-connected capacitors, cross-linked by voltage rectifiers. One set of capacitors is connected to the transformer. During the negative portion of the transformer cycle, the first capacitor in this stack C_1 becomes charged. During the positive portion, some of this charge is transferred to the first capacitor in the second set C_2. This process continues, with the transformer effectively pumping charge through the rectifier chain, until all capacitors are charged. Ultimately, the potential difference across each capacitor is $2V$, where V is the peak transformer output voltage, and the voltage across each complete set of N capacitors is $2NV$. Since the second set is referenced to ground, the voltage across it is constant except for ripple caused by power drain. To reduce ripple, alternating voltages at frequencies of several kilohertz are usually supplied to the transformer. Direct-current voltages of either sign can be obtained depending on the orientation of the rectifiers. Early power supplies, which used vacuum tube rectifiers, tended to use a high transformer voltage V and few multiplying stages. Modern supplies, which use solid-state rectifiers, have many stages of multiplication with smaller transformer voltages.

The accelerator itself is basically an ion source which injects ions into an evacuated accelerating tube. The power supply voltage is connected across the tube. Either positive or negative ions or electrons may be accelerated depending on the nature of the ion source and the polarity of the accelerating voltage. Voltage gradients along the accelerating tube may be controlled by a resistive bleeder string or by connecting conducting sections of the accelerator tube to intermediate multiplier stages. In many applications, magnetic momentum analysis of the accelerated beam is used. SEE ION SOURCES.

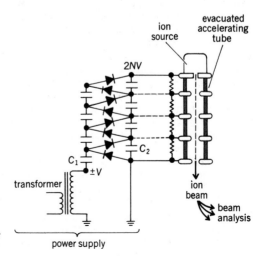

Fig. 1. Cockcroft-Walton power supply and accelerator. The multiplier circuit shown has four stages ($N = 4$).

The physical size of ordinary Cockcroft-Walton accelerators (**Fig. 2**) increases along with operating voltage, because voltage gradients along the insulating support members and in the surrounding medium must be limited to avoid electrical breakdown (sparking). The characteristically rounded corners of the structure help reduce these gradients. Air-insulated machines of the type shown have been built to produce up to 3 MV, but they are more commonly limited to 1 MV by size considerations. Air-insulated machines are easily accessible for maintenance, and their reliability in producing large ion currents has led to their use as injectors for larger accelerators and other stand-alone operations where size is not a limitation.

Fig. 2. The 750-kV Cockcroft-Walton accelerator at Fermi National Accelerator Laboratory. Protons injected by this machine into another accelerator ultimately reach energies of 500 GeV. The power supply is at left; the ion source is contained in the structure at the right from which the horizontal accelerating tube emerges. (*Fermi National Accelerator Laboratory*)

Tandem, gas-insulated accelerators. The size of a Cockcroft-Walton can be vastly reduced by housing it in a vessel filled with high-pressure insulating gas. A particularly compact version results when the power supply is connected to a tandem electrostatic accelerator. An ion source at ground potential injects negative ions into the pressurized accelerating system. The ions are accelerated to a region maintained at 1.0–3.0 MV by a power supply of the type discussed above. Within this region electrons are stripped from the ions, and the resulting positive ions continue to accelerate toward ground potential. Depending on the charge state of the positive ions, ion energies equivalent to two or more times the accelerating voltage can be obtained.

These machines are used in elemental analyses of materials through studies of characteristic x-rays produced by ion bombardment (particle-induced x-ray emission; PIXE) and in ultra-high-resolution mass spectroscopy. Radiocarbon age-dating of archeological artifacts is an example of the latter application. Ions produced from the artifact are injected into the accelerator. Of the many types of mass-14 ions injected, only ^{14}C survives the acceleration and charge-changing processes efficiently, and their numbers can therefore be determined with great precision.

Small industrial accelerators. The size of an accelerator can also be reduced if lower voltages are desired. Complete accelerator systems, including beam analysis equipment and target chamber assemblies, are commercially available in the 0.1–0.5-MV range. Hundreds of these small systems are in use. Examples of industrial applications include alteration of the electrical properties of semiconductors by ion implantation and alteration of the mechanical and thermal properties of materials by ion implantation or electron-induced radiation damage. SEE PARTICLE ACCELERATOR; RESONANCE TRANSFORMER; VAN DE GRAAFF GENERATOR.

VAN de GRAAFF GENERATOR
Harvey E. Wegner

A high-voltage electrostatic generator in which electric charge is carried on an insulated belt from ground to an insulated high-voltage terminal. The high-voltage terminal usually contains an ion source that produces protons, other types of positive ions, or electrons that can then be accelerated by the high voltage of the generator through an evacuated tube—the acceleration tube—back to ground potential. Van de Graaff generators producing positive high voltage for the acceleration of positive ions are usually used in basic research, while those producing negative high voltage for the acceleration of electrons are used in x-ray therapy, industrial radiography, various kinds of food and drug sterilization, and materials irradiation, as well as in research directed toward these goals.

Single-ended generator. Single-ended machines of the type shown in the **illustration**, rated from 1 to 6 MV, have been produced on a commercial basis. Most such machines, including support column and high-voltage terminal, are enclosed in a vessel filled with pressurized gas to prevent electrical breakdown. Most machines have used nonflammable mixtures of nitrogen and carbon dioxide and, more recently, sulfur hexafluoride, rather than pressurized air as the insulating medium to reduce fire hazards.

A motor at the base of the machine with a grounded pulley drives either a fabric or rubberized fabric belt over a second pulley that is mounted inside the high-voltage terminal and connected or grounded to it. A comblike set of points is arranged at the base of the machine to spray charge onto the belt, and a similar device removes the charge when it arrives at the high-voltage terminal.

Simple classroom demonstration machines (with no pressure vessel) have the same basic structure. The classroom machines use a simple insulated cylindrical support of plastic or composite material for the high-voltage terminal, while the research machines generally use composite glass or ceramic and metal structures.

The acceleration tube is fabricated of alternate rings of insulating glass or ceramic and

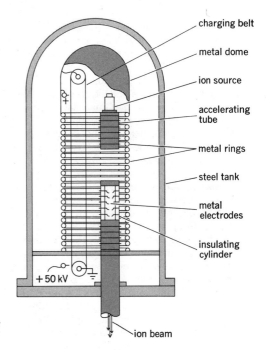

Schematic drawing of Van de Graaff generator for operation in high-pressure gas.

metal electrodes glued or bonded together to provide a hollow tubelike structure, extending from the high-voltage terminal to the base of the machine. Power for the operation of the ion source, which often requires a hot filament and an electric arc power supply, is usually supplied by a generator driven by a belt from the charging pulley. In some designs the generator is inside the charging pulley itself. The acceleration tube is evacuated by conventional vacuum pumps at the base of the machine.

The electrical voltage from the high-voltage terminal to ground is graded in a series of steps corresponding to each of the metal electrodes of the acceleration tube and the basic insulating support column itself. This gradient is provided by resistors strung together in series from the high-voltage terminal to ground, each connected to the metal electrodes and the metal rings. The resistor string then provides a stepped high voltage from the metal dome or terminal of the machine uniformly downward from terminal voltage to ground with each of the metal rings and corresponding electrodes in the tube. In this way, the injected ion from the source is accelerated by a continually accelerating electric field as it moves toward ground potential. After accelerating through the entire gradient of a machine operating at 1 MV, a charged particle has 1 MeV of energy for each unit charge.

Tandem generator. The tandem electrostatic generator, first produced for commercial sale in 1958, utilizes the basic Van de Graaff generator principle in a special form. The basic structural design is similar to that shown in Fig. 1; however, the pressure vessel is in the form of a long cylinder, and the high voltage terminal is supported in the middle of the cylinder by an insulating column structure extending from both ends of the cylinder. The charging belt operates from the high-voltage terminal to either or both ends of the cylinder, and an acceleration tube passes through from one end of the cylinder to the other so that accelerated particles can pass through the machine. The support column, high-voltage terminal, metal rings, and so forth, are all arranged concentrically about the axis of the cylindrical pressure vessel.

A negative-ion source outside one end of the cylindrical structure injects negative ions into the acceleration tube at one end of the machine, and these negative ions are then accelerated to the high-voltage terminal. There they pass through a thin foil usually made of carbon approximately 10^{-6} in. (0.02 μm) thick (or a gas stripper), and electrons are stripped away from them, leaving them as positive ions. The particles are then accelerated out of the other end of the acceleration tube just as though they were produced by an ion source in the high-voltage terminal. In this way, a 1-MV machine can provide 2-MeV protons, giving them 1-MeV acceleration as a negative ion and 1 MeV as a positive ion.

Even more advantage is gained when heavy ions of various atomic nuclei are accelerated. The negative ion again achieves 1 MeV acceleration with a 1-MV machine; however, the foil will probably strip the heavy ion to a charge state of two or three units of positive charge, which means it then achieves an additional 2 or 3 MeV of energy, reaching a total energy of 3 or 4 MeV rather than the 2 MeV achieved by a proton which has only one charge.

The tandem Van de Graaff has been widely used as a research instrument because of the convenience of having the ion source and complicated ion-producing devices of various kinds outside the accelerator at ground potential rather than inside the high-voltage terminal as in the conventional, one-stage Van de Graaff shown in the illustration. These tandem Van de Graaffs have gradually increased in size and capability from 6 MV up to 13 MV, and over 100 such machines have been sold for research throughout the world.

Large machines. The most popular large machine, first installed at Yale University in 1964, was known as the MP (Emperor). It is 18 ft (5.5 m) in diameter and 81 ft (25 m) long, and was originally warranted to produce 10 MV. With special modifications and improvements, the MP has been operated at voltages as high as 18 MV. A special version of the MP, with a larger tank diameter of 25 ft (7.6 m), at the University of Padua, Italy, operates at voltages up to 16 MV. A much larger machine, called an STU (stretched transuranium), warranted to operate at 20 MV, was offered for sale, but none was sold. However, the MP at Yale is being upgraded to (extended stretched transuranium) capability by installing the longer (100 ft or 30 m) and larger-diameter (25 ft or 7.6 m) tank required by an STU, and adding the additional components to make up the extra length of about 20 feet (6 m). This machine is expected to operate at 22 MV and, with special modifications, up to voltages as high as 25 MV.

The market for such large machines has diminished because of rapidly increasing cost.

Van de Graaff generators rated higher than 4.5 MV are no longer manufactured commercially, although lower-voltage machines will continue to be manufactured for various kinds of applications research and materials processing. However, pelletron electrostatic accelerators with terminal potentials ranging up to 20 MV continue to be manufactured commercially. They differ from the Van de Graaff's basic design only in detail (the charging belt is replaced with an insulated link chain), and have been installed in laboratories in the United States and several other countries. The vivitron, a very large 35-MV electrostatic accelerator of radically new design, dependent upon solid rather than gas insulation and incorporating seven intermediate shields between terminal and tank, is under construction in Strasbourg, France. SEE PELLETRON ACCELERATOR.

These electrostatic accelerators are largely proof against obsolescence, although the specific research programs utilizing them may change with time. Their excellent beam quality and uniquely high-energy resolution make them the accelerators of choice in a very wide spectrum of scientific areas. SEE PARTICLE ACCELERATOR.

PELLETRON ACCELERATOR
RAYMOND G. HERB

An electrostatic particle accelerator that utilizes charging chains consisting of steel cylinders joined by links of solid insulating material such as nylon. The metal cylinders are charged as they leave a pulley at ground potential, and the charge is removed as they pass over a pulley in a high-potential terminal.

Voltage range. Pelletrons range in maximum operating voltages from 200 kV to 25 MV and have beam current capabilities ranging from a few microamperes up to 0.8 milliampere. Machines to give 1 MV or more are enclosed in pressure tanks and are insulated by sulfur hexafluoride (SF_6) gas at pressures up to about 8 atm (800 kilopascals).

A large number of electrostatic accelerators of the belt-charged Van de Graaff type are in operation throughout the world with voltage capabilities of 10 MV and below. Many 10-MV Van de Graaff accelerators have been upgraded to 13 MV. Because of belt trouble above 10 MV, most upgraded machines have been converted to Pelletron charging. SEE VAN DE GRAAFF GENERATOR.

Accelerating stages. A single-stage pelletron to furnish high currents of positive ions and to furnish ions of all the noble gases has a positive ion source in the terminal. Usually ions entering the accelerating tube have a single positive charge, and these ions, when they emerge from the tube at ground potential, have an energy of 5 MeV if the terminal is at 5 MV. This arrangement can also be used to accelerate electrons if the terminal is at a negative polarity and an electron source is provided in the terminal. SEE ION SOURCES.

Commonly, two-stage (tandem) accelerators have utilized a double-ended column extending through the enclosing tank from one end to the other, with the high-potential terminal at the center. Negative ions enter at one end, are stripped of two or more electrons at the terminal, and receive a second acceleration as they proceed in a straight line through the high-energy accelerating tube to the other end of the accelerator.

An accelerator arranged like the 25-MV pelletron of the **illustration** is referred to as a folded tandem. The column extends in only one direction from the terminal, and a 180° magnet directs ions through a second accelerating tube. This arrangement requires less tank height and less building height than the straight-through tandem, but equipment in the terminal is more complex.

25-MV pelletron. The 25-MV accelerator installed at the Oak Ridge National Laboratory (see illus.) is housed in a steel tank approximately 10 m in diameter filled with SF_6 gas at a pressure of about 8 atm (800 kPa). The high-voltage terminal is supported by an insulating column made up of modules 60 cm (24 in.) high, each capable of withstanding 1 MV. Six charging chains passing over pulleys 60 cm (24 in.) in diameter extend up through the column to provide a 0.6-milliampere current to the terminal. Negative ions for this accelerator are generated in the cylindrical assembly to the lower left of the illustration, which goes to a negative voltage of 0.5 MV. Ions are formed into a beam, are deflected upward by a magnet, pass into the pelletron tank, and proceed through an evacuated accelerating tube to the terminal. Here they become positively

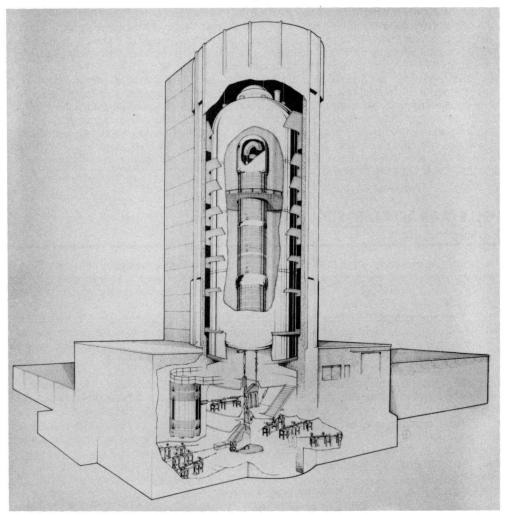

The 25-MV folded tandem pelletron at Oak Ridge National Laboratory. Service platform is stored at lower end of tank when machine is in operation.

charged as they pass through a thin foil or a small amount of gas which serves to strip off electrons; ions of medium or high atomic number may lose 15 or more electrons at the stripper. If the terminal magnet is set to select charge +15 ions of a particular species, these ions gain 15 × 25 MeV as they pass downward through the high-energy tube, and they emerge at the base with an energy of 400 MeV. If a second stripper is located one-third of the way down the high-energy tube, ions with a charge of +34 can be formed. Such ions will have an energy of 600 MeV when they emerge.

Comparison with other accelerators. Electrostatic accelerators utilize high voltage and commonly accelerate ions in only one or two stages, in contrast to machines such as cyclotrons or linear accelerators that provide high energies by a large number of relatively small energy increments. Usually, for ions of a given energy, the electrostatic machine is the most costly. Yet, electrostatic accelerators are used for more than one-half of the research work performed in nuclear physics. Their popularity is due to the quality of the ion beams. Ion currents are steady, and energy spread is very low, can be accurately determined, and easily changed. They are also used for work in atomic physics, solid-state studies, surface analysis, and in medical and industrial applications. *See* Particle accelerator.

BETATRON
Donald W. Kerst

A device by accelerating charged particles in an orbit by means of the electric field E from a slowly changing magnetic flux Φ. The electric field is given by $E = -(1/2\,\pi r_0)\,d\,\Phi/dt$ (in SI or mks units), where r_0 is the orbit radius. The name was chosen because the method was first applied to electrons. In the usual betatron both the accelerating core flux and a guiding magnetic field rise with similar time dependence, with the result that the orbit is circular. However, the orbit can have a changing radius as acceleration progresses. For the long path (usually more than 100 km), variations of axial and radial magnetic field components provide focusing forces, while space charge and space current forces due to the particle beam itself also contribute to the resulting betatron oscillations about the equilibrium orbit. In many other instances of particle beams, the term betatron oscillation is used for the particle oscillations about a beam's path.

Although there had been a number of earlier attempts to make a betatron, the complete transient theory guided the construction of a successful 2.3×10^6 eV accelerator, producing x-rays equivalent to the gamma rays from 2 g of radium, at the University of Illinois in 1940.

It can be shown that the flux change within an orbit of fixed radius must be twice that which would exist if the magnetic field were uniform. The flux can be biased backward by a fixed amount ϕ_0 to avoid iron saturation, as it was in the University of Illinois 320-MeV betatron.

The beam must be focused in both the radial direction and the axial direction (the direction perpendicular to the plane of the orbit). For radial focusing, the centripetal force F_0 required to hold a particle in a circular orbit of radius r must be supplied by the magnetic field B exerting a force F_M. At an equilibrium radius r_0, $F_C = F_M$, and if F_M decreases less rapidly than $1/r$ with increasing radius, there will be a net focusing force directed back toward r_0 when $r \neq r_0$. This focusing force will cause the radius r to undergo a simple harmonic oscillation about r_0. If the axial magnetic field B_z falls off less rapidly than $1/r$, say $B_z \sim 1/r^n$, then it can be shown that this oscillation has a frequency $\omega_r = \sqrt{1-n}\,\Omega$, where Ω is the angular velocity in the particle's orbit. Thus $n < 1$ is necessary for a stable radial oscillation giving a frequency less than the frequency of rotation.

In addition, axial focusing results from requiring the magnetic field to decrease with radius so that the lines of force are curved, giving a component of F_M on a displaced particle back toward the orbital plane, as shown in the **illustration**. It can be shown that this focusing force gives rise to an axial oscillation with frequency $W_z = \Omega\sqrt{n}$. Thus $0 < n < 1$ is necessary for complete focusing. This is the so-called weak focusing conditions for betatron oscillation in accelerators.

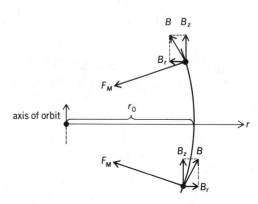

Magnetic fields and their resulting forces in the axial plane of a betatron orbit, which result in axial focusing.

By dividing the focusing structure into sectors of large and alternating gradients, $+n \gg 1$, so-called strong focusing results. This allows a small volume for the beam path, or direct-current focusing magnets 1with a finite change in orbit radius as the particle gains great energy. Such betatrons have been made.

Injection into an orbital oscillation is possible because the restoring force is proportional to B^2, which rises relatively rapidly at injection time. This increasing focusing force causes the oscillation amplitude to decrease as $1/\sqrt{B}$ and enables the particles to escape subsequent collision with a properly designed injection structure.

Collective effects from self-fields of the beam have been found important and helpful in injecting. Circulating currents of about 3 A are contained in the numerous industrial and therapeutic betatrons, although the average currents are below 10^{-7} A. Such beams have been extracted by using a magnetic shunt channel where $n = 3/4$ so that $W/\Omega = 1/2$. This field bump excites a half-integral resonance which throws the particles into the channel. *See* Particle accelerator.

ALTERNATING GRADIENT FOCUSING
Lloyd Smith

A configuration of transverse electric or magnetic fields suitable for focusing or confining a charged particle beam. Linear forces are achieved by using a four-pole (quadrupole) geometry (see **illus**.) with adjacent poles opposite in polarity. Ideally the poles are rectangular hyperbolas in cross section, but adequate linearity can be obtained even with round electrodes. According to Maxwell's equations, a magnetic or electric field which focuses in one transverse direction must be defocusing in the other, in contrast to optical lenses. However, if the polarity alternates in time or along the beam direction (alternating gradient), the net effect is to focus the beam in both directions, because the focusing effect is stronger the farther the particles are from the neutral axis. The net focusing force exceeds that which can be attained by any other external field configuration, thus the configuration is called strong focusing. With excessive focusing, the particle motion becomes unstable.

This focusing principle was discovered and developed for high-energy physics applications. It became possible to confine high-energy beams circulating in magnetic rings many kilometers (1 km = 0.6 mi) in circumference in vacuum envelopes a few centimeters (1 cm = 0.4 in.) in transverse dimensions. Thus small, and therefore economically acceptable, magnets could be used. A typical synchrotron or colliding beam storage ring consists of a lattice of bending magnets

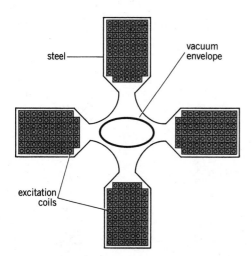

Cross-section of a typical magnetic quadrupole for a high-energy physics accelerator.

interspersed periodically with magnetic quadrupoles. Special configurations of quadrupoles bridge field-free regions necessary for the insertion of other equipment or focus the beams to submillimeter size for high-energy collisions. *See* P*article* *accelerator.*

Another application of alternating gradient focusing is in a light and compact mass filter well suited to upper atmosphere research and other applications. The device consists of four metallic rods to which dc and radio-frequency voltages specially tuned to one ion type are applied; only this ion type is transmitted to the detector. By varying the voltages, a range of masses can be scanned quickly to determine the gas composition.

PARTICLE SOURCES AND DETECTORS

Ion sources	456
Particle detector	467
Ionization chamber	470
Spark counter	476
Spark chamber	476
Cloud chamber	483
Bubble chamber	486
Junction detector	488
Crystal counter	494
Gamma-ray detectors	496
Scintillation counter	498
Liquid scintillation detector	501
Transition radiation detectors	502
Particle track etching	502
Cerenkov radiation	506
Time-projection chamber	508

ION SOURCES

ROY MIDDLETON AND JOSEPH L. MCKIBBEN

<small>J. L. McKibben wrote the first section of Polarized Ion Sources and the sections Conventional or Ground-State Source and Lamb-Shift or Metastable-Atom Source.</small>

Devices which produced positive or negative electrically charged atoms or molecules. In general, ion sources fall into three major categories: those designed for positive-ion generation, those for negative-ion generation, and a highly specialized type of source designed to produce a polarized ion beam. The positive-ion source category may further be subdivided into sources specifically designed to generate singly charged ions and those designed to produce very highly charged ions.

Desirable qualities of an ion source are large yield, high ionization efficiency, low energy spread, high brightness, and low emittance. Practical considerations such as reliability, long source life, and ease of changing ion species are also important.

Ion sources have acquired a wide variety of applications. They are used in a variety of different types of accelerators for nuclear research; have application in the field of fusion research; and are used for ion implantation, in isotope separators, in ion microprobes, as a means of rocket propulsion, in mass spectrometers, and for ion milling. SEE NUCLEAR FUSION; PARTICLE ACCELERATOR.

METHODS OF POSITIVE-ION FORMATION

The principal methods of positive-ion formation are electron impact, surface ionization, spark discharge, laser ionization, field ionization, thermal ionization, and sputtering.

Electron impact. A common method of ionizing a gas or vapor is to pass high-velocity electrons through it, with the ions being formed as a result of electron-atom collisions. Electron energies are typically a few 100 eV but in some special sources may be as high as 20 keV. An externally applied magnetic field is frequently used to cause the electrons to travel along a helical path and thereby increase the ionization efficiency. Examples of ion sources utilizing this concept are the duoplasmatron and the Penning ion source.

Surface ionization. Atoms possessing low ionization potentials can be ionized by allowing them to strike a heated surface having a high work function. Provided that the ionization potential of the atom is less than or about equal to the work function of the surface, there is a high probability that the atom will be thermally desorbed as a positive ion. The method is particularly well suited, though not entirely restricted, to producing ions of the alkali metals, all of which have ionization potentials of less than 5.4 eV. Some high work function metals are: platinum (approximately 5.7 eV), tungsten (approximately 4.5–5.5 eV), and rhenium (approximately 5 eV).

Spark discharge. There are several variations of this technique, but basically a spark is induced between two electrodes, one of which, at least, contains the element to be ionized. Generally speaking, the spark consists of a high-density, high-temperature plasma from which ions can be extracted. The spark can be produced by applying a high alternating potential between two fixed electrodes or by mechanically breaking contacting electrodes. SEE ELECTRIC SPARK.

Laser ionization. A focused beam from a high-power pulsed laser can be used to produce a small ball of dense plasma from essentially any solid, and positive ions can be extracted from this plasma. The high temperature of the plasma results in the formation of many multiply stripped ions and thus may prove a very effective method of generating highly charged positive ions.

In principle, lasers or other strong sources of electromagnetic radiation can be used to produce ions by photoionization. A photon can directly ionize an atom if its energy exceeds the ionization potential of the atom. Unfortunately, the probability of photoionization is low (the cross section is of the order of 10^{-19} cm^2), making it difficult to design efficient ion sources based on this process.

Field ionization. If an atom passes close to or gets absorbed on a very sharp point where the electric field exceeds a few times 10^{10} V/m, there is a probability that it will get ionized; the phenomenon is known as field ionization. Such large electric fields can be achieved in the vicinity of a specially sharpened tungsten needle placed close to an annular electrode, and gas or vapor passing close to the tip of the needle can be ionized.

Field emission or ionization is generally believed to be the underlying operating principle of a novel type of ion source known as the electrohydrodynamic source. In this source a conducting liquid, usually a metal, is allowed to be drawn down a fine-bore tube by capillary action. When an electric field is applied to the tip of the tube, the liquid meniscus, normally spherical, distorts and becomes conical. As the electric field is increased, the tip of the cone becomes sharper, and eventually the field at the tip becomes sufficiently large to cause field emission.

Thermal ionization. Although the term thermal ionization is ill-defined, it is generally used in the context of heating certain complex compounds, resulting in positive-ion emission. An example is the emission of lithium ions from a heated surface coated with β-eucryptite (a lithium aluminosilicate mineral). The technique has found extensive application in mass spectroscopy to produce a wide variety of ions. The sample to be ionized is usually mixed with silica gel and phosphoric acid and deposited on a rhenium filament. After a preliminary baking, the filament is introduced into the ion source and heated to the point of positive-ion emission.

Sputtering. When a solid is bombarded with energetic heavy ions, a fraction of the sputtered particles leaves the surface as ions. This fraction is usually too low for direct application in an ion source, but the sputtering process is frequently used to introduce solids into an electron impact source such as a Penning source.

METHODS OF NEGATIVE-ION FORMATION

All elements can be positively ionized, but not all form stable negative ions. For example, none of the noble gases forms negative ions. However, helium is an exception in that it does have a metastable negative ion with a lifetime of about 1 millisecond. The noble gases are not the only elements that do not form stable negative ions, but most form metastable ones with lifetimes long enough to permit acceleration. Nitrogen has an exceptionally short-lived metastable negative ion, and it is customary to accelerate either NH^- or NH_2^- molecular ions, both of which are stable.

Direct extraction. Most discharge sources, such as the duoplasmatron (**Fig. 1**), yield negative ions when the polarity of the extraction voltage is reversed. However, the yield is usually low, the electron current is high, and there are difficulties when operating them with elements other than hydrogen. These sources are now used almost exclusively to generate intense beams of hydrogen and its isotopes, and several important changes have been made to improve their performance. In 1959 it was discovered that the negative-ion-to-electron yield from a direct extraction duoplasmatron could be greatly improved if the intermediate electrode was displaced by a millimeter or so off the source axis (Fig. 1b). The introduction of cesium vapor into the discharge plasma greatly enhances the negative-ion yield. Negative hydrogen currents from such sources have been increased from a few microamperes to several tens of milliamperes.

Charge exchange. When a positive-ion beam, ranging in energy from a fraction of a keV to several tens of keV, is passed through a gas or vapor, some of the ions emerge negatively charged. At the optimum energy (depending upon the ion), and with vapors of lithium, sodium, and magnesium, the negatively charged fraction is quite high, ranging from 1 to 90%. The technique is also highly effective for the creation of metastable negative ions, such as helium.

Cesium-beam sputtering. In 1963 V. E. Krohn discovered that when a solid surface is sputtered with cesium positive ions, a surprisingly large fraction of the sputtered particles emerge as negative ions. He further noted that this fraction could be increased by almost an order of magnitude by overlaying the sputter surface with additional neutral cesium. Krohn's discovery seemed to pass unnoticed until the early 1970s, when highly efficient and versatile negative-ion sources involving cesium sputtering were developed.

Surface ionization. Just as positive ions of elements possessing low ionization potentials can be generated by thermally desorbing them from high work-function surfaces, negative ions having large electron affinities can be similarly generated on a low work-function surface. The method is particularly suited to generating negative ions of the halogens, all of which have electron affinities in excess of 3 eV. A particularly suitable low work-function surface is lanthanum hexaboride, which is reported to have a work function of about 2.6 eV.

POSITIVE-ION SOURCE CONCEPTS

Positive-ion source concepts include the duoplasmatron source for protons and the Penning ion source and ion confinement sources for multiply charged heavy ions.

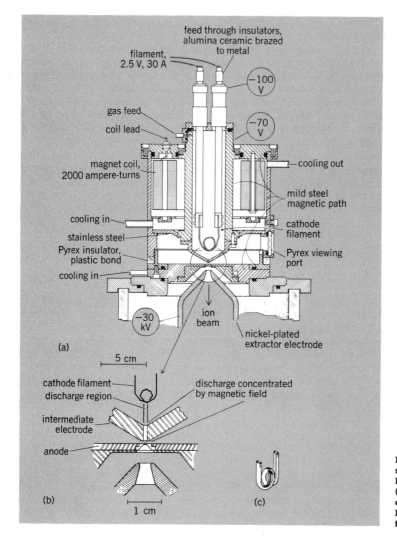

Fig. 1. Duoplasmatron ion source for accelerators. (a) Diagram of entire source. (b) Detail of discharge and extraction regions. (c) Drawing of cathode filament.

Duoplasmatron. The development of the duoplasmatron ion source (Fig. 1) in 1956 marked the beginning of the high-current era for proton sources. The duoplasmatron makes use of an arc discharge which is constricted as it passes into a very strong magnetic field shaped by iron or mild steel inserts in an intermediate electrode and anode. The beam is extracted at the point where the arc has reached a very small diameter and a very high brilliance. Sources of this type have been developed for accelerators.

Heavy-ion sources. The term heavy-ion is used to designate atoms or molecules of elements heavier than helium which have been ionized. As was mentioned earlier, ion sources that are used to generate such ions fall into two categories: those intended to form singly charged ions and those designed to produce multiply charged ions. A heavy ion can be singly ionized (one electron removed or added), can be fully stripped as in argon 18+, or can have any intermediate charge state.

Singly charged heavy ions are most frequently used in isotope separators, mass spectrographs, and ion implantation accelerators. These are much easier to generate than multiply charged ions, and frequently the experimenter has several source concepts to choose from, depending upon the application and the physical characteristics of the element to be ionized.

Multiply charged heavy ions are almost exclusively used in acclerators, particularly cyclotrons and linear accelerators. A very considerable amount of research has been devoted to developing such sources, and the bulk of this has been directed to perfecting the Penning ion source (often referred to as PIG source).

Penning ion sources. This source is based on a high-current gaseous discharge in a magnetic field with gas at a relatively low pressure (10^{-3} torr or 0.1 pascal). The source (**Fig. 2**) consists of a hollow anode chamber, cathodes at each end, a means for introducing the desired element (usually a gas), and electrodes for extracting the ions (not shown). The cathode may be heated to emit electrons, which then help to initiate the arc discharge current, creating the plasma in which the atoms are ionized. The discharge column between the cathodes (the plasma) consists of approximately equal numbers of low-energy electrons and positive ions. The electron density is much larger than can be accounted for by the primary electrons from the cathodes. The average energy of plasma electrons may range from a few volts to a few tens of volts. Electrons travel parallel to the magnetic field, are reflected from the opposite cathode, and make many traversals of the length of the hollow chamber. The electrons confined by the magnetic field and the cathode potential thereby have a high probability of making ionizing collisions with any gas present in the chamber.

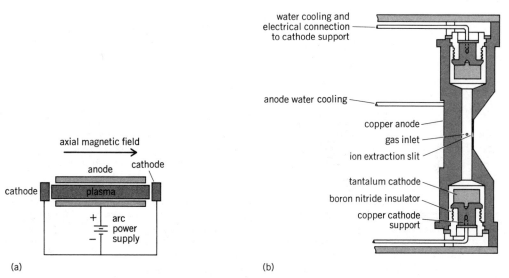

Fig. 2. Penning ion source. (a) Schematic diagram illustrating basic principles. (b) Section showing geometry. (*After J. R. J. Bennett, A review of PIG sources for multiply charged heavy ions, IEEE Trans. Nucl. Sci., NS-19(2):48–68, 1972*)

The net result of all the processes in the arc plasma is that some partially stripped atoms diffuse perpendicular to the magnetic field out of the arc, experience the field of the accelerating electrode, and are moved into the accelerator. The arc potential may be constant with time, or it may be pulsed so that ions are produced as needed by the accelerator.

High yields of charge states 1+ through 8+ (and for heavier elements, perhaps up to 12+) have been obtained for many elements of the periodic table. **Table 1** shows relative yields for selected elements. Each of the currents I^{Q+} for the charge state Q^+ has been normalized to the value for the 1+ state for each respective element. Thus for argon 5+ the observed current is three times as much as for argon 1+. For argon 8+ there is about 1% as much current as for the reference state.

Table 1. Penning heavy-ion source performance

	Charge state distributions (I^{Q+}/I^{1+})							
Element	1+	3+	4+	5+	6+	7+	8+	9+	12+
Argon	1	8.5	8	3	0.8	0.09	0.01		
Calcium	1	23	22	15	3	0.3	.035		
Krypton	1	5	5.5	6	4.5	4	2.2	0.6	
Xenon	1	7	9	9	7.5	5.5	4	1.5	0.025

The results of acceleration of heavy ions from a Penning source in a cyclotron are summarized in **Table 2** for the same elements as in Table 1. While these results are from similar sources, they are not to be compared directly because of numerous differences, including the characteristics of the cyclotron. Extracted beam currents for a number of charge states are shown in Table 2. Each beam will have its own characteristic energy, with the higher charge states having considerably higher energies.

Most studies with Penning sources were made with gases, thus greatly restricting nuclear research programs to a rather limited number of projectiles, until it was discovered that the versatility of the source could be very greatly extended by a simple but ingenuous trick. It was observed that while operating a cyclotron Penning source on a heavy noble gas, such as krypton, there was also present a fairly intense beam of highly charged copper ions. Further investigation revealed that some of the krypton ions from the source were partially accelerated and then returned back into the source where they sputtered the source body material (copper) into the discharge, resulting in a prolific beam of highly charged copper ions. Once the significance of this discovery was realized, it was a short step to introduce small pieces of various solids immediately behind the ion extraction slit (Fig. 2b) and to use the source to produce beams of a wide range of elements.

Ion-confinement sources. The yield of multiply charged heavy ions from a modern Penning source cannot be significantly improved, largely as a result of the short ion-containment time and the relatively high source pressure. Two promising sources, designed to overcome these limitations, are the electron cyclotron resonance (ECR) source and the electron-beam ion source (EBIS).

The electron cyclotron resonance source uses microwave power to heat electrons in two magnetic mirror confinement chambers which are in series with one another. Ions formed in the first chamber are allowed to drift slowly into the second chamber which has a much better vacuum, and it is here that the highly charged ions are formed. The electron energies in the source are quite high and frequently a few tens of keV. Continuous beams of several microamperes of C^{6+}, N^{7+}, Ne^{9+}, and Ar^{11+} have been extracted from such a source. The original source required a large amount of electrical power, and it is hoped that this requirement will be substantially reduced by the development of a source using superconducting magnetic coils.

Table 2. Extracted cyclotron currents from Penning source

	Extracted current for charge state, μA						
Element	3+	4+	5+	6+	7+	8+	9+	12+
Argon	3	34	1	6	1.6	1.2	0.015	
Calcium				0.3	0.7	0.012		
Krypton	0.032	1	2.2	0.15			0.020	
Xenon			0.003		0.09	0.013	0.004	0.0013

The electron-beam ion source uses an intense electron beam of 5–10 keV in a superconducting solenoid magnet to successively ionize injected gas. The ions are confined for several milliseconds in the electron beam by the radial electric field created by the electron beam and by potential barriers at the ends. After a certain confinement time, the potential barrier at one end of the solenoid is removed, and the ions are extracted. Quite high charge states have been obtained, for example, Ar^{16+}, Ar^{17+}, and Ar^{18+}, with intensities of between 10^8 and 10^{10} particles per pulse.

NEGATIVE-ION SOURCE CONCEPTS

Negative-ion source concepts will be discussed involving the methods of negative-ion formation discussed above.

Charge exchange source. Although charge exchange can be used to provide a very large variety of negative ions, it is relatively infrequently used with the exception of producing the metastable negative ion of helium. Indeed, charge exchange is the sole method of producing negative helium beams.

In a typical charge exchange source (**Fig. 3**), about 1 mA of positive helium ions is generated in a duoplasmatron ion source which is usually at ground potential. The ions are extracted by an electrode at a potential of about -20 kV, focused by an electrostatic lens, and directed through a donor canal usually containing lithium vapor. In the vapor a fraction of the positively charged helium ions sequentially picks up two electrons to form negative ions. The fraction is usually a little less than 1%, resulting in a negative helium beam of several microamperes.

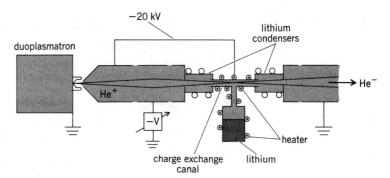

Fig. 3. Method of producing beams of helium negative ions by charge exchange in lithium vapor.

Cesium-beam sputter source. The cesium-beam sputter source is the negative-ion source most widely used on tandem accelerators. Much of its success is due to the fact that ion species can be changed rapidly, the ion output is usually high, and the source will operate for several hundreds of hours before a major cleanup is necessary.

Cesium positive ions are formed in the source (**Fig. 4**a) by passing cesium vapor through a porous disk of tungsten, heated to about 2012°F (1100°C), by the surface ionization process. These ions are accelerated through a potential difference of between 20 and 30 kV and are allowed to impinge upon a hollow cone that is either fabricated from or contains the element whose negative ion is required. Sputtering of the cesium-coated inner surface of the cone results in a large fraction of the sputtered particles leaving the surface as negative ions. An appreciable fraction of these are extracted from the rear hole of the sputter cone due to electric field penetration and are accelerated toward the ground electrode as a beam.

The sputter cones or targets can be inserted into a cooled copper wheel resembling the chamber of a revolver (Fig. 4b). Thus, by rotating the wheel, sputter cones can be rapidly changed, enabling the negative-ion species to be quickly changed. The source shown has provision for 18 different cones. In addition, negative-ion beams of gaseous elements, such as oxygen, can be formed by leaking the gas into the source and directing the flow onto a suitable getter surface such as titanium. **Table 3** lists some typical negative-ion currents obtainable from a cesium-beam sputter source.

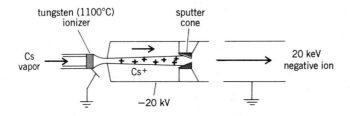

(a)

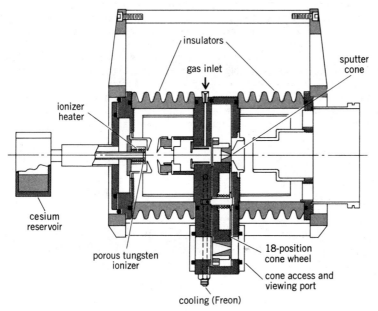

(b)

Fig. 4. Cesium-beam sputter source: (a) schematic showing operating principles; (b) section of a typical source.

Cesium-vapor Penning source. The cesium-vapor Penning source (**Fig. 5**) is a direct-extraction negative-ion source. Basically this is a conventional Penning source but with two important modifications making it suitable for negative-ion generation. The first is the introduction of a third sputter cathode, which is the source of negative ions. This cathode is made from or contains the element of interest, has a spherical face centered on the extraction aperture, and is operated at a higher negative potential than the normal cathodes. The second change involves introduction of cesium vapor into the arc chamber in addition to the support gas (usually xenon).

Table 3. Negative-ion currents from cesium-beam sputter source

Element	Negative-ion current, μA	Element	Negative-ion current, μA
Lithium	2	Sulfur	20
Boron	2	Nickel	6
Carbon	>50	Copper	3
Oxygen	>50	Gold	10
Silicon	20	Lead	0.1

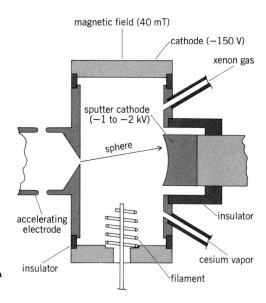

Fig. 5. Cesium-vapor Penning source for generation of negative ions.

The source operates in the normal Penning mode, and some of the cesium vapor introduced into the arc chamber becomes ionized and is accelerated toward the third sputter cathode. The negative ions that are formed as a result of sputtering are focused and accelerated toward the extraction aperture, and under the influence of the strong electric field generated by the acceleration electrode are extracted as an intense low-divergence beam. The negative-ion yield of the source is quite good and is comparable to that of the cesium-beam sputter source.

POLARIZED ION SOURCES

A polarized ion source is a device that generates ion beams in such a manner that the spins of the ions are aligned in some direction. The usual application is to inject polarized ions into a particle accelerator; however, it also has applications in atomic physics. The possible types of polarized sources are numerous, for in theory the nuclei of all kinds of atoms can be polarized, provided their spin is not zero. *See* SPIN.

The first polarized ion source was reported in 1956 by G. Clausnitzer, R. Fleischmann, and H. Schopper. The original type of source generated only positive ions but was effectively used on many types of accelerators. However, the development of the tandem Van de Graaff accelerator and its general acceptance for nuclear research created a demand for a polarized source capable of generating negative ions. This stimulated the development of the metastable-state or Lamb-shift type of polarized ion source that produces directly a high-quality negative ion beam with a high degree of polarization. The older type of source is referred to as the conventional or ground-state type of polarized ion source. Its output current of positive ions is an order of magnitude larger than the negative-ion output from the Lamb-shift source. With these two types of sources and their variants, polarized ions have been obtained from hydrogen, deuterium, tritium, helium having mass three, both isotopes of lithium, and others. The extra complication involved in producing polarized ions is such that the output is a factor of a thousand or more below the output of a moderately sized unpolarized ion source.

Conventional or ground-state source. In this type of source (**Fig. 6**) the first step consists in forming a beam of atoms in the ground state by a technique similar to that used in molecular beams. In the case of hydrogen or deuterium, this is done by dissociating the molecular gas in a glass discharge tube and allowing the atoms to escape through a nozzle. The atoms escaping at thermal energies are collimated into a well-directed beam by plates with holes or an adjustable skimmer. High-capacity diffusion pumps sweep away the large quantity of excess hydrogen or deuterium.

464 NUCLEAR AND PARTICLE PHYSICS SOURCE BOOK

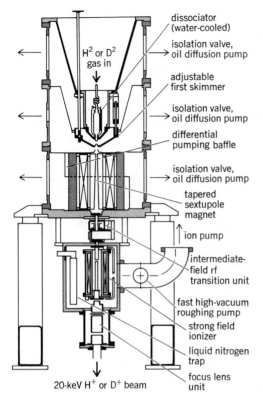

Fig. 6. Conventional or ground-state polarized ion source. (*After H. F. Glavish, Polarized ion sources, in Proceedings of the 2d Symposium on Ion Sources and Formation of Ion Beams, Berkeley, pp. IV-1-1 through IV-1-7, 1974*)

The beam is then passed along the axis of an inhomogeneous magnetic field, which is most commonly generated by a sextupole magnet. This type of magnet consists of six magnets arranged in a circular pattern with alternating polarities.

In a sextupole magnet the absolute magnitude of the field increases as the distance squared from the axis. The atoms are subjected to a radial force that is proportional to their magnetic moment times the gradient of the absolute magnitude of the field strength. In the case of a sextupole magnet the force is proportional to the first power of the distance from the axis. The sign of the force does not depend upon the direction of the magnetic lines, but only on the projection of magnetic moment along these lines of force or m_j where j is the spin value of the electron. (The atomic magnetic moment results almost entirely from the electron, since its magnetic moment is 662 times larger than that of the proton and 2156 times larger than that of the deuteron.) The result is that atoms with a positive value of m_j are subjected to a force that is directed radially inward and pass through the sextupole magnet, while atoms with a negative m_j experience a force that is directed outward and are rapidly lost from the beam. Out of the sextupole comes a beam of atomic hydrogen that is polarized with respect to the orientation of its electrons but is, as yet, unpolarized in its nuclear spin. *See Magneton.*

Since aligned nuclei rather than aligned electrons are desired, it is necessary to subject the atomic beam to other fields. Each hydrogen atom is in one of two pure states. It is possible to apply an oscillating magnetic field in combination with a dc magnetic field that will flip the sign of m_I (the projection of the nuclear spin) of one of the pure states and not the other. That aligns the spins of the nuclei but may depolarize the electrons. That does not matter, however, since they will be removed.

The final stage is to send the atomic beam into a strong solenoidal magnetic field. As the atoms from the sextupole field—having all orientations in each cross-sectional plane—enter the solenoid, they adiabatically come into alignment with the parallel lines of force within the solenoid since their m_j components of spin are conserved. In the solenoid the atoms are ionized by energetic electrons as in an arc discharge. The ionizer is actually the most difficult part of this type of polarized source to make function efficiently, even though it is conceptionally simple. The ionizer is followed by electric fields that accelerate and focus the ions to get a beam that can be accepted by the accelerator.

Lamb-shift or metastable-atom source. The polarization process in the Lamb-shift type of source (**Fig. 7**) is also performed upon atoms, in this case, metastable ones. The process is most efficient if the atoms have a velocity of approximately 10^{-3} of light rather than thermal velocity as in the case of the ground-state type of source. To get the beam, hydrogen, deuterium, or tritium can be used, but only hydrogen is discussed in this article. The hydrogen is ionized in a conventional ion source such as a duoplasmatron. The H^+ ions are then accelerated and focused into a beam at about 500 eV. The beam is passed through cesium vapor where cesium atoms donate electrons which are resonantly captured in an $n = 2$ state by the hydrogen ions.

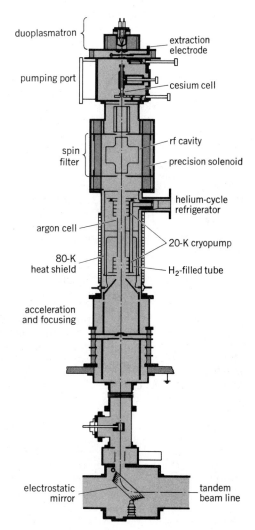

Fig. 7. Lamb-shift polarized ion source. (*After R. A. Hardekopf, Operation of the LASL polarized triton source, in Proceedings of the 4th International Symposium on Polarization Phenomena in Nuclear Reactions, Birkhäuser Verlag, Basel, 1976*)

Atoms are formed in both the 2p and the 2s states in the cesium vapor. However, those in the 2p state decay almost immediately to the ground state by the emission of a Lyman alpha photon, energy 10.15 eV. The small energy difference between the 2p and the 2s states in is the Lamb shift. The lifetime of the 2p atoms is 1.6×10^{-9} s, while the lifetime of the 2s atoms is 0.15 s because two photons must be emitted simultaneously in their decay to the ground state. Actually few 2s atoms decay by emission of two photons for they are necessarily subjected to small electric fields which mix into the 2s and 2p wave function and its tendency to decay to the ground state. To take advantage of this tendency to decay to the ground state, apparatus can be built so that those atoms having the undesired value of m_I are stimulated to decay, while those with the desired value of m_I are allowed to pass on without decay.

The polarized H^- ions are formed in argon because its atoms are capable of donating electrons to metastable atoms but have a very weak capability of forming H^- ions out of ground state atoms. The ground-state charge-changing cross section appears to be lower by a factor of about 400; however, ground-state atoms outnumber the metastable atoms at this region by a factor of 40 so that the net polarization is 90%. The remainder of the apparatus consists of electric fields that accelerate and focus the beam so it can be accepted by an accelerator.

The electron spins are polarized by applying a transverse field of about 100 kV/m while the atomic beam of metastables is passing along the axis of a solenoid at a field strength of about 57.5 millitesla. The transverse electric field couples the 2s and 2p levels through the Stark effect, and the magnetic field is just sufficient to bring the levels with $m_j = -1/2$ very close together in energy, while those with $m_j = +1/2$ have their energy separation doubled, so that 2s atoms with $m_j = +1/2$ are transmitted without loss.

There are several methods of going on to polarize the nuclei, including a device known as the spin filter. To produce the spin filter, a longitudinal electric field of about the same strength as the transverse field is added to the apparatus that polarizes the electrons, with the longitudinal field oscillating at about 1.60 GHz. The complete explanation is complicated. However, results are that if the magnetic field is adjusted so the Larmor frequency of the electron in the metastable atom is made equal to the oscillating electric field, then the lifetime of the atom for decay becomes very long exactly at resonance, yet short not far off resonance. The magnetic field that determines the Larmor frequency of the electron in the metastable atom is the sum of that due to the solenoid and that due to the proton aligned in the solenoidal field. These two fields have opposite signs in the case of $m_I = -1/2$ and it is found that the two resonances for transmission are at 54.0 mT for $m_I = +1/2$ and at 60.5 mT for $m_I = -1/2$. In the case of deuterons, there are three resonances, and they are well resolved even though $m_I = +1$ is at 56.5 mT, 0 is at 57.5 mT, and -1 is at 58.5 mT.

Colliding-beam source. In 1968 W. Haeberli proposed a new method of making polarized beams of negative hydrogen and deuterium. The most important feature of the scheme was the direct conversion of polarized neutral hydrogen (H^0) or deuterium (D^0) atoms into polarized negative ions using the reaction below.

$$H^0(\text{polarized}) + Cs^0 \rightarrow H^-(\text{polarized}) + Cs^+$$

Although conceptually very attractive, this idea presents some severe experimental difficulties. Undoubtedly the greatest of these is that if the polarized H^0 or D^0 atoms are produced at thermal energies, where production is greatly facilitated, the cross section or probability of the above reaction proceeding is extremely small. To circumvent this difficulty and to capitalize on a much higher cross section, it was proposed to accelerate a positively charged cesium (Cs) beam to an energy of about 40 keV, neutralize it by passing it through a canal containing cesium vapor, and allow this high-velocity beam of neutral cesium atoms to collide with the polarized atomic beam. Probably as a result of the experimental difficulties and the promising development of the Lamb shift source, almost a decade passed before a source was built utilizing this concept.

Such a source has been built and demonstrated to yield 2.9 μA of polarized hydrogen negative ions and 3.1 μA of deuterium ions. These currents are about five times larger than those obtainable from the best Lamb shift sources.

Bibliography. J. Arianer, The status of novel ion sources, *Nuc. Instrum. Meth.*, 39:516–521, 1985; D. Fick (ed.), *Polarization Nuclear Physics: Proceedings of a Meeting Held at Ebermannstadt, 1973*, Lecture Notes in Physics, 1974; O. B. Morgan, G. G. Kelley, and R. C. Davis, Technology of intense dc ion beams, *Rev. Sci. Instr.*, 38:467–480, 1967; *Proceedings of the International Conference on*

Multiply Charged Heavy Ion Sources and Accelerating Systems, IEEE Trans. Nucl. Sci., NS-19(2), 1972; *Proceedings of the 1975 Particle Accelerator Conference*, IEEE Trans. Nucl. Sci., NS-22(3), 1975; *Proceedings of the 2d International Conference on Ion Sources*, SGAE, Vienna, 1972; *Proceedings of the 2d Symposium on Ion Sources and Formation of Ion Beams*, Berkeley, 1974; *Proceedings of the Symposium on Ion Sources and Formation of Ion Beams*, Brookhaven National Laboratory, 1971; *Proceedings of the 3d International Symposium on Polarization Phenomena in Nuclear Reactions*, 1970.

PARTICLE DETECTOR
FRED S. GOULDING

A device used to detect and measure radiations characteristically emitted in nuclear processes, including gamma rays or x-rays, lightweight charged particles (electrons or positrons), nuclear constituents (neutrons, protons, and heavier ions), and subnuclear constituents such as mesons. The device is also known as a radiation detector. Since human senses do not respond to these types of radiation, detectors are essential tools for the discovery of radioactive minerals, for all studies of the structure of matter at the atomic, nuclear, and subnuclear levels, and for protection from the effects of radiation. They have also become important practical tools in the analysis of materials using the techniques of neutron activation and x-ray fluorescence analysis. SEE ELEMENTARY PARTICLE; NUCLEAR REACTION; NUCLEAR SPECTRA; PARTICLE ACCELERATOR; RADIOACTIVITY.

Classification by use. A convenient way to classify radiation detectors is according to their mode of use: (1) For detailed observation of individual photons or particles, a pulse detector is used to convert each such event (that is, photon or particle) into an electrical signal. (2) To measure the average rate of events, a mean-current detector, such as an ion chamber, is often used. Radiation monitoring and neutron flux measurements in reactors generally fall in this category. Sometimes, when the total number of events in a known time is to be determined, an integrating version of this detector is used, (3) Position-sensitive detectors are used to provide information on the location of particles or photons in the plane of the detector. (4) Track-imaging detectors image the whole three-dimensional structure of a particle's track. The output may be recorded by immediate electrical readout or by photographing tracks as in the bubble chamber. (5) The time when a particle passes through a detector or a photon interacts in it is measured by a timing detector. Such information is used to determine the velocity of particles and when observing the time relationship between events in more than one detector.

Ionization detectors. Any radiation-induced effect in a solid, liquid, or gas can be used in a detector. To be useful, however, the effect must be directly or indirectly interpretable in terms of either the quantity or quality (that is, type, energy, and so on) of the incident radiation or both. The ionization produced by a charged particle is the effect commonly employed.

Gas ionization detectors. In the basic type of gas ionization detector, an electric field applied between two electrodes separates and collects the electrons and positive ions produced in the gas by the radiation to be measured. Depending on the intensity of the electric field, the charge signal in the external circuit may be equal to the charge produced by the radiation, or it may be much larger. In a proportional counter, the output charge is larger than the initial charge by a factor called the gas amplification. A Geiger-Müller counter provides still larger signals, but each signal is independent of the original amount of ionization. All three types of gas ionization detectors can be used as pulse detectors, mean current detectors, or, with an indicator such as a quartz-fiber electroscope, as integrating detectors. SEE IONIZATION CHAMBER.

Position-sensitive and track-imaging detectors. Position-sensitive detectors and track-imaging detectors are nearly all based on the ionization process. Multiwire proportional chambers and spark chambers are position-sensitive adaptations of gas detectors. The signal division or time delay that occurs between the ends of an electrode made of resistive material is sometimes used to provide position sensitivity in gas and semiconductor detectors. Track-imaging detectors rely on a secondary effect of the ionization along a particle's track to reveal its structure. In Wilson cloud chambers, ionization triggers condensation along particle tracks in a supercooled vapor; bubble formation in liquids is the basis for operation of bubble chambers. A secondary effect of ionization is also employed in photographic emulsions used as radiation detectors where ionization triggers the formation of an image. SEE BUBBLE CHAMBER; CLOUD CHAMBER; SPARK CHAMBER.

Semiconductor detectors. In this type of detector, a solid replaces the gas of the previous example. The "insulating" region (depletion layer) of a reverse-biased pn junction in a semiconductor is employed. Choice of materials is very limited; very pure single crystals of silicon or germanium are presently the only fully suitable materials, although other semiconductors can be used in noncritical applications. Collection of the primary ionization is normally used, but an avalanche mode is sometimes employed in which an intense electric field causes charge multiplication. Although semiconductor devices are mostly used as pulse detectors, mean-current and integrating modes are possible when the significant leakage currents of pn junctions can be tolerated.

Since solids are approximately 1000 times denser than gases, absorption of radiation can be accomplished in relatively small volumes. A less obvious but fundamental advantage of semiconductor detectors is the fact that much less energy is required (~ 3 eV) to produce a hole-electron pair than that required (~ 30 eV) to produce an ion electron pair in gases. This results in better statistical accuracy in determining radiation energies. For this reason, semiconductor detectors have become the main tools for nuclear spectroscopy, and they have also made neutron activation analysis and x-ray fluorescence analysis of materials practical tools of great value. SEE JUNCTION DETECTOR.

Scintillation detector. In addition to producing free electrons and ions, the passage of a charged particle through matter temporarily raises electrons in the material into excited states. When these electrons fall back into their normal state, light may be emitted and detected as in the scintillation detector. The early scintillation detectors consisted of a layer of powder (zinc sulfide, for example) that, when struck by charged particles, produced light flashes, which were observed by eye and counted. The meaning of the term "scintillation detector" has changed with time to refer to the combination of a scintillator and a photomultiplier tube that converts light scintillations into signals that can be processed electronically. Various organic and inorganic crystals, plastics, liquids, and glasses are used as scintillators, each having particular virtues in regard to radiation absorption, speed, and light output. SEE LIQUID SCINTILLATION DETECTOR; SCINTILLATION COUNTER.

Neutral particles. Neutral particles, such as neutrons, cannot be detected directly by ionization. Consequently, they must be converted into charged particles by a suitable process and then observed by detecting the ionization caused by these particles. For example, high-energy neutrons produce "knock-on" protons in collisions with light nuclei, and the protons can be detected. Slow neutrons are usually detected by using a nuclear reaction in which the neutron is captured and a charged particle is emitted. For ex- ample, in boron trifluoride (BF_3) detectors, neutrons react with boron to produce alpha particles which are detected.

Other detector types. Although ionization detectors dominate the field, a number of detector types based on other radiation-induced effects are used. Notable examples are: (1) transition radiation detectors, which depend on the x-rays and light emitted when a particle passes through the interface between two media of different refractive indices; (2) track detectors, in which the damage caused by charged particles in plastic films and in minerals is revealed by etching procedures; (3) thermo- and radiophoto-luminescent detectors, which rely on the latent effects of radiation in creating traps in a material or in creating trapped charge; and (4) Cerenkov detectors, which depend on measurement of the light produced by passage of a particle whose velocity is greater than the velocity of light in the detector medium. SEE CERENKOV RADIATION; TRANSITION RADIATION DETECTORS.

Large detector systems. The very large detector systems used in relativistic heavy-ion experiments and in the detection of the products of collisions of charged particles at very high energies, typically at the intersection region of storage rings, deserve special consideration. These detectors are frequently composites of several of the basic types of detectors discussed above and are designed to provide a detailed picture of the multiple products of collisions at high energies. The complete detector system may occupy a space tens of feet in extent and involve tens or hundreds of thousands of individual signal processing channels, together with large computer recording and analysis facilities.

The time projection chamber (TPC) typifies this class of detector. This chamber, in its initial configuration, is a cylinder 6.5 ft (2 m) long and 6.5 ft (2 m) in diameter, containing a gas, with a

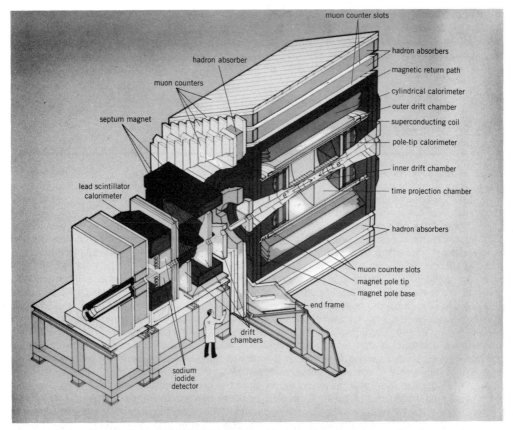

Time projection chamber designed for the PEP (Positron-Electron Projection) storage ring at Stanford University, with surrounding detection equipment and magnets.

thin-plane central electrode that splits the chamber longitudinally into two drift spaces, each 3 ft (1 m) in length. A high potential (approximately 150 kV) is applied to this electrode to cause the electrons produced by ionization along tracks to drift to the end-cap regions of the cylinder. The end caps contain arrays of a few thousand gas proportional wire chambers designed to locate the position of signals with an accuracy of a small fraction of an inch in both radius and azimuth. The signal amplitude is maintained proportional to the original ionization in the track segment. When this information is combined with the drift time information, the effect is to provide the capability to measure the amount of ionization in individual cells a small fraction of an inch in linear dimensions throughout the whole volume of the chamber. The chamber is immersed in an axial magnetic field so that the bending of the tracks can be used to provide particle momentum information. Finally, the central time projection chamber is surrounded by other detector chambers (see **illus**.) to provide information on particles leaving the time projection chamber and also to perform ray calorimetry. The time projection chamber may be thought of as a modern version of the cloud or bubble chamber with prompt electrical readout of track information.

Very complex large-area and -volume chambers are also used in heavy-ion physics experiments. Here identification of the type of ion is usually important, and this is achieved partly by measuring the ionization or energy loss at several points along the track and by combining this information with time-of-flight or magnetic bending information. Complex combinations of gas, semiconductor, and scintillation detectors are frequently used in these detector systems.

Bibliography. G. Bertolini and A. Coche, *Semiconductor Detectors,* 1968; C. W. Fabjan and T. Ludlam, Calorimetry in high-energy physics, *Annu. Rev. Nucl. Part. Sci.,* 32:335–389, 1982; February issues of *IEEE Trans. Nucl. Sci.,* 1970 to present; G. F. Knoll, *Radiation Detection and Measurement,* 1979; S. C. Loken and P. Nemethy (eds.), *Proceedings of the 1983 Division of Particles and Fields Workshop on Collider Detectors,* 1983; *Nucl. Instrum. Meth. Phys. Res.,* vol. 196, no. 1, 1982, and vol. 226, no. 1, 1984.

IONIZATION CHAMBER
William A. Lanford

An instrument for detecting ionizing radiation by measuring the amount of charge liberated by the interaction of ionizing radiation with suitable gases, liquids, or solids. These radiation detectors have played an important part in the development of modern physics and have found many applications in basic scientific research, in industry, and in medicine.

Principle of operation. While the gold leaf electroscope (**Fig. 1**) is the oldest form of ionization chamber, instruments of this type are still widely used as monitors of radiations by workers in the nuclear or radiomedical professions. In this device, two thin flexible pieces of gold leaf are suspended in a gas-filled chamber. When these are electrically charged, as in Fig. 1, the electrostatic repulsion causes the two leaves to spread apart. If ionizing radiation is incident in the gas, however, electrons are liberated from the gas atoms. These electrons then drift toward the positive charge on the gold leaf, neutralizing some of this charge. As the charge on the gold leaves decreases, the electrostatic repulsion decreases, and hence the separation between the leaves decreases. By measuring this change in separation, a measure is obtained of the amount of radiation incident on the gas volume. While this integrated measurement may be convenient for applications such as monitoring the total radiation exposure of humans, for many purposes it is useful to measure the ionization pulse produced by a single ionizing particle.

The simplest form of a pulse ionization chamber consists of two conducting electrodes in a container filled with gas (**Fig. 2**). A battery, or other power supply, maintains an electric field between the positive anode and the negative cathode. When ionizing radiation penetrates the gas in the chamber—entering, for example, through a thin gas-tight window—this radiation liberates electrons from the gas atoms leaving positively charged ions. The electric field present in the gas sweeps these electrons and ions out of the gas, the electrons going to the anode and the positive ions to the cathode.

The basic ion chamber signal consists of the current pulse observed to flow as a result of this ionization process. Because the formation of each electron-ion pair requires approximately 30

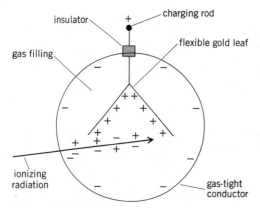

Fig. 1. Gold leaf electroscope used as a radiation detector.

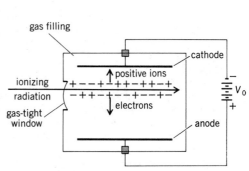

Fig. 2. Parallel-plate ionization chamber.

eV of energy on the average, this signal is proportional to the total energy deposited in the gas by the ionizing radiation.

Because the charge liberated by a single particle penetrating the chamber is small, very low-noise high-gain amplifiers are needed to measure this charge. In the early days, this was a severe problem, but such amplifiers have become readily available with the development of modern solid-state electronics.

In a chamber, such as that represented in Fig. 2, the current begins to flow as soon as the electrons and ions begin to separate under the influence of the applied electric field. The time it takes for the full current pulse to be observed depends on the drift velocity of the electrons and ions in the gas. These drift velocities are complicated functions of gas type, voltage, and chamber geometry. However, because the ions are thousands of times more massive than the electrons, the electrons always travel several orders of magnitude faster than the ions. As a result, virtually all pulse ionization chambers make use of only the relatively fast electron signal. The electron drift velocities for a few gases are given in **Fig. 3**. Using one of the most common ion chamber gases—argon with a small amount of methane—with electrode spaces of a few centimeters and voltages of a few hundred volts, the electron drift time is of order a microsec- ond, while the positive-ion drift time is of order milliseconds. By using narrow-bandpass amplifiers sensitive only to signals with rise times of order a microsecond, only the electron signals are observed.

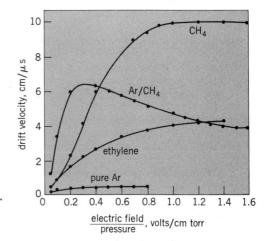

Fig. 3. Electron drift velocity in four different gases as a function of the ratio of the applied electric field strength in volts/centimeter to gas pressure in torrs. 1 torr = 133 Pa. (*After H. W. Fulbright, Ionization chambers, Nucl. Instrum. Meth., 162(1979):21–28, 1979*)

Energy spectrum. One of the most important uses of an ionization chamber is to measure the total energy of a particle or, if the particle does not stop in the ionization chamber, the energy lost by the particle in the chamber. When such an energy-sensitive application is needed, a simple chamber geometry such as that shown in Fig. 2 is not suitable because the fast electron signal charge is a function of the relative distance that the ionization even occurred from the anode and cathode. If an ionization even occurs very near the cathode, the electrons drift across the full electric potential V_0 between the chamber electrodes, and a full electron current pulse is recorded; if an ionization count occurs very near the anode, the electrons drift across a very small electric potential, and a small electron pulse is recorded. This geometrical sensitivity is a result of image charges induced by the very slowly moving positive ions. It can be shown that if the electrons drift through a potential difference ΔV, the fast electron charge pulse is $q' = (\Delta V/V_0)q$, where q is the total ionization charge liberated in the gas.

This geometrical dependence can be eliminated by introducing a Frisch grid as indicated in **Fig. 4**. This grid shields the anode from the positive ions and, hence, removes the effects of the image charges. By biasing the anode positively, relative to the grid, the electrons are pulled

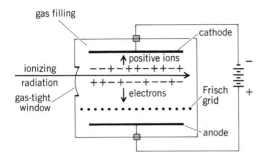

Fig. 4. Frisch grid parallel-plate ionization chamber.

through the grid and collected on the anode. Now no signal is observed on the anode until the electrons drift through the grid, but the signal charge which is then observed is the full ionization charge q.

While the ionization chamber generates only small quantities of signal charge for incident particles or photons of MeV energies, the resulting signals are nevertheless well above the noise level of modern low-noise electronic amplifiers. When the signals generated by many incident particles of the same energy are individually measured, and a histogram is plotted representing the magnitude of a signal pulse versus the total number of pulses with that magnitude, then an energy spectrum results. Such a spectrum, smoothed out, consists of an essentially gaussian distribution with standard deviation σ (**Fig. 5**). Assuming a negligible contribution from amplifier noise, it might at first sight appear that σ should correspond to the square root of the average number of electron-ion pairs produced per incident particle. In fact, σ is usually found to be less than this by a substantial amount, usually designated F, where F is the Fano factor.

It is usual to express the width of an energy distribution such as that of Fig. 5 not in terms of σ but in terms of the "full width at half maximum," usually designated FWHM, or Δ. It can be shown that, for situations in which the width of the energy spectrum is governed by statistics alone, the FWHM is given by the equation $\Delta = 2.36\sqrt{F\epsilon E}$, where ϵ is the average energy required to create an electron-ion pair and E is the energy deposited in the chamber by each incident particle or photon. Values of the Fano factor F as low as 0.1 have been observed for certain gases.

Gaseous ionization chambers. Because of the very few basic requirements needed to make an ionization chamber (namely, an appropriate gas with an electric field), a wide variety of different ionization chamber designs are possible in order to suit special applications. In addition to energy information, ionization chambers are now routinely built to give information about the position within the gas volume where the initial ionization event occurred. This information can be important not only in experiments in nuclear and high-energy physics where these position-sensitive detectors were first developed, but also in medical and industrial applications.

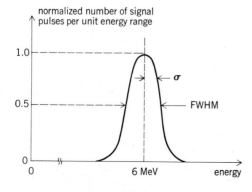

Fig. 5. Idealized energy spectrum produced by monoenergetic (6-MeV) alpha particles incident on an ideal gridded ionization chamber. The spectrum consists of a single gaussian peak with standard deviation σ.

This position-sensitivity capability results from the fact that, to a good approximation, electrons liberated in an ionizing event drift along the electric field line connecting the anode and cathode, and they drift with uniform velocity (Fig. 3). Hence, a measure of the drift time is a measure of the distance from the anode that the ionization occurred. A simple illustration of this use is a heavy-ion nuclear physics detector "telescope" used in the basic study of nuclear reactions (**Fig. 6**). Ionizing charged particles (such as ^{1}H, ^{4}He, and ^{12}C) produced in nuclear collisions enter the detector telescope through a thin gas-tight window at the left, pass through two Frisch grid ionization chambers, and then stop in a solid-state ionization detector. Measurement of the ionization produced in the gas versus the total energy of the particle as determined by the solid-state ionization detector gives sufficient information to uniquely identify the mass and atomic charge of the incident particle. Because the response of the solid-state detector is fast relative to the electron drift time, the difference in time of the signals from the solid-state detector and the anode determines the electron drift time and, hence, the distance above the grid that the particle entered the ionization chamber. This distance can be used to determine the nuclear scattering angle. Hence, a very simple device can be designed which gives several pieces of useful information. While this example illustrates the principles, very complex ionization chambers are now routinely used in heavy-ion and high-energy physics where tens or a hundred signals are recorded (using a computer) for a single ionization event. Position-sensitive heavy-ion detectors with active surfaces as large as a square meter have been developed.

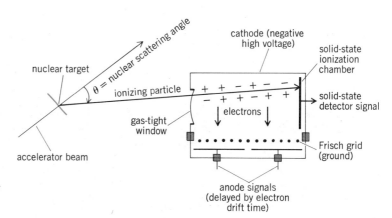

Fig. 6. Heavy-ion detector telescope used to study nuclear reactions.

Aside from applications in basic nuclear physics research, ionization chambers are widely used in other applications. Foremost among these is the use of gas ionization chambers for radiation monitoring. Portable instruments of this type usually employ a detector containing approximately 1 liter of gas, and operate by integrating the current produced by the ambient radiation. They are calibrated to read out in convenient units such as milliroentgens per hour.

Another application of ionization chambers is the use of air-filled chambers as domestic fire alarms. These employ a small ionization chamber containing a low-level radioactive source, such as ^{241}Am, which generates ionization at a constant rate, the resulting current being monitored by a small solid-state electronic amplifier. On the introduction of smoke into the chamber (which is open to the ambient air), the drifting charged ions tend to attach themselves to the smoke particles. This reduces the ionization chamber current, since the moving charge carriers are now much more massive than the initial ions and therefore exhibit correspondingly reduced mobilities. The observed reduction in ion current is used to trigger the alarm.

Another development in ion chamber usage was that of two-dimensional imaging in x-ray medical applications to replace the use of photographic plates. This imaging depends on the fact that if a large flat parallel-plate gas ionization chamber is illuminated with x-rays (perpendicular to its plane), the resulting charges will drift to the plates and thereby form an "image" in electrical charge of the point-by-point intensity of the incident x-rays. This image can be recorded xerographically by arranging for one plate to be a suitably charged insulator. This insulator is preferentially discharged by the collected ions. The resulting charge pattern is recorded by dusting the insulator with a fine powder and transferring this image to paper as in the usual xerographic technique. Alternatively, the xerographic insulator may be a photoconductor, such as selenium, which is preferentially discharged by the ionization produced in the solid material. This is then an example of a solid ionization chamber, and its action closely parallels the operation of the optical xerographic copying machines. Such x-ray imaging detectors provide exceedingly high-quality images at a dosage to the patient substantially less than when photographic plates are used.

Gaseous ionization chambers have also found application as total-energy monitors for high-energy accelerators. Such applications involve the use of a very large number of interleaved thin parallel metal plates immersed in a gas inside a large container. An incident pulse of radiation, due for example to the beam from a large accelerator, will produce a shower of radiation and ionization inside the detector. If the detector is large enough, essentially all of the incident energy will be dissipated inside the detector (mostly in the metal plates) and will produce a corresponding proportional quantity of charge in the gas. By arranging that the plates are alternately biased at a positive and negative potential, the entire device operates like a large interleaved gas ion chamber. The total collected charge is then a measure of the total energy in the initial incident pulse of radiation.

Solid ionization chambers. Ionization chambers can be made where the initial ionization occurs, not in gases, but in suitable liquids or solids. In fact, the discovery of extremely successful solid-state ionization detectors in the early 1960s temporarily diverted interest from further developments of gas-filled chambers.

In the solid-state ionization chamber (or solid-state detector) the gas filling is replaced by a large single crystal of suitably chosen solid material. In this case the incident radiation creates electron-hole pairs in the crystal, and this constitutes the signal charge. In practice, it has been found that only very few materials can be produced with a sufficiently high degree of crystalline perfection to allow this signal charge to be swept out of the crystal and collected. Although many attempts were made in this direction in the 1940s in crystal counters, using such materials as AgCl, CdS, and diamond, these were all failures due to the crystals not having adequate carrier transport properties. In the late 1950s, however, new attempts were made in this direction using single crystals of the semiconductors silicon and germanium. These were highly successful and led to detectors that revolutionized low-energy nuclear spectroscopy.

There are two important differences between solid and gas-filled ionization chambers. First, it takes much less energy to create an electron-hole pair in a solid than it does to ionize gas atoms. Hence, the intrinsic energy resolution obtainable with solid-state detectors is better than with gas counters. Gamma-ray detectors with resolutions better than 180 eV are commercially available. Second, in the case of solid semiconductors, the positive charge is carried by electron "holes" whose mobilities are similar to those of electrons. Hence, both the electrons and holes are rapidly swept away by the electric field and, as a result, no Frisch grid is needed to electrically shield the anode from the image charge effects of slow-moving positive ions as in the case of gas- or liquid-filled ionization chambers. SEE GAMMA-RAY DETECTORS.

Liquid ionization chambers. The use of a liquid in an ionization chamber combines many of the advantages of both solid and gas-filled ionization chambers; most importantly, such devices have the flexibility in design of gas chambers with the high density of solid chambers. The high density is especially important for highly penetrating particles such as gamma rays. Unfortunately, until the 1970s the difficulties of obtaining suitable high-purity liquids effectively stopped development of these detectors. During the 1970s, however, a number of groups built liquid argon ionization chambers and demonstrated their feasibility. A Frisch grid liquid argon chamber achieved a resolution of 34 keV (FWHM).

Proportional counters. If the electric field is increased beyond a certain point in a gas ionization chamber, a situation is reached in which the free electrons are able to create additional electron-ion pairs by collisions with neutral gas atoms. For this to occur, the electric field must be sufficiently high so that between collisions an electron can pick up an energy that exceeds the ionization potential of the neutral gas atoms. Under these circumstances gas multiplication, or avalanche gain, occurs, thereby providing additional signal charge from the detector.

A variety of electrode structures have been employed to provide proportional gas gain of this type. The most widely used is shown in **Fig. 7**. Here a fine central wire acts as the anode, and the avalanche gain takes place in the high field region immediately surrounding this wire. In practice, under suitable circumstances, it is possible to operate at gas gains of up to approximately 10^6.

The gas gain is a function of the bias voltage applied to the proportional counter and takes the general form shown in **Fig. 8**.

Similar avalanche multiplication effects can occur in semiconductor junction detectors, although there the situation is less favorable, and such devices have not found very widespread use except as optical detectors. SEE JUNCTION DETECTOR.

The large gas gains realizable with proportional counters have made them extremely useful for research applications involving very low-energy radiation. In addition, their flexibility in terms of geometry has made it possible to construct large-area detectors, of the order of 10 ft^2 (1 m^2), suitable for use as x-ray detectors in space. Essentially all that has been learned to date regarding x-ray astronomy has involved the use of such detectors aboard space vehicles.

Further exceedingly useful applications of gas proportional counters involve their use as position-sensitive detectors. In Fig. 7, for example, if the anode wire is grounded at both ends, then the signal charge generated at a point will split and flow to ground in the ratio of the resistance of the center wire between the point of origin and the two ends of the wire. This device therefore comprises a one-dimensional position-sensitive detector. Such devices are widely used as focal plane detectors in magnetic spectrographs. Similar position-sensitive operation can be obtained by taking account of the rise time of the signals seen at each end of the wire. Further extension of such methods allows two-dimensional detectors to be produced, a wide variety of which are under investigation for medical and other imaging uses.

The relatively large signals obtainable from gas proportional counters simplifies the requirements of the subsequent amplifiers and signal handling systems. This has made it economically feasible to employ very large arrays, of the order of thousands, of such devices in multidimensional arrays in high-energy physics experiments. By exploiting refinements of technique, it has proved possible to locate the tracks of charged particles to within a fraction of a millimeter in distances measured in meters. Such proportional counter arrays can operate at megahertz counting rates since they do not exhibit the long dead-time effects associated with spark chambers. SEE SPARK CHAMBER.

Geiger counters. If the bias voltage across a proportional counter is increased sufficiently, the device enters a new mode of operation in which the gas gain is no longer proportional

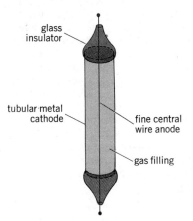

Fig. 7. Basic form of a simple single-wire gas proportional counter.

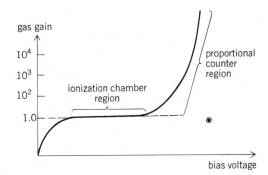

Fig. 8. Plot of gas gain versus applied voltage for a gas-filled radiation detector.

to the initial signal charge but saturates at a very large, and constant, value. This provides a very economical method of generating signals so large that they need no subsequent amplification.

The most widespread use of Geiger counters continues to be in radiation monitoring, where their large output signals simplify the readout problem. They have also found extensive use in cosmic-ray research, where again their large signals have made it feasible to use arrays of substantial numbers of detectors without excessive expenditures on signal-processing electronics. SEE PARTICLE DETECTOR.

Bibliography. D. A. Bromley (ed.), *Detectors in Nuclear Science*, 1979; G. G. Eichholz and J. W. Poston, *Principles of Nuclear Radiation Detection*, 1979; G. F. Knoll, *Radiation Detection and Measurement*, 1979; P. W. Nicholson, *Nuclear Electronics*, 1974; W. J. Price, *Nuclear Radiation Detection*, 2d ed., 1964.

SPARK COUNTER
WILLIAM B. FRETTER

A particle detector which uses the ionization produced in a gas by high-speed charged particles to trigger a spark between two electrodes. Spark counters react in a very short time (10^{-9} s) to the particle, and thus can be used for fast timing. The spark is visible and can be photographed. SEE PARTICLE DETECTOR.

The principal components of a spark counter are two plane, parallel metallic electrodes, with a gas between the electrodes consisting of a mixture of argon and an organic gas such as xylene. A potential difference of about 2000 V is placed across the electrodes, which are spaced about 0.08 in. (2 mm) apart. The function of the xylene gas is to aid in quenching the discharge which occurs between the plates when ions are produced in the gas by a charged particle passing through the counter. Although the response time of a spark counter is very fast, the counting rate is very low, since additional quenching must be provided by an electronic circuit which has a recovery time of 0.25 s. SEE SPARK CHAMBER.

Bibliography. J. W. Keuffel, Parallel-plate counters, *Rev. Sci. Instrum.*, 20:202–208, 1949; E. Segré, *Nuclei and Particles*, 2d ed., 1977.

SPARK CHAMBER
ARTHUR ROBERTS

A triggered electronic particle-detecting device whose purpose is to make visible and to locate accurately in space the tracks of charged particles. It is generally classified as wide-gap or narrow-gap; the most common form is a narrow-gap array of parallel-plate condensers, the plates of which are spaced about ⅜ in. (9.3 mm) apart, filled with a mixture of helium and neon at atmospheric pressure. A spark-chamber system requires an external initiating signal from an auxiliary particle-detection system, which triggers the application of a short-duration high-voltage pulse to the array of plates. The high-voltage pulse produces a spark discharge that follows or marks the path of the ionizing particles. If several tracks are present they are all visible, up to at least 20 (although the chamber will not resolve tracks less than about 0.04 in. or 1 mm apart). Stereo cameras usually provide accurate track location; electronic and digital methods of track location are also used. Spark chambers are characterized by somewhat lower resolution than bubble chambers, accuracy of track location ranging from moderate to excellent (the best performance exceeding that of the bubble chamber), a high rate of data collection, and a unique triggering capability that selects only those events that have passed logical selection tests to be photographed. SEE BUBBLE CHAMBER; PARTICLE DETECTOR; SPARK COUNTER.

GENERAL PROPERTIES

Figure 1 shows a schematic diagram of a spark-chamber system using cameras for data recording.

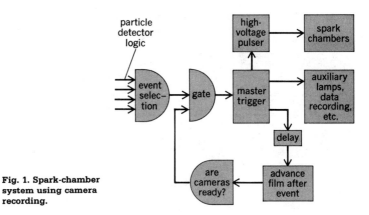

Fig. 1. Spark-chamber system using camera recording.

Characteristic times. The time during which the ionization produced by the particle can produce a good track when the high-voltage pulse is applied is called the memory time of the chamber; it may be decreased by an electric clearing field that removes electrons rapidly, or by electronegative gases like sulfur dioxide, deliberately introduced to capture electrons and provide chemical clearing. Memory times as short as 0.2 microsecond or as long as 30 microseconds can thus be obtained. If the high-voltage pulse is applied when no particle has produced a track, or after the track is cleared, no spark will occur (spurious sparks may sometimes appear; good design eliminates or minimizes them).

After the spark discharge, all the ionization produced must be removed before another event can be observed. The spark-chamber recovery time (or dead time) depends on the total energy dissipated in the spark; camera recording requires brighter sparks, hence more energy and longer dead times (usually 10–30 milliseconds). Short dead times (less than 1 millisecond) can be achieved with digital readout chambers with dim or invisible spark discharges.

Particle tracks and direction. In any spark chamber the appearance of the discharge depends upon the direction the charged particle has taken in the chamber. Because the pulsed electric field has a definite direction, the nature of the discharge is determined by the direction of the electric field as well as by the particle track. The spark chamber is therefore inherently anisotropic in its response.

In narrow-gap chambers (gaps up to about 3/4 in. or 18.75 mm), the discharge usually follows the electric field; that is, it is perpendicular to the plates. In wide-gap chambers, it follows the particle trajectory if the trajectory is not too different from the electric field direction—in practice at angles up to about 50° with the field. Whenever the particle path is more or less parallel to the plates (at right angles to the electric field), the discharge occurs as a "curtain" of streamers randomly distributed along the track with a spacing of a few millimeters.

In practice this lack of isotropy is often not a serious disadvantage. **Figure 2**, which shows a complex event in a set of narrow-gap chambers, gives a good indication of how well tracks may be followed and located no matter what their direction. A well-designed spark-chamber system can be isotropic in its particle detection efficiency.

TYPES OF SPARK CHAMBER

Spark chambers may be classified according to several different characteristics. There are wide-gap and narrow-gap chambers; sampling, projection, and track-delineating chambers; analog and digital chambers; optical, acoustic, and electronic data-readout chambers.

Sampling chamber. A sampling chamber is one that yields as output data the coordinates of a single point on the track (or tracks) in each gap. All narrow-gap chambers are of this type, including the oldest type of spark chamber, the multiplate chamber with individual gap spacings of 1/2 in. (12.5 mm) or less.

Projection chamber. A projection chamber is one in which the track of the particle is perpendicular, or nearly so, to the electric field; then the individual electrons of the track each

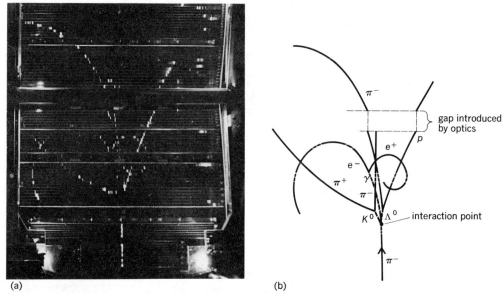

Fig. 2. Narrow-gap spark-chamber system in a magnetic field. (a) The photograph is taken through a pair of cylindrical field lenses that allow the camera to see into the narrow gaps. The large gap just above the center, where the right-hand edge appears bent, is spurious; it is introduced by the optics. (b) Interpretation of the event. A 1.2 BeV/c π^- meson enters from below and strikes an invisible target in the gap below the first full-width chamber. The interaction is $\pi^- + p \rightarrow K^0 + \Sigma^0$, where the Σ^0 decays almost instantaneously by the mode $\Sigma^0 \rightarrow \Lambda^0 + \gamma$. The ordinary decay products are neutral and hence leave no tracks until they decay or interact, K^0 decays into $\pi^+ + \pi^-$, Λ^0 into $p + \pi^-$. The gamma ray interacts to produce a positron-electron pair $e^+ - e^-$. (*After A. Roberts, Spark chambers, Encyclopaedic Dictionary of Physics, Pergamon Press, 1962–1964*)

produce a streamer across the gap, and the resulting curtain contains information only as to the projection of the track along the electric field, information on the third dimension being lost. This type of chamber is now superseded by the streamer chamber, in which the third-dimension information is preserved.

Track-delineating chamber. A track-delineating chamber is one in which the actual particle trajectory is made visible in space, just as in a cloud chamber or bubble chamber. Wide-gap and streamer chambers belong in the track-delineating category.

Conventional narrow-gap chamber. The great majority of spark-chamber experiments have used this kind of chamber, in which photographs are taken of the sparks between plates ¼ to ⅜ in. (6.25 to 9.3 mm) apart. There may or may not be a magnetic field. Under certain conditions the spark may follow the particle trajectory when it makes an angle with the plates, but in general the spark follows the electric field and coincides with the track in at most one point. Such a chamber is therefore a sampling chamber, since it samples rather than delineates the track.

When the track approaches parallelism with the plates, the number of sparks increases until finally a curtain of streamers marks the track. There is still track location information in such curtains, but frequently the track is simply not measured in those gaps.

Very narrow gaps (down to 0.08–0.12 in. or 2–3 mm) are occasionally used where particle fluxes are high, to achieve very rapid clearing (<0.2 μs) of unwanted electrons, and thus short memory times. At about ¾ in. (18.75 mm) there is a transition to wide-gap chamber operation, which is distinguished by a change from the sampling to the delineating mode. Instead of following the electric field, which is normal to the plates, the spark now follows the track, even if it makes an angle up to 40–50° with the electric field.

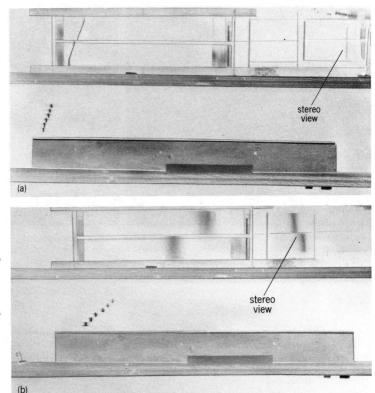

Fig. 3. Comparison between a wide-gap and a narrow-gap chamber stacked to observe the same particles. The wide-gap chamber (above) has a mirror showing a 90° stereo view at right. (a) A cosmic-ray particle making a small angle with the electric field direction. (b) A similar track, in this case making a large angle. Note the curtain discharge in the wide-gap chamber. (*From A. Roberts, Spark chambers, Encyclopaedic Dictionary of Physics, Pergamon Press, 1962–1964*)

Wide-gap chambers. These contain fewer plates, hence less matter than narrow-gap chambers, and offer less particle scattering and absorption. It is easier to "see into" a wide-gap chamber, since one is not peering down a narrow channel. The intrinsic accuracy of track location in wide-gap chambers (at least in the spark mode) is extremely good; it approaches that of the bubble chamber. The chief disadvantages are the need for a high pulse voltage and the loss of information from tracks parallel to the plates, a problem which is more serious than in a narrow-gap chamber, which restricts the loss to a small region for each track.

Figure 3 compares narrow- and wide-gap chambers, with the same particle traversing them. **Figure 4** shows a pair of 8-in. (20-cm) gaps with tracks traversing them, both without and with a magnetic field.

The streamer chamber is a development of the wide-gap chamber; the same wide-gap chamber can be operated in either the spark or the streamer mode. If the high-voltage pulse on a wide-gap chamber is made very short (about 20 to 50 nanoseconds) and very intense (2000 kV/m) a streamer discharge begins to grow along the electric field from many different electrons along the particle track. If the track is nearly parallel to the plates, the streamers are approximately at right angles to it, and grow to a length of perhaps ¼ to ½ in. (6.25 to 12.5 mm). Viewed end-on along the electric field, each streamer appears as a bright dot. The track of the particle then resembles that of a particle in a cloud chamber. The fact that any number of such tracks can be seen simultaneously makes the chamber suitable for looking at complex events.

The tracks in the streamer mode are much dimmer than those of a conventional wide-gap chamber, since much less energy goes into the discharge. Their intensity is also a complicated function of their direction; tracks along the electric field tend to break into a conventional spark mode, very much brighter than the streamers. The resulting wide light-intensity range gives rise

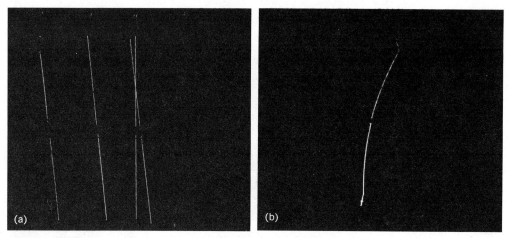

Fig. 4. Tracks in a pair of wide-gap chambers operating in the spark mode. (a) No magnetic field. (b) With magnetic field. The kinks at the end are reflections in chamber plates. (*Courtesy of K. Strauch*)

to problems in photography. **Figure 5** shows a set of tracks in a streamer chamber in a magnetic field.

Heavy spark-chamber plates. Spark-chamber plates usually contain as little matter as possible, to avoid scattering or interacting with the charged particles whose tracks are being measured. Stretched aluminum-foil plates one- or two-thousandths of an inch thick are common; aluminum-coated self-supporting low-density polyurethane foam plates are also used. Chambers in which neutral particles are to be detected use thick plates in which the neutral particles interact to produce charged particles. Thus, for gamma rays, lead plates may be used; then the electron-positron shower produced by the high-energy gamma ray is visible, and the point of interaction, the direction, and the approximate energy of the gamma ray may be measurable. By using separate chambers for measuring charged particle momentum (with thin plates) and for detecting neutral particles (thick plates) optimum conditions for both can be achieved simultaneously.

Heavy plate chambers can also be used for detecting neutrons by their nuclear interactions and for scattering protons (using carbon or aluminum) in order to measure their polarization.

DATA READOUT

Data-readout methods include digitized chambers, sonic chambers, and the use of vidicon tube and magnetic fields.

Fig. 5. Tracks in a wide-gap chamber in a magnetic field, operating in the streamer mode. (*Courtesy of R. F. Mozley*)

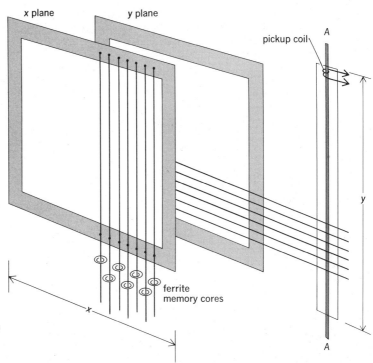

Fig. 6. Digitized spark chamber. Each plate is an array of parallel wires. One array determines the *x* coordinate of a spark, the other the *y* coordinate.

Figures 6 and 7 illustrate two of the principles described below.

Digitized chambers. In the narrow-gap chamber, the spark samples the track at one point, so that the spark coordinate may be regarded as digitized on the z axis, normal to the plates; in the plane normal to that axis it is still in analog form. The other axis may also be made to yield digitized data if the homogeneous plates of the chamber are instead replaced by arrays of parallel wires, spaced about 1 mm apart. By noting on which wire the spark occurs, the coordinate at right angles to the wire direction in the wire plane is given a digitized value. If the two wire planes are then oriented at right angles to each other (Fig. 6), all three dimensions of the spark location will be in digital form.

Of the two different readout methods illustrated, only one would be used in a real chamber. The x array uses ferrite memory cores, which are set by the spark current. They can be read out

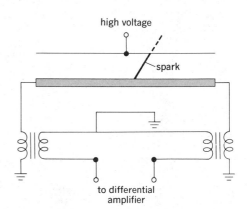

Fig. 7. Electronic determination of spark position in a narrow gap by the Charpak-Massonet current-distribution system. The spark location is determined by measuring the way in which the spark current divides between the two parallel paths to ground.

directly into a computer. The y array uses a magnetostrictive ribbon AA, in which the current pulse on the wire to which a spark has occurred produces an elastic (sonic) pulse which may be detected by a pickup coil at the end. The time it takes the pulse to arrive determines the wire at which it originated. For use in magnetic fields, the magnetostrictive ribbon is replaced by a nonmagnetic elastic ribbon. An elastic pulse produced by the mechanical shock associated with a spark from the wire to the ribbon across a tiny gap propagates down the ribbon and is detected by a piezoelectric pickup at the end (not shown).

Digitized wire chambers have assumed great importance because of the possibility of introducing their output directly into a computer. Many experiments in which the chambers were connected on-line to a computer have been run, so that the reconstruction and analysis of each event are immediately computed and made available to the experimenter. In an alternative arrangement, the data are recorded on magnetic tape, ready for later introduction into a computer, with slightly delayed analysis; this avoids tying down a computer to the experiment, and allows the use of a larger, more powerful computer.

Use of vidicon tube. An alternative method for digitizing the output of narrow-gap chambers is the use of a vidicon tube. As in all television pickup tubes, the stored image is in the form of a charge distribution on the photocathode, the amount of charge corresponding to intensity of light. The charge distribution is scanned in the conventional television raster, by an electron gun in the tube, and the output amplitude then corresponds to the light intensity. In this case, by using a uniform linear horizontal scan and measuring the position of the spark image along the scan, the position of the spark image can be digitized and recorded on magnetic tape. The chief drawback of the vidicon is its relatively limited resolution and sensitivity, compared to film recording; the convenience of immediate digitization and the elimination of film may well compensate for this disadvantage.

Sonic chambers. One of the earliest methods of locating a spark in a gap was by timing the arrival of the sound from the spark. The instant at which the spark occurs is known; hence, given the location of two microphones in each gap that receive the sound signal, and the time of arrival of the sound from the spark at each microphone, the position of the sound source is readily found. This method is difficult to apply if there is more than one spark present; it has been generally superseded by other techniques.

Electronic readout. Another method applicable to the location of a single spark in a narrow gap is the current-distribution method, shown in Fig. 7. The location of the spark (only one is allowed) is determined by observing how the spark current divides between the two available paths to ground. The transformer method shown measures the current ratio (or difference), and the sign and magnitude of the output pulse (properly normalized) determines the location of the spark (in one coordinate) to within 0.02 in. (0.5 mm). The other coordinate can be obtained from another gap.

Use of magnetic fields. Any spark chamber operates successfully in magnetic fields up to at least 20 kilogauss; the spark discharge is hardly affected. The magnetic field allows the sign and momentum of the particles to be measured by observing the curvature. The magnetic field restricts the data-readout technique somewhat: Certain types of digital readout using magnetic core storage or magnetostrictive delay lines are interfered with. It converts the system into a sort of electronic cloud chamber, with some important advantages, namely, the ability to tolerate relatively large fluxes of particles, the ability to select particular types of events by suitable triggering logic, the absence of background, ready adaptability to automatic data processing, and intrinsically higher precision.

The spark chamber is readily adaptable to complex systems in which several different kinds of chambers are used, including those with heavy plates for inducing interactions with neutral particles.

Spark-chamber photographs, if taken in a well-designed system, have certain advantages over bubble-chamber photographs when it comes to automatic processing. Many spark-chamber experiments have yielded film which has been successfully processed by semiautomatic or fully automatic film-reading data-processing systems; only a beginning in this direction has been made in bubble-chamber film, which generally contains many more background tracks and presents a much more difficult problem. Scanning can also be simpler (it can also be more difficult if the system is poorly designed). SEE CERENKOV RADIATION; SCINTILLATION COUNTER.

Bibliography. O. C. Allkofer, *Spark Chambers*, 1969; G. Charpak, L. Massonet, and J. Favier, The development of spark chamber techniques, *Prog. Nucl. Tech. Instrum.*, 1:323, 1965; J. W. Cronin, Spark chambers, in R. P. Shutt (ed.), *Bubble and Spark Chambers: Principles and Use*, vol. 1, 1967; E. Segré, *Nuclei and Particles*, 2d ed., 1977.

CLOUD CHAMBER
WILLIAM B. FRETTER

A particle detector in which the path of a fast charged particle is made visible by the formation of liquid droplets on the ions left by the particle as it passes through the gas of the chamber. Cloud chambers are used in research in nuclear physics, and can give detailed information on the particle in addition to the simple fact that the particle passed through. They can be operated at pressures below atmospheric, near atmospheric, or up to 50 atm (5000 kilopascals) depending on the application. They vary in size from a few inches in diameter up to large walk-in chambers containing many metal plates. SEE PARTICLE DETECTOR.

Supersaturated vapor is vapor whose pressure exceeds the saturation vapor pressure at the temperature prevailing. It is unstable, and condensation occurs in the presence of suitable nuclei. The interior of a cloud chamber contains a mixture of a gas such as air, argon, or helium and a vapor such as water or alcohol. When such a mixture is made supersaturated in the vapor component by cooling of the gas, the vapor tends to condense preferentially on charged atoms in the gas. Since a fast-moving charged particle produces many such ions in its passage through the gas, condensation of the vapor leaves a trail of droplets to indicate the path of the particle. When properly illuminated, these droplets are visible and may be photographed. A permanent record of the path of the particle may thus be obtained for examination at a later time.

Supersaturation methods. The supersaturation necessary for drop formation may be obtained in two ways. (1) In the Wilson cloud chamber the saturated vapor-gas mixture is suddenly expanded, causing supersaturation of the vapor at the resulting lower temperature. Usually the expansion is produced by motion of a rubber diaphragm or a piston. The amount of expansion must be carefully controlled; too little expansion will result in insufficient supersaturation to cause drop condensation, while overexpansion will give a dense fog, irrespective of the presence of ions in the gas. (2) The second method for obtaining supersaturation is used in the diffusion cloud chamber. A temperature gradient is maintained in the gas by cooling the bottom of the chamber

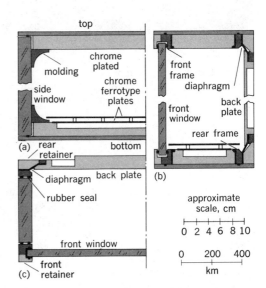

Fig. 1. Cloud chamber designed for use in a magnetic field. (*a*) Vertical section parallel to front. (*b*) Vertical section parallel to side. (*c*) Horizontal section. The back plate moves to produce the expansion. The mechanism for compressing the chamber and producing the expansion is not shown. Photograph in Fig. 2 was taken by using this cloud chamber. 1 in. = 2.5 cm.

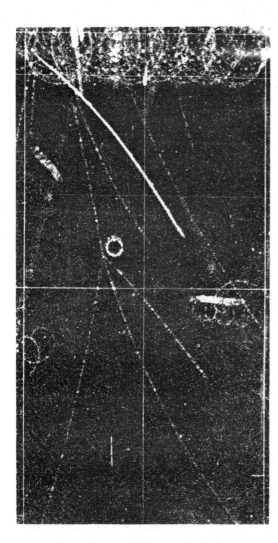

Fig. 2. Nuclear interaction produced by a cosmic-ray particle above chamber analyzed in cloud chamber placed in magnetic field. Circles at top of picture are tracks of electrons spiraling in magnetic field. Heavy track is a proton. Light track which becomes three tracks above center of chamber is a K meson, which disintegrates in flight into three π mesons; $K^+ \rightarrow \pi^+ + \pi^+ + \pi^-$. Tracks are curved because of magnetic field.

and warning the top. Vapor introduced at the top diffuses toward the bottom and cools as it goes. A certain region in the chamber becomes supersaturated, and the vapor will condense on ions if they are present. The diffusion chamber is thus continuously sensitive, as contrasted to the Wilson chamber, which is sensitive only for a fraction of a second after an expansion is made.

To overcome the short sensitive time, Wilson cloud chambers are often counter-controlled. Auxiliary particle detectors are used to signal the presence of a fast charged particle and to cause the expansion of the chamber within a few milliseconds after the passage of the particle. Wilson cloud chambers used in cosmic-ray research are usually counter-controlled because of the random time of arrival of cosmic-ray particles.

Information obtained. A cloud chamber can give information on the momentum of the particle if the chamber is placed in a magnetic field (**Figs. 1** and **2**). The curvature of the track can be measured on the photograph, and the value of the curvature, together with the magnitude of the field, gives the momentum. The sign of the charge can usually be determined also, if the direction of motion of the particle is known, since positive and negative particles curve in opposite directions. Another physical quantity that may be determined in a cloud chamber is the velocity of the particle, since the ionization produced by the charged particle depends on its velocity in a

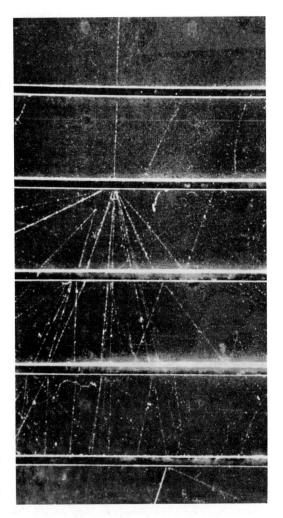

Fig. 3. Nuclear interaction produced by cosmic-ray particle occuring in a lead plate of cloud chamber. Particles produced are mesons, nucleons, and a hyperon, which traverse other lead plates in cloud chamber.

known way. With good photographic technique, individual droplets along the track are visible, and the number of these droplets is closely related to the number of ions along the track. A combination of the measurements of velocity and momentum yields a value for the mass of the particle, thus identifying it.

Cloud chambers may be used to determine the range of a particle, the distance required to bring it to rest. For low-energy particles the energy lost to the gas may be sufficient to bring a particle to rest in the chamber. To stop a high-energy particle, metal plates are inserted in the chamber with spaces between where the tracks may be seen (**Fig. 3**). A particle brought to rest in a given plate will have passed through a certain amount of material, which can be determined to give the range of the particle.

Dead time. Wilson cloud chambers have the disadvantage of a long dead time; it takes at least a minute for an expansion-type chamber to come to equilibrium and be ready for the next expansion. Diffusion chambers have no such dead time and are therefore much more suitable for use near a high-energy particle accelerator, where the particles come out in pulses only a few seconds apart. *See* BUBBLE CHAMBER.

Bibliography. American Institute of Physics, *The Cloud Chamber*, 1975; E. Segre, *Nuclei and Particles*, 2d ed., 1977.

BUBBLE CHAMBER
DONOVAN A. LJUNG

A particle detector used in elementary particle physics research. Charged particles passing through liquid contained in the chamber leave visible tracks along their paths (**Fig. 1**). These tracks are then photographed for later study. Each curved track is actually a string of small bubbles marking the path of an elementary particle. Bubble chambers vary in size, in the liquid used, and in how fast photographs can be taken. They are particularly useful because they allow direct visualization of all the charged particles involved in a high-energy interaction, and they have been used to discover many new particles, such as the rho (ρ) and omega minus (Ω^-). SEE ELEMENTARY PARTICLE.

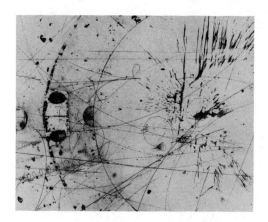

Fig. 1. Photograph of tracks produced by neutrinos in Fermilab's bubble chamber. (*Fermilab*)

Principle of operation. The bubble chamber was invented by Donald Glaser in 1952. It was known that a pure liquid in a clean glass container can be raised above its boiling temperature without its starting to boil spontaneously. The boiling must be triggered by something. For example, a broken piece of glass dropped into a superheated liquid can make boiling violently erupt. As an electrically charged particle passes through material (such as a liquid), it interacts with electrons in the material, leaving atoms ionized. Glaser discovered that in a superheated liquid this ionization energy is enough to trigger boiling along the path of the particle. The boiling bubbles grow rapidly, and so a photograph is quickly taken before the resolution of the individual tracks is destroyed. In order to make the bubble chamber sensitive to the arrival of another charged particle, the boiling must be stopped, and the liquid must be superheated again.

Operating cycle. A bubble chamber is cycled by the up and down movement of a piston (**Fig. 2**), causing pressure changes in the liquid which in turn cause a change in the liquid's boiling temperature. The piston first compresses the liquid so that the boiling point is above the temperature of the liquid. An accelerator produces the elementary particles to be studied and sends them as a beam toward the bubble chamber. Several milliseconds before the particles pass through the chamber, the piston expands the liquid, causing it to become superheated. The particles resulting from beam interaction with fluid nuclei in the chamber then produce bubbles, and about 1 ms later lights are flashed inside the chamber to record the tracks on film. Finally, the piston recompresses the liquid to stop the boiling and make the chamber ready for the next pulse of beam particles. This cycling can be faster than ten times a second for a rapid-cycling bubble chamber or as slow as four times a minute. Magnets are used to produce a magnetic field, which causes the charged particles to follow curved paths. Measurements of the curvature are then made to determine the particle's momentum. SEE PARTICLE ACCELERATOR.

Interactions and choice of liquid. Interest is focussed on the nuclear interactions that a beam particle has with the particles in the liquid. In Fig. 1, the beam particles are neutrinos

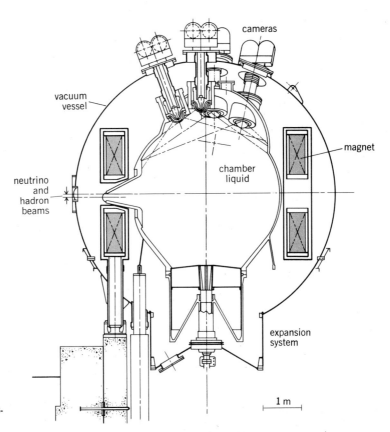

Fig. 2. Fermilab's 15-ft (4.6-m) bubble chamber.

which pass through the chamber from left to right. Neutrinos are electrically neutral, and so they do not leave tracks. When they interact, however, some of the particles produced will be charged, and the bubble chamber will record them. There are two interactions in Fig. 1: the first one with four produced particles in the upper left, and the second one in the lower center with five charged particles. A typical experiment comprises 50,000 analyzed photographs. Up to four cameras are used to photograph the tracks so that they can be reconstructed in three dimensions. Since the interactions occur with particles making up the liquid, the choice of liquid is important to the experimenter. Liquid hydrogen is used to study proton interactions because hydrogen is a simple atom made up of a single proton and electron. Likewise deuterium, with one neutron and one proton, is used to study neutron and proton interactions. Liquid hydrogen and deuterium boil at a very low temperature; thus chambers using these liquids operate at about 27 K above absolute zero.

Photons are produced in most interactions, but they are hard to study directly since they too are neutral. However, a photon can convert into an electron-positron pair, which can be seen. The chances of this conversion process occurring increase as the density or the atomic number of the liquid increases. Thus some bubble chambers are built to use a heavy liquid such as propane or Freon. These heavy-liquid bubble chambers operate at much higher temperatures, around 122°F (50°C), but otherwise operate under the same principle as hydrogen bubble chambers.

Size. The first test bubble chamber was only a few inches across. Bubble chambers have steadily become larger and are now built as large as 15 ft (4.6 m) across. The main reason for making such a larger chamber is to be able to study neutrino interactions. Neutrinos interact only very weakly with other particles, and thus one uses a large volume of liquid to increase the chance of an interaction occurring.

Hybrid systems. Bubble chambers have been combined with other particle detectors to learn even more about the interactions occurring inside the chamber. In these hybrid systems, other particle detectors such as proportional wire chambers, spark chambers, lead glass counters, and Cerenkov counters are used to further study the particles before and after they leave the bubble chamber. SEE CERENKOV RADIATION; PARTICLE DETECTOR; SPARK CHAMBER.

Bibliography. B. Barish, Experiments with neutrino beams, *Sci. Amer.*, 229(2):30–38, August 1973; D. Glaser, The bubble chamber, *Sci. Amer.*, 192(2):46–50, February 1955; D. M. Ritson (ed.), *Techniques of High Energy Physics*, pp. 87–114, 1961; R. P. Shutt (ed.), *Bubble and Spark Chambers: Principle and Use*, 2 vols., 1967.

JUNCTION DETECTOR
JAMES McKENZIE

A device in which detection of radiation takes place in or near the depletion region of a reverse-biased semiconductor junction. The electrical output pulse is linearly proportional to the energy deposited in the junction depletion layer by the incident ionizing radiation. SEE CRYSTAL COUNTER; IONIZATION CHAMBER.

Introduced into nuclear studies in 1958, the junction detector, or more generally, the nuclear semiconductor detector, revolutionized the field. In the detection of both charged particles and gamma radiation, these devices typically improved experimentally attainable energy resolutions by about two orders of magnitude over that previously attainable. To this they added unprecedented flexibility of utilization, speed of response, miniaturization, freedom from deleterious effects of extraneous electromagnetic (and often nuclear) radiation fields, low-voltage requirements, and effectively perfect linearity of output response. They are now used for a wide variety of diverse applications, from logging drill holes for uranium to examining the Shroud of Turin. They are used for general analytical applications, giving both qualitative and quantitative analysis in the microprobe and the scanning transmission electron microscopes. They are used in medicine, biology, environmental studies, and the space program. In the last category they play a very fundamental role, ranging from studies of the radiation fields in the solar system to the composition of extraterrestrial surfaces.

Fabrication of diodes. The first practical detectors were prepared by evaporating a very thin gold layer on a polished and etched wafer of *n*-type germanium (Ge). To reduce noise these devices were operated at liquid nitrogen temperature (77 K). Silicon (Si), however, with its larger band gap, 1.107 eV compared to 0.67 eV for germanium, offered the possibility of room-temperature operation. Gold-silicon surface barrier detectors and silicon *pn* junction detectors were soon developed.

Surface barrier detectors are made from wafers of *n*-type silicon semiconductor crystals. The etching and surface treatments create a thin *p* layer, and the gold contacts this layer (**Fig. 1**a). The *pn* junction silicon detectors are usually made by diffusing phosphorus about 2 micrometers into the surface of a *p*-type silicon base (Fig. 1b). Both techniques give a *pn* junction. When this junction is reverse-biased, a depletion region, or a region devoid of carriers (electrons and holes), forms mainly in the higher-resistivity base material. A high field now exists in this region, and any carriers born or generated in it are rapidly swept from the region. The requirement for detection is that the ionizing radiation must lose its energy by creating electron-hole pairs (2.96 eV/pair in germanium and 3.66 eV/pair in silicon) in the depletion region or within a carrier diffusion length of this region. Both carriers have to be collected to give an output pulse proportional to the energy of the incident particle. Electrons and holes have similar mobilities in both silicon and germanium, and although carrier trapping occurs it is not as severe as in the II-VI compounds.

Control of depletion region width. The detection of charged particles in the presence of gamma rays or higher-energy particles can be optimized by controlling the width of the depletion region. This width is a function of the reverse bias and of the resistivity of the base material.

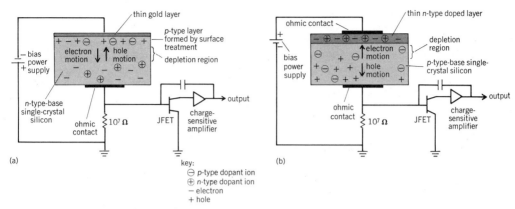

Fig. 1. Silicon junction detectors. (a) Surface barrier detector. (b) A pn junction detector. The p-type dopant ions are fixed in the crystal lattice. JFET = junction field-effect transistor.

There is a practical limit to the voltage that can be applied to a junction. Thus detectors for higher-energy or lower-mass particles (electrons) requiring wider depletion regions are made from high-resistivity material. This material occurs, by accident, during the growth of some crystals.

Lithium-drifted silicon detectors. Still wider depletion-width detectors can be made from lithium-drifted silicon. Lithium (Li) is a donor in silicon. In addition, at elevated temperatures (392°F or 200°C), the lithium ion is itself mobile. Thus when lithium is diffused in p-type silicon, a pn junction results. Reverse-biasing this junction at elevated temperatures causes the lithium ion, now appearing as a positive charge, to migrate toward the negative side. On the way it encounters an acceptor ion, negatively charged, which is fixed in the crystal lattice. The lithium ion and the acceptor ion compensate each other, and the lithium ion remains in this location. As more lithium ions drift into and across the depletion region, they compensate the acceptor ions and the region widens (**Fig. 2**). Depletion regions, or compensated regions, up to 0.8 in. (2 cm) wide have been achieved with this technique. Lithium-drifted silicon detectors can be operated at room temperature, but the larger volume gives a greater thermally generated leakage current, which degrades the resolution. The best energy resolution is obtained by operating the detectors at low temperature. However, they may be stored at room temperature.

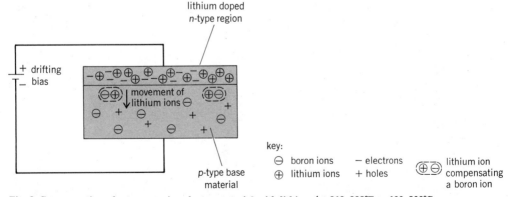

Fig. 2. Compensation of p-type semiconductor material with lithium (at 212–392°F or 100–200°C for silicon, 104–140°F or 40–60°C for germanium). Boron ions are fixed in the lattice. Lithium ions are fixed in the lattice, but at elevated temperature can be drifted under an electric field and will compensate boron ions to widen the depletion region.

Lithium-drifted silicon detectors are widely used to detect particle- or photon-induced x-rays. The resolution, when operated at 77 K, is sufficient to resolve the K x-rays for all elements higher in atomic number Z than carbon ($Z=6$). A resolution of 100 eV has been obtained at 2 keV. At the lower x-ray energies the effects of the detector window thickness and the absorption in the window of the mounting are important, and silicon is preferred for these applications. For x-rays the efficiency of a 5-mm depletion-width lithium-drifted silicon detector is about 50% at 30 keV and 5% at 60 keV. Typically these detectors have capacitances of about 2 picofarads and, to minimize noise, are operated with an optical or diode reset mechanism rather than a feedback resistor (**Fig. 3**). The detector bias is about 1000 V, and the junction field-effect transistor (JFET) gate operates at about -2.5 V. A radiation event causes a pulse of current in the detector. The amplifier drives this current i through the feedback capacitor with capacitance C and in doing so steps a voltage an amount e_{step} proportional to the charge, as given by the equation below. Each

$$e_{step} = \frac{1}{C}\int i\, dt$$

subsequent radiation event causes a voltage step. To keep the amplifier within its dynamic range the feedback capacitor must be discharged. The analyzing circuits are first gated off, and in the diode case (Fig. 3a) the reverse bias on the diode is momentarily increased to give a picoampere current pulse. The amplifier output voltage changes to allow this current to flow through the feedback capacitor, discharging it. The analyzing circuit is now gated on, and counting can resume. For the optical reset (Fig. 3b), a light is flashed on the JFET, momentarily increasing the source-to-gate leakage current and discharging the feedback capacitor. The output from the amplifier and the differentiated output for the analyzer are shown in Fig. 3c.

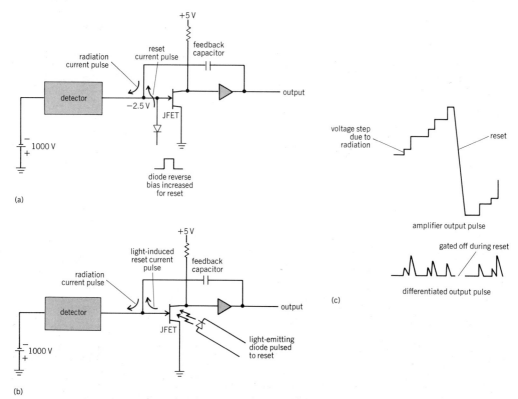

Fig. 3. Reset mechanisms for junction detectors. (**a**) Diode reset. (**b**) Optical reset. (**c**) Amplifier output and differentiated output for the pulse height analyzer.

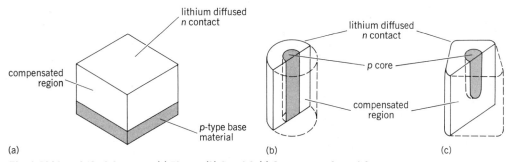

Fig. 4. Lithium-drifted detectors. (a) Planar. (b) Coaxial. (c) Open-one-end coaxial.

Lithium-drifted germanium detectors. Germanium with its higher atomic number, 32 compared with 14 for silicon, has higher radiation absorption than silicon. Lithium may also be drifted in germanium. But in germanium, lithium is mobile at room temperature and will precipitate or diffuse further if the units, after fabrication, are not kept at liquid nitrogen temperature. Lithium-drifted germanium detectors revolutionized the field of gamma-ray spectroscopy. They may be manufactured in planar, coaxial, or open-one-end coaxial geometry (**Fig. 4**).

Figure 5 compares the gamma-ray spectrum of ^{188}Os taken with a 21-cm^3 lithium-germanium detector with that from a sodium iodide (NaI) scintillator-type spectrometer which is 3 in. (7.5 cm) in diameter by 3 in. deep (27 in.^{3} or 330 cm^3). The counting efficiency of the lithium-germanium detector is lower than the scintillator, but the resolution is at least an order of magnitude better. This higher resolution often reduces the actual counting time to adequately identify a particular energy peak even with an order-of-magnitude less sensitive volume. Also, as shown in Fig. 5, the lithium-germanium detector is able to resolve more energy groups than the scintillator.

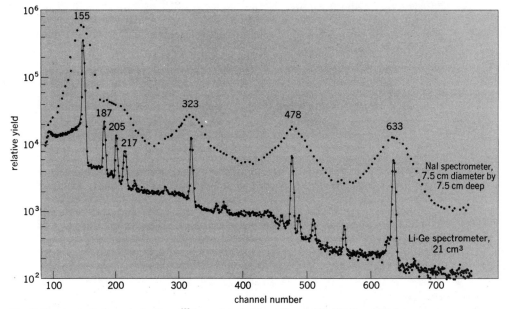

Fig. 5. Gamma-radiation spectra from ^{188}Os as detected in sodium iodide (NaI) and lithium-germanium (Li-Ge) spectrometers. 1 cm = 0.4 in.; 1 cm^3 = 6 × 10^{-2} in.^{3}

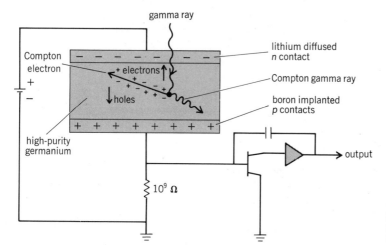

Fig. 6. High-purity germanium gamma-ray detector.

Hyperpure germanium detectors. Intrinsic or hyperpure germanium (**Fig. 6**) was grown to overcome the low-temperature-storage and the lithium-drifting problems associated with lithium-germanium. Planar detectors with up to an 0.8 in.-thick (2-cm) depletion region and coaxial detectors with 3-in.^{3} (50-cm^3) volume have been made with the material. Low-temperature processing is used in the fabrication—usually lithium diffused at 536°F (280°C) for the $n+$ contract and implanted boron for the $p+$ contract. This low-temperature processing is desirable to prevent diffusion of copper, with its subsequent charge trapping, into the germanium. Presently hyperpure germanium detectors cannot be made either as large as, nor with as high a resolution as, lithium-germanium detectors. Both types are operated at liquid nitrogen temperature, 77 K. However, the hyperpure germanium detector is easier to manufacture and can be stored at room temperature when not in use. This is a tremendous practical advantage.

Special detector configurations. Among the many other advantages of semiconductor detectors is the ease with which special detector configurations may be fabricated. One of the simple yet very important examples of this is the annular detector (**Fig. 7**), which is characteris-

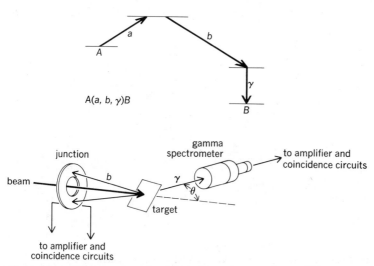

Fig. 7. Schematic of use of annular detector in nuclear reaction studies.

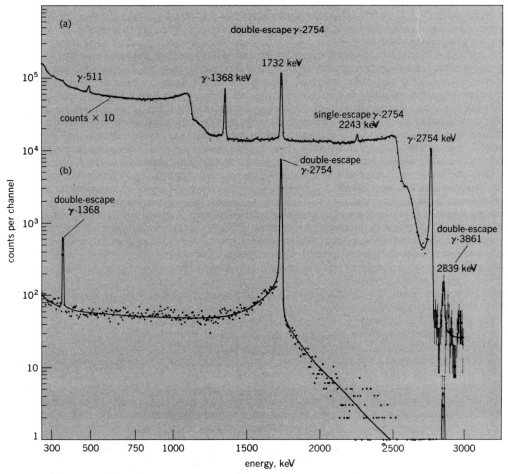

Fig. 8. Comparison of (a) direct single detector and (b) three-crystal spectrometer spectra from ^{24}Na source.

tically used to detect nuclear reaction products from a bombarded target in a tight cone around the incident beam. By examining the decay radiation in coincidence with such products, studies may be carried out only on residual nuclei which have had their spins very highly aligned in the nuclear reaction; this has been shown to provide an extremely powerful nuclear spectroscopic probe. The annular detector is extensively used in laboratories worldwide.

Composite detector systems are very readily assembled with the semiconductor devices. For example, it is standard in charged-particle detection to use a very thin detector and a very thick detector (or even two thin and one thick) in series. Multiplication of the resultant signals readily provides a characteristic identification signature for each nuclear particle species in addition to its energy. Three-crystal gamma-ray spectrometers are readily assembled, wherein only the output of the central detector is examined whenever it occurs in time coincidence with two correlated annihilation quanta escaping from the central detector. These systems essentially eliminate background from Compton scattering of other more complex electro-magnetic interactions and yield sharp single peaks for each incident photon energy (**Fig. 8**).

Similarly neutrons may be indirectly detected through examination of recoil protons from a hydrogenous radiator in the case of high-energy neutrons, or through examination of fission fragments resulting from slow neutrons incident on a fissile converter foil mounted with the semicon-

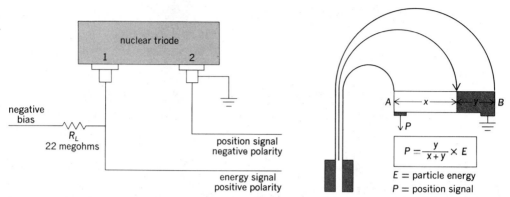

Fig. 9. Schematic of position-sensitive (nuclear triode) detector in focal plane of 180° magnetic spectrograph.

ductor detectors. (It should be noted that the response of the detectors is essentially perfectly linear all the way from electrons and photons to fission fragments.) Neutrons also may be detected and their energy spectra studied through examination of the charged products of the ($n\alpha$) reaction (where alpha particles are emitted from incident neutrons) induced in the silicon or germanium base material of the detector itself.

Fabrication of triodes. Whereas the detectors thus far discussed are electrically nothing more than diodes, it has been possible to construct equivalent triodes which have extremely important uses in that they provide not only an output which is linearly proportional to the energy deposited in them, but also a second output which in combination with the first establishes the precise location on the detector itself where the ionizing radiation was incident. This has very obvious advantages in the construction of simple systems for the measurement of angular distributions, where such position-sensitive detectors are located about a bombarded target. their most important impact, however, has been in terms of their on-line use in the focal planes of large nuclear magnetic spectrographs. Simultaneous determination of the energy and location of a particle in the focal plane, together with the momentum determination by the magnet itself, establishes unambiguously both the mass and energy of the particle, and does so instantaneously so that additional logical constraints may be imposed through a connected on-line computer—something totally impossible with the earlier photographic plate focal-plane detectors (**Fig. 9**).

A further important utilization of the nuclear triodes has followed their fabrication in an annular geometry similar to that shown in Fig. 7. With radial position sensitivity it becomes possible to correct on-line, and event by event, for the kinematic variation of particle energy with angle over the aperture of the detector. Without this correction possibility all particle group structures in the detector spectrum are smeared beyond recognition. *See* Particle detector.

Bibliography. D. A. Bromley, Detectors in nuclear science, *Nucl. Instrum. Meth.*, 162:1–8, 1979; G. T. Ewan, The solid ionization chamber, *Nucl. Instrum. Meth.*, 162:75–92, 1979; E. Laegsgaard, Position-sensitive semiconductor detectors, *Nucl. Instrum. Meth.*, 162:93–111, 1979; J. M. McKenzie, Development of the semiconductor radiation detector, *Nucl. Instrum. Meth.*, 162:49–73, 1979.

CRYSTAL COUNTER
James McKenzie

A detector of radiation in which the sensitive material is a high-resistivity crystal. The crystals are diced into wafers 0.02 to 0.8 in. (0.5 to 2 mm) in thickness and 0.006 to 0.155 in. (4 to 100 mm^2) in surface area. Various techniques are used to contact the larger surfaces. A field of about

10^4 V/cm is used to collect the charge liberated by the radiation. Crystal and junction counters are very similar, but differ fundamentally in the way the high-resistivity material is created. In a junction counter, a reverse-biased junction creates a depletion region which is the high-resistivity region. At the same time, the reverse bias gives high charge-collection field. In the crystal counter, the high resistivity comes from the basic crystalline material. It is often difficult to decide whether a particular radiation detector should be categorized as a crystal counter or a junction counter. SEE JUNCTION DETECTOR.

Crystals that have been studied as possible crystal counters include diamond, silver chloride, zinc sulfide, thallium iodide, thallium chloride, cadmium telluride, and mercuric iodide. Cadmium telluride (CdTe) and mercury iodide (HgI_2) are considered the most preferable crystals. Their absorption for gamma rays is greater than that for silicon or germanium. Also, they have larger band gaps (1.45 eV for cadmium telluride and 2.15 eV for mercury iodide)—a characteristic which lowers leakage currents, and makes possible high-resolution room-temperature detection of gamma rays.

Circuit. The basic circuit for a crystal counter is shown in the **illustration**. When a charged particle enters the crystal, it loses energy by creating electron-hole pairs. A gamma ray entering the crystal interacts with an atom of the crystal. Then, depending upon the interaction process—photoelectric, Compton, or pair production—the gamma ray will transfer all or part of its energy to an electron or to an electron-positron pair. These charged particles, in turn, dissipate their energy by creating electron-hole pairs. The number of electron-hole pairs is proportional to the energy deposited in the crystal. SEE GAMMA-RAY DETECTORS.

When an electron-hole pair is generated in the crystal, the electron and the hole induce equal and opposite charges on the electrodes (see illus.). As the electron moves toward the positive electrode, the positive charge it induces on this electrode increases uniformly. The operational amplifier provides this charge through the feed-back capacitor, which becomes charged as shown. The positive charge on the collecting electrode reaches its maximum value and is neutralized when the electron is collected. Simultaneously, the hole induces a negative charge on the collecting electrode, and this charge decreases uniformly to zero as the hole moves to the negative electrode. This decreasing negative charge on the collecting electrode is compensated by supplying positive charge from the operational amplifier and results in even more positive charge on the feedback capacitor.

Limitations and performance improvement. Only when both positive and negative carrier types are collected does a charge that is equal in magnitude to one of the carrier types appear on the feedback capacitor. This requirement of collecting both carriers places stringent demands on the crystal. That is, both carrier types should be similarly mobile to ensure an output charge signal proportional to the energy deposited in the crystal. If the carrier types are not similarly mobile, the charge signal is determined by the charge deposited and the location of that deposit. Hole trapping (or incomplete charge collection), excessive noise due to the high leakage current, the electronic noise of the preamplifier system, and the statistics of charge generation limit the resolution achieved with cadmium telluride and mercury iodide. Increasing the collecting field to minimize trapping causes an increase in the noise leakage current.

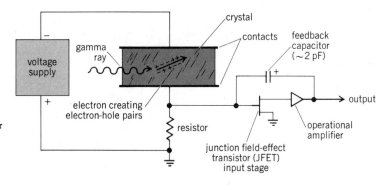

Counter circuit. Output signal depends of number of electron-hole pairs, which is proportional to energy deposited in crystal.

Other methods for reducing the effects of hole trapping include: (1) Differentiating the charge pulse. This gives a pulse which is proportional in magnitude to the number of moving carriers (in this case, electrons). (2) Searching for the spot in the crystal which gives the best resolution. (3) Introducing low-energy x-rays near the negative electrode. The electron-hole pairs are formed very near this electrode. Since the holes which become trapped are close to the negative electrode, their contribution to the total charge pulse is small.

Other methods such as cooling the detector, pulsing the bias, recording the spectra for only the first 20 s, and illuminating the crystal with infrared can be used to improve the performance.

Many of the preceding solutions may be unacceptable in the practical use of counters; however, they do show the inherent potential of good material. As methods of growing crystals improve to yield larger volumes that are highly uniform in purity, stoichiometry, and crystalline perfection, and as better detector fabrication methods are developed, crystal counters will achieve the efficiency of scintillators with the energy resolution of germanium. SEE PARTICLE DETECTOR; SCINTILLATION COUNTER.

Bibliography. D. A. Bromley (ed.), *Detectors in Nuclear Science, Nuclear Instruments and Methods*, 1979.

GAMMA-RAY DETECTORS
J. W. OLNESS

Devices for detecting gamma rays. Most detectors are capable of detecting and registering the passage of individual gamma rays, or photons. A detector which simultaneously determines the energy of a gamma ray may be properly termed a spectrometer. The term detector is frequently used as a generic name for devices that are also spectrometers. SEE GAMMA RAYS.

Detector-spectrometer devices. The two most popular devices in this category are the (thallium-activated) sodium-iodide [NaI(Tl)] detector and the (lithium-drifted) germanium [Ge(Li)] detector, which were developed in the 1950s and 1960s, respectively. These detectors are constructed from large single crystals of Ge or NaI; the activator element or dopant critical to operation of the device is written in parentheses, for example, (Tl).

Since gamma rays have no intrinsic charge or mass, both detectors depend on the fact that the three processes whereby gamma rays interact with matter (photoemission, Compton effect, and pair production) result in the production of energetic electrons whose total energy is exactly proportional to the original gamma ray energy. The electrons, in turn, lose energy by the ionization of the material (Ge or NaI), producing thereby a number of ions pairs exactly proportional to the electron energy. In both the NaI(Tl) and Ge(Li) detectors, the ionization energy is ultimately converted to a voltage-current pulse proportional to the total ionization produced by the gamma ray interaction. The output pulses from the detector (after amplification) are then analyzed by a pulse-height analyzer to provide a quantitative measure of the pulse-height spectrum.

In the ideal spectrometer, a uniform gamma ray energy should produce a uniform voltage pulse which results in a peak in the pulse-height spectrum at a pulse height proportional to the gamma ray energy. Of the three interactions, however, only the photoemission process results in a direct total conversion of gamma energy to ion energy. In the Compton process, the scattering of a gamma ray of initial energy E_γ results in a photon of lesser energy E'_γ and a broad distribution of electron energies (pulse heights), ranging from 0 to $E_\gamma/(1 + m_0c^2/2E_\gamma)$, is obtained; here m_0c^2 is the rest energy (511 keV) of the scattered electron. In the pair-production process, the gamma energy is converted into a positron and an electron pair of total kinetic energy $E_\gamma - 2m_0c^2$. The positron subsequently annihilates by pairing itself with another electron to produce two gamma rays of energy 511 keV. The lower-energy photons resulting from the Compton and pair-production processes may themselves subsequently interact by means of the first two processes. Thus, if the detector volume is large enough, the increased probability for multiple processes ensures that there also is a reasonable probability for complete conversion of gamma energy to ionization energy.

A major aim of detector development has therefore been the production of large-volume detectors which retain the essential property of high energy-resolving power. These detectors are characterized by relatively large, narrow, full-energy peaks. Because of incomplete conversion, a broad Compton distribution will remain, together with weaker subsidiary peaks (at E_γ = 511 keV and E_γ = 1022 keV) resulting from incomplete energy conversion in the pair-production process due to escape from the detector of the 511-keV annihilation radiations. SEE COMPTON EFFECT; ELECTRON-POSITRON PAIR PRODUCTION.

Ge(Li) semiconductor detectors. Illustration a is a schematic illustration of a "coaxial" Ge(Li) detector, showing the structure in terms of the n-type (n), intrinsic (i), and p-type (p) volumes. This is a single-crystal semiconductor device in which a large voltage applied in the reverse direction between the coaxial n and p surfaces produces a correspondingly large radial electric field, but a very low radial current.

Two of the three gamma ray interaction processes are shown schematically in illus. a, in which the production of ionization with the active (intrinsic) volume by the interacting electron is illustrated. The resultant distribution of pulse heights from these two processes in indicated in illus. b. The negative ionization products (electrons) are rapidly swept by the radial electric field to the n surface, whereas the positive "holes" are swept to the p surface, resulting in a current pulse exactly proportional to the total ionization produced by the interaction electron, and thus also to the initial electron energy. Assuming the detector is large enough so that the gamma ray energy is totally transferred (by means of multiple interactions) to ionizing electrons, the resultant current pulse is also exactly proportional to the gamma ray energy, and the necessary condition for a useful and efficient spectrometer is realized.

The typical large detector is approximately 6 in.^{3} (100 cm^3) in volume, roughly 2 in. (5 cm) in diameter by 2 in. (5 cm) in length. The resolution, expressed as the full width at half maximum (FWHM) of the full energy peak, is approximately 2.0 keV for 1332-keV gamma rays. The peak efficiency is about 20–30% that of a 3-in. by 3-in. (7.6-cm by 7.6-cm) NaI(Tl) detector; the ratio of peak-to-Compton counts is approximately 30:1.

Ge(Li) detectors are prepared from single crystals of germanium by coating the n surface with lithium and then diffusing the lithium ions radially inward to form the intrinsic volume; within this volume they serve the purpose of compensating for impurity atoms within the crystalline lattice. The crystal is contained in a vacuum cryostat and operated at liquid-nitrogen temperatures. In addition to the coaxial design shown in illustration a, "planar" and "trapezoidal" configurations are also used.

Two other semiconductor detectors, the lithium-drifted silicon [Si(Li)] detector and the intrinsic germanium [Ge(HP)] detector, have been widely used. Both are operated in configurations similar to that described for the Ge(Li) detector. The Si(Li) detector offers slightly improved resolution for gamma rays of energy less than 400 keV, but the detection efficiency for higher-energy gamma rays is relatively poor. Detectors using ultrapure intrinsic germanium without lithium com-

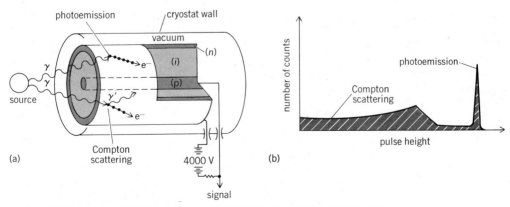

Coaxial Ge(Li) detector. (a) Schematic diagram. (b) Distribution of pulse heights from photoemission and Compton process.

pensation have also been developed. The advantages, in terms of stability and serviceable lifetime, are outweighed in most cases by the smaller size (1.2 in.^{3} or 20 cm ^{3}, as opposed to 6 in.^{3} or 100 cm ^{3}) and relatively high cost of these detectors. SEE JUNCTION DETECTOR.

NaI(Tl) detector. This is a member of the family of detectors designated scintillation detectors. Other frequently used scintillators are CsI(Tl) and various organic phosphors, both liquid and solid, of the general form $(CH_2)^n$. SEE LIQUID SCINTILLATION DETECTOR; SCINTILLATION COUNTER.

In the NaI(Tl) detector the ionization energy produced by the gamma-ray interaction is converted into a light pulse generated by the subsequent recombination of these ions. The many low-energy photons composing this light pulse are reflected onto the photocathode of a photomultiplier tube, optically coupled to the crystal, to produce finally a current pulse which is then amplified by the photomultiplier to a useful current range. The purpose of the thallium is to "shift" the resultant light spectrum into a region appropriate to the spectral response of the photocathode.

The most popular crystal configurations are cylindrical, ranging in size (diameter by height) from 2 by 2 in. to 5 by 6 in. (5 by 5 cm to 12.5 by 15 cm). Crystals as large as 12 by 12 in. (30 by 30 cm) have been employed, and a variety of smaller detectors of differing configurations have been utilized for specific purposes. The photomultiplier is optically coupled to one of the plane faces, and provides a current gain of 10^5–10^{10}; the remaining surfaces of the crystal are surrounded by a light-reflective shield. The entire assembly must be airtight, because sodium iodide is extremely hydroscopic.

Although the NaI(Tl) detector is more efficient than the Ge(Li) detector, its peak resolution is considerably poorer; for example, for 1332-keV gamma rays the resolution (FWHM) is approximately 80 keV, and the peak-Compton ratio is only about 6. It is very widely employed, however, because of its durability, relatively low cost, and high absolute peak efficiency (30–90%).

Other detectors. Gas-filled ionization chambers have also been used for gamma ray detection. The efficiencies are generally poor compared to those of NaI(Tl) or Ge (Li) detectors, and the energy-resolving capabilities are either inferior or nonexistent. But due to relatively low cost, they are still used in specialized applications, particularly for radiation monitoring for personnel safety. SEE IONIZATION CHAMBER.

Other spectrometers. Specific investigation in basic and industrial research have employed specialized spectrometers for gamma ray analysis. Among these, the crystal diffraction spectrometer has been frequently used for measurements of gamma rays up to 1 MeV in energy. The Compton spectrometer and the pair spectrometer, both of which employ magnetic analysis of interaction electrons ejected from thin metallic foils, have also been utilized for analysis of higher-energy gamma rays. SEE PARTICLE DETECTOR.

Bibliography. D. A. Bromley, *Detectors in Nuclear Science*, Nuclear Instruments and Methods, vol. 162, 1979; K. Debertin and W. B. Mann, *Gamma and X-ray Spectrometry Techniques and Applications*, 1983.

SCINTILLATION COUNTER
ROBERT HOFSTADTER

A particle or radiation detector which operates through emission of light flashes that are detected by a photosensitive device, usually a photomultiplier. The scintillation counter not only can detect the presence of a particle, gamma ray, or x-ray, but can measure the energy, or the energy loss, of the particle or radiation in the scintillating medium. The sensitive medium may be solid, liquid, or gaseous, but is usually one of the first two. The scintillation counter is one of the most widely versatile particle detectors, and is widely used in industry, scientific research, and radiation monitoring, as well as in exploration for petroleum and radioactive minerals that emit gamma rays. Many low-level radioactivity measurements are made with scintillation counters.

The scintillation phenomenon was used by E. Rutherford and colleagues, who observed the light flashes from screens of powdered zinc sulfide struck by alpha particles. The scattering of alpha particles by various foils was used by Rutherford to establish the modern notion of the nuclear atom. In modern times photomultipliers, which feed into amplifiers, have replaced the eye and have also proved to be very efficient in measuring the time of arrival of a particle, as well as its energy or energy loss. Scintillations are light pulses emitted by atoms which return to their ground states after having been raised to excited states by passing particles. Scintillations are one example of luminescent behavior, and are related to the process called fluorescence.

Operation of bulk counter. Following the innovation in 1947 of using bulk material, instead of screens of powdered luminescent materials, scintillation counters have been made of transparent crystalline materials or liquids or plastics. In order to be an efficient detector, the bulk scintillating medium must be transparent to its own luminescent radiation, and since some detectors are quite extensive, covering meters in length, the transparency must be of a high order. One face of the scintillator is placed in optical contact with the photosensitive surface of the photomultiplier, as shown in the **illustration**. In order to direct as much as possible of the light flash to the photosensitive surface, reflecting material is placed between the scintillator and the inside surface of the container.

In many cases it is necessary to collect the light from a large area and transmit it to the small surface of a photomultiplier. In this case, a "light pipe" leads the light signal from the scintillator surface to the photomultiplier with only small loss. The best light guides and light fibers are made of glass, plastic, or quartz. It is also possible to use lenses and mirrors in conjunction with scintillators and photomultipliers.

A charged particle, moving through the scintillator, leaves a trail of excited atoms which emit the characteristic luminescence of that particular material. When a particle stops in the material, all its energy may be lost in the scintillator, and therefore, after calibration with known sources, the particle's energy can be measured. When a particle passes through a scintillator, the energy loss of the particle is measured. When a gamma ray converts to charged particles in a scintillator, its energy may also be determined. When the scintillator is made of dense material and of very large dimensions, the entire energy of a very energetic particle or gamma ray may be contained within the scintillator, and again the original energy may be measured. Such is the case for energetic electrons, positrons, or gamma rays which produce electromagnetic showers in the scintillator. When the sizes of the light flashes are measured in the photomultipliers, the results are recorded in pulse-height analyzers, and then one can readily determine the energy spectra of particles in these various cases.

Characteristics. Scintillation counters have several characteristics which make them particularly useful as detectors of x-rays, gamma rays, and other nuclear and high-energy particles.

Efficiency and size. Counting efficiencies close to 100% are often not difficult to achieve, and these detectors can often be made quite thin and small and can then define a particle's position rather accurately. Arrays of small crystals can be used, and "light pipes" can be employed to bring the luminous radiation some distance away to more bulky photomultipliers. When detection of very energetic particles is desired, the scintillator can be made very large and mas-

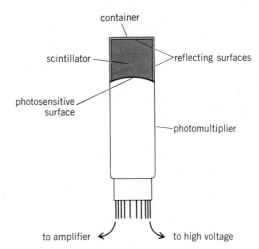

Diagram of a scintillation counter.

sive, and often requires the use of many photomultipliers "looking" at the scintillator through transparent windows.

Speed. Useful scintillation materials emit light flashes as short as 10^{-9} s, although most inorganic materials have light flashes 10 to 100 times longer. Even so, the flash is very rapid. High speed allows very fast counting and avoids pileup, which is very characteristic of older and slower particle detectors such as the Geiger counter, which has been largely superseded. The Cerenkov counter is similar to a scintillation counter, but depends only on the index of refraction of the transparent medium and the particles' velocity in the medium, and not specifically on the crystalline luminescent processes. SEE CERENKOV RADIATION.

Energy resolution. While the energy resolution is quite good in scintillation counters, their performance in this respect is markedly inferior to that of semiconductor detectors such as silicon, germanium, or lithium-drifted germanium at low temperatures. Semiconductor counters have been rather limited in size, so that they do not usually compete with the large inorganic scintillators needed in high-energy research. Often anticoincidence combinations are made between semiconductor counters and scintillation counters. SEE JUNCTION DETECTOR.

Substances used. Scintillation counter materials are generally classified into inorganic and organic types.

Inorganic scintillators. Inorganic scintillators are generally characterized by the presence of heavy elements. The most useful inorganic scintillator is sodium iodide activated with a small amount of thallium salt. The usual designation of this material is NaI(Tl). Cesium iodide is also useful, and may be activated with thallium or sodium. Other useful crystalline scintillators include calcium fluoride, barium fluoride, and cadmium tungstate. Bismuth germanate is a new and interesting scintillator. NaI(Tl) has a decay time of 2.5×10^{-7} s, and other alkali halide materials are similar. NaI(Tl) is particularly useful for detecting gamma rays because of the presence of iodine, which is a relatively heavy atom of high atomic number (53). The high atomic number is important in both the photoelectric and pair production processes which result from the interaction of a gamma ray with the base material. SEE GAMMA-RAY DETECTORS.

Organic scintillators. Materials of this type are typically naphthalene, anthracene, trans-stilbene, terphenyl, $C_{24}H_{16}N_2O_2$, and similar materials. Very successful liquid scintillators have been made by dissolving terphenyl in an organic liquid such as xylene, or in a polymeric material to make a solid plastic scintillator. Organic scintillators find special application when short decay times of the scintillation flashes are required. Scintillation decay times of 10^{-9} s are not unusual, and therefore very high counting rates, avoiding pile-up, may be achieved. Since organic materials contain hydrogen, they are also very useful in detecting neutrons. Neutrons collide with protons (hydrogen nuclei), thereby transferring their kinetic energy to charged particles. Organic scintillators are not as useful as inorganic scintillators for the detection of gamma rays, electrons, or positrons. Large volumes of liquid scintillators have been used to detect the elusive neutrinos. SEE LIQUID SCINTILLATION DETECTOR; NEUTRINO.

Coincidence counting. Scintillation counters are often used in temporal coincidence with other particle or radiation detectors. The coincidence technique eliminates or reduces the problem of local background or false events. For large counting systems recording complicated events, such as the "NaI(Tl) crystal ball" used in high-energy physics, the coincidence technique is used to make a fast trigger for desirable events and to exclude false or uninteresting events. It is also possible to use fast organic scintillators separated by a given distance to measure the time of flight (delay in time between the two pulses) of particles traversing a common pair. In this way the speed of a particle may be measured. Speeds approaching the velocity of light can easily be measured by a fast scintillator pair. Also radioactive decay times, or unstable particle decays, such as those of pions or muons, may be studied in the same scintillator by observing the incident particle pulse and later the pulse produced by the new particle or particles resulting from the decay. SEE PARTICLE DETECTOR.

Bibliography. J. B. Birks, *The Theory and Practice of Scintillation Counting*, 2d ed., 1967; F. D. Brooks, Development of organic scintillators, *Nucl. Sci.*, pp. 477-505, 1979; R. L. Heath, R. Hofstadter, and E. B. Hughes, Inorganic scintillators in detectors, *Nucl. Sci.*, pp. 431-476, 1979; C.-T. Cheng, D. L. Horrocks, and E. L. Alpen (eds.), *Liquid Scintillation Counting: Recent Applications and Development*, 2 vols., 1980.

LIQUID SCINTILLATION DETECTOR
Frank D. Brooks

A particle dectector in which the sensitive medium is a liquid scintillator. For a general discussion of scintillators as particle detectors SEE SCINTILLATION COUNTER.

Liquid and plastic scintillators are the most extensively used forms of organic scintillator. They are noted for their fast response, for the ease with which they may be formed in arbitrary shapes, large or small, and for the economy they offer in achieving large, sensitive detecting volumes. The timing precision achieved by the faster liquids and plastics is typically in the range 10^{-9} to 10^{-10} s. Liquid scintillators are used in volumes ranging from a few tenths of a cubic inch or less through several hundreds of cubic feet. The largest scintillators are used in high-energy physics to study neutrino interactions. Liquid scintillators also have the capacity to assimilate other liquids or substances and form homogeneous media, thereby providing an efficient and simple means for measuring the products of nuclear reaction or radioactive decay in those substances. SEE NEUTRINO.

Solvent and solute. A simple liquid scintil- lator might consist of 0.2 oz (5 g) of solute, such as p-terphenyl, PPO (2,5-diphenyloxazole), or PBD [2-phenyl,5-(4-biphenyl)-1,3,4,oxadiazole], dissolved in 1 quart (1 liter) of solvent, such as xylene, toluene, or benzene. The solvent, being the bulk constituent, absorbs virtually all of the energy deposited by the ionizing particle being detected in the scintillator. For efficient operation, the excitation energy imparted to the solvent molecules must transfer rapidly through the solvent to the solute molecules, which then deexcite and produce the scintillation emission. Large liquid scintillators usually incorporate an additional small quantity (0.003 02/quart or 0.1 g/liter) of secondary solute or "wavelength shifter," such as POPOP [1,4-bis-{2-(5-phenyloxazolyl)}-benzene], which captures excitation energy from the primary solute. The resulting scintillation output from the secondary solute is emitted at longer wavelengths than that of the primary solute, and is therefore more efficiently transmitted through the scintillator to the surrounding photomultiplier tubes.

Pulse shape discrimination. The response of most liquid scintillators can be improved significantly if dissolved oxygen is either removed from the solution or displaced by nitrogen or an inert gas. This procedure also endows many liquid scintillators with the ability to pulse-shape-discriminate, that is, to exhibit scintillation decay properties which depend on the type of ionizing particle causing the scintillation. Most liquid scintillators contain a high proportion of hydrogen, and therefore detect fast neutrons (energy greater than 0.1 MeV) via the proton recoils generated by internal neutron-proton scattering. Pulse shape discrimination may be used to select these proton scintillations while rejecting others, especially those caused by gamma rays. Liquid scintillators are therefore very effective in detecting and identifying neutrons, even when the background of gamma radiation is high. The response of liquid scintillators to electrons is linear, that is, proportional to electron energy. However, their response to heavier particles is nonlinear; and for different particles of the same energy, the response is smaller, the heavier the particle.

Additions of compounds and samples. Appreciable quantities of compounds of a number of elements can be incorporated in liquid scintillators (in appropriate chemical form) without degrading the scintillation performance excessively. Some examples are: boron, gadolinium, or cadmium compounds, which facilitate the detection of low-energy neutrons (less than about 0.1 MeV) through neutron capture in these elements; and tin or lead compounds, which enhance the scintillator sensitivity to low energy gamma rays or x-rays (less than about 0.1 MeV). Hexafluorobenzene may be used as a solvent, instead of benzene, to obtain a hydrogen-free, neutron-insensitive liquid detector for gamma rays. Organic liquid scintillators are extensively used in the medical, biological, and environmental fields to detect radioactive samples and tracer compounds labeled with the beta emitters ^{14}C or ^{3}H, or with other radionuclides. A carrier solution containing the sample or tracer compound is assimilated or dissolved in the liquid scintillator to form a uniform counting sample or scintillating medium with a volume of a few milliliters. Such counting samples are often processed in large batches by using sophisticated instrumentation which incorporates automatic sample changing and the capacity to carry out all the necessary data reduction automatically.

Manufacture and handling. The handling of liquid scintillators requires extreme care, because many are highly inflammable and because small concentrations of common impurities can drastically impair the scintillation performance by quenching the energy-transfer mechanism of the scintillation process. Commercial manufacturers offer a wide variety of liquid scintillators packaged in a variety of forms, for example, bottled in bulk form or encapsulated in glass or metal-and-glass containers. Encapsulated scintillators may have reflector paint applied to all surfaces other than which is to be coupled to the photomultiplier cathode. SEE PARTICLE DETECTOR.

TRANSITION RADIATION DETECTORS
WILLIAM J. WILLIS

Detectors of energetic charged particles that make use of radiation emitted as the particle crosses boundaries between regions with different indices of refraction. An energetic charged particle moving through matter momentarily polarizes the material nearby. If the particle crosses a boundary where the index of refraction changes, the change in polarization gives rise to the emission of electromagnetic transition radiation. About one photon is emitted for every 100 boundaries crossed, for transitions between air and matter of ordinary density. Transition radiation is emitted even if the velocity of the particle is less than the light velocity of a given wavelength, in contrast to Cerenkov radiation. Consequently, this radiation can take place in the x-ray region of the spectrum where there is no Cerenkov radiation, because the index of refraction is less than one. SEE CERENKOV RADIATION.

The radiation extends to frequencies greater than the plasma frequency by the factor γ = particle energy divided by particle mass. The production of x-rays requires γ equal to or greater than 1000. A threshold as high as this is difficult to achieve by other means. This fact has led to the application of this effect for the identification of high-energy particles. For example, electrons of about 10^9 eV will produce x-rays of a few kiloelectronvolts, while the threshold for pions is on the order of 10^{11} eV. The solid material should be of low atomic number, carbon or lighter, to minimize absorption of x-rays, which are emitted to close to the particle direction. The material is often in the form of foils, which must be of the order of 0.0004 in. (0.01 mm) thick to avoid destructive interference of the radiation from the two surfaces, and similarly, the foil spacing is typically 0.004 in (0.1 mm). Random assemblies of fibers or foams are almost as effective as periodic arrays of foils. Effective electron detectors have been made with several hundred foils followed by a xenon proportional chamber for x-ray detection. SEE PARTICLE DETECTOR.
Bibliography. X. Artru, G. B. Yodh, and G. Menessier, Practical theory of the multilayered transition radiation detector, *Phys. Rev. D*, 12:1289–1306, 1975; J. Cobb et al., Transition radiators for electron identification at the CERN ISR, *Nucl. Instrum. Meth.*, 140:413–427, 1977; G. M. Garibian, Transition radiation effects in particle energy losses, *Sov. Phys.—JETP*, 10:372–376, 1960.

PARTICLE TRACK ETCHING
P. BUFORD PRICE

A technique of selective chemical etching to reveal tracks of heavy nuclear particles in a wide variety of solid substances. Developed in order to see fossil particle tracks in extraterrestrial materials, the technique finds application in many fields of science and technology.

Identification of nuclear particles. An etchable track is produced if the charged particle has a sufficiently high radiation-damage rate and if the damaged region in the solid is permanently localized. Thus only highly ionizing particles are detectable; only nonconductors record tracks; and radiation-sensitive plastics can detect lighter particles than can radiation-insensitive minerals and glasses. The conical shape of the etched track depends on the ratio of the rate of etching along the track to the bulk etching rate of the solid. Careful measurements with accelerated ions of known atomic number Z and velocity $v = \beta c$ (where c = velocity of light) have shown that this ratio is an increasing function of Z/β. Measurements of the shapes of etched tracks thus serve to identify particles (**Fig. 1**). The most sensitive detector, a plastic known as

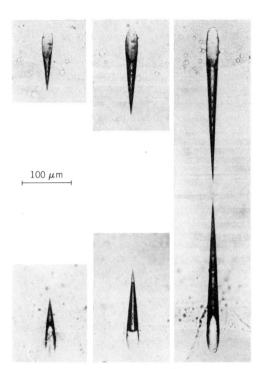

Fig. 1. Upper images show increasing radiation damage produced by a high-energy uranium nucleus in the cosmic radiation as it penetrates a thick stack of Lexan plastic sheets and slows down. The lengths of the etched cones in these three sheets, taken at intervals of several millimeters from within the stack, permit the atomic number and energy of the nucleus to be determined. Lower images show the exit points of the tracks.

CR-39, detects particles with Z as low as 1 (**Fig. 2**) provided $Z/\beta \gtrsim 10$. The clarity and contrast of the images and its high sensitivity make CR-39 a very attractive track-etch detector for nuclear physics, cosmic-ray research, element mapping, personnel neutron dosimetry, and many other applications. Minerals, on the other hand, are insensitive to particles with $Z/\beta \lesssim 150$ and are therefore useful in recording rare, very heavily ionizing, relatively low-energy nuclei with $Z \gtrsim 25$. SEE CHARGED PARTICLE BEAMS.

Solar and galactic irradiation history. The lunar surface, meteorites, and other objects exposed in space have been irradiated by charged particles from a variety of sources in the Sun and the Galaxy. The particles of lowest energy are produced by the expanding corona of the Sun—the solar wind. Arriving in prodigious numbers, but penetrating only some millionths of a centimeter, the solar wind particles quickly produce an amorphous radiation-damaged layer on

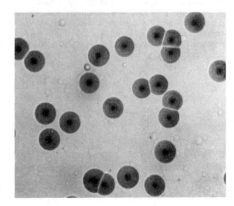

Fig. 2. Etched conical pits in a sheet of allyl diglycol carbonate (CR-39 plastic) irradiated at normal incidence with 60-MeV alpha particles. (*From P. B. Price et al., Do energetic heavy nuclei penetrate deeply into the Earth's atmosphere?, Proc. Nat. Acad. Sci., 77;44–48, 1980*)

crystalline grains. At depths from about 0.00004 to 0.04 in. or 1 micrometer to 1 mm, individual tracks produced in solar flares (sporadic, energetic outbursts on the Sun) can be resolved by electron microscopy or sometimes by optical microscopy. At depths greater than 1mm, most of the particle tracks are produced by heavy nuclei in the galactic cosmic radiation.

Comparison of fossil particle tracks in lunar rocks and meteorites with spacecraft measurements of present-day radiations has established that solar flares and galactic cosmic rays have not changed over the last 2×10^7 years—the typical time a lunar rock exists before being shattered by impacting interplanetary debris. Observations of grains in stratified lunar cores and lunar and meteorite breccias (**Fig. 3**) enable the particle track record to be extended back more than 4×10^9 years in time. Breccias, which are complex grain assemblages, often contain grains that have high solar flare track densities on their edges, indicating exposure to free space prior to breccia formation. Dating work, using spontaneous-fission tracks, indicates that some of the breccias were assembled soon after the beginning of the solar system some 4.6×10^9 years ago. The study of the time history of energetic radiations in space, using lunar samples and meteorites, elucidate various dynamic processes such as rock survival lifetimes, microerosion of rocks, and the formation and turnover rates of planetary regoliths.

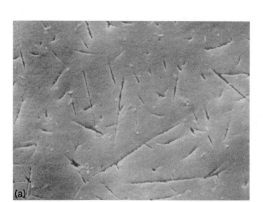 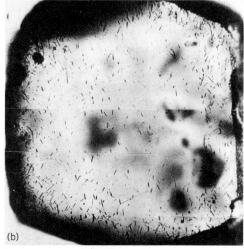

Fig. 3. Etched fossil particle tracks. (a) Tracks from spontaneous fission of ^{238}U and ^{244}Pu in a zircon crystal from a lunar breccia, seen by scanning electron microscopy (*from D. Braddy et al., Crystal chemistry of Pu and U and concordant fission track ages of lunar zircons and whitlockites, Proceedings of the 6th Lunar Science conference, pp. 3587–3600, 1975*). (b) Tracks of energetic iron nuclei from solar flares, showing a decreasing concentration from edges to center of a 150-μm-diameter olivine crystal from a carbonaceous chondritic meteorite. The irradiation occurred about 4×10^9 years ago, before the individual grains were compacted into a meteorite (*from P. B. Price et al., Track studies bearing on solar-system regoliths, Proceedings of the 6th Lunar Science Conference, pp. 3449–3469, 1975*).

Nucleosynthesis. In terrestrial crystals, which are well shielded from external radiations, the dominant source of tracks is the spontaneous fission of ^{238}U. Certain meteorites and lunar rocks (Fig. 3a) contain additional fission tracks due to the presence of ^{244}Pu (half-life, 8×10^7 years) when the crystal was formed. After corrections are made for chemical fractionation of Pu and U, the data indicate that a large spike of newly synthesized elements was produced at the time of formation of the Galaxy, followed by a period of continuous synthesis and relatively rapid mixing.

Current solar and galactic irradiation. Studies of tracks in a piece of glass from the *Surveyor 3* spacecraft after a 2.6-year exposure on the lunar surface, and of tracks in plastic detectors exposed briefly above the Earth's atmosphere in rockets, led to the surprising discovery that the Sun preferentially ejects heavy elements in its flares rather than an unbiased sample of its atmosphere. *See Sun.*

The existence of galactic cosmic rays with atomic number greater than 30 was discovered in 1966 when fossil particle tracks were first studied in meteorites. Since then many stacks of various types of plastics and nuclear emulsions up to 220 ft^2 (20 m^2) in area have been exposed in high-altitude balloons and in Skylab in order to map out the composition of the heaviest, rarest cosmic rays. Several particles heavier than uranium have been detected, indicating that cosmic rays originate in sources where synthesis has proceeded explosively beyond uranium. Exposures of giant detectors of about 1100 ft^2 (100 m^2) for a year in space are planned. Hybrid detectors using stacks of track-recording plastics to measure range, and photomultiplier tubes to measure light from Cerenkov radiation and scintillation detectors, have made it possible to determine relative abundances of the isotopes of very heavy elements in the cosmic rays. These data bear both on the history of cosmic rays and on nucleosynthesis in their sources.

Nuclear and elementary particle physics. Unique advantages of etched-track detectors are their ability to distinguish heavy-particle events in a large background of lightly ionizing radiation and their ability to detect individual rare events by a specialized technique such as electric-spark scanning or ammonia penetration through etched holes. These advantages have permitted such advances as the measurement of very long fission half-lives; the discovery of ternary fission; the determination of fission barriers; the production of numerous isotopes of several far-transuranic elements; the discovery of several light, neutron-rich nuclides such as ^{20}C at the limit of particle stability; and highly sensitive searches for magnetic monopoles, superheavy elements, and anomalously dense nuclear matter in nature and in accelerators. *See Magnetic monopoles; Nuclear fission; Supertransuranics; Transuranium elements.*

Geochronology. The spontaneous fission of ^{238}U, present as a trace-element purity, gives tracks that can be used to date terrestrial samples ranging from rocks to human artifacts. Because fission tracks are erased in a particular mineral at a well-defined temperature (for example, 212°F or 100°C in apatite), one can use the apparent fission-track ages as a function of distance from the heat source to measure the thermal (tectonic) history of regions. Examples are the rate of sea-floor spreading (about 1 cm per year) and the surprisingly rapid rate of uplift of the Alps (1.2 × 10^{-2} to 5.6 × 10^{-2} in. or 0.3–1.4 mm per year) and of the Wasatch range in Utah (4 × 10^{-3} to 1.6 × 10^{-2} in. or 0.01–0.4 mm per year). Because of its very high uranium concentration (typically a few hundred parts per million), a few tiny zircon crystals less than 4 × 10^{-3} in. (0.1 mm) in diameter suffice to give a fission track age of sedimentary volcanic ash and thus to determine absolute ages of stratigraphic boundaries. Occasional, anomalously low fission-track ages can sometimes be used to locate valuable ore bodies and even petroleum.

Geophysics and element mapping. In these applications a track detector placed next to the material being studied records the spatial distribution of certain nuclides that either spontaneously decay by charged particle emission or are induced to emit a charged particle by a suitable bombardment. A resolution of a few micrometers is easily attained. Such micromaps make it possible to identify radionuclides in atmospheric aerosols, to measure the distribution and transport of radon, thorium, and uranium, and to measure sedimentation rates at ocean floors. Thermal neutrons, readily available in a reactor, can be used to map ^{235}U (via fission), ^{10}B (via alpha particles), ^{6}Li (via tritons), ^{14}N (via protons), and several other nuclides. Deuterium, a tracer in biological studies, and lead and bismuth can also be mapped by using different irradiations.

Practical applications. Filters are produced by irradiating thin plastic sheets with fission fragments and then etching holes to the desired size. Uses include biological research, wine filtration, and virus sizing. In virus sizing a single hole is formed that separates two halves of a conducting solution. When a virus or other tiny object passes through the hole, the resistance increases drastically, and the size, shape, and speed of the object can be determined by analyzing the electric signal.

There are numerous other uses. A uranium exploration method relies on a survey of radon emanation, as measured by alpha-particle tracks in plastic detectors, to locate promising locations in which to drill. CR-39 plastic detectors are used in a Fresnel zone-plate imaging technique to

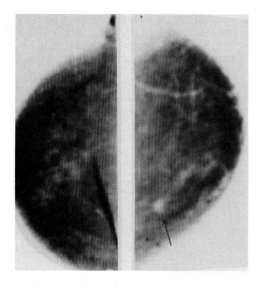

Fig. 4. Reconstructed images of right and left breasts of a patient taken by using a 3600-MeV carbon-ion beam at the Lawrence Berkeley Laboratory Bevalac accelerator. Arrow indicates 1-cm (0.4-in.) carcinoma. (*From C. A. Tobias et al., Lawrence Berkeley Laboratory*)

make high-resolution images of the thermonuclear burn region in laser fusion experiments. Plastic detectors are also used in conjunction with a beam of high-energy heavy ions to take radiographs of cancer patients that reveal details not detectable in x-rays. **Figure 4** shows reconstructed images of right and left breasts of a patient taken by using a 3600-MeV carbon-ion beam at the Lawrence Berkeley Laboratory Bevalac accelerator. A single beam pulse is passed through each breast as it is immersed in water, and the stopping points of the ions are recorded in a stack of 30 cellulose nitrate sheets, each 0.01 in. (0.025 cm) thick, located behind the water bath. After the sheets are etched, the information on each sheet is converted to digital data by a scanning system, and two images capable of revealing slight differences in density are displayed on a television screen. The patient dose for such images is about 20 to 100 millirads (200 to 1000 micrograys), considerably lower than the usual dose from x-ray mammographic imaging.

Bibliography. B. G. Cartwright, E. K. Shirk, and P. B. Price, A nuclear-track-recording polymer of unique sensitivity and resolution, *Nucl. Instrum. Meth.*, 153:457–460, 1978; R. L. Fleischer, Where do nuclear tracks lead?, *Amer. Sci.*, 67:194–203, 1979; R. L. Fleischer, P. B. Price, and R. M. Walker, *Nuclear Tracks in Solids*, 1975; R. M. Walker, Interaction of energetic nuclear particles in space with the lunar surface, *Annu. Rev. Earth Planet. Sci.*, 3:99–128, 1975.

CERENKOV RADIATION
William B. Fretter

Light emitted by a high-speed charged particle when the particle passes through a transparent, nonconducting, solid material at a speed greater than the speed of light in the material. The blue glow observed in the water of a nuclear reactor, close to the active fuel elements, is radiation of this kind. The emission of Cerenkov radiation is analogous to the emission of a shock wave by a projectile moving faster than sound, since in both cases the velocity of the object passing through the medium exceeds the velocity of the resulting wave disturbance in the medium. This radiation, first predicted by P. A. Cerenkov in 1934 and later substantiated theoretically by I. Frank and I. Tamm, is used as a signal for the indication of high-speed particles and as a means for measuring their energy in devices known as Cerenkov counters.

Direction of emission. Cerenkov radiation is emitted at a fixed angle θ to the direction of motion of the particle, such that $\cos\theta = c/nv$, where v is the speed of the particle, c is the speed of light in vacuum, and n is the index of refraction of the medium. The light forms a cone

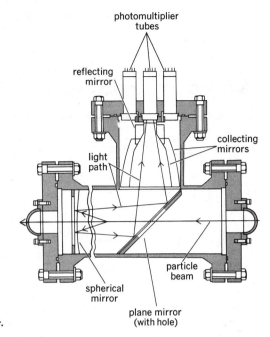

Differential gas Cerenkov counter.

of angle θ around the direction of motion. If this angle can be measured, and n is known for the medium, the speed of the particle can be determined. The light consists of all frequencies for which n is large enough to give a real value of $\cos \theta$ in the preceding equation.

Cerenkov counters. Particle detectors which utilize Cerenkov radiation are called Cerenkov counters. They are important in the detection of particles with speeds approaching that of light, such as those produced in large accelerators and in cosmic rays, and are used with photomultiplier tubes to amplify the Cerenkov radiation. These counters can emit pulses with widths of $\sim 10^{-10}$ s, and are therefore useful in time-of-flight measurements when very short times must be measured. They can also give direct information on the velocity of the passing particle. *See* PARTICLE DETECTOR.

Dielectrics such as glass, water, or clear plastic may be used in Cerenkov counters. Choice of the material depends on the velocity of the particles to be measured, since the values of n are different for the materials cited. By using two Cerenkov counters in coincidence, one after the other, with proper choice of dielectric, the combination will be sensitive to a given velocity range of particles.

The counters may be classified as nonfocusing or focusing. In the former type, the dielectric is surrounded by a light-reflecting substance except at the point where the photomultiplier is attached, and no use is made of the directional properties of the light emitted. In a focusing counter, lenses and mirrors may be used to select light emitted at a given angle and thus to give information on the velocity of the particle.

Cerenkov counters may be used as proportional counters, since the number of photons emitted in the light beam can be calculated as a function of the properties of the material, the frequency interval of the light measured, and the angle θ. Thus the number of photons which make up a certain size pulse gives information on the velocity of the particle.

Gas, notably CO_2 (see **illus**.), may also be used as the dielectric in Cerenkov counters. In such counters the intensity of light emitted is much smaller than in solid or liquid dielectric counters, but the velocity required to produce a count is much higher because of the low index of refraction of gas.

Bibliography. J. Litt and R. Meunier, Cerenkov counter technique in high-energy physics, *Annu. Rev. Nucl. Sci.*, 23:1–43, 1973; L. C. L. Yuan and C. S. Wu (eds.), *Nuclear Physics*, pp. 162–194, vol. 5A of L. Marton (ed.), *Methods of Experimental Physics*, 1961.

TIME-PROJECTION CHAMBER
DAVID R. NYGREN

An advanced particle detector for the study of ultrahigh-energy collisions of positrons and electrons developed originally at the Lawrence Berkeley Laboratory. The underlying physics of the scattering process can be studied through precise measurements of the momenta, directions, particle species, and correlations of the collision products. The time-projection chamber (TPC) provides a unique combination of capabilities for these studies and other problems in elementary particle physics by offering particle identification over a wide momentum range, and by offering high resolution of intrinsically three-dimensional spatial information for accurate event reconstruction.

The time-projection chamber concept is based on the maximum utilization of ionization information, which is deposited by high-energy charged particles traversing a gas. The ionization trail, a precise image of the particle trajectory, also contains information about the particle velocity. A strong, uniform magnetic field and a uniform electric field are generated within the time-projection chamber active volume in an exactly parallel orientation. The parallel configuration of the fields permits electrons, products of the ionization processes, to drift through the time-projection chamber gas over great distances without distortion; the parallel configuration offers a further advantage in that the diffusion of the electrons during drift can be greatly suppressed by the magnetic field, thus preserving the quality of track information. In practice, the track images are drifted on the order of 100 cm (40 in.) or more, yet with measurement precision typically better than ± 0.02 cm.

At the end of the drift volume the ionization electrons are multiplied by an avalanche process on an array of several hundred wires acting as proportional amplifiers. A highly segmented cathode plane just behind the wire array detects the avalanches and provides two-dimensional spatial coordinates in the plane of the array. The drift time provides the trajectory coordinate perpendicular to the plane of the array, hence suggesting the name time-projection chamber. The ionization density, measured precisely by the wire plane signals, offers the means to determine the particle velocity with resolution sufficient to establish the particle mass by a comparison of velocity and momentum.

Several large time-projection chambers are in operation or under construction at the premier storage ring facilities in the United States, Europe, and Japan. SEE ELECTRICAL BREAKDOWN; PARTICLE DETECTOR.

Bibliography. J. A. Macdonald (ed.), *Time Projection Chamber, TRIUMF, Vancouver, 1983*, AIP Conf. Proc. 108, 1984; R. J. Madaras and P. J. Oddone, Time-projection chambers, *Phys. Today*, 37(8):36–47, August 1984; J. N. Marx and D. R. Nygren, The time projection chamber, *Phys. Today*, 31:46–53, October 1978.

CONTRIBUTORS

CONTRIBUTORS

Adair, Dr. Robert K. Department of Physics, Yale University.

Alburger, Dr. David E. Brookhaven National Laboratory, Upton, New York.

Appelquist, Dr. Thomas. Department of Physics, Yale University.

Backstrom, Prof. Gunnar. Institute of Physics, University of Umea, Sweden.

Baltay, Dr. Charles. Department of Physics, Columbia University.

Bars, Dr. Itzhak. Department of Physics, Yale University.

Barton, Dr. Mark Q. Brookhaven National Laboratory, Upton, New York.

Batchelor, Kenneth. National Synchrotron Light Source Department, Brookhaven National Laboratory, Upton, New York.

Bertsch, Dr. George F. Department of Physics, Michigan State University.

Betts, Dr. Russell. Argonne National Laboratory, Argonne, Illinois.

Bichsel, Dr. Hans. Department of Radiology, University of Washington.

Bjorken, Prof. J. D. Stanford Linear Acceleration Center, Stanford University.

Bromley, Prof. D. Allan. Henry Ford II Professor and Director, A. W. Wright Nuclear Structure Laboratory, Yale University.

Brooks, Dr. Frank D. Department of Physics, University of Capetown, Rondebosch, South Africa.

Browne, Prof. Cornelius P. Department of Physics, University of Notre Dame.

Caughlan, Prof. Georgeanne R. Department of Physics, Montana State University.

Clark, Dr. David J. Lawrence Berkeley Laboratory, Berkeley, California.

Commins, Prof. Eugene D. Department of Physics, University of California, Berkeley.

Dalitz, Prof. Richard H. Department of Theoretical Physics, Oxford University, England.

Daniels, Prof. James M. Department of Physics, University of Toronto, Ontario, Canada.

Dover, Dr. Carl B. Department of Physics, Brookhaven National Laboratory, Upton, New York.

Duckworth, Dr. Henry E. Department of Physics, University of Manitoba, Winnipeg, Canada.

Durand, Prof. Loyal, III. Department of Physics, University of Wisconsin.

Egan, Patrick O. Department of Physics, Yale University.

Erb, Dr. K. A. Oak Ridge National Laboratory, Oak Ridge, Tennessee.

Evans, Prof. Robley D. Retired; formerly, Department of Physics, Massachusetts Institute of Technology.

Ferguson, Dr. A. John. Retired; formerly, Chalk River Nuclear Laboratories, Chalk River, Ontario, Canada.
Feshbach, Dr. Herman. Department of Physics, Massachusetts Institute of Technology.
Fidecaro, Dr. Giuseppe. EP Division, CERN, Geneva, Switzerland.
Fields, Thomas H. Physics Division, Argonne National Laboratory, Argonne, Illinois.
Firk, Dr. Frank W. K. Department of Physics, Yale University.
Fitch, Prof. Val L. Department of Physics, Princeton University.
Franzini, Dr. Paolo. Department of Physics, Columbia University.
French, Dr. J. B. Department of Physics, University of Rochester.
Fretter, Prof. William B. Department of Physics, University of California, Berkeley.

Garvey, Dr. Gerald T. Physics Division, Argonne National Laboratory, Argonne, Illinois.
Goebel, Prof. Charles J. Department of Physics, University of Wisconsin.
Goldhaber, Dr. Alfred S. Department of Physics, State University of New York, Stony Brook.
Goldhaber, Dr. Gerson. Lawrence Berkeley Laboratory, University of California, Berkeley.
Good, Dr. Roland H., Jr. Department of Physics, Pennsylvania State University.
Goulding, Dr. Fred S. Lawrence Berkeley Laboratory, Berkeley, California.
Greenberg, Prof. Jack S. Wright Nuclear Structure Laboratory, Yale University.
Greenberg, Prof. O. W. Department of Physics and Astronomy, University of Maryland.
Greiner, Dr. Walter. Institut für Theoretische Physik, Universität Frankfurt, West Germany.
Grotch, Prof. Howard. Physics Department, Pennsylvania State University.
Grunder, Hermann A. Deputy Director, General Sciences, Lawrence Berkeley Laboratory, Berkeley, California.

Hamilton, Prof. Joseph H. Department of Physics-Astronomy, Vanderbilt University.
Hanna, Prof. Stanley S. Department of Physics, Stanford University.
Harvey, Dr. John A. Oak Ridge National Laboratory, Oak Ridge, Tennessee.
Hatch, Dr. Eastman N. Department of Physics, Utah State University.
Heckman, Dr. Harry H. Lawrence Berkeley Laboratory, University of California, Berkeley.
Herb, Raymond G. President, National Electronics Corporation, Middleton, Wisconsin.
Herber, Prof. Rolfe H. Department of Chemistry, Rutgers University.
Hofstadter, Dr. Robert. Department of Physics, Stanford University.
Horen, Dr. Daniel J. Nuclear Division, Oak Ridge National Laboratory, Oak Ridge, Tennessee.
Hughes, Prof. Vernon. Department of Physics, Yale University.
Huizenga, Dr. John R. Nuclear Structure Research Laboratory, University of Rochester.
Hull, Dr. McAllister H., Jr. Department of Physics, University of New Mexico.

Iachello, Dr. Franco. Department of Physics, Yale University.

Jackiw, Prof. Roman. Department of Physics, Massachusetts Institute of Technology.
Jackson, Dr. Harold E., Jr. Argonne National Laboratory, Argonne, Illinois.
Jones, Prof. Lawrence W. Harrison M. Randell Laboratory of Physics, University of Michigan.

Kerst, Prof. Donald W. Department of Physics, University of Wisconsin.
Kim, Prof. Chung W. Department of Physics, Johns Hopkins University.
Koch, Dr. Peter M. Department of Physics, State University of New York, Stony Brook.
Koller, Dr. Noémie. Department of Physics, Rutgers University.
Kovar, Dr. Dennis G. Hahn-Meitner-Institut für Kernforschung, Berlin, West Germany.

Lach, Dr. Joseph. Fermi National Accelerator Laboratories, Batavia, Illinois.
Lanford, Prof. William A. Department of Physics, State University of New York, Albany.
Ljung, Dr. Donovan A. Fermi National Accelerator Laboratory, Batavia, Illinois.

McKenzie, Dr. James. Sandia Laboratories, Albuquerque, New Mexico.

McKibben, Dr. Joseph L. Physics Division, Los Alamos Scientific Laboratory, Los Alamos, New Mexico.
Mann, Prof. Alfred K. Department of Physics, University of Pennsylvania.
Mansouri, Freydoon. Department of Physics, Yale University.
Marshak, Dr. Harvey. National Bureau of Standards.
Middleton, Dr. Roy. Department of Physics, University of Pennsylvania.
Mohr, Prof. Peter. Department of Physics, Yale University.

Nix, Dr. Rayford J. Nuclear Theory Group Leader, Los Alamos National Laboratory, Los Alamos, New Mexico.
Nygren, Dr. David R. Physics Division, Lawrence Berkeley Laboratory, University of California, Berkeley.

Olness, Dr. John W. Brookhaven National Laboratory, Associated Universities, Inc., Upton, New York.
Overley, Dr. J. C. Department of Physics, University of Oregon.

Perl, Prof. Martin L. Stanford Linear Accelerator Center, Stanford, California.
Post, Dr. Richard F. Lawrence Livermore Laboratory, Livermore, California.
Povh, Dr. Bogdan. Max-Planck Institut für Kernphysik, Saupfercheckweg, West Germany.
Price, Dr. P. Buford, Jr. Lawrence Berkeley Laboratory, University of California, Berkeley.
Purser, Dr. Kenneth H. Ionex/HEI Corporation, Newburyport, Massachusetts.

Quigg, C. Fermi National Accelerator Laboratory, Batavia, Illinois.

Ramond, Dr. Pierre. Department of Physics, University of Florida.
Rasmussen, Dr. Norman C. Department of Nuclear Engineering, Massachusetts Institute of Technology.
Roberts, Dr. Arthur. National Accelerator Laboratory, Batavia, Illinois.
Roberts, Prof. Louis D. Department of Physics, University of North Carolina.

Rockett, Frank H. Engineering Consultant, Charlottesville, Virginia.
Rohlf, Dr. James W. Assistant Professor of Physics, High Energy Physics Laboratory, Harvard University.

Salam, Prof. Abdus. Director, International Centre for Theoretical Physics, Strada Costiera, Italy.
Samios, Dr. Nicholas P. Brookhaven National Laboratory, Upton, New York.
Sargent, Dr. Murray, III. Max-Planck Institut für Festkörperforschung, Stuttgart, West Germany.
Schumacher, Dr. Uwe. Max-Planck Institut für Plasmaphysik, Munich, West Germany.
Scott, Dr. David K. National Superconducting Cyclotron Laboratory, Michigan State University.
Seaborg, Dr. Glenn T. Lawrence Berkeley Laboratory, University of California, Berkeley.
Segrè, Emilio G. Department of Physics, University of California, Berkeley.
Sertorio, Dr. Luigi. Department of Physics, University of Rome, Italy.
Shaw, Prof. Gordon L. Department of Physics, University of California, Irvine.
Shivakumar, Dr. B. Oak Ridge National Laboratory, Oak Ridge, Tennessee.
Smith, Dr. Lloyd. Lawrence Berkeley Laboratory, Berkeley, California.
Snell, Dr. Arthur H. Associate Director, Oak Ridge National Laboratory, Oak Ridge, Tennessee.
Souder, Dr. Paul A. Department of Physics, Yale University.
Stein, Dr. Nelson. Physics Division, Los Alamos Scientific Laboratory, Los Alamos, New Mexico.
Stelson, Dr. Paul H. Physics Division, Oak Ridge National Laboratory, Oak Ridge, Tennessee.
Strauch, Dr. Karl. Physics Laboratories, Harvard University.

Taft, Dr. Horace D. Department of Physics, Yale University.
Ting, Prof. Samuel C. C. Laboratory for Nuclear Science, Massachusetts Institute of Technology.

Watson, Dr. W. W. Professor Emeritus of Physics, Yale University.

Wegner, Dr. Harvey E. Physics Division, Brookhaven National Laboratory, Upton, New York.

Wiedemann, Helmut. Stanford Synchrotron Radiation Laboratory, Stanford, California.

Wiegand, Dr. Clyde E. Lawrence Berkeley Laboratory, University of California, Berkeley.

Wilkinson, Prof. D. H. Department of Nuclear Physics, Oxford University, England.

Willis, Dr. William J. Department of Physics, Brookhaven National Laboratory, Upton, New York.

Wilson, Prof. Richard. Department of Physics, Harvard University.

Yang, Prof. Chen Ning. Institute for Theoretical Physics, State University of New York, Stony Brook.

Zeller, Prof. Michael E. Department of Physics, Yale University.

INDEX

INDEX

Asterisks indicate page references to article titles.

Abe, Y. 122
Abelson, P.H. 41
Alfvén, H. 213
Almqvist, E. 122
Alpha-particle decay 60–65
  beta-delayed 77
  Geiger-Nuttall rule 61–63
  nuclear potential barrier 63–65
Alpha particles 167–170*
  average charge 167
  delta electrons 167
  energy loss per ion pair 168
  ion pairs 167
  ionization by 167–168
  range versus energy 168–170
  special ionization 167–168
  spectrum 80
  straggling of 168
Alternating gradient focusing 452–453*
Americium 42
Analog states 26–29*
  applications 29
  isobaric spin 26
  mass 208 system 27–28
  nuclei with $N = Z$ 26–27

Analog states (*cont.*):
  nuclei with neutron excess 27–28
  width of states 28
Anders, E. 49
Anderson, C.D. 213, 216, 217, 266
Angular correlations 83–84*
  bombardment experiments 83–84
  compound nucleus interpretations 83
  perturbations 84
  radioactive decay 84
Anomalons 164*
Antimatter 197–200, 212–214*
  antibaryons 255–256
  antineutron 263–264*
  antiproton 263*
  in cosmology 213–214
  Dirac theory 212–213
  discovery of antiparticles 213
  positron 217*
Antineutron 263–264*
  neutron-antineutron oscillations 342–343*
Antiproton 263*
Antiproton atom 289

Appelquist, T. 329
Arima, A. 12
Arnbruster, P. 49
Atomic mass 7
  Garvey-Kelson mass relations 24*
  mass defect 22*
Atomic nucleus 6*
  atomic number 15*
  Coulomb excitation 113–114*
  mass number 16*
  nuclear isomerism 29–31*
  see also Nuclear moments; Nuclear radiation; Nuclear reaction; Nuclear structure; Scattering experiments (nuclei)
Atomic number 15*
Atomic weight 21–22*
  accurate mass values 21–22
  isotope 20
  mass tables 22
  relative weights 21
  unified scale 21

Bahcall, J.N. 153
Bare charm states 371–372

Baryon 208–209, 242–256*
  antibaryons 255–256
  baryon octet 245–247
  baryonic spectroscopy 255
  beta decay 247–249
  charmed 254–255
  color and quantum chromodynamics 243–244
  delta resonance 265*
  gauge invariance 353
  higher excited states 251–254
  hypercharge 370*
  hyperon 264–265*
  interaction and meson exchange 321
  interaction with matter 256
  meson see Meson
  nonleptonic decay 249
  nucleons and isospin 244–245
  proton 257–260*
  proton stability 256
  strange particles 264*
  3/2 baryon decuplet 249–251
Beauty (quarks) 280
Becquerel, H. 52
Berkelium 42–43
Beta-delayed alpha radioactivity 77
Beta-delayed neutron radioactivity 77
Beta-particle decay 65–70
  average beta energy 67
  double beta decay 69–70
  electron-capture transitions 70
  Fermi theory 67–68
  Konopinski-Uhlenbeck theory 68–69
  Kurie plots 69
  neutrinos 66–67
Beta particles 170–172*
  detection of 171
  interaction with matter 170–171
  spectrometers 171–172
  spectrum 80
Betatron 451–452*

Bethe, H.A. 152, 302, 407
Bilaniuk, O.M.P. 239
Binding energy per nucleon 22–23
Bless, A.A. 180
Bohr, A. 10
Bohr, N. 129
Boson: supersymmetry 373–376*
Bothe, W. 180
Bragg, W.H. 168
Bragg-Kleeman rule 169–170
Breit, G. 108, 180
Briggs, G.H. 168
Bromley, D.A. 122
Brossel, J. 97
Bubble chamber 486–488*
  hybrid systems 488
  interactions and choice of liquid 486–487
  operating cycle 486
  principles of operation 486
  size 487–488

Cabibbo, N. 226
Cabibbo theory: baryon beta decay 248–249
Californium 43
Carbon-nitrogen-oxygen cycles 152–154*
Cartan, E. 332
Causality 391–392*
  dispersion relations 387–388
Cerenkov, P.A. 506
Cerenkov radiation 506–508*
  Cerenkov counters 507–508
  direction of emission 506–507
Chain reaction 162–163*
Chamberlain, O. 263
Charged particle beams 183–191*
  alternating gradient focusing 452–453*
  applications 191
  atomic collisions 186–187
  biological effects 190
  bremsstrahlung 186
  electronic collisions 187–188

Charged particle beams (cont.):
  electrons 190
  interactions 185–188
  ionization 190
  observation 190–191
  particle properties 183–185
  statistics of energy loss 188–190
Charm 371–372*
  charmed baryons 254–255
  observations 371–372
  production by neutrinos 219–220
  quarks 222
  theory 371
Charmonium 371
Chiral symmetry 352
Chopin, G.R. 44
Circular accelerator 399–400, 406–423
  cyclotrons 406–414
  electron synchrotron 414–417
  heavy-ion synchrotron 422–423
  proton synchrotron 417–422
Cleland, M.R. 443
Cloud chamber 483–485*
  dead time 485
  information obtained 484–485
  supersaturation methods 483–484
Cockcroft, J.D. 445
Cockcroft-Walton accelerator 445–446*
  small industrial accelerators 446
  tandem, gas-insulated accelerators 446
Collective-effect accelerator 401
Color 372*
  baryons 243–244
  gluons 228–229
  meson 267–268
  quantum chromodynamics 324–325

Color (*cont.*):
  quarks 223
  unitary symmetry 369
Compton, A.H. 177
Compton effect 176–182*
  experimental demonstration 177
  intensity distribution 180–182
  theoretical explanation 177–180
Condon, E.U. 61
Coulomb excitation 113–114*
  experiments 113–114
  heavy-ion collisions 114
  measurement and applications 114
  theory and interpretation 114
Coulomb scattering 104–105
CPT theorem 364*
  experimental tests 364
  significance 364
  symmetry laws 351
Cronin, J. 320
Crystal counter 494–496*
  circuit 495
  limitations and performance improvement 495–496
Cunningham, B.B. 41, 42, 43
Curie, I. 52
Curie, M. 79
Curium 42
Cyclotron 406–414
  applications 414
  basic principles 406–407
  ion sources 413–414
  isochronous 410–412
  limitations 407–410
  superconducting 412–413

Dalitz plot 378–381*
  antiproton annihilation 379–380
  comparison with Goldhaber triangle 383
  equal mass representation 378–379

Dalitz plot (*cont.*):
  relativistic three-particle system 380
  unsymmetrical plot 380
Deep inelastic collisions 132–133*
  characteristic features 132
  interpretation 133
  light-particle emission 132
  reaction fragment distributions 132
Delta electrons 167, 192–193*
Delta resonance 265*
Deshpande, V.K. 239
Deuterium: deuteron 40*
Deuteron 40*
Dirac, P.A.M. 180, 182, 212, 216, 217, 238, 263, 299, 301, 317
Dispersion relations 386–391*
  causality 387–388, 391–392*
  dielectric constant and refractive index 389–390
  elementary particle theory 390–391
  forward-scattering amplitude 389–390
  pure-tone response 387
  resonance 388–389
  response function 387
Double beta decay 69–70
Duoplasmatron ion source 458
Dynamic nuclear polarization 95–99*
  laser-induced 98–99
  microwave pumping 95–97
  optical pumping 97–98
  polarization by rotation 97
Dynamitron accelerator 443–445*
Dyson, F.J. 301, 385

Einstein, A. 343
Einsteinium 43
Elastic scattering 104
Electric moment *see* Nuclear moments

Electromagnetic interaction 292–293
  gauge theory and quantum chromodynamics 323
Electromagnetism: as a gauge theory 331–332
Electron 215–216*
  charge 215
  leptons 216
  magnetic moment 216
  and matter 215–216
  muonium 285–286*
  positron 217*
  positronium 284–285*
  spin 216
Electron capture 191–192*
Electron linear accelerator 425–427
Electron-nucleon scattering 105–106
Electron-positron pair production 182–183*
Electron ring accelerator 438–441
  collective acceleration 438
  electron ring collective field 438–439
  methods of acceleration 440
  results of experiments 440–441
  ring stability 439
Electron synchrotron 414–417
  acceleration and beam control 415–416
  examples 417
  magnet system 416
  vacuum system 416–417
Electronuclear interaction 296
Electrostatic accelerator 398–399, 402–406
  applications 406
  Cockcroft-Walton accelerator 445–446*
  control and adjustment 405
  dual accelerator systems 405–406
  folded tandem design 403–404
  large machines 404–405

Electrostatic accelerator (*cont.*):
  pelletron accelerator 449–451*
  resonance transformer 441*
  tandem design 403
Electroweak interaction 294–296
  neutral currents 335–337*
  supergravity 343–344
  Weinberg-Salam model 334–335*
Element: natural isotopic composition 17–19
  transmutation 79*
  transuranium *see* Transuranium elements
Element 104 45
Element 106 45–46
Element 107 46
Element 108 46
Element 109 46
Elementary particle 196–212*
  antiparticles *see* Antimatter
  baryon *see* Baryon
  charm 371–372*
  Dalitz plot 378–381*
  delta resonance 265*
  dispersion relations 390–391
  electron *see* Electron
  Feynman diagram 299, 300–301*, 303–305
  general properties 197–200
  gluons *see* Gluons
  grand unified theories 211–212
  hadron *see* Hadron
  high-energy hadron-hadron collisions 207
  hyperon 264–265*
  instanton 333–334*
  interactions 200
  intermediate vector boson 232–236*
  J particle 277–280*
  lepton *see* Lepton
  meson *see* Meson
  multiplets 202–203
  neutrino *see* Neutrino
  neutron *see* Neutron

Elementary particle (*cont.*):
  nomenclature 203–204
  positron *see* Positron
  proton *see* Proton
  quantum chromodynamics 204–205
  quantum field theory *see* Quantum field theory
  quarks *see* Quarks
  Regge recurrences 204
  resonances 201–202
  stability 200–201
  strange particles 264*
  string model of confinement 205–207
  supermultiplets 203
  weak interaction 210–211
Energy level (quantum mechanics): sum rules 82–83*
Eskola, K. 45
Eskola, P. 45
Eta meson 271
Excited nuclear states: giant nuclear resonances 31–35*
  nuclear isomerism 29–31*
Exotic atoms: hadronic atom 287–289*
  muonium 285–286*
  pionium 286–287*
  positronium 284–285*
  strong interactions 322
Explicit symmetry breaking 373

Faddeev, L.D. 385
Fairbank, W.M. 224
Faraday, M. 294
Feinberg, G. 239
Fermi, E. 67, 317
Fermi theory (beta particles) 67–68
Fermion: masses 320
  supersymmetry 373–376*
Fermium 43–44
Feynman, R.P. 298, 300, 301, 385

Feynman diagram 299, 300–301*, 303–305
Fitch, V. 320
Flavor 370*
  unitary symmetry 369
Flerov, G.N. 44, 49
Fock, V. 331
Frank, I. 506
Fundamental interactions 292–297*
  consequences of symmetry breaking 296
  electromagnetic interaction 292–293
  electronuclear interaction 296
  electroweak interaction 294–296
  gauge interactions 294
  gauge theory *see* Gauge theory
  gravitational interaction 292
  mesons 269–270
  properties 292–294
  prospects for including gravity in unification theories 296–297
  strong interaction *see* Strong nuclear interactions
  unification 294–297
  unification of gravitation and electromagnetism 294
  weak interaction *see* Weak nuclear interactions
  Weinberg-Salam model 334–335*
  see also Grand unification theories

Gamma-ray decay: internal conversion 71–74
  internal pair formation 74
  isomeric transitions 74
  mean life for transitions 71
  radiationless transitions 74
Gamma-ray detector 496–498*
  detector-spectrometer devices 496–497

Gamma-ray detector (*cont.*):
  Ge(Li) semiconductor detectors 497–498
  miscellaneous detectors and spectrometers 498
  NaI(Tl) detector 498
Gamma rays 173–176*
  applications to nuclear research 175
  Doppler shift 175
  interaction with matter 175–176
  Mössbauer effect 84–87*
  nature of 173
  nuclear labels 174–175
  origin 173–174
  practical applications 175
  spectra 80–81
  wave-particle duality 173
Gamow, G. 317
Garvey-Kelson mass relations 24*
Gatti, R.C. 43
Gauge theory 331–333*
  electromagnetism as a gauge theory 331–332
  gauge interactions 294
  gluons 229–230
  nonabelian 332
  quantum chromodynamics 323–324
  symmetry dictates interactions 332–333
Geiger, H. 61, 104, 168, 180
Geiger-Briggs rule 168
Geiger counter 475–476
Geiger-Nuttall rule 61–63
Gell-Mann, M. 220, 228, 318, 324, 370
Georgi, H. 296, 340
Ghiorso, A. 42, 43, 44, 45, 46, 49
Giant nuclear resonances 31–35*
  E0 resonances 33–34
  E1 resonances 31–32
  E2 resonances 33
  M1 resonances 32–33

Glashow, S.L. 219, 222, 226, 227, 294, 316, 340, 342, 343
Gluons 228–232*
  charmonium lifetimes 230–232
  color 228–229
  experimental evidence 230–232
  gauge symmetry 229–230
  implications 232
  inelastic electron-proton scattering 230
  quantum chromodynamics 230
  supercritical fields 313–314
  three-jet pattern 232
  Yang-Mills theory 229
Goldhaber, M. 363
Goldhaber triangle 381–384*
  comparison with Dalitz plot 383
  example of triangle plot 383
  extension to five particles 383–384
  kinematical limits 382
  phase space distribution 382–383
Goldstone, G. 343
Goldstone, J. 319
Golfand, Y.A. 344
Grand unification theories 211–212, 340–342*
  proton stability 256
  supergravity 344
Gravitational field: supercritical fields 314–315
Gravitational interaction 292
Graviton 236–237*
Greenberg, O.W. 372
Greiner, W. 75, 120
Grodzins, L. 363
Gross, D.J. 325
Gurney, R.W. 61

Hadron 242*
  baryon *see* Baryon
  charm 371

Hadron (*cont.*):
  high-energy hadron-hadron collisions 207
  isobaric spin 365–367*
  meson *see* Meson
  strong nuclear interactions 321–323*
  structure 207–210
  unconventional 209–210
Hadronic atom 287–289*
  antiproton atoms 289
  kaonic atoms 288–289
  pionic atoms 288
  sigma-hyperonic atoms 289
  x-ray emissions 287–288
Hahnium 45
Half-life 56–57, 78–79*
Half-period *see* Half-life
Harris, J. 45
Hartree, D.R. 182
Harvey, B.G. 44
Heavy-ion linear accelerator 428
Heavy-ion sources 458–461
Heavy-ion synchrotron 422–423
  acceleration in 423
  basic considerations 422–423
  ion sources and injectors 423
  prospects 423
Heavy nuclei: analog states 27
Heisenberg, W. 301, 317, 384
Helicity 363–364*
  chiral symmetry 352
Herrmann, G. 49
Higgs, P.W. 226, 319, 324
Higgs bosons 320
  electroweak theory 295–296
Hooft, G. 't 227, 319, 324, 332
Hund, F. 115
Hydrogen: deuteron 40*
  triton 40*
Hypercharge 370*
Hyperfine structure 100–101*
  atoms and molecules 100
  isotope effect 100
  liquid and solid systems 100–101
  nuclear spin effect 100

Hypernuclei 50*
Hyperon 264–265*
  hypernuclei 50*

Iachello, F. 12, 124
Iliopoulos, J. 226
Instanton 333–334*
Intermediate vector boson 232–236*
  discovery of W particle 233
  discovery of Z particle 233
  production and detection 233
  properties of W and Z particles 233–234
  Weinberg-Salam model 334–335
Internal symmetries 348–352
  CPT theorem 351
  invariances of strong interactions 351
  isotopic spin 351
  unitary symmetry see Unitary symmetry
Ion-confinement sources 460–461
Ion sources 456–467*
  cesium-beam sputter negative-ion source concept 461
  cesium-beam sputtering for negative ions 457
  cesium-vapor Penning negative-ion source concept 462–463
  charge exchange for negative ions 457
  charge-exchange negative-ion source concept 461
  colliding-beam polarized ion source 466
  conventional polarized ion source 463–465
  direct extraction for negative ions 457
  duoplasmatron ion source 458
  electron impact method for positive ions 456

Ion sources (cont.):
  field ionization for positive ions 456–457
  heavy-ion sources 458–461
  ion-confinement sources 460–461
  Lamb-shift polarized ion source 465–466
  methods of negative-ion formation 457
  methods of positive-ion formation 456–457
  negative-ion source concepts 461–463
  Penning ion sources 459–460
  polarized 463–466
  positive-ion source concept 457–461
  spark discharge for positive ions 456
  sputtering for positive ions 457
  surface ionization for negative ions 457
  surface ionization for positive ions 456
  thermal ionization for positive ions 457
Ionization chamber 470–476*
  energy spectrum 471–472
  gaseous 472–474
  Geiger counters 475–476
  liquid 474
  principles of operation 470–471
  proportional counters 474–475
  solid 474
Ionization detector 467–468
Isobar 20*
  analog states 26–29*
Isobaric analog states see Analog states
Isobaric spin 26, 365–367*
  classification of states 366–367
  importance in reactions and decays 365–366
  similarity to spin 365

Isochronous cyclotron 410–412
Isomer: nuclear isomerism 29–31*
Isospin see Isobaric spin
Isotone 20*
Isotope 16–20*
  atomic mass 20
  isotopic abundance 16–19
  nuclear stability 16
  radioisotope 78*
  use of separated isotopes 19–20
Isotope shift 99–100*
Isotopic spin see Isobaric spin

J particle 277–280*
  discovery 277–278
  properties 278–280
James, R.A. 42
Jensen, J.H. 9
Joliot, F. 52
Junction detector 488–494*
  fabrication of diodes 488–494
  fabrication of triodes 494

Kaluza, T. 296, 343
Kaon 266, 271
Kaonic atom 288–289
Kastler, A. 97
Kennedy, J.W. 41
Kibble, T.W.B. 226
Kleeman, R. 169
Klein, O. 181, 296, 343
Koboyashi, M. 227
Konopinski, E.J. 68
Konopinski-Uhlenbeck theory (beta particles) 68–69
Kramers, H.A. 386
Kronig, R. 386
Kuehner, J.A. 122
Kurie plots (beta particles) 69
Kuz'min, V.A. 342

La Chapelle, T.J. 41
Lamb, W.E. 302

Lambda particle: hypernuclei 50*
Langer, L.M. 68
Larsh, A.E. 44
Laser: dynamic nuclear polarization 98–99
Laser-driven accelerator 401
Laser ionization for positive ions 456
Latimer, R.M. 44
Lawrence, E.O. 406
Lawrencium 44–45
Lederman, L.M. 222
Lee, T.D. 317, 360
Lefort, M. 49
Lepton 214–215*
  charm 371
  conservation 214
  decay 214–215
  electron 215–216*
  gauge invariance 353
  masses 215
  muonium 285–286*
  neutrino see Neutrino
  production 215
  Weinberg-Salam model 334–335*
Lie, S. 225, 332, 374
Likhtman, E.P. 344
Linac see Linear accelerator
Linear accelerator 399, 424–428
  electron accelerators 425–427
  heavy-ion accelerators 428
  principles of operation 424–425
Liquid scintillation detector 501–502*
  additions of compounds and samples 501
  manufacture and handling 502
  pulse shape discrimination 501
  solvent and solute 501
Local guage theories: quarks 225–226
London, F. 331
Luders, G. 351, 364

McMillan, E.M. 41
Magic numbers 25–26*
Magnetic moment see Nuclear moments
Magnetic monopoles 237–239*
  charge quantization 238
  grand unification models 238
  parity violation 237
  properties of Dirac monopoles 238
  use in searches 238–239
Magneton 37–38*
Magnusson, L.G. 41
Maiani, L. 219, 226
Marsden, E. 104
Marshak, R. 318
Maskawa, T. 227
Mass defect 22*
Mass number 16*
Mass spectroscopy: atomic mass measurement 21–22
Maxwell, J.C. 177, 294, 324, 334, 343
Mayer, M.G. 9
Mendelevium 44
Meson 207–208, 265–277*
  color and quantum chromodynamics 267–268
  discovery 266
  higher excited mesonic states 275–276
  multiplets with positive parity 274–275
  pseudoscalar mesons 270–271
  quark and antiquark interactions 269–270
  as quark-antiquark systems 268
  quark structure 266–267
  scattering 113
  strong interactions 322–323
  unitary flavor symmetries 268–269
  vector mesons 272–274
Microtron 428–430
  applications 429
  comparison with other accelerators 430

Microtron (cont.):
  design 429
  principles of operation 428–429
Microwave pumping: dynamic nuclear polarization 95–97
Millikan, R.A. 215, 224
Mills, R.L. 225, 229, 294, 319, 324, 332
Monoenergetic neutron spectrometers 91–92
Morgan, L.O. 42
Mössbauer, R.L. 84
Mössbauer effect 84–87*
  energy modulation 85
  experimental realization 85–86
  theory of 84–85
Mottelson, B.R. 10
Mullikan, R. 115
Multipole radiation 81–82*
Muon 266
  muonium 285–286*
Muonium 285–286*

Nambu, Y. 343
Ne'eman, Y. 228
Neptunium 41
Neutral currents 335–337*
  discovery and investigation 336
  electroweak interactions 335–336
  properties 336
  Weinberg-Salam model 334
Neutrino 217–220*
  basic properties 218
  beta-particle decay 66–67
  charged current interactions 219
  charm production 219–220
  neutral current interactions 219
  study of interactions 218–219
Neutron 260–263*
  antineutron 263–264*
  detection 261

Neutron (cont.):
  intrinsic properties 261–262
  magic numbers 25–26*
  in nuclei 260
  nucleon 256–257*
  penetrating power 261
  sources of free neutrons 260–261
  ultracold 262
Neutron-antineutron oscillations 342–343*
Neutron-proton scattering 106, 109–110
Neutron radioactivity 76
  beta-delayed 77
Neutron spectrometry 87–92*
  applications 92
  monoenergetic neutron spectrometers 91–92
  neutron cross sections 89–90
  neutron reactions and resonance parameters 88–89
  techniques 90–92
  time-of-flight spectrometers 90–91, 92–93*
  unbound and bound states of nuclides 87–88
Nishijima, K. 370
Nishina, Y. 181
Nobelium 44
Nonabelian gauge theory 332
Nuclear binding energy 22–23*
  mass defect 22*
Nuclear fission 154–162*
  beta-delayed spontaneous fission 78
  chain reaction 162–163*
  double-humped barrier 156
  experimental consequences of double-humped barrier 156–158
  liquid-drop model 154
  postscission phenomena 160–162
  probability for 158–159
  scission 159–160
  shell corrections 154–155
  spontaneous 74–75

Nuclear fusion 133–148*
  critical temperatures 136
  cross sections 134
  energy division 134–135
  field-reversed confinement systems 142–143
  fusion power 136–147
  magnetic confinement 137–143
  mirror machine and tandem mirror 138–142
  pellet fusion 143–147
  power plants 147–148
  properties of reactions 134–136
  reaction rates 135–136
  simple reactions 134
  thermonuclear reaction 149*
  tokamak 137–138
Nuclear gamma resonance fluorescence see Mössbauer effect
Nuclear isomerism 29–31*
  other mechanisms 30–31
  spin isomerism 29–30
Nuclear molecule 122–127*
  carbon-12 + carbon-12 system 122–124
  dipole nuclear collectivity 127
  models of molecular phenomena 124–125
  phenomena in heavier systems 125–126
  radioactive decay 126
Nuclear moments 35–37*
  effects of 35–36
  electron magnetic moment 216
  magnetic and electric moments 9
  magneton 37–38*
  measurement 36–37
Nuclear orientation 94–95*
  applications 94–95
  nuclear polarization and alignment 94
  production 94

Nuclear polarization 94
  see also Dynamic nuclear polarization
Nuclear radiation 166*
  see also Alpha particles; Beta particles; Gamma rays
Nuclear reaction 127–131*
  carbon-nitrogen-oxygen cycles 152–154*
  importance of isobaric spin 365–366
  mechanism 129–131
  nuclear cross section 128
  nuclear fission see Nuclear fission
  nuclear fusion see Nuclear fusion
  proton-proton chain 149–152*
  requirements for a reaction 128–129
  spallation reaction 163*
  spectra from 81
  studies 131
  types of nuclear interaction 127–128
Nuclear spectra 79–81*
  alpha-particle spectra 80
  beta-ray spectrum 80
  gamma-ray spectra 80–81
  from reactions 81
  spontaneous fission 80
Nuclear structure 6–15*
  analog states 26–29*
  bulk properties 6–9
  detailed properties 9
  energy-level diagram 9
  magnetic and electric moments 9
  nuclear densities and sizes 6–7
  nuclear forces 13–15
  nuclear masses 7–8
  nuclear models 9–13
  nuclei far from stability 7–8
  nuclei with few valence particles 10
  nuclei with many valence particles 10–12

Nucelar structure (*cont.*):
  shell model 9
  simple modes of excitation 13
  statistical model 12–13
Nucleon 256–257*
  binding energy 22–23
  mass number 16*
Nucleon-nucleon interaction 321–322
Nucleon-nucleus scattering 111
Nucleus *see* Atomic nucleus
Nucleus-nucleus scattering 111–113
Nuclide: isotope 16–20*
  neutron spectrometry 87–92*
  radioactivity 52–78*
Nurmia, M. 45
Nuttall, J.M. 61

Oganessian, Y.T. 45, 49
Optical pumping: dynamic nuclear polarization 97–98
Oriented nuclei 94–95

Pairing isomer 31
Parity 358–361*
  conservation 359–360
  nonconservation 360–361
Particle accelerator 398–441*
  alternating gradient focusing 452–453*
  betatron 451–452*
  circular accelerator *see* Circular accelerator
  Cockcroft-Walton accelerator 445–446*
  collective accelerators 401
  Dynamitron accelerator 443–445*
  electron ring accelerator 438–441
  electrostatic accelerator *see* Electrostatic accelerator

Particle accelerator (*cont.*):
  focusing 400
  laser-driven accelerator 401
  linear accelerator *see* Linear accelerator
  microtron 428–430
  pelletron accelerator 449–451*
  performance characteristics 401
  positron-electron storage rings 431–434
  principles of colliding-beam systems 430–431
  proton and $\bar{p}p$ storage rings 434–438
  resonance transformer 441*
  storage rings 401, 430–438
  superconducting magnets 400–401
  tandetron 441–443*
  time-varying field accelerators 399–400
  Van de Graaff generator 447–449*
Particle detector 467–470*
  bubble chamber 486–488*
  classification by use 467
  cloud chamber 483–485*
  crystal counter 494–496*
  ionization detectors 467–468
  junction detector *see* Junction detector
  large systems 468–469
  liquid scintillation detector 501–502*
  neutral particles 468
  scintillation counter 498–500
  scintillation detector 468
  spark chamber *see* Spark chamber
  spark counter 476*
  time-projection chamber 508*
  transition radiation detectors 502*
  types 468

Particle track etching 502–506*
  current solar and galactic radiation 505
  geochronology 505
  geophysics and element mapping 505
  identification of nuclear particles 502–503
  nuclear and elementary particle physics 505
  nucleosynthesis 504
  practical applications 505–506
  solar and galactic irradiation history 503–504
Parton model: quantum chromodynamics 328
Pati, J.C. 296, 301
Pauli, W. 301, 363, 364
Pellet fusion 143–147
Pelletron accelerator 449–451*
  accelerating stages 449
  comparison with other accelerators 450–451
  25-MV model 449–450
  voltage range 449
Penning ion sources 459–460
Perlman, I. 42
Phillips, L. 43
Photon 228*
Pi-mu atom *see* Pionium
Pion 266, 270–271
Pionic atom 288
Pionium 286–287*
Polikanov, S.M. 156
Politzer, H.D. 325, 329
Positron 217*
  creation in quasiatoms 119–122
  electron-positron pair production 182–183*
  observation of spontaneous positron emission 311–312
  positronium 284–285*
Positron-electron storage rings 431–434
Positronium 284–285*
  decay 285

Positronium (*cont.*):
   energy levels 284–285
Powell, C.F. 266
Price, H.C. 68
Proportional counter 474–475
Proton 257–260*
   antiproton 263*
   atomic number 15*
   magic numbers 25–26*
   nucleon 256–257*
   quark structure 258
   search for decay 258–260
   size and structure 257–258
Proton-proton chain 149–152*
   neutrino emission 151–152
   reactions 150–151
Proton-proton scattering 108, 109–110
Proton radioactivity 75–76
   beta-delayed 77–78
Proton storage rings 434–438
Proton synchrotron 417–422
   acceleration and beam dynamics 419–420
   extraction 420–421
   proton sources and injectors 418–419
   tevatron and UNK 421–422
Pseudoscalar mesons 270–271

QCD *see* Quantum chromodynamics
QED *see* Quantum electrodynamics
Quantum chromodynamics 230, 323–331*
   asymptotic freedom 325–328
   baryons 243–244
   color 324–325
   elementary particle 204–205
   experimental consequences 328
   gauge invariance 353–354
   gauge theories 323–324
   meson 267–268
   quarkonium 329–330
   quarks 227

Quantum electrodynamics 301–307*
   Feynman diagram 300–301*, 303–305
   free electromagnetic field 302–303
   free electron field 303
   general applications 302
   interaction between radiation and electrons 303
   relation to other theories 306–307
   renormalization 305–306
   vacuum decay in electrostatic fields 308–309
Quantum field theory 297–300*
   axiomatic 298
   CPT theorem 364*
   Feynman diagram 299, 300–301*
   mathematical structure 299
   nonrelativistic applications 299–300
   renormalization 337–340*
   Yukawa force 297–298
Quarkonium 329–330
Quarks 220–228*
   charm 222
   color 223, 372*
   electroweak interactions 226–227
   evidence supporting quark model 220–221
   flavor 370*
   gluons *see* Gluons
   kinds 221–222
   local gauge theories 225–226
   meson structure 266–267
   proton structure 258
   quantum chromodynamics *see* Quantum chromodynamics
   search for fundamental constituents 220
   search for quarks in stable matter 224
   searches for free quarks 223–224

Quarks (*cont.*):
   beyond the standard model 227–228
   standard model 227
   theory of properties and interactions 225–228
   understanding 224–225
   unresolved questions 225
   upsilon particle 222–223
   Weinberg-Salam model 334–335*
Quasiatom 114–122*
   delta-electron spectra 117
   electron promotion 115
   formation 114–115
   positron creation in 119–122
   quasimolecular x-ray spectra 115–117
   superheavy atoms 117–119
   superheavy quasimolecules 309

Radioactivity 52–78*
   alpha-particle decay 60–65
   beta-delayed alpha radioactivity 77
   beta-delayed neutron radioactivity 77
   beta-delayed proton radioactivity 77–78
   beta-delayed spontaneous fission 78
   beta-particle decay 65–70
   delayed particle emissions 76–78
   dual decay 55
   in Earth 58–59
   exponential decay law 55–56
   gamma-ray decay 70–74
   half-life 56–57, 78–79*
   heavy-cluster decays 75
   laboratory produced radioactive nuclei 59–60
   mean life 56
   neutron 76
   proton 75–76
   radioactive decay constant 55

Radiactivity (cont.):
  radioactive equilibrium 57–58
  radioactive series decay 56–60
  radioactive transformation series 60
  spontaneous fission 74–75
  transition rates and decay laws 55–56
  types 52–53
Radioisotope 78*
Rainwater, J. 10
Regge pole 392–396*
  hypothesis 395–396
  theory 392–395
Relative atomic weight 21
Renormalization 337–340*
  divergences in quantum field theory 337
  examples of renormalizable fields 340
  nonabelian gauge theory 332
  of parameters 338–339
  quantum electrodynamics 305–306
  regularization procedure 337–338
  renormalization group 340
Resonance transformer 441*
Richter, B. 222
Richter, J. 222
Roentgen, W. 52
Rose, M.E. 407
Rutherford, E. 79, 104
Rutherfordium 45

Salam, A. 227, 294, 316, 340, 343
Scattering experiments (nuclei) 104–113*
  Coulomb scattering by nuclei 104–105
  cross sections 104
  definitions of elastic scattering 104
  electron-nucleon scattering 105–106

Scattering experiments (nuclei) (cont.):
  intermediate-energy np and pp scattering 109–110
  low-energy np scattering 106–108
  meson scattering 113
  nucleus-nucleus scattering 111–113
  phase shifts 108–109
  potential scattering 109
  Regge pole 392–396*
  scattering matrix 384–386*
Scattering matrix 384–386*
  calculation 385–386
  definition and properties 384–385
  renormalization 337–340*
Schwinger, J.S. 294, 301, 318, 385
Scintillation counter 498–500*
  characteristics 499–500
  coincidence counting 500
  liquid scintillation detector 501–502*
  operation of bulk counter 499
  substances used 500
Scintillation detector 468
Seaborg, G.T. 41, 42, 43, 44
Segrè, E. 41, 213
Selection rules 354–358*
  angular momentum and parity rules 354–355
  conservation laws 354
  forbidden transitions 355–357
  further symmetries 357–358
  molecular 357
  parity nonconservation 360
Shape isomer 30
Shaw, R. 294
Shell model (nucleus) 9
Sigma-hyperonic atom 289
Sikkeland, T. 44
Space-time symmetries 348
Spallation reaction 163*

Spark chamber 476–483*
  characteristic times 477
  conventional narrow-gap chamber 478
  data readout 480–482
  digitized 481–482
  electronic readout 482
  general properties 476–477
  heavy spark-chamber plates 480
  particle tracks and direction 477
  projection chamber 477–478
  sampling chamber 477
  sonic chambers 482
  track-delineating chamber 478
  types 477–480
  use of magnetic fields 482
  use of vidicon tube 482
  wide-gap chambers 479–480
Spark counter 476*
Spectrometer: beta-particle 171–172
Spectroscopy: mass see Mass spectroscopy
  Mössbauer effect 84–87*
  neutron see Neutron spectrometry
Spin 363*
  dynamic nuclear polarization 95–99*
  electron 216
  helicity 363–364*
  isobaric see Isobaric spin
  nuclear orientation 94–95*
  similarity to isobaric spin 365
Spin isomer 29–30
Spontaneous symmetry breaking 373
Storage rings 430–438
Strange particles 264*
  hypernuclei 50*
  hyperon 264–265*
  unitary symmetry 367–369
Street, K., Jr. 43
String model (elementary particle) 205–207

Strong nuclear interactions
  13–15, 294, 321–323*
 exotic atoms 322
 hypernuclei 322
 interactions of mesons
   322–323
 isobaric spin 365–367*
 meson exchange and bar-
   yon interaction 321
 nucleon-nucleon interaction
   321–322
 quantum chromodynamics
   see Quantum chromody-
   namics
 range 321
Strutinsky, V.M. 156
Sudershan, E.C.G. 239, 318
Sum rules 82–83*
Sunyar, A. 363
Superconducting cyclotron
  412–413
Supercritical fields 307–315*
 application to other field
   theories 312–315
 concept of the vacuum 307
 dynamical heavy-ion collision processes 309–310
 giant nuclear systems 311
 observation of spontaneous
   positron emission
   311–312
 superheavy quasimolecules
   309
 vacuum decay in quantum
   electrodynamics 308–309
Supergravity 343–346*
 electroweak theory 343–344
 grand unification theories
   344
 models 344–345
 superunification in superspace 345–346
 superunification theories
   and supersymmetry 344
Superheavy elements see
  Supertransuranics
Supersymmetry 373–376*
 definition 373–374
 mathematical formalism 374
 in nuclear physics 374–376

Supersymmetry (cont.):
 in particle physics 374
Supertransuranics 46–50*
 nuclear stability 47–48
 production methods 48–49
 searches at accelerators
   49–50
 searches in nature 49
Symmetry breaking 372–373*
 explicit 373
 spontaneous 352–353, 373
Symmetry laws 348–354*
 chiral symmetry 352
 gauge invariance 352–354
 internal symmetries
   348–352
 selection rules see Selection
   rules
 space-time symmetries 348
 supersymmetry 373–376*
 symmetry breaking
   372–373*
 symmetry in quantum me-
   chanics 348
 unitary symmetry see
   Unitary symmetry

Table of Atomic Masses 22
Tachyon 239–240*
Talmi, I. 12
Tamm, I. 506
Tanaka, S. 239
Tandetron 441–443*
 accelerating voltage 443
 acceleration principle
   442–443
 applications 442
 reliability and safety 443
Teller, E. 317
Tevatron 421–422
Thermonuclear reaction 149*
Thomas, L.H. 410
Thompson, G.N. 42, 43, 44,
   49
Thomson, J.J. 177, 215
't Hooft, G. 227, 319, 324, 332
Time-of-flight spectrometers
  90–91, 92–93*
Time-projection chamber 508*

Time reversal invariance
  361–363*
 evidence for violation
   362–363
 quantum mechanics
   361–362
 tests 362
Time-varying field accelerator
  399–400
 see also Circular accelerator; Linear accelerator
Ting, S.C.C. 222
Tokamak 137–138
Tomonaga, S. 301
Transition radiation detectors
  502*
Transmutation 79*
Transuranium elements
  40–46*
 americium 42
 berkelium 42–43
 californium 43
 curium 42
 einsteinium 43
 element 104 45
 element 106 45–46
 elements 107–109 46
 fermium 43–44
 hahnium 45
 lawrencium 44–45
 mendelevium 44
 neptunium 41
 nobelium 44
 plutonium 41–42
 superheavy elements 46
Tritium: triton 40*
Triton 40*

Uhlenbeck, G.E. 68
Ultracold neutrons 262
Unified Scale of Atomic
  Masses 21
Unitary symmetry 351–352,
  367–369*
 charge independence 367
 color SU_3 369
 mesons 268–269
 nonrelativistic symmetries
   369

Unitary symmetry (cont.):
 SU_N symmetry 369
 SU_3 symmetry 367–369
Upsilon meson see Upsilon particles
Upsilon particles 274, 280–282*
 family of states 281–282
 heavy quarks 280–281
 potential models 281
 quarks 222–223

Valence nucleon 10
Van de Graaff generator 447–449*
 large machines 448–449
 single-ended 447–448
 tandem 448
Vector mesons 272–274
von Weiszacker, C.F. 152

W particle see Intermediate vector boson
Wahl, A.C. 41
Waller, I. 182
Wallman, J.C. 43
Walton, E. 445
Walton, J.R. 44
Ward, J.C. 294, 343
Weak nuclear interactions 210–211, 293–294, 315–321*
 early study 316–317
 Fermi theory 317
 Feynman–Gell-Mann theory 318
 mesons 270
 parity violation 317–318
 present and future problems 320
 Weinberg-Salam model 318–320
Weinberg, S. 227, 316, 340, 343
Weinberg-Salam model 334–335*
 intermediate bosons 334–335
 neutral currents 334
 predictions from 335
 standard model 335
Weinberg-Salam model (cont.):
 weak nuclear interactions 318–320
Weiszacker, C.F. von 152
Werner, L.B. 41, 42
Weyl, H. 294, 323, 331, 343, 345
Wheeler, J.A. 384
Wigner, E. 106, 369
Wilczek, F. 325
Wilson, C.T.R. 180
Wilson cloud chamber 483–485
Woods-Saxon distribution 6

Yang, C.N. 225, 229, 294, 317, 319, 324, 332, 360
Yang-Mills theory: gluons 229
 instanton 333–334*
 quantum chromodynamics 324
Yukawa, H. 229, 265, 297, 324

Z particle see Intermediate vector boson
Zweig, G. 220, 228, 324